Electromagnetic Fields of Wireless Communications: Biological and Health Effects

This book reflects contributions from experts in the biological and health effects of Radio Frequency (RF)/Microwave and Extremely Low Frequency (ELF) Electromagnetic Fields (EMFs) used in wireless communications (WC) and other technological applications. Diverse topics related to physics, biology, pathology, epidemiology, and plausible biophysical and biochemical mechanisms of WC EMFs emitted by antennas and devices are included. Discussions on the possible consequences of fifth generation (5G) mobile telephony (MT) EMFs based on available data and correlation between anthropogenic EMF exposures and various pathological conditions such as infertility, cancer, electro-hypersensitivity, organic and viral diseases, and effects on animals, plants, trees, and environment are included. It further illustrates individual and public health protection and the setting of biologically- and epidemiologically-based exposure limits.

Features:

- Covers biological and health effects, including oxidative stress, DNA damage, reproductive effects of mobile phones/antennas (2G, 3G, 4G), cordless phones, Wi-Fi, etc.
- Describes effects induced by real-life exposures by commercially available devices/antennas.
- Illustrates biophysical and biochemical mechanisms that fill the gap between recorded experimental and epidemiological findings and their explanations.
- Explores experimental and epidemiological facts and mechanisms of action. Provides explanations and protection tips.
- Transcends across physical, biological, chemical, health, epidemiological, and environmental aspects of the topic.

This book is aimed at senior undergraduate/graduate students in physics, biology, medicine, bioelectromagnetics, electromagnetic biology, non-ionizing radiation biophysics, telecommunications, electromagnetism, bioengineering, and dosimetry.

Electromagnetic Fields of Wireless Communications: Biological and Health Effects

Edited by Dimitris J. Panagopoulos

CRC Press
Taylor & Francis Group
Boca Raton London New York

CRC Press is an imprint of the
Taylor & Francis Group, an **informa** business

Designed cover image: © Shutterstock

First edition published 2023
by CRC Press
6000 Broken Sound Parkway NW, Suite 300, Boca Raton, FL 33487-2742

and by CRC Press
4 Park Square, Milton Park, Abingdon, Oxon, OX14 4RN
CRC Press is an imprint of Taylor & Francis Group, LLC

© 2023 selection and editorial matter, Dimitris J. Panagopoulos; individual chapters, the contributors

Reasonable efforts have been made to publish reliable data and information, but the author and publisher cannot assume responsibility for the validity of all materials or the consequences of their use. The authors and publishers have attempted to trace the copyright holders of all material reproduced in this publication and apologize to copyright holders if permission to publish in this form has not been obtained. If any copyright material has not been acknowledged please write and let us know so we may rectify in any future reprint.

Except as permitted under U.S. Copyright Law, no part of this book may be reprinted, reproduced, transmitted, or utilized in any form by any electronic, mechanical, or other means, now known or hereafter invented, including photocopying, microfilming, and recording, or in any information storage or retrieval system, without written permission from the publishers.

For permission to photocopy or use material electronically from this work, access www.copyright.com or contact the Copyright Clearance Center, Inc. (CCC), 222 Rosewood Drive, Danvers, MA 01923, 978-750-8400. For works that are not available on CCC please contact mpkbookspermissions@tandf.co.uk

Trademark notice: Product or corporate names may be trademarks or registered trademarks and are used only for identification and explanation without intent to infringe.

ISBN: 9781032061757 (hbk)
ISBN: 9781032061764 (pbk)
ISBN: 9781003201052 (ebk)
DOI: 10.1201/9781003201052

Typeset in Times
by Deanta Global Publishing Services, Chennai, India

Contents

The Editor vii
Contributors ix
Foreword xi

Introduction 1

PART A Physical Properties of Wireless Communication Electromagnetic Fields

Chapter 1 Defining Wireless Communication (WC) Electromagnetic Fields (EMFs):
A. Polarization Is a Principal Property of All Man-made EMFs
B. Modulation, Pulsation, and Variability Are Inherent Parameters of WC EMFs
C. Most Man-made EMF Exposures Are Non-thermal
D. Measuring Incident EMFs Is More Relevant than Specific Absorption Rate (SAR)
E. All Man-made EMFs Emit Continuous Waves, Not Photons
F. Differences from Natural EMFs. Interaction with Matter 17

Dimitris J. Panagopoulos, Andreas Karabarbounis, and Constantinos Lioliousis

PART B Biological and Health Effects of Wireless Communication Electromagnetic Fields

Chapter 2 Public Health Implications of Exposure to Wireless Communication Electromagnetic Fields 79

Anthony B. Miller

Chapter 3 Oxidative Stress Induced by Wireless Communication Electromagnetic Fields 97

Igor Yakymenko and Oleksandr Tsybulin

Chapter 4 Genotoxic Effects of Wireless Communication Electromagnetic Fields 137

Ganesh Chandra Jagetia

Chapter 5 DNA and Chromosome Damage in Human and Animal Cells Induced by Mobile Telephony Electromagnetic Fields and Other Stressors 189

Dimitris J. Panagopoulos

Chapter 6 The Impacts of Wireless Communication Electromagnetic Fields on Human Reproductive Biology 219

Kasey Miller, Kiara Harrison, Jacinta H. Martin, Brett Nixon, and Geoffry N. De Iuliis

Chapter 7 Effects of Wireless Communication Electromagnetic Fields on Human and Animal Brain Activity 275

Haitham S. Mohammed

Chapter 8 Electro-hypersensitivity as a Worldwide, Man-made Electromagnetic Pathology: A Review of the Medical Evidence 297

Dominique Belpomme and Philippe Irigaray

Chapter 9 Carcinogenic Effects of Non-thermal Exposure to Wireless Communication Electromagnetic Fields 369

Igor Yakymenko and Oleksandr Tsybulin

PART C *Effects on Wildlife and Environment*

Chapter 10 Effects of Man-made and Especially Wireless Communication Electromagnetic Fields on Wildlife 393

Alfonso Balmori

PART D *Biophysical and Biochemical Mechanisms of Action*

Chapter 11 Mechanism of Ion Forced-Oscillation and Voltage-Gated Ion Channel Dysfunction by Polarized and Coherent Electromagnetic Fields 449

Dimitris J. Panagopoulos

Chapter 12 Electromagnetic Field-induced Dysfunction of Voltage-Gated Ion Channels, Oxidative Stress, DNA Damage, and Related Pathologies 481

Dimitris J. Panagopoulos, Igor Yakymenko, and George P. Chrousos

Index 509

The Editor

Dr. Dimitris J. Panagopoulos (electromagnetic fields – biophysicist) was born in Athens, Greece, where he lives and works. He has a degree in Physics and a PhD in Biophysics both from the National and Kapodistrian University of Athens (NKUA). He completed his PhD on the Biological Effects of Electromagnetic Fields (EMFs) in 2001, and two post-doctoral studies on the biological effects of microwaves (2004) and on cell death induction by wireless communication (WC) EMFs (2006). He worked as a post-doctoral researcher and lecturer at the Department of Cell Biology and Biophysics, NKUA, (2002–2014), where he gave undergraduate and graduate lectures on radiation and EMF biophysics and performed research on the effects of various types of EMFs in experimental animals. From 2014 to 2018, he worked as a research associate at the National Centre for Scientific Research "Demokritos", Laboratory of Health Physics, Radiobiology, and Cytogenetics, researching effects of ionizing and non-ionizing radiation on human cells. Since 2018, he has been working as a researcher at the Choremeion Research Laboratory, Medical School, NKUA. His experiments were among the first that showed damaging effects of man-made EMFs on DNA and reproduction. He has also shown beneficial effects on reproduction of EMFs that mimic natural ones. His theory on the biophysical mechanism of action of EMFs on cells is considered the most valid amongst all proposed theories and is cited by more than 700 scientific publications. This theory has explained the sensing of upcoming earthquakes by animals and the sensing of upcoming thunderstorms by sensitive individuals through the action of the natural EMFs associated with these phenomena. The same theory has recently explained the induction of oxidative stress in cells by EMF exposure. Dr. Panagopoulos has shown why the specific absorption rate (SAR) is not a proper metric for non-thermal effects; why man-made (totally polarized and coherent) EMFs are damaging, while natural EMFs are vital; and why highly varying real-life exposures from mobile phones and other WC devices are significantly more damaging than simulated exposures with invariable parameters. He has also shown that genetic damage caused by WC EMFs occurs similarly in human and animal cells. Dr. Panagopoulos has also argued that photons are strictly wave-packets, not particles of light, and that man-made electromagnetic radiation does not consist of photons but of continuous "classical" polarized waves, in contrast to what has been postulated by quantum physicists for the past 100 years. He is the first or sole author in more than 40 peer-reviewed highly influential scientific publications, which have been referenced more than 1,800 times by other scientific publications and has been included in the Top 10 cited authors by the *Mutation Research* journals.

Correspondence:
Dr. Dimitris J. Panagopoulos
Email: dpanagop@biophysics.gr

Contributors

Alfonso Balmori
Environmental Department
Castilla and León,
Valladolid, Spain

Dominique Belpomme
[1]Medical Oncology Department
Paris University
Paris, France
[2]European Cancer and Environment Research Institute
Brussels, Belgium

George P. Chrousos
University Research Institute of Maternal and Child Health and Precision Medicine
UNESCO Chair on Adolescent Health Care
National and Kapodistrian University of Athens, Medical School
Aghia Sophia Children's Hospital
Athens, Greece

Geoffry N. De Iuliis
Reproductive Science Group
School of Environmental and Life Sciences
College of Engineering, Science and Environment
University of Newcastle
Callaghan, Australia

Kiara Harrison
Reproductive Science Group
School of Environmental and Life Sciences
College of Engineering, Science and Environment
University of Newcastle
Callaghan, Australia

Philippe Irigaray
European Cancer and Environment Research Institute
Brussels, Belgium

Ganesh Chandra Jagetia
Department of Zoology
Cancer and Radiation Biology Laboratory
Mizoram University
Aizawl, India

Andreas Karabarbounis
Physics Department
Section of Nuclear and Particle Physics
National and Kapodistrian University of Athens
Athens, Greece

Constantinos Lioliousis[†]
Physics Department
Section of Applied Physics
Electronics Laboratory
National and Kapodistrian University of Athens
Athens, Greece

Jacinta H. Martin
Reproductive Science Group
School of Environmental and Life Sciences
College of Engineering, Science and Environment
University of Newcastle
Callaghan, Australia

Anthony B. Miller
Dalla Lana School of Public Health
University of Toronto
Toronto, Canada

Kasey Miller
Reproductive Science Group
School of Environmental and Life Sciences
College of Engineering, Science and Environment
University of Newcastle
Callaghan, Australia

Haitham S. Mohammed
Biophysics Department
Faculty of Science
Cairo University
Giza, Egypt

Brett Nixon
Reproductive Science Group
School of Environmental and Life Sciences
College of Engineering, Science and Environment
University of Newcastle
Callaghan, Australia

Dimitris J. Panagopoulos
[1]National Center for Scientific Research "Demokritos"
Athens, Greece
[2]Choremeion Research Laboratory, Medical School
National and Kapodistrian University of Athens
Athens, Greece
[3]Electromagnetic Field-Biophysics Research Laboratory
Athens, Greece

Oleksandr Tsybulin
European Collegium
Kyiv, Ukraine

Igor Yakymenko
[1]Department of Public Health
Kyiv Medical University
Kyiv, Ukraine
[2]Department of Environmental Safety
National University of Food Technologies
Kyiv, Ukraine

---

† We regret to announce that our teacher, colleague, coworker and friend, Dr. Constantinos Lioliousis, Professor at the Department of Physics of the National and Kapodistrian University of Athens, who distinguished himself in the fields of applied physics, microwave electronics, telecommunications, and biological effects of electromagnetic fields, and contributed to this book as a coauthor in Chapter 1, a brilliant scientist and a man of ideals and integrity, passed away soon after the completion of the chapter, at his 82 years. Chapter 1 of this book, which represents his final contribution, is dedicated to his memory.

Foreword

Information and communication technology is an ever-evolving medium which has penetrated all aspects of life as we know it, accruing unprecedented societal benefits. But with those benefits come risks that need to be managed, and this book presents an exceptional, fact-based foundation for the latter.

Originally developed as a military tool, the wireless aspects of information and communication technology are complex and multi-faceted, with nuances unique to each type of mobile or cordless device, infrastructure platform, and exposure character, with effect metrics driven in large part by the sophisticated interplay of genetics and epigenetics.

To fully understand the depth and breadth of wireless technology and its health implications takes time, intelligence, and longitudinal effort. The growing scientific literature on wireless technology's biological activity alone includes thousands of peer-reviewed papers. The tasks of both learning this field and integrating the many strands of emerging knowledge portend multiple years of commitment.

In a time when opinions, informed and not, are readily available through multiple journalistic, entertainment, and social media platforms, the attainment of factual truth is elusive. But it is only factual truth that can ensure societal decision-making that is both reliable and actionable. Therein lies the value of this coalescence of professional factual thinking put forth here by Dr. Panagopoulos and his colleagues that represent a cross section of the world's top scientists and their work on the biological impacts of wireless communications technology.

This book effectively and efficiently presents the critically important science that leads to informed decisions about health, safety, and the environment. All the critical scientific aspects are considered in a learned fashion on these pages, presented by scientists who do the actual work, and who have done the heavy lifting of sorting and integrating these complexities for practical application.

This book is a necessary factual truth resource for scientists looking to informedly pursue wireless communications subject matter; for responsible employers looking to protect those in their workplaces; for regulators looking to protect health and the environment; for clinicians looking to do the best for their patients; for policy-makers looking to make informed changes to ensure public safety; and for consumers looking to balance the benefits of technology with protections for their children, families, and friends.

Read this book. Absorb its contents. Believe it. And be comfortable acting on this knowledge.

Dr. George L. Carlo
Washington, DC, USA
December 2022

Introduction

Abbreviations: B-field: magnetic field. E-field: electric field. EHS: electro-hypersensitivity. ELF: Extremely Low Frequency. EMF: electromagnetic field. EMR: electromagnetic radiation. LF: Low Frequency. MT: mobile telephony. MW: Microwaves. NR: New Radio. OS: oxidative stress. RF: Radio Frequency. ROS: reactive oxygen species. SAR: Specific Absorption Rate. ULF: Ultra Low Frequency. VGIC: voltage-gated ion channel. VLF: Very Low Frequency. WC: wireless communications. Wi-Fi: Wireless Fidelity. 1G/2G/3G/4G/5G: first/second/third/fourth/fifth generation of MT.

Static electric (E) fields are generated by (macroscopicaly) standing electric charges (actually nothing is "standing" at microscopic level), and static magnetic (B) fields are generated by direct and constant electric currents (directional movement of electric charge with a constant velocity). Only static (invariable in time) E- or B-fields can each exist alone without the coexistence of the other. But again, nothing is absolutely invariable in time, and, thus, totally static, and single E- and B-fields exist only approximately in certain occasions, such as electric fields of "isolated" charged objects or of electric batteries and magnetic fields of certain minerals (magnets).

When electric charges oscillate back and forth (as e.g., in alternating electric currents), both E- and B-fields are generated that also oscillate in phase with the charges. Thus, oscillating electric charges generate oscillating E- and B-fields simultaneously, and the frequency of the generated fields is the same as the frequency of the oscillating charges. The generated E-field oscillates in parallel to the direction of charge oscillation, while the generated B-field oscillates vertically to this direction. Due to this strong interrelation and coexistence between oscillating electric and magnetic fields, we talk about electromagnetic fields (EMFs). Oscillating electric and magnetic fields are not only generated simultaneously by oscillating electric charges but also the one reproduces the other each moment, and the two of them (vertical to each other) can propagate in space in the form of electromagnetic waves or electromagnetic radiation (EMR) vertically to both of them. Thus, electromagnetic waves are E- and B-fields oscillating with the same frequency vertically to each other and vertically to the direction of their propagation. The plane of oscillation of the E-field is called the polarization plane of the wave. The frequency of the emitted EMR is the same as the frequency of its oscillating fields and the oscillating charges that generate them. Thus, oscillating E-/B-fields not only create each other and always coexist, but they also have the unique property to reproduce and propagate themselves in the surrounding space, even in the absence of a material medium, i.e. in the vacuum. All electromagnetic waves propagate with the velocity of light, which is different in each medium. The velocity of light (and of any electromagnetic wave) in the vacuum or in the air (measured by the pioneer physicist H. Hertz to be approximately equal to ($\approx$) 3×10^8 m/s) represents an upper limit for all known velocities of any material or energetic entities (Tesla 1905; Alonso and Finn 1967; Jackson 1975).

In nature, all electric charges oscillate in all possible directions, and the generated EMFs/EMR have similarly random polarizations; in other words, they are not polarized, apart from specific occasions that are locally and partially polarized. Moreover, they do not oscillate with a unique frequency and in phase (coordinately). By contrast, electric charges in electric/electronic circuits oscillate in unique directions (determined by the geometry of the metallic conductors) and coordinately (with a unique frequency and phase), and, thus, the generated technical (man-made) EMFs/EMR are totally polarized and coherent (Panagopoulos et al. 2015a). All anthropogenic EMFs/EMR oscillate at subinfrared frequencies (0–3×10^{11} Hz) (Figure 0.1).

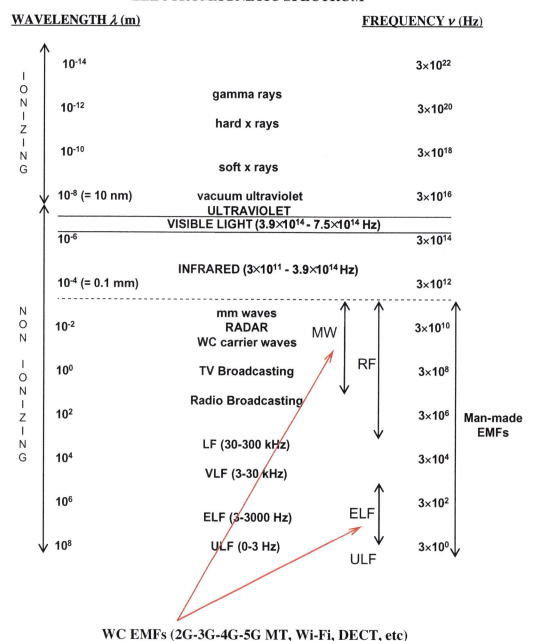

FIGURE 0.1 The electromagnetic spectrum with the ionizing, visible, infrared, and subinfrared parts. Man-made EMFs occupy the subinfrared frequency range (0–3 × 10^{11} Hz), and WC EMFs always combine MW carrier frequencies with ELF modulation and pulsation. Natural EMFs in the subinfrared part of the spectrum are cosmic microwaves, atmospheric EMFs due to lightning discharges (VLF, ELF), Schumann resonances (ELF), spontaneous ionic oscillations in cells (ULF), etc.

During the past five decades and beyond, a great amount of scientific knowledge has been accumulated regarding the biological and health effects of man-made EMFs and corresponding EMR. High-voltage power transmission lines and transformers operating at the Extremely Low Frequency (ELF) (3–3000 Hz) band, specifically at the so-called power frequency (50–60 Hz), radars, and various types of analog transmitters operating at the Radio Frequency (RF) (300 kHz–300 GHz) band and in its highest part called Microwave (MW) band (300 MHz–300 GHz) (Figure 0.1), were the first powerful man-made EMF/EMR sources that attracted the attention and concern of scientists and physicians for their biological/health effects (Persinger 1974; Presman 1977; Marino and Becker 1977; Adey 1981; 1993; Goodman et al. 1995; Puranen and Jokela 1996).

This accumulated knowledge is of particular importance today, as wireless communications (WC) have become an important part of daily life. In WC technologies, the information is conveyed by electtromagnetic waves transmitted by devices and corresponding antennas. Today's digital WC technological products include mobile phones and corresponding mobile telephony (MT) base antennas; cordless domestic phones; wireless Internet connections called Wi-Fi (Wireless Fidelity); wireless connections among electronic devices (called "Bluetooth"), etc. All digital WC devices and corresponding antennas emit MW carrier waves that are necessarily modulated and pulsed by low frequency (mostly ELF) signals in order to carry variable information and provide simultaneous service to many users. The levels of EMF emissions from WC and other technologies have increased exponentially, especially during the past 25–30 years that digital WC are in use and, similarly, the human exposure to these EMF emissions. This tremendous increase of human exposure to EMFs is an unprecedented phenomenon throughout the billions of years of biological evolution. Most importantly, as explained already, all anthropogenic EMFs differ significantly from the natural EMFs in that they are totally polarized and coherent. Therefore, living organisms are not expected to have natural defenses against anthropogenic EMFs.

While the first-generation (1G) mobile phones in the 1980s were analog and of limited use, digital MT technology since the mid-1990s has evolved fast by producing the existing second, third, and fourth generations (2G/3G/4G) of devices/antennas with each next generation transmitting increasing amounts of information/data (voice, text, pictures, video, Internet). Today the massive deployment of the New Radio (NR) 5G (fifth generation) MT/WC system around the world by the telecommunications industry, which is expected to further increase considerably the existing ambient EMF levels, has already started and is rolling out, despite serious concerns expressed by scientists (Miller et al. 2018; 2019; Hardell and Nyberg 2020; Kostoff et al. 2020; Levitt et al. 2021). At the same time, during the past 2 years, humanity was suddenly confronted by a pandemic due to a new virus. As a result, a lot of concern has been raised among scientists and the general population regarding the health and environmental consequences of a vast technological expansion that is taking place uninvited to such an extend.

Natural EMFs/EMR in the terrestrial environment (geoelectric and geomagnetic fields, atmospheric discharges, Schumann resonances, solar light, cosmic microwaves, gamma radiation, etc.) are never totally polarized and maintain relatively constant average intensities. Those that are locally polarized, to a significant degree, are static with constant polarities, such as the geoelectric and geomagnetic fields (with average intensities approximately (~) 130 V/m and ~ 0.05 mT, respectively). Similarly, static and locally polarized are the cell membrane fields (~ 10^7 V/m). During short-term changes of 20%–30% in the average constant intensities of both the environmental and the cell membrane natural EMFs, health problems and biological effects respectively are initiated (see Chapter 1 and Presman 1977; Dubrov 1978; Panagopoulos 2019). This fact suggests that the combination of polarization and variability provides a basis for EMF bioactivity (Panagopoulos 2019).

Now, all man-made EMFs produced by electric/electronic circuits are totally polarized and oscillating, and especially modern digital WC EMFs vary greatly and unpredictably at all times displaying, apart from the ELF pulsing and modulation mentioned already, significant random variability, mainly in the Ultra Low Frequency (ULF) (0–3 Hz) band, with intensity variations usually exceeding by more than 30% (and even by more than 100% in many instances) the average values because of the varying information they transmit and many other factors (see Chapter 1 and Panagopoulos 2019). These significant physical differences between natural and man-made EMFs explain their corresponding differences in the induced biological/health effects.

Natural EMFs are necessary for maintaining the health and wellbeing of all living organisms on Earth. A characteristic example is the atmospheric "Schumann" resonances that attune the brain electrical activity in all animals (Persinger 1974; Wever 1979; Cherry 2002; 2003; Panagopoulos and Chrousos 2019). By contrast, man-made EMFs have been found to produce a great number of adverse biological and health effects. These include changes in key cellular functions; oxidative stress (OS); DNA and protein damage; cell death; infertility; cancer; effects on the immune system; changes in human/animal physiology, such as brain activity; pathological symptoms referred to as electro-hypersensitivity (EHS); etc. (Adey 1981; 1993; Liburdy 1992; Walleczek 1992; Goodman et al. 1995; Santini et al. 2005; Phillips et al. 2009; Hardell and Carlberg 2009; Khurana et al. 2009; De Iuliis et al. 2009; Johansson 2009; Szmigielski 2013; Houston et al. 2016; Yakymenko et al. 2011; 2016; 2018; Mohammed et al. 2013; Balmori 2015; 2021; Gulati et al. 2016; Zothansiama et al. 2017; Miller et al 2018; 2019; Panagopoulos 2019; 2020; Irigaray et al. 2018; Belpomme and Irigaray 2020). From all anthropogenic EMF types, WC EMFs seem to be the most adversely bioactive, mainly because of their increased variability (Panagopoulos 2019).

Under the weight of accumulating scientific evidence, the International Agency for Research on Cancer (IARC), which is part of the World Health Organization (WHO), has categorized both ELF and RF (in fact WC) man-made EMFs as possibly carcinogenic to humans (IARC 2002; 2013). More recent updates on human cancer epidemiology and animal carcinogenicity studies argue that WC EMFs should be categorized as "probably carcinogenic" or "carcinogenic" (Miller et al. 2018; 2019; NTP 2018; Falcioni et al. 2018; Hardell and Nyberg 2020).

Significant scientific evidence shows that the bioactivity of WC EMFs is mainly due to their ELF/ULF components and that RF/microwave carrier signals alone, without modulation, pulsing, and variability, do not usually induce biological effects other than heating at adequately high intensities and frequencies (Bawin et al. 1975; 1978; Blackman et al. 1982; Frei et al. 1988; Walleczek 1992; Bolshakov and Alekseev 1992; Goodman et al. 1995; Penafiel et al. 1997; Creasey and Goldberg 2001; Huber et al. 2002; Betti et al. 2004; Goldsworthy 2006; Höytö et al. 2008; Franzellitti et al. 2010; Campisi et al. 2010; Mohammed et al. 2013; Panagopoulos 2019). As summarized by Goldsworthy (2006), "it is widely accepted that continuous unmodulated radio waves are of too high a frequency to give biological effects but they do become effective when pulsed or amplitude modulated at a low frequency". All endogenous physiological EMFs discovered so far within living organisms, such as the intracellular spontaneous ionic oscillations, the endogenous electric currents that control all cellular and tissue functions, or the electromagnetic signals of brain and heart activities, oscillate at low frequencies (ELF/ULF). And then, like in all forms of matter, molecular oscillations and thermal noise have frequencies in the infrared band. RF EMFs have not been detected in living organisms (Alberts et al. 1994; McGaig and Zhao 1997; Huber et al. 2002; Nuccitelli 2003; Mohammed et al. 2013). It is, thus, absolutely expected for living organisms to be more responsive to external EMFs of similar frequencies. Although this notion for the principal role of ELFs in the bioactivity of WC EMFs has long been available and repeatedly verified, many studies focus exclusively on the RF part of the WC EMFs. A most common problem in published reports on the effects of WC EMFs is that many of them refer to these EMFs simply as "RF" or "microwave", without assessing or even mentioning the inevitable coexistence of ELFs, which are actually the most bioactive (Pakhomov et al. 1998; Betskii and Lebedeva 2004; Belyaev 2005; EPRS 2020; 2021; Karipidis et al. 2021).

Recently, because of the highest microwave carrier frequencies ("mm-waves") of the 5G, certain Russian studies reporting "non-thermal effects of microwave/mm-wave EMFs" came to light. These studies were written in Russian and became known mostly from reviews in English by other Russian scientists. Three such reviews are by Pakhomov et al. (1998), Betskii and Lebedeva (2004), and Belyaev (2005). In several studies reviewed in Pakhomov et al. (1998) and Belyaev (2005), ULF/ELF and Very Low Frequency (VLF) (3–30 kHz) components were present in the form of pulsing and/or modulation/intermittence/variability, while no information on possible existence of such components was provided in the rest of the reviewed studies. Similarly, in the Betskii and Lebedeva (2004) review, information on possible existence of low-frequency components (ULF/ELF/VLF) is absent throughout the paper, but their presence was again not excluded. As it is unlikely that any microwave electronic circuit/generator is not turned on and off, even only for energy-saving reasons, the existence of ULF/ELF/VLF components, and the separate roles of the low and high frequencies in the biological effects, need to be carefully addressed in all experimental studies employing RF/microwave EMF exposures and in the related reviews in order to prevent misleading conclusions. This can be done easily and reliably in experimental studies by performing and reporting electric and magnetic field measurements in the ELF band by ELF field meters and/or spectrum analyzers. Thus, all experimental RF/microwave studies should necessarily include such measurements, and review studies should necessarily report the ELF components in the various exposures.

While one effect induced by high intensity (>0.1 mW/cm^2) and frequency (≥ 1 GHz) microwave EMFs is that of heating exposed materials and living tissues (as happens in microwave ovens with food) ("thermal effects") (Metaxas 1991; Goodman et al. 1995; Creasey and Goldberg 2001), the vast majority of the recorded biological/health effects at lower – environmentally relevant – intensities (from either RF/WC or purely ELF EMFs) are not accompanied by any significant temperature increases and, thus, have been categorized as non-thermal effects (Goodman et al. 1995; Belyaev 2005; Panagopoulos et al. 2013; Yakymenko et al. 2016). Still, the metric for RF EMF bioactivity suggested by health agencies is the Specific Absorption Rate (SAR) (IARC 2013), which, apart from the fact that it is impractical because it cannot be measured directly but has to be calculated (usually by simplistic and inaccurate methods), actually accounts only for thermal effects because the only reliable way to estimate it is by measuring temperature increases (see Chapter 1 and Gandhi et al. 2012; Panagopoulos et al. 2013).

While man-made EMFs cannot directly break chemical bonds and, thus, cause direct ionization of molecules, they are capable of inducing such effects indirectly, by triggering production of free radicals and reactive oxygen species (ROS) in the cells (De Iuliis et al. 2009; Burlaka et al. 2013; Pall 2013; Houston et al. 2016; Yakymenko et al. 2016; 2018; Zothansiama et al. 2017; Panagopoulos et al. 2021). Such species can damage any critical biomolecules, including DNA. The (over)production of ROS in cells and the consequent OS that arises can be triggered by irregular gating of voltage-gated ion channels (VGICs) in the cell membranes due to purely ELF/VLF man-made EMFs or the ULF/ELF/VLF components of the complex WC EMFs (Creasey and Goldberg 2001; Panagopoulos et al. 2002; 2015a; 2021). Today, irregular gating of VGICs in cell membranes by man-made EMFs has been verified by many experimental studies (e.g., Liburdy 1992; Piacentini et al. 2008; Cecchetto et al. 2020; Zheng et al. 2021) and presented by reviews (Pall 2013; Bertagna et al. 2021). Thus, we are dealing with mechanisms that result in chemical changes of critical biomolecules without heating the exposed biological tissues.

Although the majority of peer-reviewed published studies (more than 60%–70%) indicate effects of purely ELF man-made EMFs for field intensities down to less than a few V/m or a few µT, or of pulsed/modulated RF/WC EMFs for RF intensities down to less than 1 µW/cm^2 even for short-term exposures (Goodman et al. 1995; Santini et al. 2005; Phillips et al. 2009; Panagopoulos et al. 2010; Szmigielski 2013; Burlaka et al. 2013; Manna and Ghosh 2016; Yakymenko et al. 2011; 2016; Leach et al. 2018; Panagopoulos 2019), health authorities responsible for setting exposure guidelines in most countries have adopted limits that are thousands (and even millions, in some cases) of times higher, as set by a private, non-governmental organization (NGO) called the International Commission on Non-Ionizing Radiation Protection (ICNIRP 1998; 2010; 2020; Hardell and Carlberg 2021). These limits may provide limited protection against thermal RF effects, but certainly not against the non-thermal effects of the purely ELF man-made EMFs or the ELF components of the complex RF/WC EMFs, which actually constitute the vast majority of the reported biological and health effects.

Indicative threshold EMF/EMR intensity levels found to induce significant (non-thermal) biological/health effects and the corresponding ICNIRP (2010; 2020) limits for public exposure in the ELF and RF bands are shown in Table 0.1. It is evident that serious biological and health effects

TABLE 0.1
Threshold EMF/EMR Intensities for Indicative Biological/Health Effects and Corresponding ICNIRP Limits

Incident EMF	ICNIRP Intensity Limit (6 min Average, Local Exposure)	Threshold Intensity for Effect Initiation	Exposure Duration	Effect	References
ELF-E (CW or pulsed) (1–50 Hz)	5000 V/m	0.002 V/m	12 h	Decrease in protein synthesis rate	McLeod et al. (1987)
		0.0021 V/m	4 days	Increase in DNA synthesis rate	Cleary et al. (1988)
		10 V/m	Years	Cancer (humans)	Coghill et al. (1996)
ELF-B (50 Hz CW)	2 G (200 µT)	0.002 G (0.2 µT)	Years	Cancer (humans)	Feychting and Ahlbom (1994)
Pulsed RF (GSM) 1800 MHz	3655.6 µW/cm²	<1 µW/cm²	6 min/day, 6 days	DNA damage, cell death (fruit fly ovarian cells)	Panagopoulos et al. (2010)
Pulsed RF (GSM) 900 MHz	2014.0 µW/cm²	0.25 µW/cm²	158–360 h intermittently	OS, DNA damage, embryonic death (bird embryos)	Burlaka et al. (2013)
Pulsed RF (GSM) 1800 MHz	3655.6 µW/cm²	0.32 µW/cm²	19 days (48 s On/12 s Off)	OS, DNA damage, embryonic death (bird embryos)	Yakymenko et al. (2018)

[ICNIRP (2020) limits calculated for 900 and 1800 MHz according to formula: $0.058 f_M^{0.86}$. CW: continuous-wave. G: Gauss].

such as OS, DNA damage, cancer, etc. may occur from exposure to ELF EMFs or WC EMFs at levels thousands of times lower than the corresponding ICNIRP limits, while more subtle cellular effects may be initiated at ELF thresholds more than a million times lower than the corresponding ICNIRP limits (McLeod et al. 1987; Cleary et al. 1988; Feychting and Ahlbom 1994; Coghill et al. 1996; Panagopoulos et al. 2010; Burlaka et al. 2013; Yakymenko et al. 2018). Hence, these limits do not provide any health protection.

Because of these facts, and regardless of our remaining knowledge gaps, health complaints are increasing, especially among people residing close to antennas or high-voltage power lines, accompanied by increasing cancer rates and symptoms of unwellness (Kundi and Hutter 2009; Gulati et al. 2016; Zothansiama et al. 2017; Miller et al. 2018; 2019; Belpomme and Irigaray 2020; Lopez et al. 2021).

The situation may seem confusing with several other studies reporting no effects of ELF or pulsing/modulated RF/WC EMFs, especially studies that have employed simulated WC EMF exposures from generators, "exposure chambers", or "test" mobile phones with invariable parameters (carrier frequency, intensity, pulsations) and no modulation (ICNIRP 1998; 2020; IARC 2002; 2013; EPRS 2020; 2021; Karipidis et al. 2021). Indeed, about 50% of the studies that employ simulated EMFs do not find effects. In contrast, among studies that employ real-life exposures from commercially available devices with high variability (such as mobile or cordless phones, Wi-Fi, etc.), more than 95% find effects (Panagopoulos et al. 2015b; Yakymenko et al. 2016; Leach et al. 2018; Panagopoulos 2019; Kostoff et al. 2020).

Bioelectromagnetics is a complex scientific field, featuring an equal combination of physics and biology. This is why collaboration among experts from different areas (e.g., physicists with biologists, or medical doctors with engineers) is necessary. EMF bioeffect experiments must necessarily be carried out by scientists/teams that combine adequate knowledge in both the physical and the biological parts; otherwise, the methodology may be flawed and the conclusions misleading. The use of any devices such as generators and exposure chambers provided by companies for exposure of biological samples to simulated EMFs without knowing and measuring the physical details of the generated EMFs is a major flaw in experimental studies.

Unfortunately, conflict of interest, corruption, results depending on funding, and misleading information in scientific papers have become usual phenomena in the field (Hardell and Carlberg 2021; Leach et al. 2018). Conflict of interest is not necessarily limited to economical/professional benefits but may also include other types of personal rewards (Panagopoulos and Karabarbounis 2020; Panagopoulos 2021). It is not unusual for important findings such as those reported above to be concealed or neglected in many publications, while their consideration is necessary for further developments.

At the same time, the massive deployment of the 5G MT/WC system in order to achieve ever increasing data transmission rates and the so-called Internet of Things (IoT) is well underway despite serious concerns expressed by many expert scientists who have asked for a moratorium in 5G deployment (Hardell and Nyberg 2020), as implied by the Precautionary Principle (Harremoes 2013; Read and O'Riordan 2017; Frank 2021). Indeed, the deployment of 5G will require a huge increase in the number of base antennas, combined with potential increases in transmission power/intensity, and thousands of satellites in the atmosphere to complement the base antennas. Moreover, the increased amount of variable data transmitted by this new WC EMR type make it even more variable in intensity, waveform, frequency, etc., with inclusion of ever more variable ELF pulsations than previous types of MT/WC EMFs (Rappaport et al. 2013; Dahlman et al. 2018). Thus, 5G is expected to significantly increase public exposure and consequent health problems (Panagopoulos 2019; Hardell and Nyberg 2020; Kostoff et al. 2020; Levitt et al. 2021).

Strangely, in 2020, the ICNIRP increased the general public exposure limit for WC EMFs (2–6 GHz) averaged over 6 minutes (min) from 1000 to 4000 $\mu W/cm^2$ (from 1 to 4 mW/cm^2) instead of decreasing it (ICNIRP 1998; 2020). Also strange were the technical reports and papers referring to the characteristics of 5G that do not provide any information on the ULF/ELF/VLF components of this new WC EMF type, as if their authors are not aware of their existence (EPRS 2020; 2021;

Karipidis et al. 2021). As already mentioned, carrying out studies involving WC EMF exposures without searching the low-frequency components and attributing any observed effects to the RF/MW carrier can be very misleading. Similarly, reviewing and evaluating other studies by looking only at the RF/MW part of their EMF exposures and ignoring the low-frequency part or not examining whether the exposures are from real-life WC devices/antennas or simulated signals with fixed parameters and, thus, significantly less bioactive, as in EPRS (2020; 2021) (EPRS: European Parliamentary Research Service) and Karipidis et al. (2021), is a flawed methodology. Thus, not only are WC EMFs dangerous to life, but the evaluation of their risks by certain reviews and organizations is flawed as well. In view of the fact that the ULF/ELF/VLF EMFs are actually the most bioactive, the low frequency (ULF/ELF/VLF) pulsations of the most recent generations of WC signals such as the 4G and 5G should be in the forefront of bioelectromagnetic research in order to allow the correct evaluation of their risks.

Because of the described confusion and misinformation, many people, especially among the general public, make careless use of WC devices, utilizing cordless domestic phones and Wi-Fi at homes and workplaces for convenience instead of using wired connections and attaching the mobile phones on their heads/bodies, subjecting themselves day and night to simultaneous telephone and Wi-Fi/Bluetooth EMF emissions from their "smart" devices. Unfortunately, they also give such devices in the hands of young children or even expose their embryos during pregnancy. So far, the authorities do nothing to educate them or protect them.

When, in many cases, people realize they have become hypersensitive to man-made EMFs, their efforts to restrict unbearable symptoms, especially from WC antennas, usually lead to risky solutions, such as metal shielding in their houses and even in their clothes. Any EMF-shielding attenuates not only the detrimental anthropogenic EMFs but also the natural and absolutely vital Schumann electromagnetic resonances, which actually attune our brain activity (Persinger 1974; 2014; Wever 1979; Cherry 2002; 2003; Panagopoulos and Chrousos 2019). Therefore, such solutions should be considered only when other ways of protection are not possible, and should be applied cautiously (after careful EMF measurements) and partially (e.g., only on certain wall(s) of a house), possibly in combination with earthing, and/or scientifically tested "Schumann generators" emitting very weak signals that mimic as closely as possible the Schumann oscillations (Panagopoulos and Chrousos 2019).

It seems that humanity and science are coming to realize that the price for the comfort provided by technology and the convenience in sharing information may be compromised health, wellbeing, and natural environment, when technology is not carefully designed to respect these values and health authorities do not set safe limits.

Another particularly worrying phenomenon is research on nanobiotechnology – magnetogenetics carried out during the past 10 years. Such research is crossing sensitive boundaries of bioethics by trying to control the cellular processes via magnetic nanoparticles injected in cells and manipulated by external electromagnetic signals (Monzel et al. 2017; He et al. 2021). Such methods can not only have unpredictable, adverse effects on the cell/organism, granted that such nanoparticles are unnatural and foreign to the cells, and are, thus, dangerous for many reasons and probably toxic, but can also violate the privacy and freedom of a treated individual, who could then be monitored and manipulated remotely by electromagnetic signals. It is questionable how such research is considered acceptable in the scientific community and compatible with bioethical principles.

This book on the Biological and Health Effects of WC EMFs includes contributions from top international experts on the various areas of this important subject and is published to increase scientific knowledge, awareness, and debate that would benefit science, public health, and the environment. Expert scientists were invited to submit specific chapters on the physics, biology, pathology, epidemiology, and plausible biophysical and biochemical mechanisms of action of WC EMFs. The invitations were specific. The contributors were invited to write on topics related to their previous publications and expertise so that the book covers a minimum number of the most important topics.

Introduction

Because both RF and ELF EMFs are contained in all WC EMFs, studies on both frequency bands are examined in the chapters. Thus, the book describes effects from most types of man-made EMF exposures. In all chapters, the terms "EMF" and "EMR" (for example WC EMFs or WC EMR) are used interchangeably with equivalent meaning, as EMR is produced by temporally varying EMFs, and man-made (polarized and coherent) EMR carries net EMFs as well (Panagopoulos et al. 2015a). Moreover, because all types of WC EMFs commonly combine MW carrier waves with ELF modulation, pulsation, and ULF random variability, their biological/health effects are very similar, and, thus, they are treated similarly in the chapters, emphasizing though that newer generations of WC/MT EMFs (3G/4G/5G) are increasingly more variable and, thus, increasingly more bioactive. The terms "cell phone" and "cell tower" used occasionally in the book refer to digital mobile phones and base station antennas respectively. Thus, they have the same meaning as "mobile phones" and "MT base station antennas" (used in most cases) since all existing MT/WC systems (2G, 3G, 4G, 5G) today are digital. Digital MT uses the so called "cellular system" according to which the space is divided into areas called "cells" with one base station in each "cell".

The chapters present cutting-edge knowledge on the effects of man-made EMFs on living systems and their mechanisms. It is evident that serious effects induced by man-made and especially WC EMFs, such as genetic damage, are well documented as resulting from OS. This explains other reported pathological conditions, such as infertility or cancer. It is also evident that a most plausible biophysical/biochemical mechanism for OS induction in the cells is the dysfunction of the VGICs in the cell membranes and that the low frequency (ELF/ULF/VLF) components (pulsation, modulation, etc.) of the WC EMFs play a major role. The chapters emphasize the need for setting much tighter exposure limits and recommend prudent avoidance of exposure to man-made and especially WC EMFs, a moratorium in 5G roll-out, and urgent application of the Precautionary Principle (Harremoes 2013; Read and O'Riordan 2017; Frank 2021). Moreover, the chapters underline the need for improvement and standardization of the experimental procedures, use of real-life EMFs, and better definition of the EMF exposures by measuring all their parameters, especially the low-frequency ones.

I thank all the distinguished scientists in this book for kindly accepting my invitation, for their high-quality contributions, and their collaboration during the editing process. Inviting the chapters, editing, and shaping this book was, for me, a unique experience and a great source of combined knowledge. I also thank Dr. G.L. Carlo for writing the Forward of the book, and Dr. G. Singh from CRC for his invitation and repeated reminders to submit a book proposal.

This book will have served its purpose if it contributes toward setting scientific research in this field on a better base, leaving behind conflicts of interest and misinformation; when evidence-based discussions on the consequences of WC EMFs and possible correlations between an EMF-polluted environment and viral and other diseases are unbiased and welcomed by scientists, concerned individuals, health authorities and governments; when suggestions on individual and public health protection, and the setting of biologically and epidemiologically based exposure limits are also welcomed.

Finally, this book will have served its purpose if it contributes toward a *"real and honest science"* as Dr. Neil J. Cherry (1946–2003) would say. A science that is applicable to life and works for the benefit of humanity, not for its destruction or enslavement. A science that increases awareness on the safety of our natural environment and our planet Earth, which is in great danger because of the uncontrolled expansion of human technology and the unrestricted use of the natural resources.

We are gifted to live on this beautiful planet. We should love and respect it and live in harmony with it without destroying it. We should not disturb its natural balance by destroying the forests, changing the weather, genetically modifying our food and the natural organisms, filling the sky with thousands of satellites, and polluting the atmosphere with chemicals and artificial polarized EMFs/EMR. Instead of trying to inhabit other planets unfriendly to life, we should rather protect our home Earth, which is unique in the known universe. The balance of our planet is very fragile, and so is our existence. We all share the same home and the same future. It is our duty to protect it.

Dr. Dimitris J. Panagopoulos (Editor)

REFERENCES

Adey WR, (1981): Tissue interactions with non-ionizing electromagnetic fields. *Physiological Reviews*, 61(2), 435–514.
Adey WR, (1993): Biological effects of electromagnetic fields. *Journal of Cellular Biochemistry*, 51(4), 410–416.
Alberts B, Bray D, Lewis J, Raff M, Roberts K, Watson JD, (1994): *Molecular Biology of the Cell*. Garland Publishing, Inc., New York.
Alonso M, Finn EJ, (1967): *Fundamental University Physics* (Vol. 2). Fields and Waves, Addison-Wesley.
Balmori A, (2015): Anthropogenic radiofrequency electromagnetic fields as an emerging threat to wildlife orientation. *Science of the Total Environment*, 518–519, 58–60.
Balmori A, (2021): Electromagnetic radiation as an emerging driver factor for the decline of insects. *Science of the Total Environment*, 767, 144913.
Bawin SM, Kaczmarek LK, Adey WR, (1975): Effects of modulated VHF fields, on the central nervous system. *Annals of the New York Academy of Sciences*, 247, 74–81.
Bawin SM, Adey WR, Sabbot IM, (1978): Ionic factors in release of $^{45}Ca^{2+}$ from chick cerebral tissue by electromagnetic fields. *Proceedings of the National Academy of Sciences of the United States of America*, 75(12), 6314–6318.
Belpomme D, Irigaray P, (2020): Electrohypersensitivity as a newly identified and characterized neurologic pathological disorder: How to diagnose, treat, and prevent it. *International Journal of Molecular Sciences*, 21(6), 1915. https://doi.org/10.3390/ijms21061915.
Belyaev I, (2005): Non-thermal biological effects of microwaves. *Microwave Review*, 11(2), 13–29.
Bertagna F, Lewis R, Silva SRP, McFadden J, Jeevaratnam K, (2021): Effects of electromagnetic fields on neuronal ion channels: A systematic review. *Annals of the New York Academy of Sciences* May 4. https://doi.org/10.1111/nyas.14597.
Betskii OV, Lebedeva NN, (2004): Low-intensity millimeter waves in biology and medicine. In *Clinical Application of Bioelectromagnetic Medicine* (Vol. 2004). Marcel Decker, New York, 30–61.
Betti L, Trebbi G, Lazzarato L, Brizzi M, Calzoni GL, et al, (2004): Nonthermal microwave radiations affect the hypersensitive response of tobacco to tobacco mosaic virus. *Journal of Alternative and Complementary Medicine*, 10(6), 947–957.
Blackman CF, Benane SG, Kinney LS, Joines WT, House DE, (1982): Effects of ELF fields on calcium-ion efflux from brain tissue in vitro. *Radiation Research*, 92(3), 510–520.
Bolshakov MA, Alekseev SI, (1992): Bursting responses of Lymnea neurons to microwave radiation. *Bioelectromagnetics*, 13(2), 119–129.
Burlaka A, Tsybulin O, Sidorik E, Lukin S, Polishuk V, et al, (2013): Overproduction of free radical species in embryonic cells exposed to low intensity radiofrequency radiation. *Experimental Oncology*, 35(3), 219–225.
Campisi A, Gulino M, Acquaviva R, Bellia P, Raciti G, et al, (2010): Reactive oxygen species levels and DNA fragmentation on astrocytes in primary culture after acute exposure to low intensity microwave electromagnetic field. *Neuroscience Letters*, 473(1), 52–55.
Cecchetto C, Maschietto M, Boccaccio P, Vassanelli S, (2020): Electromagnetic field affects the voltage-dependent potassium channel Kv1.3. *Electromagnetic Biology and Medicine*, 39(4), 316–322.
Cherry NJ, (2002): *Schumann Resonances, a Plausible Biophysical Mechanism for the Human Health Effects of Solar/Geomagnetic Activity.* https://researcharchive.lincoln.ac.nz/bitstream/handle/10182/3935/90_n1_EMR_Schumann_Resonance_paper_1.pdf?sequence=1.
Cherry NJ, (2003): Human intelligence: The brain, an electromagnetic system synchronised by the Schumann Resonance signal. *Medical Hypotheses*, 60(6), 843–844.
Cleary SF, Liu LM, Graham R, Diegelmann RF, (1988): Modulation of tendon fibroplasia by exogenous electric currents. *Bioelectromagnetics*, 9(2), 183–194.
Coghill RW, Steward J, Philips A, (1996): Extra low frequency electric and magnetic fields in the bed place of children diagnosed with leukaemia: A case-control study. *European Journal of Cancer Prevention*, 5(3), 153–158.
Creasey WA, Goldberg RB, (2001): A new twist on an old mechanism for EMF bioeffects? *EMF Health Report*, 9(2), 1–11. https://mdsafetech.files.wordpress.com/2021/07/creasey-wa-goldberg-rb-2001-a-new-twist-on-an-old-mechanism-for-emf-bioeffects.pdf.
Dahlman E, Parkvall S, Skoeld J, (2018): *5G NR: The Next Generation Wireless Access Technology*. Academic Press, Elsevier, London.

De Iuliis GN, Newey RJ, King BV, Aitken RJ, (2009): Mobile phone radiation induces reactive oxygen species production and DNA damage in human spermatozoa in vitro. *PLOS ONE*, 4(7), e6446.

Dubrov AP, (1978): *The Geomagnetic Field and Life*. Plenum Press, New York.

EPRS, (2020): *Effects of 5G Wireless Communication on Human Health, European Parliamentary Research Service*. Scientific Foresight Unit (STOA), PE 646.172, March 2020.

EPRS, (2021): *Environmental Impacts of 5G. A Literature Review of Effects of Radio-Frequency Electromagnetic Field Exposure of Non-human Vertebrates, Invertebrates and Plants*, Scientific Foresight Unit (STOA), PE 690.021, June 2021.

Falcioni L, Bua L, Tibaldi E, Lauriola M, De Angelis L, et al, (2018): Report of final results regarding brain and heart tumors in Sprague-Dawley rats exposed from prenatal life until natural death to mobile phone radiofrequency field representative of a 1.8 GHz GSM base station environmental emission. *Environmental Research*, 165, 496–503.

Feychting M, Ahlbom A, (1994): Magnetic fields, leukemia and central nervous system tumors in Swedish adults residing near high - voltage power lines. *Epidemiology*, 5(5), 501–509.

Frank JW, (2021): Electromagnetic fields, 5G and health: What about the precautionary principle? *Journal of Epidemiology and Community Health*, 75(6), 562–566.

Franzellitti S, Valbonesi P, Ciancaglini N, Biondi C, Contin A, et al, (2010): Transient DNA damage induced by high-frequency electromagnetic fields (GSM 1.8 GHz) in the human trophoblast HTR-8/SVneo cell line evaluated with the alkaline comet assay. *Mutation Research*, 683(1–2), 35–42.

Frei M, Jauchem J, Heinmets F, (1988): Physiological effects of 2.8 GHz radio-frequency radiation: A comparison of pulsed and continuous-wave radiation. *Journal of Microwave Power and Electromagnetic Energy*, 23(2), 2.

Gandhi Om P, Morgan LL, De Salles AA, Han Y-Y, Herberman RB, Davis DL, (2012): Exposure limits: The underestimation of absorbed cell phone radiation, especially in children. *Electromagnetic Biology and Medicine*, 31(1), 34–51.

Goldsworthy A, (2006): Effects of electrical and electromagnetic fields on plants and related topics. In AG Volkov (Ed.), *Plant Electrophysiology–Theory & Methods*. Springer-Verlag, Berlin Heidelberg, 247–267.

Goodman EM, Greenebaum B, Marron MT, (1995): Effects of electro-magnetic fields on molecules and cells. *International Review of Cytology*, 158, 279–338.

Gulati S, Yadav A, Kumar N, Kanupriya, Aggarwal NK, et al, (2016): Effect of GSTM1 and GSTT1 polymorphisms on genetic damage in humans populations exposed to radiation From mobile towers. *Archives of Environmental Contamination and Toxicology*, 70(3), 615–625.

Hardell L, Carlberg M, (2009): Mobile phones, cordless phones and the risk for brain tumours. *International Journal of Oncology*, 35(1), 5–17.

Hardell L, Nyberg R, (2020): Appeals that matter or not on a moratorium on the deployment of the fifth generation, 5G, for microwave radiation. *Molecular and Clinical Oncology*. https://doi.org/10.3892/mco.2020.1984.

Hardell L, Carlberg M, (2021): Lost opportunities for cancer prevention: Historical evidence on early warnings with emphasis on radiofrequency radiation. *Reviews on Environmental Health*. https://doi.org/10.1515/reveh-2020-0168.

Harremoes P, Gee D, MacGarvin M, Stirling A, Keys J, et al. (Eds.), (2013): *The Precautionary Principle in the 20th Century: Late Lessons from Early Warnings*. Routledge, London.

He Y, Yi C, Zhang X, Zhao W, Yu D, (2021): Magnetic graphene oxide: Synthesis approaches, physicochemical characteristics, and biomedical applications. *TrAC Trends in Analytical Chemistry*, 136, 116191.

Houston BJ, Nixon B, King BV, De Iuliis GN, Aitken RJ, (2016): The effects of radiofrequency electromagnetic radiation on sperm function. *Reproduction*, 152(6), R263–R276.

Höytö A, Luukkonen J, Juutilainen J, Naarala J, (2008): Proliferation, oxidative stress and cell death in cells exposed to 872 MHz radiofrequency radiation and oxidants. *Radiation Research*, 170(2), 235–243.

Huber R, Treyer V, Borbely AA, Schuderer J, Gottselig JM, et al, (2002): Electromagnetic fields, such as those from mobile phones, alter regional cerebral blood flow and sleep and waking EEG. *Journal of Sleep Research*, 11(4), 289–295.

IARC, (2002): *Non-ionizing Radiation, Part 1: Static and Extremely Low-frequency (ELF) Electric and Magnetic Fields* (Vol. 80). International Agency for Research on Cancer, Lyon, France.

IARC, (2013): *Non-ionizing Radiation, Part 2: Radiofrequency Electromagnetic Fields* (Vol. 102). International Agency for Research on Cancer, Lyon, France.

ICNIRP, (1998): Guidelines for limiting exposure to time-varying electric, magnetic and electromagnetic fields (up to 300 GHz). *Health Physics*, 74, 494–522.

ICNIRP, (2010): Guidelines for limiting exposure to time-varying electric and magnetic fields (1 Hz to 100 kHz). *Health Physics*, 99(6), 818–836.

ICNIRP, (2020): Guidelines for limiting exposure to electromagnetic fields (100 kHz to 300 GHz). *Health Physics*, 118(5), 483–524.

Irigaray P, Caccamo D, Belpomme D, (2018): Oxidative stress in electrohypersensitivity self-reporting patients: Results of a prospective *in vivo* investigation with comprehensive molecular analysis. *International Journal of Molecular Medicine*, 42(4), 1885–1898.

Jackson JD, (1975): *Classical Electrodynamics*. John Wiley & Sons, Inc., New York.

Johansson O, (2009): Disturbance of the immune system by electromagnetic fields-A potentially underlying cause for cellular damage and tissue repair reduction which could lead to disease and impairment. *Pathophysiology*, 16(2–3), 157–177.

Karipidis K, Mate R, Urban D, Tinker R, Wood A, (2021): 5G mobile networks and health—A state-of-the-science review of the research into low-level RF fields above 6 GHz. *Journal of Exposure Science and Environmental Epidemiology*. https://doi.org/10.1038/s41370-021-00307-7.

Khurana VG, Teo C, Kundi M, Hardell L, Carlberg M, (2009): Cell phones and brain tumors: A review including the long-term epidemiologic data. *Surgical Neurology*, 72(3), 205–214.

Kostoff RN, Heroux P, Aschner M, Tsatsakis A, (2020): Adverse health effects of 5G mobile networking technology under real-life conditions. *Toxicology Letters*, 323, 35–40.

Kundi M, Hutter HP, (2009): Mobile phone base stations-effects on wellbeing and health. *Pathophysiology*, 16(2–3), 123–135.

Leach V, Weller S, Redmayne M, (2018): A novel database of bio-effects from non-ionizing radiation. *Reviews on Environmental Health*, 33(3), 1–8.

Levitt BB, Lai HC, Manville AM, (2021): Effects of non-ionizing electromagnetic fields on flora and fauna, part 1. Rising ambient EMF levels in the environment. *Reviews on Environmental Health*. https://doi.org/10.1515/reveh-2021-0026.

Liburdy RP, (1992): Calcium signalling in lymphocytes and ELF fields: Evidence for an electric field metric and a site of interaction involving the calcium ion channel. *FEBS Letters*, 301(1), 53–59.

López I, Félix N, Rivera M, Alonso A, Maestú C, (2021): What is the radiation before 5G? A correlation study between measurements in situ and in real time and epidemiological indicators in Vallecas, Madrid. *Environmental Research*, 194, 110734.

Manna D, Ghosh R, (2016): Effect of radiofrequency radiation in cultured mammalian cells: A review. *Electromagnetic Biology and Medicine*, 35(3), 265–301.

Marino AA, Becker RO, (1977): Biological effects of extremely low frequency electric and magnetic fields: A review. *Physiological Chemistry and Physics*, 9(2), 131–147.

McCaig CD, Zhao M, (1997): Physiological electric fields modify cell behaviour. *BioEssays*, 19(9), 819–826.

McLeod KJ, Lee RC, Ehrlich HP, (1987): Frequency dependence of electric field modulation of fibroblast protein synthesis. *Science*, 236(4807), 1465–1469.

Metaxas AC, (1991): Microwave heating. *Power Engineering Journal*, 5(5), 237–247.

Miller AB, Morgan LL, Udasin I, Davis DL, (2018): Cancer epidemiology update, following the 2011 IARC evaluation of radiofrequency electromagnetic fields (Monograph 102). *Environmental Research*, 167, 673–683.

Miller AB, Sears ME, Morgan LL, Davis DL, Hardell L, et al, (2019): Risks to health and well-being from radio-frequency radiation emitted by cell phones and other wireless devices. *Frontiers in Public Health*, 7, 223. https://doi.org/10.3389/fpubh.2019.00223.

Mohammed HS, Fahmy HM, Radwan NM, Elsayed AA, (2013): Non-thermal continuous and modulated electromagnetic radiation fields effects on sleep EEG of rats. *Journal of Advanced Research*, 4(2), 181–187.

Monzel C, Vicario C, Piehler J, Coppey M, Dahan M, (2017): Magnetic control of cellular processes using biofunctional nanoparticles. *Chemical Science*, 8(11), 7330.

NTP (National Toxicology Program), (2018): Toxicology and Carcinogenesis studies in Hsd: Sprague Dawley SD rats exposed to whole-body radio frequency radiation at a frequency (900 MHz) and modulations (GSM and CDMA) used by cell phones. *NTP TR* 595, Department of Health and Human Services, USA.

Nuccitelli R, (2003): Endogenous electric fields in embryos during development, regeneration and wound healing. *Radiation Protection Dosimetry*, 106(4), 375–383.

Pakhomov AG, Akyel Y, Pakhomova ON, Stuck BE, Murphy MR, (1998): Current state and implications of research on biological effects of millimeter waves: A review of the literature. *Bioelectromagnetics*, 19(7), 393–413.

Pall ML, (2013): Electromagnetic fields act via activation of voltage-gated calcium channels to produce beneficial or adverse effects. *Journal of Cellular and Molecular Medicine*, 17(8), 958–965.

Panagopoulos DJ, Karabarbounis A, Margaritis LH, (2002): Mechanism for action of electromagnetic fields on cells. *Biochemical and Biophysical Research Communications*, 298(1), 95–102.

Panagopoulos DJ, Chavdoula ED, Margaritis LH, (2010): Bioeffects of mobile telephony radiation in relation to its intensity or distance from the antenna. *International Journal of Radiation Biology*, 86(5), 345–357.

Panagopoulos DJ, Johansson O, Carlo GL, (2013): Evaluation of specific absorption rate as a dosimetric quantity for electromagnetic fields bioeffects. *PLOS ONE*, 8(6), e62663. https://doi.org/10.1371/journal.pone.0062663.

Panagopoulos DJ, Johansson O, Carlo GL, (2015a): Polarization: A key difference between man-made and natural electromagnetic fields, in regard to biological activity. *Scientific Reports*, 5, 14914. https://doi.org/10.1038/srep14914.

Panagopoulos DJ, Johansson O, Carlo GL, (2015b): Real versus simulated mobile phone exposures in experimental studies. *BioMed Research International*, 2015, 607053.

Panagopoulos DJ, Chrousos GP, (2019): Shielding methods and products against man-made electromagnetic fields: Protection versus risk. *Science of the Total Environment*, 667C, 255–262.

Panagopoulos DJ, (2019): Comparing DNA damage induced by mobile telephony and other types of man-made electromagnetic fields. *Mutation Research Reviews*, 781, 53–62.

Panagopoulos DJ, (2020): Comparing chromosome damage induced by mobile telephony radiation and a high caffeine dose: Effect of combination and exposure duration. *General Physiology and Biophysics*, 39(6), 531–544.

Panagopoulos DJ, Karabarbounis A, (2020): Comments on "diverse radiofrequency sensitivity and radiofrequency effects of mobile or cordless phone near fields exposure in Drosophila melanogaster". *Advances in Environmental Studies*, 4(1), 271–276.

Panagopoulos DJ, (2021): Comments on Pall's "Millimeter (MM) wave and microwave frequency radiation produce deeply penetrating effects: the biology and the physics". *Reviews on Environmental Health* 37(2), 295–297.

Panagopoulos DJ, Karabarbounis A, Yakymenko I, Chrousos GP, (2021): Mechanism of DNA damage induced by human-made electromagnetic fields. *International Journal of Oncology*, 59: 92.

Penafiel LM, Litovitz T, Krause D, Desta A, Mullins JM, (1997): Role of modulation on the effects of microwaves on ornithine decarboxylase activity in L929 cells. *Bioelectromagnetics*, 18(2), 132–141.

Persinger MA, (1974): *ELF and VLF Electromagnetic Fields*. Plenum Press, New York.

Persinger MA, (2014): Schumann Resonance frequencies found within quantitative electroencephalographic activity: Implications for earth-brain interactions. *International Letters of Chemistry, Physics and Astronomy*, 11(1), 24–32.

Phillips JL, Singh NP, Lai H, (2009): Electromagnetic fields and DNA damage. *Pathophysiology*, 16(2–3), 79–88.

Piacentini R, Ripoli C, Mezzogori D, Azzena GB, Grassi C, (2008): Extremely low-frequency electromagnetic fields promote in vitro neurogenesis via upregulation of Ca_v1-channel activity. *Journal of Cellular Physiology*, 215(1), 129–139.

Presman AS, (1977): *Electromagnetic Fields and Life*. Plenum Press, New York.

Puranen L, Jokela K, (1996): Radiation hazard assessment of pulsed microwave radars. *Journal of Microwave Power and Electromagnetic Energy*, 31(3), 165–177.

Rappaport TS, Sun S, Mayzus R, Zhao H, Azar Y, et al, (2013): Millimeter wave mobile communications for 5G cellular: It will work! *IEEE Access*, 1, 335–349. https://doi.org/10.1109/ACCESS.2013.2260813.

Read R, O'Riordan T, (2017): The precautionary principle under fire. *Environment: Science and Policy for Sustainable Development*, 59(5), 4–15.

Santini MT, Ferrante A, Rainaldi G, Indovina P, Indovina PL, (2005): Extremely low frequency (ELF) magnetic fields and apoptosis: A review. *International Journal of Radiation Biology*, 81(1), 1–11.

Szmigielski S, (2013): Reaction of the immune system to low-level RF/MW exposures. *Science of the Total Environment*, 454–455, 393–400.

Tesla N, (1905): The transmission of electrical energy without wires as a means of furthering world peace. *Electrical World and Engineer*, 7, 21–24.

Walleczek J, (1992): Electromagnetic field effects on cells of the immune system: The role of calcium signaling. *FASEB Journal*, 6(13), 3177–3185.

Wever R, (1979): *The Circadian System of Man: Results of Experiments under Temporal Isolation*. Springer-Verlag, New York.

Yakymenko I, Sidorik E, Kyrylenko S, Chekhun V, (2011): Long-term exposure to microwave radiation provokes cancer growth: Evidences from radars and mobile communication systems. *Experimental Oncology*, 33(2), 62–70.

Yakymenko I, Tsybulin O, Sidorik E, Henshel D, Kyrylenko O, et al, (2016): Oxidative mechanisms of biological activity of low-intensity radiofrequency radiation. *Electromagnetic Biology and Medicine*, 35(2), 186–202.

Yakymenko I, Burlaka A, Tsybulin I, Brieieva I, Buchynska L, et al, (2018): Oxidative and mutagenic effects of low intensity GSM 1800 MHz microwave radiation. *Experimental Oncology*, 40(4), 282–287.

Zheng Y, Xia P, Dong L, Tian L, Xiong C, (2021): Effects of modulation on sodium and potassium channel currents by extremely low frequency electromagnetic fields stimulation on hippocampal CA1 pyramidal cells. *Electromagnetic Biology and Medicine*, 17, 1–12.

Zothansiama, Zosangzuali M, Lalramdinpuii M, Jagetia GC, (2017): Impact of radiofrequency radiation on DNA damage and antioxidants in peripheral blood lymphocytes of humans residing in the vicinity of mobile phone base stations. *Electromagnetic Biology and Medicine*, 36(3), 295–305.

Part A

Physical Properties of Wireless Communication Electromagnetic Fields

1 Defining Wireless Communication (WC) Electromagnetic Fields (EMFs):

A. Polarization Is a Principal Property of All Man-made EMFs
B. Modulation, Pulsation, and Variability Are Inherent Parameters of WC EMFs
C. Most Man-made EMF Exposures Are Non-thermal
D. Measuring Incident EMFs Is More Relevant than Specific Absorption Rate (SAR)
E. All Man-made EMFs Emit Continuous Waves, Not Photons
F. Differences from Natural EMFs. Interaction with Matter

Dimitris J. Panagopoulos[1,2,3], Andreas Karabarbounis[4], and Constantinos Lioliousis[5]

[1]National Center for Scientific Research "Demokritos", Athens, Greece
[2]Choremeion Research Laboratory, Medical School, National and Kapodistrian University of Athens, Athens, Greece
[3]Electromagnetic Field-Biophysics Research Laboratory, Athens, Greece

[4]Physics Department, Section of Nuclear and Particle Physics, National and Kapodistrian University of Athens, Athens, Greece
[5]Physics Department, Section of Applied Physics, Electronics Laboratory, National and Kapodistrian University of Athens, Athens, Greece

CONTENTS

Abstract 19
1.1 Introduction 19
1.2 Polarization Is a Principal Property of All Man-made EMFs 27
    1.2.1 Defining Polarization. Why Man-made Polarized EMFs Are More Adversely Bioactive than Natural Non-polarized EMFs 27
    1.2.2 Field Intensity and Radiation Intensity 29
    1.2.3 Superposition of Non-polarized EMR/EMFs 31
    1.2.4 Superposition of Polarized and Coherent EMR/EMFs: Constructive and Destructive Interference 32
    1.2.5 Polarization Combined with Variability Is the Trigger for Biological/Health Effects 34
1.3 Modulation, Pulsation, and Variability Are Inherent Parameters of WC EMFs 35
    1.3.1 Information-Carrying WC EMFs. Combination of Frequency Bands 35
    1.3.2 Modulation, Pulsation, and Random Variability 37
1.4 Most Man-made EMF exposures Are Non-thermal 43
    1.4.1 Energy of EMF-Induced Molecular Oscillations 43
    1.4.2 Non-thermal Exposures. A New Biophysical Constant 43
    1.4.3 Thermal Exposures 45
1.5 Measuring Incident EMFs Is More Relevant than SAR 46
    1.5.1 Analysis of the SAR 46
    1.5.2 SAR Estimation Methods 49
    1.5.3 Incident EMF 50
1.6 All Man-made EMFs Emit Continuous Waves Not Photons 51
    1.6.1 Misleading Assessment of EMF Bioactivity Based on Photon Energy 51
    1.6.2 Quantized States Produce Quantized Emissions (Photons) and Line Spectra ... 52
    1.6.3 Continuous States Produce Continuous Waves and Continuous Spectra ... 53
    1.6.4 How Causality Was Abandoned in Modern Quantum Physics 55
    1.6.5 The Mathematical "Quantization" of EMF/EMR 56
    1.6.6 No Evidence of Photons at Frequencies below Infrared in Environmental Conditions 58
1.7 Differences from Natural EMFs. Interaction with Matter 60
    1.7.1 Differences between Natural and Man-made EMFs/EMR 60
    1.7.2 Basic Concepts of Interaction of EMFs/EMR with Matter 61
1.8 Discussion and Conclusions 63
References 67

Keywords: electromagnetic fields, non-ionizing electromagnetic radiation, physical properties, extremely low frequency, radio frequency, non-thermal biological effects, health effects.

Abbreviations: AM: amplitude modulation. CDMA: Code Division Multiple Access. DECT: Digitally Enhanced Cordless Telecommunications. DTX: Discontinuous Transmission Mode. EEG: electro-encephalogram. EHS: electro-hypersensitivity. ELF: Extremely Low Frequency. EMF:

electromagnetic field. EMR: electromagnetic radiation. ESR: electron spin resonance. FDTD: Finite Difference Time Domain. FM: frequency modulation. GMSK: Gaussian Minimum Shift Keying modulation. GSM: Global System for Mobile Telecommunications. LF: Low Frequency. LTE: Long-Term Evolution. MT: mobile telephony. MW: Microwaves. NMR: nuclear magnetic resonance. NR: New Radio. OS: oxidative stress. PM: phase modulation. QED: quantum electrodynamics. QEM: quantum electromagnetism. RADAR: radio detection and ranging. RF: Radio Frequency. SAR: Specific Absorption Rate. SCN: supra-chiasmatic nucleus. SD: Standard Deviation. SES: seismic electric signals. TDMA: Time Division Multiple Access. ULF: Ultra Low Frequency. UMTS: Universal Mobile Telecommunications System. VGIC: voltage-gated ion channel. VLF: Very Low Frequency. WC: wireless communications. Wi-Fi: Wireless Fidelity. WLAN: Wireless Local Area Network. 1G/2G/3G/4G/5G: first/second/third/fourth/fifth generation of MT.

ABSTRACT

All types of man-made electromagnetic fields (EMFs) and corresponding non-ionizing electromagnetic radiation (EMR) produced by electric/electronic circuits and antennas – in contrast to natural EMFs/EMR – are totally polarized and coherent. Polarized/coherent EMFs/waves can produce constructive interference and amplify their intensities at certain locations. Moreover, they induce parallel/coherent forced oscillations of charged/polar molecules – especially mobile ions – in living cells/tissues, which can trigger biological effects. The most bioactive man-made EMFs are those employed in wireless communications (WC). They are usually referred to simply as Radio Frequency (RF) or Microwave (MW) EMFs/EMR because they emit carrier signals in the RF/MW band. Yet, WC EMFs contain emissions in the Extremely Low Frequency (ELF), Ultra Low Frequency (ULF), and Very Low Frequency (VLF) bands as well in the form of modulation, pulsing, and variability. This complexity and variability of WC EMFs, combined with polarization, is what makes them even more bioactive. Man-made EMFs (including WC) at environmentally existing intensities do not induce significant heating in living tissues. The Specific Absorption Rate (SAR) was introduced by health agencies as the principal metric for the bioactivity of RF/microwave EMFs. Estimation of SAR from tissue conductivity is inaccurate, and estimation from tissue specific heat is possible only for thermal effects. Thus, SAR is of little relevance, and EMF exposures should better be defined by their incident radiation/field intensity at the included frequency bands, exposure duration, and other field parameters. The present chapter also explains that man-made EMFs/EMR, in contrast to light and ionizing electromagnetic emissions, do not consist of photons but of continuous "classical" waves and, thus, do not obey Planck's formula connecting photon energy (ϵ) with frequency (v), $\epsilon = h\ v$. Apart from polarization, man-made EMFs differ from natural EMFs in frequency bands and emission sources. Basic concepts of interaction with living tissue are discussed.

1.1 INTRODUCTION

To address the bioactivity of electromagnetic fields (EMFs) and corresponding electromagnetic radiation (EMR) emitted by wireless communication (WC) devices/antennas, we must first know their physical properties. Applying various types of EMFs in biological experiments without good knowledge of their physical properties/parameters, or without good knowledge of the exposed biological model, will most likely lead to misinterpreted effects and misleading conclusions. Thus, the aim of this chapter is to define WC EMFs, the most complex type of man-made EMFs, by analyzing and describing their various physical parameters. This may provide a basis for future studies on the biological/health effects of WC EMFs.

Among the most important parameters of EMFs, in general, are frequency bands and corresponding intensities, polarization, waveform of the emitted waves/signals, and modulation/variability with a general meaning, which may include pulsations and different types of signal variability.

Apart from the field characteristics, the exposure duration is an additional important parameter for the induced effects (Panagopoulos 2011; 2017; 2019a).

The whole part of the electromagnetic spectrum from 0 Hz (static electric and magnetic fields) up to the low limit of infrared (approximately (~) 300 GHz = 3×10^{11} Hz) is, today, mainly occupied by anthropogenic/technical/artificial/man-made EMFs. They are produced by electric/electronic circuits and antennas of human technology. Applied voltages on those circuits force all free electrons in the metallic conductors to move back and forth in phase (coherently). As a basic principle of electromagnetism summarized in the Maxwell equations, EMR is produced when electric charges are accelerating (Tesla 1905; Alonso and Finn 1967; Reitz and Milford 1967; Alexopoulos 1973; Jackson 1975; Panagopoulos 2013). In this case, we have a continuous coherent acceleration (and deceleration) of free electrons. Due to the fixed position, geometry, and orientation of the circuits/antennas, all artificial (man-made) electromagnetic emissions are totally polarized, meaning their electric and magnetic fields oscillate on single planes (while being perpendicular to each other). This makes them particularly bioactive as discussed in Section 1.2 (and originally in Panagopoulos et al. 2015a; Panagopoulos 2017).

The velocity of any electromagnetic wave is the velocity of light c, as light consists of electromagnetic wave-packets called photons. In the vacuum or in the air, $c \approx 3 \times 10^8$ m/s, as measured experimentally by Heinrich Hertz in 1888. This represents an upper limit for all known velocities (Alonso and Finn 1967; Jackson 1975; Panagopoulos 2013). The velocity of EMR/light is an absolute, universal constant independent of any reference system (Beiser 1987).

The velocity of any wave in any medium is expressed as the product of its wavelength (λ) times its frequency (ν). Accordingly, the velocity of an electromagnetic wave is:

$$c = \lambda \cdot \nu \qquad [1.1]$$

The part of the man-made electromagnetic spectrum with the highest frequencies is called Radio Frequency (RF) band (300 kHz–300 GHz). RF EMFs are produced by electromagnetic oscillation circuits/antennas and are mainly used as carriers for transmitting information. Microwaves (MWs) are called the highest part of the RF band, with frequencies (300 MHz–300 GHz) higher than those which can be reflected by the ionosphere and transmitted over long distances around the Earth. This inability to travel long distances in the atmosphere is due to their smaller wavelengths, as described by the Rayleigh law (Eq. 1.2), which declares that the intensity of scattered EMR in any material medium is inversely proportional to λ^4 (λ the wavelength of EMR) or equivalently proportional to ν^4 (ν the frequency), when the dimensions of the scattering particles are smaller than the wavelength (Alexopoulos 1966; Jackson 1975) (which is the case for man-made EMFs):

$$J_{scat} \propto \frac{1}{\lambda^4} \quad or \quad J_{scat} \propto \nu^4 \qquad [1.2]$$

Since scattering increases with increasing frequencies, penetration into a material decreases. Because MWs are unable to travel long distances, unlike the electromagnetic waves of lower frequencies, and cannot be reflected by the ionosphere to go practically everywhere, their receiving and emitting antennas need to have optical contact between them or be close to each other, as with the antennas of mobile telephony (MT). This is why, while radio and television broadcasting antennas are restricted within antenna parks on top of mountains, WC antennas are excluded from this restriction. This, in turn, shows that health concerns are not taken into account by health authorities and national/international laws. The continuous demand for increasing the amount of transmitted information by MW antennas leads to the continuous increase in the MW frequencies, and the consequent approximation toward the low limit of infrared (Lioliousis 1979; 1997; 2009; Panagopoulos 2017).

In all information-carrying electromagnetic waves there is an RF/MW frequency carrier wave and a modulation field/wave which is, in most cases, of Extremely Low Frequency (ELF) (3–3000 Hz) (mostly) or Very Low Frequency (VLF) (3–30 kHz) and includes the information to be transmitted by the carrier. The frequency and the amplitude of the modulation field/signal vary continuously, depending on the varying information this signal includes (speech, text, images, etc.). In older analog radio, television broadcasting signals, or first-generation (1G) mobile phone signals, the RF carrier was a continuous-wave amplitude modulated (AM), or frequency modulated (FM), or phase modulated (PM) by the ELF/VLF information signal. In modern digital WC, the emissions are in the form of "on/off" pulses, repeated with a namely standard frequency in the ELF/VLF band which actually varies as well. The pulses are most usually rectangular with fast rise and fall times and variable intensity. Each rectangular pulse is an "envelope" containing the RF carrier wave/signal modulated by the information signal (ELF/VLF). Radio detection and ranging (RADAR) emissions/signals are also pulsed for energy-saving reasons with "on/off" pulses, but in this case, the pulses are invariable (Puranen and Jokela 1996). The pulses are, in most cases, emitted at rates of tens, or hundreds, or thousands per second (ELF). In WC signals, the pulses are used not only for energy saving but also mainly for increasing the number of users communicating each moment with the same antenna and exchanging different types of information (speech, text, images, etc.). This is called "multiplexing". The variable pulsations in combination with the modulation and other factors create an additional variability of the final signal, which is usually in the Ultra Low Frequency (ULF) (0–3 Hz) band (Alonso and Finn 1967; Alexopoulos 1973; Jackson 1975; Schwartz 1990; Holma and Toskala 2004; Panagopoulos 2011; 2013; 2017; 2019a; Pirard and Vatovez).

A single WC device (e.g., mobile or cordless phone) emits pulses for a single user. Groups of thousands of such pulses emitted by MT base antennas, or Wireless Local Area Network (WLAN) also called Wireless Fidelity (Wi-Fi) routers for Internet access, carry the transmitted information of many users simultaneously assigning a single pulse-type, or pulse-timing, or code to each user, differing slightly in position/frequency/code from other pulse types of other users. When a "smart" mobile phone, with its multiple antennas, is simultaneously connected to telephony, Internet, or/and other devices (e.g., printers) via local ("Bluetooth") connections, different types of pulses from different antennas with different carrier/modulation/pulsing frequencies and intensities, etc., accommodate each connection, making the overall field/signal extremely complicated and unpredictably varying each moment.

Thus, WC EMF emissions, except for the RF/MW carrier signal, always include ELF/ULF (0–3000 Hz) emissions in the form of modulation, pulsing, and random variability. The intensity, frequency, and shape of these ELF/ULF components are not invariable/predictable as in non-information-carrying RF emissions (e.g., from radars or MW ovens) but unpredictably varying each moment (Pedersen 1997; Hyland 2000; 2008; Zwamborn et al. 2003; Holma and Toskala 2004; Curwen and Whalley 2008; Pirard and Vatovez). This high complexity and variability of the WC EMFs makes them significantly more bioactive than other types of man-made EMFs, as living organisms cannot adapt to unpredictably varying stressors (Panagopoulos 2019a).

A careful examination of the so-called "RF" EMF exposures employed in the vast majority of experimental EMF bioeffects studies would reveal that these were not purely RF but complex EMFs like those employed in WC and, in most cases, simulated MT EMFs, or real-life MT EMFs from commercially available mobile/cordless phones, combining RF and ELF components (Azanza et al. 2002; Panagopoulos et al. 2004; 2007a; 2007b; 2015b; 2021; Belyaev 2005; Behari 2010; Yakymenko et al. 2016; Wust et al. 2021; Bertagna et al. 2021). The combined frequency bands and variability in WC EMFs are discussed in Section 1.3 of this chapter (and originally in Panagopoulos et al. 2015b; Panagopoulos 2017; 2019a).

Living organisms have developed effective protection mechanisms against natural stress of different types (heat, cold, starvation, natural chemical toxicity, solar ultraviolet radiation, natural radioactivity, etc.). Moreover, it seems they can adapt to stressors which are predictable (invariable).

They are adapted to the presence of the significantly/locally polarized static terrestrial EMFs (geoelectric and geomagnetic field), but only when these fields are kept relatively constant despite their normal small ELF variations. When such fields vary by ~ 20% of their normal average intensities during magnetic storms taking place on Earth every several years due to increased solar activity, adverse health effects initiate in humans/animals (Presman 1977; Dubrov 1978; Panagopoulos 2013). Thus, the combination of polarization and significant ELF variability of EMF exposure is a natural trigger of biological effects (Panagopoulos 2019a). This bioactive combination in maximum levels is the case in WC EMFs. In the present chapter, we shall examine this systematically.

Man-made EMFs and corresponding non-ionizing EMR are actually very different and much more adversely bioactive than natural EMFs. Natural EMFs on Earth (such as natural light, the geoelectric and geomagnetic fields, and the atmospheric "Schumann" oscillations) are vital, as all living creatures have evolved in their presence, and no life would exist without them. The 24-hour (h) day–night periodicity of natural light attunes the central nervous system of all animals on Earth. In mammals, this is accomplished via the supra-chiasmatic nucleus (SCN) a group of neurons above the optic chiasm (Panagopoulos 2013). Atmospheric electromagnetic ELF oscillations created by lightning discharges, called Schumann resonances, play a most vital role in attuning the brain's electrical activity in all animals. It is no chance that the basic frequency of the alpha rhythms of animal/human brain oscillations (7.83 Hz) coincides with the basic frequency of the atmospheric Schumann resonances (Berger 1929; Schumann 1952; Wever 1974; 1979; Panagopoulos and Balmori 2017; Panagopoulos and Chrousos 2019). Similar vital action is exerted by the natural EMFs in all living creatures (trees, plants, etc.) (Presman 1977; Dubrov 1978; Alberts et al. 1994).

By contrast, man-made EMFs have an adverse action on living organisms, except for specific therapeutic applications when they are specifically designed to amplify/restore endogenous electric currents in cells and tissues or simulate natural exogenous EMFs such as the Schumann resonances (Wever 1974; 1979; Nuccitelli 1992; 2003; Panagopoulos 2013). Indeed, externally applied static electric fields of similar intensities and directions with endogenous fields controlling, e.g., cell proliferation, have been found to stimulate mammalian and amphibian nerve regeneration, nerve sprouting at wounds, wound healing, or spinal cord injury healing (Borgens 1988; Nuccitelli 2003; Wang and Zhao 2010). Accordingly, pulsing ELF EMFs have been found to accelerate bone regeneration and bone fracture healing in mammals (Bassett et al. 1964; Brighton et al. 1979; 1987; 1989).

Apart from the specific therapeutic effects when weak static or ELF technical EMFs mimic natural/endogenous EMFs, thousands of studies during the past five decades have indicated a variety of adverse biological effects induced in a variety of organisms/cell types by exposure to man-made EMFs, especially ELF and complex "RF" (including ELF modulation/pulsation/variability). The recorded biological and health effects range from alterations in the synthesis rates of critical biomolecules such as proteins, RNA, DNA, etc., alterations in enzymatic activity, in intracellular ionic concentrations (Ca^{+2}, Na^+, K^+, Cl^-, etc.), or in cell proliferation rates, to oxidative stress (OS), DNA and protein damage, chromosome damage, cell death, infertility, electro-hypersensitivity (EHS), and cancer (Marino and Becker 1977; Wertheimer and Leeper 1979; Adey 1981; 1993; Goodman et al. 1995; Santini et al. 2005; Diem et al. 2005; Hardell et al. 2007; 2013; Phillips et al. 2009; Khurana et al. 2009; Blackman 2009; Johansson 2009; De Iuliis et al. 2009; Yakymenko et al. 2011; 2016; 2018; Houston et al. 2016; Panagopoulos 2011; 2017; 2019a; 2019b; 2020; Panagopoulos et al. 2007a; 2007b; 2010; 2013a; Chavdoula et al. 2010; Miller et al. 2018; 2019; Belpomme and Irigaray 2020). All these reported effects are not accompanied by heating of the exposed biological tissues.

Under the weight of this accumulating evidence, especially on genotoxic effects and carcinogenicity, the International Agency for Research on Cancer (IARC), which is part of the World Health Organization (WHO), has classified both ELF and RF EMFs as possibly carcinogenic to humans (Group 2B) (IARC 2002; 2013). Recent carcinogenicity updates advocate that WC EMFs containing both RF and ELF should be categorized as "probably carcinogenic" or "carcinogenic" (Miller et al. 2018; NTP 2018; Falcioni et al. 2018; Hardell and Nyberg 2020). The field/radiation intensities and exposure durations in the majority of published man-made EMF studies are significantly smaller than those of exposures to natural EMFs in the terrestrial environment, even though in different frequency bands (Panagopoulos 2015a; 2019a).

Solar EMR intensity incident upon a human body ranges normally between 8 and 24 mW/cm^2 (depending on seasons, atmospheric conditions, geographical location, etc.) (Roller and Goldman 1968; Parsons 1993; Panagopoulos 2017), while corresponding intensity from a digital second or third, or fourth generation (2G/3G/4G) mobile phone handset upon a human head (even in contact) during a usual phone-call in "talk" mode is normally less than 0.2 mW/cm^2 (Table 1.1). Similarly, infrared radiation, from every human body at normal temperature has significantly greater incident intensities and exposure durations on any human than most artificial EMF sources (Presman 1977; Dubrov 1978; Gulyaev et al. 1995). How, then, can natural EMFs be beneficial, while man-made EMFs are detrimental?

The unique property that makes human-made EMFs so much more adversely bioactive compared to natural EMFs and natural light is polarization (combined with coherence) (Panagopoulos et al. 2015a; Panagopoulos 2017). Polarized and coherent EMFs/EMR are specifically bioactive because they can induce parallel and coherent forced oscillations of electrically charged and polar molecules which constitute the vast majority of molecules in living tissues. Moreover, they can interfere with each other and amplify their intensities at certain locations, (Panagopoulos et al. 2015a). The combination of polarization/coherence and the extreme complexity/variability of the WC EMF exposures is what makes them extremely bioactive and, thus, dangerous to all living organisms (Panagopoulos 2019a). Before the past ~120 years (and intensely the past ~25 years), living organisms had never been exposed to anything similar to man-made polarized/coherent and oscillating/pulsing EMFs and, thus, have not developed any defense mechanisms against this new unphysical type of stress.

Modulated (especially in amplitude) or pulsed RF EMFs are repeatedly found to be more bioactive than non-modulated or non-pulsing fields of the same carrier frequency and of the same average intensity (Bawin et al. 1975; 1978; Blackman et al. 1980; 1982; Lin-Liu and Adey 1982; Byus et al. 1984; 1988; Frei et al. 1988; Somosy et al. 1991; Veyret et al. 1991; Bolshakov and Alekseev 1992; Litovitz et al. 1993; Thuroczy et al. 1994; Goodman et al. 1995; Penafiel et al. 1997; Huber et al. 2002; Höytö et al. 2008; Hinrikus et al. 2008; Franzellitti et al. 2010; Campisi et al. 2010; Mohammed et al. 2013). Frei et al. (1988) found that a 2.8 GHz RF EMF pulsed at 500 Hz was significantly more effective in increasing heart rate in rats than the corresponding non-pulsed RF 2.8 GHz EMF with the same average intensity and exposure duration. Huber et al. (2002) and Mohammed et al. (2013) found that exposure to 900 MHz RF EMF pulse-modulated with various ELF pulsations induced changes in the human and rat electro-encephalograms (EEG), while the corresponding non-pulsed EMF (same RF frequency without any pulsation) with the same exposure duration did not. Similarly, Franzellitti et al. (2010) found that a 1.8 GHz RF signal amplitude-modulated by ELF pulsations induced DNA damage in cultured human trophoblast cells, while the corresponding non-modulated signal with the same exposure duration was ineffective. In all the above cases, the reported effects were not accompanied by any significant tissue heating. This significant evidence indicates that the non-thermal effects of WC EMFs on living organisms are mainly due to the included ELFs.

Moreover, ELF EMFs alone are found independently to be bioactive, as are RF EMFs modulated or pulsed by ELFs (Bawin and Adey 1976; Blackman et al. 1982; Walleczek 1992; Ma et al. 1993; Goodman et al. 1995; Azanza et al. 2002; Ivancsits et al. 2002; 2003; Santini et al. 2005; Panagopoulos et al. 2013a). Bawin and Adey (1976) found that the ELF sinusoidal signals previously used to modulate an RF carrier EMF (Bawin et al. 1975; 1978), induced alone (without the RF carrier) alterations in Ca^{2+} concentration in chicken and cat brain cells as did the modulated RF EMF, while the RF carrier alone (non-modulated) was ineffective. Azanza et al. (2002) found that the ELF pulsations employed in 2G MT at 8.3 and 217 Hz could, by themselves (without the carrier RF signal), induce changes in the spontaneous bioelectric activity of neurons. Again, in all cases, the described effects were non-thermal.

Thus, in the absence of the ELF/ULF components, the effects usually disappear, as several studies have shown (Bawin et al. 1975; 1978; Blackman et al. 1980; 1982; Goodman et al. 1995; Huber et al. 2002; Belyaev 2005; Franzellitti et al. 2010; Mohammed et al. 2013; Panagopoulos 2019a), and purely RF EMFs, without ELF pulsing or modulation, usually do not induce the above reported non-thermal effects. By contrast, ELF EMFs alone induce non-thermal effects, alike the RF EMFs modulated or pulsed by ELFs (Bawin and Adey 1976; Blackman et al. 1982; Walleczek 1992; Goodman et al. 1995; Ivancsits et al. 2002; 2003; Santini et al. 2005; Panagopoulos et al. 2013a). The fact that a variety of biological systems/living tissues respond differently to pure RF exposures than to those including ELF modulation/pulsation/variability shows that living tissue responds specifically to the ELF components of a complex RF signal containing both RF and ELF components. This is of great significance. Whether living tissue has the ability to "demodulate" the ELF components from the complex signal (Blackman et al. 1982) or these components are already independent within the signal is not the case.

The above experimental findings showing the unique ability of ELF EMFs to induce bioeffects are well explained by the "ion forced-oscillation mechanism" for irregular gating of electro-sensitive ion channels in cell membranes which predicts that pulsing EMFs are more bioactive than non-pulsing EMFs of the same other parameters and that the biological activity of any specific type of EMF is inversely proportional to its frequency and proportional to its intensity (Panagopoulos et al. 2000; 2002; 2015a; 2020; 2021).

As reported already, the above-described effects induced only in the presence of ELF EMFs are non-thermal. The only EMF exposures that cause heating (thermal effects) in living tissues are those to high frequency (of the order of GHz or higher) and high intensity/power EMR (≥ 0.1 mW/cm^2), in other words to intense RF/MW EMFs and, in this case, the presence of ELF components is not necessary. This is a well-known effect called "microwave heating" (Metaxas 1991; Walleczek 1992; Creasey and Goldberg 2001; Belyaev 2005; Panagopoulos 2011; 2017; Wust et al. 2021). Therefore, polarized ELF EMFs at environmental intensities induce non-thermal adverse effects in living organisms, while polarized intense RF EMFs induce only heating (just like the infrared or visible light) in both inanimate and living matter. Thermal and non-thermal effects and related mechanisms are analyzed in Section 1.4.

In addition, real-life, highly varying WC EMFs have been found to be more bioactive than corresponding simulated WC EMFs with invariable parameters produced by generators or "test" phones (Panagopoulos et al. 2015b; Panagopoulos 2017; 2019a; Leach et al. 2018; Kostoff et al. 2020). This shows that unpredictable, intense variability of an EMF exposure is an additional bioactive factor.

The International Commission for Non-Ionizing Radiation Protection (ICNIRP) is a private, non-governmental organization (NGO) that sets EMF exposure standards and claims that the only biological effects induced by EMFs are those due to tissue heating (thermal effects) in the case of RF EMFs, and denies any non-thermal effects (ICNIRP 1998; 2020; Hardell and Carlberg 2021). Facts show that only RF exposures with frequencies at the GHz range or higher and intensities greater than 0.1 mW/cm^2 may induce tissue heating, usually of the order of 0.1–0.3°C, and, thus, the vast majority of EMF exposures at environmentally existing intensities, mainly due to ELF EMFs alone or combined with RF, are non-thermal (Panagopoulos et al. 2013b). Yet, the thermal effects are expected to become more significant with the higher frequencies of 5G (up to 100 GHz) (Neufeld and Kuster 2018). Even though ICNIRP accepts (only) the thermal effects of RF EMFs, it has recently increased the average 6-minute (min) exposure limit for 2–6 GHz from 1 mW/cm^2 to 4 mW/cm^2 (ICNIRP 2020). Thus, not even thermal effects are prevented by the ICNIRP limits anymore.

The IARC (2002; 2013) accepts the non-thermal biological effects recorded by thousands of experimental studies at different frequencies and intensities; however, like the ICNIRP, it does not recognize that what are called simply "RF" EMFs are actually, in most cases, complex WC EMFs, including both RF and ELF/ULF components. Moreover, the IARC (2013) adopts metrics pertaining exclusively to thermal effects such as the Specific Absorption Rate (SAR) and suggests that experimental studies should be performed with simulated MT EMFs with invariable field parameters emitted by generators, while real-life WC (including MT) EMFs are highly variable. The result is that about 50% of the studies performed with simulated signals find "no effects", while more than 95% of the studies using real-life exposures from mobile phones, cordless domestic phones, Wi-Fi routers, etc., find effects (Panagopoulos et al. 2015b; Panagopoulos 2017; 2019a; Leach et al. 2018; Kostoff et al. 2020). Thus, even though the IARC accepts the non-thermal effects, it does not recognize the combined RF–ELF character of the complex WC EMFs, evaluating them simply as "RF", adopts a thermal metric for their evaluation, and overlooks the fact that it is the intense variability that makes real-life WC EMF exposures so bioactive, accepting only studies employing simulated WC exposures with no signal variability in order for the exposures to be "determined" (accurately measured). Thus, both the ICNIRP and the IARC bear great responsibility for the continuing confusion and underestimation of the health risks of WC EMFs by scientists, physicians, health authorities, and the general population.

Health agencies introduced SAR as a principal metric for the bioactivity of RF/MW EMFs/non-ionizing radiation. It expresses the rate of energy absorption (power) per unit of mass of exposed living tissue (in W/kg) in accordance with the rate of absorbed dose in the case of exposure to ionizing radiation (NCRP 1986; Coggle 1983).

But there is a significant difference between RF EMFs/EMR and ionizing electromagnetic radiation regarding their biological/health effects: The biological effects of ionizing EMR (from vacuum ultraviolet to gamma rays with frequencies ranging from $\sim 3 \times 10^{16}$ to $\sim 3 \times 10^{22}$ Hz) depend largely on the high energies of ionizing photons absorbed completely by electrons or nuclei. Such photons are capable of causing direct ionization by breaking chemical bonds, expelling electrons from atoms, or even breaking nuclei in the case of gamma rays, etc. The amounts of energy deposited in exposed single molecules, even in the softest ionizing case of vacuum ultraviolet (>10 eV $\approx$ 1.6 $\times 10^{-18}$ J), are great enough to ionize them. By contrast, the corresponding amounts of absorbed energy by single molecules (mobile ions), in the case of man-made EMF exposures, are millions of times smaller than the average thermal energy of the same molecules at human body temperature ($kT \approx 4.3 \times 10^{-21}$ J) (as analyzed in Section 1.4) and, thus, are billions of times smaller than in the softest case of ionizing exposures. In fact, in most cases, ionizing exposures are of several orders of magnitude greater photonic energy than 10 eV (vacuum ultraviolet), as x-rays have energies around 1–100 keV and gamma rays 100 keV–100 MeV (Alexopoulos 1963; Gautreau and Savin 1978; Coggle 1983; Prasad 1995). Thus, evaluating man-made EMF exposures by metrics similar to those used for ionizing radiations is neither very relevant nor useful.

As shown in Section 1.5, SAR actually accounts only for thermal effects (heating), while the effects of man-made EMFs (frequencies lower than infrared) do not usually induce any significant (or even measurable) heating in living tissues.

Moreover, there are other parameters of an EMF exposure more important for the induced biological effects, such as polarization, the field/radiation intensity at the various included frequencies (carrier, pulsing, etc.), the variability of the field, the duration, intermittence, and timing of the exposure (Diem et al. 2005; Belyaev 2005; Chavdoula et al. 2010; Panagopoulos and Margaritis 2010a; Panagopoulos et al. 2007a; 2007b; 2010; 2015a; 2015b; Panagopoulos 2017; 2019a). These important parameters are not included in the absorbed power (SAR).

Today, there are hundreds of studies that correspond specific biological effects to specific incident radiation/field intensities at different frequency bands which can be measured much more easily and reliably than SAR (see Panagopoulos et al. 2010, and reviews Adey 1981; 1993; Goodman et al. 1995; Santini et al. 2005; Phillips et al. 2009; Panagopoulos and Margaritis 2009; Manna and Gosh 2016; Leach et al. 2018; Panagopoulos 2019a). Thus, we can predict the expected effect by knowing the incident radiation/field intensity plus the other parameters of the field/exposure (Panagopoulos et al. 2013b).

Another important parameter for the definition of a particular type of EMF/EMR and consequently, for its predicted bioactivity, is whether its emitted waves are continuous waves as those described by classical electromagnetism, or discrete wave-packets (photons). It is well documented that natural light (infrared, visible, ultraviolet), x-rays, and gamma radiation are emitted in the form of discrete wave-packets (or "particles" of light) called quanta or photons, each having a discrete frequency, phase, and polarization, and its energy (ϵ) is given by the Planck formula:

$$\epsilon = h\nu \quad \text{or} \quad \epsilon = \hbar\omega \qquad [1.3]$$

($h = 6.625 \times 10^{-34}$ J·s is called Planck's constant, $\hbar = h/2\pi$, ν is the frequency of the wave-packet, and $\omega = 2\pi\nu$ is the circular frequency).

An unphysical postulate of modern quantum physics called quantum electromagnetism (QEM) or quantum electrodynamics (QED) is that not only light, x-, and gamma radiation, but also every form of EMF/EMR is quantized, i.e., consists of quanta (photons) (Panagopoulos 2015; 2018). This was established around 1925–1930 when the founders of QED/QEM (Heisenberg, Dirac, Born, and others) mathematically transformed the energy of the EMF into a Fourier series of discrete terms which were arbitrarily attributed to photons. This was not dictated or even implied by experimental facts and was based on the simplistic hypothesis that any EMF/EMR is a periodic function of time (Panagopoulos 2018).

It was already shown by Planck, Einstein, Bohr, and others that natural light is quantized, i.e., consists of photons, and the physics community considered that any form of EMF/EMR should, therefore, consist of photons. That was a flawed and arbitrary extrapolation. Technical (man-made EMFs) were still very new at that time and not explored in depth with regard to their differences from natural light or other types of natural EMR, which had not been discovered yet, such as the Schumann oscillations or the cosmic "microwaves". Possibly, the founders of QEM/QED did not mean that their "quantization" applies to every form of EMF/EMR that was to be discovered or produced in the future. But the physics community of that time and during the next decades (Feynman 1950), apart from a few exceptions, took for granted that this was the case.

The "official" opinion is that "electromagnetic signals are always composed of photons, although in the circuit domain those signals are carried as voltages and currents on wires, and the discreteness of the photon's energy is usually not evident" (Schuster et al. 2007). While they take for granted the existence of photons in man-made EMFs/EMR and especially in the RF/MW band, they admit that single MW photons have not been detected: "Verifying the single-photon output is a substantial challenge in on-chip microwave experiments. The simplest approach, that of looking for a photon each time one is created, is not currently possible; no detectors can yet resolve single MW photon events in a single shot" (Houck et al. 2007). While the alleged evidence for the existence of RF/MW photons is highly questionable, there is absolutely no evidence of photons in lower frequency bands such as VLF/LF (3–300 kHz), or ELF/ULF (0–3000 Hz). Several quantum physicists have objected to QEM/QED (Jaynes 1966; 1978; 1980; Lamb and Scully 1969; Hunter and Wadlinger 1987; Vistnes and Gjoetterud 2001; Roychoudhuri et al. 2008; Roychoudhuri 2014). Vistnes and Gjoetterud (2001) have argued that considering ELF EMFs as consisting of photons is highly misleading.

The following facts contradict the existence of photons for frequencies below infrared (0–3 × 10^{11} Hz): 1) There is no experimental proof nor explanation based on physical phenomena for the existence of such photons in environmental conditions; 2) there are no discrete lines in ELF, VLF, LF, RF antennae spectra; 3) all interactions of man-made EMFs (from ELF to RF) with matter (both biological and inanimate) are very successfully studied by classical electromagnetism; and 4) the "quantization" of the EMF was a mathematical transformation based on simplistic hypotheses.

Today, those who claim that man-made EMFs are harmless to life argue that their frequencies and, consequently, their "photon energies" are smaller than those of visible light (according to Eq. 1.3) and, thus, are unable to induce any adverse effect in living organisms (Valberg et al. 1997; Sheppard et al. 2008; Levitt et al. 2021). We shall show that this argument is flawed because: a) man-made EMFs/EMR produced by electric/electronic circuits/antennas consist of continuous/uninterrupted waves like those described by classical electromagnetism, not photons (Section 1.6);

b) man-made EMFs are totally polarized and coherent, while natural light is not (Section 1.2); c) most biological/health effects of man-made (including WC) EMFs are not accompanied by any significant heating of the exposed living tissues, in contrast to those of natural light, and are due to ELF, not to the RF or the even higher frequencies of natural light (Section 1.4); and, most importantly, d) thousands of experimental studies have already shown a plethora of effects induced by man-made EMFs which cannot be denied.

The IARC (2013) has correctly avoided mentioning the alleged "photonic" nature of man-made EMFs/EMR, noting that "the photon energy is generally referred to in the x-ray and gamma-ray regions, and also to some extent in the ultraviolet range, because the particle-like properties of the EMFs become more obvious in these spectral regions" and points out that RF EMFs are described by Maxwell's equations (classical electromagnetism) (pages 37–38).

The important differences among natural and man-made EMFs/EMR in the non-ionizing band (from 0 Hz up to ultraviolet) are summarized in Section 1.7. The differences are specified in polarization, frequency bands, and emission sources (bound versus unbound charged microparticles). The basic concepts of interaction of natural and man-made EMFs/EMR with matter, such as excitation/de-excitation and forced oscillation of charged/polar particles, are also discussed in the same section.

In Sections 1.2–1.7, the above briefly described important issues regarding the definition of man-made EMFs, and particularly WC EMFs, are specifically examined. As already noted, the purpose of this chapter is to increase knowledge, awareness, and debate among scientists on the complexity of these new types of man-made EMFs which have already overflowed the planet, exposing every living creature for the first time within the billions of years of biological evolution. Understanding the properties and complexity of these man-made EMFs and clarifying confusing diverse information is the first necessary step to understanding their impacts on life.

1.2 POLARIZATION IS A PRINCIPAL PROPERTY OF ALL MAN-MADE EMFs

1.2.1 Defining Polarization. Why Man-made Polarized EMFs Are More Adversely Bioactive than Natural Non-polarized EMFs

Here, we shall explain theoretically that the increased adverse biological action of man-made EMFs is, first of all, due to polarization, a property that only partially and occasionally exists in natural EMFs (Panagopoulos et al. 2015a; Panagopoulos 2017). Man-made EMFs are produced by electric/electronic circuits, and the corresponding EMR is emitted by the acceleration of free electrons forced to oscillate back and forth along the metallic conductors of such circuits. Because the electronic oscillations take place macroscopically in specific directions/orientations determined by the geometry/orientation of the circuit/antenna, the corresponding oscillating fields and generated waves oscillate on a single plane, and thus, they are totally polarized (in most cases linearly polarized) In contrast, natural electromagnetic emissions (cosmic microwaves/infrared, visible, ultraviolet, x-, and gamma radiation) produced by molecular/atomic/nuclear events are not polarized, and only in specific occasions may light be partially polarized. First, we must provide some definitions and equations on polarization, field intensity, wave intensity, and superposition/interference of EMFs/EMR, which will be necessary for understanding why polarized EMFs are so much more bioactive than non-polarized.

A field/wave is called linearly polarized when it oscillates on a single plane, which is called the "polarization plane". While the intensity of a non-polarized field at any point in space oscillates in every possible direction, the intensity of a linearly polarized field at any specific point oscillates on one line (Figure 1.1). Linearly polarized waves are also called "plane waves". A combination of linearly polarized fields/waves with certain phase differences among them can give circularly or elliptically polarized fields/waves. Specifically, superposition of two identical fields with a phase difference of 90° among them creates a circularly polarized field. Superposition of three identical fields with a phase difference of 120° among each two of them also creates a circularly polarized field. The same conditions with unequal amplitudes create elliptically polarized fields. Circularly and elliptically polarized 50–60 Hz sinusoidal alternating electric and magnetic fields produced by three-phase electric power transmission lines (120° phase difference among each two phases) are accused for association with cancer, while linearly polarized such fields produced in the lab are repeatedly found to induce DNA damage, cell death, infertility, alterations in DNA synthesis, and cell proliferation rates, and a variety of other adverse effects in experimental animals and cell cultures (Marino and Becker 1977; Wertheimer and Leeper 1979; Adey 1981; 1993; Schimmelpfeng and Dertinger 1993; Goodman et al. 1995; IARC 2002; Ivancsits et al. 2002; 2003; Santini et al. 2005; Phillips et al. 2009; Panagopoulos et al. 2013a).

Natural EMR/EMFs (atmospheric "Schumann" oscillations, cosmic MWs, infrared, visible light, ultraviolet, gamma rays) and several forms of artificially triggered natural electromagnetic emissions (such as from incandescent lamps, gas discharge lamps, x-rays, lasers, etc.) are not polarized, meaning that their electric and magnetic fields oscillate on any possible random plane while being perpendicular to each other. Light (infrared, visible, ultraviolet), x, and gamma rays are produced by great numbers of molecular, atomic, or nuclear transitions of random orientation and random phase difference among them (except for lasers, which are coherent). These transitions are excitations/de-excitations of molecules, atoms, or atomic nuclei. During each such transition, a single photon is emitted (Beiser 1987). Each photon (i.e., wave-packet) oscillates on a distinct random plane, and, therefore, it has a distinct different polarization. Moreover, the different photons are not produced simultaneously, but they have random phase differences among them (Panagopoulos et al. 2015a; Panagopoulos 2018). Schumann oscillations in the Earth's atmosphere are non-polarized stationary waves generated by atmospheric discharges (lightning) during thunderstorms that occur any moment on Earth. The above natural EMFs are oscillating and non-polarized.

The geoelectric and geomagnetic fields (with average intensities ~ 130 V/m and ~ 0.5 G = 0.05 mT, respectively) and the electric fields of cell membranes in all living organisms (~10^7 V/m) are locally polarized, accepting that their field lines are practically parallel among them at a certain location. These are examples of locally polarized natural EMFs. All three of them are basically static (invariable in their polarities and average intensities). There are also transient polarized signals associated with certain natural phenomena. The strongest lightning discharges from clouds to ground during thunderstorms can be considered as ~ 70% straight lines with a reasonable approximation, and thus, their emitted EMFs (called "sferics") can be considered as being ~ 70% polarized. Seismic electric signals (SES) emitted a few days or weeks prior to major earthquakes are weak, significantly polarized pulses. Both of these natural EMFs, due to their significantly polarized nature, can be sensed by sensitive animals/individuals (Panagopoulos and Balmori 2017; Panagopoulos et al. 2020), and this is probably a way for protection of the living organisms against intense natural phenomena developed during the biological evolution.

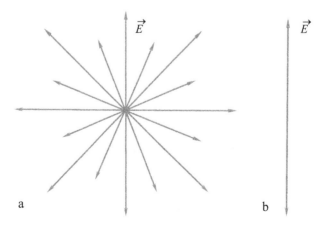

FIGURE 1.1 (a) Non-polarized field, (b) Linearly polarized field.

The effect of light interference discovered by Thomas Young in the early 1800s takes place among waves (photons) having identical polarization and frequency (Arago and Fresnel 1819; Panagopoulos 2015). In his experiments, natural light from a single source passes through two identical small slits at equal distances from the source which, in turn, become two identical coherent secondary sources, according to the Huygens principle, and the light from the two secondary sources forms standing luminous and dark parallel fringes on a screen behind the slits (Pohl 1960; Alonso and Finn 1967). As became clear from subsequent experiments in the following years, and summarized in the Arago-Fresnel experimental laws, only coherent polarized fields/waves of identical polarization and frequency are able to produce clear standing interference effects (fringes of maximum and minimum light intensity) (Arago and Fresnel 1819). [An explanation of how natural non-polarized and incoherent light in the Young experiments produces standing interference is given in Panagopoulos (2015) and is based on the fact that each single photon of natural light has a distinct polarization, frequency, and phase, though different than those of the other photons. Two parts of each single photon pass simultaneously through the two slits and then interfere with each other.] What is important, here, is that only polarized EMFs/EMR of the same polarization can produce constructive or destructive interference with each other, and amplify or cancel their intensities respectively, at the specific locations where two or more waves are superimposed on each other with the same or opposite phases. The ability of constructive or destructive interference is a unique property of polarized waves/fields with great significance in their bioactivity.

Apart from polarization, when the EMFs are in addition of the same frequency, the interference fringes are standing at certain locations (when the sources are also standing). This is called standing interference. When the polarization is fixed (e.g., vertically oriented antennas), but there are differences in frequency among the sources, the interference effects are not standing at fixed locations but, instead, change with time, creating instantaneous peaks at changing locations. Several oscillating EMFs of the same polarization, such as the fields from different antennas vertically oriented, may also produce transient constructive interference effects and instantly amplify the local field intensity at different locations. At such locations, any living organism can be instantly exposed to significantly higher intensities and become more vulnerable to the adverse action of these fields (Sangeetha et al. 2014; Panagopoulos et al. 2015a).

In addition, oscillating polarized (and coherent) EMFs/EMR (in contrast to non-polarized) have the ability to induce parallel and coherent forced oscillations on any charged/polar particles within a medium. In case the medium is biological tissue, the result is that all charged (bio)molecules will be forced to oscillate in parallel and in phase with the field. These parallel and coherent forced oscillations can trigger biological effects (Panagopoulos et al. 2000; 2002; 2015a; 2020; 2021).

Non-polarized EMR can become polarized when it passes through anisotropic media with specific molecular orientations, as are certain crystals. In fluids (gases and liquids), the molecules are randomly oriented and, macroscopically, are considered isotropic, inducing no polarization in the electromagnetic waves transmitted through them. Non-polarized and incoherent natural light can become partially polarized to a small degree after diffraction on atmospheric molecules or reflection on water, mirrors, metallic surfaces, etc. In contrast, a polarized beam cannot be unpolarized but may only be absorbed by a medium (Alonso and Finn 1967). Thus, living organisms, exposed to natural radiation throughout biological evolution, have been exposed to incoherent, partially polarized to a small degree light, under certain circumstances, but have never been exposed to totally polarized and coherent radiation, such as the EMR/EMFs of the human technology (Chen and Rao 1968; Cronin et al. 2006; Panagopoulos et al. 2015a).

1.2.2 Field Intensity and Radiation Intensity

Any harmonically oscillating physical quantity A propagating along a direction r with velocity u, is described by the classical harmonic plane wave equation:

$$A = A_o sin(\omega t - k_w r) \quad [1.4]$$

where A_o is the amplitude (max value) of the oscillating quantity, r the distance of propagation in time t, k_w (=$2\pi/\lambda$) is the wave number (λ the wavelength), and $\omega = 2\pi\nu = k_w u$ is the circular frequency of the wave (ν the frequency). The product $k_w r$ is the phase difference of the oscillation at distance r from the oscillation at the source.

The oscillating quantity A can be an elastic/mechanical disturbance transmitted in a material medium or a time-varying electric/magnetic field transmitted in any medium (including vacuum). The first is an elastic wave like the sound waves or the ripples on water. The latter is an electromagnetic wave.

Any time-varying (oscillating) electric field generates a time-varying magnetic field of the same time variations (frequency, waveform) and vice versa. The two of them constitute an electromagnetic wave. The intensities-vectors of the two fields are always vertical to each other, and both are vertical to the direction of the wave. This is described by classical electromagnetism, which is summarized in the Maxwell equations (Tesla 1905; Alonso and Finn 1967; Reitz and Milford 1967; Alexopoulos 1973; Jackson 1975; Panagopoulos 2013). Almost all electromagnetic technological applications, including WC, are based on classical electromagnetism. Electromagnetic waves do not need a material medium to accommodate their transmission and can be transmitted in the void as well due to some inherent property which is not yet entirely understood. We shall simply accept that EMFs/EMR can be transmitted by themselves in the void (and in material media) with the velocity of light c (which is smaller in the material media than in the vacuum/air depending on the permittivity of each medium).

In electromagnetic waves, the oscillating–propagating quantities are the electric and the magnetic field intensities (the electric and magnetic components of the electromagnetic wave). A plane harmonic electromagnetic wave is the simplest form of such a wave with electric (E) and magnetic field (B) intensities (vertical to each other and to the direction of propagation r) described by Eq. 1.4:

$$E = E_o sin(\omega t - k_w r) \quad [1.5]$$

$$B = B_o sin(\omega t - k_w r) \quad [1.6]$$

E_o, B_o are the amplitudes of electric and magnetic field intensities. In this case, the velocity of the wave is the velocity of light c.

The energy density (energy per unit volume) (in J/m³) of a plane harmonic EMF/EMR in a medium is connected to its electric field intensity according to the equation:

$$W = \varepsilon\varepsilon_o E^2$$

where E (in V/m) is the intensity of the electric field or the electric component of the wave in the medium, ε is the relative permittivity of the medium ($\varepsilon = 1$ in the vacuum and in the air), and $\varepsilon_o = 8.854 \times 10^{-12}$ C²/N·m² the vacuum permittivity.

The radiation intensity $\vec{J}$ in the medium (also called wave intensity, power density, or "Poynting vector") defined as the incident power per unit surface (in W/m², and more often in mW/cm², or μW/cm²) is the product of the energy density with the velocity of the wave:

$$\vec{J} = \vec{c}W = c^2 \varepsilon\varepsilon_o \vec{E} \times \vec{B} \quad [1.7]$$

For plane harmonic waves, the wave intensity becomes:

$$\vec{J} = \vec{c}\varepsilon\varepsilon_o E^2 \quad [1.8],$$

Defining Wireless Communication Electromagnetic Fields

and the average value of its magnitude is:

$$J_{ave} = \frac{1}{2} c \varepsilon \varepsilon_o E_o^2 \qquad [1.9]$$

where c is the velocity of electromagnetic waves in the medium with relative permittivity, ε (Alonso and Finn 1967). [The labeling ($\rightarrow$) on the vectors A, k_w, r, λ, u, ω, E, B, J, is omitted for simplicity in most cases.]

Equations 1.7 and 1.8 show that the wave/radiation intensity (having the direction of the wave propagation) is vertical to both the electric and the magnetic fields (1.7), and in the case of plane harmonic waves it depends upon the square of the electric field intensity (1.8) (Alonso and Finn 1967; Panagopoulos et al. 2015a).

1.2.3 Superposition of Non-polarized EMR/EMFs

Consider two incoherent, non-polarized electromagnetic beams with resultant electric components E_1, E_2, reaching a certain point, P, in space at a certain moment, t, in time. Each beam consists of a great number of individual plane harmonic waves (e.g., photons) of random but discrete polarizations and phases transmitted toward the same direction. For the sake of simplicity, let us pick two individual plane harmonic waves, one from each beam. The two vectors, $\vec{E}_1$, $\vec{E}_2$ due to the different polarizations, oscillate on different planes. Because the two beams are not polarized, the polarizations of their constituent plane harmonic elementary waves vary randomly at point P each moment. The total angle ϕ between the two vectors each moment at point P is determined by the different polarizations, plus the different phases, and varies randomly in time.

The magnitude of the resultant electric field $\vec{E}$ (electric component of the resultant electromagnetic wave) of the two elementary plane harmonic waves each moment at point P is given by the equation describing the superposition of the two vectors $\vec{E}_1$ and $\vec{E}_2$:

$$E = \sqrt{E_1^2 + E_2^2 + 2E_1 E_2 \cos\phi} \qquad [1.10]$$

E varies with time due to the temporal variations of E_1, E_2, $\cos\phi$. The average value of $\cos\phi$ is zero: $\frac{1}{2\pi}\int_0^{2\pi} \cos\phi \, d\phi = 0$, and the averages of E^2, E_1^2, and E_2^2 are $E_o^2/2$, $E_{o1}^2/2$, and $E_{o2}^2/2$, respectively (E_o, E_{o1}, and E_{o2} are the amplitudes of E, E_1, and E_2).

The magnitude of the average resultant electric field is then:

$$E_{ave} = \sqrt{\frac{1}{2}\left(E_{o1}^2 + E_{o2}^2\right)} \quad \text{or} \quad E_o^2 = E_{o1}^2 + E_{o2}^2 \; (= \text{constant})$$

and (according to Eq. 1.9):

$$J_{ave} = J_{1,ave} + J_{2,ave} \quad (= \text{constant}) \qquad [1.11]$$

Even when the two component waves have the same frequency and phase, due to the randomly changing polarizations, the result is still the same.

Thus, the total time average radiation intensity due to the superposition of two (or more) rays consisting of individual plane harmonic waves of random polarizations (natural EMR/EMFs) is the sum of the two individual average intensities, and it is constant at every point. In other words, macroscopically, there is no local variation in the resultant radiation intensity, meaning there are no locations of increased or decreased intensity (Panagopoulos et al. 2015a; Panagopoulos 2017).

Radiation Intensity Versus Field Intensity of Non-polarized EMR

Although the sum average radiation/wave intensity due to superposition of natural non-polarized rays is the sum of individual average intensities, each one depending on the square amplitude of individual electric field (Eq. 1.11), the sum electric field intensity from infinite number of individual elementary waves constituting each ray (as e.g., with natural light), at any moment, approaches zero:

$$\lim_{n \to \infty} \sum_{i=1}^{n} \vec{E}_i = \vec{E}_1 + \vec{E}_2 + \vec{E}_3 + \ldots + \vec{E}_n = 0 \qquad [1.12]$$

Let us explain this in more detail: Consider many photons of natural non-polarized light superposed on each other at a particular point in space. Let us assume, for simplicity, that these photons have equal amplitudes and are of the same frequency but have different polarizations, meaning that their electric vectors have all possible orientations forming angles among each two of them from 0° to 360°. Since all possible orientations have equal probabilities, the superposition of a large number of such equal vectors applied on the same point in space will be the sum of vectors applied on the center of a sphere with their ends equally distributed around the surface of the sphere. The sum of an infinite number of such vectors (all applied on the same point – center of the sphere – and with their ends evenly distributed at all points of the spheric surface) tends to be zero.

In other words, at any given location at any moment, the sum electric field of a great number of incident photons of random polarization tends to zero because the individual vectors are in all possible directions with equal probabilities, diminishing each other when superimposed (destructive interference of electric vectors). Similarly, for the sum magnetic field: $\lim_{n \to \infty} \sum_{i=1}^{n} \vec{B}_i = 0$

Thus, the result of superposition of a great number of incident natural waves is increased radiation intensity, but negligible electric and magnetic fields approaching zero with infinite number of individual waves/photons. Since the electric forces on charged particles depend only upon the electric and magnetic field intensities ($\vec{E}$, $\vec{B}$), and not upon the wave intensity $\vec{J}$, non-polarized (and/or incoherent) EMFs/EMR cannot induce any net forced oscillations on any charged or polar particles (e.g., biological molecules). They may only induce heat, i.e., random oscillations in all possible directions due to momentary non-zero field intensities, but this does not result to any net electric or magnetic field or to any net forced oscillation of charged/polar molecules. This is an important point of our whole reasoning.

1.2.4 SUPERPOSITION OF POLARIZED AND COHERENT EMR/EMFs: CONSTRUCTIVE AND DESTRUCTIVE INTERFERENCE

When two or more waves/fields of the same polarization and frequency are coherent, in other words, when their phase difference at the location of superposition is:

$$\varphi = 2n\pi, \text{ (with } n = 0, 1, 2, 3, \ldots) \qquad [1.13],$$

the result is constructive interference, meaning that the resultant wave has an amplitude (max intensity) equal to the sum of amplitudes of the single waves that interfere at the particular location.

When two waves of the same polarization have opposite phases at another location, in other words, when their phase difference is:

$$\varphi = (2n+1)\pi \quad [1.14],$$

then the result of their superposition is destructive interference, i.e., a wave of the same polarization but with diminished intensity (or even zero when the two amplitudes are equal).

The electrical components of two such waves (plane harmonic waves of the same polarization and frequency) reaching a certain location after having traveled different distances, r_1, and r_2, from their two coherent sources are given by the equations:

$$E_1 = E_{o1} \sin(\omega t - k_w r_1) \quad [1.15]$$

$$E_2 = E_{o2} \sin(\omega t - k_w r_2) \quad [1.16]$$

Again, the amplitude, E_o, of the resultant electric field, $\vec{E}$, (electric component of the resultant electromagnetic wave) is:

$$E_o = \sqrt{E_{o1}^2 + E_{o2}^2 + 2E_{o1}E_{o2}\cos\varphi} \quad [1.17]$$

where the phase difference among the two vectors is: $\varphi = \dfrac{2\pi}{\lambda}(r_1 - r_2)$ depending, in this case, only upon the difference in the distances traveled by the two waves.

At any location where: $\varphi = 2n\pi$, Eq. 1.17 gives:

$$E_o = \sqrt{E_{o1}^2 + E_{o2}^2 + 2E_{o1}E_{o2}} \quad (=|E_{o1} + E_{o2}|) \quad [1.18]$$

At these locations, we have constructive interference.

At any location where: $\varphi = (2n+1)\pi$, Eq. 1.17 gives:

$$E_o = \sqrt{E_{o1}^2 + E_{o2}^2 - 2E_{o1}E_{o2}} \quad (=|E_{o1} - E_{o2}|) \quad [1.19]$$

At these locations, we have destructive interference.

The intensity of the resultant wave at any location is:

$$\vec{J} = \vec{J}_1 + \vec{J}_2 \quad [1.20]$$

The amplitude of the resultant wave intensity will be, correspondingly:

$$J_o = c\varepsilon\varepsilon_o (E_{o1} + E_{o2})^2 \quad [1.21]$$

$$J_o = c\varepsilon\varepsilon_o (E_{o1} - E_{o2})^2 \quad [1.22]$$

(at the locations of constructive interference and at the locations of destructive interference, respectively).

Thus, at the locations of constructive interference, the electric field vectors of the two waves/fields are parallel and in the same direction, and both the resultant field and the resultant wave intensity are maximum (Eqs. 1.18 and 1.21).

For two identical sources ($E_{o1} = E_{o2}$): $E_o = 2E_{o1}$ and $J_o = 4\, c\varepsilon\varepsilon_o E_{o1}^2 = 4\, J_{o1}$

For N identical sources:
$$E_o = NE_{o1} \quad [1.23]$$

and:
$$J_o = N^2 J_{o1} \quad [1.24]$$

This is why a series of parallel RF/MW antennas can be used to produce high-intensity beams in certain directions (Alonso and Finn 1967), which is the case with the so-called "antenna arrays" in 5G MT technology.

At the locations of destructive interference, the electric field vectors of the two waves are anti-parallel, and thus, both the resultant field and the resultant wave intensity are minimum (Eqs. 1.19 and 1.22). For identical sources ($E_{o1} = E_{o2}$): $E = 0, J = 0$.

Thus, at the locations of constructive interference, the resultant electric field from N number of polarized coherent electromagnetic sources of the same polarization, frequency, and different intensities $E_1, E_2, \ldots, E_N$, is the sum electric field from all the individual sources (e.g., antennas):

$$E = E_1 + E_2 + E_3 + \ldots + E_N \quad [1.25]$$

The greater the number of coherent superimposed waves/fields (from the same or different sources), the higher and narrower the peaks (Alonso and Finn 1967). That situation can create very sharp peaks of wave and field intensities at certain locations that are not easily detectable by field meters where any living organism may be exposed to peak electric and magnetic field intensities.

Therefore, the difference between superposition of non-polarized and polarized electromagnetic waves/fields is that, in the first case, we have increased average radiation intensity but zeroed net fields at any location, while in the second case we have increased both radiation intensity and fields at certain locations where constructive interference occurs. This difference is of crucial importance for understanding the differences in biological activity between natural (non-polarized and incoherent) and man-made (polarized and coherent) EMFs/non-ionizing EMR.

Thus, polarized and coherent (man-made) EMFs (in contrast to non-polarized) possess a net electric and magnetic field at any point in space, apart from radiation/wave intensity, and this is the key point for their increased biological activity. They can produce interference effects increasing their intensities at certain locations and induce coherent and parallel forced oscillations/rotations on charged/polar molecules in living tissues (Panagopoulos et al. 2015a). For this reason, comparing man-made EMFs with natural EMFs, in terms of their bioactivity, is a flawed methodology, resulting in misleading conclusions.

1.2.5 Polarization Combined with Variability Is the Trigger for Biological/Health Effects

Throughout biological evolution, living organisms have been constantly exposed to the geoelectric and geomagnetic fields which, as already mentioned, are static and locally polarized with average intensities ~ 130 V/m and ~ 0.5 G (0.05 mT), respectively. While no adverse health effects are connected to normal exposure to these natural ambient fields, variations in their intensities of the order of ~ 20% during "magnetic storms" or "geomagnetic pulsations" due to increased solar activity, with an average periodicity of about 11 years and lasting for a few days or weeks, are connected with increased rates of animal/human health implications, including nervous and psychic diseases, hypertensive crises, heart attacks, cerebral accidents, and mortality (Presman 1977; Dubrov 1978; Panagopoulos 2013; 2019a).

All cells and intra-cellular organelles, such as nuclei, mitochondria, etc., are protected by cell membranes, and across all cell membranes there is an intense transmembrane static and locally polarized electric field of the order of ~ 10^7 V/m (average membrane width is ~ 10 nm and average transmembrane voltage ~ 100 mV). All physiological cellular functions are initiated and accompanied by endogenous electric currents consisting of ion flows through the cytoplasm and the cell membranes with corresponding changes in the intracellular ionic concentrations. These vital ionic currents and concentration changes are mediated by ion channel gating (opening and closing) in the cell membranes. Voltage-gated ion channels (VGICs) in all cell membranes switch between open and closed state whenever a change exceeding ~ 30% in the transmembrane voltage/field takes place. It is known that ~ 30 mV changes in the normal ~ 100 mV transmembrane voltage are required to change the status of the VGICs in cell membranes (from opened to closed and vice versa). Obviously no life, as we know it, could exist without proper functioning of ion channels (Weisenseel 1983; Liman et al. 1991; Nuccitelli 1992; 2003; Alberts et al. 1994; McGaig and Zhao 1997; Panagopoulos and Margaritis 2003; Panagopoulos 2013).

There are important similarities in the above two classes of natural EMFs, the terrestrial (geoelectric and geomagnetic) and the cell membrane fields: They are both static and almost totally polarized at any certain location. The terrestrial (geo)electric field and the cell plasma membrane electric field both have a direction vertical to the curved surface and toward its internal (earth, cell). Under normal/usual conditions, these fields do not induce any biological/health effects in the living organisms.

During magnetic storms, there are changes in the terrestrial static fields of the order of 20% of their normal intensities (electric and magnetic), and when the transmembrane electric field undergoes changes of the order of 30% of its normal value, the VGICs of the membrane get activated or deactivated (change their status from closed to opened and vice versa), ion flows are properly controlled, and physiological cellular effects are initiated (Panagopoulos 2013; 2019a).

A conclusion we can draw from these two natural phenomena is that biological and health effects initiate when polarized fields undergo changes of the order of 20%–30% of their normal intensities. Thus, these two similar natural phenomena provide an important clue for the bioactivity of EMFs in general: It is the combination of polarization and variability exceeding a threshold of about 20%–30% in normal average intensity that triggers biological and health effects.

1.3 MODULATION, PULSATION, AND VARIABILITY ARE INHERENT PARAMETERS OF WC EMFs

1.3.1 INFORMATION-CARRYING WC EMFs. COMBINATION OF FREQUENCY BANDS

WC EMFs are not simply RF/MW EMFs. They do have an RF signal, like in emissions from radars or MW ovens, but in addition, the RF carrier signal is digitally modulated, pulsed (it is included within on/off pulses), and highly variable each moment. Even when emissions from radars include on/off pulsations as well, because their power supply has to be turned on and off for energy-saving reasons, their emissions like those from MW heating devices, do not carry information, they are invariable in time and totally repetitive/predictable. In contrast, WC EMFs carry variable information (speech, text, music, images, etc.) in the form of ELF/VLF digital modulation (bits). Moreover, their pulsations are not invariable, as in radars, but are affected by many network/communication factors making the overall signal unpredictably varying in intensity, frequency, and waveform. All this creates a random variability of the final signal each moment that makes WC EMF signals totally unpredictable in their intensity and other parameters. This whole variability lies in the ELF/ULF band (0–3000 Hz) and is always present in all WC EMFs.

Indicative RF/MW (radiation intensity) and ELF (E-field and B-field) emission measurements ± Standard Deviation (SD), at different distances from Universal Mobile Telecommunication System (UMTS) and Global System for Mobile Telecommunications (GSM) 900 and 1800 mobile phones while operating in "talk" mode and under similar conditions and signal reception, are shown in Table 1.1. We note that, while UMTS (3G/4G) in the MW band is somehow lower than both GSM

TABLE 1.1
Intensity Measurements in the MW and ELF Bands above Background Levels of UMTS (3G/4G), and 2G (GSM 900, GSM 1800), According to Distance from Source (Mobile Phone)

Distance from source (cm)	UMTS Rad. Int. 1.95 GHz (µW/cm²)	UMTS ELF E-Field (V/m)	UMTS ELF B-Field (mG)	GSM 900 Rad. Int. 0.9 GHz (µW/cm²)	GSM 900 ELF E-Field (V/m)	GSM 900 ELF B-Field (mG)	GSM 1800 Rad. Int. 1.8 GHz (µW/cm²)	GSM 1800 ELF E-Field (V/m)	GSM 1800 ELF B-Field (mG)
0	232 ± 89	33 ± 8.2	3.8 ± 1.3	378 ± 59	19 ± 2.5	0.9 ± 0.15	252 ± 50	13 ± 2.1	0.6 ± 0.08
1	33 ± 10	22 ± 5.9	3.0 ± 0.7	262 ± 46	12 ± 1.7	0.7 ± 0.13	65 ± 15	6 ± 0.8	0.4 ± 0.07
10	19 ± 7.1	12 ± 3.1	1.5 ± 0.4	62 ± 20	7 ± 0.8	0.3 ± 0.05	29 ± 5	2.7 ± 0.5	0.2 ± 0.05
20	9 ± 4.1	5.3 ± 2.0	0.8 ± 0.2	32 ± 8	2.8 ± 0.4	0.2 ± 0.04	11 ± 3	0.6 ± 0.12	0.1 ± 0.02
30	6 ± 2.3	3.2 ± 1.1	0.4 ± 0.1	10 ± 2	0.7 ± 0.09	0.1 ± 0.02	7 ± 1	0.3 ± 0.06	0.06 ± 0.01
40	4 ± 1.3	2.4 ± 0.7	0.3 ± 0.1	6 ± 1	0.2 ± 0.03	0.05 ± 0.01	4 ± 0.7	0.1 ± 0.04	–
50	3 ± 1.1	1.6 ± 0.4	0.2 ± 0.05	4 ± 0.6	0.1 ± 0.02	–	2 ± 0.3	–	–
60	2.1 ± 0.8	1.0 ± 0.3	0.1 ± 0.03	2 ± 0.3	–	–	1.6 ± 0.2	–	–
70	1.8 ± 0.6	0.4 ± 0.1	0.1 ± 0.02	1.7 ± 0.2	–	–	1.3 ± 0.2	–	–
80	1.3 ± 0.4	0.1 ± 0.04	–	1.2 ± 0.2	–	–	1.1 ± 0.2	–	–
90	0.8 ± 0.3	–	–	1.0 ± 0.1	–	–	0.5 ± 0.1	–	–
100	0.5 ± 0.2	–	–	0.4 ± 0.1	–	–	0.2 ± 0.1	–	–

900 and 1800 (2G), its corresponding emissions in the ELF band are stronger. While ELF emissions from GSM 900 and 1800 mobile phones fall within the background of the stray 50 Hz fields for distances longer than 30–50 cm from the source, the corresponding UMTS emissions fall within the same background for distances longer than 70 cm. As MT base antennas are usually ~ 100 times stronger than corresponding mobile phones with similar radiation patterns in response to distance, the EMF levels in Table 1.1 correspond to base antenna emissions at ~ 100 times longer distances. For example, power density ~ 10 μW/cm² usually measured at 20–30 cm distances from mobile phones is usually measured at 20–30 m from corresponding base station antennas. After the installation of the 4G "UMTS Long Term Evolution" (LTE) system, base antennas and devices emit signals not only for MT but also for the Internet simultaneously, making EMF emission patterns even more complicated (and adversely bioactive). As noted, the EMF measurements in Table 1.1 are only indicative because they depend strongly on signal reception/availability, weather conditions, etc. Within metallic chambers (e.g., cars, elevators, etc.), mobile phone emissions can be significantly stronger.

Next, we shall explain the role of each parameter in the variability of the WC EMF signals.

1.3.2 Modulation, Pulsation, and Random Variability

Modulation

The simplest form of electromagnetic emission that can be manufactured is a single harmonic (sinusoidal) electromagnetic wave (described by Eqs. 1.4, 1.5, and 1.6). However, no information, such as voice, pictures, and other data, can be transmitted by such a signal alone. In order to convey information, the single-frequency signal – called a "carrier wave" – must be "modulated" by another signal which contains the information to be sent. Modulation of the RF signal by a signal containing the information (voice, message, pictures, video, data, etc.) is apparently the case in all WC EMF emissions. We may say that modulation is the information signal "loaded" on the RF carrier. The modulation signal is, in most cases, an ELF/VLF signal. There are three basic types of modulation, according to the physical parameters which characterize the carrier signal: Amplitude, frequency, and phase (Alexopoulos 1973; Lioliousis 1979; Schwartz 1990).

Amplitude modulation (AM) means that the amplitude (max intensity) of the carrier varies according to the modulating signal. The curve of the amplitude variations depicts the modulating signal. When the modulating signal can take any value in a given range, the AM is called analog. The AM radio broadcasting or the first-generation (1G) mobile phones are analog AM applications. When the modulating signal can only take discrete values, the modulation is called "digital". Usually, the modulating signal has a rectangular shape which takes the values "1" or "0" (binary system). With value "1", the carrier is emitted, while with value "0", it is not. This is the simplest case of digital amplitude modulation, called "OOK" (on-off keying). Another type of digital amplitude modulation is the Time Division Multiple Access (TDMA) applied in 2G (GSM) MT and in cordless domestic phones, referred to as Digitally Enhanced Cordless Telecommunications (DECT) phones (Schwartz 1990; Pedersen 1997; Tisal 1998; Pirard and Vatovez).

Frequency modulation (FM) means that the frequency of the carrier varies within a given range according to the modulating signal. Respectively, FM can be analog when the carrier frequency can take any value of the given range (such as in the older FM broadcasting) or digital when the carrier frequency can take only discrete values. FSK (frequency-shift keying) is a simple case of digital frequency modulation in which the carrier frequency can take only two values: One corresponding to 0, and the other to 1 of the modulating signal (Schwartz 1990; Pirard and Vatovez).

Phase modulation (PM) accordingly means that the phase of the carrier signal varies according to the modulating signal. As with AM and FM, it is analog when the carrier phase takes any value within a given range or digital when it takes only discrete values. A simple type of digital phase modulation is the binary phase-shift keying (BPSK) modulation for which the phase becomes 0° or 180° corresponding to 0 or 1 values of the modulating signal. Another type is the Gaussian

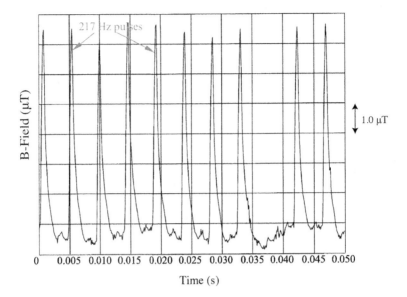

FIGURE 1.2 217 Hz pulses from a GSM mobile phone (adapted from Andersen and Pedersen 1997).

Minimum Shift Keying (GMSK) modulation applied in 2G MT and in DECT phones. GSM and DECT phones/antennas combine GMSK phase modulation with TDMA amplitude modulation (Schwartz 1990; Pedersen 1997; Tisal 1998).

In all three types of modulation, the envelope of the radiated (final) signal (amplitude, shape, and content) is modified according to the modulating signal.

Pulsation

Apart from modulation, all modern digital WC EMFs are pulsed in order to increase the density of information conveyed by the WC signal and the number of subscribers communicating simultaneously via the same antenna and occupying the same frequency band. This is called multiplexing in WC terminology. The pulses are usually (but not necessarily) rectangular with a pulse repetition rate in the ELF band, always variable in intensity and frequency, and their number increases with increasing amount of transmitted information and number of subscribers simultaneously using the same base antenna. Because the information is variable each moment (speech, text, music, images, video, Internet, etc.), and the number of users is also variable, the final signal is variable as well. For these reasons, WC EMFs are not like other RF emissions which do not carry variable information, such as pure RF signals from signal generators, or radar signals with invariable pulsations (Puranen and Jokela 1996; Pedersen 1997; Tisal 1998; Hyland 2000; 2008; Zwamborn et al. 2003; Holma and Toskala 2004; Tuor et al. 2005; Curwen and Whalley 2008; Zhou et al. 2010; Sauter 2011; Shim et al. 2013; Pirard and Vatovez; Panagopoulos 2019a).

Thus, all types of modern WC EMFs, such as from MT, DECT phones, Wi-Fi, wireless communication among electronic devices (Bluetooth), combine MW fields (with frequency usually around ~ 1–3 GHz and increasing with newer systems) as the carrier signals, with variable ELF (in most cases) fields to modulate the carrier and to increase the number of users, and the amount of transmitted information by pulsing the signals.

More specifically, 2G GSM MT EMFs, emitted by mobile phones and base antennas, except for their MW carrier signal, (900, 1800, or 1900 MHz) include a pulse repetition frequency ~ 217 Hz (Figure 1.2) plus other ELF pulsations, such as the multi-frame repetition frequency of ~ 8.34 Hz, and the Discontinuous Transmission Mode (DTX) frequency ~ 2 Hz (only in mobile phones) when the user does not speak ("listening mode"). See recorded pulsations from GSM mobile phones in Figure 1.2 and in Pedersen (1997). GSM uses the TDMA AM for the pulse amplitude, and the radiation is emitted in frames of 4.615 ms duration at a repetition rate of ~ 217 Hz. Each frame consists of eight "time slots", and each user occupies one of them. Within each time slot, the RF carrier is phase modulated by GMSK modulation (Pedersen 1997; Tisal 1998; Hyland 2000; 2008; Zwamborn et al. 2003; Tuor et al. 2005; Curwen and Whalley 2008).

3G (UMTS) MT EMFs from mobile phones and base station antennas emit a MW carrier signal at 1950–2150 MHz with basic ELF pulsations at ~ 100 Hz (frame repetition called "Time Division Duplex"), and ~ 1500 Hz (called "Adaptive Power Control"). See recorded UMTS pulsations in Figure 1.3 and in Zwamborn et al. (2003). UMTS uses the Code Division Multiple Access (CDMA) technology for multiplexing, which assigns a special code to each user (Zwamborn et al. 2003; Holma and Toskala 2004; Hyland 2008; Curwen and Whalley 2008).

The GSM (2G) and UMTS (3G) technologies are retained also in the 4G (LTE) MT system which still uses UMTS or GSM for telephony (voice) and LTE for internet connection and other applications. A newer version of 4G called VoLTE (Voice over LTE) uses the LTE system for telephony as well, being able to handle data services and voice calls concurrently. The LTE carrier frequencies (mostly 1800–2600 MHz) differ in different countries. The 100 Hz on/off (frame) pulsations of UMTS are also used in the pure LTE (4G), and there are additional 1000 Hz (subframe), 200 Hz (synchronization signals), plus other ELF synchronization and reference pulsations (Sesia et al. 2011; Sauter 2011; Shim et al. 2013). Various LTE pulsations and random signal variability are shown in Figures 1.4–1.6.

In the 5G or New Radio (NR) system which is being deployed, the carrier frequencies are extending up to 80–100 GHz with two basic frequency ranges: 1) existing MT bands ≤6 GHz, and 2) 24.25–52.6 GHz with a tendency to increase. Moreover, 5G uses new technologies such as Multiple-Input Multiple-Output (MIMO) for multi-stream transmission and high data rates, and adaptive beam-forming by use of antenna arrays (which can be used to amplify beam intensity – see Section 1.2.4 equations 1.23, 1.24). The 100 Hz and 1000 Hz pulsations (frame, subframe) are retained, and there are synchronization and reference pulsations at ~ 6–200 Hz called Synchronization Signal Blocks (SSB) (Rappaport et al. 2013; Dahlman et al. 2018).

WLAN (Wi-Fi) and Bluetooth signals used for connection to the internet and communication among devices (portable computers/laptops, "smart" phones, printers, etc.), respectively, have main carrier frequencies around 2.45 GHz (with a tendency to increase in newer devices) and pulsations at ~ 10 Hz called beacons which are synchronization signals (Figure 1.7). DECT phones and their corresponding domestic bases emit a carrier signal of around 1880 MHz with two basic ELF

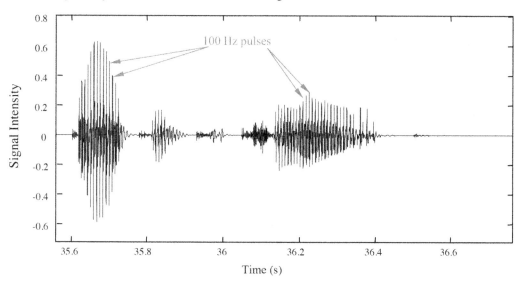

FIGURE 1.3 100 Hz "frame" pulses of a UMTS (3G/4G) mobile phone signal. Each vertical line is a pulse containing the carrier signal (adapted from Holma and Toskala 2004).

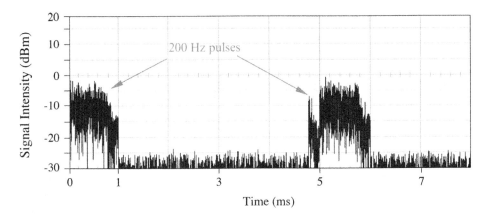

FIGURE 1.4 200 Hz pulses plus random variability in LTE (4G) signal. The variability exists also within the pulses. It seems that the synchronization signals (200 Hz) have boosted the whole corresponding subframes (subframe duration in the figure is the time among successive vertical lines) (adapted from High Performance Solutions).

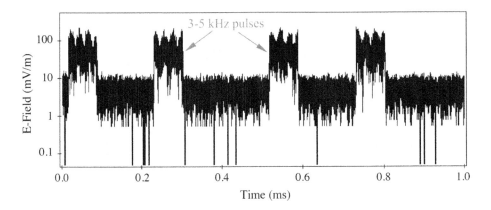

FIGURE 1.5 3–5 kHz pulses plus random pulsations from an LTE (4G) base station antenna with no traffic (adapted from Pirard and Vatovez).

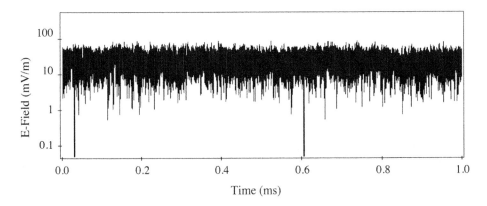

FIGURE 1.6 Random variability/pulsations of LTE (4G) base station antenna emission while communicating (downloading) (adapted from Pirard and Vatovez).

Defining Wireless Communication Electromagnetic Fields

pulsations (frame repetition at ~ 100 and an additional on/off pulsation at ~ 200 Hz) (Figure 1.8). Terrestrial Trunked Radio (TETRA) antennas/devices used by emergency services emit a carrier signal of around 400 MHz with ELF/ULF pulsations at ~ 0.98, ~ 17.64, and ~ 70.4 Hz (Pedersen 1997; Hyland 2008; Curwen and Whalley 2008; Zhou et al. 2010).

The carrier (RF) and pulsing (ELF/ULF) frequencies of GSM, UMTS, LTE, DECT, Wi-Fi/Bluetooth, and TETRA are shown in Table 1.2. Both carrier and pulsing frequencies are variable in all systems. Figures 1.2–1.8 show ELF pulsations and random variability of GSM, UMTS, LTE, Wi-Fi, and DECT signals.

Random Variability

In addition to modulation and pulsing, in all modern digital WC EMFs, the envelope (final signal) is further modified (in amplitude/intensity, pulse repetition frequency, shape, etc.) due to various

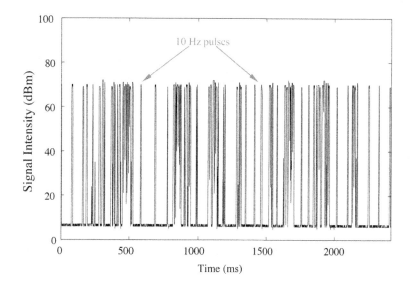

FIGURE 1.7 10 Hz pulses of WLAN (Wi-Fi) (adapted from Zhou et al. 2010).

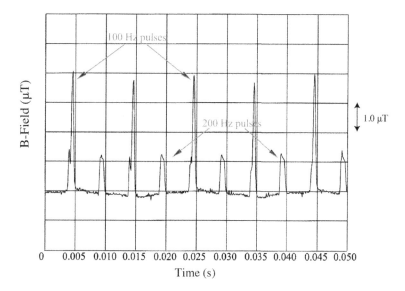

FIGURE 1.8 100 Hz and 200 Hz pulsations from a DECT phone (adapted from Andersen and Pedersen 1997).

TABLE 1.2
Basic Carrier Frequencies and ELF Pulsations of Most Common WC EMFs

WC EMF Type	Carrier Frequencies (RF)	Pulsing Frequencies (ELF/ULF)
GSM (2G MT)	900 MHz, 1800 MHz, 1900 MHz	217 Hz (frame repetition), 8.34 Hz (multi-frame repetition), 2 Hz (DTX mode)
UMTS (3G, 4G MT)	1950 MHz, 2150 MHz	100 Hz (Time Division Duplex), 1500 Hz (Adaptive Power Control)
LTE (4G MT/WC)	1.8 - 2.6 GHz (in most cases)	100 Hz (frame repetition), 1000 Hz (subframe repetition), 200 Hz (synchronization pulses)
NR (5G MT/WC)	Frequency range 1: ~0.7–6 GHz Frequency range 2: 24.25–52.6 GHz	100 Hz (frame repetition), 1000 Hz (subframe), 6–200 Hz (synchronization pulses)
DECT	1880 MHz	100 Hz (frame repetition), 200 Hz (energy saving on/off)
WLAN (Wi-Fi), Bluetooth	2450 MHz	10 Hz (beacons)
TETRA	400 MHz	17.64 Hz (frame repetition), 0.98 Hz (multi-frame repetition), 70.4 Hz (burst repetition)

physical imperfections in the electronic circuits and other parameters such as heat, noise, interference with various other electromagnetic sources, etc., plus multiple other variable physical parameters during transmission. Each moment when the number of users performing different tasks (voice, data, etc.) increases, more pulses are emitted, each one accommodating a different user or task. The final signal from both base antennas and devices depends also on additional uncontrollable parameters, such as the position of each user with respect to the base antenna, air conductivity, signal availability/reception at the specific place and time, etc. All these functions and uncontrollable parameters result in intense unpredictable variability of the final signal with variable frequency, mainly in the ELF/ULF band (see Figures 1.2–1.7). This random ELF/ULF variability is perhaps the most intense and bioactive parameter of the WC EMF emissions (in combination with the fact that the signals are totally polarized) (Holma and Toskala 2004; Panagopoulos et al. 2015b; Panagopoulos 2019a; Pirard and Vatovez).

Thus, apart from the ELF/VLF pulsing and modulation frequencies always included in the WC EMFs, during any signal transmission, there are additional continuous unpredictable changes due to the varying physical parameters, the varying information transmitted each moment, the varying number of users at various locations, environmental factors, etc. Especially with mobile phones/antennas, there are continuous sudden unexpected changes in intensity due to changes in location, number of subscribers using the network each moment, air conductivity changes, etc. These sudden unexpected changes in the final signal may exceed by 100% and even more the average intensity. Finally, for energy-saving reasons, when GSM handsets operate in DTX ("listening") mode, the average emitted power is much less (about one tenth) than when the user speaks ("speaking mode") (Pedersen 1997; Panagopoulos et al. 2004; Hyland 2008). The described final random variability of WC signals can be easily recorded by any RF field meter measuring power density in any urban environment or close to any WC device. The reading of the instrument shows continuous unexpected changes in the measured power density, usually ranging in urban environments between 0.01 and 1 $\mu W/cm^2$ and reaching ~ 10 $\mu W/cm^2$ in closer proximity to antennas. This variability lies mainly in the ULF band (0–3 Hz). The random variability of the final signal can be seen in Figures 1.3 and 1.6 for UMTS (3G/4G) mobile phones and LTE (4G) base antennas, respectively.

Due to the above inherent variability of all WC EMFs, any EMF/EMR measurements can only be representative for average or peak values. The variability becomes more intense in the near field of the emitting devices/antennas (Panagopoulos et al. 2016; Panagopoulos and Karabarbounis 2020). For this reason, health organizations such as the IARC (2013) have recommended that experimental studies on the effects of WC EMFs should be performed with invariable simulated signals emitted by generators or test phones. But exactly because of this inherent variability, it is impossible to simulate the real emissions by use of invariable emissions of fixed parameters (such as fixed intensity, frequency, and pulsation), and when such simulated EMFs are used in experiments, they are significantly less bioactive than real-life WC EMFs. Thus, the simulated signals are very different and much less effective in inducing adverse biological/health effects (Panagopoulos et al. 2015b; Panagopoulos 2017; 2019a; Pall 2018; Leach et al. 2018; Kostoff et al. 2020). Even though the measurements of real WC signals can only be representative, there is actually no need for "exact" measurements. Average and peak measurements are enough to predict bioactivity (Panagopoulos et al. 2016; Panagopoulos and Karabarbounis 2020).

In fact, health agencies', including the IARC, acceptance of simulated exposures with fixed parameters for studying the effects of WC EMFs and the exclusion of the studies having used real-life exposures is one of the most serious flaws in the evaluation of WC EMF bioactivity by these agencies, resulting in the underestimation of the adverse effects (Panagopoulos et al. 2015b). As a result, about 50% of the experimental studies having employed simulated WC signals (in line with IARC's recommendation) do not find any effects, while more than 95% of the studies employing real-life WC exposures from commercially available devices or antennas find effects (Panagopoulos et al. 2015b; Panagopoulos 2017; 2019a; Gulati et al. 2016; Zothansiama et al. 2017; Leach et al. 2018; Kostoff et al. 2020).

1.4 MOST MAN-MADE EMF EXPOSURES ARE NON-THERMAL

1.4.1 Energy of EMF-Induced Molecular Oscillations

In living tissue, most (bio)molecules are polar, (meaning they have a positive side and a negative side separated by some distance of atomic/molecular dimensions, as e.g., water molecules) or carry a net electric charge. Thus, any man-made (polarized) oscillating EMF (and corresponding EMR) induces a forced oscillation on each of these charged/polar molecules and transfers to each of them a tiny part of its energy. This forced oscillation is linear in the case of molecules bearing a net electric charge or rotational in the case of polar molecules.

It seems that a main mechanism of action for both ELF and purely RF man-made (polarized) EMFs is this forced oscillation/rotation of charged/polar particles (Metaxas 1991; Panagopoulos et al. 2000; 2002; 2015a; 2020; 2021).

This induced oscillation will be of greatest amplitude on the smallest (and lightest) mobile particles which carry a net electric charge, i.e., the mobile ("free") ions that exist in large concentrations in all types of cells and extracellular aqueous solutions determining practically all cellular/biological functions (Alberts et al. 1994; Panagopoulos and Margaritis 2003). The induced oscillation will be much smaller on the polar water molecules and even of negligible amplitude on the much larger polar biological macromolecules such as proteins, lipids, nucleic acids, etc., which are, in most cases, bound with other molecules.

The amount of energy absorbed by a single mobile ion in biological tissue will manifest itself as kinetic energy of the forced oscillation induced on that particle. The maximum kinetic energy of such an oscillation is:

$$\in (\max) = \frac{1}{2} m_i u_o^2 \qquad [1.26]$$

where, m_i is the ion mass (e.g., for Na$^+$ ions $m_i \cong 3.8 \times 10^{-26}$ kg), and u_o is the particle's maximum velocity acquired by the forced oscillation.

1.4.2 Non-thermal Exposures. A New Biophysical Constant

Significant experiments in the mid-1970s with, what was at the time, a novel technique called "patch-clamp" allowed the measurement of ion currents through open ion channels in cell membranes

(Neher and Sakmann 1992; Stryer 1996). This technique is widely used today in the study of ion channels (Cecchetto et al. 2020; Zheng et al. 2021). It was found that the electric current through an open sodium channel is of the order of 4×10^{-12} A when the transmembrane voltage is around 100 mV. That means 2.5×10^7 Na$^+$ ions per s flow through an open channel. Taking the channel's length equal to the membrane's width $\cong 10$ nm $= 10^{-8}$ m and accepting that the ions pass through the channel in single file (Palmer 1986; Panagopoulos et al. 2000), we find that the transit time of every Na$^+$ ion through the Na$^+$ channel is $\sim 0.4 \times 10^{-7}$ s, and thus, the ion velocity through the channel is: $u = 2.5 \times 10^7 \times 10^{-8}$ m/s $\Rightarrow u = 0.25$ m/s (see also Chapter 11).

Considering that this velocity is acquired under the force of the transmembrane electric field, which is a huge field ($\sim 10^7$ V/m), any other velocity acquired by any charged particle/molecule within biological tissue due to any externally applied EMF will normally be several orders of magnitude smaller than that. Thus, we can reasonably accept that this ion drift velocity through an open ion channel represents an upper limit for the maximum velocity an ion can acquire within living tissue. Indeed, the velocity of an oscillating ion, according to the ion forced oscillation mechanism, is found for all frequencies and for all possible field intensities of environmentally existing polarized EMFs to be much smaller than 0.25 m/s (see Chapter 11 and Panagopoulos et al. 2021). Thus, the max ion velocity in biological tissue is:

$$u_o = 0.25 \text{ m/s} \quad [1.27]$$

This maximum velocity (and corresponding kinetic energy) of the mobile ion was calculated independently of any externally applied EMF, and it is similar for any living system because cells in most organisms (e.g. in all animals) have identical cell membranes and ion channels. It, thus, represents a biophysical constant which is important for electromagnetic interactions in living tissues.

From Eq. 1.26, we get that the maximum kinetic energy corresponding to u_o, is: $\epsilon(\max) \approx 1.2 \times 10^{-27}$ J. This is respectively an upper limit for the energy that may be absorbed by a single sodium ion due to the interaction with an applied EMF (which is usually several orders of magnitude smaller).

The Thermal Energy

The average kinetic energy of a mobile ion (and of any free molecule) of mass, m_i, and velocity u_{kT} due to thermal motion for tissue temperature T is (Alexopoulos 1962; Mandl 1988; Panagopoulos et al. 2013b):

$$\epsilon_{kT} = \frac{1}{2} m_i u_{kT}^2 = \frac{3}{2} kT \quad [1.28]$$

which gives:

$$u_{kT} = \sqrt{\frac{3kT}{m_i}} \quad [1.29]$$

(T the tissue absolute temperature in K, and $k = 1.381 \times 10^{-23}$ J·K^{-1} the Boltzmann's constant). For Na$^+$ ions ($m_i \cong 3.8 \times 10^{-26}$ kg) and $T = 310$ K (human body temperature 37°C) we get: $\epsilon_{kT} \cong 6.4 \times 10^{-21}$ J, and $u_{kT} \cong 0.58 \times 10^3$ m/s.

It follows that the thermal velocity and energy of a sodium ion in living tissue at human body temperature are $\sim 2.3 \times 10^3$ times and $\sim 5.3 \times 10^6$ times greater, respectively, than the maximum velocity and kinetic energy that could ever be acquired by this ion due to any expected applied EMF. In fact, as explained, the differences are several orders of magnitude greater in the case of environmental EMF exposures. This result is in agreement with experimental studies showing that the vast majority of recorded EMF bioeffects are non-thermal (Carpenter and Livstone 1968; Adey 1981; 1993; Gründler 1992; Kwee and Raskmark 1998; Velizarov et al. 1999; Panagopoulos et al. 2007a; 2007b; 2010; and reviews Walleczek 1992; Goodman et al. 1995; Creasey and Goldberg 2001; Belyaev 2005; Panagopoulos and Margaritis 2009; Phillips et al. 2009; Behari 2010; Panagopoulos 2011; 2017; 2019a; 2019b; 2020;

Defining Wireless Communication Electromagnetic Fields   **45**

Wust et al. 2021). Moreover, the above result is in agreement with the suggested mechanism of action of EMFs on cells (Panagopoulos et al. 2000; 2002; 2015; 2020; 2021). Thus, environmental EMF exposures (even in today's EMF-polluted environment) do not normally result in increasing tissue temperature.

1.4.3 Thermal Exposures

Naturally, heating of any material occurs when the absorbed radiation has a frequency close to the infrared band (~ $3 \times 10^{11} – 3 \times 10^{14}$ Hz). This comes from the fact that the emission and absorption spectrum of a "black body" has a peak mainly in the infrared and, secondarily, in the visible band of the electromagnetic spectrum. According to Kirchhoff's theorem, any material body of temperature T absorbs and emits radiation at the same frequencies/wavelengths as a "black body" at the same temperature (Alexopoulos 1962; Alonso and Finn 1967; Panagopoulos and Margaritis 2003).

Heating of materials occurs also by artificial exposures to MWs of high intensity/power (≥ 0.1 mW/cm^2) and frequency (≥ 1 GHz), such as in MW ovens, which emit MW EMR at 2.45 GHz with a power of ~ 1000 W focused within the metal cavity of the oven. This is a well-established phenomenon in physics called "microwave heating" (Metaxas 1991; Clark et al. 2000; Olaniyi 2017). Man-made MW radiations used in WC and other applications with frequencies 1–10 GHz may start inducing slight temperature increases in living tissue when their power density increases more than ~ 0.1 mW/cm^2 (Panagopoulos and Margaritis 2003; Panagopoulos et al. 2013b). Environmentally existing MW exposures mainly due to mobile/cordless phones and corresponding antennas, Wi-Fi, wireless connections (Bluetooth), etc., range between 0.001 µW/cm^2 and ~200 µW/cm^2 (very close to mobile phones) (Panagopoulos et al. 2010; Panagopoulos 2017; 2019b; Wongkasem 2021).

The induction of small temperature increases of the order of 0.15–0.3°C has been reported after exposure of biological samples (*Caenorhabditis elegans*) to continuous-wave 1 W, 1 GHz emitted by a generator within an exposure chamber (Dawe et al. 2006). In real exposure conditions, a GSM mobile phone in "talk" mode at 0–1 cm distance (0.2–0.3 mW/cm^2, 0.9, or 1.8 GHz) was not found to induce heating at a 0.05°C level within the mass of food for fruit flies in exposed glass vials (Panagopoulos et al. 2004; 2007a; 2007b; 2010). Similar non-thermal findings are also presented by many studies referenced above (Carpenter and Livstone 1968; Kwee and Raskmark 1998; Velizarov et al. 1999; Belyaev 2005; Wust et al. 2021). A UMTS mobile phone at 1–2 cm distance in "talk" mode (~ 0.1 mW/cm^2, ~1.95 GHz) was found to increase the temperature in 5.6 ml blood cultures after 25 min exposure by 0.1–0.2 °C (Panagopoulos 2019b; 2020). Human exposures from base station antennas at distances ≥ 10 m are normally of significantly lower power densities than a mobile phone at 0–1 cm proximity. Thus, in most cases, man-made EMFs at environmentally existing levels are unlikely to induce significant temperature increases in biological tissue, not even at the level of 0.1–0.3°C; however, newer WC technologies and especially 5G with higher MW frequencies and intensities may do (Neufeld and Kuster 2018; Thielens et al. 2018; 2020).

In order for the EMF exposures to cause heating, they should be millions of times more powerful than most environmental ELF EMFs and significantly more intense than environmentally existing RF EMFs, such as, for example, the ELF fields in close proximity to high-voltage/power transformers or power lines or the RF fields within a MW oven focusing all of its radiating power within its cavity. GSM (2G), UMTS (3G/4G), or LTE (4G) mobile phones (with average radiating power ~ 0.1–1 W) at a few cm distance or more, or even a corresponding base station antenna (~ 10–100 W) distributing their power in all directions within wide angles, would not cause any heating apart from 0–0.3°C when used in contact or very close proximity during "talk" mode (or video calling) and after several min of exposure.

The mechanism of heating biological tissues is as follows: Due to friction during the induced forced oscillation of the charged/polar molecules (and especially mobile ions), a part of the particle's kinetic energy is converted to heat. The damping coefficient of electrolytes increases (conductivity decreases) with higher (MW) frequencies (Chandra and Bagchi 2000). This results in increased friction of the oscillating molecules and slight tissue heating which may become significant for increasing frequency and power. While with 2G, 3G, 4G mobile phones (v ~ 1–2 GHz), the heating effect, even with the device in close proximity to the body, ranges from 0°C to 0.2°C; newer WC radiation types, with increasing frequencies and especially 5G combining significantly higher frequencies (up to 80–100 GHz) and denser radiation beams of anticipated greater intensity, may produce significant thermal effects in addition to the already existing non-thermal induced by the ELF pulsation, modulation, and variability (Neufeld and Kuster 2018; Thielens et al. 2018; 2020; Panagopoulos 2020; Wongkasem 2021). Thus, RF/MW EMF exposures with frequencies approaching infrared and with high enough power density (≥ 0.1 mW/cm^2) may cause tissue heating.

The absorbed power per unit volume can be written according to tissue specific conductivity (σ) and electric field intensity (E) as, (see Section 1.5, Eq. 1.34):

$$\frac{dP}{dV} = \sigma E^2 \qquad [1.30]$$

As the specific conductivity of tissue depends on the frequency v of the field, the absorbed power P by living tissue will also depend on frequency. In MW heating, the absorbed power by a material (e.g., living tissue) per unit volume dP/dV increases with increasing wave/field frequency v, the dielectric loss factor ε' of the material, and the electric field within the material E according to the equation (Metaxas 1991; Clark et al. 2000; Olaniyi 2017):

$$\frac{dP}{dV} = 2\pi v \varepsilon_o \varepsilon' E^2 \qquad [1.31]$$

Thus, the MW heating effect increases as the EMF frequency (v) increases approaching the low limit of infrared, and as the EMF power density (depending on E^2 according to Eq. 1.8) increases, resulting in measurable heating. Apart from the forced oscillation of charged/polar molecules, the MW heating effect seems to be related with some kind of not yet fully explored resonant absorption mechanism when the MW EMF frequency approaches the low limit of infrared (and accordingly the wavelength reduces to a few mm – "mm-waves"). The more the EMF frequency approaches infrared and the EMF power density increases, the more significant becomes the effect, resulting in measurable heating. This is probably related to the natural phenomenon expressed by Kirchhoff's law that any material body absorbs EMR at the same wavelengths/frequencies at which this body emits electromagnetic radiation. These wavelengths/frequencies for all bodies are mainly in the infrared and, secondarily, in the visible part of the electromagnetic spectrum, as described above (Alexopoulos 1962; Panagopoulos and Margaritis 2003).

5G MT employs higher MW carrier frequencies (called mm-waves) in order to accomplish higher quality of simulations (data transfer). But with higher frequencies, the heating of exposed living tissues increases (Eq. 1.31), while penetration through different materials (e.g., air, buildings, etc.) decreases (Eq. 1.2). In order to overcome the low penetration, the number of antennas must be significantly increased, and the intensity of the emissions as well. Under such conditions, thermal effects in exposed humans cannot be excluded in addition to the already existing non-thermal effects. Studies have theoretically predicted the induction of significant thermal effects (Neufeld and Kuster 2018; Thielens et al. 2018; 2020). These facts further justify the concerns expressed by the scientific community against the installation of 5G (Hardell and Nyberg 2020; Kostoff et al. 2020; Panagopoulos 2020).

1.5 MEASURING INCIDENT EMFs IS MORE RELEVANT THAN SAR

1.5.1 Analysis of the SAR

SAR (in W/kg) is defined as the incremental power dP absorbed by an incremental mass of tissue dm contained in a volume element dV of a given density $\rho = dm/dV$ (in kg/m^3) (NCRP 1986):

$$SAR = \frac{dP}{dm} \qquad [1.32]$$

Defining Wireless Communication Electromagnetic Fields

Eq. 1.32 can be expressed according to tissue conductivity, density, and internal electric field, or according to tissue specific heat and temperature increase.

SAR According to Tissue Conductivity

By use of the Ohm's law:

$$j = \sigma E \quad [1.33]$$

where j is the electric current density (in A/m²) within the tissue due to the internal electric field E, and σ is the tissue specific conductivity (in S/m) and assuming certain quantities to be constant within the tissue, Eq. 1.32, after operations, becomes:

$$SAR = \frac{\sigma \cdot E^2}{\rho} \quad [1.34]$$

which is equivalent to Eq. 1.30.

Eq. 1.34 is frequently reported in papers for defining and estimating SAR, but its derivation is never described or considered. Actually, Eq. 1.34 cannot be derived unless certain physical quantities are assumed to be constant. This, of course, is a simplification that minimizes its validity. To address these issues, we must see how this formula is derived.

Derivation of Eq. 1.34

Neglecting thermal losses, the absorbed electric power dP can be expressed as the power of an electric current i (generated within the tissue by the applied EMF) flowing vertically across an area S, $dP = d\Psi \cdot i$, where $d\Psi$ is an incremental voltage induced by the EMF exposure. Then, Eq. 1.32 becomes: $SAR = d\Psi \cdot i / dm$, which can be written as: $SAR = \frac{d\Psi \cdot i \cdot S}{dm \cdot S}$, or $SAR = \frac{d\Psi \cdot j \cdot S}{dm}$, where $j = i/S$ is the current density across the area S. Since $d\Psi = E\, dr$, where E is the generated electric field within the tissue and dr is a displacement of electric charge as part of the current i, we get: $SAR = \frac{E \cdot dr \cdot j \cdot S}{dm}$. Considering that $dr \cdot S$ is the volume dV defined by the area S and the charge displacement dr containing tissue mass dm, and dm/dV is the tissue density ρ, assuming it is constant within the volume dV, the previous relation becomes $SAR = \frac{j \cdot E}{\rho}$, and replacing j with σE due to Ohm's law (Eq. 1.33), we reach the desired formula (Eq. 1.34): $SAR = \frac{\sigma \cdot E^2}{\rho}$.

It is obvious that in the above operations, the quantities i, j, S, E, ρ, and σ were assumed to be constant within the incremental volume dV, and, moreover, it is obvious that Eq. 1.34 refers to this volume only. In any other volume outside dV, SAR has a different value and must be calculated separately. By applying Eq. 1.34 to the whole volume of an animal, organ (e.g., eye), a group of organs (e.g., head), or even a single cell, it is assumed that j, E, ρ, and σ are constant within those volumes. This, of course, is an oversimplification, as every organ or group of organs consists of many different tissues, and all the above quantities vary significantly between different tissues and even within a single type of tissue and within a single cell (Panagopoulos et al. 2013b).

In particular, specific conductivity varies significantly among different tissues. For example, at a frequency of 1 GHz, specific conductivity in different tissues of the human body varies from about 0.04 S/m (bone marrow) to about 2.45 S/m (cerebrospinal fluid). Even within a single cell, specific conductivity can have huge variations from 10^{-7} S/m in the cell membrane to 0.5–1 S/m in the cytoplasm (Foster and Schwan 1989; Fear and Stuchly 1998).

In addition, the available data on tissue conductivity are collected from measurements on dead animals (Schwan 1957; 1963; Gabriel et al. 1996a; 1996b). The variations become significantly greater in live animals. Conductivity values in the ELF band of up to ~ 300% higher than those previously reported by Schwan (1957; 1963) were measured in porcine organs of just sacrificed animals. The differences from the previously known corresponding conductivity values were attributed to the fact that the organs were still alive and filled with blood during the measurements in contrast to the previous studies which were performed on dead organs. It was found that within an hour from animal sacrifice, the conductivities of different organs/tissues decreased by up to 50% of their original values in the alive animal (Spottorno et al. 2008; 2012), which is absolutely reasonable. These findings raise serious questions about the validity of tissue conductivity data measured before and their dependence on frequency. Moreover, the conductivity of the various organs – especially of the brain – in all animals changes with age. The conductivity of a young child's brain is almost double the conductivity of an adult's brain, resulting in almost double radiation absorption and SAR (Peyman et al. 2001; Christ et al. 2010).

Finally, human tissue density varies from about 900 kg/m³ (fat) to about 1200 kg/m³ (tumor) between different soft tissue types and reaches a value of about 1800 kg/m³ for bones (Gabriel et al. 1996b).

Thereby, Eq. 1.34 provides a poor expression/definition of SAR because of the large variations of the related quantities, and any estimating method for SAR based on Eq. 1.34 includes very large uncertainty. Eq. 1.34 actually applies only within incremental volumes dV significantly smaller than single cells. Applying Eq. 1.34 on whole organs (e.g., heart, spleen, eye, etc.), groups of organs (e.g., head), or on whole animals by using average conductivity, density, and internal field values can be very misleading, as it grossly underestimates the local microscopic variations of these parameters which determine the potential biological effects.

SAR According to Tissue Specific Heat

For a homogeneous medium (thus, neglecting density and tissue-type variations) with specific heat c_h, [in J/(kg·K)] (thus, neglecting also any variations in specific heat) and by use of a form of the calorimetry law,

$$\frac{dQ}{dt} = mc_h \cdot \frac{dT}{dt} \qquad [1.35]$$

Eq. 1.32 becomes:

$$SAR = c_h \cdot \frac{dT}{dt} \qquad [1.36]$$

where dQ/dt is the radiation power transformed into an amount of heat dQ within the tissue mass m, producing a temperature increase dT during an incremental time interval dt.

For a measurable temperature increase δT during a measurable time interval δt, Eq. 1.36 would be written as:

$$SAR = c_h \cdot \frac{\delta T}{\delta t} \qquad [1.37]$$

Since variations in specific heat within living tissue are much smaller than corresponding variations in conductivity (Gabriel et al. 1996a; 1996b; Haemmerich et al. 2005), resulting in much more uniform temperature than electric field distribution, Eq. 1.37 provides a better way for SAR estimation and, therefore, definition (Panagopoulos et al. 2013b).

In addition, while differences in internal electric field intensity are retained during the whole exposure period as they depend on tissue permittivity, which has large variations even within a single cell, differences in temperature between different parts of a tissue or organ disappear a short time after the beginning of a constant exposure, and temperature gets evenly distributed within a whole organ or even body. Moreover, while tissue conductivity and permittivity/internal electric field are reported to change significantly with different frequencies of the applied EMF/EMR (Gabriel et al. 1996a; 1996b), specific heat is independent from the applied field and depends only on tissue properties. In case of exposure to WC EMFs, which include several different frequencies (carrier, pulsing, modulation), conductivity and internal field intensity become even more variable, and their accurate estimation even more complicated, while specific heat is constant.

Even if we consider a single frequency and additionally neglect internal field intensity and density differences, spatial conductivity variations alone result in considerably greater variability of SAR when calculated by Eq. 1.34 than by Eq. 1.37. For example, most organs/parts of the human/animal body contain both muscle and fat tissues. While at 1 GHz muscle specific conductivity (~ 1.006 S/m) is about 1,760% higher than fat specific conductivity (~ 0.054 S/m), muscle specific heat (~ 3.5 kJ/kg·K) is only 56% higher than fat specific heat (~ 2.3 kJ/kg·K). This would result to a ~ 1,700% larger variability in the SAR of this specific organ or part of the animal body when estimated by Eq. 1.34 than when estimated by Eq. 1.37. At lower frequencies, conductivity variations increase considerably, resulting in an even larger variability in the SAR calculation, while specific heat has the same value. For example, at 10 MHz, the above difference in SAR variability (~ 1,700%) between Eq. 1.34 and Eq. 1.37 becomes ~ 2,125% (or 21.25 times greater according to Eq. 1.34 than according to Eq. 1.37) (Leonard et al. 1984; IEEE 2002). If we add variations in internal electric field intensity and tissue density we may have hundreds of times greater variability in SAR values according to Eq. 1.34 than according to Eq. 1.37. Thus, while variation in SAR calculation according to Eq. 1.37 is restricted to measurement errors and the assumption that c_h has the same value throughout the tissue, which somehow can be tolerated, corresponding variation in SAR according to Eq. 1.34 includes similar errors plus tenths or even hundreds of times greater variability. This shows that the only way to reliably estimate SAR is by macroscopically measuring the temperature increases – if any – within biological tissue according to Eq. 1.37 (Panagopoulos et al. 2013b).

In fact, Eqs. 1.36 and 1.37 are also inaccurate, as it is assumed that all power absorbed by the exposed biological tissue is converted into heat, which, of course, is not true either. In the non-thermal effects, the power absorbed by mobile ions that are forced to oscillate in phase with the external field can be converted to gate electrosensitive ion channels (VGICs) by exerting electric forces on their channel sensors (Panagopoulos et al. 2000; 2002; 2015a; 2020; 2021). But, as we showed that the absorbed energy of these forced oscillations is more than millions of times smaller than the thermal energy of the same particles, once we have measurable heating, we may assume that this is by far greater than any other non-thermal energy absorption.

From the above analysis, it follows that SAR actually applies only to thermal effects, and it actually expresses the rate by which electromagnetic energy from an incident electromagnetic wave/field is converted into heat within living tissue. But, as already explained (Section 1.4), man-made electromagnetic fields at environmentally existing intensities do not normally induce measurable heating within exposed living tissue. Thus, SAR is not a proper metric to describe the biological activity of man-made electromagnetic fields at environmental intensities.

1.5.2 SAR Estimation Methods

SAR is estimated by 1) insertion of micro-antennas/probes into the tissue to detect the internal electric field. Assuming the conductivity and the density of the tissue to be constant, SAR is computed by Eq. 1.34; 2) insertion of miniature thermal probes into the tissue to detect changes δT in the temperature caused by the exposure during a time interval δt, assuming the tissue is homogeneous with known specific heat. Then SAR is computed by Eq. 1.37; 3) numerical modeling, such as the Finite Difference Time Domain (FDTD) method, simulating the spatial distribution of the absorbed energy within an object and computing SAR by Eq. 1.34 (Moulder et al. 1999).

Apart from the disadvantage of the first method regarding oversimplification of Eq. 1.34, in both the first and second methods, the insertion of needles/probes in living tissue disturbs its physiological function and distorts its physical properties in unpredictable ways. Moreover, in the case of live animals, it causes trauma and pain. Such methods are improper to be used in live animals and may only be used in *in vitro* experiments with cell cultures.

Numerical modeling, such as the FDTD method, which is considered the best, numerically simulates the energy absorption within the tissue by use of special computer programs, dividing the tissue volume into cubes (voxels), and assigning each of them certain values of conductivity, permittivity, and density. Then SAR is (again) computed by Eq. 1.34. Because conductivity, permittivity, and density are assumed to be constant within each voxel, this method, like the first one, is a simplification. This explains why earlier SAR estimates used in the currently accepted criteria for whole body average SAR (ICNIRP 1998; 2020) are questioned by more recent and more accurate FDTD calculations (Wang et al. 2006; Flyckt et al. 2007; Gandhi et al. 2012).

All methods of simulation, no matter how much improved, will always be highly simplified compared to the complexity of living tissue because they can never take into account the countless microscopic variations in its physical parameters. Modeling living tissue by attributing average dielectric values in whole animals or organs has been a method applied by engineers treating living tissue as an inanimate material. Such methods highly underestimate the potential biological effects which depend on significant variations of dielectric properties at microscopic level and are not taken into account by average values. Unfortunately, such simplistic methods continue to dominate in EMF dosimetry (ICNIRP 1998; 2020; Behari 2010; IARC 2013; Wongkasem 2021).

In conclusion, all the existing methods for SAR estimation, especially those based on Eq. 1.34, have serious deficiencies. Actually, only the second method, which is based on measurable tissue heating, is reliable to be applied only in cell cultures. Finally, all methods for SAR estimation are highly complicated and time-consuming, so that SAR cannot be readily measured/calculated by use of the equipment of an ordinary EMF laboratory. In other words, SAR is not only a flawed metric but impractical as well.

1.5.3 Incident EMF

A more precise and practical EMF exposure metric than SAR is the incident radiation/field intensities on the surface of the exposed biological tissue at the various frequency bands (RF, ELF, VLF, etc.) plus the additional physical parameters of the field and the exposure which can readily and accurately be known, such as pulse and carrier frequency, exposure duration, modulation, waveform, etc., as measured by reliable radiation/field meters, frequency meters, and spectrum analyzers. (Panagopoulos et al. 2013b)

As already explained, today there are thousands of studies corresponding specific biological effects to specific radiation/field intensities at the different frequency bands plus the other exposure parameters. Therefore, one can approximately predict the biological effect by knowing these field/exposure parameters, which can be readily and objectively measured. An example of different GSM intensities inducing DNA fragmentation in fruit fly ovarian cells for 6 min exposure is found in Panagopoulos et al. (2010).

Any inaccuracy in the intensity measurement, especially of the highly variable WC EMFs, and especially in the near fields, would be further increased in a corresponding SAR estimation. More specifically, if the electric field intensity E varies significantly, the corresponding SAR value depending on E^2 (according to Eq. 1.34) will include the square of this variation plus the variation in the conductivity and density of the biological tissue. Moreover, the SAR will refer to the absorbed field, which introduces an additional error in its estimation than in the incident field which is directly measured by any reliable instrument.

Intensity measurements of incident WC EMFs, and especially in the near field, may, indeed, include errors due to the described increased variability (see Section 1.3) and even possible capacitive coupling between the antenna/device and the sensor of the field meter. The error can be effectively minimized by a) using near-field probes that are now available in the market, b) increasing the number of measurements and reporting average intensity and SD and even excluding certain unrealistically high measurements which could possibly be attributed to capacitive coupling. This provides a representative estimation of the incident field. "Accurate" estimation of the instant intensity of WC EMFs, especially in the near field, has no meaning, as these EMFs are highly varying any moment due to the reasons described above (and in Panagopoulos et al. 2015b; 2016; Panagopoulos and Karabarbounis 2020). Similarly, accurately calculating the SAR or internal fields within organisms exposed by WC antennas/devices and especially in the near fields is actually impossible and introduces significantly greater errors than measuring the incident fields. While average and peak intensity values can be representatively measured, SAR corresponding values still carry the flaws described above (Sections 1.5.1 and 1.5.2).

For taking into account possible field distortion by the exposed object due to possible resonance phenomena and areas of increased radiation absorption, although such phenomena are not expected to cause any significant alterations, radiation/field intensity measurements should be carried out both in the presence and in absence of the object and in different locations corresponding to different parts of its surface. If the measured values in the presence and in the absence of the object are significantly different, both sets of measurements should be reported.

Certainly, due to the usually encountered non-linearity in the response of living organisms to different environmental stimuli and especially EMFs, not even radiation/field intensity (along with the rest field parameters) is expected to be precise predictor of the expected biological effect at all frequency/intensity areas. Non-linear effects in which the dose-response relation is not a straight line, such as intensity or frequency "windows" reported occasionally in the EMF bioeffects literature (Bawin et al. 1975; 1978; Blackmann et al. 1980; Liboff 2003; Panagopoulos et al. 2010; Panagopoulos and Margaritis 2010b), cannot be predicted by either intensity or SAR dosimetry. At the very least, radiation/field intensity can be readily and more accurately measured than SAR in any case.

As there is today overwhelming evidence on the intense adverse biological activity of man-made EMFs, and especially WC EMFs, with detrimental effects on human/animal health and the natural environment, the need for fast and reliable EMF monitoring has become necessary on a regular basis, especially at residential, social, and working places. EMF measurements should be readily performed by EMF laboratories around the world by proper use of reliable field/radiation meters and spectrum analyzers. Today, such instruments are widely available in the market, relatively cheap, and easily used by qualified and experienced scientists/engineers and trained individuals. EMF dosimetry should not be based on complicated, time-consuming, and largely inaccurate methods of SAR estimation. [The problems with the *SAR* versus incident EMFs were originally analyzed in Panagopoulos et al. 2013b.]

1.6 ALL MAN-MADE EMFs EMIT CONTINUOUS WAVES NOT PHOTONS

1.6.1 MISLEADING ASSESSMENT OF EMF BIOACTIVITY BASED ON PHOTON ENERGY

The physics community has accepted that all EMFs (including man-made) and all corresponding types of EMR consist of photons (Alonso and Finn 1967; Beiser 1987; Walleczek 1992; Valberg et al. 1997; Pall 2013; Levitt et al. 2021). According to this postulate, man-made EMFs having frequencies in the subinfrared range ($0-3 \times 10^{11}$ Hz) cannot induce any biological effects because their "photon energies" (according to Planck's law – Eq. 1.3) are lower than those of natural light (Valberg et al. 1997; ICNIRP 1998; 2020; Balzano and Sheppard 2003), which is not harmful at normal intensities but vital. But then, what about the thousands of studies showing a plethora of biological and health effects at man-made frequencies? Is it possible that all these findings corroborating each other are wrong and should be ruled out? The experimental/epidemiological findings are scientific facts and cannot be ruled out. Therefore, obviously, the hypothesis must be ruled out. The hypothesis, here, is that all EMFs (including man-made) consist of photons, and, thus, any effect is due to photon absorption. This hypothesis was not based on experimental facts in the case of man-made EMFs, but on the mathematical "quantization of the EMF" performed by the founders of QED. We shall show here that this hypothesis is flawed in the case of man-made EMFs because spectral data show otherwise, and because the mathematical "quantization" offered by QED/QEM was actually based on the simplistic assumption that all EMFs are periodic in time.

1.6.2 Quantized States Produce Quantized Emissions (Photons) and Line Spectra

There is an intrinsic property of matter that the energy of its elementary particles in a bound state can only take discrete values; in other words, their energy is quantized. It is actually a hypothesis that atoms bound in molecules, electrons bound in atoms, and nucleons bound in nuclei are in perpetual periodic motions at stationary states (discrete energy levels) without emitting radiation, despite their accelerated motion. Radiation is only emitted during transitions from one discrete energy level to another. Such transitions are very fast (of the order of ~ 10^{-9} s for electrons), and, during this time, a wave-packet of certain frequency, phase, polarization, and length (~ 30 cm) is emitted/absorbed (Alexopoulos 1963; 1966; Beiser 1987; Panagopoulos 2015; 2018). These nanosecond wave-packets are the photons. Photons produced by specific transitions (molecular/atomic/nuclear) have discrete frequencies and, thus, give discrete lines in molecular/atomic/nuclear emission spectra. It is well-known in physics that individual sources of quantized emissions (molecules/atoms/nuclei) produce spectra with discrete lines (Herzberg 1944; 1950; Alexopoulos 1963; 1966; Klimov 1975; Burcham and Jobes 1995).

This general hypothesis for the quantization of the electronic energies in all atoms was made by Bohr (1913a; 1913b; 1914; 1928) in the study of the hydrogen atom and was extended by Wilson (1915) and Sommerfeld (1916) for any periodic motion in a single-electron atom. The Bohr-Sommerfeld-Wilson quantization rules allowed the calculation of the stationary energy levels in the hydrogen atom and in single-electron ions, which really corresponded to the observed frequency lines of the atomic spectra. This fact proved correct Bohr's hypothesis for the energy quantization of electrons bound in atoms, and soon it was found that similar quantization rules apply to all bound micro-particles not only in atoms but also in molecules and nuclei (Gautreau and Savin 1978; Beiser 1987).

The energy quantization of all molecules, atoms, and nuclei explains their stability and this, in turn, explains the stability of matter. The quantization implies that bound micro-particles in molecules/atoms/nuclei cannot spontaneously jump from one stationary state to another, as that would require the absorption/emission of energy amounts corresponding to the energy differences between different stationary states. If bound electrons' energies were to take not only discrete values, the electrons would constantly lose energy due to their acceleration around the nuclei (with consequent emission of EMR), and, inevitably, they would collapse and fall on the nuclei. In such a case, no matter would exist in the form of the chemical elements we know (Panagopoulos 2018). A direct consequence of this is that molecules/atoms/nuclei emit and absorb only discrete amounts of energy (photons) corresponding to transitions between discrete energy levels. It was found that the energy differences between such levels in molecules/atoms/nuclei correspond to frequency bands from infrared to gamma rays. More specifically, transitions between different molecular oscillation energy levels correspond to the emission/absorption of photons in the infrared band; electronic transitions in atoms correspond to photons in the visible, ultraviolet; and x-ray bands; and nuclear transitions correspond to photons in the gamma-ray band (Gautreau and Savin 1978; Beiser 1987). Moreover, these quantized transitions correspond to discrete frequencies, and this is why all molecular, atomic, and nuclear spectra are line spectra consisting of discrete lines (Herzberg 1944; 1950; Alexopoulos 1963; 1966).

No transitions were found to correspond to photon energies/frequencies below infrared except for the rare case of photons in the RF/MW band emitted after artificial excitation and/or in the presence of a strong static magnetic field (usually of the order of ~ 0.1–1 T), as in the Stern-Gerlach experiment, in the nuclear magnetic resonance (NMR) spectroscopy, the electron spin resonance (ESR) spectroscopy, and the maser MW amplifiers (see Section 1.6.6). But such strong static magnetic fields do not exist in the environment. The intensity of the terrestrial static magnetic field is ~ 0.5 G = 0.5 × 10^{-4} T, which is much smaller (~ 2000–20,000 times) than the magnetic field in NMR/ESR spectroscopy (Panagopoulos 2018).

Cosmic MW radiation is known to be of originally higher frequency (infrared/visible) which reaches the Earth reduced due to the Doppler effect taking place because of the cosmic expansion (Durrer 2008; Panagopoulos 2018). Thus, cosmic MWs, indeed, consist of photons, but they are not actually MWs. They are infrared radiation shifted toward lower frequencies. Moreover, they are not polarized or coherent in contrast to man-made MWs which are totally polarized and coherent. Thereby, the argument that living organisms on Earth have always been exposed to MW radiation of cosmic origin does not stand.

In conclusion, all quantized (photonic) emissions occurring spontaneously in our natural and daily environments correspond to discrete frequencies which are, in all cases, higher than the low limit of infrared.

1.6.3 Continuous States Produce Continuous Waves and Continuous Spectra

In contrast to the time-finite emissions from bound micro-particles, free charged particles emit EMR continuously during acceleration, as predicted by classical electromagnetism (Alonso and Finn 1967; Alexopoulos 1973; Jackson 1975). A continuous emission generates continuous waves of length increasing with the duration of the emission. This is fundamentally different from a time-finite quantized emission. It is obvious that such a continuous emission cannot correspond to discrete energy/frequency transitions but to a continuous range of energies/frequencies.

The intensity J of EMR (in the vacuum or in the air) emitted by an accelerating particle of charge q, with non-relativistic velocity (as is the case with free electrons accelerating in the metallic conductors of all electric/electronic circuits/antennas), at any angle θ with the direction of motion, and at distance r from the charged particle, is described by the equation

$$J(\theta) = \frac{q^2 a^2}{16\pi^2 \varepsilon_o c^3 r^2} \cdot \sin^2 \theta \qquad [1.38]$$

where a is the acceleration/retardation of the charged particle, ε_o is the vacuum permittivity, and c is the speed of light in the vacuum/air (Alonso and Finn 1967; Panagopoulos 2018).

The frequency range of the emitted radiation is determined by the curves in the free electron trajectories, which, in turn, are determined by the frequency and amplitude of the applied alternating voltage, the electron velocity, and the collision parameters with the ions of the metal. This frequency range extends within a narrow band around the main frequency of the applied voltage (Jackson 1975). When direct (non-alternating) voltages are applied in the circuit, the frequency of the emission is determined by the velocity and the collision parameters only.

Thus, radiation emitted by accelerating/decelerating free electrons in circuits/antennas, or ions and electrons in air discharges, etc., depends upon the square of the acceleration/deceleration a^2. Because the acceleration a can take any possible value (within a range determined by the applied forces), the emitted radiation can also take any possible value within a corresponding range. The emission is not time-finite, and the emitted electromagnetic waves do not consist of discrete wave-packets of finite length but of continuous waves like those described by classical electromagnetism, containing a continuous range of frequencies around the main frequency of the applied voltage in the circuit.

The continuous part of x-ray spectra emitted by retarding free electrons impinging on a metallic surface consists of continuous "classical" waves not photons. Parts of the energy of the continuous waves are absorbed by inner bound electrons in the metal atoms, which get excited to higher energy states, and emit discrete frequencies by de-excitation providing the discrete spectral lines in the final x-ray spectra. The discrete lines correspond to photons, while the continuous part of the spectra corresponds to continuous waves.

Ionic oscillations discovered in all living cells with ULF frequencies of the order of 0.01–0.2 Hz are continuous oscillations and, thus, emit continuous waves not photons. Similarly, atmospheric discharges in the VLF and ELF bands and their resulting Schumann resonances are continuous emissions of accelerating charges (electrons, ions) taking place for as long as the discharge lasts. Finally, all forms of EMR produced by all man-made electric/electronic circuits (e.g., power lines, antennas, etc.) are continuous emissions of accelerating free electrons within the metallic conductors (Panagopoulos 2013; 2018; Panagopoulos and Balmori 2017).

The continuous emission spectra of "black body" radiation, sunlight, light from lamps, hot solid bodies, x-rays, RF/MW antennas, cosmic MWs, and atmospheric discharges are, at least in part, due to accelerating free charged particles such as free electrons/ions (and even neutral molecules in thermal motion) existing in all the above EMR sources and not exclusively to quantized transitions (photons).

Any continuous emission spectrum may be attributed either to a) acceleration of unbound charged microparticles such as free electrons/ions accelerated by an applied electric field and uncharged particles in thermal motion, or b) to transitions of bound microparticles corresponding to a continuous range of photon frequencies, resulting in a seemingly continuous spectrum that even a spectrum analyzer with the highest resolution cannot discriminate the individual spectral lines; or c) to a combination of both a and b cases.

One could, undoubtedly, clarify whether a certain continuous emission spectrum is due to accelerating free microparticles or quantized transitions (photons) of a continuous frequency range were it, indeed, possible to detect discrete photons from the emission source, but it is not. Single photons have not been detected, in spite of opposite claims. In fact, what are really detected are "clicks" in photomultipliers (detectors). Each "click" represents the emission of a discrete photoelectron, and this is interpreted as corresponding to the absorption of a discrete photon (Roychoudhuri and Tirfessa 2008). But highly accurate photon counting experiments have more recently shown that actually the simultaneous detection of multiple photons ("multiple units of $h\nu$") is necessary for the emission of a single photoelectron, and, thus, the production of a single "click" on a detector does not correspond to the detection of a single photon (Panarella 2008). Thus, in reality, single photons have not been detected, in spite of the widely spread impression for the opposite (Roychoudhuri et al. 2008; Roychoudhuri 2014). Since photoelectron emission could also, hypothetically, be triggered by partial absorption of a (divisible) continuous wave, there is no way to verify beyond any doubt the existence of photons by use of photomultipliers.

Therefore, we cannot undoubtedly verify the existence of photons in the continuous spectra, and it is actually only the line spectra that show the existence of photons with discrete frequencies emitted by bound microparticles. As for the continuous spectra of free electron emissions in all man-made EMFs, similarly, there is no proof nor any indication that they consist of photons.

A single charged free microparticle accelerating in the vacuum due to an alternating applied voltage may move periodically, and then its emission spectrum would (theoretically) contain only discrete lines/frequencies. But in electric/electronic circuits, we do not have a single microparticle accelerating in the vacuum due to a perfectly alternating applied voltage. Instead, we have innumerable microparticles (free electrons) moving not periodically (even in case of a perfectly alternating field), with each one's individual period/frequency slightly differing from all others' due to the chaotic friction forces which are different for each individual microparticle, plus their random thermal motion, which is also different for each one. This is why EMR produced by accelerating free charged microparticles gives continuous emission spectra and why all antennae spectra are continuous spectra (Panagopoulos 2018).

In conclusion, bound charged microparticles produce photons with discrete frequencies/energies and line spectra, while free charged microparticles produce continuous waves and continuous spectra. This distinction is fundamental for understanding the arguments presented here.

1.6.4 How Causality Was Abandoned in Modern Quantum Physics

According to the Fourier theory, all periodic motions of any frequency can be represented as a sum of discrete harmonic oscillations with a basic frequency v and its harmonic frequencies $2v$, $3v$, ..., etc. Non-periodic functions/motions may also be developed into Fourier series, but, in this case, the Fourier series do not approach the initial functions. One of the three Dirichlet conditions in order for a Fourier series to approach the initial function is that the initial function must be periodic. Therefore, non-periodic undulations cannot be represented as a sum of discrete harmonic terms. Except for the Fourier series, any continuous and integrable function, periodic or not, can be transformed by the Fourier integral/transform into another continuous function consisting of infinite number of (non-discrete) harmonic terms (Stephenson 1973; Spiegel 1974; Panagopoulos 2018). But a continuous function with no discrete terms cannot be considered as "quantized". It is most strange that these simple facts were overlooked by the founders of QEM/QED and their successors. Let us see briefly how this happened.

De Broglie (1924) ascribed a wavelength (λ) on elementary particles. such as electrons, accepting that they possess a wave-like nature and called these waves "matter-waves":

$$\lambda = \frac{h}{mu} \quad \text{or} \quad k = \frac{p}{\hbar} \qquad [1.39]$$

(m, u, and p are the particle's mass, velocity, and momentum, respectively, and k is the wave number). His hypothesis was soon confirmed experimentally when it was shown that electrons produce diffraction patterns just like x-rays (Davisson and Germer 1927).

In his attempt to find an equation to describe the energy of the electronic "matter-waves" in the many-electron atoms, Schroedinger (1926) took the classical wave-function,

$$\xi(r, t) = e^{i(\omega t - kr)} \qquad [1.40]$$

which describes a plane harmonic wave of circular frequency $\omega = 2\pi v$ (v the frequency) and wave number $k = 2\pi/\lambda$ (λ the wavelength) at distance r from its source along the direction of propagation [i is the imaginary unit ($i^2 = -1$)] (Alonso and Finn 1967). [The fact that Eq. 1.40 describes a plane harmonic wave comes from the Euler formula, $e^{i\theta} = cos\theta + isin\theta$ (Stephenson 1973), with the convention that physical quantities are obtained by taking the real parts of complex quantities (Jackson 1975). Thus, a physical wave described by Eq. 1.40 depends solely upon $cos(\omega t - kr)$ (which is a harmonic function of time) and, therefore, it is a plane harmonic wave.]

Then he substituted ω and k by their corresponding quantum mechanical expressions (derived directly from Planck's and De Broglie's Eqs. 1.3 and 1.39, respectively, $\omega = \epsilon/\hbar$ and $k = p/\hbar$) and derived what he called the quantum mechanical "wave-function" in direct analogy with the classical wave-function (Schroedinger 1926; Tarasov 1980; Trachanas 1981):

$$\psi(r, t) = e^{i(\epsilon t - pr)/\hbar} \qquad [1.41]$$

The square of the wave-function ψ^2 supposedly describes the probability for the electron to be found at distance r from the nucleus at a given time. That was arbitrarily accepted also in analogy with classical wave physics in which the square of the oscillating quantity (wave-function) is proportional to the energy density of the wave. Thus, Schroedinger identified the energy density of the matter wave associated with the electron at a specific location around the nucleus, as the probability of finding the electron at this location (Panagopoulos 2018).

Differentiating Eq. 1.41 with respect to r and t, he found the operator $-i\hbar(\partial/\partial r)$ corresponding to the momentum and the operator $i\hbar(\partial/\partial t)$ corresponding to the energy of the particle, respectively.

In classical physics, the total (conserved) energy value ∈ of a particle with mass m and momentum p moving with potential energy $V(r)$ is the sum of its kinetic and potential energy. The equation $\epsilon = p^2/2m + V(r)$ expresses the energy conservation law.

Since the wave-function $\psi(r, t)$ was introduced to represent the wave associated with the particle under study, Schroedinger demanded *a priori* that it must satisfy the equation:

$$\epsilon \psi = \left(\frac{p^2}{2m}\right)\psi + V(r)\psi \qquad [1.42]$$

Substituting in the last equation the energy and momentum by their corresponding operators, we get:

$$i\hbar\left(\frac{\partial \psi}{\partial t}\right) = -\left(\frac{\hbar^2}{2m}\right)\left(\frac{\partial^2 \psi}{\partial r^2}\right) + V(r)\psi \qquad [1.43]$$

In three dimensions, the equation, describing the energy conservation law for a "matter-wave", becomes:

$$i\hbar\frac{\partial \psi}{\partial t} = -\frac{\hbar^2}{2m}\nabla^2\psi + V(r)\psi \qquad [1.44]$$

($\nabla^2 = \frac{\partial^2}{\partial x^2} + \frac{\partial^2}{\partial y^2} + \frac{\partial^2}{\partial z^2}$ is the Laplace operator)

Despite the arbitrary assumptions made by Schroedinger in order to derive his equation (Eq. 1.44), there was a causal reasoning in his methodology up to this point, and this is probably the reason why this equation seems to really work in describing the electronic states in atoms. But causality was abandoned in the next step.

Although the Schroedinger equation was originally written to describe the energy of electrons bound in atoms, since they were described by harmonic wave functions with quantized energy, it was arbitrarily extrapolated for the case of a free electron/particle with zero potential energy [$V(r) = 0$] when it was also written by Schroedinger himself as,

$$i\hbar\frac{\partial \psi}{\partial t} = -\frac{\hbar^2}{2m}\nabla^2\psi \qquad [1.45]$$

But in such a case, how can a harmonic wave-function (Eq. 1.41) with quantized energy be attributed to a free particle? By doing this, it was automatically accepted that any free particle can only have discrete energy values by itself, even when it is not in periodic motion. That was an unphysical extrapolation, and the start of a wrong direction that was to be followed. Causality was ruined by this step.

This unphysical extrapolation made by Schroedinger (1926) was blindly followed by Klein, Gordon, Dirac, Heisenberg, and everybody else at that time, when they all adopted this equation to describe a free particle (!), and this was surprisingly accepted by everyone else in the quantum physics community until today without any objections (Panagopoulos 2018).

1.6.5 The Mathematical "Quantization" of EMF/EMR

The reasoning of "quantization" of an EMF/EMR is described by Dirac (1927): "Resolving the radiation into its Fourier components, we can consider the energy and phase of each of the components to be dynamical variables describing the radiation field". But according to the Fourier theory, a non-periodic function (such as any random emission of radiation) cannot be represented/approached by Fourier components.

Let us see in brief how Heisenberg, Born, Jordan, Pauli, and Dirac "quantized" the EMF/EMR, starting from its classical description by Maxwell's equations. In the vacuum or the air and considering the free fields (without electric charges or currents), Maxwell's equations (Alonso and Finn 1967) are written as:

$$\nabla \cdot E = 0 \qquad [1.46]$$

$$\nabla \cdot B = 0 \qquad [1.47]$$

$$\nabla \times E = -\frac{\partial B}{\partial t} \qquad [1.48]$$

$$\nabla \times B = \varepsilon_o \mu_o \frac{\partial E}{\partial t} \qquad [1.49]$$

They introduced a vector potential $A(r, t)$ which should, *a priori*, satisfy both the constraint $\nabla \cdot A = 0$ and the equations:

$$B = \nabla \times A \qquad [1.50]$$

$$E = -\frac{\partial A}{\partial t} \qquad [1.51]$$

[vector labeling ($\rightarrow$) on E, B, A, is omitted for simplicity]

After such a fabrication, substituting Eqs. 1.50 and 1.51 into Eq. 1.49, it comes that $A(r, t)$ satisfies the classical wave equation (to be transmitted along the direction r with velocity c, just like the electric and magnetic components of an electromagnetic undulation):

$$\frac{\partial^2 A(r,t)}{\partial t^2} = c^2 \nabla^2 A(r,t) \qquad [1.52]$$

with $c = \frac{1}{\sqrt{\varepsilon_o \mu_o}}$ the velocity of the electromagnetic wave.

Then they demanded the vector potential to be a periodic function of time and separated it into a sum of two conjugate complex terms:

$$A(r,t) = A^{(+)}(r,t) + A^{(-)}(r,t) \qquad [1.53]$$

where $A^{(+)}(r, t)$ contains all amplitudes which vary as $e^{-i\omega t}$ for $\omega > 0$, and $A^{(-)}(r, t)$ contains all amplitudes which vary as $e^{i\omega t}$, and $A^{(-)} = A^{(+)*}$. Thus, they fabricated $A(r, t)$ in such a way that a) satisfies the wave equation and b) is a periodic function of time and, thus, contains only harmonically varying terms.

Since they accepted that $A(r, t)$ is periodic in time, they developed its terms into Fourier series of harmonic terms, according to the Fourier theorem, with a set of vector "mode" functions $u_k(r)$ satisfying the wave equation, $(\nabla^2 + \omega_k^2/c^2) u_k(r) = 0$ for harmonic waves, corresponding to the frequencies ω_k, describing the field restricted in a volume V in space (with c_k the Fourier coefficients):

$$A^{(+)}(r,t) = \sum_k c_k u_k(r) e^{-i\omega_k t} \qquad [1.54]$$

Finally, after additional arbitrary requirements and operations, the vector potential is transformed as:

$$A(r,t) = \sum_k \left(\hbar/2\omega_k \varepsilon_o\right)^{1/2} \left[a_k u_k(r) e^{-i\omega_k t} + a_k^\dagger u_k^*(r) e^{i\omega_k t}\right] \qquad [1.55]$$

The Fourier amplitudes a_k, and $a_k^\dagger$, were arbitrarily chosen to mutually adjoint operators which satisfy the commutation relations: $[a_k, a_{k'}] = [a_k^\dagger, a_{k'}^\dagger] = 0$, $[a_k, a_{k'}^\dagger] = \delta_{kk'}$
[$\delta_{kk'} = 1$ for $k = k'$, and 0 for $k \neq k'$ (the "Kronecker's delta" function)]

Replacing $A(r,t)$ into Eq. 1.51, the electric field becomes:

$$E(r,t) = i\sum_k \left(\hbar\omega_k/2\varepsilon_o\right)^{1/2} \left[a_k u_k(r) e^{-i\omega_k t} - a_k^\dagger u_k^*(r) e^{i\omega_k t}\right] \qquad [1.56]$$

and a similar expression is found for the magnetic field. These finally transform the Hamiltonian (total energy) of the EMF as:

$$H = \sum_k \hbar\omega_k \left(a_k^\dagger a_k + \frac{1}{2}\right) \qquad [1.57]$$

[For details see Mandel and Wolf (1995), Walls and Milburn (2008), and Panagopoulos (2018)]

Eq. 1.57 represents the total energy of the EMF as the sum of the number of photons in each mode $a_k^\dagger a_k$, multiplied by the photon energy in this mode $\hbar\omega_k$, plus $\frac{1}{2}\hbar\omega_k$ representing the energy of the "vacuum fluctuations" in each mode.

Thus, the famous "EMF quantization" is nothing more than mathematically transforming a periodic EMF into a sum of discrete terms by use of the Fourier series. But in nature, most forms of EMFs are not periodic and cannot be approximated as such. Finally, the fact that they mathematically transformed a periodic EMF into a sum of discrete terms does not mean that these terms represent photons. There should be facts supporting this "quantization", and such facts do not exist for man-made EMFs or for the other EMF continuous emissions with frequencies below infrared described above (see Section 1.6.3).

1.6.6 No Evidence of Photons at Frequencies below Infrared in Environmental Conditions

Let us now examine what are referred to as "microwave photons" and the ways they are generated (originally discussed in Panagopoulos 2018). As for lower frequency bands (lower RF, LF, VLF, ELF, ULF), there is not even a mention in the physics literature regarding actual evidence of photon existence.

MW Generators

MWs are produced artificially by generators such as the magnetron, the klystron, and the masers (Lioliousis 1979). The magnetron and the klystron produce electron beams emitted by a cathode and directed to pass through a series of positively charged metal cavities called "cavity resonators". The frequency of the produced oscillations in the electron beam is determined by the cavities' dimensions and the beam's speed. Such MW generators are used in radars and in MW ovens. The produced MWs last for as long as the electrons accelerate within the cavities and are, thus, continuous/uninterrupted waves. They are produced by unbound electrons accelerated by an applied voltage just like in every electric/electronic circuit, and there is no reason to assume that they are quantized. There are no time-finite emissions to correspond to quanta (photons).

In the case of masers (microwave amplification by stimulated emission of radiation), the continuous MWs produced by a klystron or magnetron are amplified by MW photons produced by some paramagnetic material, such as NH_3 or crystals such as silicon (Si), after excitation by the continuous MWs and in the presence of a strong static and spatially inhomogeneous magnetic field with intensity of the order of ~ 1T, like in the Stern-Gerlach experiment. This, indeed, describes conditions of photon production in the RF/MW band. It is related to the splitting of spin energy levels of uncoupled electrons or nucleons within a strong static magnetic field B (~ 1T) (Gautreau and Savin 1978), which is the underlying effect in the ESR and NMR spectroscopies. Uncoupled particles may jump between the two separated spin levels with corresponding emission/absorption of photons in the MW band. Thus, such photons may exist under the specific conditions.

But such strong static magnetic fields (~ 1T) do not exist in human environments. Moreover, the production of MW photons cannot take place without excitation by the artificial (continuous) MWs. Thus, we do not expect to have MW photons due to this mechanism in environmental conditions.

Atomic Transitions in the RF/MW Band

There are atomic transitions due to the hyperfine splitting of electronic energy levels in atoms, corresponding to photon energy in the RF/MW band (typically of a few GHz). The hyperfine splitting is due to the interaction of the nuclear magnetic moment with the electron magnetic moment. The function of "atomic clocks" is based on this effect. Such hyperfine transitions do not occur naturally/spontaneously and need to be excited artificially. In atomic clocks, excitation of cesium atoms is achieved by periodic laser signals in a chamber at superconductive conditions (extremely low temperature very close to absolute zero, -273°C). By de-excitation, the cesium atoms emit photons of precise MW frequency. Other ways to excite MW transitions in atoms involve magnetic resonance by an externally applied magnetic field (of the order of ~ 0.1–1 T) and artificial MW radiation (see above). The resulting magnetic resonance is observed by changing the frequency and magnitude of the applied RF field. Again, such conditions do not exist environmentally, and the described hyperfine transitions do not occur naturally (Major 2014; Kraus et al. 2014; Panagopoulos 2018).

MWs Produced by "Qubits"

In practice, the devices that are currently being developed to produce MW "photons" need to be operated at temperatures below 0.1K (or -272.9°C) (Houck et al. 2007; Inomata et al. 2016). Until recently, this would have meant using cryostats with liquid helium for cooling, which is generally not possible in conditions outside of research labs. Rapid progress in cryogenics has already produced dry mechanical systems that only require a source of electricity to operate (Radebaugh 2009), but still, such conditions do not exist environmentally.

Recent claims that MW/RF photons can be generated in electronic circuits also involve superconductive/cryogenic conditions. The so-called "microwave photons", generated by special MW oscillation circuits, called quantum bits ("qubits"), are manifested as electromagnetic pulses. Qubits are integrated micro-circuits made by lithography and containing capacitors (C) and inductors/coils (L) forming LC harmonic oscillators. They are the basic units of the so-called "quantum computers" (Houck et al. 2007). A large amplitude trigger pulse generated by a conventional MW pulse generator in the "in" port excites the qubit which, a few tens of nanoseconds later, decays into the "out" port by emitting a second pulse which is interpreted as a MW "photon". With the circuit resistance approaching to zero in superconductive conditions, the generated pulses (interpreted as "photons") are practically harmonic (Houck et al. 2007; Schuster et al. 2007; Clarke and Wilhelm 2008).

Thus, the so-called MW "photons" emitted by qubits are not quantized transitions of bound micro-particles but pulses of a continuous carrier wave at MW frequency produced by the LC artificial micro-circuits. Even if we interpreted these artificial MW pulses as photons, which is definitely not the case, they could not exist in the environment (without superconductive/cryogenic conditions and without artificial excitation) (Panagopoulos 2018).

In conclusion, all present day "quantum" MW emitters a) need to be triggered by artificial pulses and b) are cooled down to extremely low temperatures (Houck et al. 2007; Kraus et al. 2014).

Antennae Spectra

If man-made EMR types were indeed quantized, according to QEM/QED hypothesis, then all antennae emission spectra anywhere in the whole band below infrared ($0-3 \times 10^{11}$ Hz) would be line spectra consisting of discrete lines corresponding to the basic carrier frequency emitted by the antenna and its harmonics plus the modulation frequencies. Although spectra may be very complicated, and discrete lines may broaden due to a variety of reasons, as is usually the case in molecular, x-ray, and gamma-ray spectra, acquiring these spectra with increased resolution reveals their discrete lines. In contrast, all antennae emission spectra do not display discrete lines regardless of resolution, but they do display continuous frequency bands around the main emission frequencies. This is because, even though macroscopically the free electron cloud in the antenna circuit may perform a periodic motion at a certain carrier frequency v, the motion of each individual free electron is not periodic due to the chaotic friction forces which are different for each individual free electron plus the individual random thermal motion, as explained. The result is that, instead of an individual emitted frequency, we have a continuous range of frequencies $\pm \Delta v$ around the carrier frequency v of the alternating voltage applied on the antenna circuit. In other words, instead of single lines, we have continuous frequency bands with peaks on the main frequencies (Panagopoulos 2018).

Thus, antennae spectra are continuous spectra, even though antennas in most (almost all) cases emit a periodic carrier signal, and this is an additional indication that all man-made EMR types do not consist of photons but of continuous waves.

In conclusion, there is actually no evidence showing photon existence at frequencies below infrared, in environmental conditions, or showing that man-made MW radiation types transmitted by WC antennas/devices, radars, satellites, etc., consist of photons.

1.7 DIFFERENCES FROM NATURAL EMFs. INTERACTION WITH MATTER

1.7.1 Differences between Natural and Man-made EMFs/EMR

Many people, including scientists, are not aware of the differences in the physical properties and the consequent differences in biological activity between natural and man-made EMFs/EMR, coming to the erroneous conclusion that since natural light, which is of significantly higher intensity and frequency, does not induce adverse health effects, man-made EMFs/EMR should not induce adverse effects either. Let us summarize the differences between natural and man-made EMFs/EMR which were analyzed in the previous sections (and originally in Panagopoulos 2018).

A. *Polarization*: All man-made EMFs/EMR emitted by circuits/antennas are totally polarized (and coherent), in most cases linearly polarized, oscillating on a certain plane determined by the orientation/geometry of the antenna/circuit. By contrast, natural EMFs/EMR (such as Schumann resonances, cosmic MW, infrared, visible, ultraviolet, gamma) are never totally polarized (nor coherent) and may only be partially polarized in a small degree under certain conditions. Exceptions are the geomagnetic and geoelectric fields and the cell membrane electric fields, which are locally polarized but static.

B. *Frequency bands*: Man-made EMFs/EMR occupy the lower frequency bands, from 0 Hz up to the low limit of infrared ($\sim 3 \times 10^{11}$ Hz). Natural EMFs/EMR occupy the higher frequency bands of the electromagnetic spectrum, from infrared to gamma rays (3×10^{11}–3×10^{22} Hz). Exceptions include a) the VLF/ELF EMFs of atmospheric discharges (lightning) and consequent Schumann resonances; b) the geoelectric, geomagnetic, and cell membrane electric fields which are basically static with ELF variations (Presman 1977; Dubrov 1978; Panagopoulos 2013; Panagopoulos and Balmori 2017); c) the preseismic ULF/ELF/VLF pulsations (including SES) recorded a few days or weeks before major earthquakes (Panagopoulos et al. 2020); and d) the ULF ionic oscillations in all living cells.

C. *Bound versus unbound emission sources*: Natural EMR is produced by time-finite transitions (excitations/de-excitations) of bound charged microparticles (i.e., atoms/ions, electrons, or nucleons, in molecules, atoms and nuclei respectively), between quantized energy levels, and for this reason it consists of time-finite wave-packets (photons). By contrast, man-made EMR types (and the above-mentioned exceptions of the atmospheric/terrestrial/biological natural ULF/ELF/VLF EMFs), are produced by continuous (uninterrupted) acceleration of free electrons/ions due to an applied EMF, and for this reason they consist of continuous "classical" waves.

The above fundamental differences indicate that man-made EMFs should not be confused or compared with the natural ones without addressing these differences, and they should not be evaluated for their biological activity by the same criteria (Panagopoulos and Margaritis 2003; Panagopoulos 2011; 2013; 2018).

1.7.2 Basic Concepts of Interaction of EMFs/EMR with Matter

Natural EMR (from infrared to gamma) passing through inanimate matter can be absorbed by bound charged atoms/ions in molecules (infrared), electrons in atoms (visible, ultraviolet, x), or nucleons in nuclei (gamma) in all materials and by free electrons in metals. The main mechanisms of interaction are:

A. *Excitations*: They take place when the frequency of the radiation is close to the frequencies of the molecular/atomic/nuclear spectra in the corresponding bands. Bound charged atoms and electrons absorb the necessary amount of energy in order to jump to a higher stationary energy level. The excited molecules/atoms/nuclei are unstable, re-emit the absorbed energy in the form of time-finite emissions (photons) in random directions, and get back to their initial energy levels.

B. *Ionizations*: For higher frequencies (vacuum ultraviolet, x-rays, gamma rays) the absorbed energy is adequate to ionize the atoms by expelling electrons and even excite or break nuclei (in the case of gamma radiation). These are known effects of ionizing radiations (Alexopoulos 1963; Klimov 1975; Gautreau and Savin 1978; Beiser 1987; Burcham and Jobes 1995).

C. *Forced oscillations*: Bound charged atoms and electrons in all materials and free electrons in metals are forced to oscillate at the frequency of the radiation in addition to their initial motions. The energy of the forced oscillation is subtracted from the radiation and re-emitted by the charged particles in all directions. This causes scattering of the initial waves (Alonso and Finn 1967; Alexopoulos 1963; Klimov 1975; Panagopoulos 2018). In all cases, the initial EMR is left with the same frequency but reduced intensity.

Man-made EMR has several orders of magnitude lower frequency than the frequencies of the molecular/atomic/nuclear spectra (ranging from the infrared to the gamma-ray band), and thus, it is not expected to induce excitations or forced oscillations on bound microparticles and certainly not ionizations.

Forced-oscillation of free electron clouds on metallic surfaces is the mechanism by which metals absorb man-made EMFs/EMR. In this case, the absorption is so intense as to practically eliminate EMR in the interior of the metallic object and shield other objects behind the metallic surface (e.g., "Faraday cage"). This is how metals can insulate space from EMFs/EMR (Alexopoulos 1973; Panagopoulos 2018; Panagopoulos and Chrousos 2019).

The situation is different when the continuous polarized waves of man-made EMFs/EMR pass through living tissue. Living tissue consists of biological cells, and in all types of cells (and in the extracellular fluids), except for the bound electrons in atoms/molecules, there are trillions of mobile ions, water polar molecules, and polar macromolecules. The vast majority of biological molecules such as proteins, lipids, nucleic acids, etc., are either polar or carry a net electric charge (Alberts et al. 1994; Stryer 1996). Therefore, except for the above mechanisms of energy loss on bound electrons, there are induced forced oscillations on every charged or polar molecule of the biological tissue (as described in Section 1.4). These forced oscillations of ions and polar (macro)molecules absorb much more energy than the induced oscillations on the bound electrons of the biological molecules because the masses of the charged/polar particles are now several orders of magnitude (more than 10^4 times) bigger. The forced oscillations induced by man-made EMFs/EMR in biological tissue are parallel and coherent oscillations since, as explained, these fields are totally polarized and coherent.

The induced oscillations will be most intense on the mobile ions which carry a net electric charge and have smaller mass and higher mobility than other charged or polar molecules (Alberts et al. 1994; Panagopoulos 2013). The induced oscillations will be much smaller or even negligible on the polar macromolecules that do not carry a net electric charge, they have much greater masses, and they are usually chemically bound to other molecules. Forced oscillations of mobile ions can trigger biological effects (Panagopoulos et al. 2000; 2002; 2015; 2020; 2021).

After induction of forced oscillations by the continuous polarized waves on the charged/polar molecules of living tissue and consequent abstraction of energy from the initial wave, the remaining wave continues its way through the tissue with the same frequency but reduced amplitude/intensity. After countless numbers of such events, depending on the tissue's mass, density, and the number of polar/charged molecules, any remaining wave leaves the tissue scattered and with reduced amplitude/intensity (Panagopoulos et al. 2013b).

The wave intensity J (as in the simplest case of a plane harmonic electromagnetic wave described by Eq. 1.7) decreases with decreasing amplitude/intensity E of the oscillating field/wave within the tissue after interaction with the charged/polar molecules. Thus, the amplitude and energy of each individual continuous wave decrease.

The energy loss of the man-made electromagnetic waves may be manifested as heating of the exposed material (e.g., MW heating) without any frequency reduction as, e.g., in the Compton effect. Information-carrying MWs do not change their frequency when passing through matter, but they can cause heating when they have sufficient intensity and frequency (MWs in the GHz range with intensity ≥ 0.1 mW/cm^2).

Thus, man-made EMF/EMR types lose energy not by losing a number of photons absorbed by the medium or by decreasing their frequency as in the Compton effect (by getting absorbed and giving rise to scattered photons of decreased frequency). This might explain why MW radiation can cause greater temperature increases than ionizing radiation when absorbed by matter, although it has considerably lower frequency. Ionizing radiation is quantized (photonic) and described by Planck's equation (Eq. 1.3) in terms of its energy, while man-made radiation (including MWs) consists of continuous waves, and described by Eq. 1.7, in which the energy loss is not dependent on quantized (all or nothing) absorption but on partial absorption from a continuous/uninterrupted wave, inducing a continuous forced oscillation on charged/polar particles. In this case, the energy loss transformed into heat may be greater, even though the frequency is several orders of magnitude smaller.

Natural non-ionizing quantized EMR (infrared, visible light) also decreases in intensity (number of photons) when passing through biological matter by causing forced oscillations on charged/polar particles. But these oscillations are in random directions (each photon oscillates on a different plane) and not coherent. For this reason, they only cause heating (increase in molecular random thermal motion) which is tolerated by living organisms if it is not excessive. Important adverse biological effects and cancer may be caused by (natural quantized) ionizing radiations through the breakage of chemical bonds in biological molecules. Thus, the mechanisms of interaction with living tissue are quite different between quantized and not quantized EMR, even though they may finally result in the same effects (e.g., genetic damage, cell death, cancer, etc.).

1.8 DISCUSSION AND CONCLUSIONS

In this chapter we described the physical properties that characterize WC EMFs. Some of these properties (polarization/coherence, non-thermal energies, and emission of continuous waves instead of photons) account not only for WC EMFs but for all types of man-made EMFs. The combination of polarization/coherence with the intense variability of the WC signals, the combination of different frequency bands, and the ULF/ELF components in the form of pulsing, modulation, and random variability, are specific properties of the WC EMFs. Although WC EMFs are usually referred to in the literature simply as "RF" EMFs, this is not only inaccurate but also misleading, as these fields/radiations necessarily combine RF carrier signals with ELF/VLF modulation and pulsing plus ELF/ULF random variability. These ELF/ULF components are the most bioactive, not the RF carrier, which is usually responsible only for heating.

We explained the property of polarization which (combined with coherence) is inherent in all technical/artificial/man-made EMF/EMR emissions, including those of WC. We showed how this property is necessary for the induction of biological effects through the phenomena of constructive interference and most importantly the induced forced oscillations on every charged particle in biological tissue and especially mobile ions. We showed that the biological effects of man-made EMFs arise from their unique property of being totally polarized (and coherent) capable of inducing parallel and coherent forced oscillations/rotations on charged/polar molecules which are the vast majority of molecules in living tissue.

We underscored that polarization alone is not enough for the induction of biological effects but low frequency (ULF/ELF/VLF) variability of the EMF exposure is also necessary. In a comparison study, 36 min total exposure to real-life GSM (2G) EMF emitted by a mobile phone induced DNA damage in fruit fly ovarian cells in a much higher degree than 120 h total exposure to 50 Hz alternating EMF significantly stronger than those of high-voltage power lines. The crucial difference between the two exposures was found to be the intense variability of the real-life GSM EMF (Panagopoulos 2019a). The importance of field variability, especially in intensity, is also indicated by the recorded health effects in human populations during magnetic storms, the nerve impulses which are voltage changes in the membranes of nerve cells, and the gating of VGICs in all cell membranes. These effects do not occur while the static polarized terrestrial or cell membrane fields retain their regular field intensities but initiate once their intensities undergo changes of the order of 20%–30% of their regular values. This bioactive variability lies mainly in the ELF/ULF band. In addition, a plethora of experimental findings show the increased ability of ELF/ULF man-made (polarized) EMFs to induce biological effects.

We noted the similarity between the terrestrial fields and the cell membrane fields. They are both locally polarized and static and normally not bioactive. Effects are triggered whenever changes of ~ 20%–30% of their regular field intensities occur. This observation is important for the explanation of the biological/health effects of EMFs in general and shows that polarization, combined with variability, is the trigger for EMF bioeffects (Panagopoulos 2019a).

We explained that all WC EMFs necessarily contain ULF/ELF/VLF components in the form of modulation, pulsing, and random variability, and thus, they combine polarization with ELF/ULF variations. Although information regarding the ELF pulsations of WC EMFs (especially of LTE, 5G, and Wi-Fi) is limited in the literature and not easily accessible for reasons unknown to us, we provided measurements of the ELF components (Table 1.1), and we showed pulsations of the most common forms of such emissions, such as GSM (2G MT), UMTS (3G/4G MT), LTE (4G), DECT, and Wi-Fi/Bluetooth, (Figures 1.2–1.8) collected from the available specialized studies on this topic. The difficulty in finding information in the literature regarding the ELF pulsations of WC EMFs (summarized in Table 1.2), in spite of the fact that the pulsing character of these EMFs/radiations is their most important technical feature and their most bioactive component, shows the degree of misinformation prevailing today in science.

In a recent review of studies of the European Parliamentary Research Service (EPRS 2021) (authored by Thielens and reviewed by Vacha and Vian) regarding environmental impacts of 5G, there is no mention of pulsations or any other ELF components, and the only examined frequency band of the radiation is the carrier (MW) frequency. Moreover, the importance of the inherent variability of the real WC exposures in inducing biological/health effects is not even mentioned, and studies are criticized for having used real-life emissions from mobile phones for the exposures, which, as explained, is the only realistic exposure method (Panagopoulos et al. 2015b; 2016; Panagopoulos 2017; 2019a; Leach et al. 2018; Kostoff et al. 2020). Thus, the most important parameters of WC EMFs (low frequency components, variability) were completely ignored. They criticized the real-life exposures and the EMF measurements in our and others' studies, based on Verschaeve (2014) and do not mention our published comments on Verschaeve's paper (Panagopoulos et al. 2016). Reproducing the criticism expressed in a paper without referring to the peer-reviewed published response to this criticism is a major flaw. Verschaeve is known for attempting to discredit every study that has found effects from man-made EMFs. His "arguments" collapsed in our comments (Panagopoulos et al. 2016). As a result, he did not comment on our studies again (Verschaeve 2017). Now EPRS (2021) reproduce Verschaeve's (2014) "arguments" as if they were not rebutted. This is not a way for science to move forward.

Another recent review of 107 experimental and 31 epidemiological studies with "RF" EMFs above 6 GHz (in order to assess bioactivity of 5G) by members of the Australian Radiation Protection and Nuclear Safety Agency again makes no mention of pulsations or any other ELF components in the 5G or in the examined studies, and no mention whether there is any similarity of the signals produced by generators in the studies with those of the 5G apart from the carrier frequency. Although most of the reviewed studies had reported genotoxic and various other effects, the authors of the review found "no confirmed evidence" of adverse effects on human health and criticized the studies for not being "independently replicated" and for employing "low quality methods of exposure assessment and control" (Karipidis et al. 2021). The same authors also made a "meta-analysis" of the same 107 experimental studies and found that the studies "do not confirm an association between low-level mm-waves and biological effects" (Wood et al. 2021). They also estimated the "effect size" (an arbitrary measure of bioactivity) among studies that reported "continuous wave" and "modulated" "RF" EMFs and found "non-significant difference". But the "effect size" of the studies reporting modulation was found to be almost double (4.3 ± 1.6) than that of the studies reporting "continuous wave" (2.2 ± 0.6), and it is strange how this difference was reported as "non-significant". Moreover, as explained in the present chapter and in Panagopoulos (2021), it is unlikely that any MW generator does not contain on/off pulsations, even only for energy-saving reasons, as in radars. Even the onset and removal of an EMF exposure alone may produce the greatest effects (Goodman et al. 1995).

The fact that these two publications and the EPRS (2021) ignore the presence of ELF components and whether the reviewed studies employed simulated signals or real-life WC signals, shows that they are not reliable for investigating the health issues of these types of EMFs. Such publications attempt to present 5G radiation as harmless to health and environment, which is clearly not the case.

A part of the scientific community believes that the ELF/ULF components of WC EMFs do not exist independently of the RF carrier and need to be "demodulated" in order to affect living organisms (Goldsworthy 2006; Sheppard et al. 2008; Wust et al. 2021). Demodulation of a modulated RF signal is accomplished by "non-linear" electronic elements in the RF receivers in electronics, such as diodes, transistors, etc. (Alexopoulos 1973; Schwartz 1990). Studies have clearly shown that the

ELF elements exist and can be recorded independently of the RF carrier, as shown in Section 1.3.2 (Pedersen 1997; Holma and Toskala 2004; Zhou et al. 2010; Pirard and Vatovez). "Demodulated" or not, the fact is that both ELF meters and living organisms detect them and are affected by them. This is why modulated and pulsed RF EMFs by ELF are shown by plethora of studies (cited in the Introduction of this chapter) to be bioactive, while the corresponding non-modulated and non-pulsed signals are not.

We analyzed the physics of non-thermal effects of man-made EMFs in biological tissue, which constitute the vast majority of effects at environmental conditions and the physics of thermal effects (the known phenomenon of MW heating). We calculated the velocity of an ion passing through an open channel in a cell membrane (Eq. 1.27), which represents an upper limit for any velocity of a mobile ion in living tissue under the influence of an applied EMF. This velocity is of major importance for the estimation of physical effects in living cells (see also Chapter 11) and represents a biophysical constant. We calculated the corresponding maximum kinetic energy and compared it with the average thermal energy of the same particle. We showed that this upper limit energy of an ion is millions of times smaller than the average thermal energy of the same particle, and this explains why the vast majority of the recorded biological/health effects of man-made EMF exposures are non-thermal. The available evidence shows that these non-thermal effects are due to the ELF EMFs included in almost all artificial EMFs in combination with their totally polarized and coherent character.

In recent publications, Wust et al. (2020; 2021) (Table 1.1 in both papers) provide ion velocities though opened channels about four orders of magnitude smaller (~ 10,000 times). They estimated these as being due to an applied RF field supposedly "rectified" by the membrane and superimposed to the transmembrane field. But how can an externally applied field be rectified by a cell membrane? Ions (both positive and negative) flow in and out of the membrane through the channels all the time. If the membranes were "rectifiers", they would only allow ion flows in one direction. They "estimated" this "rectified" voltage to be of the order of 1 μV while the transmembrane voltage is ~ 100 mV. This is completely hypothetical and not based on measurements (in contrast to Eq. 1.27). Moreover, it can be very misleading, as readers may think that the ion velocities through open channels may be of such magnitude.

Recently, due to the higher MW frequencies ("mm-waves") included in 5G, certain Russian studies came to light reporting "non-thermal effects of MW/mm-wave EMFs". Three reviews of such studies in English are Pakhomov et al. (1998), Betskii and Lebedeva (2004), and Belyaev (2005). In several studies reviewed in Pakhomov et al. (1998), and in Belyaev (2005), ULF/ELF, and VLF components were reported to be present in the form of pulsing, and/or modulation/intermittence/variability, while for the rest of the reviewed studies, no information on possible existence of such components was provided, and thus, their presence is not excluded. In the Betskii and Lebedeva (2004) review paper, information on the possible existence of low frequency components (ULF/ELF/VLF) is totally missing throughout the paper, and thus, their presence is again not excluded. Since, as explained, it is unlikely that any MW electronic circuit/generator is not turned on and off even only for energy-saving reasons, the existence of ULF/ELF/VLF components and the separate roles of the low and high frequencies in the biological effects need to be carefully investigated in order to prevent misleading conclusions. In this context, speaking of "non-thermal MW effects" without having clarified whether these effects are indeed due to the MWs or to their low frequency components can be very misleading. Systematic attempts by Gandhi and coworkers to reproduce "non-thermal biological effects" induced by pure MW carrier signals without modulation or pulsations as reported by Russian and German researchers were unsuccessful, and only thermal effects could be elicited by such exposures at higher power densities (Bush et al. 1981; Stensaas et al. 1981; Gandhi 1983; Furia et al. 1986). Wust et al. (2020; 2021) also speak of "non-thermal effects of RF fields" without reporting any measurements in the low frequencies (ULF/ELF/VLF) for the emissions of the device they used. Speaking of "RF" effects without having explored the possible coexistence of low frequencies (which unfortunately is the common case in many publications) is very misleading.

As reported earlier in this chapter, in most of the studies which compared a pulsed and/or modulated complex RF EMF with the same EMF without pulsation/modulation, it was found that it was the low frequency (ULF/ELF/VLF) pulsation/modulation and not the carrier alone that produced the non-thermal biological effects. As correctly summarized by Goldsworthy (2006), "Radio waves can also give biological effects, but only if they are pulsed or amplitude modulated at biologically active low frequencies". These facts are fully explained by the ion forced oscillation mechanism (Panagopoulos et al. 2000; 2002; 2015a; 2020; 2021), and there is no corresponding mechanism to explain non-thermal effects by high frequencies (RF/MW) alone (see Chapter 11).

Polarized and coherent ELF EMFs induce parallel and coherent forced oscillations on any charged/polar particle with energy well below the thermal level. The oscillating ions exert forces on the sensors of electrosensitive ion channels (VGICs) in cell membranes causing their irregular opening or closing with consequent disruption of the intracellular ionic concentrations and the electrochemical balance in all types of cells. This biophysical mechanism, known as "ion forced-oscillation mechanism" (described in Chapter 11 of this book), provides the basis for the explanation of the non-thermal effects of all man-made EMFs (Panagopoulos et al. 2000; 2002; 2015a; 2020; 2021). Today, the unique ability of ELF polarized EMFs to irregularly gate VGICs is widely recognized, verifying the aforementioned mechanism (Liburdy 1992; Walleczeck 1992; Pall 2013; Ceccetto et al. 2020; Zheng et al. 2021; Bertagna et al. 2021). Because of these unique properties of the man-made EMFs, EMF exposure by a mobile phone with average intensity ~ 10 μW/cm^2 on a human body may initiate adverse non-thermal biological effects, while ~ 10 mW/cm^2 (1000 times stronger) solar EMR with significantly longer exposure during the day does not (Panagopoulos 2017; Panagopoulos et al. 2015a).

While the vast majority of EMF-induced recorded bioeffects are non-thermal, heating increases with increasing RF frequency, as shown by Eq. 1.31, and may become significant with the higher frequencies employed in 5G MT technology. As, at the same time, penetration of the EMR decreases with increasing frequency (Eq. 1.2), it will likely become necessary to increase the intensity of the 5G signals in addition to the installation of huge number of additional base stations, antennas, and satellites. The existence of antenna arrays in 5G technology provides the ability of stronger and focused radiation/field beams (Eqs. 1.23–1.24). It is noteworthy that just before the massive deployment of 5G, the ICNIRP (2020) increased the limit for 6 min average exposure at 2–6 GHz from 1 to 4 mW/cm^2 (ICNIRP 1998; 2020; Panagopoulos 2020). While the older limit (1 mW/cm^2) provided limited protection against heating, the new one does not. A combination of non-thermal and thermal biological effects can be far more dangerous than non-thermal effects alone.

We discussed how WC emissions should be better described according to incident EMF than according to SAR. The argument that we need to know the power absorbed by the tissue in order to predict the biological effect has been disproven by the plethora of published peer-reviewed experimental studies, which correspond specific field/radiation intensities, frequencies, exposure durations, etc., to specific biological effects. For example, we know that WC EMF exposure with intensities ≥1 μW/cm^2 may initiate biological effects within minutes, and the effects increase with increasing intensity and exposure duration (Panagopoulos et al. 2004; 2007a; 2007b; 2010; Panagopoulos and Margaritis 2010a). We do not need to calculate the SAR by complicated methods to know this. We can predict the effect by knowing the incident radiation intensity, frequency, exposure duration, etc. We showed that a) when SAR is estimated from tissue conductivity and internal electric field, important microscopic variations in tissue conductivity are overlooked, and b) when SAR is estimated from tissue specific heat and increased temperature is significantly more accurate, but most environmental EMF exposures do not cause measurable tissue heating. Moreover, this method cannot be used in experiments with live animals, as needles/thermal probes need to be inserted, but only in experiments with cell cultures. Thus, SAR is rendered useless for the majority of EMF exposures which are non-thermal and for those involving live animals. Although, at higher MW frequencies of newer technologies (≥ 2 GHz) and high intensities (≥ 0.1 mW/cm^2) (such as 3G, 4G), there may be temperature increases at 0.1–0.3°C level (which will likely become more significant with the 5G) the biological effect of man-made EMFs is determined by field parameters not directly (or at all) included in SAR such as polarization, frequency, pulsing, modulation, variability, exposure duration, etc. Moreover, the biological effect depends on microscopic power absorption by specific biomolecules (e.g., DNA), which is not easy to estimate. Thus, SAR is of very limited value to describe bioactivity of EMF exposures. Instead, the incident radiation/field intensity at the included frequency bands should be reported along with the other field parameters, the exposure duration, variability (SD) of the measured intensity values, etc. SAR may be used complementarily in experiments with cell cultures exposed to high frequency/power MWs causing measurable heating (Panagopoulos et al. 2013b). Marino et al. (2016) have expressed similar views: "To provide an objective basis for follow-up studies, the power density of the incident radiation, which was the independent variable in the study, was characterized by direct measurement rather than by employing model-dependent dosimetry parameters, such as the specific absorption rate".

Similarly, Baker et al. (2004), even though they explored thermal effects in magnetic resonance (MR) imaging, concluded: "using *SAR* to guide MR safety recommendations for neuro-stimulation systems or other similar implants across different MR systems is unreliable and, therefore, potentially dangerous. Better, more universal, measures are required in order to ensure patient safety".

We analyzed the important issue of whether man-made EMFs/EMR consist of photons or continuous "classical" waves and the mathematical "quantization" of the EMF/EMR by the founders of QED/QEM. We showed that the mathematical "quantization" was based on the simplistic assumption that any EMF is periodic in time, allowing them to transform it into a Fourier series of discrete terms. The discrete terms were then interpreted as the "photons" of the EMF/EMR. But any random EMF is not periodic in time and, thus, cannot be transformed by application of the Fourier series. This simplistic approach started by Schroedinger, who used a harmonic wave-function to describe a free particle (Eq. 1.45). By application of the Fourier integral (Spiegel 1974), a randomly varying EMF could be theoretically transformed into a continuous of an infinite number of (non-discrete) harmonic oscillators. But this is not a "quantization". Thus, the argument that man-made (including WC) EMFs cannot induce any biological/health effects due to their small "photon energy" collapses simply because there are no such "photons", and this is in agreement with the thousands of experimental and epidemiological studies showing a vast number of adverse effects on a variety of organisms/tissues/cells.

Finally, we summarized the important differences between natural and man-made EMFs/EMR which imply that these two categories of EMFs should not be evaluated by the same criteria for their bioactivity. General concepts for the interaction of both natural and man-made EMFs/EMR with inanimate matter and biological tissue were discussed as well. We hope the presented chapter is useful in clarifying important aspects of the physical properties of man-made EMFs and, in particular, WC EMFs, which, in turn, determine their increased adverse biological activity and explain the plethora of experimental and epidemiological findings. We hope the present chapter forms a basis for the systematic study of WC EMFs and the health risks associated with exposures to these EMFs.

"This work is valuable to the society. Among many other details, it correctly identifies that 'low photon energy' must not be used to justify that microwaves are benign to living organisms. That is an irresponsible scientific thinking".
(Dr. Chandrasekhar Roychoudhuri, Photonics Laboratory, Physics Department, University of Connecticut, US)

REFERENCES

Adey WR, (1981). Tissue interactions with non-ionizing electromagnetic fields. *Physiol Rev.* 61:435–514.

Adey WR, (1993). Biological effects of electromagnetic fields. *J Cell Biochem.* 51:410–416.

Alberts B, Bray D, Lewis J, Raff M, Roberts K, Watson JD, (1994). *Molecular Biology of the Cell.* Garland Publishing, Inc., New York.

Alexopoulos CD, (1962). *Heat.* Papazisis Publ., Athens [Αλεξόπουλος ΚΔ, Θερμότης, Εκδ. Παπαζήση, Αθήνα 1962].

Alexopoulos CD, (1963). *Atomic and Nuclear Physics.* Papazisis Publ., Athens [Αλεξόπουλος ΚΔ, Ατομική και Πυρηνική φυσική, Εκδ. Παπαζήση, Αθήνα 1963].

Alexopoulos CD, (1966). *Optics.* Papazisis Publ., Athens [Αλεξόπουλος ΚΔ, Οπτική, Εκδ. Παπαζήση, Αθήνα 1966].

Alexopoulos CD, (1973). *Electricity.* Papazisis Publ., Athens [Αλεξόπουλος ΚΔ, Ηλεκτρισμός, Εκδ. Παπαζήση, Αθήνα 1973].

Alonso M, Finn EJ, (1967). *Fundamental University Physics, Vol. 2: Fields and Waves.* Addison-Wesley, USA.

Andersen JB, Pedersen GF, (1997). The technology of mobile telephone systems relevant for risk assessment. *Radiat Prot Dosimetry.* 3–4(72):249–257.

Arago DFJ, Fresnel AJ, (1819). On the action of rays of polarized light upon each other. *Ann Chim Phys.* 2:288–304.

Azanza MJ, Perez Bruzon RN, Lederer D, et al, (2002). Reversibility of the effects induced on the spontaneous bioelectric activity of neurons under exposure to 8.3 and 217.0 Hz low intensity magnetic fields. 2nd Int. Workshop Biol. Effects of EMFs, Rhodes, Grece, 651–659.

Baker KB, Tkach JA, Nyenhuis JA, Phillips M, Shellock FG, et al, (2004). Evaluation of specific absorption rate as a dosimeter of MRI-related implant heating. *J Magn Reson Imaging.* 20:315–320.

Balzano Q, Sheppard A, (2003). RF nonlinear interactions in living cells I: nonequilibrium thermodynamic theory. *Bioelectromagnetics.* 24:473–482.

Bassett CAL, Pawluk RJ, Becker RO, (1964). Effect of electric currents on bone *in vivo. Nature.* 204:652–654.

Bawin SM, Kaczmarek LK, Adey WR, (1975). Effects of modulated VMF fields, on the central nervous system. *Ann NY Acad Sci.* 247:74–81.

Bawin SM, Adey WR, Sabbot IM, (1978). Ionic factors in release of $^{45}Ca^{2+}$ from chick cerebral tissue by electromagnetic fields. *Proc Natl Acad Sci USA.* 75:6314–6318.

Bawin SM, Adey WR, (1976). Sensitivity of calcium binding in cerebral tissue to weak environmental electric fields oscillating at low frequency. *Proc Natl Acad Sci USA.* 73:1999–2003.

Behari J, (2010). Biological responses of mobile phone frequency exposure. *Ind J Exp Biol.* 48:959–981.

Beiser A, (1987). *Concepts of Modern Physics.* McGraw-Hill, Inc., New York.

Belpomme D, Irigaray P, (2020). Electrohypersensitivity as a newly identified and characterized neurologic pathological disorder: how to diagnose, treat, and prevent it. *Int J Mol Sci.* 21:1915; doi:10.3390/ijms21061915

Belyaev I, (2005). Non-thermal biological effects of microwaves. *Microwave Rev.* 11(2):13–29.

Berger H, (1929). Ueber das Elektrenkephalogramm des Menschen (On the human electroencephalogram). *Archiv f. Psychiatrie u. Nervenkrankheiten.* 87:527–570.

Bertagna F, Lewis R, Silva SRP, McFadden J, Jeevaratnam K, (2021). Effects of electromagnetic fields on neuronal ion channels: a systematic review. *Ann NY Acad Sci.* 1499(1):82–103.

Betskii OV, Lebedeva NN, (2004). Low-intensity millimeter waves in biology and medicine. In *Clinical Application of Bioelectromagnetic Medicine.* Marcel Decker, New York, 30–61.

Blackman CF., Benane SG, Elder JA, House DE, Lampe JA, Faulk JM, (1980). Induction of calcium - ion efflux from brain tissue by radiofrequency radiation: effect of sample number and modulation frequency on the power - density window. *Bioelectromagnetics.* 1:35–43.

Blackman CF, Benane SG, Kinney LS, Joines WT, House DE, (1982). Effects of ELF fields on calcium-ion efflux from brain tissue in vitro. *Radiat Res.* 92(3):510–20.

Blackman C, (2009). Cell phone radiation: evidence from ELF and RF studies supporting more inclusive risk identification and assessment. *Pathophysiology.* 16:205–16.

Bohr N, (1913a). On the constitution of atoms and molecules, part I. *Philos Mag.* 26:1–24.

Bohr N, (1913b). On the constitution of atoms and molecules, part II systems containing only a single nucleus. *Philos Mag.* 26:476–502.

Bohr N, (1914). The spectra of helium and hydrogen. *Nature.* 92:231–232.

Bohr N, (1928). The quantum postulate and the recent development of atomic theory. *Nature* 121:580–590.

Bolshakov MA, Alekseev SI, (1992). Bursting responses of Lymnea neurons to microwave radiation. *Bioelectromagnetics.* 13(2):119–129.

Borgens RB, (1988). Stimulation of neuronal regeneration and development by steady electrical fields. *Advances in Neurology.* 47; *Functional Recovery in Neurological Disease*, SG Waxman (Ed), Raven Press, New York.

Brighton CT, Friedenberg ZB, Black J, (1979). Evaluation of the use of constant direct current in the treatment of non-union. In CT Brighton (Ed), *Electrical Properties of Bone and Cartilage.* Plenum Press, New York, 519–545.

Brighton CT, McClusky WP, (1987). Response of cultured bone cells to capacitively coupled electrical field: inhibition of cAMP response to parathyroid hormone. *J Orthop Res.* 6:567–571.

Brighton CT, Jensen L, Pollack SR, Tolin BS, Clark CC, (1989). Proliferative and synthetic response of bovine growth plate chondrocytes to various capacitively coupled electrical fields. *J Orthop Res.* 7:759–765.

Burcham WE, Jobes M, (1995). *Nuclear and Particle Physics.* Prentice Hall, England.

Bush LG, Hill DW, Riazi A, Stensaas LJ, Partlow LM, Gandhi OP, (1981). Effects of millimeter wave radiation on monolayer cell cultures III: a search for frequency-specific effects on protein synthesis. *Bioelectromagnetics.* 2:151–160.

Byus CV, Lundak RL, Fletcher RM, Adey WR, (1984). Alterations in protein Kinase activity following exposure of cultured lymphocytes to modulated microwave fields. *Bioelectromagnetics (N.Y.).* 5:341–351.

Byus CV, Kartum K, Pieper SE, Adey WR, (1988). Ornithine decarboxylase activity in liver cells is enhanced by low-level amplitude modulated microwave fields. *Cancer Res.* 48:4222–4226.

Campisi A, Gulino M, Acquaviva R, Bellia P, Raciti G, et al., (2010). Reactive oxygen species levels and DNA fragmentation on astrocytes in primary culture after acute exposure to low intensity microwave electromagnetic field. *Neurosci Lett.* 473(1):52–55.

Carpenter RL, Livstone EM, (1968). Evidence for nonthermal effects of microwave radiation: abnormal development of irradiated insect pupae. *IEEE Trans Microwave Theory Tech.* 19(2):173–178.

Cecchetto C, Maschietto M, Boccaccio P, Vassanelli S, (2020). Electromagnetic field affects the voltage-dependent potassium channel Kv1.3. *Electromagn Biol Med.* 39(4):316–322.

Chandra A, Bagchi B, (2000). Frequency dependence of ionic conductivity of electrolyte solutions. *J Chem Phys.* 112:1876–1886.

Chavdoula ED, Panagopoulos DJ, Margaritis LH, (2010). Comparison of biological effects between continuous and intermittent exposure to GSM-900 MHz mobile phone radiation. Detection of apoptotic cell death features. *Mut Res.* 700:51–61.

Chen HS, Rao CRN, (1968). Polarization of light on reflection by some natural surfaces. *Brit J Appl Phys.* 1:1191–1200.

Christ A, Gosselin M-C, Christopoulou M, et al. (2010). Age-dependent tissue-specific exposure of cell phone users. *Phys Med Biol.* 55:1767–1783.

Clark DE, Folz DC, West JK, (2000). Processing materials with microwave energy. *Adv Mater Sci Eng.* 287:153–158.

Clarke J, Wilhelm FK, (2008). Superconducting quantum bits. *Nature.* 453:1031–1042.

Coggle JE, (1983). *Biological Effects of Radiation.* Taylor & Francis.

Creasey WA, Goldberg RB, (2001). A new twist on an old mechanism for EMF bioeffects?, *EMF Health Rep.* 9(2):1–11. https://www.emfsa.co.za/research-and-studies/creasey-wa-goldberg-rb-2001-a-new-twist-on-an-old-mechanism-for-emf/

Cronin TW, Warrant EJ, Greiner B, (2006). Celestial polarization patterns during twilight. *Applied Optics.* 22:5582–5589.

Curwen P, Whalley J, (2008). *Mobile* Communications in the 21st century, In AC Harper, RV Buress (Eds), *Mobile Telephones: Networks, Applications and Performance.* Nova Science Publishers, New York, 29–75.

Dahlman E, Parkvall S, Skoeld J, (2018). *5G NR: The Next Generation Wireless Access Technology.* Academic Press, Elsevier, London.

Davisson C, Germer L, (1927). Reflection of electrons by a crystal of nickel. *Nature.* 119:558–560.

Dawe AS, Smith B, Thomas DW, Greedy S, Vasic N, et al., (2006). A small temperature rise may contribute towards the apparent induction by microwaves of heat-shock gene expression in the nematode *Caenorhabditis elegans. Bioelectromagnetics.* 27(2):88–97.

De Broglie L, (1924). *Recherches sur la théorie des quanta*, doctoral thesis, Paris.

De Iuliis GN, Newey RJ, King BV, Aitken RJ, (2009). Mobile phone radiation induces reactive oxygen species production and DNA damage in human spermatozoa in vitro. *PLoS One.* 4(7):e6446.

Diem E, Schwarz C, Adlkofer F, Jahn O, Rudiger H., (2005). Non-thermal DNA breakage by mobile-phone radiation (1800 MHz) in human fibroblasts and in transformed GFSH-R17 rat granulosa cells in vitro. *Mutat Res.* 583(2):178–83.

Dirac PAM. (1927). The quantum theory of the emission and absorption of radiation. *Proc R Soc London.* 114:243–256.

Dubrov AP, (1978). *The Geomagnetic Field and Life.* Plenum Press, New York.

Durrer R, (2008). *The Cosmic Microwave Background.* Cambridge Univ. Press, Cambridge.

EPRS, (2021). Environmental impacts of 5G. A literature review of effects of radio-frequency electromagnetic field exposure of non-human vertebrates, invertebrates and plants. Scientific Foresight Unit (STOA), PE 690.021, June 2021.

Falcioni L, Bua L, Tibaldi E, et al. (2018). Report of final results regarding brain and heart tumors in Sprague-Dawley rats exposed from prenatal life until natural death to mobile phone radiofrequency field representative of a 1.8GHz GSM base station environmental emission. *Environ Res.* 165:496–503.

Fear EC, Stuchly MA, (1998). A novel equivalent circuit model for gap connected cells. *Phys Med Biol.* 43:1439–1448.

Feynman RP, (1950). Mathematical formulation of the quantum theory of electromagnetic interaction. *Phys Rev.* 80:440–457.

Flyckt VM, Raaymakers BW, Kroeze H, Lagendijk JJ, (2007). Calculation of SAR and temperature rise in a high-resolution vascularized model of the human eye and orbit when exposed to a dipole antenna at 900, 1500 and 1800 MHz. *Phys Med Biol.* 52(10):2691–701.

Foster KR, Schwan HP, (1989). Dielectric properties of tissues and biological materials: a critical review, *Crit Rev Biomed Eng.* 17(1):25–103.

Franzellitti S, Valbonesi P, Ciancaglini N, Biondi C, Contin A, et al., (2010). Transient DNA damage induced by high-frequency electromagnetic fields (GSM 1.8 GHz) in the human trophoblast HTR-8/SVneo cell line evaluated with the alkaline comet assay. *Mutat Res.* 683(1–2):35–42.

Frei M., Jauchem J, Heinmets F, (1988). Physiological effects of 2.8 GHz radio-frequency radiation: a comparison of pulsed and continuous-wave radiation. *J Microwave Power Electromagn.* 23:2.

Furia L, Hill DW, Gandhi OP, (1986). Effect of millimeter wave irradiation on growth of saccaromyces cerevisiae. *IEEE Trans Biomed Eng.* 33:993–999.

Gabriel S, Lau RW, Gabriel C, (1996a). The dielectric properties of biological tissues: II. measurements in the frequency range 10 Hz to 20 GHz. *Phys Med Biol.* 41:2251–2269.

Gabriel S, Lau RW, Gabriel C, (1996b). The dielectric properties of biological tissues: III. parametric models for the dielectric spectrum of tissues. *Phys Med Biol.* 41:2271–93.

Gandhi OP, (1983). Some basic properties of biological tissues for potential biomedical applications of millimeter waves. *J Microwave Power.* 18:295–304.

Gandhi OP, Morgan LL, de Salles AA, Han Y-Y, Herberman RB, Davis DL, (2012). Exposure limits: the underestimation of absorbed cell phone radiation, especially in children. *Electromagn Biol Med.* 31(1):34–51.

Gautreau R, Savin W, (1978). *Theory and Problems of Modern Physics.* McGraw-Hill, New York.

Goldsworthy A, (2006). Effects of electrical and electromagnetic fields on plants and related topics, In Volkov (Ed), *Plant Electrophysiology–Theory & Methods.* Springer-Verlag, Berlin Heidelberg.

Goodman EM, Greenebaum B, Marron MT, (1995). Effects of electro- magnetic fields on molecules and cells. *Int Rev Cytol.* 158:279–338.

Gründler W, (1992). Intensity- and frequency-dependent effects of microwaves on cell growth rates. *Bioelectrochem. Bioenerg.* 27:361–365,

Gulati S, Yadav A, Kumar N, Kanupriya ANK, et al, (2016). Effect of GSTM1 and GSTT1 polymorphisms on genetic damage in humans populations exposed to radiation from mobile towers. *Arch Environ Contam Toxicol.* 70(3):615–625.

Gulyaev YuV, Markov AG, Koreneva LG, Zakharov PV. (1995). Dynamical infrared thermography in humans. *Eng Med Biol Mag IEEE.* 14:766–771.

Haemmerich D, Schutt DJ, dos Santos I, Webster JG, Mahvi DM, (2005). Measurement of temperature-dependent specific heat of biological tissues. *Physiol. Meas.* 26(1):59–67.

Hardell L, Carlberg M, Söderqvist F, Mild KH, Morgan LL. (2007). Long-term use of cellular phones and brain tumours: increased risk associated with use for > or =10 years. *Occup Environ Med.* 64(9):626–32. Review.

Hardell L, Carlberg M, Hansson Mild K, (2013). Use of mobile phones and cordless phones is associated with increased risk for glioma and acoustic neuroma. *Pathophysiology.* 20:85–110.

Hardell L, Nyberg R, (2020). Appeals that matter or not on a moratorium on the deployment of the fifth generation, 5G, for microwave radiation. *Mol Clin Oncol.* doi:10.3892/mco.2020.1984

Hardell L, Carlberg M, (2021). Lost opportunities for cancer prevention: historical evidence on early warnings with emphasis on radiofrequency radiation. *Rev Environ Health.* doi:10.1515/reveh-2020-0168. Online ahead of print.

Herzberg G, (1944). *Atomic Spectra and Atomic Structure.* Dover publications, USA.

Herzberg G, (1950). *Molecular Spectra and Molecular Structure.* D Van Nostrand company Inc, USA.

High performance solutions for peak and average power measurements. https://emin.com.mm/high-performance-solutions-for-peak-and-average-power-measurements-myanmar-83633/pr.html.

Hinrikus H, Bachmann M, Lass J, Tomson R, Tuulik V, (2008). Effect of 7, 14 and 21 Hz modulated 450 MHz microwave radiation on human electroencephalographic rhythms. *Int J Radiat Biol.* 84(1):69–79.

Holma H, Toskala A, (2004). *WCDMA for UMTS, Radio Access for Third Generation Mobile Communications.* John Wiley & Sons Inc., Chichester, England.

Houck AA, Schuster DI, Gambetta JM, Schreier JA, Johnson BR, Chow JM, Frunzio L., Majer J., Devoret MH, Girvin SM, Schoelkopf RJ, (2007). Generating single microwave photons in a circuit. *Nature* 449:328–331.

Houston BJ, Nixon B, King BV, De Iuliis GN, Aitken RJ, (2016). The effects of radiofrequency electromagnetic radiation on sperm function. *Reproduction.* 152(6):R263–R276.

Höytö A, Luukkonen J, Juutilainen J, Naarala J, (2008). Proliferation, oxidative stress and cell death in cells exposed to 872 MHz radiofrequency radiation and oxidants. *Radiat Res.* 170(2):235–243.

Huber R, Treyer V, Borbely AA, Schuderer J, Gottselig JM, Landolt HP, Werth E, Berthold T, Kuster N, Buck A, Achermann P, (2002). Electromagnetic fields, such as those from mobile phones, alter regional cerebral blood flow and sleep and waking EEG. *J Sleep Res.* 11(4):289–95.

Hunter G, Wadlinger RLP, (1987). Physical photons: theory, experiment, interpretation. In *Quantum Uncertainties: Recent and Future Experiments and Interpretations: Proceedings of tile NATO Workshop,* University of Bridgeport, CT, 1986, NATO ASI Series B, Vol.162, Plenum Press.

Hyland GJ, (2000). Physics and biology of mobile telephony. *Lancet.* 356:1833–36.

Hyland GJ, (2008). Physical basis of adverse and therapeutic effects of low intensity microwave radiation. *Ind J Exp Biol.* 46:403–419.

IARC, (2002). Non-ionizing radiation, part 1: static and extremely low-frequency (ELF) electric and magnetic fields. Vol. 80. World Health Organization.

IARC, (2013). Non-ionizing radiation, part 2: radiofrequency electromagnetic fields. Vol. 102. Lyon, France.

ICNIRP, (1998). Guidelines for limiting exposure to time-varying electric, magnetic and electromagnetic fields (up to 300GHz). *Health Phys.* 74:494–522.

ICNIRP, (2020). Guidelines for limiting exposure to electromagnetic fields (100 kHz to 300GHz). *Health Phys.* [Published ahead of print].

IEEE, (2002). IEEE recommended practice for measurements and computations of radio frequency electromagnetic fields with respect to human exposure to such fields, 100 kHz–300 GHz, IEEE Std C95.3™-2002 (R2008).

Inomata K., Lin Z., Koshino K., Oliver WD, Tsai JS, Yamamoto T., Nakamura Y, (2016). Single microwave-photon detector using an artificial [Lambda]-type three-level system. *Nat. Commun.* 7.

Ivancsits S, Diem E, Pilger A, Rüdiger HW, Jahn, (2002). Induction of DNA strand breaks by intermittent exposure to extremely-low-frequency electromagnetic fields in human diploid fibroblasts. *Mutat Res.* 519(1–2):1–13.

Ivancsits S, Diem E, Jahn O, Rüdiger HW, (2003). Intermittent extremely low frequency electromagnetic fields cause DNA damage in a dose-dependent way. *Int Arch Occup Environ Health.* 76(6):431–6.

Jackson JD, (1975). *Classical Electrodynamics.* John Wiley & Sons, Inc, New York.

Jaynes ET, (1966). Is QED necessary?. In L Mandel, E Wolf (Eds), *Proceedings of the Second Rochester Conference on Coherence and Quantum Optics.* Plenum, New York, 21.

Jaynes ET, (1978). Electrodynamics today. In L Mandel, E Wolf (Eds), *Coherence and Quantum Optics IV.* Plenum Press, New York, 495.

Jaynes ET, (1980). *Quantum Beats.* http://bayes.wustl.edu/etj/articles/ quantum. beats.pdf.

Johansson O, (2009). Disturbance of the immune system by electromagnetic fields-A potentially underlying cause for cellular damage and tissue repair reduction which could lead to disease and impairment. *Pathophysiology.* 16:157–77.

Karipidis K, Mate R, Urban D, Tinker R, Wood A, (2021). 5G mobile networks and health—a state-of-the-science review of the research into low-level RF fields above 6 GHz. *J Exposure Sci Environ Epidemiol.* https://doi.org/10.1038/s41370-021-00307-7

Khurana VG, Teo C, Kundi M, Hardell L, Carlberg M, (2009). Cell phones and brain tumors: a review including the long-term epidemiologic data. *Surg Neurol.* 72(3):205–14.

Klimov A., (1975). *Nuclear Physics and Nuclear Reactors*. Mir Publishers, Moscow.

Kostoff RN, Heroux P, Aschner M, Tsatsakis A, (2020). Adverse health effects of 5G mobile networking technology under real-life conditions. *Toxicol Lett.* 323:35–40.

Kraus H., Soltamov VA, Riedel D., Väth S., Fuchs F., Sperlich A., Baranov PG, Dyakonov V., Astakhov GV, (2014). Room-temperature quantum microwave emitters based on spin defects in silicon carbide. *Nat Phys*.10:157–162.

Kwee S, Raskmark P, (1998). Changes in cell proliferation due to environmental non-ionizing radiation 2. Microwave radiation. *Bioelectrochem Bioenerg.* 44:251–255.

Lamb WE, Scully MO, (1969). The photoelectric effect without photons. In *Polarization, Matter and Radiation*. Presses Universitaires de France, Paris, 363–369.

Leach V, Weller S, Redmayne M, (2018). A novel database of bio-effects from non-ionizing radiation. *Rev Environ Health.* 33(3):1–8.

Leonard J, Foster K, Athey TW, (1984). Thermal properties of tissue equivalent phantom materials. *IEEE Trans Biomed Eng.* 31:533–536.

Levitt BB,. Lai HC, Manville AM, (2021). Effects of non-ionizing electromagnetic fields on flora and fauna, part 1. Rising ambient EMF levels in the environment. *Rev Environ Health*. doi.org/10.1515/reveh-2021-0026

Liboff AR, (2003). Ion cyclotron resonance in biological systems: experimental evidence. In Stavroulakis P (Ed), *Biological Effects of Electromagnetic Fields*, Springer, Berlin, 76–113.

Liburdy RP, (1992). Calcium signalling in lymphocytes and ELF fields: Evidence for an electric field metric and a site of interaction involving the calcium ion channel. *FEBS Lett.* 301: 53–59.

Liman ER, Hess P, Weaver F, Koren G, (1991). Voltage sensing residues in the S4 region of a mammalian K+ channel. *Nature.* 353:752–756.

Lin-Liu S, Adey WR, (1982). Low frequency amplitude modulated microwave fields change calcium efflux rates from synaptosomes. *Bioelectromagnetics.* 3(3):309–22.6.

Lioliousis C, (1979). *Microwaves*, Athens. [Λιολιούσης Κ, Μικροκύματα, Αθήνα 1979]

Lioliousis C, (1997). *Biological Effects of Electromagnetic Radiation*. Diavlos, Athens. [Λιολιούσης Κ, Βιολογικές επιδράσεις της ηλεκτρομαγνητικής ακτινοβολίας, Δίαυλος, Αθήνα 1997]

Lioliousis C, (2009). *Mobile Phone and Health*. Diavlos, Athens. [Λιολιούσης Κ, Κινητό τηλέφωνο και υγεία, Δίαυλος, Αθήνα 2009]

Litovitz TA, Krause D, Penafiel M, Elson EC, Mullins JM, (1993). The role of coherence time in the effect of microwaves on ornithine decarboxylase activity. *Bioelectromagnetics.* 14:395–403.

Ma TH, Chu KC. (1993). Effect of the extremely low frequency (ELF) electromagnetic field (EMF) on developing embryos of the fruit fly (Drosophila melanogaster L.). *Mutat Res*, 303(1):35–9.

Major FG, (2014). The classical atomic clocks. In *Quo Vadis: Evolution of Modern Navigation*. Springer, New York, NY, 151–180.

Manna D, Ghosh R, (2016). Effect of radiofrequency radiation in cultured mammalian cells: a review. *Electromagn Biol Med.* 35(3):265–301.

Mandel L, Wolf E. (1995). *Optical Coherence and Quantum Optics*. Cambridge University Press.

Mandl F, (1988). *Statistical Physics*. Wiley, Chichester, 2nd edition.

Marino AA, Becker RO, (1977). Biological effects of extremely low frequency electric and magnetic fields: a review. *Physiol Chem Phys.* 9:131–147.

Marino AA, Kim PY, Frilot C, (2016). Trigeminal neurons detect cellphone radiation: thermal or nonthermal is not the question. *Electromagn Biol Med.* 36(2):123–131.

McCaig CD, Zhao M, (1997). Physiological electric fields modify cell behaviour. *Bioessays.* 19(9):819–826.

Metaxas AC, (1991). Microwave heating. *Power Eng.* 5(5):237–247.

Miller AB, Morgan LL, Udasin I, Davis DL, (2018). Cancer epidemiology update, following the 2011 IARC evaluation of radiofrequency electromagnetic fields (Monograph 102). *Environ Res.* 167:673–683.

Miller AB, Sears ME, Morgan LL, Davis DL, Hardell L, et al, (2019). Risks to health and well-being from radio-frequency radiation emitted by cell phones and other wireless devices. *Front Public Health.* 7:223. doi:10.3389/fpubh.2019.00223.

Mohammed HS, Fahmy HM, Radwan NM, Elsayed AA, (2013). Non-thermal continuous and modulated electromagnetic radiation fields effects on sleep EEG of rats. *J Adv Res.* 4(2):181–7.

Moulder JE, Erdreich LS, Malyapa RS, Merritt J, Pickard WF et al, (1999). Cell phones and cancer. what is the evidence for a connection?. *Radiat Res.* 151:513–531. 60.

NCRP, (1986). Biological effects and exposure criteria for radiofrequency electromagnetic fields. Properties, quantities and units, biophysical interaction and measurements. National Council on Radiation Protection and Measurements, Report No 86, Bethesda, MD.

Neher E, Sakmann B, (1992). The patch clamp technique. *Sci Am.* 266:28–35.

Neufeld E, Kuster N, (2018). Systematic derivation of safety limits for time-varying 5G radiofrequency exposure based on analytical models and thermal dose. *Health Phys.* 115(6):705–711.

NTP (National Toxicology Program), (2018) Toxicology and carcinogenesis studies in HSD: sprague dawley SD rats exposed to whole-body radio frequency radiation at a frequency (900 MHz) and modulations (GSM and CDMA) used by cell phones. NTP TR 595, Department of Health and Human Services, USA.

Nuccitelli R, (1992). Endogenous ionic currents and DC electric fields in multicellular animal tissues. *Bioelectromagnetics.* Suppl 1:147–157.

Nuccitelli R, (2003). Endogenous electric fields in embryos during development, regeneration and wound healing. *Radiat Prot Dosimetry*.106(4):375–83.

Olaniyi IJ, (2017). Microwave heating in food processing. *BAOJ Nutr.* 3:027.

Pakhomov AG, Akyel Y, Pakhomova ON, Stuck BE, Murphy MR, (1998). Current state and implications of research on biological effects of millimeter waves: a review of the literature. *Bioelectromagnetics.* 19:393–413.

Pall ML, (2013). Electromagnetic fields act via activation of voltage-gated calcium channels to produce beneficial or adverse effects. *J Cell Mol Med.* 17(8):958–965.

Pall ML, (2018). Wi-Fi is an important threat to human health. *Environ Res.* 164:405–416.

Palmer LG, (1986) *New Insights into Cell and Membrane Transport Processes.* G Poste, ST Crooke, Eds., Plenum, New York, 331.

Panagopoulos DJ, Messini N, Karabarbounis A, Philippetis AL, Margaritis LH, (2000). A mechanism for action of oscillating electric fields on cells. *Biochem Biophys Res Commun.* 272:634–640.

Panagopoulos DJ, Karabarbounis A, Margaritis LH, (2002). Mechanism for action of electromagnetic fields on cells. *Biochem Biophys Res Commun.* 298(1):95–102.

Panagopoulos DJ, Margaritis LH, (2003). Theoretical considerations for the biological effects of electromagnetic fields. In P Stavroulakis (Ed), *Biological Effects of Electromagnetic Fields.* Springer, Berlin, 5–33.

Panagopoulos DJ, Karabarbounis A, Margaritis LH, (2004). Effect of GSM 900-MHz mobile phone radiation on the reproductive capacity of drosophila melanogaster. *Electromagn Biol Med.* 23(1):29–43.

Panagopoulos DJ, Chavdoula ED, Nezis IP, Margaritis LH, (2007a). Cell death induced by GSM 900MHz and DCS 1800MHz mobile telephony radiation. *Mutat Res.* 626:69–78.

Panagopoulos DJ, Chavdoula ED, Karabarbounis A, Margaritis LH, (2007b). Comparison of bioactivity between GSM 900 MHz and DCS 1800 MHz mobile telephony radiation. *Electromagn Biol Med.* 26(1):33–44.

Panagopoulos DJ, Margaritis LH, (2009). Biological and health effects of mobile telephony radiations. *Int J Med Biol Front.* 15(1/2):33–76.

Panagopoulos DJ, Chavdoula ED, Margaritis LH, (2010). Bioeffects of mobile telephony radiation in relation to its intensity or distance from the antenna. *Int J Radiat Biol.* 86(5):345–357.

Panagopoulos DJ, Margaritis LH, (2010a). The effect of exposure duration on the biological activity of mobile telephony radiation. *Mutat Res.* 699(1/2):17–22.

Panagopoulos DJ, Margaritis LH, (2010b). The identification of an intensity "window" on the bioeffects of mobile telephony radiation. *Int J Rad Biol.* 86(5):358–366.

Panagopoulos DJ, (2011). Biological impacts, action mechanisms, dosimetry and protection issues of mobile telephony radiation. In MC Barnes, NP Meyers (Eds), *Mobile Phones: Technology, Networks and User Issues.* Nova Science Publishers, Inc., New York, 1–54.

Panagopoulos DJ, (2013). Electromagnetic interaction between environmental fields and living systems determines health and well-being. In MH Kwang and SO Yoon (Eds), *Electromagnetic Fields: Principles, Engineering Applications and Biophysical Effects.* Nova Science Publishers, New York, 87–130.

Panagopoulos DJ, Karabarbounis A, Lioliousis C, (2013a). ELF alternating magnetic field decreases reproduction by DNA damage induction. *Cell Biochem Biophys.* 67:703–716.

Panagopoulos DJ, Johansson O, Carlo GL, (2013b). Evaluation of specific absorption rate as a dosimetric quantity for electromagnetic fields bioeffects. *PLoS One.* 8(6):e62663. doi:10.1371/journal.pone.0062663.

Panagopoulos DJ, Johansson O, Carlo GL, (2015a). Polarization: a key difference between man-made and natural electromagnetic fields, in regard to biological activity. *Sci Rep.* 5:14914

Panagopoulos DJ, Johansson O, Carlo GL, (2015b). Real versus simulated mobile phone exposures in experimental studies. *BioMed Res Int.* 2015:607053.

Panagopoulos DJ, (2015). Considering photons as spatially confined wave-packets. In A Reimer (Ed), *Horizons in World Physics.* Vol. 285, Nova Science Publishers, New York.

Panagopoulos DJ, Cammaerts MC, Favre D, Balmori A, (2016). Comments on environmental impact of radiofrequency fields from mobile phone base stations. *Crit Rev Environ Sci Technol.* 46(9):885–903.

Panagopoulos DJ, Balmori A, (2017). On the biophysical mechanism of sensing atmospheric discharges by living organisms. *Sci Total Environ.* 599–600(2017):2026–2034.

Panagopoulos DJ, (2017). Mobile telephony radiation effects on insect ovarian cells. the necessity for real exposures bioactivity assessment. the key role of polarization, and the ion forced-oscillation mechanism. In CD Geddes (Ed), *Microwave Effects on DNA and Proteins.* Springer, Cham, Switzerland, 1–48.

Panagopoulos DJ, (2018). Man-made electromagnetic radiation is not quantized. In A Reimer (Ed), *Horizons in World Physics.* Vol. 296, Nova Science Publishers, New York.

Panagopoulos DJ, Chrousos GP, (2019). Shielding methods and products against man-made electromagnetic fields: protection versus risk. *Sci Total Environ.* 667C:255–262.

Panagopoulos DJ, (2019a). Comparing DNA damage induced by mobile telephony and other types of man-made electromagnetic fields. *Mutat Res Rev.* 781:53–62.

Panagopoulos DJ, (2019b). Chromosome damage in human cells induced by UMTS mobile telephony radiation. *Gen Physiol Biophys.* 38:445–454

Panagopoulos DJ, (2020). Comparing chromosome damage induced by mobile telephony radiation and a high caffeine dose: effect of combination and exposure duration, *Gen Physiol Biophys.* 39:531–544.

Panagopoulos DJ, Balmori A, Chrousos GP, (2020). On the biophysical mechanism of sensing upcoming earthquakes by animals. *Sci Total Environ.* 717(2020):136989.

Panagopoulos DJ, Karabarbounis A, (2020). Comments on "diverse radiofrequency sensitivity and radiofrequency effects of mobile or cordless phone near fields exposure in drosophila melanogaster. *Adv Environ Stud.* 4(1):271–276.

Panagopoulos DJ, Karabarbounis A, Yakymenko I, Chrousos GP, (2021). Human-made electromagnetic fields: ion forced-oscillation and voltage-gated ion channel dysfunction: oxidative stress and DNA damage. *Int J Oncol*, 59: 92.

Panarella E, (2008). Single photons have not been detected: the alternative photon clump model. In C Roychoudhuri, AF Kracklauer, K Creath (Ed), *The Nature of Light. What Is a Photon?* CRC Press, Taylor & Francis, USA.

Parsons KC, (1993). *Human Thermal Environments.* Taylor and Francis, London.

Pedersen GF, (1997). Amplitude modulated RF fields stemming from a GSM/DCS-1800 phone. *Wireless Networks.* 3:489–498.

Penafiel LM, Litovitz T, Krause D, Desta A, Mullins MJ, (1997). Role of modulation on the effect of microwaves on ornithine decarboxylase activity in L929 cells. *Bioelectromagnetics.* 18(2):132–41.

Peyman A, Rezazadeh AA, Gabriel C, (2001). Changes in the dielectric properties of rat tissue as a function of age at microwave frequencies. *Phys Med Biol.* 46(6):1617–1629.

Phillips JL, Singh NP, Lai H., (2009). Electromagnetic fields and DNA damage. *Pathophysiology.* 16(2–3):79–88.

Pirard W, Vatovez B. Study of pulsed character of radiation emitted by wireless telecommunication systems. Institut scientifique de service public, Liège, Belgium. https://www.issep.be/wp-content/uploads/7IWSBEEMF_B-Vatovez_W-Pirard.pdf.

Pohl R, (1960). Discovery of interference by Thomas Young. *Am J Phys.* 28:530.

Prasad KN, (1995). *Handbook of Radiobiology.* CRC Press, Boca Raton, USA, 2nd edition.

Presman AS, (1977). *Electromagnetic Fields and Life.* Plenum Press, New York.

Puranen L, Jokela K, (1996). Radiation hazard assessment of pulsed microwave radars. *J Microwave Power Electromagn Energy.* 31(3):165–177.

Radebaugh R, (2009). Cryocoolers: the state of the art and recent developments. *J Phys Conden Matt.* 21(16):164219.

Rappaport, TS, Sun, S, Mayzus, R, et al. (2013). Millimeter wave mobile communications for 5G cellular: it will work! *IEEE Access.* 1:335–349.

Reitz JR, Milford FJ, (1967). *Foundations of Electromagnetic Theory.* Addison-Wesley Publishing Company, Inc, Boston, MA.

Roller WL, Goldman RF, (1968). Prediction of solar heat load on man. *J Appl Physiol.* 25:717–721.

Roychoudhuri C., Kracklauer AF, Creath K, (2008). *The Nature of Light. What Is a Photon?* CRC Press, Taylor & Francis, USA.

Roychoudhuri C., Tirfessa N, (2008). Do we count indivisible photons or discrete quantum events experienced by detectors?. In Roychoudhuri, Kracklauer, Creath (Eds), *The Nature of Light. What Is a Photon?.* CRC Press, Taylor & Francis, USA.

Roychoudhuri C., (2014). *Causal Physics – Photon Model by Non-Interaction of Waves.* CRC Press, Taylor and Francis, Boca Raton, USA.

Sangeetha M, Purushothaman BM, Suresh Babu S, (2014). Estimating cell phone signal intensity and identifying radiation hotspot area for tirunel veli taluk using RS and GIS. *Int J Res Eng Technol.* 3:412–418.

Santini MT, Ferrante A, Rainaldi G, Indovina P, Indovina PL, (2005). Extremely low frequency (ELF) magnetic fields and apoptosis: a review. *Int J Radiat Biol.* 81(1):1–11.

Sauter M, (2011). *From GSM to LTE: An Introduction to Mobile Networks and Mobile Broadband.* John Wiley & Sons, Chichester, UK.

Schimmelpfeng J, Dertinger H, (1993). The action of 50Hz magnetic and electric fields upon cell proliferation and cyclic AMP content of cultured mammalian cells. *Bioelectrochem. Bioenerg.* 30:143–150.

Schroedinger E, (1926). An undulatory theory of the mechanics of atoms and molecules. *Physical Review.* 28(6):1049–1070.

Schumann WO, (1952). Uber die strahlunglosen eigenschwingungen einer leitenden Kugel, die von einer Luftschicht und einer Ionospharenhulle umgeben ist (On the characteristic oscillations of a conducting sphere which is surrounded by an air layer and an ionospheric shell). *Zeitschrift Naturforschung.* 7A:149–154.

Schuster DI, Houck AA, Schreier JA, Wallraff A, Gambetta JM, Blais A, Frunzio L, Majer J, Johnson B, Devoret MH, Girvin SM, Schoelkopf RJ, (2007). Resolving photon number states in a superconducting circuit. *Nature.* 445:515–518.

Schwan HP, (1957). Electrical properties of tissues and cell suspensions. *Adv Phys Med Biol.* 5:147–209.

Schwan HP, (1963). Determination of biological impedances. In Nastuk WL (Ed), *Physical Techniques in Biological Research*, Vol. 6. New York: Academic Press, 323–407.

Schwartz M, (1990). *Information Transmission, Modulation, and Noise*, McGraw-Hill, New York, 4th edition.

Sesia S, Toufik I, Baker M (Eds), (2011). *LTE – The UMTS Long Term Evolution.* John Wiley & Sons Ltd., West Sussex, UK.

Sheppard AR, Swicord ML, Balzano Q, (2008). Quantitative evaluations of mechanisms of radiofrequency interactions with biological molecules and processes. *Health Phys.* 93(4):365–396.

Shim Y, Lee I, Park S, (2013). The impact of LTE UE on audio devices. *ETRI J.* 35(2):332–335.

Sommerfeld A, (1916). Zur Quantentheorie der Spektrallinien. *Ann D Phys.* 51:1.

Somosy Z, Thuroczy G, Kubasova T, Kovacs J, Szabo LD, (1991). Effects of modulated and continuous microwave irradiation on the morphology and cell surface negative charge of 3T3 fibroblasts. *Scanning Microsc.* 5(4):1145–1155.

Spiegel MR, (1974). *Fourier Analysis with Applications to Boundary Value Problems.* McGraw-Hill, New York.

Spottorno J, Multigner M, Rivero G, Alvarez L, de la Venta J, Santos M, (2008). Time dependence of electrical bioimpedance on porcine liver and kidney under a 50 Hz ac current. *Phys Med Biol.* 53:1701–1713.

Spottorno J, Multigner M, Rivero G, Alvarez L, de la Venta J, Santos M, (2012). In vivo measurements of electrical conductivity of porcine organs at low frequency: new method of measurement. *Bioelectromagnetics.* 33(7):612–9.

Stensaas LJ, Partlow LM, Bush LG, Iversen PL, Hill DW, Hagmann MJ, Gandhi OP, (1981). Effects of millimeter-wave radiation on monolayer cell cultures. II. Scanning and transmission electron microscopy. *Bioelecytromagnetics.* 2(2):141–150.

Stephenson G, (1973). *Mathematical Methods for Science Students.* Longman Group, London, 2nd edition.

Stryer L, (1996). *Biochemistry.* W.H. Freeman and Co, New York, 4th edition.

Tarasov LV, (1980). *Basic Concepts of Quantum Mechanics.* Mir Publishers, Moscow.

Tesla N, (1905). The transmission of electrical energy without wires as a means of furthering world peace. *Electrical World and Engineer.* 7:21–24.

Thielens A, Bell D, Mortimore DB, Greco MK, Martens L, Joseph W, (2018). Exposure of insects to radio-frequency electromagnetic fields from 2 to 120 GHz. *Sci Rep.* 8:3924. doi:10.1038/s41598-018-22271-3.

Thielens A, Greco MK, Verloock L, Martens L, Joseph W, (2020). Radio-frequency electromagnetic field exposure of western honey bees. *Sci Rep.* 10:461. doi:org/10.1038/s41598-019-56948-0.

Thuroczy G, Kubinyi G, Bodo M, Bakos J, Szabo LD, (1994). Simultaneous response of brain electrical activity (EEG) and cerebral circulation (REG) to microwave exposure in rats. *Rev Environ Health.* 10(2):135–148.

Tisal J, (1998). *GSM Cellular Radio Telephony.* J.Wiley & Sons, West Sussex, England.

Trachanas SL, (1981). *Quantum Mechanics.* Athens. [Τραχανάς Σ.Λ., «Κβαντομηχανική», Εκδ. Σύγχρονες Επιστήμες, Αθήνα 1981].

Tuor M, Ebert S, Schuderer J, Kuster N, (2005). Assessment of ELF exposure from GSM handsets and development of an optimized RF/ELF exposure setup for studies of human volunteers. BAG Reg. No. 2.23.02.-18/02.001778, IT'IS Foundation.

Valberg PA, Kavet R, Rafferty CN, (1997). Can low-level 50/60Hz electric and magnetic fields cause biological effects? *Rad Res.* 148:2–21.

Velizarov S, Raskmark P, Kwee S, (1999). The effects of radiofrequency fields on cell proliferation are non-thermal. *Bioelectrochem Bioenerg.* 48:177–180.

Verschaeve L, (2014). Environmental impact of radiofrequency fields from mobile phone base stations. *Crit Rev Environ Sci Technol.* 44:1313–1369.

Verschaeve L, (2017). Misleading scientific papers on health effects from wireless communication devices. In CD Geddes (Ed), *Microwave Effects on DNA and Proteins*, Springer, Cham, Switzerland.

Veyret B, Bouthet C, Deschaux P, de Seze R, Geffard M, et al., (1991). Antibody responses of mice exposed to low-power microwaves under combined, pulse-and-amplitude modulation. *Bioelectromagnetics.* 2(1):47–56.

Vistnes AI, Gjoetterud K, (2001). Why arguments based on photon energy may be highly misleading for power line frequency electromagnetic fields. *Bioelectromagnetics.* 22:200–204.

Walleczek J, (1992). Electromagnetic field effects on cells of the immune system: the role of calcium signaling. *FASEB J.* 6: 3177–3185.

Walls DF, Milburn GL (2008). *Quantum Optics*, Springer.

Wang ET, Zhao M, (2010). Regulation of tissue repair and regeneration by electric fields. *Chin J Traumatol.* 13(1):55–61.

Wang J., Fujiwara O., Kodera S., Watanabe S., (2006). FDTD calculation of whole-body average SAR in adult and child models for frequencies from 30 MHz to 3 GHz. *Phys Med Biol.* 51(17):4119–27.

Weisenseel MH, (1983). Control of differentiation and growth by endogenous electric currents. In W Hoppe, W Lohmann, H Markl, H Ziegler (Eds), *Biophysics*, Springer–Verlag, Berlin, 460–465.

Wertheimer N, Leeper E, (1979). Electrical wiring configurations and childhood cancer. *Am J Epidemiol.* 109.

Wever R, (1974). ELF effects on human circadian rhythms. In MA Persinger (Ed), *ELF and VLF Electromagnetic Fields*. Plenum Press, New York.

Wever R, (1979). *The Circadian System of Man: Results of Experiments under Temporal Isolation*. Springer-Verlag, New York.

Wilson W, (1915). The quantum theory of radiation and line spectra. *Phil Mag* 29:795–802.

Wongkasem N, (2021). Electromagnetic pollution alert: microwave radiation and absorption in human organs and tissues. *Electromag Biol Med.* Feb 10:1–18. doi:10.1080/15368378.2021.1874976. Online ahead of print.

Wood A, Mate R, Karipidis K, (2021). Meta-analysis of in vitro and in vivo studies of the biological effects of low-level millimetre waves. *J Exposure Sci Environ Epidemiol.* https://doi.org/10.1038/s41370-021-00307-7.

Wust P, Kortüm B, Strauss U, et al, (2020). Non-thermal effects of radiofrequency electromagnetic fields. *Sci Rep.* 10(1):13488. doi:10.1038/s41598-020-69561-3.

Wust P, Stein U, Ghadjar P, (2021). Non-thermal membrane effects of electromagnetic fields and therapeutic applications in oncology. *Int J Hyperthermia.* 38(1):715–731.

Yakymenko I, Sidorik E, Kyrylenko S, Chekhun V, (2011). Long-term exposure to microwave radiation provokes cancer growth: evidences from radars and mobile communication systems. *Exp Oncol.* 33(2):62–70.

Yakymenko I, Tsybulin O, Sidorik E, Henshel D, Kyrylenko O, et al, (2016). Oxidative mechanisms of biological activity of low-intensity radiofrequency radiation. *Electromagn Biol Med.* 35(2):186–202.

Yakymenko I, Burlaka A, Tsybulin I, Brieieva I, Buchynska L, et al, (2018). Oxidative and mutagenic effects of low intensity GSM 1800 MHz microwave radiation. *Exp Oncol.* 40(4):282–287.

Zheng Y, Xia P, Dong L, Tian L, Xiong C, (2021). Effects of modulation on sodium and potassium channel currents by extremely low frequency electromagnetic fields stimulation on hippocampal CA1 pyramidal cells. *Electromagn Biol Med.* 17:1–12.

Zhou R, Xiong Y, Xing G, Sun L, Ma J, (2010). ZiFi: wireless LAN Discovery via ZigBee Interference Signatures *MobiCom'10*, September 20–24, Chicago, IL.

Zothansiama, ZM, Lalramdinpuii M, Jagetia GC, (2017). Impact of radiofrequency radiation on DNA damage and antioxidants in peripheral blood lymphocytes of humans residing in the vicinity of mobile phone base stations. *Electromagn Biol Med.* 36(3):295–305.

Zwamborn APM, Vossen SHJA, van Leersum BJAM, Ouwens MA, Mäkel WN. (2003). Effects of global communication system radio-frequency fields on well being and cognitive functions of human subjects with and without subjective complaints. *FEL-03-C148*. TNO Physics and Electronics Laboratory, The Hague, the Netherlands. http://home.tiscali.be/milieugezondheid/dossiers/gsm/TNO_rapport_Nederland_sept_2003.pdf.

Part B

Biological and Health Effects of Wireless Communication Electromagnetic Fields

2 Public Health Implications of Exposure to Wireless Communication Electromagnetic Fields

Anthony B. Miller

Dalla Lana School of Public Health, University of Toronto, Toronto, Canada

CONTENTS

Abstract 79
2.1 Introduction 80
2.2 Human and Animal Carcinogenicity Studies 81
2.3 Children and Reproduction 84
2.4 Related Observations, Implications, and Strengths of Current Evidence 85
2.5 Discussion 86
    2.5.1 Challenges to Research from Rapid Technological Advances 86
    2.5.2 Exposure Limits and Gaps in Applying Current Evidence 87
    2.5.3 Policy Recommendations Based on the Evidence to Date 88
    2.5.4 Application to the General Public 89
Funding 90
References 90

Keywords: electromagnetic fields, wireless communication radiation, mobile phones, cordless phones, health effects, brain cancers, glioma, glioblastoma, vestibular Schwannoma, epidemiology, children, reproduction

Abbreviations: ADHD: attention-deficit hyperactivity disorder. CI: Confidence Interval. DDT: dichloro-diphenyl-trichloroethane. DECT: Digitally Enhanced Cordless Telecommunications. EHS: electro-hypersensitivity. ELF: Extremely Low Frequency. EMF: electromagnetic field. EMR: electromagnetic radiation. IARC: International Agency for Research on Cancer. ICNIRP: International Commission on Non-Ionizing Radiation Protection. IoT: Internet of Things. GSM: Global System for Mobile Telecommunications. LTE: Long Term Evolution. MT: mobile telephony. NR: New Radio. NTP: US National Toxicology Program. OR: Odds Ratio. RF: Radio Frequency. SAR: Specific Absorption Rate. UMTS: Universal Mobile Telecommunications System. WC: wireless communication. WCD: wireless communication devices. Wi-Fi: Wireless Fidelity. 2G/3G/4G/5G: second/third/fourth/fifth generation of MT.

ABSTRACT

Anthropogenic electromagnetic fields (EMFs) and corresponding electromagnetic radiation (EMR) exposure has long been a concern for the public, policy makers, and health researchers. Beginning

DOI: 10.1201/9781003201052-5

with radar during World War II, human exposure to Radio Frequency (RF) radiation, and to modulated RF wireless communication (WC) EMFs/EMR has grown substantially over time. In 2011, a working group of the International Agency for Research on Cancer (IARC) reviewed the published literature and categorized WC EMR, termed as RF radiation, as a "possible" (Group 2B) human carcinogen. A broad range of adverse human health effects associated with WC EMFs/EMR have been reported since the IARC review. In addition, two large-scale carcinogenicity studies in rodents exposed to levels of WC EMR that mimic lifetime human exposures have shown significantly increased rates of Schwannomas and malignant gliomas, as well as chromosomal DNA damage. Of particular concern are the effects of WC EMR exposure on the developing brain in children. Compared with an adult male, a mobile phone held against the head of a child exposes deeper brain structures to greater radiation doses per unit volume, and the young, thin skull's bone marrow absorbs a roughly tenfold higher local dose. Experimental and observational studies also suggest that men who keep mobile phones in their trouser pockets have significantly lower sperm counts and significantly impaired sperm motility and morphology, including mitochondrial DNA damage as well as an increased risk of colon cancer. Pending an updated IARC working group review, current knowledge provides justification for governments, public health authorities, and physicians/allied health professionals to warn the population that having a cell phone next to the body is harmful, and to support measures to reduce all exposures to WC EMFs/EMR to as low as reasonably achievable.

2.1 INTRODUCTION

We live in a world that relies heavily on technology. Whether for personal use or work, wireless communication devices (WCDs), such as cell (digital mobile) phones, cordless domestic phones (DECT: Digitally Enhanced Cordless Telecommunications), internet connections (Wi-Fi: Wireless Fidelity), Bluetooth, etc., are commonly used around the world, and exposure to wireless communication (WC) electromagnetic fields (EMFs) and corresponding electromagnetic radiation (EMR) is widespread, including in public spaces (Carlberg et al., 2019; Hardell et al., 2018). All types of WC EMFs/EMR combine Radio Frequency (RF) (300 kHz–300 GHz) carrier waves with Extremely Low Frequency (ELF) (3–3000 Hz) pulsing and modulation of the signals (Hyland 2000; 2008; Zwamborn et al. 2003; Huber et al. 2005; Tuor et al. 2005; Panagopoulos 2019).

In this chapter, I address the current scientific evidence on health risks from exposure to WC EMR, which is in the non-ionizing frequency range. I focus on human health effects but also note evidence that WC EMFs/EMR can cause biological, physiological, and/or morphological effects on insects (bees, fruit flies, etc.), birds, amphibia, plants, and trees (Panagopoulos et al. 2004; Balmori 2009; 2014; Sharma and Kumar 2010; Waldmann-Selsam et al. 2016; Halgamuge 2017; Odemer and Odemer 2019). Because EMR is always produced by time-varying EMFs, and because all types of anthropogenic (polarized) EMR (including WC EMR) carry net electric and magnetic fields (Panagopoulos et al. 2015), the terms WC EMFs and WC EMR are used similarly throughout the chapter.

There are a diversity of opinions on the potential adverse effects of WC EMFs/EMR exposure from cell phones and other WCDs, including DECT phones and Wi-Fi. The paradigmatic approach of cancer epidemiology, which considers the body of epidemiological, toxicological, and mechanistic/cellular evidence when assessing causality, is applied.

Since 1998, the International Commission on Non-Ionizing Radiation Protection (ICNIRP) has maintained that no evidence of adverse biological effects of WC EMFs/EMR exist in either RF or ELF bands other than tissue heating at exposures above prescribed thresholds (ICNIRP, 1998; 2018; 2020).

In contrast, in 2011, an expert working group of the International Agency for Research on Cancer (IARC) categorized "RF EMFs" (actually WC EMFs/EMR) emitted by mobile phones and other WCDs as a Group 2B ("possible") human carcinogen (IARC 2013), a category that also includes lead and the insecticide DDT (dichloro-diphenyl-trichloroethane). ELF magnetic fields were already categorized by IARC (2002) as a Group 2B human carcinogen as well. Both ELF and RF EMFs are parts of all types of WC EMFs/EMR.

2.2 HUMAN AND ANIMAL CARCINOGENICITY STUDIES

Since the IARC categorizations, new science has emerged, both human and animal, confirming that WC EMFs/EMR cause cancer.

The human evidence comprises three large case–control studies[*] on mobile phone exposure and brain tumors, including the highly malignant gliomas and acoustic neuromas (tumors of the vestibular/hearing nerve), which demonstrated significantly elevated risks that tended to increase with increasing latency (time from first exposure), increasing cumulative duration of use, ipsilateral phone use (on the same side of the head), and earlier age at first exposure (Miller et al. 2018). These important case–control studies are:

1. The multi-country Interphone study which found a twofold increased risk of glioma after 10 or more years of regular use of mobile phones, with a dose–response relationship (Interphone Study Group 2010).
2. Several studies by Hardell and his colleagues in Sweden (one of the first countries to introduce mobile phones) showing two- to fivefold increased risk of glioma after prolonged use, especially when exposure began early in life (Hardell and Carlberg 2015). Analyses that pooled the findings from a number of studies (pooled analyses) by the Hardell group that examined the risks of glioma and acoustic neuroma stratified by age at first exposure to mobile phones and cordless (DECT) phones found the highest risks, or highest Odds Ratio[†] (OR), among those first exposed before the age of 20 (Hardell and Carlberg 2015; Hardell et al. 2013a; 2013b). For glioma, first use of cell (mobile) phones before the age of 20 years resulted in an OR of 1.8 with 95% Confidence Interval[‡] (CI) (CI: 1.2–2.8). For ipsilateral use, the OR was 2.3 (CI: 1.3–4.2); contralateral use (on both sides of the head) was 1.9 (CI: 0.9–3.7). Use of cordless phones before the age of 20 yielded OR 2.3 (CI: 1.4–3.9), ipsilateral OR 3.1 (CI: 1.6–6.3) and contralateral use OR 1.5 (CI: 0.6–3.8) (Hardell and Carlberg, 2015).
3. A large study in France (CERENAT study), which found a fivefold increased risk of glioma after 5 or more years of use of mobile phones (Coureau et al. 2014).

These studies all show that the longer the exposure, and the earlier the exposure begins in life, the greater the risk. Although an increased risk of glioma was not reported from a cohort study[§] in the UK (with some potential errors in classification of exposure), there was a doubling of risk of acoustic neuroma with 10 or more years of mobile phone use (Benson et al. 2013), as was also found in a case–control study by Hardell et al. (2013a), though not by Moon et al. (2014) from Korea. However, a case–control study of brain tumors in adolescents, using operator records for exposure in Nordic countries, found more than a doubling of risk after 2.8 years since initial subscription for mobile phone use (Aydin et al. 2011).

---

[*] Case–control studies are used to identify factors that may contribute to a medical condition by comparing subjects who have that condition/disease (the "cases") with patients who do not have the condition/disease but are otherwise similar (the "controls"), preferably drawn from the same population as the cases.
[†] The OR represents the probability of an outcome due to a particular cause compared to the probability of the outcome in the absence of that cause.
[‡] 95% CI is the interval with 0.95 probability that the outcome (cancer) is due to the examined parameter. When both numbers of the 95% CI are ≥1.0, the increasing risk is statistically significant. When one of the numbers is <1.0, the increasing risk is not statistically significant.
[§] A cohort study is a longitudinal study that samples a cohort (a group of people who share a defining characteristic), typically those that were at risk of the same exposure in a defined period.

West et al. (2013) reported four "extraordinary" multi-focal breast cancers that arose directly under the antennae of the cell (mobile) phones habitually carried within the bra, on the sternal side of the breast (the opposite of the norm). Case reports can point to major unrecognized hazards and avenues for further investigation, although they do not usually provide direct causal evidence.

The incidence of parotid or salivary gland tumors has tripled in Israel (Czerninski et al. 2011), and a rise in the incidence of glioblastoma in the temporal and frontal regions of the brain has been reported in the United Kingdom (UK) (Philips et al. 2018), while the incidence of neuro-epithelial brain cancers has significantly increased in children, adolescents, and young adults from birth to 24 years of age in the United States (US) (Gittleman et al. 2015; Ostrom et al. 2016; 2018; 2020).

Although Karipidis et al. (2018) and Nilsson et al. (2019) found no evidence of an increased incidence of gliomas in recent years in Australia and Sweden, respectively, Karipidis et al. (2018) reported only on brain tumor data for ages 20–59, and Nilsson et al. (2019) failed to include data for high grade glioma (the most malignant form of brain cancer).

In contrast, others have reported evidence that increases in specific types of brain tumors seen in laboratory studies are occurring in Britain and the US: As noted above, the incidence of neuro-epithelial brain cancers has significantly increased in the US (Gittleman et al. 2015; Ostrom et al. 2016; 2018; 2020), while a sustained and statistically significant rise in glioblastomas across all ages has been described in the UK (Phillips et al. 2018).

The incidence of several brain tumors is increasing at statistically significant rates, according to the 2010–2017 *Central Brain Tumor Registry of the U.S.* (CBTRUS) data set (Ostrom et al. 2016; 2018; 2020). For example:

1. There was a significant increase in the incidence of radiographically diagnosed tumors of the pituitary from 2006–2012 with no significant change in incidence from 2012–2015.
2. Meningioma (tumors of the meninges, the lining of the brain) rates have increased in all age groups from 15 through 85 or more years.
3. Nerve sheath tumor (Schwannoma, a usually benign tumor composed of Schwann cells, which normally produce the insulating myelin sheath covering peripheral nerves) rates have increased in all age groups from 20 through 84 years.
4. Vestibular (hearing nerve) Schwannoma rates, as a percentage of nerve sheath tumors, have also increased from 58% in 2004 to 95% in 2010–2014.

Epidemiological evidence was subsequently reviewed and incorporated in a meta-analysis (a combined analysis of all relevant studies) by Röösli et al. (2019). They concluded that, overall, epidemiological evidence does not suggest increased brain or salivary gland tumor risk with mobile phone use, although the authors admitted that some uncertainty remains regarding long latency periods (>15 years), rare brain tumor subtypes, and mobile phone usage during childhood. Of concern is that these analyses included cohort studies with poor exposure classification (as exposure was only characterized at the onset of the study and could not reflect subsequent changes) (Söderqvist et al. 2012).

In epidemiological studies, recall bias (an error caused by differences in the accuracy or completeness of the recollections recalled by study participants regarding events or experiences from the past) can play a substantial role in the attenuation of risks (OR) toward the null. However, an analysis of data from one large multi-center case–control study of WC EMF/EMR exposure did not find that recall bias was an issue (Vrijheid et al. 2006). In another multi-country study, it was found that young people can recall phone use moderately well, with recall depending on the amount of phone use and participants' characteristics (Goedhart et al. 2018). With less rigorous querying of exposure, prospective cohort studies are, unfortunately, vulnerable to exposure misclassification and imprecision in identifying risk from rare events to the point that negative results from such studies are almost inevitable (Brzozek et al. 2018; Miller et al. 2018). This is because cohort studies are impractical for rare diseases.

To take brain cancer as an example of why cohort studies are impractical, brain cancer incidence should be, according to IARC (2021), about seven cases per 100,000 (10^5) persons per year. Even if as many as 100,000 people are enrolled in the study, if there is no risk associated with exposure to WC EMR, after 10 years (given the long latency times of brain cancer, this is the minimum time to expect a risk), there would be seven brain cancers per year. If there was a doubled risk, there would be 14 cases per year. Whether or not such an increased risk would be detected would depend on how long the cohort was followed after the end of the latent period and how efficiently the use or non-use of WCDs was determined, both initially and during the follow-up period.

Another example of disparate results from studies of different design focuses on the prognosis for patients with gliomas, depending upon mobile phone use: A Swedish study found lower survival in patients with glioblastoma associated with long-term use of mobile and cordless (DECT) phones (Carlberg and Hardell 2014; 2017). Olsson et al. (2019), however, reported no indication of reduced survival among glioblastoma patients in Denmark, Finland, and Sweden with a history of mobile phone use (ever regular use, time since start of regular use, cumulative call time overall, or in the last 12 months) relative to no use or non-regular use. Notably, Olsson et al. (2019) differed from Carlberg and Hardell (2014) in that the study did not include use of cordless phones, used a shorter latency time, and excluded patients older than 69 years. Furthermore, a major shortcoming was that patients with the worst prognosis were excluded, and inoperable cases were also excluded, all of which would increase the risk estimate.

Tumor promotion by exposure to WC EMFs/EMR below the ICNIRP exposure limits for humans in exposed mice were reported by Lerchl et al. (2015). No clear dose–response relationship was evident. Lerchl et al. (2015) hypothesized that metabolic changes are responsible for the effects observed.

Two large-scale toxicological (animal carcinogenicity) studies support the human evidence – as do modeling, cellular, and DNA studies – identifying vulnerable subgroups of the population:

1. The US National Toxicology Program (NTP) (National Toxicology Program 2018a; 2018b) has reported significantly increased incidence of glioma and malignant Schwannoma (mostly on the nerves of the heart, but also additional organs) in rat and mouse carcinogenicity studies with exposure to levels of second generation (2G) and third generation (3G) simulated mobile telephony (MT) radiation that did not significantly heat tissue. Multiple organs of the exposed animals (e.g., brain, heart) also had evidence of DNA damage. Although these findings have been dismissed by the ICNIRP (2018) for reasons including lack of blinding and longer survival of exposed rats compared to controls, one of the key originators of the NTP study has refuted the criticisms (Melnick 2019).
2. A study by Italy's Ramazzini Institute has evaluated lifespan environmental exposure of rodents to WC EMR, as generated by simulated 2G Global System for Mobile Telecommunications (GSM) MT radiation (1.8 GHz plus 8 and 217 Hz pulsations). Although the exposure levels were significantly lower than those in the NTP study, statistically significant increases in Schwannomas of the heart in male rodents exposed to the highest dose and Schwann-cell hyperplasia in the heart in male and female rodents were observed (Falcioni et al. 2018). A non-statistically significant increase in malignant glial tumors in female rodents also was detected. These findings with far-field exposure (not close to the antenna) to WC EMFs/EMR are consistent with and reinforce the results of the NTP study on near-field exposure (close to the antenna). Both reported an increase in the incidence of tumors of the brain and heart in WC EMF-exposed Sprague-Dawley rats, which are tumors of the same histological type as those observed in some epidemiological studies on cell phone users.

Co-carcinogenicity of WC EMFs/EMR was also demonstrated by Soffritti and Giuliani (2019) who examined both power-line frequency (50–60 Hz) ELF magnetic fields as well as 1.8 GHz modulated

WC EMR. They found that exposure to sinusoidal 50 Hz magnetic fields combined with acute exposure to gamma radiation or to chronic administration of formaldehyde in drinking water induced a significantly increased incidence of malignant tumors in male and female Sprague-Dawley rats. In the same report, preliminary results indicate higher incidence of malignant Schwannoma of the heart after exposure to WC EMR in male rats. Given the ubiquity of many of these co-carcinogens, this provides further evidence to support the recommendation to reduce the public's exposure to WC EMFs/EMR to as low as is reasonably achievable.

A review of ELF EMFs carcinogenicity studies conducted by a working group convened by IARC (2002) concluded that the evidence of carcinogenicity of ELF EMFs, mainly from power transmission line studies, was limited, largely because it had not yet been possible to perform adequate animal experiments capable of mimicking the human exposure situation, and they placed ELF magnetic fields in Category 2B, possibly carcinogenic to humans (Wertheimer and Leeper 1979; Savitz et al. 1988; Coleman et al. 1989; Feychting and Ahlbom 1993; 1994; 1995; Miller et al. 1996; Coghill et al. 1996; Green et al. 1999a; 1999b; Villeneuve et al. 2000a; 2000b; Ahlbom et al. 2000; Greenland et al. 2000). The IARC group acknowledged that the evidence for risk in humans was greatest for childhood leukemia.

I and my co-authors conducted several studies in Canada in which we found an increased risk of childhood leukemia from exposure to ELF (60 Hz) magnetic fields (Green et al., 1999a; 1999b), an increased risk of leukemia in those occupationally exposed to ELF (60 Hz) electric fields (Miller et al. 1996; Villeneuve et al. 2000b), and an increased risk of non-Hodgkin lymphoma in workers occupationally exposed to high levels of ELF (60 Hz) electric fields (Villeneuve et al. 2000a).

In a study of four groups of men, of which one group did not use mobile phones, it was found that DNA damage indicators in hair follicle cells in the ear canal were higher in the WC EMF/EMR exposure groups than in the control subjects. In addition, DNA damage increased with the daily duration of exposure (Akdag et al. 2018).

Many profess that WC EMFs/EMR cannot be carcinogenic, as it has insufficient energy to cause direct DNA damage (Moulder et al. 1999; 2005; Adair 2003). In a review, Vijayalaxmi and Prihoda (2019) found that some studies suggested significantly increased damage in cells exposed to WC EMR compared to unexposed and/or sham-exposed control cells, others did not. Unfortunately, however, in grading the evidence, these authors failed to consider baseline DNA status or the fact that genotoxicity has been poorly predicted using tissue culture studies (Corvi and Madia 2017). Additionally, funding, which constitutes a strong source of bias in this field of inquiry, was not considered (Huss et al. 2007). Moreover, it has been documented that WC EMFs/EMR induce oxidative stress in exposed cells, which, in turn, can cause DNA damage (Yakymenko et al. 2016).

2.3 CHILDREN AND REPRODUCTION

As a result of rapid growth rates and the greater vulnerability of developing nervous systems, the long-term risks to children from WC EMF/EMR exposure from mobile phones and other WCDs are expected to be greater than those to adults (Redmayne et al. 2013). By analogy with other carcinogens, longer opportunities for exposure due to earlier use of mobile phones and other WCDs could be associated with greater cancer risks in later life.

Modeling of energy absorption can be an indicator of potential exposure to WC EMR. A study modeling the exposure of children 3–14 years of age to WC EMR has indicated that a mobile phone held against the head of a child exposes deeper brain structures to roughly double the radiation doses (including fluctuating/oscillating electrical and magnetic fields) per unit volume than in adults, and the marrow in the young, thin skull absorbs a roughly tenfold higher local dose than in the skull of an adult male (Fernández et al. 2018). Thus, pediatric populations are among the most vulnerable to WC EMF/EMR exposure.

The increasing use of mobile phones in children, which can be regarded as a form of addictive behavior (De-Sola Gutiérrez et al. 2016), has been shown to be associated with emotional and behavioral disorders. Divan et al. (2008) studied 13,000 mothers and children and found that prenatal exposure to cell phones was associated with behavioral problems and hyperactivity in children. A subsequent Danish study of 24,499 children found a 23% increased risk of emotional and behavioral difficulties at the age of 11 among children whose mothers reported any cell phone use for their children at the age of 7, compared to children whose mothers reported no use at the age of 7 (Sudan et al. 2016). A cross-sectional study of 4,524 US children aged 8–11 years from 20 study sites indicated that shorter screen time and longer sleep periods independently improved child cognition, with maximum benefits achieved with low screen time and age-appropriate sleep times (Walsh et al. 2018). Similarly, a cohort study of Swiss adolescents suggested a potential adverse effect of WC EMR on cognitive functions that involve brain regions mostly exposed during mobile phone use (Foerster et al. 2018). Sage and Burgio (2018) posit that epigenetic drivers and DNA damage underlie adverse effects of wireless devices on childhood development.

WC EMF/EMR exposure occurs in the context of other exposures, both beneficial (e.g., nutrition) and adverse (e.g., toxicants or stress). Two studies identified that WC EMFs/EMR potentiated the adverse effects of lead on neuro-development, with higher maternal use of mobile phones during pregnancy in 1,198 mother–child pairs (Choi et al. 2017) and attention-deficit hyperactivity disorder (ADHD) with higher cell phone use and higher blood lead levels, in 2,422 elementary school children (Byun et al., 2013).

A study of MT base station tower settings adjacent to school buildings has found that high exposure of male students to WC EMR from these towers was associated with delayed fine and gross motor skills, spatial working memory, and attention in adolescent students compared with students who were exposed to low exposure of WC EMR (Meo et al. 2018). A recent prospective cohort study showed a potential adverse effect of WC EMR brain dose on adolescents' cognitive functions, including spatial memory that involve brain regions exposed during cell phone use (Foerster et al. 2018).

In a review, Pall (2016) concluded that various non-thermal microwave EMF exposures produce diverse neuro-psychiatric effects. Both animal research (Deniz et al. 2017; Eghlidospour et al. 2017; Aldad et al. 2012) and human studies of brain activity research (Huber et al. 2002; Huber et al. 2005; Volkow et al. 2011; Kostoff and Lau 2013) indicate potential roles of WC EMFs/EMR in these outcomes.

Male fertility has been addressed in cross-sectional studies in men. Associations between keeping cell phones in trouser pockets and lower sperm quantity and quality have been reported (Adams et al. 2014). Both *in vivo* and *in vitro* studies with human sperm confirm adverse effects of WC EMR on the testicular proteome and other indicators of male reproductive health (Wdowiak et al. 2007; Agarwal et al. 2008; De Iuliis et al. 2009; Adams et al. 2014; Houston et al. 2016), including infertility (Kesari et al. 2018). Rago et al. (2013) found significantly altered sperm DNA fragmentation in subjects who use mobile phones for more than 4 hours a day (h/day) and, in particular, those who carry the device in the trousers' pockets. In a cohort study, Zhang et al. (2016) found that cell phone use may negatively affect sperm quality in men by decreasing the semen volume, sperm concentration, or sperm count, thus, impairing male fertility. Gautam et al. (2019) studied the effect of 3G Universal Mobile Telecommunications System (UMTS) mobile phone radiation (1.9–2.2 GHz plus 100 and 1500 Hz pulsations) on the reproductive system of male Wistar rats. They found that exposure to mobile phone radiation induces oxidative stress in the rats that may lead to alteration in sperm parameters affecting their fertility.

2.4 RELATED OBSERVATIONS, IMPLICATIONS, AND STRENGTHS OF CURRENT EVIDENCE

An extensive review of numerous published studies confirms non-thermally induced biological effects or damage (e.g., oxidative stress, damaged DNA, altered gene expression and protein synthesis, breakdown of the blood–brain barrier) from exposure to WC EMFs/EMR (Salford et al. 2003;

BioInitiative Working Group 2012; Yakymenko et al. 2016; Panagopoulos 2019), as well as adverse chronic health effects from long-term exposure (Belyaev 2010). Biological effects of typical population exposures to WC EMFs/EMR are largely attributed to fluctuating polarized electrical and magnetic fields (Ying et al. 2014; Panagopoulos et al. 2015; Barnes and Greenebaum 2016).

Indeed, an increasing number of people have developed constellations of symptoms attributed to exposure to WC EMFs/EMR (e.g., headaches, fatigue, appetite loss, insomnia), a syndrome termed microwave sickness or electro-hypersensitivity (EHS) (Belyaev et al. 2016, Heuser and Heuser 2017, Belpomme et al. 2018).

Causal inference is supported by consistency between epidemiological studies of the effects of WC EMFs/EMR on induction of human cancer, especially glioma and vestibular Schwannomas, and evidence from animal studies (Miller et al. 2018). The combined weight of the evidence linking WC EMFs/EMR to public health risks includes a broad array of findings: experimental biological evidence of non-thermal effects of WC EMFs/EMR, concordance of evidence regarding carcinogenicity of WC EMFs/EMR, human evidence of male reproductive damage, human and animal evidence of developmental harms, and limited human and animal evidence of potentiation of effects of chemical toxicants. Thus, diverse independent evidence of a potentially troubling and escalating problem warrants policy intervention.

2.5 DISCUSSION

2.5.1 Challenges to Research from Rapid Technological Advances

Advances in WC EMF/EMR-related technologies have been and continue to be rapid. Changes in carrier frequencies and the growing complexity of modulation/pulsing technologies can quickly render "yesterday's" technologies obsolete. This rapid obsolescence restricts the amount of data on human WC EMF/EMR exposure to particular frequencies and modulations and related health outcomes that can be collected during the lifespan of the technology in question. Certainly, there is something common in all types of modern digital WC EMFs/EMR. As noted in the Introduction of this chapter, they all combine RF carrier waves with ELF pulsations and modulation, and in addition, they all display random variability of the final signals (Panagopoulos 2019).

Epidemiological studies with adequate statistical power must be based upon large numbers of participants with sufficient latency and intensity of exposure to specific technologies. Therefore, a lack of epidemiological evidence does not necessarily indicate an absence of effect but, rather, an inability to study an exposure for the length of time necessary, with an adequate sample size and unexposed comparators, to draw clear conclusions. For example, no case–control study has been published on fourth generation (4G) Long-Term Evolution (LTE) MT EMFs/EMR (1.8–2.6 GHz carrier frequencies plus 100 Hz and other pulsations), even though this type of WC EMF/EMR was introduced in 2010 and had achieved a 39% market share worldwide by 2018 (Anonymous 2018).

With this absence of human evidence, governments must require large-scale animal studies (or other appropriate studies of indicators of carcinogenicity and other adverse health effects) to determine whether the newest modulation technologies incur risks prior to their release into the marketplace. Governments should also investigate short-term impacts, such as insomnia, memory loss, changes in reaction time, hearing, and vision, especially those that can occur in children and adolescents, whose use of WCDs has grown exponentially within the past few years.

The telecom industry's fifth generation (5G) New Radio (NR) wireless service will require the placement of many times more base antennae/antenna towers close to all recipients of the service because solid structures, rain, and foliage block the associated millimeter wave[*] WC EMR (Rappaport et al. 2013). Carrier frequency bands for 5G are separated into two different frequency

---

[*] Millimeter waves are microwaves with wavelengths 1–10 mm, or frequencies 30–300 GHz

ranges. Frequency range 1 (FR1) includes sub-6 GHz frequency bands, some of which are bands traditionally used by previous standards but has been extended to cover potential new spectrum offerings from approximately (~) 0.7 GHz to ~6 GHz. Frequency range 2 (FR2) includes higher frequency bands from 24.25 GHz to 52.6 GHz. Bands in FR2 are largely of millimeter wavelength. These have a shorter range but a higher available bandwidth than bands in the FR1. NR 5G technology is being developed as it is also being deployed, with antenna arrays for directional – steerable beam-forming, operating at higher power than previous technologies. 5G is not stand-alone – it will operate and interface with other (including 2G, 3G, and 4G) frequencies and modulations to enable diverse devices under continual development for the "Internet of Things" (IoT), driverless vehicles, and more (Rappaport et al. 2013).

Novel 5G technology is being rolled out in several densely populated cities, although potential chronic health or environmental impacts have not been evaluated and are not being followed. Higher carrier frequencies (shorter wavelength) associated with 5G do not penetrate the body as deeply as frequencies from older technologies, but the low frequency pulsations do. Moreover, the effects may be systemic (at whole organism level) (Beltzalel et al. 2018; Russell 2018). The range and magnitude of potential impacts of 5G technologies are under-researched, although important biological outcomes have been reported with millimeter wavelength exposure. These include oxidative stress and altered gene expression, effects on skin, and systemic effects such as on immune function (Szmigielski 2013; Yakymenko et al. 2016; Russell, 2018). *In vivo* studies reporting resonance with human sweat ducts (Beltzalel et al. 2018), acceleration of bacterial and viral replication, and other endpoints indicate the potential for novel as well as more commonly recognized biological impacts of this range of frequencies and highlight the need for research before population-wide continuous exposures. While information on the carrier frequencies of 5G technology are available in the related technical literature, there is no information regarding the lower frequency components (pulsations, modulations) of this new type of WC EMFs/EMR.

2.5.2 Exposure Limits and Gaps in Applying Current Evidence

Current exposure limits are based on an assumption that the only adverse health effect from WC EMFs/EMR is heating from short-term (acute), time-averaged exposures (Federal Communication Commission 2013). Unfortunately, in some countries, notably the US, scientific evidence of the potential hazards of WC EMFs/EMR has been largely dismissed (Alster 2015). Findings of carcinogenicity, infertility, and cell damage occurring at daily exposure levels – within current limits – indicate that existing exposure standards are not sufficiently protective of public health. Evidence of carcinogenicity alone, such as that from the NTP study, should be sufficient to recognize that current exposure limits are inadequate.

Public health authorities in many jurisdictions have not yet incorporated the latest science from the US NTP or other groups. Many cite 28-year-old guidelines by the Institute of Electrical and Electronic Engineers (IEEE), which claimed that "Research on the effects of chronic exposure and speculations on the biological significance of non-thermal interactions have not yet resulted in any meaningful basis for alteration of the standard" (IEEE 1991). The US Federal Communications Commission (FCC) adopted the IEEE C95.1 1991 standard in 1996.

Conversely, some organizations have taken specific actions to reduce exposure to citizens (Environmental Health Trust 2018), including testing and recalling phones that exceed current exposure limits.

While we do not know how risks to individuals from using cell phones may be offset by the benefits to public health of being able to summon timely health, fire, and police emergency services, the findings reported above underscore the importance of evaluating potential adverse health effects from WC EMF/EMR exposure and taking pragmatic, practical actions to minimize exposure.

Miller et al. (2018) proposed the following considerations to address gaps in the current body of evidence:

A. As many claim that we should, by now, be seeing an increase in the incidence of brain tumors if WC EMR causes them, ignoring the increases in brain tumors summarized above, a detailed evaluation of age-specific, location-specific trends in the incidence of gliomas in many countries is warranted.
B. Studies should be designed to yield the strongest evidence, most efficiently. 1) Population-based case–control designs can be more statistically powerful to determine relationships with rare outcomes, such as glioma, than cohort studies. Such studies should explore the relationship between intensity of radiation, energy absorption (Specific Absorption Rate – SAR*), duration of exposure, and adverse outcomes, especially brain cancer, cardiomyopathies and abnormal cardiac rhythms, hematologic malignancies, and thyroid cancer. 2) Cohort studies are inefficient in the study of rare outcomes with long latencies, such as glioma, because of cost-considerations relating to the follow-up required of very large cohorts needed for the study of rare outcomes. In addition, without continual resource-consuming follow-up at frequent intervals, it is not possible to ascertain ongoing information about changing technologies, uses (e.g., phoning versus texting or accessing the Internet), and/or exposures. 3) Cross-sectional studies comparing high-, medium-, and low-exposure persons may yield hypothesis-generating information about a range of outcomes relating to memory, vision, hearing, reaction time, pain, fertility, and sleep patterns.
C. Exposure assessment is poor in this field, with very little fine-grained detail as to frequencies and modulations, doses and dose rates, and peak exposures, particularly over the long term. Solutions such as wearable meters and phone apps have not yet been incorporated in large-scale research.
D. Systematic reviews on the topic could use existing databases of research reports, such as the one created by Oceania Radiofrequency Science Advisory Association (Leach et al., 2018) or EMF Portal (2018), to facilitate literature searches.
E. Studies should be conducted to determine appropriate locations for installation of antennae and other broadcasting systems; these studies should include examination of biomarkers of inflammation, genotoxicity, and other health indicators in persons who live at different radiuses around these installations. This is difficult to study in the general population because many people's greatest exposure arises from their personal devices.
F. Further work should be undertaken to determine the distance that wireless technology antennae should be kept away from humans to ensure acceptable levels of safety, distinguishing among a broad range of sources (e.g., MT base antennas, cell phones, cordless phones, Wi-Fi, Bluetooth devices, etc.), recognizing that exposures fall with increasing distance. The effective radiated power from MT base antennas (cell towers) needs to be regularly measured and monitored.

2.5.3 Policy Recommendations Based on the Evidence to Date

Miller et al. (2018) noted that, at the time of writing, a total of 32 countries or governmental bodies within these countries† have issued policies and health recommendations concerning exposure to WC EMR (Environmental Health Trust, 2018). Two US states, Connecticut and Massachusetts, have issued advisories to limit exposure to WC EMR (Connecticut Department of Public Health

---

* SAR values (in W/kg) should be adjusted for the age of the child, even though the usefulness of SAR has been questioned for non-thermal effects (Panagopoulos et al. 2013).
† Argentina, Australia, Austria, Belgium, Canada, Chile, Cyprus, Denmark, European Environmental Agency, European Parliament, Finland, France, French Polynesia, Germany, Greece, Italy, India, Ireland, Israel, Namibia, New Zealand, Poland, Romania, Russia, Singapore, Spain, Switzerland, Taiwan, Tanzania, Turkey, the UK, and the US.

– CDPH 2017a; 2017b; Massachusetts 2017), and the Worcester Massachusetts Public Schools (2017) voted to post precautionary guidelines on Wi-Fi radiation on its website. In France, Wi-Fi has been removed from pre-schools and ordered to be shut off in elementary schools when not in use, and children aged 16 years or younger are banned from bringing cell phones to school (Samuel 2018). Because the national test agency found nine out of ten phones exceeded permissible radiation limits, France is also recalling several million phones.

Miller et al. (2018) made the following recommendations:

1. Governmental and institutional support of data collection and analysis to monitor potential links between EMFs/EMR associated with wireless technology and cancers as well as effects on sperm, heart, nervous system, sleep, vision, and hearing and effects on children.
2. Further dissemination of information regarding the potential health risk from WC EMR exposures. Information that is in WCDs and manuals is necessary to respect users' *right to know*. Cautionary statements and protective measures should be posted on packaging and at points of sale. Governments should follow the practice of France, Israel, and Belgium and mandate labeling, as for tobacco and alcohol.
3. Regulations should require that any WCD that could be used or carried directly against the skin (e.g., a mobile phone) or in close proximity (e.g., a device being used on the lap of a small child) be tested appropriately as used, and that this information be prominently displayed at point of sale, on packaging, and both on the exterior and within the device.
4. IARC should convene a new working group to update the categorization of WC EMFs/EMR, including current scientific findings that highlight, in particular, risks of subsequent cancers to youngsters.
5. The World Health Organization (WHO) should complete its long-standing WC EMR systematic review project, using strong modern scientific methods. National and regional public health authorities similarly need to update their understanding and to provide adequate precautionary guidance for the public to minimize potential health risks.
6. Emerging human evidence is confirming animal evidence of developmental problems with WC EMR exposure during pregnancy. WC EMF/EMR sources should be avoided and distanced from expectant mothers, as recommended by physicians and scientists (refer to: babysafeproject.org).
7. Other countries should follow France, limiting WC EMF/EMR exposure in children younger than 16 years of age.
8. MT base antennas (cell towers) should be distanced from homes, daycare centers, schools, and places frequented by pregnant women, men who wish to father healthy children, and the young.

Specific examples of how the health policy recommendations above, invoking the Precautionary Principle, might be practically applied to protect public health, were provided by Miller et al. (2018) in the annex to their paper.

2.5.4 Application to the General Public

Unfortunately, the public is not educated on the potential risks from WC EMFs/EMR: a) There is enormous ignorance over the adverse health effects of WC EMR (e.g., mobile phones, "smart" phones with internet access, tablets, driverless cars, etc.), b) warning signs are in small print or hidden to such an extent that they are useless, c) manufacturers of products that depend upon WC EMR for functioning ignore the risks. But they do so at their peril – they cannot obtain insurance coverage against the risks of WC EMR

The telecom industry promotes the view that WC EMR is harmless. This they do largely through ICNIRP. This organization is based in Germany and is largely financed by the telecom industry

(Hardell and Carlberg 2021). In 2020, the ICNIRP released guidelines for exposure to WC EMFs/EMR. Commenting on these guidelines, Hardell and Carlberg (2021) stated "Harmful effects on human health and the environment at levels below the guidelines are downplayed although evidence is steadily increasing. Only thermal (heating) effects are acknowledged and therefore form the basis for the guidelines". They point out that

> ICNIRP is not representative of the scientific community since it does not include representatives from scientists that agree there is evidence of harmful effects at levels well below ICNIRPs limits although these scientists are in majority in the scientific community.

It is reassuring that an IARC advisory committee recently recommended that WC EMFs/EMR should be re-reviewed with high priority. An extensive literature search will be conducted for relevant peer-reviewed publications, members (and chair) of a working group will be selected by the IARC director and the head of the Monographs program, the members of the working group will be given specific tasks, and then will meet for 8 days in Lyon to reach a conclusion on the carcinogenicity of WC EMFs/EMR. That was declared in an email by IARC to those (including myself) who had asked for such a revision (personal communication).

I and many other scientists now believe that WC EMFs/EMR should be categorized as a Class 1 Human Carcinogen, in the same category as cigarette smoking, asbestos exposure, and x-rays (Miller et al. 2018; 2019; Hardell and Nyberg 2020; Hardell and Carlberg 2020). Government standards must be changed to reflect this. WC EMFs/EMR are now ubiquitous, and those who use cell phones, "smart" phones, tablets, Wi-Fi, and other WCDs are increasing the risk of cancer in their bodies, especially after prolonged exposure or exposure beginning in childhood. We emphasized that MT antenna towers should be distanced from homes, daycare centers, and schools.

Even if the risk of WC EMR per individual is low, WC EMR is now widely distributed and could become a major public health problem, especially if the planned introduction of 5G proceeds. If 5G is rolled out, we can expect to see an increase in all of the conditions associated with exposure to WC EMR. A moratorium on the roll-out of 5G is essential.

Paul Heroux of McGill University (personal communication, 2021) has suggested that 5G and the IoT is a Trojan horse, with millions of mini-cell towers soon to be installed every 150 m in our neighborhoods, which will invade the privacy of every home. Optical fibers are safer, healthier, and faster. With optical fibers, everyone could enjoy a communication speed ultimately 10,000 times faster than wireless, less vulnerable to hacking, and harmless to the health of humans and other species.

FUNDING

No funding was required for the preparation of this manuscript.

REFERENCES

Adair, R.K. (2003). Biophysical limits on athermal effects of RF and microwave radiation. *Bioelectromagnetics* 24(1):39–48.

Adams, J.A., Galloway, T.S., Mondal, D., Esteves, S.C., Mathews, F. (2014). Effect of mobile telephones on sperm quality: A systematic review and meta-analysis. *Environ. Int.* 70:106–112.

Agarwal, A., Deepinder, F., Sharma, R.K., Ranga, G., Li, J. (2008). Effect of cell phone usage on semen analysis in men attending infertility clinic: An observational study. *Fertil. Steril.* 89(1):124–128.

Ahlbom, A., Day, N., Feychting, M., et al. (2000). A pooled analysis of magnetic fields and childhood leukaemia. *Br. J. Cancer* 83(5):692–698.

Akdag, M., Dasdag, S., Canturk, F., Akdag, M.Z. (2018). Exposure to non-ionizing electromagnetic fields emitted from mobile phones induced DNA damage in human ear canal hair follicle cells. *Electromagn. Biol. Med.* 37(2):66–75.

Aldad, T.S., Gan, G., Gao, X.B., Taylor, H.S. (2012). Fetal radiofrequency radiation exposure from 800–1900 MHz-rated cellular telephones affects neurodevelopment and behavior in mice. *Sci. Rep.* 2:312. https://doi.org/10.1038/srep00312.

Alster, N. (2015). Captured agency: How the federal communications commission is dominated by the industries it presumably regulates. Edmond J. Safra Center for Ethics Harvard University, 124 Mount Auburn Street, Suite 520 N. Cambridge, MA 02138 USA.

Anonymous. (2018). LTE achieves 39% market share worldwide. *Microw. J.*, June 25, 2018. http://www.microwavejournal.com/articles/30603-lte-achieves (accessed September 29 2018).

Aydin, D., Feychting, M., Schüz, J., et al. (2011). Mobile phone use and brain tumors in children and adolescents: A multicenter case-control study. *J. Natl Cancer Inst.* 103(16):1264–1276.

Balmori, A. (2009). Electromagnetic pollution from phone masts: Effects on wildlife. *Pathophysiology* 16(2–3):191–199.

Balmori, A. (2014). Electrosmog and species conservation. *Sci. Total Environ.* 496:314–316.

Barnes, F., Greenebaum, B. (2016). Some effects of weak magnetic fields on biological systems: RF fields can change radical concentrations and cancer cell growth rates. *IEEE Power Electron. Mag.* 3(1):60–68.

Belpomme, D., Hardell, L., Belyaev, I., Burgio, E., Carpenter, D.O. (2018). Thermal and non-thermal health effects of low intensity non-ionizing radiation: An international perspective. *Environ. Pollut.* https://doi.org/10.1016/j.envpol.2018.07.019.

Beltzalel, N., Ben Ishai, P., Feldman, Y. (2018). The human skin as a sub-THz receiver - Does 5G pose a danger to it or not? *Environ. Res.* 163:208–216.

Belyaev, I. (2010). Dependence of non–thermal biological effects of microwaves on physical and biological variables: Implications for reproducibility and safety standards. In L. Giuliani, M. Soffritti (Eds.), *European J. Oncol.—Library non–thermal effects and mechanisms of interaction between electromagnetic fields and living matter, 5*, Ramazzini Institute, Bologna, Italy, pp. 187–218 (An ICEMS Monograph).

Belyaev, I., Dean, A., Eger, H., et al. (2016). EUROPAEM EMF guideline. 2016 for the prevention, diagnosis and treatment of EMF-related health problems and illnesses. https://doi.org/10.1515/reveh-2016-0011.

Benson, V.S., Pirie, K., Schüz, J., et al.; Million Women Study Collaborators. (2013). Mobile phone use and risk of brain neoplasms and other cancers: Prospective study. *Int. J. Epidemiol.* 42(3):792–802.

BioInitiative Working Group. (2012). A rationale for biologically-based exposure standards for low-intensity electromagnetic radiation. *BioInitiative*. https://www.bioinitiative.org/.

Brzozek, C., Benke, K.K., Zeleke, B.M., et al. (2018). Radiofrequency electromagnetic radiation and memory performance: Sources of uncertainty in epidemiological cohort studies. *Int. J. Environ. Res. Public Health* 15(4):592. https://doi.org/10.3390/ijerph15040592.

Byun, Y.H., Ha, M., Kwon, H.J., et al. (2013). Mobile phone use, blood lead levels, and attention deficit hyperactivity symptoms in children: A longitudinal study. *PLOS ONE* 8(3):e59742. https://doi.org/10.1371/journal.pone.0059742.

Carlberg, M., Hardell, L. (2014). Decreased survival of glioma patients with astrocytoma grade IV (glioblastoma multiforme) associated with long-term use of mobile and cordless phones. *Int. J. Environ. Res. Public Health* 11(10):10790–10805.

Carlberg, M., Hardell, L. (2017). Evaluation of mobile phone and cordless phone use and glioma risk using the Bradford Hill viewpoints from 1965 on association or causation. *BioMed Res. Int.* 2017:9218486.

Carlberg, M., Hedendahl, L., Koppel, T., Hardell, L. (2019). High ambient radiofrequency radiation in Stockholm city, Sweden. *Oncol. Lett.* 17(2):1777–1783. https://doi.org/10.3892/ol.2018.9789.

CDPH. (2017a). Connecticut department of public health: Cell phones: Questions and answers about safety. https://portal.ct.gov/-/media/Departments-and-Agencies/DPH/dph/ environmental_health/eoha/Toxicology_Risk_Assessment/050815CellPhonesFINALpdf.pdf?la=en.

CDPH. (2017b). Guidelines on how to reduce exposure to radio frequency energy from cell phones. https://www.cdph.ca.gov/Programs/OPA/Pages/NR17-086.aspx.

Choi, K.H., Ha, M., Ha, E.H., et al. (2017). Neurodevelopment for the first three years following prenatal mobile phone use, radio frequency radiation and lead exposure. *Environ. Res.* 156:810–817.

Coghill, R.W., Steward, J., Philips, A. (1996). Extra low frequency electric and magnetic fields in the bed place of children diagnosed with leukaemia: A case-control study. *Eur. J. Cancer Prev.* 5(3):153–158.

Coleman, M.P., Bell, C.M., Taylor, H.L., Primic-Zakelj, M. (1989). Leukaemia and residence near electricity transmission equipment: A case-control study. *Br. J. Cancer* 60(5):793–798.

Corvi, R., Madia, F. (2017). In vitro genotoxicity testing – Can the performance be enhanced? *Food Chem. Toxicol.* 106(B):600–608.

Coureau, G., Bouvier, G., Lebailly, P., et al. (2014). Mobile phone use and brain tumours in the CERENAT case-control study. *Occup. Environ. Med.* 71(7):514–522.

Czerninski, R., Zini, A., Sgan-Cohen, H.D. (2011). Risk of parotid malignant tumors in Israel (1970–2006). *Epidemiology* 22(1):130–131. https://doi.org/10.1097/EDE.0b013e3181feb9f0.

De Iuliis, G.N., Newey, R.J., King, B.V., Aitken, R.J. (2009). Mobile phone radiation induces reactive oxygen species production and DNA damage in human spermatozoa in vitro. *PLOS ONE* 4(7):e6446.

Deniz, O.G., Suleyman, K., Mustafa, B.S., et al. (2017). Effects of short and long term electromagnetic fields exposure on the human hippocampus. *J. Microsc. Ultrastruct.* 5(4):191–197. https://doi.org/10.1016/j.jmau.2017.07.001.

De-Sola Gutiérrez, J., Rodríguez de Fonseca, F., Rubio, G. (2016). Cell-phone addiction: A review. *Front. Psychiatry* 7:175. https://doi.org/10.3389/fpsyt.2016.00175; https://www.ncbi.nlm.nih.gov/pmc/articles/PMC5076301/.

Divan, H.A., Kheifets, L., Obel, C., Olsen, J. (2008). Prenatal and postnatal exposure to cell phone use and behavioral problems in children. *Epidemiology* 19(4):523–529. https://doi.org/10.1097/EDE.0b013e318175dd47.

Eghlidospour, M., Amir, G., Seyyed, M.J.M., Hassan, A. (2017). Effects of radiofrequency exposure emitted from a GSM mobile phone on proliferation, differentiation, and apoptosis of neural stem cells. *Anat. Cell Biol.* 50(2):115–123.

EMF-Portal of the RWTH Aachen University. (2018). https://www.emf-portal.org/en.

Environmental Health Trust. (2018). Database of worldwide policies on cell phones, wireless and health. https://ehtrust.org/policy/international-policy-actions-on-wireless/.

Falcioni, L., Bua, L., Tibaldi, E., et al. (2018). Report of final results regarding brain and heart tumors in Sprague-Dawley rats exposed from prenatal life until natural death to mobile phone radiofrequency field representative of a 1.8 GHz GSM base station environmental emission. *Environ. Res.* 17(1):50. https://doi.org/10.1186/s12940-018-0394-x.

Federal Communication Commission. (2013). Radio frequency safety 13–39 section 112, page 37 first report and order March 29, 2013. https://apps.fcc.gov/edocs_public/attachmatch/FCC-13-39A1.pdf.

Fernández, C., de Salles, A.A., Sears, M.E., Morris, R.D., Davis, D.L. (2018). Absorption of wireless radiation in the child versus adult brain and eye from cell phone conversation or virtual reality. *Environ. Res.* 167:694–699. https://doi.org/10.1016/j.envres.2018.05.013.

Feychting, M., Ahlbom, A. (1993). Magnetic fields and cancer in children residing near Swedish high - Voltage power lines. *Am. J. Epidemiol.* 138(7):467–81.

Feychting, M., Ahlbom, A. (1994). Magnetic fields, leukemia and central nervous system tumors in Swedish adults residing near high - Voltage power lines. *Epidemiology* 5(5):501–9.

Feychting, M., Ahlbom, A. (1995). Childhood leukemia and residential exposure to weak extremely low frequency magnetic fields. *Environ. Health Perspect.* 103(suppl 2):59–62.

Foerster, M., Thielens, A., Joseph, W., Eeftens, M., Röösli, M. (2018). A prospective cohort study of adolescents' memory performance and individual brain dose of microwave radiation from wireless communication. *Environ. Health Perspect.* 126(7):077007. https://doi.org/10.1289/EHP2427.

Gautam, R., Singh, K.V., Nirala, J., et al. (2019). Oxidative stress-mediated alterations on sperm parameters in male Wistar rats exposed to 3G mobile phone radiation. *Andrologia* 51(3):e13201. https://doi.org/10.1111/and.13201.

Gittleman, H.R., Ostrom, Q.T., Rouse, C.D., et al. (2015). Trends in central nervous system tumor incidence relative to other common cancers in adults, adolescents, and children in the United States, 2000 to 2010. *Cancer* 121(1):102–112. https://doi.org/10.1002/cncr.29015.

Goedhart, G., van Wel, L., Langer, C.E., et al. (2018). Recall of mobile phone usage and laterality in young people: The multinational mobi-expo study. *Environ. Res.* 165:150–157. https://doi.org/10.1016/j.envres.2018.04.018.

Green, L.M., Miller, A.B., Villeneuve, P., et al. (1999a). A case-control study of childhood leukemia in Southern Ontario, Canada, and exposure to magnetic fields in residences. *Int. J. Cancer* 82(2):161–170.

Green, L.M., Miller, A.B., Agnew, D.A., et al. (1999b). Childhood leukemia and personal monitoring of residential exposures to electric and magnetic fields in Ontario Canada. *Cancer Causes Control* 10(3):233–243.

Greenland, S., Sheppard, A.R., Kaune, W.T., Poole, C., Kelsh, M.A. (2000). A pooled analysis of magnetic fields, wire codes, and childhood leukemia: Childhood leukemia-EMF Study Group. *Epidemiology* 11(6):624–634.

Halgamuge, M.N. (2017). Review: Weak radiofrequency radiation exposure from mobile phone radiation on plants. *Electromagn. Biol. Med.* 36(2):213–235.

Hardell, L., Carlberg, M., Söderqvist, F., Kjell, H.M. (2013a). Pooled analysis of case-control studies on acoustic neuroma diagnosed 1997–2003 and 2007–2009 and use of mobile and cordless phones. *Int. J. Oncol.* 43(4):1036–1044.

Hardell, L., Carlberg, M., Gee, D. (2013b). Mobile phone use and brain tumour risk: Early warnings, early actions? Chapter 21. In *Late lessons from early warnings, part 2. European Environment Agency*, Copenhagen, Denmark. https://www.eea.europa.eu/publications/late-lessons-2/late-lessons-chapters/late-lessons-ii-chapter-21/view.

Hardell, L., Carlberg, M. (2015). Mobile phone and cordless phone use and the risk for glioma - Analysis of pooled case-control studies in Sweden, 1997–2003 and 2007–2009. *Pathophysiology* 22(1):1–13.

Hardell, L., Carlberg, M., Hedendahl, L.K. (2018). Radiofrequency radiation from nearby base stations gives high levels in an apartment in Stockholm, Sweden: A case report. *Oncol. Lett.* 15(5):7871–7883. https://doi.org/10.3892/ol.2018.8285.

Hardell, L., Nyberg, R. (2020). Appeals that matter or not on a moratorium on the deployment of the fifth generation, 5G, for microwave radiation. *Mol. Clin. Oncol.* https://doi.org/10.3892/mco.2020.1984.

Hardell, L., Carlberg, M. (2020). Health risks from radiofrequency radiation, including 5G, should be assessed by experts with no conflicts of interest. *Oncol. Lett.* 20(4):15.

Hardell, L., Carlberg, M. (2021). Lost opportunities for cancer prevention: Historical evidence on early warnings with emphasis on radiofrequency radiation. *Rev. Environ. Health.* https://doi.org/10.1515/reveh-2020-0168.

Heuser, G., Heuser, S.A. (2017). Functional brain MRI in patients complaining of electrohypersensitivity after long term exposure to electromagnetic fields. *Rev. Environ. Health* 32(3):291–299. https://doi.org/10.1515/reveh-2017-0014.

Houston, B.J., Nixon, B., King, B.V., De Iuliis, G.N., Aitken, R.J. (2016). The effects of radiofrequency electromagnetic radiation on sperm function. *Reproduction* 152(6):R263–R276.

Huber, R., Treyer, V., Borbély, A.A., et al. (2002). Electromagnetic fields, such as those from mobile phones, alter regional cerebral blood flow and sleep and waking EEG. *J. Sleep Res.* 11(4):289–295.

Huber, R., Treyer, V., Schuderer, J., et al. (2005). Exposure to pulse-modulated radio frequency electromagnetic fields affects regional cerebral blood flow. *Eur. J. Neurosci.* 21(4):1000–1006.

Huss, A., Egger, M., Hug, K., Huwiler-Müntener, K., Röösli, M. (2007). Source of funding and results of studies of health effects of mobile phone use: Systematic review of experimental studies. *Environ. Health Perspect.* 115(1):1–4.

Hyland, G.J. (2000). Physics and biology of mobile telephony. *Lancet* 356(9244):1833–1836.

Hyland, G.J. (2008). Physical basis of adverse and therapeutic effects of low intensity microwave radiation. *Indian J. Exp. Biol.* 46(5):403–419.

IARC. (2002). Non-ionizing radiation, Part 1: Static and extremely low-frequency (ELF) electric and magnetic fields. *IARC Monogr. Eval. Carcinog. Risks Hum.* 80:1–395.

IARC. (2013). IARC Monographs on the evaluation of carcinogenic risks to humans. Non-ionizing radiation, Part 2: Radiofrequency Electromagnetic fields. *IARC Monogr. Eval. Carcinog. Risks Hum.* 102(Pt 2):1–460.

IARC. (2021). *Cancer incidence in five continents*, Volume XI. International Agency for Research on Cancer, Lyon, France.

ICNIRP. (1998). Guidelines for limiting exposure to time-varying electric, magnetic, and electromagnetic fields (up to 300 GHz). International commission on non-ionizing radiation protection. *Health Phys.* 74(4):494–522.

ICNIRP. (2018). ICNIRP note on recent animal carcinogenesis studies. Munich, Germany. https://www.icnirp.org/cms/upload/publications/ICNIRP note.pdf.

ICNIRP. (2020). Guidelines for limiting exposure to electromagnetic fields (100 kHz to 300GHz). *Health Phys.* [Published ahead of print].

IEEE. (1991). IEEE c95.1 IEEE standard for safety levels with respect to human exposure to radio frequency electromagnetic fields, 3 kHZ to 300 GHz. https://ieeexplore.ieee.org/document/1626482/.

Interphone Study Group. (2010). Brain tumour risk in relation to mobile telephone use: Results of the INTERPHONE international case-control study. *Int. J. Epidemiol.* 39(3):675–694.

Karipidis, K., Elwood, M., Benke, G., et al. (2018). Mobile phone use and incidence of brain tumour histological types, grading or anatomical location: A population-based ecological study. *BMJ Open* 8(12):e024489. https://doi.org/10.1136/bmjopen-2018-024489.

Kesari, K.K., Agarwal, A., Henkel, R. (2018). Radiations and male fertility. *Reprod. Biol. Endocrinol.* 16(1):118. https://doi.org/10.1186/s12958-018-0431-1.

Kostoff, R.N., Lau, C.G.Y. (2013). Combined biological and health effects of electromagnetic fields and other agents in the published literature. *Technol. Forecast Soc. Change* 80(7):1331–1349.

Leach, V., Weller, S., Redmayne, M. (2018). Database of bio-effects from non-ionizing radiation: A novel database of bio-effects from non-ionizing radiation. *Rev. Environ. Health* 33(3):273–280. https://doi.org/10.1515/reveh-2018-0017; https://www.ncbi.nlm.nih.gov/pubmed/29874195.

Lerchl, A., Klose, M., Grote, K., et al. (2015). Tumor promotion by exposure to radiofrequency electromagnetic fields below exposure limits for humans. *Biochem. Biophys. Res. Commun.* 459(4):585–590.

Massachusetts, United States of America. (2017). Legislative update on bills on wireless and health. https://ehtrust.org/massachusetts-2017-bills-wireless-health/.

Melnick, R.L. (2019). Commentary on the utility of the national toxicology program study on cellphone radiofrequency radiation data for assessing human health risks despite unfounded criticisms aimed at minimizing the findings of adverse health effects. *Environ. Res.* 168:1–6.

Meo, S.A., Almahmoud, M., Alsultan, Q., et al. (2018). Mobile phone base station tower settings adjacent to school buildings: Impact on students' cognitive health. *Am. J. Mens Health.* 7:1557988318816914. https://doi.org/10.1177/1557988318816914.

Miller, A.B., To, T., Agnew, D.A., Wall, C., Green, L.M. (1996). Leukemia following occupational exposure to 60-Hz electric and magnetic fields among Ontario electric utility workers. *Am. J. Epidemiol.* 144(2):150–160.

Miller, A.B., Morgan, L.L., Udasin, I., Davis, D.L. (2018). Cancer epidemiology update, following the 2011 IARC evaluation of radiofrequency electromagnetic fields (monograph 102). *Environ. Res.* 167:673–683.

Miller, A.B., Sears, M.E., Morgan, L.L., et al. (2019). Risks to health and well-being from radio-frequency radiation emitted by cell phones and other wireless devices. *Front. Public Health* 7:223. https://doi.org/10.3389/fpubh.2019.00223.

Moon, I.S., Kim, B.G., Kim, J., Lee, J.D., Lee, W.S. (2014). Association between vestibular schwannomas and mobile phone use. *Tumour Biol.* 35(1):581–587. https://doi.org/10.1007/s13277-013-1081-8.

Moulder, J.E., Erdreich, L.S., Malyapa, R.S., et al. (1999). Cell phones and cancer: What is the evidence for a connection? *Radiat. Res.* 151(5):513–531.

Moulder, J.E., Foster, K.R., Erdreich, L.S., McNamee, J.P. (2005). Mobile phones mobile phone base stations and cancer: A review. *Int. J. Radiat. Biol.* 81(3):189–203.

National Toxicology Program. (2018a). NTP technical report on the toxicology and carcinogenesis studies in Hsd:Sprague-Dawley SD rats exposed to whole-body radio frequency radiation at a frequency (900 MHz) and modulations (GSM and CDMA) used by cell phones. *NTP TR* 595.

National Toxicology Program. (2018b). NTP technical report on the toxicology and carcinogenesis studies in B6C3F1/N mice exposed to whole-body radio frequency radiation at a frequency (1800 MHz) and modulations (GSM and CDMA) used by cell phones. *NTP TR* 596. https://ntp.niehs.nih.gov/ntp/about_ntp/trpanel/2018/march/tr596peerdraft.pdf.

Nilsson, J., Järås, J., Henriksson, R., et al. (2019). No evidence for increased brain tumour incidence in the Swedish national cancer register between years 1980–2012. *Anticancer Res.* 39(2):791–796. https://doi.org/10.21873/anticanres.13176.

Odemer, R., Odemer, F. (2019). Effects of radiofrequency electromagnetic radiation (RF-EMF) on honey bee queen development and mating success. *Sci. Total Environ.* 661:553–562. https://doi.org/10.1016/j.scitotenv.2019.01.154.

Olsson, A., Bouaoun, L., Auvinen, A., et al. (2019). Survival of glioma patients in relation to mobile phone use in Denmark, Finland and Sweden. *J. Neurooncol.* 141(1):139–149. https://doi.org/10.1007/s11060-018-03019-5.

Ostrom, Q.T., Gittleman, H., de Blank, P.M., et al. (2016). Adolescent and young adult primary brain and central nervous system tumors diagnosed in the United States in 2008–2012. *Neuro-Oncology* 18(suppl_1):1–50. https://doi.org/10.1093/neuonc/nov297.

Ostrom, Q.T., Gittleman, H., Truitt, G., et al. (2018). CBTRUS statistical report: Primary brain and other central nervous system tumors diagnosed in the United States in 2011–2015. *Neuro-Oncology* 20(suppl 4):1–86. https://doi.org/10.1093/neuonc/noy131.

Ostrom, Q.T., Patil, N., Cioffi, G., et al. (2020). CBTRUS statistical report. Primary brain and other central nervous system tumors diagnosed in the United States in 2013–2017. *Neuro-Oncology* 22(12 suppl 2):iv1–iv96. https://doi.org/10.1093/neuonc/noaa200.

Pall, M.L. (2016). Microwave frequency electromagnetic fields (EMFs) produce widespread neuropsychiatric effects including depression. *J. Chem. Neuroanat.* 75(Pt B):43–51. https://doi.org/10.1016/j.jchemneu.2015.08.001.

Panagopoulos, D.J., Karabarbounis, A., Margaritis, L.H. (2004). Effect of GSM 900-MHz mobile phone radiation on the reproductive capacity of Drosophila melanogaster. *Electromagn. Biol. Med.* 23(1):29–43.

Panagopoulos, D.J., Johansson, O., Carlo, G.L. (2013). Evaluation of specific absorption rate as a dosimetric quantity for electromagnetic fields bioeffects. *PLOS ONE* 8(6):e62663. https://doi.org/10.1371/journal.pone.0062663.

Panagopoulos, D.J., Johansson, O., Carlo, G.L. (2015). Polarization: A key difference between man-made and natural electromagnetic fields, in regard to biological activity. *Sci. Rep.* 5:14914. https://doi.org/10.1038/srep14914.

Panagopoulos, D.J. (2019). Comparing DNA damage induced by mobile telephony and other types of man-made electromagnetic fields. *Mutat. Res. Rev. Mutat. Res.* 781:53–62.

Philips, A., Henshaw, D.L., Lamburn, G., O'Carrol, L.M.J. (2018). Brain tumours: Rise in Glioblastoma multiforme incidence in England 1995–2015 suggests an adverse environmental or lifestyle factor. *J. Public Health Environ.*:7910754. https://doi.org/10.1155/2018/7910754.

Rago, R., Salacone, P., Caponecchia, L., et al. (2013). The semen quality of the mobile phone users. *J. Endocrinol. Invest.* 36(11):970–974. https://doi.org/10.3275/8996.

Rappaport, T.S., Sun, S., Mayzus, R., et al. (2013). Millimeter wave mobile communications for 5G cellular: It will work! *IEEE Access* 1:335–349. https://doi.org/10.1109/ACCESS.2013.2260813.

Redmayne, M., Smith, E., Abramson, M.J. (2013). The relationship between adolescents' well-being and their wireless phone use: A cross-sectional study. *Environ. Health* 12:90. https://doi.org/10.1186/1476-069X-12-90.

Röösli, M., Lagorio, S., Schoemaker, M.J., Schüz, J., Feychting, M. (2019). Brain and salivary gland tumors and mobile phone use: Evaluating the evidence from various epidemiological study designs. *Annu. Rev. Public Health*; January 11. https://doi.org/10.1146/annurev-publhealth-040218-044037.

Russell, C.L. (2018). 5G wireless telecommunications expansion: Public health and environmental implications. *Environ. Res.* 165:484–495.

Sage, C., Burgio, E. (2018). Electromagnetic fields, pulsed radiofrequency radiation, and epigenetics: How wireless technologies may affect childhood development. *Child Dev.* 89(1):129–136. https://doi.org/10.1111/cdev.12824.

Salford, L.G., Brun, A.E., Eberhardt, J.L., Marmgren, L., Persson, B.R. (2003). Nerve cell damage in mammalian brain after exposure to microwaves from GSM mobile phones. *Environ. Health Perspect.* 111(7):881–883.

Samuel, H. (2018). The telegraph. France to impose total ban on mobile phones in schools. https://www.telegraph.co.uk/news/2017/12/11/france-impose-total-ban-mobile-phones-schools/.

Savitz, D.A., Wachtel, H., Barnes, F., John, E.M., Tvrdik, J.G. (1988). Case-control study of childhood cancer and exposure to 60Hz magnetic fields. *Am. J. Epidemiol.* 128(1):21–38.

Sharma, V.P., Kumar, N.R. (2010). Changes in honeybee behaviour and biology under the influence of cellphone radiations. *Curr. Sci.* 98:1376–1378.

Szmigielski, S. (2013). Reaction of the immune system to low-level RF/MW exposures. *Sci. Total Environ.* 454–455:393–400.

Söderqvist, F., Carlberg, M., Hardell, L. (2012). Review of four publications on the Danish cohort study on mobile phone subscribers and risk of brain tumours. *Rev. Environ. Health* 27(1):51–58.

Soffritti, M., Giuliani, L. (2019). The carcinogenic potential of non-ionizing radiations: The cases of S-50 Hz MF and 1.8 GHz GSM radiofrequency radiation. *Basic Clin. Pharmacol. Toxicol.* February 24. https://doi.org/10.1111/bcpt.13215.

Sudan, M., Olsen, J., Arah, O.A., Obel, C., Kheifets, L. (2016). Prospective cohort analysis of cellphone use and emotional and behavioural difficulties in children. *J. Epidemiol. Community Health.* https://doi.org/10.1136/jech-2016-207419.

Tuor, M., Ebert, S., Schuderer, J., Kuster, N. (2005). Assessment of ELF exposure from GSM handsets and development of an optimized RF/ELF exposure setup for studies of human volunteers, BAG Reg. No. 2.23.02.-18/02.001778, IT'IS Foundation.

Vijayalaxmi, Prihoda, T.J. (2019). Comprehensive review of quality of publications and meta-analysis of genetic damage in mammalian cells exposed to non-ionizing radiofrequency fields. *Radiat. Res.* 191(1):20–30. https://doi.org/10.1667/RR15117.1.

Villeneuve, P.J., Agnew, D.A., Miller, A.B., Corey, P.N. (2000a). Non-Hodgkin's lymphoma among electric utility workers in Ontario: The evaluation of alternate indices of exposure to 60 Hz electric and magnetic fields. *Occup. Environ. Med.* 57(4):249–257.

Villeneuve, P.J., Agnew, D.A., Miller, A.B., Corey, P.N., Purdham, J.T. (2000b). Leukemia in electric utility workers: The evaluation of alternative indices of exposure to 60 Hz electric and magnetic fields. *Am. J. Ind. Med.* 37(6):607–617.

Volkow, N.D., Tomasi, D., Wang, G.-J., et al. (2011). Effects of cell phone radiofrequency signal exposure on brain glucose metabolism. *JAMA* 305(8):808–813.

Vrijheid, M., Deltour, I., Krewski, D., Sanchez, M., Cardis, E. (2006). The effects of recall errors and of selection bias in epidemiologic studies of mobile phone use and cancer risk. *J. Expo. Sci. Environ. Epidemiol.*:1–14. https://doi.org/10.1038/sj.jes.7500509.

Waldmann-Selsam, C., Balmori-de la Plante, A., Breunig, H., Balmori, A. (2016). Radiofrequency radiation injures trees around mobile phone base stations. *Sci. Total Environ.* 572:554–569. https://doi.org/10.1016/j.scitotenv.2016.08.045.

Walsh, J.J., Barnes, J.D., Cameron, J.D., et al. (2018). Associations between 24 hour movement behaviours and global cognition in US children: A cross-sectional observational study. *The Lancet Child Adolesc. Health*, September 27, 2018. https://doi.org/10.1016/S2352-4642(18)30278-5 (accessed September 29, 2018).

Wdowiak, A., Wdowiak, L., Wiktor, H. (2007). Evaluation of the effect of using mobile phones on male fertility. *Ann. Agric. Environ. Med.* 14(1):169–172.

Wertheimer, N., Leeper, E. (1979). Electrical wiring configurations and childhood cancer. *Am. J. Epidemiol.* 109(3):273–84.

West, J.G., Kapoor, N.S., Liao, S.-Y., et al. (2013). Multifocal breast cancer in young women with prolonged contact between their breasts and their cellular phones. *Case Rep. Med.* 2013:354682.

Worcester Massachusetts Public Schools. (2017). Worcester school committee precautionary option on radiofrequency exposure. http://wpsweb.com/sites/default/files/www/school_safety/radio_ frequency.pdf.

Ying, L., Héroux, P. (2014). Extra-low-frequency magnetic fields alter cancer cells through metabolic restriction. *Electromagn. Biol. Med.* 33(4):264–275. https://doi.org/10.3109/15368378.2013.817334.

Yakymenko, I., Tsybulin, O., Sidorik, E., et al. (2016). Oxidative mechanisms of biological activity of low-intensity radiofrequency radiation. *Electromagn. Biol. Med.* 35(2):186–202.

Zhang, G., Yan, H., Chen, Q., et al. (2016). Effects of cell phone use on semen parameters: Results from the MARHCS cohort study in Chongqing, China. *Environ. Int.* 91:116–121. https://doi.org/10.1016/j.envint.2016.02.028.

Zwamborn, A.P.M., Vossen, S.H.J.A., van Leersum, B.J.A.M., et al. (2003). *Effects of global communication system radio-frequency fields on well being and cognitive functions of human subjects with and without subjective complaints.* FEL-03-C148. The Hague, the Netherlands: TNO Physics and Electronics Laboratory. http://home.tiscali.be/milieugezondheid/dossiers/gsm/TNO_rapport_Nederland_sept_2003.pdf.

3 Oxidative Stress Induced by Wireless Communication Electromagnetic Fields

Igor Yakymenko[1,2] and Oleksandr Tsybulin[3]

[1]Department of Public Health, Kyiv Medical University, Kyiv, Ukraine
[2]Department of Environmental Safety, National University of Food Technologies, Kyiv, Ukraine
[3]European Collegium, Kyiv, Ukraine

CONTENTS

Abstract 98
3.1 Introduction 98
    3.1.1 Man-made Electromagnetic Fields (EMFs) and Health Effects 98
    3.1.2 ELF, RF, and WC Man-made EMFs 99
    3.1.3 Exposure Limits for ELF and RF EMFs 101
3.2 Biophysical Effects of Man-made EMFs in Living Cells 102
3.3 Oxidative Stress induced by Man-made EMFs 104
    3.3.1 Experimental Findings 104
    3.3.2 ROS Sources in Cells 107
3.4 Oxidative Damage of DNA under EMF Exposure 119
3.5 Discussion and Conclusions 120
    3.5.1 Non-thermal Man-made EMF Exposure and Cellular Signaling 120
    3.5.2 OS and Non-carcinogenic Health Effects of Man-made EMFs 121
    3.5.3 OS and Carcinogenic Potential of Man-made EMFs 122
    3.5.4 Conclusions 123
Acknowledgment 123
References 123

Keywords: extremely low frequency, radio frequency, microwaves, reactive oxygen species, oxidative stress, free radicals, cellular signaling, cancer, electromagnetic hypersensitivity

Abbreviations: CAPE: caffeic acid phenethyl ester. CAT: catalase. CDMA: Code Division Multiple Access. DECT: Digitally Enhanced Cordless Telecommunications. ELF: Extremely Low Frequency. EMF: electromagnetic field. EMR: electromagnetic radiation. ETC: electron transport chain. GR: glutathione reductase. GSH: reduced glutathione. GPx: glutathione peroxidase. GSM: Global System for Mobile Telecommunications. GSSG: oxidized glutathione. GST: glutathione *S*-transferase. IgE: immunoglobulin E. LPO: lipid peroxidation. LTE: Long Term Evolution. MDA: malondialdehyde. MPO: myeloperoxidase. MT: mobile telephony. MW: microwaves. NF-κB: nuclear factor kappa B. NO_x: nitric oxides. NOS: nitric oxide synthase. iNOS: inducible nitric oxide synthase. eNOS: endothelial nitric oxide synthase. NTT: nitrotyrosine. ODC: ornithine decarboxylase. OS: oxidative stress. OSI: oxidative stress index. PO: protein oxidation. RF: Radio Frequency. ROS: reactive oxygen species. SAR: Specific Absorption Rate. SOD: superoxide dismutase. TAC: total antioxidative capacity. TBARs: thiobarbituric acid-reactive substances. TBHP: t-butyl hydroperoxide. TDMA: Time Division Multiple Access. TOS: total oxidant status. TTG: total thiol group. UMTS: Universal Mobile Telecommunications System. VGIC: voltage-gated ion channel. WC: wireless communication. Wi-Fi: Wireless Fidelity. XO: xanthine oxidase. 2G/3G/4G/5G: second/third/fourth/fifth generation of MT/WC. 8-OH-dG: 8-hydroxy-2′-deoxyguanosine.

ABSTRACT

This chapter describes experimental data on oxidative effects induced by man-made electromagnetic fields (EMFs) and corresponding electromagnetic radiation (EMR) in living cells. Analysis of the currently available peer-reviewed scientific literature reveals important molecular effects induced by non-thermal exposures to man-made EMFs, especially wireless communication (WC) EMFs, in living cells. They include significant activation of key cellular pathways generating oxidative stress (OS) by reactive oxygen species (ROS), activation of peroxidation, oxidative damage of DNA, and changes in activities of antioxidant enzymes. Critically important features of man-made EMFs, compared to natural EMFs, are their totally polarized and coherent character and, in the case of WC EMFs, combined frequency bands and sophisticated modulation. These features provide these types of EMFs/EMR with the unique and unexpected capacity of inducing biological effects such as pronounced oxidative effects in exposed living cells. It is indicative that among 131 analyzed peer-reviewed studies dealing with oxidative effects of non-thermal Radio Frequency (RF) EMFs, mostly pulsed/modulated by Extremely Low Frequencies (ELF), 124 (95%) confirmed statistically significant oxidative effects on various types of biological systems. And among 39 analyzed studies on oxidative effects of purely ELF EMFs, 36 of them (92%) also revealed significant oxidative effects of the exposure. The wide pathogenic potential of induced ROS and their involvement in cell signaling explains a range of biological/health effects of non-thermal man-made EMF exposures, which includes both carcinogenic and non-carcinogenic pathologies. In conclusion, our analysis demonstrates that a) man-made EMFs, and especially those employed in WC combining both RF and ELF components, is a pronounced oxidative agent for living cells with high pathogenic potential; and b) the OS induced by man-made EMF exposures should be recognized as one of the primary mechanisms of biological activity of this new environmental agent.

3.1 INTRODUCTION

3.1.1 Man-made Electromagnetic Fields (EMFs) and Health Effects

Starting from global electrification more than a hundred years ago, the population of industrialized countries became chronically exposed to man-made electromagnetic fields (EMFs), in this case, mostly to Extremely Low Frequency (ELF) EMFs (3–3000 Hz), and Ultra Low Frequency (ULF) EMFs (0–3 Hz). Recently, the intensive development of wireless communication (WC) technologies, mainly including mobile phones and corresponding base antennas, cordless domestic [Digitally Enhanced Cordless Telecommunications (DECT)] phones, internet wireless connections called Wi-Fi (Wireless Fidelity), etc., during the past three decades led to a new dramatic increase of background EMFs and corresponding electromagnetic radiation (EMR) in the human environment and stimulated a new surge of research on the biological effects of man-made EMFs/EMR. The EMFs/EMR emitted by WC devices/antennas necessarily combine Radio Frequency (RF) (300 kHz–300 GHz) carrier waves, which are modulated by ELF signals containing the transmitted information, and pulsed by ELF pulsations for multiplexing purposes (simultaneous connection of increasing number of users). Therefore, WC EMFs/EMR are a combination of ELF and RF signals (Pedersen 1997; Hyland 2000; 2008; Panagopoulos 2019). The part of the RF band with the highest frequencies (300 MHz–300 GHz) is also called microwaves (MWs).

A series of epidemiological studies which have shown increased risk of some types of cancer in "heavy" users of WC devices, such as mobile or DECT phones, is a proof of a reasonable social concern (Hardell et al. 2007; 2011; Sadetzki et al. 2008; Yakymenko et al. 2011; Sato et al. 2011; Miller et al. 2019). In addition to that, some studies indicate that long-term exposure of humans to WC EMFs can cause various non-cancerous disorders, e.g., headache, fatigue, depression, tinnitus, skin irritation, hormonal disorders, and other conditions (Abdel-Rassoul et al. 2007; Belpomme and Irigaray 2020; Buchner and Eger 2011; Chu et al. 2011; Johansson 2006; Santini et al. 2002; Yakymenko et al. 2011). In addition, convincing studies on the hazardous effects of WC EMFs in human germ cells have been published (Agarwal et al. 2009; De Iuliis et al. 2009). Most importantly, genetic damage in animal/human cells induced by WC EMFs has been repeatedly and reliably documented by several research groups (Lai and Singh 1996; 1997; Diem et al. 2005; Burlaka et al. 2013; Panagopoulos 2019; 2020; Yakymenko et al. 2018).

Also, the significantly increased levels of background man-made EMFs are obviously a potential stress factor for entire ecosystems. Recently, a clear demonstration has been provided for forest ecosystems (Ozel et al. 2021). Depending on the distance from a mobile telephony (MT) base station antenna as a source of WC EMFs (from 100 m to 800 m), the flowers and cones in *Pinus brutia* trees (a species of pine) decreased by 11 and 7 times, respectively compared to unexposed trees of the same species at longer distances from the MT antenna.

All the abovementioned studies dealt with the effects of WC EMFs/EMR at non-thermal intensities. This means that the intensity (power density) of radiation was below observable thermal effects (heating) in biological tissues and below EMF exposure limits set by the International Commission on Non-Ionizing Radiation Protection (ICNIRP 1998) (see Section 3.1.3). A number of studies on metabolic changes in living cells under non-thermal man-made EMF exposure were carried out, and comprehensive reviews were published (Belyaev 2010; Consales et al. 2012; Desai et al. 2009; Yakymenko et al. 2011; 2016). In the present work, we analyze molecular effects of man-made non-thermal EMF exposure in living cells and model biosystems, with a special emphasis on oxidative stress (OS) effects. It might seem paradoxical that, despite being non-ionizing, man-made ELF and RF/WC EMFs can induce significant damage in biological molecules. This becomes possible by activation of free radical processes and overproduction of reactive oxygen species (ROS) in living cells. The analysis of these findings may provide a general picture of the health effects of the already ubiquitous and still ever-increasing ambient levels of WC EMFs.

Generally, OS represents an imbalance between production of ROS in cells and tissues and the ability of a biological system to detoxify them. Many pathologies and medical conditions, including cancer, are closely related with chronic OS. In turn, ROS are molecules that contain oxygen and are highly reactive. Most, but not all of them, are free radicals, i.e., molecules or atoms with one unpaired electron on the outer orbit. Cells always produce some level of ROS because of normal metabolic processes and have some specific enzyme-antioxidants for their neutralization. For example, antioxidant enzymes include superoxide dismutase (SOD), catalase (CAT), glutathione peroxidase (GPx), and others. Also, some non-enzymatic antioxidants, such as vitamin E and vitamin C, are essential in protection of cells against OS. We have different indicators of OS in living cells, e.g., malondialdehyde (MDA) or thiobarbituric acid-reactive substances (TBARs) are indicators of an increased level of lipid peroxidation (LPO), which is an indicator of OS itself, and e.g., an increased level of 8-hydroxy-2′-deoxyguanosine (8-OH-dG) is a clear indicator of oxidative damage of DNA.

3.1.2 ELF, RF, AND WC MAN-MADE EMFs

ELF EMFs occupy the lower part of the electromagnetic spectrum in the frequency range of 3–3000 Hz. ELF EMFs result from electrically charged particles and are usually associated with the generation, distribution, and use of electricity at the frequency of 50 Hz or 60 Hz. Electric fields are measured in Volts per meter (V/m). Magnetic fields arise from the motion of electric charges (i.e., a current), and they are expressed in subunits of Tesla (T), such as milli-Tesla (mT), micro-Tesla (μT), or nano-Tesla (nT). A practical unit called Gauss (G) is also commonly used (1G = 10^{-4} T) along with its subunit milli-Gauss (mG). Close to certain electrical appliances at home, the magnetic field values can be up to a few hundred nano-Teslas (or up to a few mG), and the electric field values up to 100 V/m (usually several V/m). Underneath high-voltage power lines, magnetic fields can be about 20–50 µT (0.2–0.5 G), and electric fields can be several thousand Volts per meter (several kV/m) (WHO 2007).

RF is the part of the electromagnetic spectrum with frequencies of 300 kHz–300 GHz. Like ELF EMFs, RF EMFs/EMR are also classified as non-ionizing, which means that they do not carry sufficient energy for the ionization of atoms and molecules. The part of RF EMFs/EMR with the highest frequencies (300 MHz–300 GHz), referred to as MWs, can potentially generate the highest thermal effects in the absorbing matter.

The main parameters of RF EMFs/EMR are i) frequency (in Hz), ii) intensity or power density of radiation (in W/m^2 or $\mu W/cm^2$), iii) its modulated or non-modulated nature, and iv) continuous or discontinuous (pulsed) pattern of radiation. For the absorbed RF energy by living tissues, the metric of Specific Absorption Rate (SAR) is used (in W/kg).

A common digital technology using RF/MW and ELF EMFs for MT/WC is the Global System for Mobile Telecommunications (GSM) or second-generation (2G) MT system, which utilizes carrier frequencies at about 900, 1800, or 1900 MHz. This radiation is amplitude- and phase-modulated, with a main pulse ("frame") repetition frequency at 217 Hz and additional pulsations at 2 and 8 Hz. The Time Division Multiple Access (TDMA) standard is employed for multiplexing. This provides a single pulse from the pulse-packet (frame) to each user with the pulse rate of 217 Hz (Pedersen 1997; Hyland 2000; 2008). The third-generation (3G) MT system is called Universal Mobile Telecommunications System (UMTS). The carrier frequency in UMTS is between 1920 MHz and 2170 MHz, with pulsing frequencies at 100 and 1500 Hz. In the UMTS, the Code Division Multiple Access (CDMA) standard is employed according to which a special code is assigned to each user. The Long Term Evolution (LTE) or fourth-generation (4G) MT/WC system is the upgrade to UMTS/CDMA with pulsing frequencies at 100 Hz (frame repetition), and 1000 Hz (sub-frame repetition). The carrier frequency bands allocated for LTE are different (up to 2600 MHz) in different countries around the world (see Chapter 1, and Hyland 2000; 2008; Ericsson 2009; Shim et al. 2013; Torrieri 2018).

The most recent, currently under deployment, fifth-generation (5G) MT/WC system uses much higher carrier frequencies (up to 100 GHz). The 5G system, apart from ELF pulsations for multiplexing, reference, and synchronization, will also use very new supportive technology, including phased arrays for directional beam forming and so-called multiple input/multiple output (MIMO) technology to enable higher data transmission capacity (see Chapter 1). Since this new standard has not gone through any reliable assessment regarding its biological effects and health risks, it represents a huge global challenge against the Precautionary Principle (Frank 2021).

It is important to underline, again, that all WC technologies utilize pulsed and modulated RF EMFs/EMR. This actually means that exposure of biological systems to any type of WC EMFs represents simultaneous exposure to both RF/MW and ELF EMFs. Also, all types of man-made EMFs, in contrast to natural EMFs, are totally polarized and coherent, and this characteristic significantly increases their biological activity (Panagopoulos et al. 2015).

"Non-thermal" RF EMF/EMR refers to radiation intensities which do not induce significant thermal effects in exposed biological tissues. Accordingly, any intensity of RF EMF/EMR under the ICNIRP (1998) limits can be referred to as non-thermal. From a practical approach, non-thermal EMR refers to radiation which does not increase the temperature of exposed biological tissues by 1°C. This is actually too much for biological systems in most cases. In this paper, we will only discuss effects of non-thermal EMF-exposures.

The ambient levels of WC EMFs/EMR in terms of power density in most urban (and even more in natural/rural) environments are (before the advent of 5G) below 0.01 $\mu W/cm^2$. The corresponding levels closer to MT base stations (≤100 m) are significantly higher (of the order of 1 $\mu W/cm^2$ or higher), and close to WC devices (mobile phones, DECT phones, Wi-Fi), especially mobile phones, the levels may exceed 100 µW/cm² (Panagopoulos et al. 2010; Panagopoulos 2020; Carpenter 2013; Yakymenko et al. 2018).

3.1.3 Exposure Limits for ELF and RF EMFs

The international recommendations on safety limits of non-ionizing EMFs/EMR adopted by health authorities and governments are given in the document titled "Guidelines for Limiting Exposure to Time-Varying Electric, Magnetic, and Electromagnetic Fields (up to 300 GHz)" published by ICNIRP (1998). The document provides recommended safety limits in all bands of the man-made EMF spectrum, both for occupational and general public exposure. The section of the guidelines called "basis for limiting exposure" is important for understanding the flaws of this document. The document directly states that

> "induction of cancer from long-term EMF exposure was not considered to be established, and so these guidelines are based on short-term, immediate health effects such as stimulation of peripheral nerves and muscles, shocks and burns caused by touching conducting objects, and elevated tissue temperatures resulting from absorption of energy during exposure to EMF".

For ELF EMFs (at power line frequencies 50–60 Hz), the ICNIRP (1998) limits for magnetic and electric field intensities were 0.5 mT (5G) and 10 kV/m for occupational exposure and 0.1 mT (1G) and 5 kV/m respectively for the public. Although these limits were already obsolete, as significant adverse biological effects are known to take place at several orders of magnitude weaker ELF EMFs (Goodman et al. 1995; Panagopoulos et al. 2021), the ICNIRP revised its limits for EMFs up to 100 kHz in 2010, and the 0.1 mT limit for the public was raised to 0.2 mT (2G) (ICNIRP 2010).

For RF EMFs, according to the ICNIRP (1998) document, the limits are given in terms of certain parameters of the exposure: 1) The SAR (in W/kg), which indicates the amount of EMR energy absorbed per unit mass of human tissue per second, and 2) the power density or intensity of the incident radiation in W/m² (or mW/cm² or µW/cm²), which indicates the amount of electromagnetic energy incident upon a unit surface per second.

The SAR limit for the general public for RF EMF exposure in the ICNIRP guidelines is 2 W/kg (for head and trunk). This limit is accepted by the telecom industry as mandatory for every commercially available mobile phone over the world, and the value of SAR of each mobile phone model must be indicated in the technical specifications of the device. Unfortunately, SAR is not only inconvenient but also unreliable as a metric for the assessment of non-thermal biological effects (Panagopoulos et al. 2013). Moreover, only simplified models of the adult human head are currently used by the industry for calculation of the SAR, while real SAR values depend on geometry, conductivity, and other tissue details, and it was shown that SAR is much higher for a child's head than for an adult's head (Christ et al. 2010; de Salles et al. 2006; Gandhi et al. 1996).

The power density, or intensity of the EMR, is a much more direct and simple metric compared to the SAR, although it does not estimate the amount of energy of EMR that is absorbed by the exposed tissue. The occupational exposure limits in the RF/MW band according to ICNIRP (1998) were set to 10–50 W/m² (or 1–5 mW/cm²) average values for 6 min of exposure. The corresponding public exposure limits were set to 2–10 W/m² (or 0.2–1 mW/cm²) depending on the frequency (averaged over 6 min). For example, for 900 MHz, the ICNIRP (1998) public exposure limit was set to 0.45 mW/cm², and for 2–6 GHz, it was set to 1 mW/cm² (ICNIRP 1998). Surprisingly, the most recent ICNIRP update (ICNIRP 2020) raised the 2–6 GHz 6 min average limit for public exposure from 1 mW/cm² to 4 mW/cm², which does not anymore prevent not even the thermal effects.

It is important to note that the ICNIRP recommendations have no legal validity, as they are only recommendations. Each country has its own national legislation for electromagnetic safety, and national safety limits are different in different countries. Some countries, such as the United States (US) and many European countries, actually adapted their national EMF/EMR safety limits to the

ICNIRP recommendations. Other countries have rather different national limits from those of the ICNIRP guidelines. For example, for 900 MHz (e.g., GSM), the public exposure limits in different countries are: in US and Germany 450 µW/cm^2; in Italy, Russia, Poland, and China 10 µW/cm^2; and in Switzerland 4 µW/cm^2 (Chekhun et al. 2014). As we can see, some countries have significantly more stringent national safety limits than others. Such national positions are, in part, explained by long-term national research in the field of electromagnetic biology and experience in studying the non-thermal biological effects of man-made EMFs, as, e.g., in the former USSR. On the other hand, some countries, such as Switzerland or Italy, stay closer to the Precautionary Principle, which recommends caution, pausing, and review before allowing new technologies that may prove disastrous (Read and O'Riordan 2017).

3.2 BIOPHYSICAL EFFECTS OF MAN-MADE EMFs IN LIVING CELLS

RF, especially MW EMR, can produce significant thermal effects in matter due to interaction with charged particles, e.g., ions or dipole molecules, which are forced to oscillate due to an applied oscillating EMF. People utilize the heating capacity of RF/MW EMR for warming food in the microwave oven. The thermal effects become more evident with increasing intensity and frequency of the MW radiation and are mostly negligible for intensities < 100 µW/cm^2 (for frequencies up to 2 GHz). On the other hand, the energy of RF/MW EMR is insufficient not only for the ionization of molecules but also for the excitation of orbital electrons (as done by visible light). That is why non-thermal RF EMR was historically considered as biologically inert, and RF EMR with increased intensity was expected to produce only thermal effects. Nevertheless, clear and ample evidence of non-thermal biological effects induced by RF EMFs discovered during the past 50 years has stimulated research on the physical mechanisms of such effects.

The ability of non-thermal MW EMF exposures to dissociate water molecules has been demonstrated in model experiments (Vaks et al. 1994). In that study, MW EMF of 10 GHz with radiated power 30 mW produced a significant level of H_2O_2 in deionized water (and also in $MgSO_4$ solution) under stable temperature conditions. According to the authors, a kinetic excitation of liquid water associates $(H_2O)_n$ (ionic clusters) upon the absorption of MWs leads to subsequent viscous energy losses due to friction between moving clusters of water molecules. It results in partial irreversible decomposition of water, including breaks of intra-molecular bonds (H-OH) due to a mechanochemical reaction and generation of H•, OH•, H$^+$, and OH$^-$ groups. Among these, the hydroxyl free radical (OH•) is the most aggressive form of ROS, which can break any chemical bond in surrounding molecules (Halliwell 2007). The authors assessed that this type of mechanochemical transformation in water could be due to only 10^{-8}–10^{-4} relative parts of the total MW energy absorbed (meaning that only a tiny amount of the absorbed energy actually results in water decomposition). But since water molecules are ubiquitous in living cells, even a subtle chance for dissociation of water molecules under non-thermal levels of RF EMR exposure could have a profound effect on tissue homeostasis. It is of note here that one OH• radical can initiate irreversible peroxidation of many hundreds of macromolecules, e.g., lipid molecules in cell membranes (Halliwell 1991).

A biophysical model of a forced-vibration/oscillation of free ions on the surface of a cell membrane due to external oscillating EMFs has been proposed (Panagopoulos et al. 2000; 2002; 2015; 2021) It is important that the proposed ion forced-oscillation mechanism explains how exposure to polarized and coherent (man-made) ELF EMFs of even very low intensities can disrupt cell's electrochemical balance (the intra-cellular concentrations of dissolved/mobile ions) by irregular gating of voltage-gated ion channels (VGICs) in the cell membranes.

In line with the above-proposed mechanism, the so-called "calcium effects" under modulated RF EMF exposure in living cells have been demonstrated (Dutta et al. 1989; Paulraj et al. 1999; Rao et al. 2008), which include a significant increase in intra-cellular Ca^{2+}. Taking into account that calcium is an ubiquitous regulator of cellular metabolism, these data point to a possibility that even very weak EMFs can affect multiple Ca^{2+}-dependent signaling cascades.

A "moving charge interaction" hypothesis was proposed for ELF EMFs (Blank and Soo 2001). The authors suggested a possible qualitative explanation for the activation of genes and the synthesis of stress proteins under EMF exposure due to a possible interaction of the field with moving electrons in DNA (Blank and Soo 2001; Goodman and Blank 2002). They also found that ELF EMFs increased electron transfer rates in cytochrome oxidase accelerated the active ion transport in the Na^+/K^+-ATPase function and accelerated the Belousov–Zhabotinsky reaction[*] in homogeneous solutions (Blank and Soo 2003).

The ability of weak magnetic fields to trigger the onset and offset of evoked potentials was demonstrated (Marino et al. 2009). Effectiveness of a rapid magnetic stimulus (0.2 ms) led the authors to conclude a direct interaction between the field and ion channels in the cell plasma membrane.

Increase of free radical lifetimes in living cells due to electron spin recombination of free radical pairs presumably induced by magnetic field exposure has also been proposed (Brocklehurst and McLauchlan 1996; Georgiou 2010).

A significant effect of non-thermal RF EMR exposure on ferritin[†] has been reported (Céspedes and Ueno 2009). Non-thermal RF EMF exposure of ferritin solution significantly reduced, up to threefold, iron chelation with ferrozine. The authors argued that the RF magnetic component plays a principal role in the observed effect and that this effect is clearly non-thermal. The hypothetical non-thermal mechanism of interaction of RF magnetic fields with ferritin results in the reduction of iron chelates into the ferritin cage. The authors underlined the potential role of ferritin malfunction for oxidative processes in living cells due to participation of Fe^{2+} ions in the Fenton reaction, which produces hydroxyl radicals. In this respect, it is interesting to point to the results of an *in vitro* study with RF EMF exposure of rat lymphocytes treated with iron ions (Zmyślony et al. 2004). Although RF EMF exposure (930 MHz) did not induce detectable intra-cellular ROS production, the same exposure in the presence of $FeCl_2$ in the lymphocyte suspensions induced a significant production of ROS.

Another set of studies indicates a possibility of changes in protein conformation under RF EMF exposure. Exposure to 2.45 MHz EMF at non-thermal levels accelerated conformational changes in β-lactoglobulin through excitation of so-called collective intrinsic modes in the protein (Bohr and Bohr, 2000a; 2000b), which suggests a principal ability of the EMF exposure to modulate the nonrandom collective movements of entire protein domains. Similarly, a frequency-dependent effect on intrinsic flexibility in insulin structure due to applied oscillating electric field was demonstrated (Budi et al. 2007). Moreover, the macromolecular structure of the cytoskeleton was significantly altered in fibroblasts of Chinese hamsters after exposure to GSM EMF (Pavicic and Trosic, 2010). A 3 h exposure of fibroblasts to the pulsed (975 MHz) EMF led to significant changes in the structure of microtubules and actin microfilaments, which both are polar cytoskeleton structures, while non-polar vimentin filaments reportedly stayed unaffected. Severe damage to the actin cytoskeleton has also been found to accompany DNA fragmentation in fruit fly ovarian cells after only a few minutes of exposure to GSM mobile phone EMF (Chavdoula et al. 2010). Taking into account the extensive regulatory potential of the actin cytoskeleton on cell homeostasis, these data could obviously add to the nature of the biological effects of WC EMFs.

It was shown that the activity of ornithine decarboxylase (ODC) significantly changes under non-thermal pulsed RF EMF exposure with ELF modulation/pulsations (Byus et al. 1988; Hoyto et al. 2007; Litovitz et al. 1993; Litovitz et al. 1997; Paulraj et al. 1999; Penafiel et al. 1997). It is interesting that the effects depended on the ELF frequency of the amplitude modulation (in the range of 16–600 Hz) under the same MW carrier frequency (835 MHz) (Penafiel et al. 1997).

Overall, these data show that non-thermal man-made EMFs, and especially WC EMFs combining ELF and RF, can interact with particular charges, molecules, and cellular structures and, in this way, can potentially induce substantial modulatory effects in living cells.

---

[*] Oxidation of organic compounds by bromic acid catalyzed by metal ions in acidic water solutions.
[†] An iron cage protein present in most living organisms from bacteria to humans.

3.3 OXIDATIVE STRESS INDUCED BY MAN-MADE EMFs

3.3.1 Experimental Findings

Our brief analysis of peer-reviewed experimental papers on oxidative effects of ELF man-made EMFs using a scholar.google.com search has revealed that, among 39 selected relevant peer-reviewed experimental studies, 36 (92%) of them demonstrated statistically significant oxidative effects in different biological models under non-thermal exposure to ELF EMFs (Tables 3.1 and 3.2). Most research has been carried out on laboratory rats and mice. ELF EMF intensity of 100 µT (magnetic field) with 30 min exposure duration provided significant changes in pro-oxidant/antioxidant balance in the exposed biological models (Abyaneh 2018; Karimi et al. 2019). Along with ROS overproduction and depletion of antioxidant enzyme activities, exposure to technical ELF EMFs resulted in oxidative damage of DNA (Yokus et al. 2005; 2008), which is obviously the expected and most critical result of OS due to the EMF exposure. Along with these experimental data, a few papers demonstrated clear oxidative effects of ELF man-made EMFs in exposed human population, e.g., among exposed technicians and electric installation workers (Tiwari et al. 2015; Kunt et al. 2016; Hosseinabadi et al. 2020; 2021).

Similarly, based on the analysis of available peer-reviewed publications, we identified 131 experimental studies in biological models which investigated oxidative effects of non-thermal RF/WC EMF exposures. Many of them used modulated RF/MW EMFs from WC devices (WC EMFs). Of these 131 studies, 124 studies (95%) demonstrated significant oxidative effects induced by non-thermal RF/WC EMF exposure (Tables 3.3–3.5), while seven studies demonstrated insignificant changes (Table 3.6). The total number includes 25 *in vitro* studies, 95 studies in animals, three studies in plants, and eight studies in humans. The majority of research was carried out in laboratory rats and mice. Six out of eight studies in humans were positive. From the *in vitro* studies, 24 were positive (96%), including four studies in human spermatozoa and two studies in human blood cells, all of which were positive. Interestingly enough, from 31 studies on oxidative effects of non-thermal RF/WC EMF exposures published during the past 5 years (from 2017), all (100%) were positive.

Most of the RF/WC EMF studies utilized RF exposure in the MW range, which, in most cases, included ELF components (pulsation/modulation) and many used commercial or "test"* mobile phones, Wi-Fi devices, or MT base stations as sources of WC EMFs. The power densities of the RF EMR applied in positive studies varied from 0.1 µW/cm^2 (Oksay et al., 2014) to 680 µW/cm^2 (Jelodar et al. 2013b), and SAR values varied from 3 µW/kg (0.25 µW/cm^2) (Burlaka et al., 2013) to the ICNIRP recommended limit of 2 W/kg (Xu et al. 2010; Naziroglu et al. 2012). Exposure times in positive studies varied from 5 min (Friedman et al., 2007) to 12.5 years, 29.6 h/month (Hamzany et al. 2013).

The most frequently used indexes of OS analyzed in both ELF and RF/WC EMF studies were ROS production, levels of LPO/MDA, protein oxidation (PO), nitric oxides (NO$_x$), reduced glutathione (GSH), and activity of antioxidant enzymes such as SOD, CAT, or GPx.

It is important to emphasize a strict non-thermal character of ROS overproduction under EMF exposure described in the cited papers. As low as 0.1 µW/cm^2 in intensity of modulated RF/WC EMF and 0.3 µW/kg SAR were demonstrated to be effective in inducing significant OS in living cells (Burlaka et al. 2013; Oksay et al., 2014). These data are particularly important, as the existing international safety limits on ELF and RF EMF/EMR exposure are based solely on the thermal effects and only restrict RF EMR intensity to 450–4000 µW/cm^2 and SAR to 2 W/kg (ICNIRP 1998; 2020). Moreover, several studies that employed high (thermal) intensities of RF EMR could not reveal oxidative effects (Luukkonen et al. 2009; Hong et al. 2012; Kang et al. 2013; Jeong et al. 2018), which may point to the variety of molecular mechanisms for different radiation/field intensities.

---

* "Test" mobile phones emit pulsing signals like those of the commercially available handsets but are invariable/predictable and non-modulated. Thus, their signals are similar to simulated ones emitted by generators.

TABLE 3.1
Studies with Positive Findings on OS Induced by Man-made ELF EMF Exposure

S. No	Biological System Exposed	EMF Exposure	Recorded Effects[a]	Reference
1.	Lab mice	ELF EMF 60 Hz, 12 G, for 3 h	Lucigenin-amplified t-butyl hydroperoxide (TBHP)-initiated brain homogenates chemiluminescence was increased and brain SOD activity was increased	(Lee et al., 2004)
2.	Lab rats	ELF EMF 50 Hz, 0.97 mT, for 100 days	Increased levels of 8-OH-dG and TBARs after 50 and 100 days	(Yokus et al., 2005)
3.	Helix aspersa snails	ELF EMF 50 Hz, 0.5, 2.5, 10, and 50 µT, up to 2 months	Changes in CAT, glutathione reductase (GR), and the overall capability to neutralize peroxyl radicals	(Regoli et al., 2005)
4.	Lab rats	ELF EMF 50 Hz, 0.5 mT, for 7 days	Increased superoxide radical contents in all brain structures and increased production of nitric oxide in the frontal cortex	(Jelenković et al., 2006)
5.	Neuroblastoma cells	ELF EMF 50 Hz, 1.0 mT	Reduced cell tolerance to oxidative attacks	(Falone et al., 2007)
6.	Guinea pigs	ELF EMF 50 Hz, 1, 2, and 3 mT, 4 h/day and 8 h/day, for 5 days	Increased MDA levels and NO_x levels. Decreased myelo-peroxidase (MPO) activity in liver	(Canseven et al., 2008)
7.	Lab mice	ELF EMF 50 Hz, for 30 days	Increased glutathione S-transferase (GST) activity and LPO level in the liver. Decreased hepatic gluthathione content	(Hashish et al., 2008)
8.	Lab rats	ELF EMF 50 Hz, 100 and 500 µT, 2 h/day, for 10 months	Increased 8-OH-dG in DNA by 100 µT	(Yokus et al., 2008)
9.	Lab rats	ELF EMF 50 Hz, 1 mT, 4 h/day, for 45 days	Increased 3-nitrotyrosine (3-NT) levels in exposed female rats (a deteriorative effect on cellular proteins)	(Erdal et al., 2008)
10.	Human keratinocyte cells	ELF EMF, 50 Hz, 1 mT	Increased inducible nitric oxide synthase (iNOS) and endothelial NOS (eNOS) expression levels and increased NO production	(Patruno et al., 2010)
11.	Lab rats	ELF EMF 40 Hz, 7 mT, 30 min/day or 60 min/day, for 14 days (commonly applied in magnetotherapy)	Increase in TBARS and H_2O_2 concentration in heart (by the 60 min/day exposure)	(Goraca et al., 2010)
12.	Lab rats	ELF EMF, 50 Hz, 100 and 500 µT, 2 h/day, for 10 months	Increased MDA, total oxidant status (TOS) and oxidative stress index (OSI) values. Decreased total antioxidative capacity (TAC) and CAT activity (by the 500 µT exposure)	(Akdag et al., 2010)
13.	C2C12 muscle cells	ELF EMF, 50 Hz, 1 µT–1 mT	Increased ROS production in myoblasts and myotubes, increased CAT and GPx activities, and altered intra-cellular Ca^{2+} homeostasis	(Morabito et al., 2010)

(Continued)

TABLE 3.1 (CONTINUED)
Studies with Positive Findings on OS Induced by Man-made ELF EMF Exposure

S. No	Biological System Exposed	EMF Exposure	Recorded Effects[a]	Reference
14.	Lab mice	ELF EMF 60 Hz, 2.3 mT, for 3 h	Increased MDA level and hydroxyl radical in cerebellum	(Chu et al., 2011)
15.	Lab rats	Pulsed square-wave ELF magnetic field with consecutive four pulse trains of 1, 10, 20, and 40 Hz, 1.5 mT, 1 h/day, for 30 days	Increased levels of MDA and SOD in liver	(Emre et al., 2011)
16.	Lab rats	ELF EMF 40 Hz, 7 mT, 30 min/day, or 60 min/day, for 10 days	Exposure of 30 min/day increased LPO, and 60 min/day increased free –SH groups in the brain	(Ciejka et al., 2011)
17.	Lab rats	ELF EMF 60 Hz, 2 mT, for 5 days	Increased NO levels in the cerebral cortex, striatum, and hippocampus	(Cho et al., 2012)
18.	Lab mice	ELF EMF 50 Hz, 0.1 or 1 mT, 4 h/day, for 12 consecutive weeks	Increased MDA levels. Decreased CAT, GPx, and TAC levels in hippocampus and striatum with 1 mT	(Cui et al., 2012)
19.	Lab rats	ELF EMF, 60 Hz, 2.4 mT, for 2 h	Reduction in CAT and SOD activities	(Martínez-Sámano et al., 2012)
20.	Gerbils (rodents)	ELF EMF 50 Hz; 0.1, 0.25, and 0.5 mT, for 7 days	Increased OS in all tested brain regions. The effect was correlated with intensity and was higher in middle-aged gerbils	(Selaković et al., 2013)
21.	Lab mice	ELF EMF 50 Hz, 8 mT, for 28 days	Increased levels of MDA, ROS, nitric oxide, and nitric oxide synthase (NOS). Decreased activities of GPx, CAT, and SOD	(Duan et al., 2013)
22.	Lab mice	ELF EMF 2 mT, 4 h/day, for 8 weeks	Decrease in SOD activity and increase in the levels of MDA in brain	(Deng et al., 2013)
23.	Lab rats	ELF EMF 50 Hz, 50 and 100 µT, for 90 days continuously	Altered reduced glutathione and oxidized glutathione (GSH/GSSG) levels in a dose-dependent manner in all the regions of the brain	(Manikonda et al., 2014)
24.	Gerbils/Rodents	ELF EMF 50 Hz, 0.5 mT, for 7 days continuously	The exposure increased OS (indexes of nitric oxide, superoxide anion, SOD) to a greater extent than ischemia immediately after cessation of exposure	(Rauš Balind et al., 2014)
25.	Human SH-SYSY cells	ELF EMF 50 Hz, 1 mT, for 24 h	Elevated expression of NOS, increased levels of superoxide radical ($O_2^{\bullet-}$), and compensatory changes in enzymatic kinetic parameters related to cytochrome P450 and CAT activity	(Reale et al., 2014)
26.	Human osteosarcoma MG-63 cells	ELF EMF 50 Hz, 1 mT, for 1, 2, and 3 h	Decreased cell viability, inhibited cell growth, induced cell apoptosis, and increased the level of ROS	(Yang, Ye 2015)

S. No	Biological System Exposed	EMF Exposure	Recorded Effects[a]	Reference
27.	Lab mice	ELF EMF 8 mT, 4 h/day, for 28 consecutive days	The exposure increased OS (indexes of MDA, SOD, CAT, GPx, GR, and GST in the blood and cerebral cortex)	(Luo et al., 2016)
28.	Lab mice	ELF EMF 50 Hz, 200 µT, for 30 days	Increased MDA and nitric oxide levels. Decreased CAT and SOD activities	(Gao et al., 2017)
29.	Seeds of *Lepidium sativum* plant	ELF EMF 60 Hz, 3.7 mT, for 30 or 60 min	Increased peroxidation of membrane lipids, activities of SOD, CAT, and ascorbate peroxidase	(Abyaneh, 2018)
30.	*C elegans* worm eggs	ELF EMF 50 Hz, 3 mT, usually 48 h (until reaching the L4 stage)	Elevation of ROS levels accompanied with the depression of total antioxidant capacity	(Sun et al., 2018)
31.	Male lab rats	ELF EMF 50 Hz, 10 kV/m, 4.3 pT, for 22 h/day, for 28 days	Decreased activities of antioxidant enzymes in brain	(Budziosz et al., 2018)
32.	*C. elegans* worms	ELF EMF 50 Hz, 3 mT	Elevated ROS levels, upregulated TAC proteins, decreased ROS-TAC score	(Sun et al., 2019)
33.	Lab rats	ELF EMF 50 Hz, 0.7 mT, 2 h/day	Increase in plasma protein carbonyl (PCO), metHb, and hemichrome concentrations	(Seif et al., 2019)
34.	Male lab rats	ELF EMF 1 µT, 100 µT, 500 µT, and 2000 µT, 2 h/day for 60 days	Increased MDA concentration (by 100 and 500 µT exposure)	(Karimi et al., 2019)
35.	Lab rats	ELF EMF 50 Hz, 2.5 mT, duty cycle 40%, 1 h/day for 14 days	Reduced iNOS and ROS levels and upregulated the expression of CAT and SOD	(Wang et al., 2019)[b]
36.	Pregnant lab rats	ELF EMF 50 Hz, 500 µT, 24 h/day for 21 days	Increased MDA concentration and decreased GSH levels in blood plasma of rat offspring	(Guleken, 2021)

S. No: study number. [a]All effects were statistically significant ($p < 0.05$ or less in all cases) as compared to control or sham-exposed groups. [b]This paper demonstrates antioxidant effects of a particular regime of ELF EMF exposure, which needs further investigation.

3.3.2 ROS Sources in Cells

The mechanism of ROS overproduction in living cells under non-thermal levels of ELF EMFs, as was indicated above, may well be the dysfunction/irregular gating of VGICs in cell membranes under this type of exposure (Panagopoulos et al. 2000; 2002; 2015; 2021). It is important to underline that this mechanism may be applied both to purely ELF EMFs and to RF/MW EMFs modulated by ELF components, which is the case for all WC EMFs.

NADH oxidase, a plasma membrane enzyme, was suggested as a primary mediator of RF EMF interaction with living cells (Friedman et al. 2007). Using purified membranes from HeLa cells, the authors experimentally proved that exposure to "mobile phone irradiation" of 875 MHz from a "frequency generator",[*] 200 µW/cm^2 for 5 or 10 min, significantly (almost threefold)

---

[*] The authors do not specify whether the signal emitted by the generator was continuous-wave or pulsed.

TABLE 3.2
Studies with Non-significant OS Findings due to ELF EMF Exposure

S. No	Biological System Exposed	EMF Exposure	Indexes Analyzed	Reference
1.	Lab rats	ELF EMF 50 Hz, 0.25 mT, 3 h/day, for 14 days	TAC was decreased, and TOS and OSI levels were increased, but the changes were not statistically significant	(Sert and Deniz, 2011)
2.	MCF10A human breast epithelial cells	ELF EMF 60 Hz, 1 mT, for 4 h	Intra-cellular ROS level, SOD activity, and reduced versus oxidized glutathione (GSH/GSSG) ratio	(Hong et al., 2012)
3.	Male lab rats	ELF EMF 50 Hz, 100 and 500 µT, 2 h/day, 7 days/week, for 10 months	Levels of CAT, MDA, MPO, TAC, TOS, and OSI	(Akdag et al., 2013)

increased the activity of NADH oxidase. NADH oxidase serves as electron donor that catalyzes the one-electron reduction of oxygen into superoxide radical ($O_2^{\bullet -}$), which is a powerful ROS. Traditionally, this enzyme has been known for its role in induction of an oxidative burst in phagocytes as a part of the immune response. Yet later, the existence of non-phagocytic NAD(P)H oxidases was revealed in various types of cells, including fibroblasts, vascular and cardiac cells, etc. (Griendling et al. 2000). Obviously, the presence of a superoxide-generating enzyme in many types of non-phagocytic cells points to the considerable regulatory roles of ROS in living cells. On the other hand, the ability of non-thermal man-made EMF exposure to modulate the activity of the NADH oxidase automatically makes such an exposure a notable and potentially dangerous factor for cell metabolism. Notably, Friedman et al. (2007) pointed out that the NADH oxidase is different from the superoxide-generating NADPH oxidase, which is also found in plasma membranes (Low et al. 2012).

The other powerful source of ROS in cells is the mitochondrial electron transport chain (ETC), which can generate superoxide radicals due to breakdowns in electron transport (Inoue et al. 2003). A significant inverse correlation between mitochondrial membrane potential and ROS levels in living cells was found (Wang et al. 2003). As the authors underlined, such a relationship could be due to two mutually interconnected phenomena: ROS causing damage to the mitochondrial membrane and the damaged mitochondrial membrane causing increased ROS production.

It was demonstrated that the generation of ROS by the mitochondrial pathway can be activated under simulated mobile phone radiation exposure in human spermatozoa (De Iuliis et al. 2009). The authors revealed a dose-dependent effect of 1.8 GHz RF EMF exposure on ROS production in spermatozoa, particularly in their mitochondria. The significantly increased level of total ROS in spermatozoa was detected with a SAR value of 1 W/kg, which is below the exposure limits recommended by ICNIRP (1998; 2020) and accepted in many countries. Similarly, it was demonstrated in our laboratory that the exposure of quail embryos *in ovo* to mobile phone EMF from a commercially available handset (GSM 900 MHz, 0.25 µW/cm^2) during the initial days of embryogenesis resulted in overproduction of superoxide and nitric oxide free radicals in mitochondria of embryonic cells (Burlaka et al. 2013). Three possible sites of generation of superoxide in ETC have been suggested: the ETC complex I (Inoue et al. 2003), complex II (Liu et al. 2002), and complex III (Guzy and Schumacker 2006). On the model of mouse spermatogonia GC1 and spermatocytes GC2 cell lines under 1.8 GHz, 0.15 and 1.5 W/kg exposure, complex III of the ETC was identified as the potential source of electrons producing ROS (Houston et al. 2018).

TABLE 3.3
Studies with Positive Findings on OS Induced by RF/WC EMF Exposure of Cells *In Vitro*

S. No	Biological System Exposed	RF/WC EMF Exposure	Recorded Effects[a]	Reference
1.	Rat lymphocytes	930 MHz, 500 µW/cm^2, SAR = 1.5 W/kg, for 5 or 15 min	Increased ROS level in exposed FeCl$_2$-treated cells compared with unexposed FeCl$_2$-treated cells	(Zmyślony et al., 2004)
2.	Human blood cells	Continuous wave or simulated GSM 1800 MHz signal, SAR = 2 W/kg, for 30 or 45 min of continuous or intermittent (5 min ON/ 5 min OFF) exposure	ROS changes after continuous or intermittent GSM exposure. Continuous wave had no effect	(Lantow, et al., 2006)
3.	Human mono mac 6 and K562 cells	Continuous wave, GSM (simulated) 1800 MHz speaking only, GSM hearing only, GSM talk, SARs of 0.5, 1.0, 1.5, and 2.0 W/kg	ROS changes by the GSM signal at 2 W/kg	(Lantow, et al., 2006)
4.	HeLa cell plasma membranes	875 MHz, 200 µW/cm^2, for 5 and 10 min	Increased NADH oxidase activity	(Friedman et al., 2007)
5.	Human spermatozoa	1.8 GHz simulated mobile phone EMF, SAR = 0.4–27.5 W/kg	Increased ROS levels	(De Iuliis et al., 2009)
6.	Human spermatozoa	Mobile phone EMF, in "talk" mode, for 1 h	Increase in ROS level; decrease in sperm motility and viability	(Agarwal et al., 2009)
7.	Rat astroglial cells	900 MHz (continuous or pulsed), electric field 10 V/m, for 5, 10, 20 min	Increase in ROS levels and DNA fragmentation after exposure to the pulsed RF EMF for 20 min. Continuous wave RF exposure had no effect	(Campisi et al., 2010)
8.	Primary cultured neurons	1800 MHz, pulsed, SAR = 2 W/kg, for 24 h	Increased 8-OH-dG	(Xu et al., 2010)
9.	HL-60 cells	2450 MHz, pulsed, SAR = 0.1–2.5 W/kg, for 1, 2, 12, or 24 h	Increased LPO levels with all exposure times	(Naziroglu, et al., 2012)
10.	Neuronal cells and human fibroblasts	27.12 MHz, pulsed, electric field 41 V/m, 2 min prior to lipopolysaccharide administration or for 15 min	Increased level of nitric oxide (NO•) free radical	(Pilla, 2012)
11.	Human blood mononuclear cells	900 MHz, SAR = 0.4 W/kg, for 1–8 h	Increased apoptosis induced through the mitochondrial pathway and mediated by ROS and caspase-3	(Lu et al., 2012)
12.	Mouse GC-2 cells	GSM 1800 MHz EMF, SAR = 1, 2 W/kg, 5 min ON/10 min OFF, for 24 h	Increased ROS by the stronger (2 W/kg) exposure	(Liu et al., 2013)
13.	Human lens epithelial cells	1800 MHz, SAR = 2, 3, or 4 W/kg	Increased ROS and MDA levels	(Ni et al., 2013)
14.	Mouse embryonic fibroblasts (NIH/3T3)	1800 MHz GSM EMF "talk" mode, SAR = 2 W/kg, intermittent exposure (5 min ON/10 min OFF), for 30 min to 8 h	Increased intra-cellular ROS levels	(Hou et al., 2014)

(Continued)

TABLE 3.3 (CONTINUED)
Studies with Positive Findings on OS Induced by RF/WC EMF Exposure of Cells *In Vitro*

S. No	Biological System Exposed	RF/WC EMF Exposure	Recorded Effects[a]	Reference
15.	Cancer cell cultures	900 MHz modulated/pulsed EMF, SAR = 0.36 W/kg, for 1 h	Induced apoptosis effects through OS. Selenium counteracted the effects of EMF exposure	(Kahya et al., 2014)
16.	V79 cells	1800 MHz, SAR = 1.6 W/kg, for 10, 30, or 60 min	Increased ROS level after 10 min of exposure. Decreased ROS level after 30 min exposure, indicating activation of antioxidant defense mechanism	(Marjanovic et al., 2014)
17.	HEK293T cells	940 MHz, SAR = 0.09 W/kg, for 15, 30, 45, 60, or 90 min	Increased ROS with the 30 min exposure; increased CAT and SOD activities and elevation of GSH during the 45 min exposure	(Sefidbakht et al., 2014)
18.	Human embryonic kidney cells (HEK293)	2.45 GHz EMF, output 15 V as constant voltage, and 3.3 V as variable voltage, 8 V/m	Increased MDA levels. Decreased SOD and GPx activities	(Özsobacı et al., 2018)
19.	Human semen	2.45 GHz Wi-Fi EMF, 1.0–2.5 W/kg, for 45 and 90 min	Increased ROS levels with increasing exposure duration. GPx levels decreased, 8-OHdG level increased	(Ding et al., 2018)
20.	Mouse spermatogonia GC1, and spermatocyte GC2 cell lines, epididymal spermatozoa	1.8 GHz, 0.15 and 1.5 W/kg, for 4 h	Increased generation of mitochondrial ROS, identifying complex III of the ETC as the potential source of electrons producing ROS	(Houston et al., 2018)
21.	Human neuroblastoma cells (SH-SY5Y)	1800 MHz, modulated EMF, SAR 1.6 W/kg, for 10, 30, or 60 min	Increased ROS levels were observed for every exposure time; 60 min exposure induced lipid and protein damage	(Marjanovic et al., 2018)
22.	Mouse preantral follicles	GSM 1900 MHz mobile phone, SAR 0.77–0.88 W/kg, for 60 min	Decreased SOD, GPx, CAT activity, and TAC level. Increased MDA levels	(Koohestani et al., 2019)
23.	Mouse MC3T3-E1 cells	2.45 GHz Wi-Fi EMF, 100 mW and 500 mW, SAR 0.1671 W/kg and 0.8356 W/kg, for 0–180 min	Increased ROS and GSH levels by 90 min exposure	(Wang et al., 2020)
24.	Human kidney embryonic cells (HEK293)	2.45 GHz EMF, output 15 V as constant voltage, and 3.3 V as variable voltage, 8 V/m, for 1 h	Increased MDA level and decreased SOD activity	(Özsobacı et al., 2020)

[a] All effects were statistically significant ($p < 0.05$ in all cases) as compared to control or sham-exposed groups.

In addition to the well-established role of the mitochondria in energy metabolism, regulation of cell death is a second major function of these organelles. This, in turn, is linked to their role as the powerful intra-cellular source of ROS. Mitochondria-generated ROS play an important role in the release of cytochrome c and other pro-apoptotic proteins, which can trigger caspase activation and apoptosis (Ott et al. 2007). Several reports have indicated activation of apoptosis due to non-thermal

TABLE 3.4
Studies with Positive Findings on OS Induced by RF/WC EMF Exposure of Animals and Plants *In Vivo*

S. No	Biological System Exposed	RF/WC EMF Exposure	Recorded Effects[a]	Reference
1.	Rat whole body	2450 MHz, pulsed EMF, 2 mW/cm^2, SAR = 1.2 W/kg	Melatonin or spin-trap compound blocked the DNA strand breaks induced by EMF exposure in rat brain cells	(Lai and Singh, 1997)
2.	Rat whole body	900 MHz, modulated/pulsed, 30 min/day, for 10 days	Increased MDA and hydroxyproline levels and activities of CAT and GPx. Decreased SOD activity in skin. Melatonin treatment reversed effect	(Ayata et al., 2004)
3.	Rat whole body	GSM 900 MHz mobile phone EMF, SAR = 0.52 W/kg, 20 min/day, for 1 month	Increased MDA concentration in brain tissue	(Suleyman et al., 2004)
4.	Rat whole body	GSM 900 MHz, from mobile phone, 1 h/day, for 7 days	Increased MDA, and NO• levels. Decreased SOD and GPx activities in the brain. The effects were prevented by administration of Ginkgo biloba extract	(Ilhan et al., 2004)
5.	Rat whole body	900 MHz continuous wave from generator, 1.04 mW/cm^2, 30 min/day, for 10 days	Increased MDA level in renal tissue. Reduced SOD, CAT, and GPx activities	(Oktem et al., 2005)
6.	Rat whole body	GSM 900 MHz mobile phone EMF, 30 min/day, for 10 days	Increased MDA and NO levels in heart tissue. Reduced SOD, CAT, and GPx activities. Caffeic acid phenethyl ester (CAPE) treatment reversed the effects	(Ozguner et al., 2005a)
7.	Rat whole body	GSM 900 MHz mobile phone EMF	Increased MDA and NO levels in renal tissue. Reduced SOD, CAT, and GPx activities. CAPE treatment reversed the effects	(Ozguner et al., 2005b)
8.	Rat whole body	GSM 900 MHz, from mobile phone	Increased NO and MDA retinal levels. Decreased SOD, GPx, and CAT activities	(Ozguner et al., 2006)
9.	Rat whole body	900 MHz modulated/pulsed, 1.04 mW/cm^2	Increased LPO levels in brain cortex and hippocampus. Melatonin reduced the effect in hippocampus	(Koylu et al., 2006)
10.	Rat whole body	GSM 900 MHz mobile phone EMF, 30 min/day, for 30 days	Increased MDA levels and apoptosis in endometrial tissue	(Oral et al., 2006)
11.	Rat whole body	GSM 945 MHz, 367 µW/cm^2, SAR=11.3 mW/kg	Increased MDA level and SOD activity. Decreased GSH concentration	(Yurekli et al., 2006)
12.	Guinea pig whole body	GSM 890–915 MHz, from mobile phone, SAR = 0.95 w/kg, 12 h/day, for 30 days (11 h 45 min stand-by and 15 min speaking mode)	MDA level increased, GSH level and CAT activity decreased in the brain. MDA, vitamins A, D$_3$, and E levels, and CAT activity increased, and GSH level decreased in the blood	(Meral et al., 2007)
13.	Rat whole body	900 MHz pulsed, SAR 0.016–4 W/kg, 30 min/day, for 30 days	Increased endometrial NO• and MDA levels. Decreased endometrial SOD, CAT, and GPx activities	(Guney et al., 2007)
14.	Rat whole body	Mobile phone EMF, for 4 weeks	Increased MDA level and CAT activity in corneal tissue. Decreased SOD activity. Increased MDA level in lens tissues	(Balci et al., 2007)

(Continued)

TABLE 3.4 (CONTINUED)
Studies with Positive Findings on OS Induced by RF/WC EMF Exposure of Animals and Plants *In Vivo*

S. No	Biological System Exposed	RF/WC EMF Exposure	Recorded Effects[a]	Reference
15.	Plant Lemna minor (duckweed)	400–900 MHz, 10, 23, 41, or 120 V/m, for 2 or 4 h	Increased LPO and H_2O_2 content. Increased CAT activity. Decreased pyrogallol peroxidase	(Tkalec et al., 2007)
16.	Rat whole body	900 MHz EMF by "test" mobile phone, E = 9.9–18.4 V/m, B = 4.7–8.7 µT, SAR = 0.043–0.135 W/kg, for 20, 40, and 60 days	Increased MDA and carbonyl group concentration in brain tissue. Decreased CAT activity, and increased xanthine oxidase (XO) activity	(Sokolovic et al., 2008)
17.	Rat whole body	900 MHz, 78 µW/cm², 2 h/days, for 10 months	Increased MDA levels and total oxidative status in liver tissue	(Dasdag et al., 2008)
18.	Head of rats	900 MHz pulsed from MP, 2 h/day, for 10 months	Increased antioxidant capacity and CAT activity in brain	(Dasdag et al., 2009)
19.	Rat whole body	900/1800 MHz, GSM, 1 h/day, for 28 days	Increased LPO and decreased GSH content in the testis and epididymis	(Mailankot et al., 2009)
20.	Rat whole body	900 MHz modulated/pulsed, 1.04 mW/cm²	Increased XO, CAT activities, and LPO level in liver. XO, CAT activities and LPO levels were decreased by caffeic acid phenethyl ester (CAPE) administration	(Koyu et al., 2009)
21.	Plant (mung bean) whole body	GSM 900 MHz, mobile phone EMF, 8.55 µW/cm², for 0.5, 1, 2, or 4 h	Increased MDA level, H_2O_2 accumulation, and root oxidizability. Increased activities of SOD, CAT, ascorbate peroxidases, guaiacol peroxidases, and GSH reductases in roots	(Sharma et al., 2009)
22.	Rabbit whole body	GSM 1800 MHz simulated EMF, 15 min/day, for a week	Increased MDA and ferrous oxidation (indication of lipid peroxidation) in the liver	(Tomruk et al., 2010)
23.	Male rat whole body	Mobile phone EMF, SAR=0.9 W/kg, 2 h/day, for 35 days	Reduction in protein kinase activity, decrease in sperm count, and increase in apoptosis	(Kesari et al., 2010)
24.	Rat whole body	GSM mobile phone 900 MHz (pulsed), SAR=0.9 W/kg, 2 h/day, for 45 days	Increased ROS levels. Decrease SOD and GPx activities. Decreased pineal melatonin levels	(Kesari et al., 2011)
25.	Rat whole body	900 MHz, pulsed, modulated, SAR = 1.2 W/kg, 20 min/day, for 3 weeks	Increased MDA and NOx levels, and decreased GSH levels in liver, lung, testis, and heart tissues	(Esmekaya et al., 2011)
26.	Guinea pig whole body	GSM 900 or 1800 MHz from MT base station antennas, 1.50 mW/kg for the heart, 4 h/day, for 20 days	Changes in plasma oxidant status levels by either 900 or 1800 MHz. NO• level changed by GSM 900 MHz compared to the control animals	(Çenesíz et al., 2011)
27.	Rat partial body	2450 MHz, pulsed EMF, SAR = 0.1 W/kg, 1 h/day, for 28 days	Increased LPO level. Decreased vitamin A, vitamin C, and vitamin E concentrations	(Turker et al., 2011)
28.	Quail embryo *in ovo*	GSM 900 MHz EMF, from mobile phone, 0.024–0.21 µW/cm² intermittently, for 14 days	Increased level of TBARS in brains and livers of the hatched birds	(Tsybulin et al., 2012)

S. No	Biological System Exposed	RF/WC EMF Exposure	Recorded Effects[a]	Reference
29.	Plant (mung bean) whole body	GSM 900 MHz mobile phone EMF	Increased MDA, hydrogen peroxide, and proline content in hypocotyls	(Singh et al., 2012)
30.	Rat whole body	Mobile phone EMF, 15, 30, or 60 min/day, for 2 weeks	Increased levels of LPO and CAT activities in serum and testicular tissue. Decreased GSH and GPx levels in total serum and testicular tissue	(Al-Damegh, 2012)
31.	Rat whole body	1800 MHz, modulated/pulsed RFR, SAR = 0.4 W/kg, 1 h/day, for 3 weeks	Increased protein oxidation (PO) in brain tissue and NO• in serum. Garlic administration reduced PO	(Avci et al., 2012)
32.	Rabbit infant whole body	GSM 1800 MHz, 18 mW/kg 15 min/day, for 7 days (females) or 14 days (males)	Increased LPO levels in the liver of both females and males. Increased liver 8-OH-dG levels in females	(Guler et al., 2012)
33.	Head of rats	900 MHz pulsed from MP, 2 h/day, for 10 months	Increased protein carbonyl (an indicator of protein oxidation) level in the brain of exposed rats	(Dasdag et al., 2012)
34.	Rat whole body	Mobile phone EMF, SAR = 0.9 W/kg, 3 × 10, 3 × 30, or 3 × 60 min, for 20 days during gestation	3 × 30 and 3 × 60 min exposures increased MDA level and decreased SOD and GPx activities	(Jing et al., 2012)
35.	Rat whole body	900 MHz EMF, 30 min/day, for 10 days	Increased MDA levels. Decreased SOD, CAT, and GPx activities. Melatonin treatment reversed the effects	(Kerman and Senol, 2012)
36.	Rat whole body	Simulated GSM 1800 MHz from a generator, electric field 15–20 V/m, for 2 h	Increased 8-OH-dG levels in urine	(Khalil et al., 2012)
37.	Rabbit whole body (non-pregnant and pregnant)	Simulated GSM 1800 MHz, 15 min/day, for 7 days	Changes in creatine kinases levels	(Kismali et al., 2012)
38.	Rat whole body	2450 MHz with 217 Hz pulses, 1 mW/m^2 (0.1 µW/cm^2), 60 min/day, for 30 days	Increased LPO, cell viability, and cytosolic Ca^{2+} in dorsal root ganglion neurons	(Naziroglu, Celik, et al., 2012)
39.	Rat whole body	900 or 1800 MHz, 170 µW/cm^2, SAR = 0.6 mW/kg, 2 h/day, 5 days/week, for 30 days	Increased LPO and PO levels. Decreased GSH levels	(Megha et al., 2012)
40.	Mouse whole body	2450 MHz EMF from generator, 33.5 µW/cm^2, SAR = 23 mW/kg, 2 h/day, for 45 days	Increase in ROS, decrease in NO and antioxidant enzymes activities	(Shahin et al., 2013)

(Continued)

TABLE 3.4 (CONTINUED)
Studies with Positive Findings on OS Induced by RF/WC EMF Exposure of Animals and Plants *In Vivo*

S. No	Biological System Exposed	RF/WC EMF Exposure	Recorded Effects[a]	Reference
41.	Rat whole body	900 MHz EMF by "test" mobile phone, $E = 9.9–18.4$ V/m, $B = 4.7–8.7$ µT, SAR $= 0.043–0.135$ W/kg, 4 h/day, for 29; 40 or 60 days	Increased LPO and PO levels. Increased CAT and XO activities and the number of apoptotic cells in thymus tissue	(Sokolovic et al., 2013)
42.	Earthworm whole body	900 MHz EMF, amplitude modulated at 1 kHz, 30–3800 µW/cm², SAR $= 0.13–9.33$ mW/kg, for 2 h	Increased protein carbonyl content with all exposures above 30 µW/cm². Increased MDA level at 140 µW/cm²	(Tkalec et al., 2013)
43.	Rat whole body	2450 MHz, with 50 Hz modulation, 210 µW/cm², SAR $= 0.14$ W/kg, 2 h/day, for 45 days	Increased MDA and ROS levels in testis. Melatonin prevented the OS and reversed the effects	(Meena et al., 2013)
44.	Male rat whole body	GSM 1800 MHz mobile phone EMF at calling or stand-by modes	Induced OS in both calling and stand-by exposures	(Koc et al., 2013)
45.	Rat whole body	900 MHz from MW generator SMC 100 (continuous), 2 h/day, 5 days a week, for 30 days	Increased LPO and PO levels	(Deshmukh et al., 2013)
46.	Rat whole body	850–950 MHz, modulated/pulsed SAR $= 1.08$ W/kg, 1 h/day, for 3 weeks	Increased NO• in serum and MDA and PO in brain. Administration of garlic extract diminished the effects.	(Bilgici et al., 2013)
47.	Rat whole body	2450 MHz, pulsed (Wi-Fi) EMF, SAR $= 0.143$ W/kg, 60 min/day, for 30 days	Increased LPO level melatonin administration prevented the effect	(Aynali et al., 2013)
48.	Rat whole body	GSM 1800 MHz EMF, for 15 min	Reduced antioxidant capacity both in healthy animals and in those with paw inflammation	(Bodera et al., 2013)
49.	Quail embryo *in ovo*	GSM 900 MHz, 0.25 µW/cm², SAR $= 3$ µW/kg, intermittent exposure (48 s ON/12 sec OFF), for 158–360 h	Overproduction of superoxide ($O_2^{•-}$) and NO• free radicals. Increased TBARS and 8-OH-dG levels. Decreased SOD and CAT activities.	(Burlaka et al., 2013)
50.	Rat whole body	900 MHz, pulsed 680 µW/cm², 4 h/day, for 45 days	Increased MDA concentration and decreased SOD, GPx, and CAT activities in rat eyes. Vitamin C administration prevented the effects	(Jelodar et al., 2013a)
51.	Rat whole body	900 MHz, pulsed 680 µW/cm² daily, for 45 days	Increased MDA level and decreased antioxidant enzymes activity in rat testis	(Jelodar et al., 2013b)
52.	Drosophila whole body	DECT phone base EMF, 1880–1900 MHz, 2.7 V/m, SAR $= 0.009$ W/kg, for 0.5–96 h	Increased ROS levels in male and female bodies and ROS in ovaries	(Manta et al., 2013)

S. No	Biological System Exposed	RF/WC EMF Exposure	Recorded Effects[a]	Reference
53.	Rabbit whole body	1800 MHz, pulsed EMF, 15 min/day, for 7 days in pregnant animals and for 7 or 14 days in infants	Increased LPO in the group exposed prenatally	(Ozgur et al., 2013)
54.	Rat whole body	GSM 900, or 1800 MHz, or Wi-Fi 2450 MHz, pulsed, 12 µW/cm^2, 60 min/day during gestation and 6 weeks following birth	Increased LPO in the kidney and testis, and decreased level of GSH and total antioxidant status, at the age of six weeks	(Özorak et al., 2013)
55.	Male rat whole body	465 MHz or 2.45 GHz pulsed, 6.0 mW/cm^2, for 28 days	Increased superoxide ($O_2^{\bullet-}$) production; decreased activity of NADH-ubiquinone oxidoreductase complex in liver, cardiac, and aorta tissues after 28 days of exposure	(Burlaka et al., 2014)
56.	Rat whole body	WC EMF from MT base station	Decreased GPx, SOD, and CAT activities. Increased MDA level; Vitamin C reduced the effect	(Akbari et al., 2014)
57.	Pregnant rats and offspring	900 or 1800 MHz pulsed EMF, 1 h/day during pregnancy and neonatal development	Decreased brain and liver GPx activities, selenium concentrations in the brain, and liver vitamin A and β-carotene concentrations in the newborns	(Çetin et al., 2014)
58.	Rat whole body	2450 MHz pulsed, 3.68 V/m, 1 h/day, for 30 days	Increased 8-OH-dG level in both plasma and brain tissue. Increased PO level only in plasma. Garlic prevented the increase of 8-OH-dG in brain and PO in the plasma	(Gürler et al., 2014)
59.	Rat whole body	UMTS (3G) mobile phone 2115 MHz, SAR = 0.26 W/kg, 2 h/day, for 60 days	Increased ROS levels, DNA damage, and apoptosis	(Kesari et al., 2014)
60.	Rat whole body	3G (UMTS) mobile phone 1910.5 MHz EMF 2 h/day, for 60 days (6 days a week)	Increased LPO level and induced DNA damage in sperm cells	(Kumar et al., 2014)
61.	Rat whole body	GSM 900 MHz simulated mobile phone EMF, 4 h/day, for 12 days	Increased MDA and Nrf2 protein, and decreased SOD and GSH in the liver	(Luo et al., 2014)
62.	Rat whole body	900 MHz from MT base antenna tower, 24 h/day, for 8 weeks	Reduced SOD and CAT activities in blood. Sesame oil reversed the effect	(Marzook et al., 2014)
63.	Rat whole body	GSM 900 MHz mobile phone (simulated MT) EMF, SAR = 1.13 W/kg, 2 h/day, for 60 days	Increased conjugated dienes, protein carbonyls, total oxidant status, and OS index. Reduced total antioxidant capacity levels	(Motawi et al., 2014)
64.	Rat whole body	2450 MHz, pulsed, 0.1 µW/cm^2, SAR = 0.1 W/kg, 1 h/day, for 30 days	Increased LPO levels. Melatonin treatment reversed the effect	(Oksay et al., 2014)
65.	Male mouse whole body	1800 MHz, 208 µW/cm^2, 30 or 120 min/day, for 30 days	Decreased CAT and GPx activities and increased MDA level in cerebrum. Selenium nanoparticles reversed the effects	(Qin et al., 2014)

(Continued)

TABLE 3.4 (CONTINUED)
Studies with Positive Findings on OS Induced by RF/WC EMF Exposure of Animals and Plants *In Vivo*

S. No	Biological System Exposed	RF/WC EMF Exposure	Recorded Effects[a]	Reference
66.	Rat whole body	GSM 900 MHz mobile phone EMF, 1 h/day, for 60 days	Increased MDA levels and decreased total antioxidant capacity levels in brain, liver, and kidneys. The effects were corrected 30 days after withdrawal of exposure	(Ragy, 2014)
67.	Rat whole body	GSM 900 MHz mobile phone EMF, 4 h/day, for 15 days	Change in levels of antioxidant enzymes and non-enzymatic antioxidants and increase in LPO	(Saikhedkar et al., 2014)
68.	Rat whole body	2450 MHz, Wi-Fi EMF, 60 min/day, for 30 days	Decreased GPx activity. GPx activity and GSH values increased after melatonin treatment	(Tök et al., 2014)
69.	Pregnant rat whole body	900 MHz, 13.7 V/m, 50 µW/cm^2, emitted by RF generator, 1 h/day, for 13–21 days of pregnancy	Increased MDA, SOD, and CAT values. Decreased GSH values in exposed pups	(Türedi et al., 2014)
70.	Lab rats	900 MHz pulsed at 217 Hz, 0.02 mW/cm^2, SAR 1.245 W/kg, for 3 months	Induced OS in the hippocampus and striatum	(Ahmed et al., 2017)
71.	Lab rats	Mobile phone EMF (parameters are not indicated), for 28 days	Increased serum levels of MDA and decreased SOD activity	(Oyewopo et al., 2017)
72.	Male lab rats	2.45 GHz, pulsed at 217 Hz, EMF emitted by generator, 0.1–45.5 V/m, SAR 50 mW/kg, for 30 days	Increased tissue MDA, total oxidant status, and OSI. Decreased total antioxidant status	(Topsakal et al., 2017)
73.	Pregnant female lab mice	GSM (900 or 1800 MHz) EMF from mobile phone 2 h/day, for 20 days during pregnancy	Increased MDA levels, decreased total thiol groups (TTG), SOD, and CAT in the tissues of dams and their offspring	(Bahreyni Toossi et al., 2018)
74.	Lab rats	900 MHz or 1800 MHz pulsed, 0.1 W/kg, 60 min/5 days of the week during 1 year	Increased intra-cellular calcium (Ca^{2+}), ROS production, mitochondrial membrane depolarization, apoptosis, and caspase 3 and 9 activities	(Ertilav et al., 2018)
75.	Lab rats	Wi-Fi 2.45 GHz EMF, 30 cm from the antenna, 24 h/day, for 10 weeks	Decreased total antioxidant capacity of plasma. Decreased CAT, GPx, and SOD activities. Increased GST activity	(Kamali et al., 2018)
76.	Female lab rats	2.45 GHz Wi-Fi EMF, 0.8–2.27 mW/kg for different tissues, 1 h/day, for 30 days	Increased TOS and OSI levels in ovarian tissues	(Saygin et al., 2018)
77.	Male lab rats	2.45 GHz Wi-Fi EMF, 30 cm from the antenna, 4 h/day, for 45 days	Increased LPO and decreased GSH level, SOD, and GPx activities in the pancreas	(Masoumi et al., 2018)
78.	Male lab rats	Continuous-wave 900 MHz, 1 h/day during adolescence	Changes in OS biomarkers and in morphology of the rat testis	(Hancı et al., 2018)

S. No	Biological System Exposed	RF/WC EMF Exposure	Recorded Effects[a]	Reference
79.	Male lab mice	GSM 900 MHz, 2.7 W/m^2, for 3 h twice/day, for 35 days	Excess free radical generation; extensive DNA damage in germ cells	(Pandey and Giri, 2018)
80.	Quail embryos	GSM 1800 MHz from mobile phone, 0.32 μW/cm^2, 48 s ON, 12 s OFF, for 5 days before and 14 days through the incubation	Increased superoxide anion and nitrogen oxide generation rates, and oxidative damages of DNA (increased levels of 8-OH-dG)	(Yakymenko et al., 2018)
81.	Lab mice	GSM 900 MHz mobile phone EMF, 145.1 μW/cm^2, 1 h/day, for 28 days	Reduced activities of brain antioxidant enzymes. Increased lipoperoxidation. Induced impairment in learning and memory	(Akefe et al., 2019)
82.	Male lab rats	950 MHz continuous-wave, 1 mW/cm^2, SAR 0.238 and 0.372 W/kg, 1 h thrice a week, for seven weeks	Increased MDA and nitric oxide (NO•), and decreased SOD, CAT, and GPx activities in liver and brain	(Shedid et al., 2019)
83.	Male lab rats	GSM 900 MHz EMF from mobile phone, 24 h/day, for 28 days	Increased hepatic levels of MDA and erythroid 2–related factor 2 (Nrf-2). Increased SOD and CAT activities	(Ismaiil et al., 2019)
84.	Male lab rats	3G (UMTS) EMF from mobile phone, 2 h/day, for 45 days	Increased ROS and LPO levels. Decreased sperm count, and induced alterations in sperm tail morphology	(Gautam et al., 2019)
85.	Male lab rats	Simulated GSM (900 or 1800 MHz), or UMTS (2100 MHz) MT EMF (0.08 W/kg, 0.05 W/kg, and 0.034 W/kg), 2 h/day, for 6 months	Increased MDA, 8-OH-dG, and TOS	(Alkis et al., 2019)
86.	Lab rats	900 MHz GSM, 7.737 μW/m^2, 1 h, 2 h, or 4 h, for 90 days	Increased MDA level. Decreased SOD and CAT levels	(Sharma and Shukla, 2020)
87.	Lab rats	GSM 915 MHz EMF from mobile phone, or 2450 MHz from Wi-Fi, for 1 month	Increased protein carbonylation (PC), nitric oxide (NO•) and MDA; reduced GSH, GSH-Px, TAC, SOD, and CAT in brain	(Asl et al., 2020)
88.	Lab rats	Multiple transceiver GSM mobile phone (900/1800/1900 MHz), 10 min/day, for 6 weeks	Decreased GPx, SOD, and MDA levels in the serum, heart, and brain. Increased nitric oxide (NO$_x$) activities in the brain and heart	(Usman et al., 2020)
89.	Male lab rats	1800 MHz CW, 0.433 W/kg, 4 h/5 days/week, for 90 days	Alteration in GSH cycle regulating enzymes such as GR, GPx, GST, and glucose-6-phosphate dehydrogenase (G6PDH). DNA damage in brain neurons	(Sharma et al., 2020)
90.	Lab mice	MT base station 1800 MHz EMF, 37.54 mW/m^2, for 12 h or 24 h/day, for 45 days	Increased MDA levels in brain	(Zosangzuali et al., 2021)

[a] All effects were statistically significant ($p < 0.05$ in all cases) as compared to control or sham-exposed groups.

TABLE 3.5
Studies with Positive Findings on OS Induced by RF/WC EMF Exposure of Humans

S. No	Biological System Exposed	RF/WC EMF Exposure	Recorded Effects[a]	Reference
1.	Human male body	Mobile phone in a pocket in stand-by position for 1, 2, or 4 h single exposure	Increased plasma LPO level. Decreased SOD and GPx activities in erythrocytes	(Moustafa et al., 2001)
2.	Human whole body	3, 5.5, or 9.4 GHz radar pulsed EMF	Increased MDA level, decreased GSH level	(Garaj-Vrhovac et al., 2011)
3.	Human head/whole body	Mobile phone EMF, mean exposure time 29.6 h/month, for 12.5 years	Increased all salivary OS indices	(Hamzany et al., 2013)
4.	Human male head	GSM 1800 MHz mobile phone EMF, SAR = 1.09 W/kg, for 15 and 30 min	Increased SOD activity in saliva	(Abu Khadra et al., 2014)
5.	Humans	Mobile phone use and presence of Wi-Fi modem in houses	Increase in glutathione S-transferase (GST) activity in the blood, proportional to measured exposure levels	(Akkam et al., 2020)
6.	Humans (students)	Mobile phone use	Increased MDA and decreased SOD levels in serum	(Shaheen et al., 2021)

[a] All effects were statistically significant ($p < 0.05$ in all cases) as compared to control or sham-exposed groups.

TABLE 3.6
Studies that Reported No Significant OS Effects of RF/WC EMF Exposure

S. No	Biological System Exposed	RF/WC EMF Exposure	Indexes analyzed	Reference
1.	Rat whole body	Mobile phone EMF, SAR = 0.52 W/kg, 20 min/day, for 1 month	No alteration in MDA concentration	(Dasdag et al., 2003)
2.	Mammalian cells in vitro	835.62 MHz (frequency-modulated continuous-wave), or 847.74 MHz, CDMA, SAR = 0.8 W/kg, for 20–22 h	In both cases, EMF exposure did not alter OS parameters	(Hook et al., 2004)
3.	Rat whole body	GSM 800–1800 MHz, from mobile phone	No changes in lipid and protein damage or in non-enzymatic antioxidant defense in frontal cortex or hippocampus	(Ferreira et al., 2006)
4.	Rat whole body	3G (UMTS) mobile phone EMF, "standardized daily dose", for 20 days	No difference in GPx and CAT activity in eye tissues or MDA and GSH levels in blood	(Demirel et al., 2012)
5.	Human head/whole body	Mobile phone EMF 1800 MHz, 1.09 W/kg, (talking mode) for 15 or 30 min	No relationship found between exposure and changes in the salivary oxidant/antioxidant profile	(Khalil et al., 2013)
6.	Human head/whole body	Mobile phone EMF, <60 min to >300 min/month, for 5 or more years	No difference in the saliva of the exposed parotid gland from the saliva of the opposite gland of each individual	(de Souza et al., 2014)

EMF exposure. In human epidermoid cancer KB cells, 1950 MHz RF EMF induced time-dependent apoptosis (45% after 3 h) that was paralleled by a 2.5-fold decrease in the expression of ras and Raf-1 genes and the activity of ras and Erk-1/2 genes (Caraglia et al. 2005). Primary cultured neurons and astrocytes exposed to GSM 1900 MHz WC EMF by a commercially available mobile phone device for 2 h demonstrated up-regulation of caspase-2, caspase-6, and Asc*. Up-regulation in neurons occurred in both "on" (during a call) and "stand-by" modes but, in astrocytes, only in the "on" mode (Zhao et al. 2007).

It is important that some studies directly pointed to the induction of free radicals (especially the superoxide radical $O_2\bullet^-$) as a primary reaction of living cells to man-made EMF exposure (Friedman et al. 2007; Burlaka et al. 2013; Yakymenko et al. 2018). As we pointed out earlier, direct activation of NADH oxidase (Friedman et al., 2007) and the mitochondrial pathway of superoxide radical overproduction (Burlaka et al. 2013; De Iuliis et al. 2009; Yakymenko et al. 2018) have been experimentally proven. In addition, a significant overproduction of nitric oxide (NO•) was revealed in some studies (Avci et al. 2012; Bilgici et al. 2013; Burlaka et al. 2013), although it is unclear whether an induction of expression of nitric oxide synthases (NOS) or direct activation of the existing NOS molecules took place. It is, however, clear that significantly increased levels of these free radical species (superoxide and nitrogen oxide) in cells due to EMF exposure result in an activation of peroxidation and repression of activities of key antioxidant enzymes.

Many studies have demonstrated the effectiveness of different antioxidants to overcome oxidative effects caused by man-made EMF exposure, which confirms an important role of OS in the biological effects of non-thermal EMF exposures. Such protective effects have been reported for melatonin (Lai and Singh 1997; Ayata et al. 2004; Oktem et al. 2005; Ozguner et al. 2006; Sokolovic et al. 2008; Pandey and Giri, 2018), vitamins E and C (Oral et al. 2006; Jelodar et al. 2013a; Wang et al. 2020), caffeic acid phenethyl ester (Ozguner et al. 2006), selenium (Özsobacı et al. 2018), zinc (Özsobacı et al. 2020), L-carnitine (Turker et al. 2011), garlic (Avci et al. 2012; Bilgici et al. 2013), green tea extract (Ahmed et al. 2017), dietary supplements/vitamins E and C (Shaheen et al. 2021), rosmarinic acid (Asl et al. 2020), gallic acid (Topsakal et al., 2017), and light-emitting diode (LED) red light application (Tsybulin et al. 2016).

Overall, the analysis of the published peer-reviewed literature persuasively proves that non-thermal exposure to man-made ELF and WC EMFs leads to ROS/free radical overproduction and results in significant OS in living cells.

3.4 OXIDATIVE DAMAGE OF DNA UNDER EMF EXPOSURE

Many experimental studies have already investigated the mutagenic effects of ELF or RF (mostly WC) EMF exposure, and most of them have revealed significant effects, see, for example, reviews (Ruediger 2009; Phillips et al. 2009). There is a substantial number of studies which have demonstrated the formation of micronuclei (Garaj-Vrhovac et al. 1992; Tice et al. 2002; Zotti-Martelli et al. 2005) or structural anomalies of metaphase chromosomes (Kerbacher et al. 1990; Garson et al. 1991; Maes et al. 2000; Panagopoulos 2020) in living cells due to non-thermal RF/WC EMF exposure. The majority of studies on the mutagenic effects of ELF or RF/WC EMFs successfully applied a comet assay approach (Baohong et al. 2005; Belyaev et al., 2006; Diem et al., 2005; Kim et al., 2008; Lai and Singh, 1996; Liu et al., 2013a; Tsybulin et al. 2013; Yakymenko et al. 2018).

A recent study on *Drosophila melanogaster* under 900 MHz, 1800 MHz, and 2100 MHz EMF exposure is very demonstrative (Gunes et al. 2021). They applied the wing somatic mutation and recombination test (SMART), which is based on the observation of genetic changes occurring in the trichomes of the wings of *Drosophila* mutant clones under the microscope. It was observed that the number of mutant clones was statistically increased by all exposures. Previous studies on *D. melanogaster* have consistently shown DNA fragmentation in reproductive cells by only a few minutes of

---

* Apoptosis associated speck-like protein containing a caspase recruitment domain.

daily exposure during the first days of adult life to an active mobile phone (GSM 900 or 1800 MHz) in "talk" mode (Panagopoulos et al. 2007; 2010; Panagopoulos 2012). These were the first studies that demonstrated DNA damage induced by real-life WC EMF exposure on live animals. The DNA fragmentation, demonstrated by the TUNEL (Terminal deoxynucleotide transferase dUTP Nick End Labeling) assay, was shown to be accompanied by actin cytoskeleton damage (Chavdoula et al. 2010).

Particular studies have identified 8-OH-dG as a specific marker of oxidative damage to the DNA, (Yokus et al. 2005; Yokus et al. 2008; De Iuliis et al. 2009; Xu et al. 2010; Guler et al. 2012; Khalil et al. 2012; Burlaka et al. 2013; Ding et al. 2018; Alkis et al. 2019; Yakymenko et al. 2018). It is important that both ELF and RF/MW EMFs (in most cases WC EMFs) were demonstrated to cause such effects. The level of 8-OH-dG in human spermatozoa was shown to be significantly increased after *in vitro* exposure to non-thermal simulated mobile phone EMF exposure (De Iuliis et al., 2009). Likewise, we demonstrated that the exposure of quail embryos *in ovo* to real-life mobile phone GSM 900 MHz and 1800 MHz EMFs at 0.25 µW/cm^2 and 0.32 µW/cm^2, respectively, for a few days was sufficient for a significant, two- to threefold, increase of 8-OH-dG level in embryonic cells (Burlaka et al. 2013; Yakymenko et al. 2018). And, for example, significant oxidative damage of DNA (8-OH-dG increased levels) was detected in laboratory rats after long-term (100 days or longer) exposure to 50 Hz ELF EMFs, 100 µT or higher (Yokus et al. 2005; 2008).

As ELF and RF/WC EMFs at non-thermal environmentally existing intensities do not have enough energy to ionize DNA molecules, and as overproduction of ROS in living cells due to either ELF or RF/WC EMF exposures is reliably documented, it is clear that we have an indirect action of these EMF/EMR types in the form of oxidative damage to DNA. The most aggressive form of ROS, which is able to affect the DNA molecule, is the hydroxyl radical (OH•) (Halliwell 2007). The hydroxyl radicals are generated in cells by the Fenton reaction (Valko et al. 2006):

$$Fe^{2+} + H_2O_2 \rightarrow Fe^{3+} + OH\bullet + OH^-$$

and by the Haber–Weiss reaction:

$$O_2\bullet^- + H_2O_2 \rightarrow O_2 + OH\bullet + OH^-$$

The "fuel" for these reactions is the superoxide radical ($O_2\bullet^-$), which can be converted to hydrogen peroxide (H_2O_2). The superoxide radical can be generated in living cells under non-thermal levels of ELF/RF/WC EMFs, as was demonstrated above.

On the other hand, increased concentration of nitric oxide (NO•), in addition to the superoxide radical in the EMF-exposed cells can lead to the formation of another aggressive form of ROS, peroxynitrite (ONOO$^-$), which can also cause DNA damage (Valko et al. 2006).

Thus, due to overproduction of ROS and impairment of the antioxidant system in living cells under man-made ELF or RF/WC EMF exposure, the oxidative damage of DNA is the inevitable result, especially in cases of repeated/sustained and/or intense exposure.

3.5 DISCUSSION AND CONCLUSIONS

3.5.1 Non-thermal Man-made EMF Exposure and Cellular Signaling

As we discussed above, overproduction of ROS in living cells due to non-thermal man-made EMF exposure can induce harmful effects via oxidation of biological macromolecules. But, in addition to that, ROS are an intrinsic part of the cellular signaling cascades (Forman et al. 2014; Moloney and Cotter 2018; Sies and Jones 2020). For example, hydrogen peroxide appears as a second messenger both in insulin signaling and in growth factor-induced signaling cascades (Sies 2014). ROS are implicated in the biochemical mechanism of ethanol oxidation and other metabolic processes

(Oshino et al. 1975) and are also required for initiation of wound repair and elimination of infectious micro-organisms in cells (Enyedi and Niethammer 2013). In addition, ROS, at relatively low concentrations, can modulate inflammation via activation of the nuclear factor kappa B (NFκB) pathway[*] (Hayden and Ghosh 2011). Therefore, even subtle exposure to EMFs resulting in the generation of hardly detectable quantities of ROS/free radicals can have significant biological consequences.

We could ascertain the signaling effects of moderate levels of ROS from our experiments in quail embryos exposed by a commercial mobile phone. We demonstrated that the prolonged exposures of embryos *in ovo* led to significant repression in their development (Tsybulin et al. 2013), which was concomitant with significant overproduction of superoxide ($O_2^{\bullet-}$) and nitric oxide (NO•) radicals, increased rates of LPO, and oxidative damage of DNA (Burlaka et al. 2013; Tsybulin et al. 2012; Yakymenko et al. 2018). Notably, shorter exposure durations, in combination with significantly lower intensity levels (0.2–03 $\mu W/cm^2$) than those of a usual mobile phone exposure in "talk" mode, led to enhancement in embryonic development (Tsybulin et al. 2012; 2013). We demonstrated the effects of shorter exposure durations also at the molecular level. Indeed, after short-term exposure to GSM EMF, DNA comets in embryonic cells were significantly shorter than in the control non-exposed embryos, pointing to the activation of mechanisms maintaining the integrity of DNA. The "beneficial" consequences of irradiation could be explained by the so-called hormesis effect (Calabrese 2008). One could hypothesize, however, that the "beneficial" effects of irradiation could be explained by the signaling action of free radicals induced at levels below the damaging concentrations. Obviously, any seemingly beneficial effect of any external environmental impact should be considered with caution and careful evaluation of the long-term consequences, otherwise it may be misleading.

Altogether, the EMF-induced OS effects discussed above provide us with a clear warning sign of the adverse health effects of non-thermal man-made EMF exposure, which could be evoked both by oxidative damages in living cells and/or by disturbed cellular signaling.

3.5.2 OS and Non-carcinogenic Health Effects of Man-made EMFs

Health effects of ELF EMFs have been detected in some industry workers, e.g., among exposed technicians and electrical utility workers, in the form of headaches, anxiety, depression, and sleep disorders (Pourabdian et al. 2009; Hosseinabadi et al. 2019). In these studies, increased exposure to ELF EMFs had direct and significant effects, such as increased stress, depression, and anxiety. Moreover, sleep quality in highly exposed technicians was significantly lower than in other groups (Hosseinabadi et al. 2019).

Also, during the past two to three decades, a new medical condition, electrohypersensitivity (EHS), in which people suffer due to intense man-made EMF exposure, has been described (Johansson 2006; Belpomme and Irigaray 2020). Typically, these persons suffer from skin- and mucosa-related symptoms (itching, pain, heat sensation) or heart and nervous system disorders after chronic exposure to computer monitors, mobile or DECT phones, and other electromagnetic devices. This disorder is rapidly growing among the general population: starting from 0.06% of the total population in 1985, this category now includes as much as 9%–11% of the European population (Hallberg and Oberfeld 2006). In Sweden, EHS has been officially recognized as a health impairment. Today, EHS is seen as a new serious challenge both for public health and for medical professionals (Hedendahl et al. 2015).

In line with that, a high percentage, up to 18%–43% of young people, has been described to suffer from headache/earache during or after mobile phone conversations (Chu et al. 2011; Yakymenko et al. 2011). Likewise, a number of psychophysical and preclinical disorders, including fatigue,

---

[*] A signaling pathway that links pathogenic signals with cellular danger signals, helping the cell's resistance to invading pathogens.

irritation, headache, sleep disorders, and hormonal imbalances, were detected in high percentages of people living nearby MT base station antennas (Santini et al. 2002; Buchner and Eger 2011).

An allergy reaction to modulated RF EMFs in humans has been confirmed by a significant increase in the number of mast cells in the skin of persons under chronic exposure to EMF emissions from TV and computer screens (Johansson et al. 2001). Likewise, a higher number of degranulated mast cells in the dermis of EHS persons has been detected (Johansson 2006). In turn, the activated mast cells can release histamine and other mediators of such reactions, which include allergic hypersensitivity, itching, dermatoses, etc. Importantly, an implication of ROS in allergic reactions is rather clear today. For example, in the case of airway allergic inflammation, the lung cells generate superoxides in nanomolar concentrations following antigen challenges (Nagata 2005). Then, mast cells generate ROS following aggregation of FcεRI, a high-affinity receptor for immunoglobulin E (IgE) (Okayama 2005). In addition, pollen NADPH oxidases rapidly increase the level of ROS in lung epithelium cells (Boldogh et al. 2005), and the removal of pollen NADPH oxidases from the challenge material reduced the antigen-induced allergic airway inflammation.

Thus, it seems plausible that EHS-like conditions can be attributed partially or significantly to ROS overproduction in living cells due to EMF exposure. Indeed, recent studies confirm this hypothesis. For example, it was recently demonstrated that 80% of patients with EHS present one, two, or three detectable OS biomarkers in their peripheral blood (Belpomme and Irigaray 2020). And 30%–50% of EHS self-reporting patients presented statistically significant increases in TBARs, MDA, GSSG, and nitrotyrosine (NTT) mean plasmatic level values in comparison with normal values obtained in healthy controls ($P < 0.0001$) (Irigaray et al. 2018).

3.5.3 OS and Carcinogenic Potential of Man-made EMFs

The first epidemiological evidence on carcinogenic effects of man-made EMFs was published a few decades ago (Wertheimer and Leeper 1979). The data revealed a statistically significant increase of childhood leukemia in Colorado, US in a population living near high-voltage power lines and were, thus, exposed to man-made ELF EMFs. Later, the same researchers demonstrated increased percentages of some other types of cancer as well, including breast cancer and cancer of the central nervous system in adults under increased ELF EMF exposure (Wertheimer and Leeper 1982). Recent meta-analysis on childhood cancer due to ELF EMF exposure (33 studies including 186,000 participants) demonstrated significant associations between exposure to ELF magnetic fields and childhood leukemia and a dose-response relationship of the effect (Seomun et al. 2021).

Moreover, in recent years, a number of epidemiological studies have indicated a significant increase in the incidence of various types of tumors among long-term or "heavy" users of mobile or cordless phones, see, for example, reviews (Yakymenko et al. 2011; Miller et al. 2019). Briefly, reports have pointed to the increased risk in brain tumors (Hardell et al. 2007; Hardell and Carlberg 2009; Cardis et al. 2010), acoustic neuroma (Hardell et al., 2005; Sato et al., 2011), tumors of parotid glands (Sadetzki et al. 2008), seminomas (Hardell et al. 2007), melanomas (Hardell et al. 2011), and lymphomas (Hardell et al. 2005) in these cohorts of people. Also, a significant increase in tumor incidence among people living nearby MT base stations has also been reported (Eger et al. 2004; Wolf and Wolf 2007). Similarly, experimental evidence of cancer expansion in rodents caused by long-term non-thermal RF/WC EMF exposure has been published (Szmigielski et al. 1982; Chou et al. 1992; Repacholi et al. 1997; Toler et al. 1997). In addition, activation of ODC has been detected in pulsed and continuous wave RF EMF-exposed cells (Hoyto et al. 2007). ODC is involved in the processes of cell growth and differentiation, and its activity is increased in tumor cells. Although overexpression of ODC is not sufficient for tumorigenic transformation, an increased activity of this enzyme has been shown to promote the development of tumors in pre-tumor cells (Clifford et al. 1995).

Significant overproduction of ROS leads to OS in living cells, induces oxidative damage of DNA, and can cause malignant transformation (Halliwell and Whiteman 2004; Valko et al. 2007). It is known that, in addition to mutagenic effects, ROS play an important role as second messengers for intra-cellular signaling cascades which can also induce oncogenic transformation (Valko et al. 2006). Earlier we hypothesized (Burlaka et al. 2013) that low-intensity RF/WC EMF exposure leads to dysfunction of mitochondria, which results in overproduction of superoxide and nitric oxide radicals and, subsequently, to ROS-mediated mutagenesis. Moreover, it is well established that OS is associated with carcinogenesis; for instance, the OS elicited by membrane-type 1 matrix metalloproteinase is implicated in both the pathogenesis and progression of prostate cancer (Nguyen et al. 2011). Similarly, a progressive elevation in mitochondrial ROS production (chronic ROS elevation) under both hypoxia and/or low glucose, which leads to stabilization of cells via increased HIF-2alpha expression, can eventually result in malignant transformation (Ralph et al. 2010). These data, together with the strong experimental evidence on activation of NADH oxidase under man-made EMF exposure (Friedman et al. 2007), suggest that man-made EMFs at non-thermal intensities constitute a significant stress factor for living cells, inducing (chronic) OS which may potentially result in cancer.

3.5.4 Conclusions

Analysis of the available literature on the biological effects of man-made non-thermal EMF exposures leads to a firm conclusion that this physical agent is a powerful oxidative stressor for living cells. The oxidative efficiency of non-thermal man-made EMF exposures can be mediated via direct effects on water molecule clusters or irregular gating of VGICs via changes in activities of key ROS-generating systems, including mitochondria and non-phagocytic NADH oxidases, induction of conformational changes in biologically important macromolecules, etc. In turn, a broad biological potential of ROS and free radicals, including their mutagenic effects and their signaling regulatory potential, makes man-made non-thermal EMF exposure, especially from modern WC systems, a significant hazard to human health. We suggest minimizing the intensity and duration of WC EMF exposure and taking a precautionary approach in the application of WC technology in everyday life.

ACKNOWLEDGMENT

I.Ya. work was supported by the Erasmus+ Jean Monnet Projects 620627-EPP-1-2020-1-UA-EPPJMO-CoE and 611278-EPP-1-2019-1-UA-EPPJMO-SUPPA and by the Fulbright Small Grants Program.

REFERENCES

Abdel-Rassoul, G., El-Fateh, O. A., Salem, M. A., et al. (2007). Neurobehavioral effects among inhabitants around mobile phone base stations. *Neurotoxicology* 28(2):434–440.

Abu Khadra, K. M., Khalil, A. M., Abu Samak, M., et al. (2014). Evaluation of selected biochemical parameters in the saliva of young males using mobile phones. *Electromagnetic Biology and Medicine*, 34(1):1–5.

Abyaneh, E. B. (2018). Low frequency electromagnetic field induced oxidative stress in Lepidium sativum L. *Iranian Journal of Science and Technology, Transactions A: Science* 42(3):1419–1426.

Agarwal, A., Desai, N. R., Makker, K., et al. (2009). Effects of radiofrequency electromagnetic waves (RF-EMW) from cellular phones on human ejaculated semen: An in vitro pilot study. *Fertility and Sterility* 92(4):1318–1325.

Ahmed, N. A., Radwan, N. M., Aboul Ezz, H. S., Salama, N. A. (2017). The antioxidant effect of green tea mega EGCG against electromagnetic radiation-induced oxidative stress in the hippocampus and striatum of rats. *Electromagnetic Biology and Medicine* 36(1):63–73.

Akbari, A., Jelodar, G., Nazifi, S. (2014). Vitamin C protects rat cerebellum and encephalon from oxidative stress following exposure to radiofrequency wave generated by BTS antenna mobile. *Toxicology Mechanisms and Methods*, 24(5):347–352.

Akdag, M. Z., Dasdag, S., Ulukaya, E., et al. (2010). Effects of extremely low-frequency magnetic field on caspase activities and oxidative stress values in rat brain. *Biological Trace Element Research* 138(1):238–249.

Akdag, M. Z., Dasdag, S., Uzunlar, A. K., et al. (2013). Can safe and long-term exposure to extremely low frequency (50 Hz) magnetic fields affect apoptosis, reproduction, and oxidative stress? *International Journal of Radiation Biology* 89(12):1053–1060.

Akefe, I. O., Yusuf, I. L., Adegoke, V. A. (2019). C-glycosyl flavonoid orientin alleviates learning and memory impairment by radiofrequency electromagnetic radiation in mice via improving antioxidant defence mechanism. *Asian Pacific Journal of Tropical Biomedicine* 9(12):518.

Akkam, Y., A Al-Taani, A., Ayasreh, S., Akkam, N. (2020). Correlation of blood oxidative stress parameters to indoor radiofrequency radiation: A cross sectional study in Jordan. *International Journal of Environmental Research and Public Health* 17(13):4673.

Al-Damegh, M. A. (2012). Rat testicular impairment induced by electromagnetic radiation from a conventional cellular telephone and the protective effects of the antioxidants vitamins C and E. *Clinics* 67(7):785–792.

Alkis, M. E., Bilgin, H. M., Akpolat, V., et al. (2019). Effect of 900-, 1800-, and 2100-MHz radiofrequency radiation on DNA and oxidative stress in brain. *Electromagnetic Biology and Medicine* 38(1):32–47.

Asl, J. F., Goudarzi, M., Shoghi, H. (2020). The radio-protective effect of rosmarinic acid against mobile phone and wi-fi radiation-induced oxidative stress in the brains of rats. *Pharmacological Reports* 72(4):857–866.

Avci, B., Akar, A., Bilgici, B., et al. (2012). Oxidative stress induced by 1.8 GHz radio frequency electromagnetic radiation and effects of garlic extract in rats. *International Journal of Radiation Biology* 88(11):799–805.

Ayata, A., Mollaoglu, H., Yilmaz, H. R., et al. (2004). Oxidative stress-mediated skin damage in an experimental mobile phone model can be prevented by melatonin. *Journal of Dermatology* 31(11):878–883.

Aynali, G., Naziroglu, M., Celik, O., et al. (2013). Modulation of wireless (2.45 GHz)-induced oxidative toxicity in laryngotracheal mucosa of rat by melatonin. *European Archives of Oto-Rhino-Laryngology: Official Journal of the European Federation of Oto-Rhino-Laryngological Societies* 270(5):1695–1700.

Bagheri Hosseinabadi, M., Khanjani, N., Ebrahimi, M. H., Haji, B., Abdolahfard, M. (2019). The effect of chronic exposure to extremely low-frequency electromagnetic fields on sleep quality, stress, depression and anxiety. *Electromagnetic Biology and Medicine* 38(1):96–101.

Bagheri Hosseinabadi, M., Khanjani, N., Norouzi, P., et al. (2021). Oxidative stress associated with long term occupational exposure to extremely low frequency electric and magnetic fields. *Work* 68(2), 379–386.

Bahreyni Toossi, M. H., Sadeghnia, H. R., Mohammad Mahdizadeh Feyzabadi, M., et al. (2018). Exposure to mobile phone (900–1800 MHz) during pregnancy: Tissue oxidative stress after childbirth. *The Journal of Maternal-Fetal and Neonatal Medicine* 31(10):1298–1303.

Balci, M., Devrim, E., Durak, I. (2007). Effects of mobile phones on oxidant/antioxidant balance in cornea and lens of rats. *Current Eye Research* 32(1):21–25.

Baohong, W., Jiliang, H., Lifen, J., et al. (2005). Studying the synergistic damage effects induced by 1.8 GHz radiofrequency field radiation (RFR) with four chemical mutagens on human lymphocyte DNA using comet assay in vitro. *Mutation Research* 578(1–2):149–157.

Belpomme, D., Irigaray, P. (2020). Electrohypersensitivity as a newly identified and characterized neurologic pathological disorder: How to diagnose, treat, and prevent it. *International Journal of Molecular Sciences* 21(6):1915.

Belyaev, I. (2010). Dependence of non-thermal biological effects of microwaves on physical and biological variables: Implications for reproducibility and safety standards. *European Journal of Oncology Library* 5:187–217.

Belyaev, I. Y., Koch, C. B., Terenius, O., et al. (2006). Exposure of rat brain to 915 MHz GSM microwaves induces changes in gene expression but not double stranded DNA breaks or effects on chromatin conformation. *Bioelectromagnetics* 27(4):295–306.

Bilgici, B., Akar, A., Avci, B., et al. (2013). Effect of 900 MHz radiofrequency radiation on oxidative stress in rat brain and serum. *Electromagnetic Biology and Medicine* 32(1):20–29.

Blank, M., Soo, L. (2001). Electromagnetic acceleration of electron transfer reactions. *Journal of Cellular Biochemistry* 81(2):278–283.

Blank, M., Soo, L. (2003). Electromagnetic acceleration of the Belousov-Zhabotinski reaction. *Bioelectrochemistry* 61(1–2):93–97.

Bodera, P., Stankiewicz, W., Zawada, K., et al. (2013). Changes in antioxidant capacity of blood due to mutual action of electromagnetic field (1800 MHz) and opioid drug (tramadol) in animal model of persistent inflammatory state. *Pharmacological Reports: PR* 65(2):421–428.

Bohr, H., Bohr, J. (2000a). Microwave-enhanced folding and denaturation of globular proteins. *Physical Review: Part E* 61(4):4310–4314.

Bohr, H., Bohr, J. (2000b). Microwave enhanced kinetics observed in ORD studies of a protein. *Bioelectromagnetics* 21(1):68–72.

Boldogh, I., Bacsi, A., Choudhury, B. K., et al. (2005). ROS generated by pollen NADPH oxidase provide a signal that augments antigen-induced allergic airway inflammation. *The Journal of Clinical Investigation* 115(8):2169–2179.

Brocklehurst, B., McLauchlan, K. A. (1996). Free radical mechanism for the effects of environmental electromagnetic fields on biological systems. *International Journal of Radiation Biology* 69(1):3–24.

Buchner, K., Eger, H. (2011). Changes of clinically important neurotransmitters under the influence of modulated RF fields—A long-term study under real-life conditions. *Umwelt - Medizin-Gesellschaft* 24(1):44–57.

Budi, A., Legge, F. S., Treutlein, H., Yarovsky, I. (2007). Effect of frequency on insulin response to electric field stress. *The Journal of Physical Chemistry: Part B* 111(20):5748–5756.

Budziosz, J., Stanek, A., Sieroń, A., et al. (2018). Effects of low-frequency electromagnetic field on oxidative stress in selected structures of the central nervous system. *Oxidative Medicine and Cellular Longevity*, 2018, 1427412.

Burlaka, A., Tsybulin, O., Sidorik, E., et al. (2013). Overproduction of free radical species in embryonic cells exposed to low intensity radiofrequency radiation. *Experimental Oncology* 35(3):219–225.

Burlaka, A., Selyuk, M., Gafurov, M., et al. (2014). Changes in mitochondrial functioning with electromagnetic radiation of ultra high frequency as revealed by electron paramagnetic resonance methods. *International Journal of Radiation Biology* 90(5):357–362.

Byus, C. V., Kartun, K., Pieper, S., Adey, W. R. (1988). Increased ornithine decarboxylase activity in cultured cells exposed to low energy modulated microwave fields and phorbol ester tumor promoters. *Cancer Research* 48(15):4222–4226.

Calabrese, E. J. (2008). Hormesis: Why it is important to toxicology and toxicologists. *Environmental Toxicology and Chemistry* 27(7):1451–1474.

Campisi, A., Gulino, M., Acquaviva, R., et al. (2010). Reactive oxygen species levels and DNA fragmentation on astrocytes in primary culture after acute exposure to low intensity microwave electromagnetic field. *Neuroscience Letters* 473(1):52–55.

Canseven, A. G., Coskun, S., Seyhan, N. (2008). Effects of various extremely low frequency magnetic fields on the free radical processes, natural antioxidant system and respiratory burst system activities in the heart and liver tissues, *Indian Journal of Biochemistry & Biophysics*, 45:326–331.

Caraglia, M., Marra, M., Mancinelli, F., et al. (2005). Electromagnetic fields at mobile phone frequency induce apoptosis and inactivation of the multi-chaperone complex in human epidermoid cancer cells. *Journal of Cellular Physiology* 204(2):539–548.

Cardis, E., Deltour, I., Vrijheid, M., et al. (2010). Brain tumour risk in relation to mobile telephone use: Results of the INTERPHONE international case-control study. *International Journal of Epidemiology* 39(3):675–694.

Carpenter, D. O. (2013). Human disease resulting from exposure to electromagnetic fields. *Reviews on Environmental Health* 28(4):159–172.

Çenesiz, M., Atakişi, O., Akar, A., et al. (2011). Effects of 900 and 1800 MHz electromagnetic field application on electrocardiogram, nitric oxide, total antioxidant capacity, total oxidant capacity, total protein, albumin and globulin levels in guinea pigs. *Kafkas Üniversitesi Veteriner Fakültesi Dergisi* 17(3):357–362.

Céspedes, O., Ueno, S. (2009). Effects of radio frequency magnetic fields on iron release from cage proteins. *Bioelectromagnetics* 30(5):336–342.

Çetin, H., Naziroglu, M., Çelik, Ö., et al. (2014). Liver antioxidant stores protect the brain from electromagnetic radiation (900 and 1800 MHz)-induced oxidative stress in rats during pregnancy and the development of offspring. *The Journal of Maternal-Fetal and Neonatal Medicine* (Publised online):1–6.

Chavdoula, E. D., Panagopoulos, D. J., Margaritis, L. H. (2010). Comparison of biological effects between continuous and intermittent exposure to GSM-900 MHz mobile phone radiation: Detection of apoptotic cell death features. *Mutation Research* 700(1–2):51–61.

Chekhun, V., Yakymenko, I., Sidorik, E., et al. (2014). Current state of international and national public safety limits for radiofrequency radiation. *Scientific Journal of the Ministry of Health of Ukraine* 1:57–64.

Cho, S. I., Nam, Y. S., Chu, L. Y., et al. (2012). Extremely low-frequency magnetic fields modulate nitric oxide signaling in rat brain. *Bioelectromagnetics* 33(7):568–574.

Chou, C. K., Guy, A. W., Kunz, L. L., et al. (1992). Long-term, low-level microwave irradiation of rats. *Bioelectromagnetics* 13(6):469–496.

Christ, A., Gosselin, M. C., Christopoulou, M., Kühn, S., Kuster, N. (2010). Age-dependent tissue-specific exposure of cell phone users. *Physics in Medicine and Biology* 55(7):1767–1783.

Chu, L. Y., Lee, J. H., Nam, Y. S., et al. (2011). Extremely low frequency magnetic field induces oxidative stress in mouse cerebellum. *General Physiology and Biophysics* 30(4):415–421.

Chu, M. K., Song, H. G., Kim, C., Lee, B. C. (2011). Clinical features of headache associated with mobile phone use: A cross-sectional study in university students. *BMC Neurology* 11:115.

Ciejka, E., Kleniewska, P., Skibska, B., Goraca, A. (2011). Effects of extremely low frequency magnetic field on oxidative balance in brain of rats. *Journal of Physiology and Pharmacology* 62(6):657.

Clifford, A., Morgan, D., Yuspa, S. H., Soler, A. P., Gilmour, S. (1995). Role of ornithine decarboxylase in epidermal tumorigenesis. *Cancer Research* 55(8):1680–1686.

Consales, C., Merla, C., Marino, C., Benassi, B. (2012). Electromagnetic fields, oxidative stress, and neurodegeneration. *International Journal of Cell Biology* 2012:683897.

Cui, Y., Ge, Z., Rizak, J. D., et al. (2012). Deficits in water maze performance and oxidative stress in the hippocampus and striatum induced by extremely low frequency magnetic field exposure. *PLOS ONE* 7(5):e32196.

Dasdag, S., Zulkuf Akdag, M., Aksen, F., et al. (2003). Whole body exposure of rats to microwaves emitted from a cell phone does not affect the testes. *Bioelectromagnetics* 24(3):182–188.

Dasdag, S., Bilgin, H., Akdag, M. Z., Celik, H., Aksen, F. (2008). Effect of long term mobile phone exposure on oxidative-antioxidative processes and nitric oxide in rats. *Biotechnology and Biotechnological Equipment* 22(4):992–997.

Dasdag, S., Akdag, M. Z., Ulukaya, E., Uzunlar, A. K., Ocak, A. R. (2009). Effect of mobile phone exposure on apoptotic glial cells and status of oxidative stress in rat brain. *Electromagnetic Biology and Medicine* 28(4):342–354.

Dasdag, S., Akdag, M. Z., Kizil, G., et al. (2012). Effect of 900 MHz radio frequency radiation on beta amyloid protein, protein carbonyl, and malondialdehyde in the brain. *Electromagnetic Biology and Medicine* 31(1):67–74.

De Iuliis, G. N., Newey, R. J., King, B. V., Aitken, R. J. (2009). Mobile phone radiation induces reactive oxygen species production and DNA damage in human spermatozoa in vitro. *PLOS ONE* 4(7):e6446.

De Salles, A. A., Bulla, G., Rodriguez, C. E. (2006). Electromagnetic absorption in the head of adults and children due to mobile phone operation close to the head. *Electromagnetic Biology and Medicine* 25(4):349–360.

De Souza, F. T., Silva, J. F., Ferreira, E. F., et al. (2014). Cell phone use and parotid salivary gland alterations: No molecular evidence. *Cancer Epidemiology, Biomarkers and Prevention* (Published online):1357.

Demirel, S., Doganay, S., Turkoz, Y., et al. (2012). Effects of third generation mobile phone-emitted electromagnetic radiation on oxidative stress parameters in eye tissue and blood of rats. *Cutaneous and Ocular Toxicology* 31(2):89–94.

Deng, Y., Zhang, Y., Jia, S., et al. (2013). Effects of aluminum and extremely low frequency electromagnetic radiation on oxidative stress and memory in brain of mice. *Biological Trace Element Research* 156(1):243–252.

Desai, N. R., Kesari, K. K., Agarwal, A. (2009). Pathophysiology of cell phone radiation: Oxidative stress and carcinogenesis with focus on male reproductive system. *Reproductive Biology and Endocrinology: RB and E* 7:114.

Deshmukh, P. S., Banerjee, B. D., Abegaonkar, M. P., et al. (2013). Effect of low level microwave radiation exposure on cognitive function and oxidative stress in rats. *Indian Journal of Biochemistry and Biophysics* 50(2):114–119.

Diem, E., Schwarz, C., Adlkofer, F., Jahn, O., Rüdiger, H. (2005). Non-thermal DNA breakage by mobile-phone radiation (1800MHz) in human fibroblasts and in transformed GFSH-R17 rat granulosa cells in vitro. *Mutation Research/Genetic Toxicology and Environmental Mutagenesis* 583(2):178–183.

Ding, S.-S., Sun, P., Zhang, Z., et al. (2018). Moderate dose of trolox preventing the deleterious effects of wi-fi radiation on spermatozoa in vitro through reduction of oxidative stress damage. *Chinese Medical Journal (English Edition)* 131(4):402.

Duan, Y., Wang, Z., Zhang, H., et al. (2013). The preventive effect of lotus seedpod procyanidins on cognitive impairment and oxidative damage induced by extremely low frequency electromagnetic field exposure. *Food and Function* 4(8):1252–1262. https://doi.org/10.1039/C3FO60116A.

Dutta, S. K., Ghosh, B., Blackman, C. F. (1989). Radiofrequency radiation-induced calcium ion efflux enhancement from human and other neuroblastoma cells in culture. *Bioelectromagnetics* 10(2):197–202.

Eger, H., Hagen, K., Lucas, B., et al. (2004). Influence of the proximity of mobile phone base stations on the incidence of cancer. *Environmental Medicine Society* 17:273–356.

Emre, M., Cetiner, S., Zencir, S., et al. (2011). Oxidative stress and apoptosis in relation to exposure to magnetic field. *Cell Biochemistry and Biophysics* 59(2):71–77.

Enyedi, B., Niethammer, P. (2013). H_2O_2: A chemoattractant? *Methods in Enzymology* 528:237–255.

Erdal, N., Gürgül, S., Tamer, L., Ayaz, L. (2008). Effects of long-term exposure of extremely low frequency magnetic field on oxidative/nitrosative stress in rat liver. *Journal of Radiation Research* 49(2):181–187.

Ericsson, (2009). LTE – an introduction. White Paper. 2009.

Ertilav, K., Uslusoy, F., Ataizi, S., Nazıroğlu, M. (2018). Long term exposure to cell phone frequencies (900 and 1800 MHz) induces apoptosis, mitochondrial oxidative stress and TRPV1 channel activation in the hippocampus and dorsal root ganglion of rats. *Metabolic Brain Disease* 33(3):753–763.

Esmekaya, M. A., Ozer, C., Seyhan, N. (2011). 900 MHz pulse-modulated radiofrequency radiation induces oxidative stress on heart, lung, testis and liver tissues. *General Physiology and Biophysics* 30(1):84–89.

Falone, S., Grossi, M. R., Cinque, B., et al. (2007). Fifty hertz extremely low-frequency electromagnetic field causes changes in redox and differentiative status in neuroblastoma cells. *The International Journal of Biochemistry and Cell Biology* 39(11):2093–2106.

Ferreira, A. R., Bonatto, F., de Bittencourt Pasquali, M. A., et al. (2006). Oxidative stress effects on the central nervous system of rats after acute exposure to ultra high frequency electromagnetic fields. *Bioelectromagnetics* 27(6):487–493.

Forman, H. J., Ursini, F., Maiorino, M. (2014). An overview of mechanisms of redox signaling. *Journal of Molecular and Cellular Cardiology*, 73:2–9.

Frank, J. W. (2021). Electromagnetic fields, 5G and health: What about the precautionary principle? *Journal of Epidemiology and Community Health* 75(6):562–566.

Friedman, J., Kraus, S., Hauptman, Y., Schiff, Y., Seger, R. (2007). Mechanism of short-term ERK activation by electromagnetic fields at mobile phone frequencies. *Biochemical Journal* 405(3):559–568.

Gandhi, O., Lazzi, G., Furse, C. (1996). Electromagnetic absorption in the human head and neck for mobile telephones at 835 and 1900 MHz. *IEEE Transactions on Microwave Theory and Techniques* 44(10):1884–1897.

Gao, Q.-H., Cai, Q., Fan, Y. (2017). Beneficial effect of catechin and epicatechin on cognitive impairment and oxidative stress induced by extremely low frequency electromagnetic field. *Journal of Food Biochemistry* 41(6):e12416.

Garaj-Vrhovac, V., Fucic, A., Horvat, D. (1992). The correlation between the frequency of micronuclei and specific chromosome aberrations in human lymphocytes exposed to microwave radiation in vitro. *Mutation Research* 281(3):181–186.

Garaj-Vrhovac, V., Gajski, G., Pažanin, S., et al. (2011). Assessment of cytogenetic damage and oxidative stress in personnel occupationally exposed to the pulsed microwave radiation of marine radar equipment. *International Journal of Hygiene and Environmental Health* 214(1):59–65.

Garson, O. M., McRobert, T. L., Campbell, L. J., Hocking, B. A., Gordon, I. (1991). A chromosomal study of workers with long-term exposure to radio-frequency radiation. *The Medical Journal of Australia* 155(5):289–292.

Gautam, R., Singh, K. V., Nirala, J., et al. (2019). Oxidative stress-mediated alterations on sperm parameters in male Wistar rats exposed to 3G mobile phone radiation. *Andrologia* 51(3):e13201.

Georgiou, C. D. (2010). Oxidative stress-induced biological damage by low-level EMFs: Mechanism of free radical pair electron spin-polarization and biochemical amplification. *European Journal of Oncology - Library* 5:63–113.

Goodman, E. M., Greenebaum, B., Marron, M. T. (1995). Effects of electro-magnetic fields on molecules and cells. *International Review of Cytology* 158:279–338.

Goodman, R., Blank, M. (2002). Insights into electromagnetic interaction mechanisms. *Journal of Cellular Physiology* 192(1):16–22.

Goraca, A., Ciejka, E., Piechota, A. (2010). Effects of extremely low frequency magnetic field on the parameters of oxidative stress in heart. *Journal of Physiology and Pharmacology* 61(3):333.

Griendling, K. K., Sorescu, D., Ushio-Fukai, M. (2000). NAD(P)H oxidase: Role in cardiovascular biology and disease. *Circulation Research* 86(5):494–501.

Guleken, Z. (2021). Chronic low-frequency electromagnetic field exposure before and after neonatal life induces changes on blood oxidative parameters of rat offspring. *Annals of Medical Research* 28(2):361–365.

Guler, G., Tomruk, A., Ozgur, E., et al. (2012). The effect of radiofrequency radiation on DNA and lipid damage in female and male infant rabbits. *International Journal of Radiation Biology* 88(4):367–373.

Gunes, M., Ates, K., Yalcin, B., et al. (2021). An evaluation of the genotoxic effects of electromagnetic radiation at 900 MHz, 1800 MHz, and 2100 MHz frequencies with a SMART assay in *Drosophila melanogaster*. *Electromagnetic Biology and Medicine*, 40(2):254-263.

Guney, M., Ozguner, F., Oral, B., Karahan, N., Mungan, T. (2007). 900 MHz radiofrequency-induced histopathologic changes and oxidative stress in rat endometrium: Protection by vitamins E and C. *Toxicology and Industrial Health* 23(7):411–420.

Gürler, H. Ş., Bilgici, B., Akar, A. K., et al. (2014). Increased DNA oxidation (8-OHdG) and protein oxidation (AOPP) by low level electromagnetic field (2.45 GHz) in rat brain and protective effect of garlic. *International Journal of Radiation Biology*, 90(10):892-896. .

Guzy, R. D., Schumacker, P. T. (2006). Oxygen sensing by mitochondria at complex III: The paradox of increased reactive oxygen species during hypoxia. *Experimental Physiology* 91(5):807–819.

Hallberg, O., Oberfeld, G. (2006). Letter to the editor: Will we all become electrosensitive? [Letter]. *Electromagnetic Biology and Medicine* 25(3):189–191.

Halliwell, B. (1991). Reactive oxygen species in living systems: Source, biochemistry, and role in human disease [Review]. *American Journal of Medicine* 91(3C):14S–22S.

Halliwell, B., Whiteman, M. (2004). Measuring reactive species and oxidative damage in vivo and in cell culture: How should you do it and what do the results mean? *British Journal of Pharmacology* 142(2):231–255.

Halliwell, B. (2007). Biochemistry of oxidative stress [Review]. *Biochemical Society Transactions* 35(Pt 5):1147–1150.

Hamzany, Y., Feinmesser, R., Shpitzer, T., et al. (2013). Is human saliva an indicator of the adverse health effects of using mobile phones? *Antioxidants and Redox Signaling* 18(6):622–627.

Hancı, H., Kerimoğlu, G., Mercantepe, T., Odacı, E. (2018). Changes in testicular morphology and oxidative stress biomarkers in 60-day-old Sprague Dawley rats following exposure to continuous 900-MHz electromagnetic field for 1 ha day throughout adolescence. *Reproductive Toxicology* 81:71–78.

Hardell, L., Carlberg, M., Hansson Mild, K. (2005). Case-control study on cellular and cordless telephones and the risk for acoustic neuroma or meningioma in patients diagnosed 2000–2003. *Neuroepidemiology* 25(3):120–128.

Hardell, L., Eriksson, M., Carlberg, M., Sundström, C., Mild, K. H. (2005). Use of cellular or cordless telephones and the risk for non-Hodgkin's lymphoma. *International Archives of Occupational and Environmental Health* 78(8):625–632.

Hardell, L., Carlberg, M., Ohlson, C. G., et al. (2007). Use of cellular and cordless telephones and risk of testicular cancer. *International Journal of Andrology* 30(2):115–122.

Hardell, L., Carlberg, M., Soderqvist, F., Mild, K. H., Morgan, L. L. (2007). Long-term use of cellular phones and brain tumours: Increased risk associated with use for > or =10 years. *Occupational and Environmental Medicine* 64(9):626–632.

Hardell, L., Carlberg, M. (2009). Mobile phones, cordless phones and the risk for brain tumours. *International Journal of Oncology* 35(1):5–17.

Hardell, L., Carlberg, M., Hansson Mild, K., Eriksson, M. (2011). Case-control study on the use of mobile and cordless phones and the risk for malignant melanoma in the head and neck region. *Pathophysiology* 18(4):325–333.

Hashish, A. H., El-Missiry, M. A., Abdelkader, H. I., Abou-Saleh, R. H. (2008). Assessment of biological changes of continuous whole body exposure to static magnetic field and extremely low frequency electromagnetic fields in mice. *Ecotoxicology and Environment Safety* 71(3):895–902.

Hayden, M. S., Ghosh, S. (2011). NF-kappa B in immunobiology. *Cell Research* 21(2):223–244.

Hedendahl, L., Carlberg, M., Hardell, L. (2015). Electromagnetic hypersensitivity–an increasing challenge to the medical profession. *Reviews on Environmental Health* 30(4):209–215.

Hong, M.-N., Han, N.-K., Lee, H.-C., et al. (2012). Extremely low frequency magnetic fields do not elicit oxidative stress in MCF10A cells. *Journal of Radiation Research* 53(1):79–86.

Hong, M. N., Kim, B. C., Ko, Y. G., et al. (2012). Effects of 837 and 1950 MHz radiofrequency radiation exposure alone or combined on oxidative stress in MCF10A cells. *Bioelectromagnetics* 33(7):604–611.

Hook, G. J., Spitz, D. R., Sim, J. E., et al. (2004). Evaluation of parameters of oxidative stress after in vitro exposure to FMCW- and CDMA-modulated radiofrequency radiation fields. *Radiation Research* 162(5):497–504.

Hosseinabadi, M. B., Khanjani, N., Ebrahimi, M. H., Mousavi, S. H., Nazarkhani, F. (2020). Investigating the effects of exposure to extremely low frequency electromagnetic fields on job burnout syndrome and the severity of depression; the role of oxidative stress. *Journal of Occupational Health* 62(1):e12136.

Hou, Q., Wang, M., Wu, S., et al. (2014). Oxidative changes and apoptosis induced by 1800-MHz electromagnetic radiation in NIH/3T3 cells. *Electromagnetic Biology and Medicine* (Published online):1–8.

Houston, B. J., Nixon, B., King, B. V., Aitken, R. J., De Iuliis, G. N. (2018). Probing the origins of 1,800 MHz radio frequency electromagnetic radiation induced damage in mouse immortalized germ cells and spermatozoa in vitro. *Frontiers in Public Health* 6:270.

Hoyto, A., Juutilainen, J., Naarala, J. (2007). Ornithine decarboxylase activity is affected in primary astrocytes but not in secondary cell lines exposed to 872 MHz RF radiation. *International Journal of Radiation Biology* 83(6):367–374.

Hyland, G. J. (2000). Physics and biology of mobile telephony. *Lancet* 356(9244):1833–1836.

Hyland, G. J. (2008). Physical basis of adverse and therapeutic effects of low intensity microwave radiation. *Indian Journal of Experimental Biology* 46(5):403–419.

ICNIRP. (1998). Guidelines for limiting exposure to time-varying elecrtic, magnetic and electromagnetic fields (up to 300 GHz). *Health Physics* 74(4):494–522.

ICNIRP. (2010). Guidelines for limiting exposure to time-varying electric and magnetic fields (1 Hz to 100 kHz). *Health Physics* 99(6):818–836.

ICNIRP. (2020). Guidelines for limiting exposure to electromagnetic fields (100 kHz to 300 GHz). *Health Physics* 118(5):483–524.

Ilhan, A., Gurel, A., Armutcu, F., et al. (2004). Ginkgo biloba prevents mobile phone-induced oxidative stress in rat brain. *Clinica Chimica Acta* 340(1–2):153–162.

Inoue, M., Sato, E. F., Nishikawa, M., et al. (2003). Mitochondrial generation of reactive oxygen species and its role in aerobic life. *Current Medicinal Chemistry* 10(23):2495–2505.

Irigaray, P., Caccamo, D., Belpomme, D. (2018). Oxidative stress in electrohypersensitivity selfreporting patients: Results of a prospective in vivo investigation with comprehensive molecular analysis. *International Journal of Molecular Medicine* 42(4):1885–1898.

Ismaiil, L. A., Joumaa, W. H., Moustafa, M. E. (2019). The impact of exposure of diabetic rats to 900 MHz electromagnetic radiation emitted from mobile phone antenna on hepatic oxidative stress. *Electromagnetic Biology and Medicine* 38(4):287–296.

Jelenković, A., Janać, B., Pešić, V., et al. (2006). Effects of extremely low-frequency magnetic field in the brain of rats. *Brain Research Bulletin* 68(5):355–360.

Jelodar, G., Akbari, A., Nazifi, S. (2013a). The prophylactic effect of vitamin C on oxidative stress indexes in rat eyes following exposure to radiofrequency wave generated by a BTS antenna model. *International Journal of Radiation Biology* 89(2):128–131.

Jelodar, G., Nazifi, S., Akbari, A. (2013b). The prophylactic effect of vitamin C on induced oxidative stress in rat testis following exposure to 900 MHz radio frequency wave generated by a BTS antenna model. *Electromagnetic Biology and Medicine* 32(3):409–416.

Jeong, Y. J., Son, Y., Han, N.-K., et al. (2018). Impact of long-term RF-EMF on oxidative stress and neuroinflammation in aging brains of C57BL/6 mice. *International Journal of Molecular Sciences* 19(7):2103.

Jing, J., Yuhua, Z., Xiao-qian, Y., et al. (2012). The influence of microwave radiation from cellular phone on fetal rat brain. *Electromagnetic Biology and Medicine* 31(1):57–66.

Johansson, O. (2006). Electrohypersensitivity: State-of-the-art of a functional impairment. *Electromagnetic Biology and Medicine* 25(4):245–258.

Johansson, O., Gangi, S., Liang, Y., et al. (2001). Cutaneous mast cells are altered in normal healthy volunteers sitting in front of ordinary TVs/PCs – Results from open-field provocation experiments. *Journal of Cutaneous Pathology* 28(10):513–519.

Kahya, M. C., Nazıroğlu, M., Çiğ, B. (2014). Selenium reduces mobile phone (900 MHz)-induced oxidative stress, mitochondrial function, and apoptosis in breast cancer cells. *Biological Trace Element Research* 160(2):285–293.

Kamali, K., Taravati, A., Sayyadi, S., Gharib, F. Z., Maftoon, H. (2018). Evidence of oxidative stress after continuous exposure to wi-fi radiation in rat model. *Environmental Science and Pollution Research International* 25(35):35396–35403.

Kang, K. A., Lee, H. C., Lee, J. J., et al. (2013). Effects of combined radiofrequency radiation exposure on levels of reactive oxygen species in neuronal cells. Journal of Radiation Research, 55(2):265–276.

Karimi, S. A., Salehi, I., Shykhi, T., Zare, S., Komaki, A. (2019). Effects of exposure to extremely low-frequency electromagnetic fields on spatial and passive avoidance learning and memory, anxiety-like behavior and oxidative stress in male rats. *Behavioural Brain Research* 359:630–638.

Kerbacher, J. J., Meltz, M. L., Erwin, D. N. (1990). Influence of radiofrequency radiation on chromosome aberrations in CHO cells and its interaction with DNA-damaging agents. *Radiation Research* 123(3):311–319.

Kerman, M., Senol, N. (2012). Oxidative stress in hippocampus induced by 900 MHz electromagnetic field emitting mobile phone: Protection by melatonin. *Biomedical Research* 23(1):147–151.

Kesari, K. K., Kumar, S., Behari, J. (2010). Mobile phone usage and male infertility in Wistar rats. *Indian Journal of Experimental Biology* 48(10):987–992.

Kesari, K. K., Kumar, S., Behari, J. (2011). 900-MHz microwave radiation promotes oxidation in rat brain. *Electromagnetic Biology and Medicine* 30(4):219–234.

Kesari, K. K., Meena, R., Nirala, J., Kumar, J., Verma, H. N. (2014). Effect of 3G cell phone exposure with computer controlled 2-D stepper motor on non-thermal activation of the hsp27/p38MAPK stress pathway in rat brain. *Cell Biochemistry and Biophysics* 68(2):347–358.

Khalil, A. M., Gagaa, M. H., Alshamali, A. M. (2012). 8-Oxo-7, 8-dihydro-2′-deoxyguanosine as a biomarker of DNA damage by mobile phone radiation. *Human and Experimental Toxicology* 31(7):734–740.

Khalil, A. M., Abu Khadra, K. M., Aljaberi, A. M., Gagaa, M. H., Issa, H. S. (2013). Assessment of oxidant/antioxidant status in saliva of cell phone users. *Electromagnetic Biology and Medicine*, 33(2):92–97.

Kim, J. Y., Hong, S. Y., Lee, Y. M., et al. (2008). In vitro assessment of clastogenicity of mobile-phone radiation (835 MHz) using the alkaline comet assay and chromosomal aberration test. *Environmental Toxicology* 23(3):319–327.

Kismali, G., Ozgur, E., Guler, G., et al. (2012). The influence of 1800 MHz GSM-like signals on blood chemistry and oxidative stress in non-pregnant and pregnant rabbits. *International Journal of Radiation Biology* 88(5):414–419.

Koc, A., Unal, D., Cimentepe, E. (2013). The effects of antioxidants on testicular apoptosis and oxidative stress produced by cell phones. *Turkish Journal of Medical Sciences* 43:131–137.

Koohestani, N. V., Zavareh, S., Lashkarbolouki, T., Azimipour, F. (2019). Exposure to cell phone induce oxidative stress in mice preantral follicles during in vitro cultivation: An experimental study. *International Journal of Reproductive Biomedicine* 17(9):637.

Koylu, H., Mollaoglu, H., Ozguner, F., Naziroglu, M., Delibas, N. (2006). Melatonin modulates 900 MHz microwave-induced lipid peroxidation changes in rat brain. *Toxicology and Industryial Health* 22(5):211–216.

Koyu, A., Ozguner, F., Yilmaz, H., et al. (2009). The protective effect of caffeic acid phenethyl ester (CAPE) on oxidative stress in rat liver exposed to the 900 MHz electromagnetic field. *Toxicology and Industryial Health* 25(6):429–434.

Kumar, S., Nirala, J. P., Behari, J., Paulraj, R. (2014). Effect of electromagnetic irradiation produced by 3G mobile phone on male rat reproductive system in a simulated scenario. *Indian Journal of Experimental Biology* 52(9):890–897.

Kunt, H., Şentürk, İ., Gönül, Y., et al. (2016). Effects of electromagnetic radiation exposure on bone mineral density, thyroid, and oxidative stress index in electrical workers. *OncoTargets and Therapy* 9:745.

Lai, H., Singh, N. P. (1996). Single- and double-strand DNA breaks in rat brain cells after acute exposure to radiofrequency electromagnetic radiation. *International Journal of Radiation Biology* 69(4):513–521.

Lai, H., Singh, N. P. (1997). Melatonin and a spin-trap compound block radiofrequency electromagnetic radiation-induced DNA strand breaks in rat brain cells. *Bioelectromagnetics* 18(6):446–454.

Lantow, M., Lupke, M., Frahm, J., et al. (2006). ROS release and Hsp70 expression after exposure to 1,800 MHz radiofrequency electromagnetic fields in primary human monocytes and lymphocytes. *Radiation and Environmental Biophysics* 45(1):55–62.

Lantow, M., Schuderer, J., Hartwig, C., Simkó, M. (2006). Free radical release and HSP70 expression in two human immune-relevant cell lines after exposure to 1800 MHz radiofrequency radiation. *Radiation Research* 165(1):88–94.

Lee, B.-C., Johng, H.-M., Lim, J.-K., et al. (2004). Effects of extremely low frequency magnetic field on the antioxidant defense system in mouse brain: A chemiluminescence study. *Journal of Photochemistry and Photobiology, Part B: Biology* 73(1):43–48.

Litovitz, T. A., Krause, D., Penafiel, M., Elson, E. C., Mullins, J. M. (1993). The role of coherence time in the effect of microwaves on ornithine decarboxylase activity. *Bioelectromagnetics* 14(5):395–403.

Litovitz, T. A., Penafiel, L. M., Farrel, J. M., et al. (1997). Bioeffects induced by exposure to microwaves are mitigated by superposition of ELF noise. *Bioelectromagnetics* 18(6):422–430.

Liu, C., Duan, W., Xu, S., et al. (2013a). Exposure to 1800 MHz radiofrequency electromagnetic radiation induces oxidative DNA base damage in a mouse spermatocyte-derived cell line. *Toxicology Letters* 218(1):2–9.

Liu, C., Gao, P., Xu, S.-C., et al. (2013b). Mobile phone radiation induces mode-dependent DNA damage in a mouse spermatocyte-derived cell line: A protective role of melatonin. *International Journal of Radiation Biology* 89(11):993–1001.

Liu, Y., Fiskum, G., Schubert, D. (2002). Generation of reactive oxygen species by the mitochondrial electron transport chain. *Journal of Neurochemistry* 80(5):780–787.

Low, H., Crane, F. L., Morre, D. J. (2012). Putting together a plasma membrane NADH oxidase: A tale of three laboratories. *International Journal of Biochemistry and Cell Biology* 44(11):1834–1838.

Luo, X., Chen, M., Duan, Y., et al. (2016). Chemoprotective action of lotus seedpod procyanidins on oxidative stress in mice induced by extremely low-frequency electromagnetic field exposure. *Biomedicine and Pharmacotherapy* 82:640–648.

Lu, Y. S., Huang, B. T., Huang, Y. X. (2012). Reactive oxygen species formation and apoptosis in human peripheral blood mononuclear cell induced by 900 MHz mobile phone radiation. *Oxidative Medicine and Cellular Longevity* 2012:740280.

Luo, Y.-P., Ma, H.-R., Chen, J.-W., Li, J. J., Li, C. X. (2014). Effect of American ginseng Capsule on the liver oxidative injury and the Nrf2 protein expression in rats exposed by electromagnetic radiation of frequency of cell phone. *Zhongguo Zhong Xi Yi Jie He Za Zhi Zhongguo Zhongxiyi Jiehe Zazhi = Chinese Journal of Integrated Traditional and Western Medicine / Zhongguo Zhong Xi Yi Jie He Xue Hui, Zhongguo Zhong Yi Yan Jiu Yuan Zhu Ban* 34(5):575–580.

Luukkonen, J., Hakulinen, P., Maki-Paakkanen, J., Juutilainen, J., Naarala, J. (2009). Enhancement of chemically induced reactive oxygen species production and DNA damage in human SH-SY5Y neuroblastoma cells by 872 MHz radiofrequency radiation. *Mutation Research* 662(1–2):54–58.

Maes, A., Collier, M., Verschaeve, L. (2000). Cytogenetic investigations on microwaves emitted by a 455.7 MHz car phone. *Folia Biologica* 46(5):175–180.

Mailankot, M., Kunnath, A. P., Jayalekshmi, H., Koduru, B., Valsalan, R. (2009). Radio frequency electromagnetic radiation (RF-EMR) from GSM (0.9/1.8GHz) mobile phones induces oxidative stress and reduces sperm motility in rats. *Clinics* 64(6):561–565.

Manikonda, P. K., Rajendra, P., Devendranath, D., et al. (2014). Extremely low frequency magnetic fields induce oxidative stress in rat brain. *General Physiology and Biophysics* 33(1):81–90.

Manta, A. K., Stravopodis, D. J., Papassideri, I. S., Margaritis, L. H. (2013). Reactive oxygen species elevation and recovery in Drosophila bodies and ovaries following short-term and long-term exposure to DECT base EMF. *Electromagnetic Biology and Medicine* 33(2):118–131.

Marino, A. A., Carrubba, S., Frilot, C., Chesson, A. L. (2009). Evidence that transduction of electromagnetic field is mediated by a force receptor. *Neuroscience Letters* 452(2):119–123.

Marjanovic, A. M., Pavicic, I., Trosic, I. (2014). Cell oxidation–reduction imbalance after modulated radiofrequency radiation. *Electromagnetic Biology and Medicine* (Published online):1–6.

Marjanovic, A. M., Pavicic, I., Trosic, I. (2018). Oxidative stress response in SH-SY5Y cells exposed to short-term 1800 MHz radiofrequency radiation. *Journal of Environmental Science and Health, Part A* 53(2):132–138.

Martínez-Sámano, J., Torres-Durán, P. V., Juárez-Oropeza, M. A., Verdugo-Díaz, L. (2012). Effect of acute extremely low frequency electromagnetic field exposure on the antioxidant status and lipid levels in rat brain. *Archives of Medical Research* 43(3):183–189.

Marzook, E. A., Abd El Moneim, A. E., Elhadary, A. A. (2014). Protective role of sesame oil against mobile base station-induced oxidative stress. *Journal of Radiation Research and Applied Sciences* 7(1):1–6.

Masoumi, A., Karbalaei, N., Mortazavi, S., Shabani, M. (2018). Radiofrequency radiation emitted from wi-fi (2.4 GHz) causes impaired insulin secretion and increased oxidative stress in rat pancreatic islets. *International Journal of Radiation Biology* 94(9):850–857.

Meena, R., Kumari, K., Kumar, J., et al. (2013). Therapeutic approaches of melatonin in microwave radiations-induced oxidative stress-mediated toxicity on male fertility pattern of Wistar rats. *Electromagnetic Biology and Medicine* 33(2):81–91.

Megha, K., Deshmukh, P. S., Banerjee, B. D., Tripathi, A. K., Abegaonkar, M. P. (2012). Microwave radiation induced oxidative stress, cognitive impairment and inflammation in brain of Fischer rats. *Indian Journal of Experimental Biology* 50(12):889–896.

Meral, I., Mert, H., Mert, N., et al. (2007). Effects of 900-MHz electromagnetic field emitted from cellular phone on brain oxidative stress and some vitamin levels of guinea pigs. *Brain Research* 1169:120–124.

Miller, A. B., Sears, M., Hardell, L., et al. (2019). Risks to health and well-being from radio-frequency radiation emitted by cell phones and other wireless devices. *Frontiers in Public Health* 7:223.

Moloney, J. N., Cotter, T. G. (2018). ROS signalling in the biology of cancer. *Seminars in cell & developmental biology* 80:50–64.

Morabito, C., Rovetta, F., Bizzarri, M., et al. (2010). Modulation of redox status and calcium handling by extremely low frequency electromagnetic fields in C2C12 muscle cells: A real-time, single-cell approach. *Free Radical Biology and Medicine* 48(4):579–589.

Motawi, T., Darwish, H., Moustafa, Y., Labib, M. M. (2014). Biochemical modifications and neuronal damage in brain of young and adult rats after long-term exposure to mobile phone radiations. *Cell Biochemistry and Biophysics* 70(2):845–855.

Moustafa, Y. M., Moustafa, R. M., Belacy, A., Abou-El-Ela, S. H., Ali, F. M. (2001). Effects of acute exposure to the radiofrequency fields of cellular phones on plasma lipid peroxide and antioxidase activities in human erythrocytes. *Journal of Pharmaceutical and Biomedical Analysis* 26(4):605–608.

Nagata, M. (2005). Inflammatory cells and oxygen radicals. *Current Drug Targets: Inflammation and Allergy* 4(4):503–504.

Naziroglu, M., Celik, O., Ozgul, C., et al. (2012). Melatonin modulates wireless (2.45 GHz)-induced oxidative injury through TRPM2 and voltage gated Ca(2+) channels in brain and dorsal root ganglion in rat. *Physiology and Behavior* 105(3):683–692.

Naziroglu, M., Cig, B., Dogan, S., et al. (2012). 2.45-Gz wireless devices induce oxidative stress and proliferation through cytosolic Ca(2)(+) influx in human leukemia cancer cells. *International Journal of Radiation Biology* 88(6):449–456.

Nguyen, H. L., Zucker, S., Zarrabi, K., et al. (2011). Oxidative stress and prostate cancer progression are elicited by membrane-type 1 matrix metalloproteinase. *Molecular Cancer Research* 9(10):1305–1318.

Ni, S., Yu, Y., Zhang, Y., et al. (2013). Study of oxidative stress in human lens epithelial cells exposed to 1.8 GHz radiofrequency fields. *PLOS ONE* 8(8):e72370.

Okayama, Y. (2005). Oxidative stress in allergic and inflammatory skin diseases. *Current Drug Targets: Inflammation and Allergy* 4(4):517–519.

Oksay, T., Nazıroğlu, M., Doğan, S., et al. (2014). Protective effects of melatonin against oxidative injury in rat testis induced by wireless (2.45 GHz) devices. *Andrologia* 46(1):65–72.

Oktem, F., Ozguner, F., Mollaoglu, H., Koyu, A., Uz, E. (2005). Oxidative damage in the kidney induced by 900-MHz-emitted mobile phone: Protection by melatonin. *Archives of Medical Research* 36(4):350–355.

Oral, B., Guney, M., Ozguner, F., et al. (2006). Endometrial apoptosis induced by a 900-MHz mobile phone: Preventive effects of vitamins E and C. *Advances in Therapy* 23(6):957–973.

Oshino, N., Jamieson, D., Sugano, T., Chance, B. (1975). Optical measurement of catalase-hydrogen peroxide intermediate (Compound-I) in liver of anesthetized rats and its implication to hydrogen-peroxide production insitu. *Biochemical Journal* 146(1):67–77.

Ott, M., Gogvadze, V., Orrenius, S., Zhivotovsky, B. (2007). Mitochondria, oxidative stress and cell death. *Apoptosis* 12(5):913–922.

Oyewopo, A., Olaniyi, S., Oyewopo, C., Jimoh, A. T. (2017). Radiofrequency electromagnetic radiation from cell phone causes defective testicular function in male Wistar rats. *Andrologia* 49(10):e12772.

Ozel, H. B., Cetin, M., Sevik, H., et al. (2021). The effects of base station as an electromagnetic radiation source on flower and cone yield and germination percentage in Pinus brutia Ten. *Biologia Futura*, 72(3):359-365.

Ozguner, F., Altinbas, A., Ozaydin, M., et al. (2005a). Mobile phone-induced myocardial oxidative stress: Protection by a novel antioxidant agent caffeic acid phenethyl ester. *Toxicology and Industryial Health* 21(9):223–230.

Ozguner, F., Oktem, F., Ayata, A., Koyu, A., Yilmaz, H. R. (2005b). A novel antioxidant agent caffeic acid phenethyl ester prevents long-term mobile phone exposure-induced renal impairment in rat. Prognostic value of malondialdehyde, N-acetyl-beta-D-glucosaminidase and nitric oxide determination. *Molecular and Cellular Biochemistry* 277(1–2):73–80.

Ozguner, F., Bardak, Y., Comlekci, S. (2006). Protective effects of melatonin and caffeic acid phenethyl ester against retinal oxidative stress in long-term use of mobile phone: A comparative study. *Molecular and Cellular Biochemistry* 282(1–2):83–88.

Ozgur, E., Kismali, G., Guler, G., et al. (2013). Effects of prenatal and postnatal exposure to GSM-like radiofrequency on blood chemistry and oxidative stress in infant rabbits, an experimental study. *Cell Biochemistry and Biophysics* 67(2):743–751.

Özorak, A., Nazıroğlu, M., Çelik, Ö., et al. (2013). Wi-fi (2.45 GHz)- and mobile phone (900 and 1800 MHz)-induced risks on oxidative stress and elements in kidney and testis of rats during pregnancy and the development of offspring. *Biological Trace Element Research* 156(1–3):221–229.

Özsobacı, N. P., Ergün, D. D., Durmuş, S., et al. (2018). Selenium supplementation ameliorates electromagnetic field-induced oxidative stress in the HEK293 cells. *Journal of Trace Elements in Medicine and Biology: Organ of the Society for Minerals and Trace Elements* 50:572–579.

Özsobacı, N. P., Ergün, D. D., Tunçdemir, M., Özçelik, D. (2020). Protective effects of zinc on 2.45 GHz electromagnetic radiation-induced oxidative stress and apoptosis in HEK293 cells. *Biological Trace Element Research* 194(2):368–378.

Panagopoulos, D. J., Messini, N., Karabarbounis, A., Philippetis, A. L., Margaritis, L. H. (2000). A mechanism for action of oscillating electric fields on cells. *Biochemical and Biophysical Research Communications* 272(3):634–640.

Panagopoulos, D. J., Karabarbounis, A., Margaritis, L. H. (2002). Mechanism for action of electromagnetic fields on cells. *Biochemical and Biophysical Research Communications* 298(1):95–102.

Panagopoulos, D. J., Chavdoula, E. D., Nezis, I. P., Margaritis, L. H. (2007). Cell death induced by GSM 900-MHz and DCS 1800-MHz mobile telephony radiation. *Mutation Research/Genetic Toxicology and Environmental Mutagenesis* 626(1–2):69–78.

Panagopoulos, D. J., Chavdoula, E. D., Margaritis, L. H. (2010). Bioeffects of mobile telephony radiation in relation to its intensity or distance from the antenna. *International Journal of Radiation Biology* 86(5):345–357.

Panagopoulos, D. J. (2012). Effect of microwave exposure on the ovarian development of *Drosophila melanogaster*. *Cell Biochemistry and Biophysics* 63(2):121–132.

Panagopoulos, D., Johansson, O., Carlo, G. (2013). Evaluation of specific absorption rate as a dosimetric quantity for electromagnetic fields bioeffects. *PLOS ONE* 8(6):e62663.

Panagopoulos, D., Johansson, O., Carlo, G. (2015). Polarization: A key difference between man-made and natural electromagnetic fields, in regard to biological activity. *Scientific Reports* 5(1):1–10.

Panagopoulos, D. J. (2019). Comparing DNA damage induced by mobile telephony and other types of man-made electromagnetic fields. *Mutation Research Reviews* 781:53–62.

Panagopoulos, D. J. (2020). Comparing chromosome damage induced by mobile telephony radiation and a high caffeine dose: Effect of combination and exposure Duration. *General Physiology and Biophysics* 39(6):531–544.

Panagopoulos, D. J., Karabarbounis, A., Yakymenko, I., Chrousos, G. P. (2021). Mechanism of DNA damage induced by human-made electromagnetic fields. *International Journal of Oncology* 59(5):92.

Pandey, N., Giri, S. (2018). Melatonin attenuates radiofrequency radiation (900 MHz)-induced oxidative stress, DNA damage and cell cycle arrest in germ cells of male Swiss albino mice. *Toxicology and Industrial Health* 34(5):315–327.

Patruno, A., Amerio, P., Pesce, M., et al. (2010). Extremely low frequency electromagnetic fields modulate expression of inducible nitric oxide synthase, endothelial nitric oxide synthase and cyclooxygenase-2 in the human keratinocyte cell line HaCat: Potential therapeutic effects in wound healing. *British Journal of Dermatology* 162(2):258–266.

Paulraj, R., Behari, J., Rao, A. R. (1999). Effect of amplitude modulated RF radiation on calcium ion efflux and ODC activity in chronically exposed rat brain. *Indian Journal of Biochemistry and Biophysics* 36(5):337–340.

Pavicic, I., Trosic, I. (2010). *Interaction of GSM modulated RF radiation and macromolecular cytoskeleton structures*. Paper presented at the 6th International Workshop on Biological Effects of Electromagnetic Fields.

Pedersen, G. F. (1997). Amplitude modulated RF fields stemming from a GSM/DCS-1800 phone. *Wireless Networks* 3(6):489–498.

Penafiel, L. M., Litovitz, T., Krause, D., et al. (1997). Role of modulation on the effect of microwaves on ornithine decarboxylase activity in L929 cells. *Bioelectromagnetics: Journal of the Bioelectromagnetics Society, The Society for Physical Regulation in Biology and Medicine, The European Bioelectromagnetics Association* 18(2):132–141.

Phillips, J. L., Singh, N. P., Lai, H. (2009). Electromagnetic fields and DNA damage. *Pathophysiology* 16(2–3):79–88.

Pilla, A. A. (2012). Electromagnetic fields instantaneously modulate nitric oxide signaling in challenged biological systems. *Biochemical and Biophysical Research Communications* 426(3):330–333.

Pourabdian, S., Golshiri, P., Habibi, E. (2009). Anxiety disorder in exposed employment to extremely low frequency electromagnetic fields (ELF-EMF) in steel industry. *Journal of Military Medicine* 10(4):299–302.

Qin, F., Yuan, H., Nie, J., Cao, Y., Tong, J. (2014). Effects of nano-selenium on cognition performance of mice exposed in 1800 MHz radiofrequency fields. *Wei Sheng Yan Jiu = Journal of Hygiene Research* 43(1):16–21.

Ragy, M. M. (2014). Effect of exposure and withdrawal of 900-MHz-electromagnetic waves on brain, kidney and liver oxidative stress and some biochemical parameters in male rats. *Electromagnetic Biology and Medicine* (Published online):1–6.

Ralph, S. J., Rodríguez-Enríquez, S., Neuzil, J., Saavedra, E., Moreno-Sánchez, R. (2010). The causes of cancer revisited: "mitochondrial malignancy" and ROS-induced oncogenic transformation – Why mitochondria are targets for cancer therapy. *Molecular Aspects of Medicine* 31(2):145–170.

Rao, V. S., Titushkin, I. A., Moros, E. G., et al. (2008). Nonthermal effects of radiofrequency-field exposure on calcium dynamics in stem cell-derived neuronal cells: Elucidation of calcium pathways. *Radiation Research* 169(3):319–329.

Rauš Balind, S., Selaković, V., Radenović, L., Prolić, Z., Janać, B. (2014). Extremely low frequency magnetic field (50 Hz, 0.5 mT) reduces oxidative stress in the brain of gerbils submitted to global cerebral ischemia. *PLOS ONE* 9(2):e88921.

Read, R., O'Riordan, T. (2017). The precautionary principle under fire. *Environment: Science and Policy for Sustainable Development* 59(5):4–15.

Reale, M., Kamal, M. A., Patruno, A., et al. (2014). Neuronal cellular responses to extremely low frequency electromagnetic field exposure: Implications regarding oxidative stress and neurodegeneration. *PLOS ONE* 9(8):e104973.

Regoli, F., Gorbi, S., Machella, N., et al. (2005). Pro-oxidant effects of extremely low frequency electromagnetic fields in the land snail Helix aspersa. *Free Radical Biology and Medicine* 39(12):1620–1628.

Repacholi, M. H., Basten, A., Gebski, V., et al. (1997). Lymphomas in E mu-Pim1 transgenic mice exposed to pulsed 900 MHz electromagnetic fields. *Radiation Research* 147(5):631–640.

Ruediger, H. W. (2009). Genotoxic effects of radiofrequency electromagnetic fields. *Pathophysiology* 16(2–3):89–102.

Sadetzki, S., Chetrit, A., Jarus-Hakak, A., et al. (2008). Cellular phone use and risk of benign and malignant parotid gland tumors–A nationwide case-control study. *American Journal of Epidemiology* 167(4):457–467.

Saikhedkar, N., Bhatnagar, M., Jain, A., et al. (2014). Effects of mobile phone radiation (900 MHz radiofrequency) on structure and functions of rat brain. *Neurological Research* 36(12):1072–1079.

Santini, R., Santini, P., Danze, J. M., Le Ruz, P., Seigne, M. (2002). Study of the health of people living in the vicinity of mobile phone base stations: 1. Influences of distance and sex. *Pathologie biologie* 50(6):369–373.

Sato, Y., Akiba, S., Kubo, O., Yamaguchi, N. (2011). A case-case study of mobile phone use and acoustic neuroma risk in Japan. *Bioelectromagnetics* 32(2):85–93.

Saygin, M., Ozmen, O., Erol, O., et al. (2018). The impact of electromagnetic radiation (2.45 GHz, wi-fi) on the female reproductive system: The role of vitamin C. *Toxicology and Industrial Health* 34(9):620–630.

Sefidbakht, Y., Moosavi-Movahedi, A. A., Hosseinkhani, S., et al. (2014). Effects of 940 MHz EMF on bioluminescence and oxidative response of stable luciferase producing HEK cells. *Photochemical and Photobiological Sciences* 13(7):1082–1092. https://doi.org/10.1039/C3PP50451D.

Seif, F., Reza Bayatiani, M., Ansarihadipour, H., Habibi, G., Sadelaji, S. (2019). Protective properties of Myrtus communis extract against oxidative effects of extremely low-frequency magnetic fields on rat plasma and hemoglobin. *International Journal of Radiation Biology* 95(2):215–224.

Selaković, V., Rauš Balind, S., Radenović, L., Prolić, Z., Janać, B. (2013). Age-dependent effects of ELF-MF on oxidative stress in the brain of Mongolian gerbils. *Cell Biochemistry and Biophysics* 66(3):513–521.

Seomun, G., Lee, J., Park, J. (2021). Exposure to extremely low-frequency magnetic fields and childhood cancer: A systematic review and meta-analysis. *PLOS ONE* 16(5):e0251628.

Sert, C., Deniz, M. (2011). Total antioxidant capacity, total oxidant status and oxidative stress index in rats exposed to extremely low frequency magnetic field. *Asian Journal of Chemistry* 23(5):1925.

Shaheen, W., Amer, N. M., Hafez, S. F., et al. (2021). Effect of antioxidants intake on oxidative stress among mobile phone users. *Egyptian Journal of Chemistry* 64(7):3903–3912.

Shahin, S., Singh, V. P., Shukla, R. K., et al. (2013). 2.45 GHz microwave irradiation-induced oxidative stress affects implantation or pregnancy in mice, Mus musculus. *Applied Biochemistry and Biotechnology* 169(5):1727–1751.

Sharma, A., Shrivastava, S., Shukla, S. (2020). Exposure of radiofrequency electromagnetic radiation on biochemical and pathological alterations. *Neurology India* 68(5):1092.

Sharma, S., Shukla, S. (2020). Effect of electromagnetic radiation on redox status, acetylcholine esterase activity and cellular damage contributing to the diminution of the brain working memory in rats. *Journal of Chemical Neuroanatomy* 106:101784.

Sharma, V. P., Singh, H. P., Kohli, R. K., Batish, D. R. (2009). Mobile phone radiation inhibits Vigna radiata (mung bean) root growth by inducing oxidative stress. *The Science of the Total Environment* 407(21):5543–5547.

Shedid, S. M., El-Tawill, G. A., Algeda, F. R., El-Fatih, N., Eltahawy, N. (2019). The impact of 950 MHz electromagnetic radiation on the brain and liver of rats and the role of garlic treatment. *Egyptian Journal of Radiation Sciences and Applications* 32(1):51–60.

Shim, Y., Lee, I., Park, S. (2013). The impact of LTE UE on audio devices. *ETRI Journal* 35(2):332–335.

Sies, H. (2014). Role of metabolic H_2O_2 generation: Redox signalling and oxidative stress. *Journal of Biological Chemistry*, 289(13):8735–8741.

Sies, H., Jones, D. P. (2020). Reactive oxygen species (ROS) as pleiotropic physiological signalling agents. *Nature Reviews. Molecular Cell Biology* 21(7):363–383.

Singh, H. P., Sharma, V. P., Batish, D. R., Kohli, R. K. (2012). Cell phone electromagnetic field radiations affect rhizogenesis through impairment of biochemical processes. *Environmental Monitoring and Assessment* 184(4):1813–1821.

Sokolovic, D., Djindjic, B., Nikolic, J., et al. (2008). Melatonin reduces oxidative stress induced by chronic exposure of microwave radiation from mobile phones in rat brain. *Journal of Radiation Research (Tokyo)* 49(6):579–586.

Sokolovic, D., Djordjevic, B., Kocic, G., et al. (2013). Melatonin protects rat thymus against oxidative stress caused by exposure to microwaves and modulates proliferation/apoptosis of thymocytes. *General Physiology and Biophysics* 32(1):79–90.

Suleyman, D., Zulkuf, M., A., Feyzan, A., Buyukbayram, H. (2004). Does 900 MHz GSM mobile phone exposure affect rat brain? *Electromagnetic Biology and Medicine* 23(3):201–214.

Sun, Y.-Y., Wang, Y.-H., Li, Z.-H., et al. (2019). Extremely low frequency electromagnetic radiation enhanced energy metabolism and induced oxidative stress in Caenorhabditis elegans. *Sheng Li Xue Bao: [Acta Physiologica Sinica]* 71(3):388–394.

Sun, Y., Shi, Z., Wang, Y., et al. (2018). Coupling of oxidative stress responses to tricarboxylic acid cycle and prostaglandin E2 alterations in Caenorhabditis elegans under extremely low-frequency electromagnetic field. *International Journal of Radiation Biology* 94(12):1159–1166.

Szmigielski, S., Szudzinski, A., Pietraszek, A., et al. (1982). Accelerated development of spontaneous and benzopyrene-induced skin cancer in mice exposed to 2450-MHz microwave radiation. *Bioelectromagnetics* 3(2):179–191.

Tice, R. R., Hook, G. G., Donner, M., McRee, D. I., Guy, A. W. (2002). Genotoxicity of radiofrequency signals. I. Investigation of DNA damage and micronuclei induction in cultured human blood cells. *Bioelectromagnetics* 23(2):113–126.

Tiwari, R., Lakshmi, N., Bhargava, S., Ahuja, Y. R. (2015). Epinephrine, DNA integrity and oxidative stress in workers exposed to extremely low-frequency electromagnetic fields (ELF-EMFs) at 132 kV substations. *Electromagnetic Biology and Medicine* 34(1):56–62.

Tkalec, M., Malaric, K., Pevalek-Kozlina, B. (2007). Exposure to radiofrequency radiation induces oxidative stress in duckweed Lemna minor L. *Science of the Total Environment* 388(1–3):78–89.

Tkalec, M., Stambuk, A., Srut, M., et al. (2013). Oxidative and genotoxic effects of 900 MHz electromagnetic fields in the earthworm Eisenia fetida. *Ecotoxicology and Environment Safety* 90:7–12.

Tök, L., Nazıroğlu, M., Doğan, S., Kahya, M. C., Tök, O. (2014). Effects of melatonin on wi-fi-induced oxidative stress in lens of rats. *Indian Journal of Ophthalmology* 62(1):12–15.

Toler, J. C., Shelton, W. W., Frei, M. R., Merritt, J. H., Stedham, M. A. (1997). Long-term, low-level exposure of mice prone to mammary tumors to 435 MHz radiofrequency radiation. *Radiation Research* 148(3):227–234.

Tomruk, A., Guler, G., Dincel, A. S. (2010). The influence of 1800 MHz GSM-like signals on hepatic oxidative DNA and lipid damage in nonpregnant, pregnant, and newly born rabbits. *Cell Biochemistry and Biophysics* 56(1):39–47.

Topsakal, S., Ozmen, O., Cicek, E., Comlekci, S. (2017). The ameliorative effect of gallic acid on pancreas lesions induced by 2.45 GHz electromagnetic radiation (wi-fi) in young rats. *Journal of Radiation Research and Applied Sciences* 10(3):233–240.

Torrieri, D. (2018). *Principles of spread-spectrum communication systems*, 4th ed. Springer.

Tsybulin, O., Sidorik, E., Kyrylenko, S., Henshel, D., Yakymenko, I. (2012). GSM 900 MHz microwave radiation affects embryo development of Japanese quails. *Electromagnetic Biology and Medicine* 31(1):75–86.

Tsybulin, O., Sidorik, E., Brieieva, O., et al. (2013). GSM 900 MHz cellular phone radiation can either stimulate or depress early embryogenesis in Japanese quails depending on the duration of exposure. *International Journal of Radiation Biology* 89(9):756–763.

Tsybulin, O., Sidorik, E., Kyrylenko, S., Yakymenko, I. (2016). Monochromatic red light of LED protects embryonic cells from oxidative stress caused by radiofrequency radiation. *Oxidants and Antioxidants in Medical Science* 5(1):21–27.

Türedi, S., Hancı, H., Topal, Z., et al. (2014). The effects of prenatal exposure to a 900-MHz electromagnetic field on the 21-day-old male rat heart. *Electromagnetic Biology and Medicine* (Published online):1–8.

Turker, Y., Naziroglu, M., Gumral, N., et al. (2011). Selenium and L-carnitine reduce oxidative stress in the heart of rat induced by 2.45-GHz radiation from wireless devices. *Biological Trace Element Research* 143(3):1640–1650.

Usman, J. D., Isyaku, U. M., Magaji, R. A., et al. (2020). Assessment of electromagnetic fields, vibration and sound exposure effects from multiple transceiver mobile phones on oxidative stress levels in serum, brain and heart tissue. *Scientific African* 7:e00271.

Vaks, V. L., Domrachev, G. A., Rodygin, Y. L., Selivanovskii, D. A., Spivak, E. I. (1994). Dissociation of water by microwave radiation. *Radiophysics and Quantum Electronics* 37(1):85–88.

Valko, M., Rhodes, C. J., Moncol, J., Izakovic, M., Mazur, M. (2006). Free radicals, metals and antioxidants in oxidative stress-induced cancer. *Chemico-Biological Interactions* 160(1):1–40.

Valko, M., Leibfritz, D., Moncol, J., et al. (2007). Free radicals and antioxidants in normal physiological functions and human disease. *International Journal of Biochemistry and Cell Biology* 39(1):44–84.

Wang, C., Liu, Y., Wang, Y., et al. (2019). Lowfrequency pulsed electromagnetic field promotes functional recovery, reduces inflammation and oxidative stress, and enhances HSP70 expression following spinal cord injury. *Molecular Medicine Reports* 19(3):1687–1693.

Wang, M., Yang, G., Li, Y., et al. (2020). Protective role of vitamin C in wi-fi induced oxidative stress in MC3T3-E1 cells in vitro. *Applied Computational Electromagnetics Society Journal*, 35(5):587–94.

Wang, X., Sharma, R. K., Gupta, A., et al. (2003). Alterations in mitochondria membrane potential and oxidative stress in infertile men: A prospective observational study. *Fertility and Sterility* 80:844–850.

Wertheimer, N., Leeper, E. (1979). Electrical wiring configurations and childhood cancer. *American Journal of Epidemiology* 109(3):273–284.

Wertheimer, N., Leeper, E. (1982). Adult cancer related to electrical wires near the home. *International Journal of Epidemiology* 11(4):345–355.

WHO. (2007). Electromagnetic fields and public health. https://www.who.int/teams/environment-climate-change-and-health/radiation-and-health/non-ionizing/elff.

Wolf, R., Wolf, D. (2007). Increased incidence of cancer near a cell-phone transmitted station. In F. Columbus (Ed.), *Trends in cancer prevention* (pp. 1–8). Nova Science Publishers, Inc., New York, USA.

Xu, S., Zhou, Z., Zhang, L., et al. (2010). Exposure to 1800 MHz radiofrequency radiation induces oxidative damage to mitochondrial DNA in primary cultured neurons. *Brain Research* 1311:189–196.

Yakymenko, I., Sidorik, E., Kyrylenko, S., Chekhun, V. (2011). Long-term exposure to microwave radiation provokes cancer growth: Evidences from radars and mobile communication systems. *Experimental Oncology* 33(2):62–70.

Yakymenko, I., Sidorik, E., Tsybulin, O., et al. (2011). Potential risks of microwaves from mobile phones for youth health. *Environmental and Health* 56(1):48–51.

Yakymenko, I., Tsybulin, O., Sidorik, E., et al. (2016). Oxidative mechanisms of biological activity of low-intensity radiofrequency radiation. *Electromagnetic Biology and Medicine* 35(2):186–202.

Yakymenko, I., Burlaka, A., Tsybulin, I., et al. (2018). Oxidative and mutagenic effects of low intensity GSM 1800 MHz microwave radiation. *Experimental Oncology* 40(4):282–287.

Yang, M. L., Ye, Z. M. (2015). Extremely low frequency electromagnetic field induces apoptosis of osteosarcoma cells via oxidative stress. *Journal of Zhejiang University (Medical Sciences)* 44(3):323–328.

Yokus, B., Cakir, D. U., Akdag, M. Z., Sert, C., Mete, N. (2005). Oxidative DNA damage in rats exposed to extremely low frequency electro magnetic fields. *Free Radical Research* 39(3):317–323.

Yokus, B., Akdag, M. Z., Dasdag, S., Cakir, D. U., Kizil, M. (2008). Extremely low frequency magnetic fields cause oxidative DNA damage in rats. *International Journal of Radiation Biology* 84(10):789–795.

Yurekli, A. I., Ozkan, M., Kalkan, T., et al. (2006). GSM base station electromagnetic radiation and oxidative stress in rats. *Electromagnetic Biology and Medicine* 25(3):177–188.

Zhao, T. Y., Zou, S. P., Knapp, P. E. (2007). Exposure to cell phone radiation up-regulates apoptosis genes in primary cultures of neurons and astrocytes. *Neuroscience Letters* 412(1):34–38.

Zmyślony, M., Politanski, P., Rajkowska, E., Szymczak, W., Jajte, J. (2004). Acute exposure to 930 MHz CW electromagnetic radiation in vitro affects reactive oxygen species level in rat lymphocytes treated by iron ions. *Bioelectromagnetics* 25(5):324–328.

Zosangzuali, M., Lalremruati, M., Lalmuansangi, C., et al. (2021). Effects of radiofrequency electromagnetic radiation emitted from a mobile phone base station on the redox homeostasis in different organs of Swiss albino mice. *Electromagnetic Biology and Medicine*, 40(3): 393–407.

Zotti-Martelli, L., Peccatori, M., Maggini, V., Ballardin, M., Barale, R. (2005). Individual responsiveness to induction of micronuclei in human lymphocytes after exposure in vitro to 1800-MHz microwave radiation. *Mutation Research* 582(1–2):42–52.

4 Genotoxic Effects of Wireless Communication Electromagnetic Fields

Ganesh Chandra Jagetia

Department of Zoology, Cancer and Radiation Biology
Laboratory, Mizoram University, Aizawl, India

CONTENTS

Abstract 138
4.1 Introduction 138
    4.1.1 Man-made Electromagnetic Fields (EMFs) and Wireless Communications (WC) 138
    4.1.2 WC EMFs and Reported Biological and Health Effects 140
4.2 Genotoxic Effects of WC EMFs 142
    4.2.1 RF/WC EMF Studies 142
        4.2.1.1 Human Cell (In Vitro or In Vivo) Studies 142
        4.2.1.2 Animal Cell (In Vitro or In Vivo) Studies 146
    4.2.2 ELF EMF and Static Field Studies 150
        4.2.2.1 Human Cell (In Vitro or In Vivo) Studies 150
        4.2.2.2 Animal Cell (In Vitro or In Vivo) Studies 152
4.3 Studies with Negative Findings 164
4.4 Discussion and Conclusions 169
Appendix 1: Functions of Genes/Proteins Mentioned in the Chapter 172
Acknowledgments 174
References 174

Keywords: electromagnetic fields; electromagnetic radiation; wireless communications; radio frequency, extremely low frequency; micronuclei; chromosome; DNA damage; health effects

Abbreviations: AMPK: adenosine monophosphate-activated protein kinase. AOPP: advanced oxidation protein product. ATM: ataxia telangiectasia mutated. ATP: adenosine triphosphate. AVTD: anomalous viscosity time dependencies. Bax: Bcl2-associated X. Bcl: B-cell lymphoma. 53BP1: p53 binding protein 1. CAT: catalase. CDMA: Code Division Multiple Access. Chk: checkpoint protein. DCS: Digital Cellular System. DECT: Digitally Enhanced Cordless Telecommunications. DNA: deoxyribonucleic acid. DNAPK: DNA-dependent *protein* kinase. DSB: double-strand break. ELF: Extremely Low Frequency. EMF: electromagnetic field. EMR: electromagnetic radiation. FDMA: Frequency Division Multiple Access. GADD: growth arrest and DNA damage inducible. GBM: glioblastoma multiforme. GPx: glutathione peroxidase. GSH: glutathione. GSM: Global System for Mobile Telecommunications. GSSG: oxidized glutathione. GST: glutathione-*S*-transferase. H2AX: histone family member X. HPBLs: human peripheral blood lymphocytes. HSP: heat shock protein. ICNIRP: International Commission for Non-Ionizing Radiation Protection. iDEN:

Integrated Digital Enhanced Network. IL6: interleukin-6. LC3: microtubule associated protein 1A/1B-light chain 3. LDH: lactate dehydrogenase. LIG: ligase. LPO: lipid peroxidation. LTE: Long Term Evolution. MAPK: mitogen-activated protein kinase. MT: mobile telephony. mt: mitochondrion. mTOR: mammalian target of rapamycin. NCE: normochromatic erythrocyte. NFκB: nuclear factor kappa B. ODC: ornithine decarboxylase. OS: oxidative stress. PARP: poly(ADP ribose)polymerase. PCE: polychromatic erythrocytes. RAD: radiation sensitive. RF: Radio Frequency. ROS: reactive oxygen species. SAR: Specific Absorption Rate. SCE: sister chromatid exchanges. SOD: superoxide dismutase. SSB: single-strand break. STAT: signal transducer and activator of transcription. TDMA: Time Division Multiple Access. TNF: tumor necrosis factor. TUNEL: terminal deoxynucleotide transferase-mediated dUTP nick end-labeling. ULF: Ultra Low Frequency. UMTS: Universal Mobile Telecommunications System. VGCC: voltage-gated calcium channel. VGIC: voltage-gated ion channel. WC: wireless communication. Wi-Fi: Wireless Fidelity. XRCC: x-ray repair cross complementing protein. 1G/2G/3G/4G/5G: first/second/third/fourth/fifth generation of MT. 8-OH-dG: 8-hydroxy-2′-deoxyguanosine.

ABSTRACT

The tremendous development of wireless communications (WC) technology during the past 30 years has transformed telecommunications and popularized mobile phones so much that, today, their number exceeds the global population. In addition to electromagnetic fields (EMFs) and corresponding electromagnetic radiation (EMR) from natural sources like sun, cosmos, atmospheric discharges, etc., humans are exposed to man-made EMFs/EMR, especially at the Extremely Low Frequency (ELF) and the Radio Frequency (RF)/microwave bands. EMFs/EMR emitted by WC devices, such as mobile phones and corresponding antennas, contain RF carrier signals which are pulsed and modulated by ELF signals. We call these complex emissions WC EMFs. WC EMFs have generated great concern in the scientific community and the public, as they have been reported to cause headache, fatigue, tinnitus (microwave hearing), concentration problems, depression, memory loss, sleep, and hormonal disorders as short-term effects and even infertility and cancer as the long-term effects. This chapter has been written after collecting information from various search engines, including Google Scholar, PubMed, SciFinder, Science Direct, and other websites on the internet. The chapter focuses on the genotoxic cellular effects of WC EMFs on cultured cells, humans, and animals. Since WC EMFs combine both RF and ELF, in this chapter, both RF/WC and purely ELF man-made EMF studies are reviewed. Most studies conducted on the genotoxic effects of ELF or RF/WC EMFs have resulted in positive findings. Many human and animal studies have demonstrated that ELF or RF/WC man-made EMFs increased the frequency of micronuclei and induced chromosome aberrations or DNA damage, including single- and double-strand breaks. It has also been demonstrated that these EMFs trigger reactive oxygen species (ROS) formation, and changes in gene expression, particularly in genes involved in signal transduction, cytoskeleton formation, and cellular metabolism.

4.1 INTRODUCTION

4.1.1 Man-made Electromagnetic Fields (EMFs) and Wireless Communications (WC)

Wireless communication (WC) was discovered in 1899 by Guglielmo Marconi, who sent the first telegraphic message from a ship docked at the New York Harbor to the Twin Lights in Highlands, New Jersey. Thereafter, the technology of WC had a fast and continuous development, and numerous new inventions since have paved the way for mobile telephony (MT) and the Internet. The first commercially available mobile phone was launched in October 1983, and since then, the MT technology has advanced tremendously with the addition of new features over time. The twenty-first century can be regarded as the age of mobile phones (Bento 2016; Harris and Cooper 2019). Throughout the world, the number of mobile phones has greatly increased during the past 25 years, becoming approximately 1.5 times the world's human population at the time of this writing, and it will continue to grow with time (Anonymous, 2022). Mobile phones act as a two-way radio; i.e., they can receive and send information (voice, messages, pictures, videos, internet). This is accomplished through the emission of electromagnetic fields (EMFs) and corresponding electromagnetic radiation (EMR) (Aly and Crum 2016; Zothansiama et al. 2017).

All types of EMFs (natural or man-made) are self-propagating in the form of EMR produced by the vibration of charged particles. In the case of man-made EMFs/EMR, the charged particles are free electrons in the metallic conductors of the electric/electronic circuits. The electromagnetic waves (EMR) possess both electric and magnetic components which oscillate in phase normally (at an angle of 90°) to each other and to the direction of propagation of the waves. They propagate in material media and in the vacuum with the speed of light.

Man-made EMFs have frequencies from 0 Hz to 300 GHz (which is the low limit of infrared) and are non-ionizing, meaning they cannot directly break chemical bonds or cause ionization of atoms or molecules. The lower part of the frequency spectrum is the Extremely Low Frequency (ELF) band (3–3000 Hz), which is of great importance, as it includes the electric power transmission EMFs (50–60 Hz) which are reported to induce significant biological/health effects. The lowest part, 0–3 Hz, is called Ultra Low Frequency (ULF) band. The frequencies of the spontaneous intra-cellular ionic oscillations in all living cells are in the ULF band. All physiological endogenous biological activities (nerve impulses, brain waves, etc.) are in the ULF/ELF ranges. The higher part of the frequency spectrum of man-made EMFs is called Radio Frequency (RF) band (300 kHz–300 GHz), and it is also of great importance, as it is employed in radio and television broadcasting and the WCs. The highest part of RF is called microwave band (300 MHz–300 GHz). Radars, MT, and other types of digital WC, such as wireless internet, employ carrier frequencies in this band. The newer WC systems, such as the fifth generation (5G) of MT/WC, which is now under deployment, employ increasingly higher frequencies in the microwave band (up to 100 GHz). The EMFs emitted by all modern digital WC devices and corresponding antennas contain RF carrier signals, and have a "pulsed" character, meaning that the RF carrier wave is emitted in the form of on/off pulses ("frames") which are repeated at ELF (in most cases) rates. Moreover, the RF/microwave carrier wave is at the same time modulated by ELF (in most cases) signals, which include the transmitted information. Thus, all modern digital WC emissions combine RF/microwave carrier waves which are modulated by ELF signals and pulsed at ELF rates (Pedersen 1997; Zwamborn et al. 2003; Hyland 2000, 2008; Panagopoulos 2019a). Man-made EMFs, and especially pulsed WC EMFs, seem to pose additional health risks even when the exposures are well below the limits set by the International Commission for Non-Ionizing Radiation Protection (ICNIRP) (1998; 2010; 2020; Martynyuk et al. 2016; Liu et al. 2019).

Exposure assessments and limits are provided in the ELF/ULF bands in terms of electric and magnetic field intensity (in V/m, and usually mT, respectively), and, in the RF/microwave band in terms of incident power per unit surface (radiation intensity or power density in mW/cm^2 or $\mu W/cm^2$) or in terms of absorbed power per unit mass of tissue called Specific Absorption Rate (SAR) in W/kg.

WC EMFs are those emitted by mobile phones and corresponding MT base antennas, domestic cordless [Digitally Enhanced Cordless Telecommunications (DECT)] phones, wireless internet connections [Wireless Fidelity (Wi-Fi)], etc. The exposure to WC EMFs emitted by MT base station antennas and antennae "towers" (with many base station antennas) depends on the distance from a base station, the number of base stations per square kilometer, the distance between them, the intensity or power density of the radiation transmitted by the base stations, the number of active mobile phones/km^2, the call traffic, atmospheric conductivity, and many other variable factors. Similarly, the EMF exposure from mobile phones depends on the distance from the devices, the signal reception from the base station antennae (when the received signal is low the output power increases), the specific position, etc. The EMF exposure from MT/WC antennas and devices is usually much higher in the cities because of the greater number of users and the higher density of base stations to accommodate the network connectivity of a large number of users. The power density of the EMR emitted from a base station antenna within a city usually exceeds 1000 $\mu W/cm^2$ at a few meters distance from the antennas, is usually 10–100 $\mu W/cm^2$ within 100 m distance from the base stations (depending on the antenna's output power, number of antennas emitting simultaneously, etc.), and reduces almost linearly at longer distances. However, the levels may be more complicated at places where MT base stations are closely located (Haumann et al. 2002; Frei et al. 2009; Panagopoulos et al. 2010; Carpenter 2013; Aly and Crum 2016).

Apart from the first-generation (1G) MT system which was analog and of limited use, today's MT/WC technology uses the second-generation (2G) system or Global System for Mobile Telecommunications (GSM) that employs the Time-Division Multiple Access (TDMA) pulsing system for multiplexing (simultaneous service for many users), and the third/fourth generation (3G/4G) Universal Mobile Telecommunications System (UMTS) that employs the Code-Division Multiple Access (CDMA) pulsing/multiplexing system. The carrier frequencies of GSM and UMTS are in the frequency range between 900 and 2200 MHz, with pulsations mainly at 2, 8, and 217 Hz (GSM) or 100 and 1500 Hz (UMTS). The purely 4G Long Term Evolution (LTE) system employs carrier frequencies up to 2600 MHz (which differ among different countries), similar pulsations with UMTS, and additional pulsations as reference and synchronization signals. As mentioned already, apart from MT, other powerful sources of WC EMFs are DECT phones (1880 MHz carrier, with 100 and 200 Hz pulsations) and Wi-Fi (usually 2450 MHz carrier, with 10 Hz pulsations). They all combine RF and ELF emissions (Pedersen 1997; Zwamborn et al. 2003; Hyland 2000, 2008; Sesia et al. 2011; Panagopoulos 2019a; 2020). Apart from the WC antennas/devices, ELF exposure arises mainly from electric power transmission lines (50–60 Hz), transformers, and various electric devices at residential and working environments (Gosselin et al. 2013; Moon 2020; Paolucci et al. 2020; López et al. 2021).

All man-made EMFs are actually very different from the natural EMF emissions from sun, cosmos, geomagnetic/geoelectric fields, atmospheric oscillations, and other natural EMF sources because all man-made EMFs are totally polarized and coherent and, for this reason, detrimental to living organisms (Panagopoulos et al. 2015a; Panagopoulos 2019a). Moreover, especially WC EMFs are highly variable each moment in their intensity and other parameters, and this makes them even more bioactive (Panagopoulos 2019a). WC technology is developing fast, and the research on the adverse effects of WC EMF exposure on human health is not easy to keep pace with the changing technology (Russell 2018), but as already explained, all types of WC EMFs have very similar characteristics, and also their effects are similar in different animals and cell types (Yakymenko et al. 2016; Panagopoulos 2019a).

4.1.2 WC EMFs and Reported Biological and Health Effects

The increasing use of mobile phones for ever longer durations, the spread of MT base station antennas/towers to facilitate communication between mobile phones, and exposure to WC EMFs from other electronic sources such as DECT and Wi-Fi have raised concerns regarding their adverse health effects on human health and the environment, as it has become clear that, today, man-made EMFs represent one of the most important types of environmental pollution encountered daily, along with air and water pollution, chemicals, pesticides, etc. As early as 2000–2001, mobile phone users and individuals residing close to MT base station antennas/towers started complaining of a variety of non-specific symptoms, including headache, sleep problems, fatigue, tinnitus, problems with concentration, depression, memory impairment, skin irritation, cognitive function problems, cardiovascular effects, and hormonal disorders. The terms "microwave sickness" and electro-hypersensitivity (EHS) have been introduced to describe these symptoms, which are connected with WC EMF exposure and oxidative stress (OS) induced in living tissues by overproduction of reactive oxygen species (ROS) in the cells (Santini et al. 2002; Navarro et al. 2003; Hutter et al. 2006; Abdel-Rassoul et al. 2007; Augner et al. 2009; Yakymenko et al. 2016; Deniz et al. 2017; Belpomme and Irigaray 2020; López et al. 2021). In addition, WC EMF exposure has been occasionally reported to cause muscle and nerve stimulation, shock or burn after touching conducting objects, hallucinations, dry eyes, De Quervain's tenosynovitis, nomophobia, computer vision syndrome, weakness in thumbs, rigidity of hands and wrist, stiff neck, auditory and sleep disturbances, insomnia, hallucinations, reduced self-confidence, anxiety, stress, and mobile phone addiction disorders (Parasuraman et al. 2017; Amjad et al. 2020; Al-Khlaiwi et al. 2020).

It is known that the exposure of living organisms to high power/intensity RF EMFs leads to an increase in tissue temperature (heating), which is known as the "thermal effect" (Dilli 2021). The microwave oven for heating food is a well-known example of the thermal effect of high-power microwaves. Yet, the vast majority of ELF or WC EMF exposures at environmentally relevant levels are not found to induce any heating in exposed living tissues, and the recorded effects are known as "non-thermal effects" (Goodman et al. 1995; Belyaev et al. 2005; Panagopoulos et al. 2013a; Yakymenko et al. 2016, 2018; López et al. 2021). Some of the reported non-thermal biological and health effects of WC EMF exposure, in addition to the aforementioned EHS symptoms, include cellular genotoxic effects (which will be extensively described in the present chapter), ROS overproduction, and OS (which results in genotoxicity when it is not controlled by the cells) (Yakymenko et al. 2016), dysfunction of the immune system (Szmigielski 2013), increased permeability of the blood–brain barrier (Salford et al. 2003), changes in cell membrane permeability (Aly and Crum 2016), and cancer (Hardell et al. 2007; Hardell and Carlberg 2009; Miller et al. 2018). The induction of such effects confirms that WC EMF exposure is not safe for humans and living organisms.

A study conducted in Sweden between 2007 and 2009 on mobile and cordless (DECT) phone users of either sex between the age of 18 and 75 years revealed an increased risk of malignant brain tumors, and ipsilateral use (on one side of the head) was associated with a higher risk of tumors than contralateral use (on both sides). Increased daily use by an individual has a double to quadruple risk of brain tumors (Hardell et al. 2013). The incidence of neuro-epithelial brain tumors in children, adolescents, and young adults in the age group from birth to 24 years has increased between 2010 and 2017 in the United States (US). The glioblastoma multiforme (GBM) (probably the most aggressive type of brain cancer) increased among all ages in the UK, and schwannomas (nerve sheath tumors) increased in the age group from 20 to 85 years (Miller et al. 2019). This indicates that there is a close relationship between WC EMF exposure and the induction of brain tumors.

The EMF exposure from mobile phones and MT base station antennas has increased the spontaneous abortions in pregnant women and the percentage of low birth weights (Mahmoudabadi et al. 2015). The use of mobile phones for more than 30 minutes (min) per day reduced fetal growth and infant birth weight (Lu et al. 2017; Boileau et al. 2020). An epidemiological study on 55,507 women from Denmark, the Netherlands, Spain, and South Korea who used mobile phones during pregnancy showed an association between preterm delivery and mobile phone use (Tsarna et al. 2019).

The exposure of mouse follicles to 33 or 50 Hz ELF EMF reduced the developmental capacity of the follicles *in vitro* (Cecconi et al. 2000). Exposure of female rats to ELF 30 Hz sinusoidal magnetic field for 2 hours a day (h/day) for 10 weeks decreased the levels of the follicle stimulating hormone in the proestrus and of progesterone in the estrus phase. Moreover, the number of primordial follicles in the ovaries was lower in the exposed animals than in the controls (Alekperov et al. 2019). This is in agreement with the decreased ovary size and number of developing follicles reported for rats (Gul et al. 2009) and fruit flies (Panagopoulos 2012) exposed to mobile phone EMFs and attributed to cell death in ovarian cells (Panagopoulos 2012).

Excessive exposure to WC EMF for a longer duration adversely affected male and female reproduction (Dilli 2021). The exposure of human semen samples to WC EMF from mobile phones or laptop computers connected to internet via Wi-Fi reduced semen quality, sperm motility, and viability (Erogul et al. 2006; Avendaño et al. 2012; Okechukwu 2020). Several other studies had repeatedly shown decreased sperm motility and viability along with OS induction in humans and other mammals due to mobile phone EMF exposure (Agarwal et al. 2009; De Iuliis et al. 2009; Mailankot et al. 2009; Veerachari and Vasan 2012). However, a recent study conducted in Denmark and the US with 751 and 2,349 human volunteers, respectively, who kept their mobile phones in their pant pockets did not find any adverse effect on semen quality (Hatch et al. 2021).

The studies reporting adverse effects of WC EMF exposure on human/animal health significantly outnumber the studies reporting no effects. This fact further verifies that WC EMF emissions are not safe (Yakymenko et al. 2016; Miller et al. 2018, 2019; Panagopoulos et al. 2021; López et al. 2021). The time living cells take to respond to WC EMF exposure is usually a few min, indicating that the health risks posed by WC EMF exposures to human/animal health and to the environment are alarming (Hardell et al. 2002; Blank and Goodman 2009, 2011; Miller et al. 2018, 2019).

Although the effects of WC EMFs may seem contradictory, as several studies have reported no effects (negative findings), a closer examination reveals that, while studies that employ simulated WC EMFs with fixed parameters emitted by generators find effects in about 50% of the cases, studies that employ real-life WC EMF exposures (from mobile phones, DECT phones, MT base antennas, Wi-Fi, etc.) with high signal variability, demonstrate effects in more than 95% of the cases (Panagopoulos et al. 2015b; Panagopoulos 2019a). Moreover, it has been shown that funding by industry exerts influence on the reported results, and supported studies report positive findings less frequently than those that are not supported (Leach et al. 2018), which is a clear indication of conflict of interest. Overall, about 70% of the peer-reviewed published studies have found effects (Manna and Ghosh 2016; Yakymenko et al. 2016; Leach et al. 2018; Panagopoulos et al. 2015b; Panagopoulos 2019a). The number of studies carried out on man-made EMF bioeffects is enormous, and the main focus, here, is to review the genotoxic effects of WC EMFs on humans and animals, at both whole organism and cellular levels. Since WC EMFs include both RF and ELF EMFs, I have reviewed studies investigating genotoxic effects involving exposures in either RF/WC and/or ELF bands.

Since various antioxidants play key roles in EMF-induced cellular genotoxic effects discussed in this chapter, a few general words on their function should be provided. The ROS, apart from their reactions with various biological molecules, interact with cell lipids abstracting hydrogen from their hydrocarbon constituents and adding oxygen, generating peroxyl radicals and hydroperoxides. This is called lipid peroxidation (LPO). The initial products of LPO, in turn, initiate various pathways leading to cell death. The antioxidant enzymes, such as catalase (CAT) and glutathione peroxidase (GPx), detoxify by catalyzing the reduction of hydrogen peroxide (H_2O_2) into oxygen and water. Glutathione-S-transferase (GST) and GPx also detoxify peroxides of lipids in concert with reduced glutathione (GSH), whereas superoxide dismutase (SOD) detoxifies by dismutating superoxide radical ($O_2\bullet^-$) into molecular oxygen and hydroxyl radicals (OH•) into hydrogen peroxide. GSH is a non-enzymatic antioxidant produced by cells which removes free radicals and detoxifies electrophilic compounds and peroxides via catalysis by GST and GPx. GSH is converted to oxidized glutathione (GSSG) after reaction with electrophiles and catalysis by GPx.

4.2 GENOTOXIC EFFECTS OF WC EMFs

4.2.1 RF/WC EMF Studies

4.2.1.1 Human Cell (In Vitro or In Vivo) Studies

Human lymphocytes exposed in vitro: Exposure of human peripheral blood lymphocytes (HPBLs) to 7.7 GHz RF EMF (0.5, 10, and 30 mW/cm^2) for 10, 30, and 60 min was found to increase the frequency of micronuclei[*] and chromosomal aberrations, acentric fragments, dicentric and ring chromosomes, and the effect increased with increasing power density and exposure duration (Garaj-Vrhovac et al. 1992). The acentric fragments, dicentric chromosomes, and micronuclei increased in the HPBLs exposed to 2450 MHz (pulsed at 50 Hz) EMF (SAR 75 W/kg) for 30 and 120 min (Maes et al. 1993). The HPBLs exposed to a 9 GHz non-modulated continuous-wave RF EMF (SAR 90 W/kg, incident power 190 mW) or the same carrier modulated in amplitude by 50 Hz sine waves for 10 min showed an increase in micronuclei formation (D'Ambrosio et al. 1995). Application of 2.45 and 7.7 GHz (10, 20, and 30 mW/cm^2) RF EMF to HPBLs for 15, 30, and 60 min increased the frequency of micronuclei after the 30 and 60 min exposures at the power density of 30 mW/cm^2, while the 15 min exposure did not induce a statistically significant effect (Zotti-Martelli et al. 2000). The exposure of HPBLs to a non-modulated continuous-wave 1.748 GHz (SAR 5 W/kg) RF EMF or the same carrier phase modulated by Gaussian Minimum Shift Keying (GMSK) mode (GSM modulation) at SAR 2.25 ± 0.87 W/kg for 15 min increased the micronuclei frequency after the GMSK-modulated exposure, while

---

[*] The existence of micronuclei is a sign of damaged DNA and fragmented chromosomes.

the non-modulated exposure was ineffective (D'Ambrosio et al. 2002). The frequency of micronucleated lymphocytes increased after exposure of HPBLs to either CDMA or TDMA types of mobile phone signal modulations for 3 or 24 hours (h) at SAR 5 or 10 W/kg (Tice et al. 2002). The exposure of HPBLs to 830 MHz (SAR 1.6–8.8 W/kg) RF EMF for 72 h resulted in an increase in aneuploidy (abnormal number of chromosomes) with increasing SAR (Mashevich et al. 2003). Exposure of HPBLs to 1800 MHz (5 and 10 mW/cm^2) RF EMF continuously for 60, 120, and 180 min increased the frequency of micronuclei in the lymphocytes. The number of micronuclei varied among the individual donors and between the first and the repeat experiment for each donor after 3 months, with a significant variation in the lymphocyte proliferation (Zotti-Martelli et al. 2005). It should be noted that some of the intensity/SAR values used in the above early studies are considerably higher than those existing environmentally and higher than the ICNIRP (2020) limits for public exposure to WC EMFs (4 mW/cm^2 or SAR 2 W/kg). Such intensity/SAR values are indicative of exposures very close to radar or MT base station antennas where thermal effects are not excluded.

HPBLs from one donor exposed to 900 and 905 MHz GSM EMF (0.125–2 W power output) SAR 5.4 mW/kg for 30 min exhibited increased chromosome condensation, whereas exposure to 895, 905, and 915 MHz for 1 h showed a similar effect in HPBLs from other donors (Sarimov et al. 2004). Exposure of HPBLs for 2 h to 915 MHz simulated GSM from a "test" phone (SAR 37 mW/kg) induced chromatin condensation (measured by anomalous viscosity time dependencies (AVTD)), which is a sign of cell death, and inhibited tumor-suppressor p53-binding protein 1 (53BP1) in normal subjects as well in subjects reporting hypersensitivity to EMFs (Belyaev et al. 2005). The exposure of HPBLs to 890 and 915 MHz simulated GSM EMF (SAR 37 mW/kg) for 1 h led to a significant chromatin condensation followed by the suppression of 53BP1 and increased γH2AX foci formation[*] (Markovà et al. 2005). Exposure of HPBLs to an 800 MHz (SAR 2.9 or 4.1 W/kg) continuous-wave RF EMF for 72 h significantly elevated chromosome aneuploidy (abnormal number of chromosomes) (Mazor et al. 2008). An increase in aberrant cells and chromosome exchanges were detected in the HPBLs exposed for 24 h to 1.95 GHz (SAR 0.5 and 2.0 W/kg) simulated UMTS (3G) MT EMF (Manti et al. 2008).

Exposure of HPBLs for 1 h to 905 MHz or 915 MHz simulated GSM EMF (SAR 37 mW/kg) or 1947.4 MHz simulated UMTS EMF (SAR 40 mW/kg) led to a decline in 53BP1 synthesis, an increase in γH2AX DNA repair foci, and an increased chromatin condensation (Belyaev et al. 2009). Application of combined a 3 T static magnetic field and 128 MHz EMF (SAR < 3 W/kg) from a magnetic resonance imaging (MRI) scanner for 22, 45, 67, and 89 min to HPBLs significantly increased the frequency of DNA single-strand breaks (SSB), micronuclei, aberrant cells, and both chromosome and chromatid deletions (breaks) and exchanges (Lee et al. 2011).

HPBLs subjected to repetitive 1 h exposure to 916 MHz (20 dBm, 0.1 W or 30 dBm, 1 W) simulated GSM EMF for 1–8 h, with a gap of 1 h between each 1 h exposure, showed an increase in chromosome aberrations depending on the total duration of the exposure. The maximum increase in chromosomal aberrations was detected for 7 h total exposure at a power of 1 W (Shah et al. 2015). Exposure of HPBLs to 1923, 1947.47, and 1977 MHz UTMS EMF from 3G mobile phones for 1 and 3 h led to a significant rise in DNA damage (Gulati et al. 2020).

The frequency of chromatid gaps and chromatid breaks increased in HPBLs from six healthy donors activated for mitosis and exposed *in vitro* during the G$_2$ or early M-phase of the mitotic cycle to UMTS EMF from a mobile phone in "talk" mode for 15 min at 1 cm distance (average peak power density 92 ± 27 μW/cm^2). The number of chromatid-type aberrations increased in the exposed blood samples from 100% up to 275% with regard to the control blood samples (Panagopoulos 2019b). In another set of experiments, the effect of UMTS MT EMF was compared with the effect of a high caffeine dose (approximately (~) 290 times higher than the permitted single dose for an adult human) on the HPBLs of the same healthy subjects. It was found that the UMTS exposure, ~136 times below the ICNIRP (2020) limits, induced even more aberrations than the high caffeine dose of ~290 times above the corresponding caffeine limits. The combined effect of UMTS and caffeine

---

[*] A sign of DNA double-strand breaks.

significantly exceeded the sum of individual effects in most subjects and increased with increasing duration of exposure to the UMTS EMF (Panagopoulos 2020).

Human exposure in vivo: Occupational exposure of 24- to 45-year-old male individuals to RF EMR with frequencies ranging from 0.2 MHz to 26 GHz and power densities ranging from 0 to 20 mW/cm^2 for 15 years increased micronuclei frequency in their peripheral blood lymphocytes (Fučić et al. 1992). The frequency of micronuclei increased significantly in the HPBLs of 12 individuals occupationally exposed to 1250–1350 MHz (10 μW/cm^2–20 mW/cm^2) RF EMF for 10–19 years (Garaj-Vrhovac 1999). A significant rise in the dicentric and ring chromosomes was detected in the HPBLs of 24 individuals who had been using GSM 900 MHz mobile phones for 2 years (Gadhia et al. 2003). The HPBLs from 26 air traffic controllers and 24 engineers occupationally exposed to airport radar EMFs for 8–27 years showed a significant rise in aberrant cells, chromatid breaks/gaps, hypodiploid and hyperdiploid cells[*] (Othman et al. 2003). The formation of micronuclei and induction of DNA damage (assessed by the comet assay method) increased in the HPBLs of 24 individuals who were using mobile phones for 1–5 years (Gandhi and Anita 2005). Occupational exposure of 10 individuals (45–56 years old) to 1250–1350 MHz EMF (10 μW/cm^2–20 mW/cm^2) from a radar antenna for 24.3 years significantly increased DNA damage (comet assay tail length) and chromatid breaks in their peripheral blood lymphocytes (Garaj-Vrhovac and Orescanin 2009).

Exposure to GSM 900 MHz mobile phone EMF for 15 or 30 min significantly increased the DNA SSB in the hair root cells of all eight mobile phone users who participated in the study, and the damage was greater for 30 minute use than for 15 minute use (Çam and Seyhan 2012).

The exfoliated buccal mucosa cells from 85 mobile phone users (SAR 0.34–0.95 W/kg) after 2.35 years average time of use showed a significant rise in the micronuclei formation (Yadav and Sharma 2008). Micronuclei formation in exfoliated buccal mucosa cells was found to be significantly increased in men and women 19–33 years old who had used mobile phones for 4–14 years for 10–1260 minutes per week. The frequency of micronucleated cells was higher in individuals who talked on their mobile phones for more than 60 min per week in comparison to those who talked less (Daroit et al. 2015). The frequency of micronuclei was found to be significantly increased in the exfoliated buccal mucosa cells of "high mobile phone users" (who had been using the devices for more than 5 years and more than 10 h/week) compared to "low mobile phone users" (who had been using the devices for less than 5 years and less than 3 h/week), from a total of 95 3G/4G (CDMA) and 55 2G (GSM) mobile phone users. The number of micronuclei was significantly correlated with the degree of mobile phone use ("high" users had a significantly higher number of micronuclei) (Banerjee et al. 2016). The DNA damage (assessed by the comet assay) increased significantly in the hair follicle cells of the ear canal of 56 men using mobile phones (SAR 0.45–0.97 W/kg) for 0–30, 30–60, or >60 min/day for 10 years, at degrees increasing with the daily duration of exposure (Akdag et al. 2018). The use of mobile phones by individuals for more than 2–3 h/day for a minimum of 4 years resulted in a significant increase in the frequency of micronuclei in their buccal mucosa cells (Rashmi et al. 2020). Mobile phone (SAR 0.244–1.552 W/kg) use by 45 male and 55 female students (18–30 years old) for 5 to 10 years for up to 30 min or up to 60 min per day increased DNA damage (comet tail length) in their buccal mucosa cells. The DNA damage was higher in the group of heavier use. DNA fragmentation was also found to be increased non-significantly in the buccal mucosa cells, as assessed by the TUNEL (terminal deoxynucleotide transferase-mediated dUTP nick end-labeling) assay, in individuals using mobile phones for 30–60 min/day compared to those using mobile phones for less than 30 min/day (Khalil et al. 2020).

Exposure from MT base antennas: 18 males and 22 females residing (≥5 years) in the vicinity of MT base stations (within 80 m) transmitting at 900 MHz and 1800 MHz GSM EMFs/EMR (average power density 0.5 ± 0.02 μW/cm^2) showed a significant rise in micronuclei formation and LPO and a decrease in GSH concentration, CAT, and SOD activities in their peripheral blood lymphocytes,

---

[*] Hypodiploid and hyperdiploid are cells with a lower or higher number of chromosomes than the physiological 46 diploid human chromosomes.

with respect to control population residing more than 300 m away from the base station antennas (average power density 0.0035 ± 0.0002 µW/cm^2) (Zothansiama et al. 2017).

The buccal mucosa cells of 116 individuals living close to GSM 1800 MHz MT base stations (within a distance of 150 m) (highest power density measured 1.22 µW/cm^2) for 12–13 h/day for 8–9 years showed an increased frequency of micronuclei and DNA damage (comet tail moment) that increased with increasing power density compared to 106 healthy control individuals residing at distances > 800 m. The frequencies of micronuclei were higher in the individuals exposed for ≥9 years than in those exposed for less than 9 years. The genetic damage did not correlate with GST-mu 1 (GSTMI) and GST-theta 1 (GSTTI)* genes polymorphism† (Gulati et al. 2016). Exposed individuals showed a decline in manganese superoxide dismutase (Mn-SOD) and CAT activities and a rise in LPO in their serum compared to control subjects (Gulati et al. 2018). The frequency of micronuclei was found to be significantly increased in the buccal mucosa cells of 108 individuals (5–50 years of age) residing within 10–25 meters of MT base stations for 5 years. The frequency of micronuclei was higher in the younger (5–20 years old) participants than in the adults (21–50 years old) and was higher in females than in males. Power density levels of the WC EMF exposure were not reported in the study (Thamilselvan et al. 2021).

Other types of human cells in vitro: The intermittent (5 min on/10 min off) or continuous exposure of human diploid skin fibroblasts ES1 and rat granulosa GFSH-R17 cells to two different emission modes (GSM "basic" or GSM "talk")‡ of GSM 1800 MHz (SAR 1.2 or 2 W/kg) simulated mobile phone EMF for 4, 16, or 24 h induced DNA SSBs and double-strand breaks (DSBs) in both cell types and by both emission modes. Moreover, the intermittent exposure induced a stronger effect than the continuous exposure (Diem etal. 2005). Exposure of human lens epithelial (SRA01/04) cells to 1.8 GHz (217 Hz amplitude-modulated) simulated GSM EMF with SAR 1, 2, 3, or 4 W/kg for 2 h (5 min on/10 min off) significantly elevated ROS and DNA SSBs (at SAR 3–4 W/kg) but did not increase γH2AX foci (Yao et al. 2008). A rise in DNA fragmentation, accompanied by ROS formation and expression of 8-hydroxy-2′-deoxyguanosine (8-OH-dG)§ has been reported in human spermatozoa exposed to 1.8 GHz (SAR 0.4–27.5 W/kg) RF EMF for 16 h. The effect increased with increasing SAR (De Iuliis et al. 2009). Cultured human buccal epithelial cells exposed to simulated GSM 900 MHz MT EMF (SAR 0.5–1.1 W/kg) for 1, 2, 5, 10, 15, 30, and 60 min *in vitro* increased the heterochromatin¶ granules (indicating lower transcription activity) depending on the duration of exposure (Skamrova et al. 2013).

Exposure of Chinese hamster lung (CHL) cells, primary rat astrocytes, human amniotic epithelial cells, SRA01/04, human skin fibroblasts (HSFs), and human umbilical vein endothelial cells (HUVECs) to simulated GSM 1800 MHz EMF (SAR 3.0 W/kg) for 1 or 24 h intermittently (5 min on/10 min off) increased the γH2AX foci formation in CHL and HSFs cells only, whereas the other types of cultured cells did not show any detectable response (Xu et al. 2013). A significant elevation in the chromosome fragility, chromatid breaks and gaps, and chromosome condensation with increasing exposure time was observed in the human amniocytes (FCs) exposed to simulated GSM 900 (268.4 mW/m^2 = 26.84 µW/cm^2) and 1800 MHz (520 mW/m^2 = 52 µW/cm^2) EMFs (SAR 1.3 W/kg) continuously for 12 days for 3, 6, or 12 h daily (Uslu et al. 2019).

Exposure of MOLT-4 cells to Integrated Digital Enhanced Network (iDEN) mobile phone EMF 813.5 MHz (8 mW/cm^2, SAR 24 mW/kg) for 1, 1.67, and 10.67 h, increased the DNA damage at 2 or 21 h post-exposure, whereas exposure of MOLT-4 cells to TDMA (8 mW/cm^2, SAR 26 mW/

---

* GST family genes are involved in the detoxification of carcinogens, therapeutic drugs, environmental toxins, and OS products by conjugating with glutathione in humans.
† Gene polymorphism is variants of a gene's DNA sequence.
‡ GSM "basic" refers to GSM emission while speaking on the phone, and GSM "talk" refers to combined speaking and listening emission. In GSM, intensity increases during speaking.
§ A marker for oxidative damage to DNA.
¶ Heterochromatin is a more condensed part of the eukaryotic chromatin involved in gene expression silencing and restricting DNA replication and repair. In the nucleus, heterochromatin remains separately from euchromatin and is localized toward the nuclear periphery and surrounding the nucleolus.

kg) EMF decreased DNA damage at 2 h post-exposure (Phillips et al. 1998). The ROS and 8-OH-dG DNA adduct[*] formation increased significantly in human leukemia HL60 cells exposed to 900 MHz (120 µW/cm^2) EMF for 4 h/day for 5 days. Moreover, the exposure reduced mitochondrial (mt) transcription factor A, mtDNA polymerase-γ, mtDNA transcripts, mtDNA copy number, and ATP (Sun et al. 2017). U87 (p53 proficient) glioblastoma cells and U251 (p53 deficient) glioblastoma cells exposed to UMTS 1950 MHz MT EMF (SAR 0.25, 0.5, and 1 W/kg) intermittently (5 min on/10 min off) for 16 h displayed DNA damage at 0.5 and 1 W/kg (Al-Serori et al. 2018).

4.2.1.2 Animal Cell (In Vitro or In Vivo) Studies

Mammalian cells exposed in vitro: Primary newborn rat cortical neurons exposed to simulated GSM 1800 MHz (pulsed at 217 Hz) EMF (SAR 2 W/kg) for 24 h intermittently (5 min on/10 min off) showed a significant rise in ROS production, 8-OH-dG DNA adducts in the mitochondria, and a decline in the mtDNA copy number and mtDNA transcript (Xu et al. 2010). Intermittent exposure (5 min on/10 min off) of mouse spermatocyte-derived cell line to 1800 MHz GSM (SAR 1, 2, or 4 W/kg) EMF in "talk" mode for 24 h increased ROS generation, DNA base damage (as determined by formamidopyrimidine comet assay), and 8-OH-dG DNA adducts (flow cytometry assay), especially at 4 W/kg (Liu et al. 2013a). Exposure of mouse spermatocyte-derived GC2 cells to GSM 900 MHz EMF from a mobile phone intermittently (for 1 min every 20 min) for 24 h, during different operation modes – ["dialing" mode (with average power density 12.20 ± 1.84 µW/cm^2), "dialed" mode (with average power density 11.69 ± 2.17 µW/cm^2), or "listening" mode (with average power density 5.90 ± 1.16 µW/cm^2)], – resulted in DNA damage as estimated by the comet assay with all modes. The DNA damage was higher with the dialing and dialed modes than with listening mode in consistency with the higher power density (Liu et al. 2013b). Intermittent exposure (5 min on/10 min off) of mouse spermatogonia derived GC2 cell line to 1800 MHz EMF simulated GSM in "talk" mode (SAR 1, 2, and 4 W/kg) for 24 h increased DNA base damage (studied by alkaline comet assay) at 4 W/kg SAR (Duan et al. 2015). Exposure of mouse spermatogonial GC1 cells, spermatocyte GC2 cells, and epididymal spermatozoa, *in vitro* to continuous wave 1.8 GHz (SAR 0.15 or 1.5 W/kg) RF EMF for 4 h resulted in a significant rise in ROS formation in GC1 and GC2 cells. The higher SAR exposure induced DNA fragmentation (assessed by the comet assay) due to oxidative damage (as determined by the 8-OH-dG oxidative DNA damage assay) in the spermatozoa (Houston et al. 2018). The intermittent exposure of GC2cells to 1800 MHz simulated GSM EMF (SAR 4 W/kg) for 24 h (5 min on/10 min off) induced DNA damage (DNA SSBs and DSBs), ROS, and autophagic cell death (Li et al. 2018).

Rat astrocytes exposed to 900 MHz 0.26 W/m^2 (26 µW/cm^2) EMF, either non-modulated continuous-wave or amplitude modulated, by a 50 Hz sinusoidal signal for 5, 10, or 20 min, exhibited a significant rise in ROS generation and DNA base damage only after the modulated 20 min exposure, while the continuous wave RF alone was ineffective (Campisi et al. 2010). Mouse brain cells exposed to 10.7 GHz (SAR 0.725 W/kg) EMF for 6 h/day for 3 days exhibited an 11-fold rise in the micronuclei and a twofold increase in apoptotic cells. Moreover, the exposure reduced the expression of signal transducer and activator of transcription (STAT-3) gene (Karaca et al. 2012). Mouse neuroblastoma neuro-2a cells exposed to GSM 900 MHz EMF ("talk" signal) for 24 h (SAR 0, 0.5, 1, or 2 W/kg) exhibited a significant rise in ROS generation, DNA base damage, DNA SSBs, and DNA DSBs with the 2 W/kg exposure (Wang et al. 2015). The DNA strand breaks (comet tail) increased significantly in canine blood and HPBLs exposed to 123.9 kHz EMF or to 250.8 kHz EMF (0.1 or 0.79 mT) for 20 h (Brech et al. 2019).

Exposure of rat kangaroo RH16 cells to 2450 MHz (SAR 15.2 ± 1.82 W/kg) EMF for 320 days (50 passages)[†] increased chromosomal aberrations (isochromatid breaks, dicentrics, rings, gaps, and number of polyploid cells[‡])after 125 days of exposure or more (Yao 1982). Exposure of V79 cells

---

[*] From the four DNA bases, guanine is the most sensitive to ROS attacks. The hydroxyl free radicals (OH•) interact with guanine to form 8-OH-dG adducts that serve as a biomarker of oxidative DNA damage.

[†] Passage is the term used for the subculture of the cells taken from the original vessel by detaching with trypsin and transferring the cells to a new vessel containing fresh media for further processing.

[‡] Cells with more than one pair of (homologous) chromosomes are called polyploid.

to 7.7 GHz EMF (30 mW/cm^2) for 15, 30, and 60 min inhibited DNA synthesis and induced DNA damage (structure alterations). EMF exposure also induced chromatid and chromosome breaks, acentric fragments, and dicentric chromosomes with 15 and 30 min exposure, whereas ring chromosomes, symmetrical interchanges, and tetraploid* V79 cells were detected with 60 min exposure (Garaj-Vrhovac et al. 1990). Power density-dependent decline in the colony formation of V79 cells exposed to 7.7 GHz (0.5, 10, and 30 mW/cm^2) EMF for 15, 30, or 60 min was observed. The exposure to 30 mW/cm^2 induced micronuclei formation, chromosome breaks, dicentrics, and acentric fragments to a degree increasing with the exposure duration (Garaj-Vrhovac et al. 1991). A significant increase in micronuclei formation was detected in CHO-K1 cells after 18 h exposure to 2.45 GHz EMF (SAR 25, 100, or 78 W/kg) with the higher SAR exposures (78 and 100 W/kg) (Koyama et al. 2003).

Mammals exposed in vivo: Whole body exposure of Sprague Dawley rats to 2450 MHz pulsed (500 Hz pulse repetition) EMF (1 or 2 mW/cm^2, SAR 0.6 and 1.2 W/kg), or continuous-wave (2 mW/cm^2, SAR 1.2 W/kg) for 2 h, increased DNA SSBs in the hippocampus and other parts of the rat brain (Lai and Singh 1995). In another study, whole body exposure of Sprague Dawley rats to 2450 MHz pulsed (500 Hz) or continuous wave (2 mW/cm^2, SAR 1.2 w/kg) for 2 h, increased the DNA SSBs and DNA DSBs in the rat brain (Lai and Singh 1996). Latvian cows staying near to and in front of the Skrunda Radio Station (installed at Venta River Valley, Skrunda, Latvia) transmitting at 154–164 MHz (peak power 1.25 MW from each of the two antennas) for a minimum of 2 years had a significant rise in micronuclei in their peripheral blood erythrocytes (Balode 1996). Separate exposure from four different mobile phone handsets (SAR 0.9–1.18 W/kg) 3 h twice daily (with 30 min gap) for 18 weeks reduced the fertilizing capacity of rat spermatozoa and induced cell death in sperm cells (Yan et al. 2007). Exposure of rats to 200 MHz pulsed EMF 50 W/m^2 (5 mW/cm^2) for 1 h/day for 30 days decreased sperm quality, reduced serum testosterone level, and increased apoptosis in testes due to increased Bax/Bcl2 expression ratio and increased cleaved caspase 3 and caspase 3 activities† (Guo et al. 2019).

A significant increase in 8-OH-dG DNA adducts and LPO was observed in the brains of non-pregnant and pregnant New Zealand white rabbits exposed to a simulated 1800 MHz GSM EMF for 15 min/day for 7 days. No effect was detected in the newborn rabbits exposed *in utero* (Guler et al. 2010). Exposure of pregnant and newborn infant rabbits in similar conditions to 1800 MHz GSM-like EMF triggered apoptosis in the brain, eyes, heart, lung, liver, spleen, and kidneys studied by the TUNEL assay (Guler et al. 2011). The 8-OH-dG DNA adduct formation and LPO increased in rabbits exposed to 1800 MHz GSM-like EMF 15 min/day *in utero*, during Days 15–22 of gestation, and also exposed after one month of birth (Güler et al. 2012). A change in caspase 1, caspase 3, and caspase 9 activities was reported in infant rabbits exposed to 1800 MHz GSM-like EMF (SAR 18 mW/kg) *in utero* during Days 15–22 of gestation for 15 min/day (Meral et al.2016).

Whole body exposure of Wistar rats to 2.45 GHz continuous wave EMF (5–10 mW/cm^2, SAR 1–2 W/kg) 2 h/day for 30 days significantly increased the number of micronuclei in their peripheral blood polychromatic erythrocytes (PCEs) after the eighth day of exposure (Trosic et al. 2002). Similarly, whole body exposure of Wistar rats to continuous wave 2.45 GHz EMF (5–10 mW/cm^2, SAR 1.25 ± 0.36 W/kg) for 4, 16, 30, or 60 h total exposure (2 h/day) increased the frequency of micronuclei in their bone marrow PCEs significantly on Days 8 (16 h) and 15 (30 h) of exposure (Trosic et al. 2004). The frequency of micronuclei increased in the bone marrow PCEs of rats exposed to continuous wave 910 MHz (SAR 0.42 W/Kg) EMF for 2 h/day for 30 consecutive days (Demsia et al. 2004). Sprague Dawley rats exposed to 2.45 GHz (SAR 0.48, 0.95, 1.43, 1.91, 2.39, 2.90, 3.40, 3.80, or 4.30 W/kg) EMF from a generator for 8 days, showed an elevation in the DNA SSBs in the blood leukocytes and in brain, lung, and spleen cells (Aweda et al. 2010).

Exposure of Wistar rats to 10 GHz continuous-wave EMF (214 µW/cm^2, SAR 14 mW/kg), or to 50 GHz (0.86 µW/cm^2, SAR 0.8 mW/kg,) continuous-wave EMF 2 h/day for 45 days resulted (in both

---

* Tetraploid are cells with four copies of each chromosome.
† Bax gene triggers apoptosis, whereas Bcl2 gene suppresses apoptosis. Caspases (cysteine-aspartic proteases) are proteolytic enzymes that are involved in apoptosis, and caspase 3 executes apoptosis.

cases) in an increase in ROS and micronuclei formation in the PCEs accompanied by a reduction in the PCE/NCE (normochromatic erythrocyte) ratio (probably showing that many PCE cells were driven to apoptosis) and reduction in SOD, CAT, and GPx activities in the serum (Kumar et al. 2010). Microwave irradiation of Fischer rats to 900 MHz (SAR 5.953×10^{-4} W/kg), 1800 MHz (SAR 5.835×10^{-4} W/kg), or 2450 MHz (SAR 6.672×10^{-4} W/kg) EMF for 2 h/day, 5 days a week for 30 days, increased the DNA strand breaks (comet tail) in the rat brain (Deshmukh et al. 2013). Wistar rats exposed to 2.45 GHz (0.21 mW/cm^2, SAR 0.14 W/Kg) EMF 2 h/day for 45 days exhibited a significant rise in DNA damage, ROS, xanthine oxidase, protein carbonyl contents, LPO, and apoptosis, followed by a decline in lactate dehydrogenase (LDH) and testosterone levels in the testes (Meena et al. 2014). Exposure of male Wistar rats to pulsed 2.45 GHz EMF with 217 Hz pulsations (SAR 0.02 W/kg, 0.104 μW/cm^2) for 1 h/day for 30 days significantly increased the 8-OH-dG DNA adducts in the brain and in the blood and increased advanced oxidation protein products (AOPP) in the blood (Gürler et al. 2014).

Exposure of pregnant rats during gestation Days 0–20 to 1800 MHz GSM-like EMF for 2 h/day for 1, 2, or 3 weeks reduced the number of live embryos and increased the number of dead and resorbed embryos, malformed, and edematous fetuses. Specifically, the third week of exposure resulted in congestion, hematoma, malformation, short tail, growth retardation in the embryos, and reduced GPx in the pregnant rat serum (Alchalabi et al. 2016). Whole body exposure of Sprague Dawley rats to GSM 1800 MHz (SAR 0.974 W/kg) EMF for 2 h/day for 15, 30, or 60 days resulted in a significant rise in the serum 8-OH-dG DNA adducts and LPO with the 30 and 60 days exposure. The number of micronuclei increased with 60 days of exposure accompanied by a significant decline in the GPx with 30 and 60 days of exposure (Alchalabi et al. 2017). Sprague Dawley rats exposed to 900 MHz (SAR 0.0845 W/kg), 1800 MHz (SAR 0.04563 W/kg), or 2100 MHz (SAR 0.03957 W/kg) simulated GSM mobile phone EMFs for 2 h/day for 6 months showed higher DNA damage, oxidative damage, and increased LPO in the frontal lobe of the brain. The study concluded that EMFs emitted by mobile phones may cause oxidative damage, induce an increase in LPO, and oxidative DNA damage in the frontal lobe of the rat brain tissues (Alkis et al. 2019).

Exposure of male Wistar rats to 1800 MHz EMF (11.6 μW/m^2, SAR 0.433 W/kg) for 4 h/day, 5 days a week for 90 days, increased DNA damage (assessed by the comet assay), the inflammatory index interleukin-6 (IL6), and the tumor necrosis factor (TNF-α) in the rat brain. The EMF exposure decreased the GSH, GST, GPx, glucose-6-phosphate dehydrogenase,* and acetylcholinesterase levels, followed by an increase in GSSG in the rat brain, indicating OS (Sharma et al. 2020).

Exposure of Wistar rats to 2115 MHz simulated (UMTS) mobile phone EMF (SAR 1.15 W/kg, 1 mW/cm^2) for 8 h increased ROS, LPO, DNA damage, and reduced cell proliferation in the rat brain (Singh et al. 2020). The exposure of Sprague Dawley rats to 1800 MHz (SAR 0.62 W/kg, 0.127 mW/cm^2) or 2100 MHz (SAR 0.2 W/kg, at 0.038 mW/cm^2) GSM-like pulsed EMFs for 2 h/day for 7 months significantly increased 8-OH-dG DNA adducts, DNA strand breaks (increased comet tail/moment intensity), LPO, and total oxidants and reduced the total antioxidants in the rat liver (Alkis et al. 2021).

The formation of micronuclei increased significantly in PCEs of 2 days old newborn rats exposed *in utero* to 834 MHz (SAR 0.55–1.23 W/kg, 0.2 and 0.4 mW/cm^2) simulated mobile phone EMF from a "test" mobile phone for 8.5 h/day until birth (Ferreira et al. 2006). Exposure of sexually mature and immature rats to 900 MHz simulated GSM (RF electric field strength 19.6 V/m, or SAR 0.37, and 16.9 V/m or SAR 0.49 W/kg) EMF for 2 h/day for 45 days increased chromosome aberrations and micronuclei formation and reduced mitotic index and PCE/NCE ratio (Şekeroğlu et al. 2012). A similar effect was observed for SAR 0.38–78 W/kg or 28.1 ± 4.8 V/m RF electric field strength, and SAR 0.31–52 W/kg, or 20.0 ± 3.2 V/m for immature and mature rats respectively (Şekeroğlu et al. 2013).

Exposure of pregnant Sprague Dawley rats to 900 MHz EMF (26.5 μW/cm^2) during gestation Days 13–21 for 1 h daily, significantly elevated the apoptotic index, LPO, and 8-OH-dG DNA adducts in the testes of the 21-day-old progeny rats that were exposed *in utero* (Hanci et al. 2013). Sprague Dawley rats exposed to 1800 MHz simulated GSM EMF (SAR 1.0 W/kg, 15–20 V/m RF

---

* Important enzyme that protects red blood cells from destruction and neurons from ROS attack.

electric field strength) for 2 h showed a significant rise in the 8-OH-dG DNA adducts in their urine at 0.5 h post-exposure that reached a maximum at 1 h post-exposure and declined at 2 h post-exposure (Khalil et al. 2012). Exposure of differentiated pheochromocytoma (PC12) cells and Wistar rats to 2.856 GHz (average power density 30 mW/cm^2) EMF for 5 and 15 min, respectively, led to DNA fragmentation with consequent apoptosis in PC12 cells and a significant loss in mitochondrial membrane potential (an additional apoptotic event). The EMF exposure increased the ratio of apoptosis-related proteins Bax/Bcl-2 and increased the expression of cleaved caspase-3, the cytochrome-*c* release, and the cleavage of caspase cascade substrate PARP* in both PC12 cells and hippocampal neurons of the rats 6 h post-exposure. Thus, the EMF exposure induced neural cell apoptosis via the classical mitochondria-dependent caspase-3 pathway (Zuo et al. 2014). Sprague Dawley rats exposed to 900 MHz GSM- or CDMA-pulsed (simulated 2G or 3G MT/WC EMF, respectively), (SAR 0, 1.5, 3, or 6 W/kg) *in utero*, from gestation Day 5 until 14 weeks after weaning (a total exposure time of 19 weeks) 18 h/day exhibited increased DNA damage in the hippocampus of the male rats after the CDMA (3G) exposure. The exposure of 5-week-old mice for 14 weeks to 1900 MHz, CDMA- (3G) or GSM-pulsed (2G) EMF (SAR 0, 3.5, 5, or 10 W/kg), elevated DNA damage in the brain cortex of male mice and in the leukocytes of female mice. A significant increase in micronuclei was also observed in bone marrow PCEs of mice exposed to 10 W/kg CDMA (Smith-Roe et al. 2020). The results demonstrate that both MT types (2G, 3G) cause DNA damage but also suggest that 3G may be even more bioactive than 2G, as was also reported by D'Silva et al. (2017) (see below).

The exposure of CBA/CAY male mice to 2.5 GHz EMF (SAR 0.05, 0.5, 10, or 20 W/kg) for 30 min/day, 6 days a week, for 2 weeks increased the yield of chromosomal translocations in the meiotic metaphase with increasing SAR (Manikowska-Czerska et al. 1985). Exposure of Swiss albino mice to 2.45 GHz (1 mW/cm^2) continuous wave RF EMF for 2 h/day for 120, 150, and 200 days resulted in DNA rearrangement (mutations) in the brain and testes with all exposure durations (Sarkar et al. 1994). In another study, pKZ1 mice exposed to pulsed 900 MHz (with 217 Hz pulse repetition rate) simulated GSM EMF for 30 min/day for 1, 5, or 25 days exhibited an increase in DNA inversion events† in the spleens of mice exposed for 25 days (Sykes et al. 2001). Exposure of CD1 male mice to 900 MHz (SAR 90 mW/kg) simulated GSM EMF for 12 h/day for 7 days increased mitochondrial DNA damage and β-globin locus in epididymal sperms (Aitken et al. 2005). Exposure of Swiss albino male mice to 902.4 MHz simulated GSM EMF (0.250 W) (SAR 0.0054 W/kg and 0.0516 W/kg or power densities 28.4 and 271.7 μW/cm^2 respectively) for 3 h twice daily for 35 days resulted in a significant increase in DNA damage (assessed by the comet assay) in the spermatocytes and a significant increase in lipid peroxides in the testes. The GSH and SOD declined significantly after EMF exposure in the mouse testes (Pandey and Giri 2018). The whole body exposure of mice to 905 MHz (SAR 2.2 W/kg) RF EMF for 12 h/day for 1, 3, or 5 weeks caused an increase in ROS generation, DNA fragmentation, and 8-OH-dG DNA adducts in the spermatozoa, but it did not inhibit fertilization competence of the spermatozoa (Houston et al. 2019).

Bird embryos: Exposure of chick embryos to 2G (GSM) or 3G (UMTS) MT EMFs by a 2G/3G mobile phone device (SAR 0.310 W/kg) for 75 min every 24 h for 12 days increased DNA DSBs in the liver at 9–12 days after the beginning of exposure. The MT EMFs increased karyorrhexis in the embryo liver. The histological examination of chick embryo liver after the MT EMF exposure revealed marked dilation of sinusoids and hemorrhage, whereas the hepatocytes showed cytoplasmic vacuolization and nucleomegaly (enlarged nuclei). The UMTS (3G) exposure was found to be even more genotoxic than GSM (2G) under identical conditions and from the same device (D'Silva et al. 2017). The same research group also reported increased DNA damage in the chick embryo brains under similar conditions to their previous study with 2G and 3G MT EMF emitted by a mobile phone. Again, the effects were more pronounced with the UMTS (3G) than with the GSM (2G) exposure from the same device (D'Silva et al. 2021).

---

* The function of various genes and related proteins is explained in Appendix 1.
† DNA inversion is when a section of the DNA breaks away and reattaches in reversed order.

Exposure of Japanese quail embryos to GSM 900 MHz mobile phone EMF (0.25 μW/cm^2) intermittently (48 s on/12 s off) for 158–360 h, increased nitric oxide and superoxide free radicals in the embryos. Moreover, the EMF exposure significantly increased LPO and 8-OH-dG DNA adducts followed by a decline in CAT and SOD activities in the embryos (Burlaka et al. 2013). Exposure of Japanese quail embryos *in ovo* to GSM 1800 MHz (0.32 μW/cm^2) EMF from a mobile phone for 19 days (before and during incubation) (48 s on/12 s off) significantly increased DNA SSBs and DSBs (studied by comet assay) and doubled the 8-OH-dG DNA adducts in 1-day-old embryos. The mobile phone EMF exposure doubled the SOD activity and the nitric oxide levels and significantly increased LPO in the brain, heart, and liver cells in the 10-day-old embryos (Yakymenko et al. 2018).

Drosophila melanogaster: The death of follicle cells, nurse cells, and oocytes was induced in a great number of egg chambers (up to 60%) due to DNA fragmentation (TUNEL assay) in newly eclosed female fruit flies exposed to GSM 900 MHz (0.402 ± 0.054 mW/cm^2) or GSM 1800 MHz (0.288 ± 0.038 mW/cm^2) mobile phone EMF in "talk" mode for 6 min/day once daily, for 6 days. The flies were dissected on the sixth day 1 h after the last exposure. The DNA fragmentation was observed for the first time not only at the two most sensitive stages of oogenesis (germarium and Stages 7–8) but in all developmental stages and in all three types of egg chamber cells (the nurse cells, the follicle cells, and the oocyte). The DNA fragmentation and cell death in the reproductive cells explained the severe reproductive decline found in previous experiments by the same authors to be induced after similar exposure to these EMFs (Panagopoulos et al. 2007). In another series of experiments by the same group, the flies were exposed to the same types of GSM EMFs at different distances from the mobile phone simultaneously (0, 1, 10, 20, 30, 40, 50, 60, 70, 80, 90, and 100 cm) with corresponding power densities ranging from 0.378 ± 0.059 mW/cm^2 (GSM 900) or 0.252 ± 0.050 mW/cm^2 (GSM1800) to less than 0.001 mW/cm^2 (1 μW/cm^2). It was shown that DNA fragmentation and reproductive decline with 6 min exposures start from the weakest exposure (less than 1 μW/cm^2 or 100 cm distance) and become maximum at 0 cm distance from the mobile phone and also within a "window" 20–30 cm from the device (at power density ~ 10 μW/cm^2) (Panagopoulos et al. 2010).

The daily exposure of *Drosophila melanogaster* to GSM 900 MHz EMF from a mobile phone (0.35 ± 0.07 mW/cm^2) for 6 min daily (continuously or intermittently) for 6 days induced DNA fragmentation in the egg chamber cells during early and mid-oogenesis, reduced fecundity, and induced damage in the actin–cytoskeleton network of the egg chambers. The effect of continuous and intermittent exposures with a gap of 10 min between each exposure was equally effective for the same total duration, whereas longer intervals between the exposures allowed some degree of recovery from the effects of the GSM exposure (Chavdoula et al. 2010). Similarly, a few min daily exposure of newly eclosed virgin female flies with developing ovaries at different developmental stages to an active GSM mobile phone in "talk" mode increased DNA fragmentation in ovarian cells, leading to retardation of the ovarian growth in the young virgin female flies (Panagopoulos 2012).

4.2.2 ELF EMF and Static Field Studies

4.2.2.1 Human Cell (In Vitro or In Vivo) Studies

Human cells in vitro: Exposure of HPBLs to 50 Hz sinusoidal magnetic field (15 μT) for 2 h significantly increased chromatin condensation (measured by AVTD) and inhibited tumor-suppressor p53-binding protein 1 (53BP1) in normal subjects as well as in subjects reporting hypersensitivity to EMFs (Belyaev et al. 2005). HPBLs exposed to 50 Hz electric pulses 4×10^5 V/m for 120 min showed a significant rise in DNA damage (determined by the comet assay) (Delimaris et al. 2006). Exposure of HPBLs from six female and six male volunteers to 50 Hz ELF magnetic field for 1 h at 2, 3, 5, 7, or 10 mT induced DNA damage which increased with increasing magnetic field strength. The induced DNA damage was higher in females than in males (Ahuja et al. 1997, 1999). The HPBLs from 29 male electric power plant utility workers of average age 36.03 ± 3.15 years, exposed to 50/60 Hz (4–50 μT) EMF for 9.86 ± 2.99 years, showed increased DNA damage (assessed by the comet assay) compared to the control population (Zendehdel et al. 2019).

Intermittent (15 s on/15 s off, or 2 s on/20 s off) exposure of cultured human amniotic cells to 50 Hz (30 µT sine wave signal) ELF EMF for 72 h significantly increased the frequency of chromatid and chromosome aberrations (Nordenson et al. 1994). Exposure of human diploid skin fibroblasts to continuous or intermittent 50 Hz sinusoidal (0.02–2 mT) ELF EMF for 24 h (1, 3, 5, 10, 15, or 25 min on/10 min off) increased DNA SSBs and DSBs (evaluated by alkaline and neutral comet assays), depending on the field strength. The intermittent exposure of 5 min on and 10 min off was the most effective in triggering DNA strand breaks (Ivancsits et al. 2002). The exposure of human diploid skin fibroblasts (ES-1, IH-I, and KE-1) to intermittent (5 min on/10 min off) 50 Hz sinusoidal (0.02–2 mT) ELF EMF for 1–24 h increased DNA SSBs and DNA DSBs. At 15 h after termination of the exposure, the DNA damage (comet tail factors) started returning to normal levels and this process was completed within 9 h, showing a repair capability of the cells in the absence of the disturbance. The effect manifested at magnetic field strength as low as 35 µT and increased with increasing field strength, being well below proposed ICNIRP (1998) guidelines. The effect also increased with increasing exposure time, reaching a maximum by 15–19 h of exposure (Ivancsits et al. 2003). Human fibroblasts, human melanocytes, human lymphocytes, human monocytes, human skeletal muscle cells, and rat granulosa cells exposed to 50 Hz sinusoidal (1 mT) ELF EMF for 1–24 h (5 min on/10 min off) showed a differential response. The human fibroblasts, human melanocytes, and rat granulosa cells showed an increase in the DNA SSBs and DNA DSBs, whereas the human lymphocytes, human monocytes, and human skeletal muscle cells did not show that much of an increase, indicating that all cells do not respond equally to the same stressor (Ivancsits et al. 2005).

Exposure of ES-1 human diploid fibroblasts to 50 Hz (1 mT) sinusoidal ELF EMF (5 min on/10 min off) for 2–24 h resulted in a significant rise in the acentric fragments, chromosome gaps and breaks, ring, and dicentric chromosomes. The frequency of micronuclei increased with exposure time and a significant difference was observed after 10 h of exposure (Winker et al. 2005). Significant induction of DNA fragmentation (evaluated by the comet assay) was reported in HR-1d, ES-1, MRC-5 human primary fibroblasts that were exposed intermittently (5 min on/10 min off) to 50 Hz (1 mT) ELF EMF for 15 h (Focke et al. 2010). Lipopolysaccharide[*] (LPS)-stimulated (for mitosis) RAW-246 macrophage cells exposed to 50 Hz sinusoidal (0.5 mT) ELF EMF for 24 h showed increased DNA SSBs (comet assay), whereas cells of the same type that were not stimulated for mitosis did not show any significant effect (Nakayama et al. 2016).

Repetitive exposure of IMR90 (human lung fibroblasts) and HeLa cells to 60 Hz ELF EMF (6 mT) for 30 min/day for 3 days significantly increased γH2AX foci (which denote DSBs) and checkpoint kinase 2 (Chk-2) protein, which is critical for protection against DNA damage (and possible mutations) by activating apoptosis through the p38 gene activation[†] (Kim et al. 2010). IMR90 and HeLa cells exposed to 60 Hz (7 mT) ELF magnetic field for 0, 10, 20, 30, 60, and 120 min showed a rise in the DNA DSBs (neutral comet and γH2AX assays) after 30 min of exposure in both cell types (Kim et al. 2012). Exposure of GBM U87 cells to 10, 50, or 100 Hz pulsed EMF with pulses of varying amplitude of 50 ± 10 G (10 and 50 Hz) or 100 ± 15 G (50 and 100 Hz) for 2, 4, and 24 h resulted in a decline in cyclin D1 and a rise in caspase activation and p53 expression after 24 h exposure, suggesting that the proliferation and apoptosis of human GBM cells are influenced by exposure to ELF EMFs (Akbarnejad et al. 2017).

Exposure of human SH-SY5Y neuroblastoma cells to a 50 Hz (0.1 mT) ELF EMF for 24 h increased significantly the number of micronuclei at 8 and 15 days after the exposure. The ELF EMF exposure also increased mitochondrial activity on Day 8 and LPO level on Day 15 after the exposure. The OS processes (ROS production, decrease in GSH level, and increase in mitochondrial superoxide level) were induced immediately after the exposure (Luukkonen et al. 2014). Human neuroblastoma SH-SY5Y cells exposed to 50 Hz (0.1 mT) ELF EMF for 24 h and observed for 45 days after the exposure revealed a significant rise in the frequency of micronuclei at 14 and 30 days

---

[*] A bacterial endotoxin which stimulates mitosis.
[†] The p38 pathway is the major signaling cascade of the mitogen-activated protein kinase (MAPK) signaling pathway that controls apoptosis and the release of cytokines by macrophages and neutrophils. It is activated in response to various stress stimuli.

but not 45 days after the exposure. The LPO level was decreased at 30 and 45 days after the exposure (Kesari et al. 2015). Exposure of SH-SY5Y cells to 50 Hz (0.5 mT) ELF EMF for 3 h increased the expression of 64 genes and decreased the expression of 40 genes, whereas increasing the power to 1 mT resulted in the increased expression (upregulation) of 86 genes and decreased expression (downregulation) of 65 genes (Rezaie-Tavirani et al. 2017).

Human breast adenocarcinoma MCF-7 cells and SH-SY5Y cells exposed to 50 Hz (0.25 and 0.50 mT) ELF EMF (5 min on/5 min off for 30 min, or 15 min on/15 min off for 30 min, or 30 min on continuously) exhibited downregulation of mRNA levels of seven genes (GADD45A, XRCC1, XRCC4, Ku70, Ku80, DNA-PKc, and LIG4) involved in DNA repair pathways, as measured by quantitative real-time PCR (Sanie-Jahromi et al. 2016). Human leukemia HL-60 cells, Rat-I fibroblasts, and WI-38 diploid fibroblasts exposed to 50 Hz (0.5–1.0 mT) ELF EMF for 24–72 h, showed a dose-dependent rise in DNA strand breaks and 8-OH-dG DNA adducts in all cell lines when examined at 24 and 72h post-exposure. The exposure led to an increase in cell proliferation rates, percentage (number) of S-phase cells undergoing cell division in all cell lines after 12 and 48 h of exposure, and increase in ROS formation 3–24 h post-exposure in Rat-I fibroblasts. The ELF EMF exposure also increased the activation of NFκB-related proteins (an indication of OS), which was maximum at 6 h after exposure and remained high up to 12 h and 24 h after exposure, compared to control cells (Wolf et al. 2005). Exposure of lung cancer A549 cells, gastric cancer SGC7901 cells, pancreatic cancer PANC-1, and G401 nephroblastoma cells to 50 Hz ELF EMF with average magnetic field intensity 5.1 mT for 2 h/day for 3 consecutive days inhibited tumor cell growth. In addition, the exposure increased ROS levels, apoptosis, the expression of cleaved caspase-3, and PARP (poly-ADP ribose-polymerase),* and the levels of LPO, DNA SSBs, and DNA DSBs (γH2AX foci) at 1, 2, and 3 days post-exposure. The ELF EMF also stimulated DNA repair pathways by elevating the expression of DNA-PKcs, LIG4, RAD9B, and BMI1 genes in A459 and G401 cells at Day 2 post-exposure (Yuan et al. 2020).

4.2.2.2 Animal Cell (In Vitro or In Vivo) Studies

Mammalian cells in vitro: Mouse m5S cells exposed to 5, 50, and 400 mT 60 Hz ELF magnetic field for 42 h showed a dose-dependent rise in sister chromatid exchanges with most significant increase observed in 400 mT (Yaguchi et al. 1999). Exposure of HaCaT cells to 60 Hz (1.5 mT) ELF for 96 h reduced cell proliferation by upregulating the cell-cycle-related gene CDKN1A and downregulating CDC25B, CDC20, CDC2, CCNA2, and CCNB1 genes, causing cell cycle arrest in G_1/G_0 phase of the cell cycle followed by increased phosphorylation of ataxia telangiectasia mutated (ATM) and Chk2 genes (Huang et al. 2014). Mouse spermatocytes-derived GC2 cells intermittently (5 min on/10 min off) exposed to 50 Hz ELF EMF (1, 2, or 3 mT) for 24 h exhibited a significant degree of DNA damage, especially with 3 mT ELF (γH2AX foci method) (Duan et al. 2015). Similarly, the intermittent (5 min on/10 min off) exposure of mouse spermatocytes-derived GC2 cells to 50 Hz (1, 2, or 3 mT) ELF EMF for 72 h was found to affect DNA methylation and gene expression (Liu et al. 2015). Exposure of mouse GC-1 and GC-2 cells to 2, 50, and 120 Hz ELF EMF continuously or intermittently (1s on/1s off) for 2 h induced superoxide and nitric oxide free radicals, NFκB expression, apoptosis, p53 activation, and downregulation of p21, Bcl2 genes, and caspase 3 activation, especially at 2 Hz (Solek et al. 2017). Exposure of Chinese hamster lung (CHL) cells to 50 Hz (0.4 mT) sinusoidal ELF EMF for 30 min or 24 h significantly increased LC3-II protein expression, autophagosome formation, and apoptosis, and induced a non-significant γH2AX foci formation (Shen et al. 2016).

Mammals exposed in vivo: Sprague Dawley rats exposed to 60 Hz ELF EMF with magnetic field strength of 0.1, 0.25, or 0.5 mT for 2 h showed a dose-dependent increase in DNA strand breaks (comet assay) at 4 h post-exposure in the brain cells. DNA SSBs were detected at all field intensities, whereas DNA DSBs could only be observed after exposure to 0.25 and 0.5 mT (Lai and Singh 1997). Similarly, the exposure of Sprague Dawley rats to 60 Hz (0.5 mT) ELF EMF for 2 h also induced DNA-protein cross-links† in the brain cells in the presence of proteinase-K (Singh and Lai 1998). The DNA SSBs

---

* a DNA repair enzyme.
† DNA-protein crosslinks are formed when a nucleotide residue in the DNA forms a covalent bond with a protein, and this binding affects DNA replication and gene expression.

and DSBs increased significantly in the brain cells of male Sprague Dawley rats exposed to 60 Hz (0.01 mT) ELF EMF for 24 or 48 h. The effect was greater with the 48 h exposure (Lai and Singh 2004). Wistar rats exposed to 50 Hz ELF EMF (average magnetic field intensity 0.97 ± 0.136 mT) for 50 and 100 days exhibited a significant elevation in the 8-OH-dG DNA adducts in the leukocytes and LPO in the blood plasma with much higher elevation at 100 days (Yokus et al. 2005). Adult male Wistar rats exposed to 128 mT static magnetic field for 1 h/day for 30 days showed increased 8-OH-dG DNA adduct formation and subsequent decline in CAT, GPx, mitochondrial SOD, and increase in LPO in the testes (Amara et al. 2006). Exposure of Wistar rats to 50 Hz (1 mT) ELF EMF for 4 h/day for 1 day or 45 days led to a significant rise in micronuclei in PCEs in the bone marrow and reduced mitotic index and PCE/NCE ratio significantly in the rats exposed for 45 days (Erdal et al. 2007). Whole body exposure of newborn Wistar rats (10 days old) to 50 Hz (0.5 mT) EMF ELF for 24 h daily for 30 days significantly induced DNA damage (assessed by comet assay) and increased LPO followed by a reduced SOD activity in the brain. The ELF EMF exposure also significantly increased micronuclei formation and mitotic index in the bone marrow cells of neonates (Rageh et al. 2012).

Exposure of CBA/Ca mice to 50 Hz (8 µT) ELF EMF for 11, 20, and 32 days resulted in significant DNA damage (assessed by the comet assay) in the brain cells. The ELF EMF exposure also resulted in a decline in the PCE/NCE ratio, number of leucocytes, and ornithine decarboxylase (ODC) activity in the testes (Svedenstål et al. 1999). Exposure of male mice to 50 Hz ELF EMF (1.5 mT) for 8 weeks induced unscheduled DNA synthesis,[*] mitochondrial DNA synthesis, and *in situ* nick translation[†] with DNA polymerase-I in the epithelial cells of the choroid plexus in the ventricles of the brain, indicating genotoxicity (Schmitz et al. 2004). CD1 mice exposed to 50 Hz 6.5 G (0.65 mT) ELF EMF continuously during the fetal period and until 3-day old neonates for 24 h/day for a total of 21 days exhibited an increase in micronuclei in the peripheral blood PCEs of the newborn mice but not in the adult mice (Udroiu et al. 2008). Exposure of CD1 mice to 50 Hz (1 mT) ELF EMF for 15 h/day for 1 or 7 days induced DNA damage (assessed by comet tail moment) in the cortex, hippocampus, and cerebellum of the brain and activation of heat shock protein (HSP-70) gene. The DNA damage showed repair in mice sacrificed 24 h after a 7-day exposure, whereas the HSP-70 expression remained unchanged (Mariucci et al. 2010).

Exposure of male CD1 mice to 50 Hz (0.1, 0.2, 1, or 2 mT) ELF EMF for 15 h/day for 7 days immediately induced DNA SSBs and DSBs in the cortex-striatum, hippocampus, and cerebellum of the brain with 1 and 2 mT. The exposure to 0.1 mT also increased the expression of HSP-70 in the hippocampus of the mouse brain (Villarini et al. 2013). Swiss albino mice exposed to 50 Hz, 2 G (200 ± 20 µT) ELF EMF (12 h on/12 h off) for 7, 14, 21, and 28 days exhibited a significantly increased number of micronuclei in the PCEs of the bone marrow (Alcaraz et al. 2014). Exposure of CD1 pregnant female mice and their embryos to 50 Hz (65 µT) ELF EMF from the twelfth day post-conception 24 h/day for a total of 30 days (including the neonatal period) resulted in an increase in the micronuclei in peripheral blood PCEs at birth and 11, 42, and 142 days after birth (Udroiu et al. 2015).

Drosophila: The exposure of newly eclosed adult *Drosophila* flies to 50 Hz alternating ELF EMF with magnetic field strength of 1 G (0.1 mT), 11 G (1.1 mT), or 21 G (2.1 mT) and corresponding induced electric field of 0.13, 1.43, and 2.72 V/m continuously for 5 days (120 h) triggered DNA fragmentation at the two most sensitive developmental stages (checkpoints) of oogenesis, the germarium, and the mid-oogenesis checkpoint (Stages 7–8 of oogenesis), studied by the TUNEL assay, and decreased reproduction. Both effects were dose-dependent, and they were more correlated with the induced electric than with the magnetic component of the EMF. The DNA fragmentation was observed in all three types of egg chamber cells (nurse cells, follicle cells, and in the oocyte) (Panagopoulos et al. 2013b).

The studies indicating that RF/WC EMFs or pure ELF EMFs induce genotoxicity in humans and various animal/cell systems are listed in Table 4.1.

---

[*] Unscheduled DNA synthesis repairs nucleotide excision in a single DNA strand and indicates that DNA has suffered damage.

[†] Nick translation is the process of replacing cold nucleoside triphosphates in a double-stranded DNA molecule with radioactive or fluorescence labeled nucleosides during DNA replication.

TABLE 4.1
Studies Reporting Genotoxic Effects of RF/WC or ELF EMFs in Various Biological Systems
A. RF/WC EMF Studies

S. No.	Biological System	EMF Exposure	Effect	Reference
1.	Rat kangaroo RH16 cells	2450 MHz, microwave generator, 320 days	Increased isochromatid breaks, dicentrics, rings, gaps, and polyploid cells	Yao et al. 1982
2.	CBA/CAY mice	2.5 GHz microwave generator, 30 min/day 6 days/week for 2 weeks	Increased translocations in meiotic metaphase	Manikowska-Czerska et al. 1985
3.	V79 cells	7.7 GHz, microwave generator, 15, 30, and 60 min	Decreased DNA synthesis, alteration in DNA structure, Increased chromatid and chromosome breaks, acentric fragments and dicentric chromosomes, ring chromosomes, symmetrical interchanges, and tetraploidy	Garaj-Vrhovac et al. 1990
4.	V79 cells	7.7 GHz, microwave generator, 15, 30, and 60 min	Increased micronuclei, chromosome breaks, dicentrics, and acentric fragments	Garaj-Vrhovac et al. 1991
5.	HPBLs	7.7 GHz, microwave generator, 10, 30, and 60 min	Increased micronuclei chromatid breaks, chromosome breaks, acentric fragments, and dicentrics	Garaj-Vrhovac et al. 1992
6.	V79 cells	7.7 GHz, microwave generator, 15, 30, and 60 min	Increased micronuclei, chromosome breaks, dicentrics, and acentric fragments	Garaj-Vrhovac et al. 1992
7.	HPBLs of occupational workers	0.2 MHz to 26 GHz, microwave generator, 15 years	Increased micronuclei	Fučić et al. 1992
8.	HPBLs	2450 MHz, microwave generator, 30 and 120 min	Increased acentric fragments, dicentric chromosomes and micronuclei	Maes et al. 1993
9.	Swiss albino mice	2.45 GHz, microwave generator, 2h/day for 120,150, or 200 days	Increased DNA rearrangement	Sarkar et al. 1994
10.	HPBLs	9 GHz, microwave generator, 10 min	Increased micronuclei	D'Ambrosio et al. 1995
11.	Rat brain	2450 MHz microwave generator, 2 h	Increased DNA SSBs	Lai and Singh 1995
12.	Rat brain	2450 MHz, microwave generator, 2 h	Increased DNA SSBs and DNA DSBs	Lai and Singh 1996
13.	Cows (blood)	154-164 MHz from radar station, 2 years	Increased micronuclei	Balode 1996
14.	MOLT-4 cells	813.5 MHz, iDEN mobile phone, 1, 1.67, or 10.67 h	Increased DNA damage	Phillips et al. 1998
15.	HPBLs of occupational workers	1250–1350 MHz, microwave generator, 10–19 years	Increased micronuclei	Garaj-Vrhovac, 1999

S. No.	Biological System	EMF Exposure	Effect	Reference
16.	HPBLs	2.45 and 7.7 GHz microwave generator, 15, 30, and 60 min	Increased micronuclei	Zotti-Martelli et al. 2000
17.	pKZ1 mice (spleen)	900 MHz simulated GSM, 30 min/day for 1, 5, and 25 days	Increased DNA inversion events	Sykes et al. 2001
18.	HPBLs	1.748 GHz, GSM pulsation, 15 min	Increased micronuclei	D'Ambrosio et al. 2002
19.	HPBLs	837 or 1909.8 MHz CDMA or TDMA or GSM or PCS, 3 or 24 h, SAR 5, and 10 W/kg	Increased micronuclei	Tice et al. 2002
20.	Wistar rats	2.45 GHz, microwave generator, 2 h/day for 30 days	Increased micronuclei	Trosic et al. 2002
21.	HPBLs	830 MHz continuous wave RF generator, 72 h	Increased aneuploidy of chromosome 17	Mashevich et al. 2003
22.	HPBLs of mobile phone users	935–960 MHz, GSM mobile phone, 2 years	Increased dicentric and ring chromosomes	Gadhia et al. 2003
23.	HPBLs of occupational workers	60.43–105.7% of ANSI* limits, airport radar, 8–27 years	Increased aberrant cells, chromatid breaks/gaps, hypodiploids, and hyperdiploids	Othman et al. 2003
24.	CHO-K1 cells	2.45 GHz, microwave generator, 18 h	Increased micronuclei	Koyama et al. 2003
25.	HPBLs	900 and 905 MHz, GSM mobile phone, 30 min and 1 h	Increased chromosome condensation	Sarimov et al. 2004
26.	Wistar rats	2.45 GHz, microwave generator, 2 h/day, 7 days/week for 4, 16, 30, and 60 h	Increased micronuclei	Trosic et al. 2004
27.	Rat bone marrow	910 MHz, RF generator, 2h/day for 30 days	Increased micronuclei	Demsia et al. 2004
28.	HPBLs	1800 MHz, microwave generator, 60, 120, and 180 min	Increased micronuclei with variation among individual donors	Zotti-Martelli et al. 2005
29.	HPBLs	915 MHz, simulated GSM, 2 h	Increased chromatin condensation and decreased 53BP1	Belyaev et al. 2005
30.	HPBLs	890 MHz, 915 MHz, simulated GSM, 1 h	Increased chromatid condensation and decreased 53BP1 and γH2AX	Markovà et al. 2005
31.	HPBLs of mobile phone users	800–2000 MHz, mobile phone use, 1–5 years	Increased micronuclei and DNA damage	Gandhi and Anita 2005
32.	ES1 fibroblasts GFSH-R17	1800 MHz, simulated GSM "basic", or GSM "talk" modes, 4/16/24 h	Increased DNA SSBs and DNA DSBs	Diem et al. 2005

(Continued)

TABLE 4.1 (CONTINUED)
Studies Reporting Genotoxic Effects of RF/WC or ELF EMFs in Various Biological Systems

S. No.	Biological System	EMF Exposure	Effect	Reference
33.	CD1 mice (sperm)	900 MHz, simulated GSM, 12 h/day for 7 days	Increased mitochondrial DNA damage and damaged β-globin locus	Aitken et al. 2005
34.	Exposed rat embryos, (examined neonate rats blood)	834 MHz, mobile phone, 8.30 h/day during embryogenesis until birth	Increased micronuclei	Ferreira et al. 2006
35.	Rats	900 MHz, GSM mobile phone, 3 h twice daily (with 30 min gap) for 18 weeks	Induced sperm cell death	Yan et al. 2007
36.	*Drosophila melanogaster*	900 or 1800 MHz, GSM mobile phone, 6 min/day for 6 days	Increased DNA fragmentation in follicle cells, nurse cells, and oocyte. Cell death in ovarian cells and decreased reproduction	Panagopoulos et al. 2007
37.	HPBLs	800 MHz, microwave generator, 72 h	Increased aneuploidy of 1, 10, 11, and 17 chromosomes	Mazor et al. 2008
38.	HPBLs	1.95 GHz UMTS, 24 h	Increased aberrant cells and chromosome exchanges	Manti et al. 2008
39.	Buccal mucosa cells of mobile phone users	Mobile phone, SAR 0.34 - 0.95 W/kg, 2.35 years	Increased micronuclei	Yadav and Sharma 2008
40.	Human lens epithelial SRA01/04 cells	1.8 GHz, simulated GSM, 2 h (5 min on/10 min off)	Increased DNA SSBs and ROS	Yao et al. 2008
41.	HPBLs	905 MHz, 915 MHz simulated GSM, 1947.4 MHz simulated UTMS, 1 h	Increased chromatin condensation and decreased 53BP1 and γH2AX repair foci	Belyaev et al. 2009
42.	HPBLs of occupational workers	1250-1350 MHz, radar antenna, 24.3 years	Increased DNA damage and chromatid breaks	Garaj-Vrhovac and Orescanin 2009
43.	Human spermatozoa	1.8 GHz, RF generator, 16 h	Increased ROS, DNA fragmentation, and 8-OH-dG DNA adducts	De Iuliis et al. 2009
44.	Primary rat neurons	1800 MHz, simulated GSM, 24 h (5 min on/10 min off)	Increased ROS and 8-OH-dG DNA adducts. Decreased mtDNA transcripts and mtDNA copy number	Xu et al. 2010
45.	Rat astrocytes	900 MHz, continuous wave, or amplitude modulated, 20 min	Increased ROS and DNA damage only by the modulated EMF	Campisi et al. 2010

S. No.	Biological System	EMF Exposure	Effect	Reference
46.	Sprague Dawley rat brain, lung, spleen, and leucocytes	2.45 GHz, microwave generator, 8 days	Increased DNA SSBs	Aweda et al. 2010
47.	Wistar rats	10 or 50 GHz, RF generator, 2 h/day for 45 days	Increased ROS, micronuclei reduced OD, GPx, and CAT	Kumar et al. 2010
48.	*Drosophila melanogaster*	900 or 1800 GSM mobile phone, 6 min/day for 6 days, at distances 0–100 cm	Dose-dependent DNA fragmentation and reduced reproduction for power densities down to <1 $\mu W/cm^2$	Panagopoulos et al. 2010
49.	*Drosophila melanogaster* females	900 MHz GSM mobile phone, 6 min continuous or intermittent exposure 1 min × 6, 2 min × 3, 3 min × 2, and 2 min × 3	Increased DNA fragmentation and apoptosis in ovarian cells by both continuous or intermittent exposure and actin–cytoskeleton damage	Chavdoula et al. 2010
50.	Rabbit (pregnant and new born infants)	1800 MHz, simulated GSM, 15 min/day for 7 days	Increased apoptosis in brain, eyes, heart, lung, liver, spleen, and kidneys	Guler et al. 2011
51.	Human hair root cells	900 MHz, GSM mobile phone, 15–30 min	Increased DNA SSBs	Çam and Seyhan 2012
52.	Mouse brain cells	10.715 GHz, microwave generator, 6 h for 3 day	Increased micronuclei, apoptosis, and reduced STAT3	Karaca et al. 2012
53.	Rabbit (pregnant)	1800 MHz, simulated GSM, 15 min/day for 7 days	Increased 8-OH-dG DNA adducts, LPO	Güler et al. 2012
54.	Mature and immature rats	900 MHz simulated GSM, 2h/day for 45 days	Increased chromosome aberrations and micronuclei. Reduced mitotic index and PCE/NCE ratio	Şekeroğlu et al. 2012
55.	Sprague Dawley rat urine	1800 MHz, simulated GSM, 2 h	Increased 8-OH-dG DNA adducts	Khalil et al. 2012
56.	*Drosophila melanogaster* virgin females	900 MHz, GSM mobile phone, 1 × 6 min every 10 h during the first 51 h of the newly emerged adult flies	Increased DNA fragmentation in ovarian cells of virgin females. Decreased development of the ovaries	Panagopoulos 2012
57.	Human buccal epithelium cells	900 MHz, GSM mobile phone, 1, 2, 5, 10, 15, 30, and 60 min	Increased heterochromatin granules	Skamrova et al. 2013
58.	CHL cells, human fibroblasts	1800 MHz, simulated GSM, 1 or 24 h	Increased γH2AX foci	Xu et al. 2013

(Continued)

TABLE 4.1 (CONTINUED)
Studies Reporting Genotoxic Effects of RF/WC or ELF EMFs in Various Biological Systems

S. No.	Biological System	EMF Exposure	Effect	Reference
59.	Mouse spermatocytes	1800 MHz, GSM mobile phone, 24 h	Increased ROS, DNA base damage, and 8-OH-dG DNA adducts	Liu et al. 2013a
60.	Mouse GC2 cells	900 MHz, GSM mobile phone, 1 min every 20 min for 24 h	Increased DNA damage	Liu et al. 2013b
61.	Fisher rats brain	900, 1800, or 2450 MHz, microwave generator, 2 h/day 5 days/week for 30 days	Increased DNA strand breaks	Deshmukh et al. 2013
62.	Mature and immature rats	900 MHz, simulated GSM, 2h/day for 45 days	Increased chromosome aberrations, and micronuclei. Reduced mitotic index and PCE/NCE ratio	Şekeroğlu et al. 2013
63.	Sprague Dawley rats 21 days old neonates (testes)	900 MHz, RF generator, 1 h	Increased apoptotic index, LPO, and 8-OH-dG DNA adducts	Hanci et al. 2013
64.	Japanese quail embryos	900 MHz, GSM mobile phone, 48 s on/12 s off, for 158–360 h	Increased nitric oxide and superoxide free radicals, LPO, and 8 OH-dG DNA adducts	Burlaka et al. 2013
65.	PC12 cells and Wistar rat hippocampus	2.856 GHz, RF generator, 5 and 15 min	Reduced mitochondrial membrane potential, increased apoptosis, and DNA fragmentation in PC12 cells. Increased Bax/Bcl-2 ratio, cytochrome-c release, cleaved caspase-3, and PARP in PC12 and rat hippocampus	Zuo et al. 2014
66.	Wistar rat testes	2.45 GHz, microwave generator, 2 h/day for 45 days	Increased DNA damage, ROS, xanthine oxidase, protein carbonyl contents, LPO, apoptosis, and reduced LDH-X and testosterone	Meena et al. 2014
67.	Wistar rats	1800 MHz, microwave generator, 1 h/day for 30 days	Increased 8-OH-dG DNA adducts and AOPP	Gürler et al. 2014
68.	HPBLs	916 MHz, simulated GSM, 1–8 h (1 h gap)	Increased chromosome aberrations	Shah et al. 2015
69.	Buccal mucosa cells of mobile phone users	Mobile phone (SAR 2 w/kg), 10–128.6 min/week for 4–14 years	Increased micronuclei	Daroit et al. 2015
70.	Mouse GC2 cells	1800 MHz, simulated GSM talk signal, 24 h (5 min on/10 min off)	Increased DNA base damage	Duan et al. 2015

S. No.	Biological System	EMF Exposure	Effect	Reference
71.	Mouse neuroblastoma neuro-2a cells	900 MHz GSM, 24 h	Increased ROS, DNA base damage, DNA SSBs, and DNA DSBs	Wang et al. 2015
72.	Buccal mucosa cells of mobile phone users	UMTS/GSM mobile phones, 3–10 h/week up to 5 years	Increased micronuclei	Banerjee et al. 2016
73.	Buccal mucosa cells HPBLs	1800 MHz, MT base station, 12–13 h/day for 8–9 years	Increased micronuclei and DNA damage	Gulati et al. 2016
74.	Rabbit (pregnant)	1800 MHz, simulated GSM, 15 min/day for 7 or 14 days	Increased caspase 1, caspase 3, and caspase 9	Meral et al. 2016
75.	Rats (pregnant)	1800 MHz, simulated GSM, 2 h/day for 1–2 weeks	Dead and resorbed embryos, hematomas, malformed, and edematous fetuses	Alchalabi et al. 2016
76.	HPBLs of males and females	900 MHz, 1800 MHz, MT base stations, up to 5 years	Increased micronuclei and lipid peroxidation and reduced GSH, catalase, and SOD	Zothansiama et al. 2017
77.	HL60 cells	900 MHz, RF generator, 4 h/day for 5 days	Increased ROS, 8-OH-dG DNA adducts. Reduced mt-transcription factor A, mtDNA polymerase γ, mtDNA transcripts mtDNA copy number, and ATP	Sun et al. 2017
78.	Sprague Dawley rats (ovaries)	1800 MHz, simulated GSM, 2 h/day 7 days/week for 15, 30, or 60 days	Increased 8-OH-dG DNA adducts, micronuclei, and LPO and reduced GPx	Alchalabi et al. 2017
79.	Chick embryo (liver)	2G, or 3G mobile phone, SAR 0.310 W/kg, 75 min/12 h for 12 days	Increased DNA DSBs and induced dilation of sinusoids, hemorrhage, cytoplasmic vacuolization, and nucleomegaly of hepatocytes	D'Silva et al. 2017
80.	Ear hair follicle cells of male mobile phone users	Mobile phone use, SAR 0.45–0.97 W/kg, 30–60 min/day for 10 years	Increased DNA damage	Akdag et al. 2018
81.	Humans (serum)	1800 MHz, MT base stations, 9 years	Decreased SOD, and CAT, Increased LPO	Gulati et al. 2018
82.	U87 and U251 glioblastoma cells	1950 MHz UTMS, 5 min on/10 min off for 16 h	Increased DNA damage	Al-Serori et al. 2018

(*Continued*)

TABLE 4.1 (CONTINUED)
Studies Reporting Genotoxic Effects of RF/WC or ELF EMFs in Various Biological Systems

S. No.	Biological System	EMF Exposure	Effect	Reference
83.	Mouse spermatogonia GC1, spermatocyte GC2 cells, spermatozoa	1.8 GHz, RF generator, 3–4 h	Increased ROS, DNA fragmentation, and 8-OH-dG DNA adducts	Houston et al. 2018
84.	Mouse GC2 cells	1800 MHz, simulated GSM, 24 h (5 min on/10 min off)	Increased DNA SSBs, DSBs, ROS, LC3-II, and pAMPK and reduced p62 and p-mTOR	Li et al. 2018
85.	Swiss albino mice (testes)	902.4 MHz, simulated GSM, 3 h twice/day for 35 days	Increased DNA damage and LPO. Decreased GSH, and SOD	Pandey and Giri 2018
86.	Quail embryos	1800 MHz, GSM mobile phone, 5 days before and 14 days after eclosion (48 s on/12 s off)	Increased DNA SSBs, DSBs, and 8-OH-dG DNA adducts, Increased LPO, SOD, and NO in brain, heart and liver	Yakymenko et al. 2018
87.	HPBLs	3G (UTMS) mobile phone, 15 min	Increased chromatid gaps and breaks	Panagopoulos 2019a
88.	Human amniocytes	900 and 1800 MHz simulated GSM, 3, 6, and 12 h/day for 12 days	Increased chromosome fragility, chromatid gap, chromatid breaks, and chromosome condensation	Uslu et al. 2019
89.	Canine blood, and HPBLs	123.9 kHz or 250.8 kHz, RF generator, (0.1 or 0.79 mT) 20 h	Increased DNA strand breaks	Brech et al. 2019
90.	Rats	200 MHz, RF generator, 1 h/day for 30 days	Decreased serum testosterone. Increased Bax/Bcl2 ratio, cleaved caspase 3, and caspase 3 in testes	Guo et al. 2019
91.	Sprague Dawley rat brain	900, 1800, or 2100 MHz GSM, 2 h/day for 6 months	Increased DNA damage, 8-OH-dG DNA adducts, total oxidants, oxidant index, nitric oxide, and LPO. Reduced total antioxidants	Alkis et al. 2019
92.	Mouse spermatozoa	905 MHz, RF generator, 12 h/day for 1, 3, or 5 weeks	Increased ROS, DNA fragmentation, and 8-OH-dG DNA adducts	Houston et al. 2019
93.	HPBLs	1923 MHz, 1947.47 MHz, 1977 MHz 3 G (UTMS) mobile phone, 1–3 h	Increased DNA damage	Gulati et al. 2020
94.	HPBLs	3G (UTMS) mobile phone and caffeine, 5, 15, and 25 min	Increased chromatid gaps and breaks compared to those of a high caffeine dose. Increased effect with combination and increasing exposure duration	Panagopoulos 2020
95.	Buccal mucosa cells of mobile phone users	Mobile phone use, 2–3 h/day up to 4 years	Increased micronuclei	Rashmi et al. 2020

S. No.	Biological System	EMF Exposure	Effect	Reference
96.	Buccal mucosa cells of mobile phone users	Mobile phone use, SAR 0.244–1.552 W/kg, 5–10 years	Increased DNA damage	Khalil et al. 2020
97.	Wistar rats (brain)	1800 MHz, microwave generator, 4 h/day 5 days/week for 90 days	Increased DNA damage, IL6, TNF-α, and GSSG and decreased GSH, GST, GPx, glucose-6-phosphate dehydrogenase, and acetylcholinesterase	Sharma et al. 2020
98.	Wistar rats brain	2115 MHz, simulated mobile phone EMF, 8 h	Increased ROS, LPO, and DNA damage. Reduced cell proliferation	Singh et al. 2020
99.	Wistar rat *in utero* and adult mice (brain cortex and leucocytes)	900 or 1900 MHz, CDMA (3G)- or GSM (2G)-modulated, 14 weeks	Increased DNA damage	Smith-Roe et al. 2020
100.	Humans (buccal mucosa cells)	MT base stations, 5 years	Increased micronuclei	Thamilselvan et al. 2021
101.	Sprague Dawley rats (liver)	1800 or 2100 MHz, RF generator, 2/day for 7 months	Increased 8-OH-dG DNA adducts, DNA strand breaks, LPO, total oxidants, and oxidant index and decreased total antioxidants	Alkis et al. 2021
102.	Chick embryo (brain)	2G and 3G mobile phone, SAR 0.310 W/kg, 75 min/12 h for 12 days	Increased DNA damage	D'Silva et al. 2021
B. ELF EMF Studies				
103.	Human amniotic cells	50 Hz (30 µT) sine wave EMF, 72 h (2 or 15s on/15 or 20s off)	Increased chromatid and chromosome aberrations	Nordenson et al. 1994
104.	HPBLs (females)	50 Hz (2, 3, 5, 7, and 10 mT) magnetic field, 1 h	Increased DNA damage	Ahuja et al. 1997
105.	Sprague Dawley rat brain cells	60 Hz (0.1, 0.25, and 0.5 mT) EMF, 2 h	Increased DNA SSBs and DSBs	Lai and Singh 1997
106.	Sprague Dawley rat brain cells	60 Hz (0.5 mT) EMF, 2 h	Increased DNA-protein cross-links	Singh and Lai 1998
107.	HPBLs (males)	50 Hz (2, 3, 5, 7, and 10 mT) magnetic field, 1 h	Increased DNA damage	Ahuja et al. 1999
108.	Mouse m5S cells	5, 50, and 400 mT EMF, 42 h	Increased SCE	Yaguchi et al. 1999

(Continued)

TABLE 4.1 (CONTINUED)
Studies Reporting Genotoxic Effects of RF/WC or ELF EMFs in Various Biological Systems

S. No.	Biological System	EMF Exposure	Effect	Reference
109.	CBA/Ca mice brain cells	50 Hz EMF, power transmission lines, 11, 20, and 32 days	Increased DNA damage and reduced PCE/NCE ratio, leucocytes, and ODC	Svedenstål et al. 1999
110.	Human fibroblasts	50 Hz EMF, 24 h (1, 3, 5, 10, 15, or 25 min on/10 min off)	Increased DNA SSBs and DSBs	Ivancsits et al. 2002
111.	Human fibroblasts	50 Hz EMF, 1–24 h (5 min on/10 min off)	Increased DNA SSBs and DNA DSBs	Ivancsits et al. 2003
112.	Sprague Dawley rat brain cells	60 Hz (0.01 mT) EMF, 24 or 48 h	Increased DNA SSBs and DSBs (48 h exposure led to higher rise)	Lai and Singh 2004
113.	Mice	50 Hz (1.5 mT) EMF, 8 weeks	Increased unscheduled DNA synthesis and activity of DNA polymerase-I	Schmitz et al. 2004
114.	HPBLS	50 Hz, 15 µT EMF, 2 h	Increased chromatin condensation and decreased 53BP1 expression	Belyaev et al. 2005
115.	Human fibroblasts and melanocytes Rat granulosa cells	50 Hz EMF, 24 h	Increased DNA SSBs and DNA DSBs	Ivancsits et al. 2005
116.	ES-1 human diploid fibroblasts	50 Hz EMF, 2–24 h (5 min on/10 min off)	Increased acentric fragments, chromosome gaps and breaks, ring and dicentric chromosomes, and micronuclei	Winker et al. 2005
117.	HL-60 cells, Rat-I fibroblasts WI-38 diploid fibroblasts	50-Hz EMF, 24–72 h	Increased DNA strand breaks, 8-OH-dG DNA adducts, S-phase fraction, and ROS (Rat-I cells).	Wolf et al. 2005
118.	Wistar rats	50 Hz EMF, 50 or 100 days	Increased 8-OH-dG DNA adducts and LPO	Yokus et al. 2005
119.	HPBLs	50 Hz electric pulses, 120 min	Increased DNA damage	Delimaris et al. 2006
120.	Wistar rats	128 mT static magnetic field, 1 h/day for 30 days	Increased 8-OH-dG DNA adducts, LPO, and reduced GPx, CAT, and SOD	Amara et al. 2006
121.	Wistar rats bone marrow	50 Hz (1 mT) EMF, 4 h/day for 45 days	Increased micronuclei. Decreased mitotic index and PCE/NCE ratio	Erdal et al. 2007
122.	CD1 mice (embryos and neonates)	50 Hz (650 µT) EMF, 24h/day for 21 days	Increased micronuclei in neonates	Udroiu et al. 2008
123.	Human primary fibroblasts HR-1d, ES-1, MRC-5 cells	50 Hz EMF, 15 h (5 min on/10 min off)	Increased DNA fragmentation	Focke et al. 2010

S. No.	Biological System	EMF Exposure	Effect	Reference
124.	IMR90 HeLa	60 Hz (6 mT) magnetic field, 30 min/day for 3 days	Increased γH2AX, Chk-2, and p38	Kim et al. 2010
125.	CD1 mice (brain)	50 Hz (1 mT) EMF, 15 h/day for 1 or 7 days	Increased DNA damage (tail moment)	Mariucci et al. 2010
126.	IMR90 HeLa cells	60 Hz (7 mT) magnetic field, 10, 20, 30, 60, or 120 min	Increased γH2AX	Kim et al. 2012
127.	Neonate Wistar rats	50 Hz (0.5 mT) EMF, 24/day 30 days	Increased DNA damage and LPO and decreased SOD in the brain. Increased micronuclei and mitotic index in the bone marrow	Rageh et al. 2012
128.	CD1 mice (brain)	50 Hz EMF (0.1, 0.2, 1, or 2 mT), 15 h/day for 7 days	Increased DNA SSBs and DSBs immediately after exposure (1 and 2 mT) and increased HSP-70 expression (hippocampus)	Villarini et al. 2013
129.	*Drosophila melanogaster*	50 Hz EMF (1, 11, or 21 G), 5 days continuous exposure (120 h)	Increased DNA fragmentation in ovarian cells	Panagopoulos et al. 2013b
130.	SH-SY5Y cells	50 Hz (100 µT) magnetic field, 24 h	Increased micronuclei, ROS, mitochondrial activity, and LPO	Luukkonen et al. 2014
131.	HaCaT cells	60 Hz (1.5 mT) EMF, 96 h	Decreased cell proliferation, arrest of cells in G_1/G_0 phase. Decreased expression of CDC25B, CDC20, CDC2, CCNA2, and CCNBI genes. Increased activation of CDKN1A, ATM, and Chk2	Huang et al. 2014
132.	Mice	50 Hz EMF, 12 h/day for 7, 14, 21, and 28 days	Increased micronuclei	Alcaraz et al. 2014
133.	SH-SY5Y cells	50 Hz (100 µT) EMF, 24 h	Increased micronuclei and decreased LPO	Kesari et al. 2015
134.	Mouse spermatocytes	50 Hz EMF, 24 h	Increased DNA damage and γH2AX foci	Duan et al. 2015
135.	CD1 mouse fetus and neonates	50 Hz (65 µT) EMF, 24h/day for 30 days	Increased micronuclei	Udroiu et al. 2015
136.	Mouse spermatocytes	50 Hz EMF, 72 h	Altered DNA methylation and gene expression	Liu et al. 2015
137.	RAW -246 macrophages	50 Hz (0.5 mT) EMF, 24 h	Increased DNA SSBs	Nakayama et al. 2016
138.	MCF-7 and SH-SY5Y cells	50 Hz EMF, 30 min (5 min on/5 min off, 15 min on/15 min off, 30 min continuous)	Reduced the mRNA of GADD45A, XRCC1, XRCC4, Ku80 (XRCC5), Ku70 (XRCC6), and LIG4	Sanie-Jahromi et al. 2016

(Continued)

TABLE 4.1 (CONTINUED)
Studies Reporting Genotoxic Effects of RF/WC or ELF EMFs in Various Biological Systems

S. No.	Biological System	EMF Exposure	Effect	Reference
139.	CHL lung cells	50 Hz EMF, 30 min or 24 h	Increased LC3-II, autophagosome, apoptosis, and γH2AX foci	Shen et al. 2016
140.	U87 cells	10 Hz/50 Hz/100 Hz EMF, 2, 4, and 24 h	Decreased cyclin D1. Increased caspase activity and p53	Akbarnejad et al. 2017
141.	SH-SY5Y cells	50 Hz (0.5 or 1 mT) EMF, 3 h	Altered gene expression	Rezaie-Tavirani et al. 2017
142.	Mouse GC1 and GC2 cells	2, 50, 120 Hz EMF, 2 h	Increased superoxide, nitric oxide radicals, NFκB, apoptosis, and p53. Reduced, p21, activated caspase3, and Bcl2	Solek et al. 2017
143.	HPBLs from 29 males occupationally exposed	50/60 Hz (4–50 μT), electric power plant, 9.86 years	Increased DNA damage	Zendehdel et al. 2019
144.	A549 cells, SGC7901 cells, PANC-1 cells, G401 cells	50 Hz EMF, 2 h/day for 3 days	Increased ROS, apoptosis, cleaved caspase-3, PARP, LPO, DNA SSBs and DSBs, DNA-PKCs, LIG4, RAD9B, and BM11	Yuan et al. 2020

* American National Standard Institute.

4.3 STUDIES WITH NEGATIVE FINDINGS

At the same time, several other studies, looking at similar endpoints as those described above, have found no effects due to RF/WC EMF exposure a) in human lymphocytes exposed *in vitro* or *in vivo* (Lloyd et al. 1984; Garson et al. 1991; Maes et al. 1993; 1997; 2000a; 2001; 2006; Antonopoulos et al. 1997; McNamee et al. 2002b; 2003; Tice et al. 2002; McNamee et al. 2002a; Zeni et al. 2003; 2005; 2007; 2008; Stronati et al. 2006; Ros-Llor et al. 2012; Waldmann et al. 2013; de Oliveira et al. 2017), b) in other types of human/animal cells (Kerbacher et al. 1990; Malyapa et al. 1997a; 1997b; Li et al. 2001; Vijayalaxmi et al. 2001; Bisht et al. 2002; Port et al. 2003; Koyama et al. 2016; 2019; Lagroye et al. 2004a; 2004b; Hook et al. 2004; Koyama et al. 2004; Komatsubara et al. 2005; Sakuma et al. 2006; Speit et al. 2007; Bourthoumieu et al. 2010; Gurbuz et al. 2010; 2014; 2015; Kumar et al. 2015; Mizuno et al. 2015; Herrala et al. 2018; Su et al. 2018; Regalbuto et al. 2020), and c) in rodents and other animals exposed *in vivo* (Vijayalaxmi et al. 1997; 1999; 2004; Ono et al. 2004; Görlitz et al. 2005; Juutilainen et al. 2007; Ziemann et al. 2009; Jeong et al. 2018). Similarly, with pure ELF exposures, several studies have reported no effect on similar endpoints in HPBLs, human and other mammalian cells, and rodents (Antonopoulos et al. 1995; Svedenstål and Johanson 1998; Maes et al. 2000b; Stronati et al. 2004; Testa et al. 2004; Reddy et al. 2010; Zhu et al. 2016; Su et al. 2017; Ross et al. 2018; Sun et al. 2018; Wang et al. 2019; Lv et al. 2021). Table 4.2 lists the above studies that reported no genotoxic effects in various animals and cell types.

TABLE 4.2
List of Studies Reporting No Genotoxic Effects of RF/WC or ELF EMF Exposure in Various Biological Systems

A: RF/WC EMF Studies

S. No.	Biological System	EMF Exposure	Effect Studied	Reference
1.	HPBLs	2.45 GHz EMF, 20 min	Chromosome aberrations	Lloyd et al. 1984
2.	CHO	2450 MHz, microwave generator, 2h	Chromosome aberrations	Kerbacher et al. 1990
3.	L5178Y cells	2.45 GHz pulsed, microwave generator, 4 h	Mutation	Meltz et al. 1990
4.	HPBLs from linemen	400 kHz - 20 GHz EMF, 6 years	Chromosome aberrations	Garson et al. 1991
5.	HPBLs	900 MHz GSM-like, 935.2 MHz GSM-like, 455.7 MHz EMF, 30 and 120 min	Sister chromatid exchanges, cell kinetics, chromatid/chromosome breaks and gaps, dicentrics, and acentric fragments	Maes et al. 1997; 2000a; 2000b; 2001
6.	HPBLs	380, 900, 1800 MHz, simulated GSM	Sister chromatid exchange	Antonopoulos et al. 1997
7.	U87MG cells	835.62 MHz EMF, 24 h	DNA damage	Malyapa et al. 1997a; 1997b
8.	C3H/HeJ mice	2450 MHz, RF generator, 20 h/day for 18 months	Micronuclei	Vijayalaxmi et al. 1997
9.	CF1 mice	Wideband RF EMF, generator, 15 min	Micronuclei	Vijayalaxmi et al. 1999
10.	C3H 10T½ cells	835.62/847.74 MHz CDMA pulsed, 2, 4, or 24 h	DNA damage	Li et al. 2001
11.	Sprague Dawley rats	2450 MHz, RF generator, 24 h	Micronuclei	Vijayalaxmi et al. 2001
12.	HPBLs	1.9 GHz pulsed EMF, 2 h or 24 h	Micronuclei, proliferation index	McNamee et al. 2002a; 2002b; 2003
13.	HPBLs	837 MHz simulated GSM, 1909.8 MHz simulated UMTS, 3 h	DNA damage and micronuclei	Tice et al. 2002
14.	C3H 10T½ cells	835.62/847.74 MHz simulated FDMA/CDMA, 3, 8, 16, or 24 h	Micronuclei	Bisht et al. 2002
15.	HPBLs	900 MHZ simulated GSM, 44 h (6 min on/3 h off)	Micronuclei and cell cycle	Zeni et al. 2003
16.	HL-60 cells	RF EMF 25 times higher than the ICNIRP limits for occupational exposure, 24, 48, and 72h	Micronuclei, apoptosis, abnormal cells, and differential gene expression	Port et al. 2003
17.	Molt-4 cells	813.56 MHz iDEN 836.55 MHz TDMA mobile phones, up to 24 h	DNA damage and apoptosis	Hook et al. 2004
18.	BALB/c mice	42.2 GHz modulated EMF, 30 min/day for 3 days	Micronuclei	Vijayalaxmi et al. 2004
19.	LacZ-transgenic Muta™ pregnant mouse	2.45 GHz, RF generator, 16 h/day, up to 15 days	Mutant frequency	Ono et al. 2004

(Continued)

TABLE 4.2 (CONTINUED)
List of Studies Reporting No Genotoxic Effects of RF/WC or ELF EMF Exposure in Various Biological Systems

S. No.	Biological System	EMF Exposure	Effect Studied	Reference
20.	CHO-K1	2.45 GHz, sinusoidal CW RF EMF, 2 h	Micronuclei	Koyama et al. 2004
21.	Sprague Dawley rats	2450 MHz, microwave generator, 2 h	DNA damage	Lagroye et al. 2004a
22.	C3H 10T½ cells	2450 MHz EMF, 2 h	DNA–protein cross-links	Lagroye et al. 2004b
23.	HPBLs	900 MHz, simulated GSM EMF, 2 h	DNA damage	Zeni et al. 2005
24.	m5S cells	2.45 GHz, continuous wave RF EMF, 2 h	Chromosome aberrations	Komatsubara et al. 2005
25.	B6C3F1 mice	902 MHz simulated GSM/1747 MHz simulated GSM, 1 week	Micronuclei	Görlitz et al. 2005
26.	HPBLs	900 MHz GSM mobile phone, 24 h	Micronuclei	Scarfì et al. 2006
27.	HPBLs	935 MHz simulated GSM, 24 h	DNA strand breaks, micronuclei, chromosome aberrations, and nuclear division index	Stronati et al. 2006
28.	HPBLs	1950 MHz UTMS mobile phone, 24 h	DNA damage	Sannino et al. 2006
29.	HPBLs from subjects occupationally exposed to RF/WC EMFs	450 or 900 MHz EMF, 1 h/day for 2–3 years	Chromosome aberrations, DNA damage, and sister chromatid exchanges	Maes et al. 2006
30.	A172 cells	2.1425 GHz continuous wave, or WCDMA mobile phone, 2 or 24 h	DNA strand breaks	Sakuma et al. 2006
31.	ES1, V79 cells	1800 MHz continuous wave, or GSM pulsation, 1, 4, or 24 h (5 min on/10 min off)	DNA damage and micronuclei	Speit et al. 2007
32.	HPBLs	120–130 GHz pulsed microwaves, 20 min	DNA damage and cell cycle	Zeni et al. 2007
33.	CBA/S mice	902.5 MHz continuous wave, or 902.4 MHz simulated mobile phone EMF, 1.5 h/day, 5 days per week, for 78 weeks	Micronuclei	Juutilainen et al. 2007
34.	HPBLs	1900 MHz, simulated UMTS, 24–48 h, (6 min on/2 h off)	DNA strand breaks/alkali labile sites, and micronuclei	Zeni et al. 2008
35.	B6C3F1	902 MHz GSM-like, or 1747 MHz GSM-like EMF, 2 h/day, 5 days per week, for 2 years	Micronuclei	Ziemann et al. 2009
36.	Human amniotic fluid cells	900 MHz simulated GSM 24 h	Karyotype	Bourthoumieu et al. 2010
37.	Wistar rats	1800 MHz simulated GSM, 20 min/day, 5 days/week, 1 month	Micronuclei	Gurbuz et al. 2010

S. No.	Biological System	EMF Exposure	Effect Studied	Reference
38.	Exfoliated buccal mucosa cells of mobile phone users	Mobile phones, 0.8 to 10 h/week for 3–18 years	Difference in micronuclei between the "exposed side" and the other side of the users' heads	Ros-Llor et al. 2012
39.	HPBLs	1800 MHz simulated GSM, 28 h (5 min on/10 min off)	Chromosomal aberrations, micronuclei, and sister chromatid exchanges	Waldmann et al. 2013
40.	Wistar rats	1800/2100 MHz simulated GSM, 30 min/day, 6 days/week for 1–2 months	Micronuclei	Gurbuz et al. 2014
41.	Wistar rats (diabetic)	2100 MHz simulated GSM, 30 min/day, 5 days/week for 1 month	Micronuclei	Gurbuz et al. 2015
42.	WI38VA13 cells	12.5 MHz EMF, 2 or 24 h	DNA damage, micronuclei, HPRT mutation, and cell cycle	Mizuno et al. 2015
43.	Rat bone marrow lymphoblasts	900 or 1800 MHz simulated GSM, 120 min	DNA strand breaks and alkali labile sites	Kumar et al. 2015
44.	HCE-T/SRA01/4 cells	60 GHz, RF generator, 24 h	DNA damage, micronuclei, heat shock proteins	Koyama et al. 2016
45.	Exfoliated buccal mucosa cells of mobile phone users	Mobile phone	Micronuclei	de Oliveira et al. 2017
46.	A172, U251, and SH-SY5Y cells	1800 MHz simulated GSM, 1, 6, or 24 h (5 min on/10 min off)	γH2AX and cell viability	Su et al. 2017a
47.	Astrocytes, micro-glia cortical neurons of newborn rat	1800 MHz simulated GSM, 1, 2, 7, or 8 days	DNA damage (γH2AX foci) and cytokine expression	Su et al. 2018
48.	Rat primary astrocytes	872 MHz simulated GSM, 24 h	DNA damage and micronuclei	Herrala et al. 2018
49.	C57BL/6 mice brain	1950 MHz simulated 3G/4G mobile phone EMF, 2 h/day, 5 days/week for 8 months	8-OH-dG DNA adducts, p53, p21, γH2AX foci, Bax, apoptosis and cleaved caspase-3, and PARP	Jeong et al. 2018
50.	HCE-T/SRA01/4 cells	40 GHz, RF generator, 24 h	Micronuclei, DNA strand breaks, HSP-27, 70, and 90 expressions	Koyama et al. 2019
51.	Human fibroblasts	2.45 GHz continuous wave, or amplitude modulated, 2 h	γH2AX/53BP1 foci, micronuclei, expression of various genes and cell cycle kinetics	Regalbuto et al. 2020

(Continued)

TABLE 4.2 (CONTINUED)
List of Studies Reporting No Genotoxic Effects of RF/WC or ELF EMF Exposure in Various Biological Systems

S. No.	Biological System	EMF Exposure	Effect Studied	Reference
B: ELF EMF Studies				
52.	HBPLs	50 Hz EMF, 24 h	Sister chromatid exchange	Antonopoulos et al. 1995
53.	CBA/S, CBA/Ca mice	50 Hz EMF, 90 days	Micronuclei	Svedenstål and Johanson 1998
54.	HBPLs	50 Hz EMF, 48 h	Chromosome aberrations and sister chromatid exchanges	Maes et al. 2000b
55.	Human whole blood	50 Hz EMF, 2 h	Chromosome aberrations and DNA strand breaks	Stronati et al. 2004
56.	HPBLs	50 Hz EMF, 48 h	Chromosome aberrations, micronuclei, and DNA damage	Testa et al. 2004
57.	CD1	1000 Hz pulsed EMF 8 weeks	Micronuclei	Reddy et al. 2010
58.	Human lens epithelial (SRA01/04)	50 Hz EMF, 2, 6, 12, 24, or 48 h	γH2AX foci and DNA fragmentation	Zhu et al. 2016
59.	Neurogenic tumor cells and primary cultured neurogenic cells from rats (astrocytes, micro-glia, cortical neurons)	50 Hz EMF, 24 h	γH2AX foci, cell cycle, cell proliferation, and cell viability	Su et al. 2017b
60.	hMSC	5 Hz EMF, 20 min/day, 3 times/week for 2 weeks	Karyotype, cell viability, and cell proliferation	Ross et al. 2018
61.	$Atm+/+$ and $Atm-/-$ MEF cells	50 Hz EMF, 24 h (15 min/1 h)	γH2AX foci, DNA fragmentation, cell cycle progression, and cell viability	Sun et al. 2018
62.	AC16 human ventricular cardiomyocytes, Sprague Dawley rats	50 Hz EMF, 1 h continuously or 75 min intermittently (15 min field on/15 min field off) (cells) 50 Hz EMF, 15h/day for 7 days (rats)	DNA damage, ROS G_0/G_1, S or G_2/M phases p53 and HSP-70	Wang et al. 2019
63.	Human amnion epithelial cells (FLs), human umbilical vein endothelial cells (HUVECs)	50 Hz EMF, 15 min/h) for 24 h	γH2AX foci	Lv et al. 2021

4.4 DISCUSSION AND CONCLUSIONS

All man-made EMFs are totally polarized and coherent, in contrast to the natural EMFs, as they are generated by electric/electronic circuits/antennas in which free electrons oscillate in specific directions and coherently within the metallic conductors (Panagopoulos et al. 2015a; Panagopoulos 2019a). Accordingly, WC EMFs are totally polarized and coherent and highly variable in time, including both low (ELF) and high (RF) frequencies, and they produce intense biological effects including genotoxicity as reported by the plethora of studies described in this chapter and listed in Table 4.1. Although there are probably more studies with both positive and negative findings, it is clear that the negative reports on the genotoxic effects of man-made, and especially WC EMF exposures are significantly less than the studies with positive findings. Moreover, it has been shown that studies may be influenced by funding in favor of negative findings (Leach et al. 2018; Hardell and Carlberg 2020). It has been shown that industry-funded studies, more often than not, find "no effect", while studies not funded, or studies supported by institutional funding commonly reveal "effects" (Leach et al. 2018).

Thus, the reports on the genotoxic effects of ELF and MT/WC EMFs from various sources such as WC devices and antennas, high-voltage power lines, etc., significantly outnumber the corresponding negative reports, indicating the potential of man-made and especially WC EMFs to affect cell function and induce adverse effects in human and animal health. The world is moving on to the use of 5G MT/WC technology, the adverse effects of which are still not known, but they are reasonably expected to be even more intense than those of existing previous technologies due to the increasing complexity and variability of the emissions (Panagopoulos 2019a). It is now well established that long use of mobile phones causes brain cancer (Hardell and Carlberg 2021), which is in agreement with the great number of genotoxic and cellular effects reviewed in this chapter.

All the effects described in this chapter (apart, perhaps, from those induced by very high-intensity/SAR exposures that exceeded ICNIRP limits in a few studies) are non-thermal effects, as there has not been any reported rise in the temperature of the exposed living tissues during the examined EMF exposures.

It has been argued that SAR does not account for the microscopic effects on exposed cells and living tissues, which are particularly important. Indeed, the effects of applied man-made EMFs are mainly due to their interaction with various specific biomolecules, such as the free ions or the ion channel proteins, and the power density/field intensity plus the other physical parameters of the EMFs (included frequencies, waveform, etc.) aptly express such effects. Moreover, the SAR has been shown to be a reliable metric only for thermal effects and in addition is impractical, as it is not directly measured but has to be calculated. Therefore, it is appropriate to express the non-thermal effects of RF/WC EMFs in terms of power density (in $\mu W/cm^2$ or mW/cm^2) instead of SAR (Baker et al. 2004; Panagopoulos et al. 2013a; Popov and Shevchenko 2019). One weak point in many RF/WC EMF studies with either positive or negative findings is that they report only the SAR instead of reporting power density and complementarily the SAR.

Another, perhaps even more important, weak point in many RF/WC EMF studies is that they do not specify the existence or non-existence of ELF pulsing/modulation components along with the RF carrier signal and report only the carrier frequency of the EMF exposure while, as described in the present chapter, in most cases, the RF carrier is accompanied by ELF components. Several studies report that they have applied continuous-wave RF exposure from generators, but they do not provide measurements verifying that there are no ELF emissions from the exposure system. Many studies refer to "RF" effects from WC systems, while both RF and ELF are always present in all WC EMF exposures. Similarly, many studies investigating purely ELF EMF effects focus only on the magnetic component and do not report the coexisting electric component which may be even more responsible for the observed effects (Coghill et al. 1996).

Finally, many experimental studies use generators with stabilized and invariable signals instead of real-life devices to assess the effects of real-life WC EMFs, which are highly variable. These

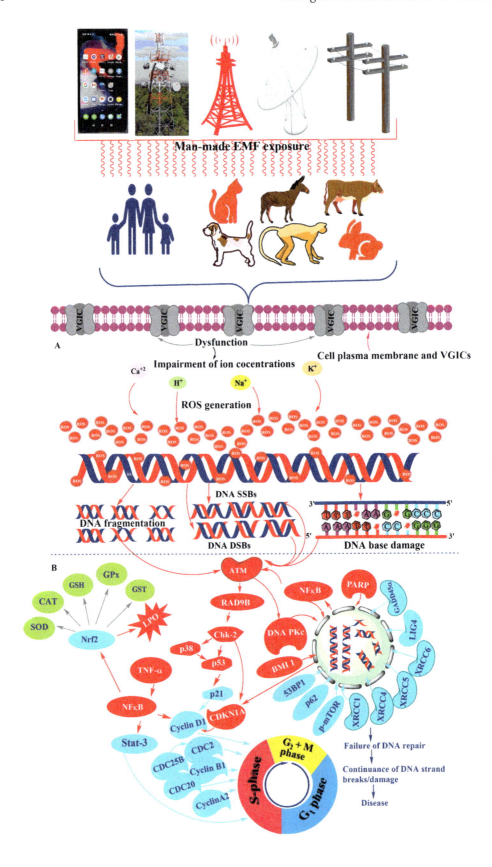

FIGURE 4.1 Possible mechanism for the genotoxic action of man-made EMFs. **A. VGIC dysfunction-ROS generation-DNA damage:** Polarized/coherent man-made EMFs (especially WC EMFs) cause ROS generation by dysfunction of VGICs leading to the impairment of intracellular ion concentrations, which, in turn, activates enzymes responsible for ROS production (Panagopoulos et al 2021). ROS damage DNA. **B. Continuation of DNA damage leading to disease:** ROS and the induced DNA damage stimulate upregulation of certain genes (red color) and downregulation of other genes (cyan color) accompanied by reduction in the levels of antioxidants (light green color), causing continuance of DNA damage and failure of DNA repair, leading to disease (infertility, aging, mutations, cancer, etc.).

weaknesses indicate a need for better definition of the physical parameters of the applied EMFs in the studies. Thus, intensity measurements should be provided both in the RF and ELF bands, all existing frequencies should be specified by proper spectrum analysis, the type of EMF (continuous-wave/pulsed/modulated), the EMF source (whether it is a real-life WC EMF or simulated), and the exposure duration should be provided. Only then we can draw safer conclusions on the mechanisms involved in the recorded biological and health effects.

The detailed mechanism by which man-made EMFs induce genotoxicity may not yet be completely revealed, but what looks as the most probable scenario is depicted in Figure 4.1. One of the principal facts involved in the action of EMFs seems to be the generation of ROS as reported by several investigations (Wolf et al. 2005; Yao et al. 2008; De Iuliis et al. 2009; Xu et al. 2010; Kumar et al. 2010; Liu et al. 2013a; Luukkonen et al. 2014; Meena et al. 2014; Wang et al. 2015; Yakymenko et al. 2016; Hui et al. 2017; Sun et al. 2017; Houston et al. 2018; Li et al. 2018; Yuan et al. 2020; Panagopoulos et al. 2021).

But how can ROS be generated by man-made EMF exposure? It has been shown that ROS generation in cells can be induced by impairment of intracellular ionic concentrations after dysfunction of voltage-gated ion channels (VGICs) in the cell membranes. The man-made EMFs are totally polarized and coherent and force charged/polar molecules and free ions to oscillate in parallel lines and in phase with the applied EMFs (Panagopoulos et al. 2015a; Panagopoulos et al. 2021). The electrochemical homeostasis of the cell is disrupted because of irregular gating of electro-sensitive ion channels (VGICs) in cell membranes caused by additive electrostatic forces exerted by the parallel ionic oscillations on the voltage-sensors of these channels. These changes cause depolarization of the cell membrane and trigger ROS formation (Panagopoulos et al. 2000, 2002, 2015a, 2021; Pall 2013; Ullrich and Apell 2021).

Direct DNA damage by specifically potent ROS such as the hydroxyl free radical (OH•) or peroxynitrite (ONOO-) can occur after the dysfunction of the VGICs in the cell membranes (Panagopoulos et al. 2021). The EMF-induced ROS overproduction and consequent OS in the cells, apart from direct action on DNA and other critically important biological molecules, reduce the levels of antioxidant enzymes/factors such as GSH, CAT, SOD, GST, and GPx and increase LPO, leading to intensification of the direct genotoxic effects in various systems, as recorded by many experimental studies described in this chapter and reported by several reviews (Amara et al. 2006; Kumar et al. 2010; Yakymenko et al. 2016, 2018; Alchalabi et al. 2017; Zothansiama et al. 2017; Gulati et al. 2018; Pandey and Giri 2018). At the molecular level, the EMF-induced ROS overproduction causes activation of a variety of apoptotic factors (genes and related proteins) such as NFκB, TNF-α, HSP 70, p53, ATM, Chk-2, p38, LC3-II, pAMPK, PARP, LIG4, Bax/Bcl-2 ratio, DNA-PKcs, RAD9B, BMI1, cytochrome-c, and cleaved caspase-3 (see Appendix 1 for the functions of these proteins/genes) (Wolf et al. 2005; Kim et al. 2010; Villarini et al. 2013; Zuo et al. 2014; Huang et al. 2014; Solek et al. 2017; Li et al. 2018; López-Furelos et al. 2018; Yuan et al. 2020). The inhibition of cyclin D1, CDC25B, CDC20, CDC2, CCNA2, and CCNB1 and upregulation of p53, ATM, Chk2, and CDKN1A genes arrest the cells in the G_0/G_1 phase of the cell cycle to allow repair of the DNA damage (Huang et al. 2014; Akbarnejad et al. 2017). However, the downregulation of p21, p62, p-mTOR, GADD45A, XRCC1, XRCC4, Ku80 (XRCC5), Ku70 (XRCC6), caspase 3, and Bcl2 can inhibit DNA base/SSBs/DSBs repair, intensifying the genotoxic and mutagenic effects of EMFs, which may lead eventually to cell death or cell senescence or otherwise to pathological conditions including cancer (Zuo et al. 2014; Sanie-Jahromi et al. 2016; Shen et al. 2016; Solek et al. 2017; Li et al. 2018). Thus, it seems that it is an avalanche of cellular effects, starting with dysfunction of VGICs, ROS generation, direct DNA damage by the ROS, reduction of antioxidant factors (GSH, CAT, SOD, GST, GPx, etc.), and activation of apoptotic genes, etc. that carry on, intensify, and finally establish the genotoxic effects.

The number of studies showing that man-made and especially WC EMF exposures are genotoxic to humans and animals has increased significantly, and there is no possibility that so many studies are wrong. The genotoxic effects of ELF and WC EMFs have been assessed in various forms, such as various types of chromosome aberrations, chromatin condensation, micronuclei formation, DNA SSBs and DSBs, fragmentation, adduct formation, and base damages. Man-made EMF exposures at environmentally encountered levels can trigger the formation of ROS, which seems to be responsible for all the genotoxic and mutagenic effects of man-made EMFs reported in the literature. It is now fairly known in the scientific community of the EMF bioeffects that repeated EMF exposures exert an adverse effect on human health and cause mutagenesis and cancer in humans and animals. Human exposure to EMFs is increasing because of the ever-increasing use of electromagnetic applications, mobile phones, and WC technology.

While adults are responsible for knowing the dangers of careless and unrestricted use of WC technology on themselves, children who are unable to know are engrossed in making unnecessary use of mobile phones and other WC devices, playing video games, searching the Internet, using Facebook, and various other applications through wireless internet access (Wi-Fi) of their "smart" phones. The general public is unaware of the severe adverse effects of WC EMFs, and there is an urgent need to be educated. Therefore, governments should make it mandatory for the manufacturers to insert warnings on mobile phones and other WC devices stating that the "use of mobile phones is injurious to health and causes cancer". The best way to reduce human exposure to WC EMFs is to follow the ALARA (as low as reasonably achievable) practice (Yeung 2019; Bryant 2021) and the Precautionary Principle (Harremoes et al. 2013) as is done with ionizing radiations. It will be purposeful to conduct more epidemiological studies by collecting samples from mobile phone users of all age groups to assess the adverse effects of MT and also other WC EMFs. In the future, experiments in the laboratory should be conducted in such a way that reproduce the actual conditions encountered by humans when using WC technology as closely as possible and avoid the weaknesses in today's literature discussed above in order to collect the right information to assess the adverse impact of man-made and especially WC EMFs.

APPENDIX 1: FUNCTIONS OF GENES/PROTEINS MENTIONED IN THE CHAPTER

AMPK (AMP-activated protein kinase) acts as a sensor of the energy status of cells and ensures survival during times of cellular metabolic stress.

ATM (ataxia telangiectasia mutated) is a tumor-suppressor gene that regulates cell division and coordinates DNA repair.

Bax/Bcl2 (Bcl2-associated X/B-cell lymphoma) belongs to the family of Bcl2 gene. The Bax triggers apoptosis, whereas the Bcl2 suppresses apoptosis and increases cell proliferation.

BMI1 (polycomb group protein) is involved in chromatin modification, stem cell function, DNA damage repair, and mitochondrial bioenergetics.

CCNA2 or cyclin A2 binds and activates cyclin-dependent kinase 2 and promotes transition of cells through G_1/S and G_2/M phases of the cell division cycle.

CCNB1 or cyclin B1 encodes G_2/mitotic-specific cyclin-B1, which complexes with p34 (CDC2) to control the G_2/M transition phase of the cell cycle.

CDC2 (cell division cycle) gene initiates both the mitotic S-phase and M-phase and premeiotic DNA synthesis and meiosis II.

CDC20 (cell division cycle) appears to act as a regulatory protein interacting with many other proteins at multiple points in the cell cycle. It is involved in nuclear movement prior to anaphase and chromosome separation.

CDC25B (cell division cycle) is a member of the CDC25 family of phosphatases. CDC25B activates the cyclin-dependent kinase CDC2 by removing two phosphate groups, and it is required for entry into mitosis. CDC25B shuttles between the nucleus and the cytoplasm due to nuclear localization and nuclear export signals.

CDKN1A (cell division kinase) is implicated in the regulation of cell growth and cell response to DNA damage. It inhibits cell cycle progression in G_1 phase by binding to G_1 cyclin-CDK complexes and to PCNA antigen and may also induce G_2 arrest.

Chk 2 (Check point kinase) is a tumor-suppressor gene that regulates cell division and has the ability to prevent cells from dividing too rapidly or in an uncontrolled manner.

DNAPK (DNA-dependent protein kinase) senses DNA damage and plays a critical role in DNA DSB repair, V(D)J recombination,[*] and apoptosis.

GADD45 (growth arrest and DNA damage-inducible 45) regulates DNA repair, cell cycle, senescence, and genotoxic stress.

GST family genes are involved in the detoxification of carcinogens, therapeutic drugs, environmental toxins, and OS products by conjugating with glutathione in humans.

HSP 70 (heat shock protein) is induced as a stress response, and it carries out protein folding.

Ku70 (XRCC6) activates DNA-binding, helicase, hydrolase, and lyase and plays a crucial role in DNA damage, DNA recombination, and DNA repair.

Ku80 (XRCC5) activates DNA-binding, helicase, and hydrolase and plays a crucial role in DNA damage, DNA recombination, and DNA repair.

LC3-II microtubule associated protein 1A/1B light chain 3B is involved in autophagy of the cells.

LIG 4 (ligase) encodes ligase involved in DNA SSBs repair.

mTOR (mammalian target of rapamycin) regulates cell growth, cell proliferation, cell motility, cell survival protein synthesis, autophagy, and transcription.

NFκB is nuclear transcription factor that regulates inflammatory responses, and its expression is increased after DNA damage.

PARP (poly-adenosine diphosphate ribose-polymerase) protein acts as a first responder that detects DNA damage and facilitates DNA repair.

p21 (cyclin-dependent kinase inhibitor 1 or CDK-interacting protein 1), is a major target of p53 activity and is associated with linking DNA damage to cell cycle arrest.

p38 is a major signaling factor of the mitogen-activated protein kinase (MAPK) signaling pathway, which plays an essential role in regulating many cellular processes, including inflammation, cell survival, cell differentiation, cell proliferation, cell migration, and cell death.

p53 is a tumor-suppressor gene which regulates cell cycle and helps in DNA repair and apoptosis induction.

p53-binding protein 1 (53BP1) is involved in DNA DSB repair.

p62 is involved in autophagy of cells.

RAD9B (RAD9 checkpoint clamp component B) is involved in delaying cells in the G_1 phase

TNF-α (tumor necrosis factor) is an inflammatory cytokine produced by macrophages/monocytes during acute inflammation and is responsible for a diverse range of signaling events within cells, leading to necrosis or apoptosis.

XRCC1 (X-ray repair cross complementing) is involved in DNA single-strand repair.

XRCC4 carries out non-homologous DNA end joining (NHEJ) in coordination with ligase 4.

---

[*] V(D)J (variable, diversity, and joining) is the process of somatic recombination that occurs only in developing lymphocytes during the early stages of T and B cell maturation in which a pair of segments introduces DSBs adjacent to each segment, deletes (or inverts) the intervening DNA, and ligates the segments together.

ACKNOWLEDGMENTS

The author is thankful to his wife Mrs. Mangla Jagetia for her encouragement and patience during the writing of this manuscript, and acknowledges the financial grant received from the University Grant's Commission, India vide grant No. F4-10/2010(BSR).

REFERENCES

Abdel-Rassoul G, El-Fateh OA, Salem MA, et al (2007) Neurobehavioral effects among inhabitants around mobile phone base stations. *Neurotoxicology* 28(2):434–440. https://doi.org/10.1016/j.neuro.2006.07.012.

Agarwal A, Desai NR, Makker K, et al (2009) Effects of radiofrequency electromagnetic waves (RF-EMW) from cellular phones on human ejaculated semen: An in vitro pilot study. *Fertil Steril* 92(4):1318–1325. https://doi.org/10.1016/j.fertnstert.2008.08.022.

Ahuja YR, Bhargava A, Sircar S, et al (1997) Comet assay to evaluate DNA damage caused by magnetic fields. *Proc Int Conf Electromagn Interf Compat* 273–276. https://doi.org/10.1109/icemic.1997.669812.

Ahuja YR, Vijayashree B, Saran R, et al (1999) In vitro effects of low-level, low-frequency electromagnetic fields on DNA damage in human leucocytes by comet assay. *Indian J Biochem Biophys* 36(5):318–322.

Aitken RJ, Bennetts LE, Sawyer D, Wiklendt AM, King BV (2005) Impact of radio frequency electromagnetic radiation on DNA integrity in the male germline. *Int J Androl* 28(3):171–179. https://doi.org/10.1111/j.1365-2605.2005.00531.x.

Akbarnejad Z, Eskandary H, Vergallo C, et al (2017) Effects of extremely low-frequency pulsed electromagnetic fields (ELF-PEMFs) on glioblastoma cells (U87). *Electromagn Biol Med* 36(3):238–247. https://doi.org/10.1080/15368378.2016.1251452.

Akdag M, Dasdag S, Canturk F, Akdag MZ (2018) Exposure to non-ionizing electromagnetic fields emitted from mobile phones induced DNA damage in human ear canal hair follicle cells. *Electromagn Biol Med* 37(2):66–75. https://doi.org/10.1080/15368378.2018.1463246.

Al-Khlaiwi TM, Habib SS, Meo SA, Alqhtani MS, Ogailan AA (2020) The association of smart mobile phone usage with cognitive function impairment in Saudi adult population. *Pak J Med Sci* 36(7):1628–1633. https://doi.org/10.12669/PJMS.36.7.2826.

Al-Serori H, Ferk F, Kundi M, et al (2018) Mobile phone specific electromagnetic fields induce transient DNA damage and nucleotide excision repair in serum-deprived human glioblastoma cells. *PLOS ONE* 13(4):e0193677. https://doi.org/10.1371/journal.pone.0193677.

Alcaraz M, Olmos E, Alcaraz-Saura M, Achel DG, Castillo J (2014) Effect of long-term 50 Hz magnetic field exposure on the micronucleated polychromatic erythrocytes of mice. *Electromagn Biol Med* 33(1):51–57. https://doi.org/10.3109/15368378.2013.783851.

Alchalabi ASH, Aklilu E, Aziz AR, et al (2016) Different periods of intrauterine exposure to electromagnetic field: Influence on female rats' fertility, prenatal and postnatal development. *Asian Pac J Reprod* 5(1):14–23. https://doi.org/10.1016/J.APJR.2015.12.003.

Alchalabi ASH, Rahim H, AbdulMalek MF, et al (2017) Micronuclei formation and 8-hydroxy-2-deoxyguanosine enzyme detection in ovarian tissues after radiofrequency exposure at 1800 MHz in adult Spragueâ€"Dawley Rats. *HAYATI J Biosci* 24:79–79. https://doi.org/10.4308/ hjb.24.2.79.

Alekperov S, Suetov A, Efremov V, Kimstach AN, Lavrenenok LV (2019) The effect of electromagnetic fields of extremely low frequency 30 Hz on rat ovaries. *Bull Exp Biol Med* 166(5):704–707. https://doi.org/10.1007/S10517-019-04422-2.

Alkis ME, Bilgin HM, Akpolat V, et al (2019) Effect of 900-, 1800-, and 2100-MHz radiofrequency radiation on DNA and oxidative stress in brain. *Electromagn Biol Med* 38(1):32–47. https://doi.org/10.1080/15368378.2019.1567526.

Alkis ME, Akdag MZ, Dasdag S (2021) Effects of low-intensity microwave radiation on oxidant-antioxidant parameters and DNA damage in the liver of rats. *Bioelectromagnetics* 42(1):76–85. https://doi.org/10.1002/bem.22315.

Aly A, Crum R (2016) Research review for possible relation between mobile phone radiation and brain tumor. *Int J Inf Technol Converg Serv* 6:1–16. https://doi.org/10.5121/ijitcs.2016.6401.

Amara S, Abdelmelek H, Garrel C, et al (2006) Effects of subchronic exposure to static magnetic field on testicular function in rats. *Arch Med Res* 37(8):947–952. https://doi.org/10.1016/j.arcmed.2006.06.004.

Amjad F, Farooq MN, Batool R, Irshad A (2020) Frequency of wrist pain and its associated risk factors in students using mobile phones. *Pak J Med Sci* 36(4):746–749.

Anonymous (2022) How many smartphones are in the world? https://www.bankmycell.com/blog/how-many-phones-are-in-the-world#sources

Antonopoulos A, Yang B, Stamm A, Heller WD, Obe G (1995) Cytological effects of 50 Hz electromagnetic fields on human lymphocytes in vitro. *Mutat Res Lett* 346(3):151–157. https://doi.org/10.1016/0165-7992(95)90047-0.

Antonopoulos A, Eisenbrandt H, Obe G (1997) Effects of high-frequency electromagnetic fields on human lymphocytes in vitro. *Mutat Res - Genet Toxicol Environ Mutagen* 395(2–3):209–214. https://doi.org/10.1016/S1383-5718(97)00173-3.

Augner C, Florian M, Pauser G, Oberfeld G, Hacker GW (2009) GSM base stations: Short-term effects on well-being. *Bioelectromagnetics* 30(1):73–80. https://doi.org/10.1002/bem.20447.

Avendaño C, Mata A, Sanchez Sarmiento C, Doncel G (2012) Use of laptop computers connected to internet through Wi-fi decreases human sperm motility and increases sperm DNA fragmentation. *Fertil Steril* 97(1):39–45.e2. https://doi.org/10.1016/J.FERTNSTERT.2011.10.012.

Aweda MA, Usikalu MR, Wan JH, et al (2010) Genotoxic effects of low 2.45 GHz microwave radiation exposures on Sprague Dawley rats. *Int J Genet Mol Biol* 2:189–197. https://doi.org/10.5897/IJGMB.9000028.

Baker KB, Tkach JA, Nyenhuis JA, et al (2004) Evaluation of specific absorption rate as a dosimeter of MRI-related implant heating. *J Magn Reson Imaging* 20(2):315–320. https://doi.org/10.1002/jmri.20103.

Balode Z (1996) Assessment of radio-frequency electromagnetic radiation by the micronucleus test in bovine peripheral erythrocytes. *Sci Total Environ* 180(1):81–85. https://doi.org/10.1016/0048-9697(95)04923-1.

Banerjee S, Singh NN, Sreedhar G, Mukherjee S (2016) Analysis of the genotoxic effects of mobile phone radiation using buccal micronucleus assay: A comparative evaluation. *J Clin Diagn Res* 10(3):82–85. https://doi.org/10.7860/JCDR/2016/17592.7505.

Belpomme D, Irigaray P (2020) Electrohypersensitivity as a newly identified and characterized neurologic pathological disorder: How to diagnose, treat, and prevent it. *Int J Mol Sci* 21(6):1915. https://doi.org/10.3390/ijms21061915.

Belyaev IY, Hillert L, Protopopova M, et al (2005) 915 MHz microwaves and 50 Hz magnetic field affect chromatin conformation and 53BP1 foci in human lymphocytes from hypersensitive and healthy persons. *Bioelectromagnetics* 26(3):173–184. https://doi.org/10.1002/bem.20103.

Belyaev IY, Markova E, Hillert L, Malmgren LO, Persson BR (2009) Microwaves from UMTS/GSM mobile phones induce long-lasting inhibition of 53BP1/γ-H2AX DNARepair foci in human lymphocytes. *Bioelectromagnetics* 30(2):129–141. https://doi.org/10.1002/bem.20445.

Bento N (2016) Calling for change? Innovation, diffusion, and the energy impacts of global mobile telephony. *Energy Res Soc Sci* 21:84–100. https://doi.org/10.1016/j.erss.2016.06.016.

Bisht K, Moros E, Straube W, Baty JD, Roti Roti JL (2002) The effect of 835.62 MHz FDMA or 847.74 MHz CDMA modulated radiofrequency radiation on the induction of micronuclei in C3H 10T(1/2) cells. *Radiat Res* 157(5):506–515. https://doi.org/10.1667/0033-7587(2002)157[0506:teomfo]2.0.co;2.

Blank M, Goodman R (2009) Electromagnetic fields stress living cells. *Pathophysiology* 16(2–3):71–78. https://doi.org/10.1016/j.pathophys.2009.01.006.

Blank M, Goodman R (2011) DNA is a fractal antenna in electromagnetic fields. *Int J Radiat Biol* 87(4):409–415. https://doi.org/10.3109/09553002.2011.538130.

Boileau N, Margueritte F, Gauthier T, et al (2020) Mobile phone use during pregnancy: Which association with fetal growth? *J Gynecol Obstet Hum Reprod* 49(8):101852. https://doi.org/10.1016/J.JOGOH.2020.101852.

Bourthoumieu S, Joubert V, Marin B, et al (2010) Cytogenetic studies in human cells exposed in vitro to GSM-900 MHz radiofrequency radiation using R-banded karyotyping. *Radiat Res* 174(6):712–718. https://doi.org/10.1667/RR2137.1.

Brech A, Kubinyi G, Németh Z, et al (2019) Genotoxic effects of intermediate frequency magnetic fields on blood leukocytes in vitro. *Mutat Res - Genet Toxicol Environ Mutagen* 845:403060. https://doi.org/10.1016/j.mrgentox.2019.05.016.

Bryant PA (2021) Communicating radiation risk: The role of public engagement in reaching ALARA. *J Radiol Prot* 41(2):s1–S8. https://doi.org/10.1088/1361-6498/abd348.

Burlaka A, Tsybulin O, Sidorik E, et al (2013) Overproduction of free radical species in embryonic cells exposed to low intensity radiofrequency radiation. *Exp Oncol* 35(3):219–225.

Çam ST, Seyhan N (2012) Single-strand DNA breaks in human hair root cells exposed to mobile phone radiation. *Int J Radiat Biol* 88(5):420–424. https://doi.org/10.3109/09553002.2012.666005.

Campisi A, Gulino M, Acquaviva R, et al (2010) Reactive oxygen species levels and DNA fragmentation on astrocytes in primary culture after acute exposure to low intensity microwave electromagnetic field. *Neurosci Lett* 473(1):52–55. https://doi.org/10.1016/j.neulet.2010.02.018.

Carpenter DO (2013) Human disease resulting from exposure to electromagnetic fields. *Rev Environ Health* 28(4):159–172. https://doi.org/10.1515/reveh-2013-0016.

Cecconi S, Gualtieri G, Di Bartolomeo A, et al (2000) Evaluation of the effects of extremely low frequency electromagnetic fields on mammalian follicle development. *Hum Reprod* 15(11):2319–2325. https://doi.org/10.1093/HUMREP/15.11.2319.

Chavdoula ED, Panagopoulos DJ, Margaritis LH (2010) Comparison of biological effects between continuous and intermittent exposure to GSM-900-MHz mobile phone radiation: Detection of apoptotic cell-death features. *Mutat Res - Genet Toxicol Environ Mutagen* 700(1–2):51–61. https://doi.org/10.1016/j.mrgentox.2010.05.008.

Coghill RW, Steward J, Philips A (1996) Extra low frequency electric and magnetic fields in the bedplace of children diagnosed with leukaemia: A case-control study. *Eur J Cancer Prev* 5(3):153–158. https://doi.org/10.1097/00008469-199606000-0000.

D'Ambrosio G, Lioi MB, Massa R, Scarfi MR, Zeni O (1995) Genotoxic effects of amplitude-modulated microwaves on human lymphocytes exposed in vitro under controlled conditions. *Electromagn Biol Med* 14(3):157–164. https://doi.org/10.3109/15368379509030726.

D'Ambrosio G, Massa R, Scarfl MR, Zeni O (2002) Cytogenetic damage in human lymphocytes following GMSK phase modulated microwave exposure. *Bioelectromagnetics* 23(1):7–13. https://doi.org/10.1002/bem.93.

D'Silva MH, Swer RT, Anbalagan J, Rajesh B (2017) Effect of radiofrequency radiation emitted from 2G and 3G cell phone on developing liver of chick embryo – A comparative study. *J Clin Diagn Res* 11(7):AC05–AC09. https://doi.org/10.7860/JCDR/2017/26360.10275.

D'Silva MH, Swer RT, Anbalagan J, Bhargavan R (2021) Assessment of DNA damage in chick embryo brains exposed to 2G and 3G cell phone radiation using alkaline comet assay technique. *J Clin Diagn Res* 15:AC01–AC04. https://doi.org/10.7860/jcdr/2021/47115.14441.

Daroit NB, Visioli F, Magnusson AS, Vieira GR, Rados PV (2015) Cell phone radiation effects on cytogenetic abnormalities of oral mucosal cells. *Braz Oral Res* 29:1–8. https://doi.org/10.1590/1807-3107BOR-2015.vol29.0114.

De Iuliis GN, Newey RJ, King BV, Aitken RJ (2009) Mobile phone radiation induces reactive oxygen species production and DNA damage in human spermatozoa in vitro. *PLOS ONE* 4(7):e6446. https://doi.org/10.1371/journal.pone.0006446.

de Oliveira FM, Carmona AM, Ladeira C (2017) Is mobile phone radiation genotoxic? An analysis of micronucleus frequency in exfoliated buccal cells. *Mutat Res - Genet Toxicol Environ Mutagen* 822:41–46. https://doi.org/10.1016/j.mrgentox.2017.08.001.

Delimaris J, Tsilimigaki S, Messini-Nicolaki N, Ziros E, Piperakis SM (2006) Effects of pulsed electric fields on DNA of human lymphocytes. *Cell Biol Toxicol* 22(6):409–415. https://doi.org/10.1007/s10565-006-0105-1.

Demsia G, Vlastos D, Matthopoulos DP (2004) Effect of 910-MHz electromagnetic field on rat bone marrow. *Sci World J* 4(Suppl 2):48–54. https://doi.org/10.1100/tsw.2004.178.

Deniz O, Kaplan S, Selçuk M, et al (2017) Effects of short and long term electromagnetic fields exposure on the human hippocampus. *J Microsc Ultrastruct* 5(4):191–197. https://doi.org/10.1016/J.JMAU.2017.07.001.

Deshmukh PS, Megha K, Banerjee BD, et al (2013) Detection of low level microwave radiation induced deoxyribonucleic acid damage vis-à-vis genotoxicity in brain of Fischer rats. *Toxicol Int* 20(1):19–24. https://doi.org/10.4103/0971-6580.111549.

Diem E, Schwarz C, Adlkofer F, Jahn O, Rüdiger H (2005) Non-thermal DNA breakage by mobile-phone radiation (1800 MHz) in human fibroblasts and in transformed GFSH-R17 rat granulosa cells in vitro. *Mutat Res - Genet Toxicol Environ Mutagen* 583(2):178–183. https://doi.org/10.1016/j.mrgentox.2005.03.006.

Dilli R (2021) Implications of mmWave radiation on human health: State of the art threshold levels. *IEEE Access* 9:13009–13021. https://doi.org/10.1109/ACCESS.2021.3052387.

Duan W, Liu C, Zhang L, et al (2015) Comparison of the genotoxic effects induced by 50 Hz extremely low-frequency electromagnetic fields and 1800 MHz radiofrequency electromagnetic fields in GC-2 cells. *Radiat Res* 183(3):305–314. https://doi.org/10.1667/RR13851.1.

Erdal N, Gürgül S, Çelik A (2007) Cytogenetic effects of extremely low frequency magnetic field on Wistar rat bone marrow. *Mutat Res - Genet Toxicol Environ Mutagen* 630(1–2):69–77. https://doi.org/10.1016/j.mrgentox.2007.03.001.

Erogul O, Oztas E, Yildirim I, et al (2006) Effects of electromagnetic radiation from a cellular phone on human sperm motility: An in vitro study. *Arch Med Res* 37(7):840–843. https://doi.org/10.1016/J.ARCMED.2006.05.003.

Ferreira AR, Knakievicz T, de Bittencourt Pasquali MA, et al (2006) Ultra high frequency-electromagnetic field irradiation during pregnancy leads to an increase in erythrocytes micronuclei incidence in rat offspring. *Life Sci* 80(1):43–50. https://doi.org/10.1016/j.lfs.2006.08.018.

Focke F, Schuermann D, Kuster N, Schär P (2010) DNA fragmentation in human fibroblasts under extremely low frequency electromagnetic field exposure. *Mutat Res - Fundam Mol Mech Mutagen* 683(1–2):74–83. https://doi.org/10.1016/j.mrfmmm.2009.10.012.

Frei P, Mohler E, Bürgi A, et al (2009) A prediction model for personal radio frequency electromagnetic field exposure. *Sci Total Environ* 408(1):102–108. https://doi.org/10.1016/j.scitotenv.2009.09.023.

Fučić A, Garaj-Vrhovac V, Škara M, Dimitrovič B (1992) X-rays, microwaves and vinyl chloride monomer: Their clastogenic and aneugenic activity, using the micronucleus assay on human lymphocytes. *Mutat Res Lett* 282(4):265–271. https://doi.org/10.1016/0165-7992(92)90133-3.

Gadhia PK, Shah T, Mistry A, Pithawala M, Tamakuwala D (2003) A preliminary study to assess possible chromosomal damage among users of digital mobile phones. *Electromagn Biol Med* 22(2–3):149–159. https://doi.org/10.1081/ JBC-120024624.

Gandhi G, Anita (2005) Genetic damage in mobile phone users: Some preliminary findings. *Indian J Hum Genet* 11(2):99–104. https://doi.org/10.4103/0971-6866.16810.

Garaj-Vrhovac V, Horvat D, Koren Z (1990) The effect of microwave radiation on the cell genome. *Mutat Res Lett* 243(2):87–93. https://doi.org/10.1016/0165-7992(90)90028-I.

Garaj-Vrhovac V, Horvat D, Koren Z (1991) The relationship between colony-forming ability, chromosome aberrations and incidence of micronuclei in V79 Chinese hamster cells exposed to microwave radiation. *Mutat Res Lett* 263(3):143–149. https://doi.org/10.1016/0165-7992(91)90054-8.

Garaj-Vrhovac V, Fučić A, Horvat D (1992) The correlation between the frequency of micronuclei and specific chromosome aberrations in human lymphocytes exposed to microwave radiation in vitro. *Mutat Res Lett* 281(3):181–186. https://doi.org/10.1016/0165-7992(92)90006-4.

Garaj-Vrhovac V (1999) Micronucleus assay and lymphocyte mitotic activity in risk assessment of occupational exposure to microwave radiation. *Chemosphere* 39(13):2301–2312. https://doi.org/10.1016/S0045-6535(99)00139-3.

Garaj-Vrhovac V, Orescanin V (2009) Assessment of DNA sensitivity in peripheral blood leukocytes after occupational exposure to microwave radiation: The alkaline comet assay and chromatid breakage assay. *Cell Biol Toxicol* 25(1):33–43. https://doi.org/10.1007/s10565-008-9060-3.

Garson O, McRobert T, Campbell L, Hocking BA, Gordon I (1991) A chromosomal study of workers with long-term exposure to radio-frequency radiation. *Med J Aust* 155(5):289–292. https://doi.org/10.5694/J.1326-5377.1991.TB142282.X.

Görlitz B, Müller M, Ebert S, et al (2005) Effects of 1-week and 6-week exposure to GSM/DCS radiofrequency radiation on micronucleus formation in B6C3F1 mice. *Radiat Res* 164(4 Pt 1):431–439. https://doi.org/10.1667/RR3440.1.

Goodman EM, Greenebaum B, Marron MT, (1995): Effects of electromagnetic fields on molecules and cells. *International Rev. Cytol.* 158, 279–338.

Gosselin M-C, Kühn S, Kuster N (2013) Experimental and numerical assessment of low-frequency current distributions from UMTS and GSM mobile phones. *Phys Med Biol* 58(23):8339. https://doi.org/10.1088/0031-9155/58/23/8339.

Gul A, Celebi H, Uğraş S (2009) The effects of microwave emitted by cellular phones on ovarian follicles in rats. *Arch Gynecol Obstet* 280(5):729–733. https://doi.org/10.1007/s00404-009-0972-9.

Gulati S, Yadav A, Kumar N, et al (2016) Effect of GSTM1 and GSTT1 polymorphisms on genetic damage in humans populations exposed to radiation from mobile towers. *Arch Environ Contam Toxicol* 70(3):615–625. https://doi.org/10.1007/s00244-015-0195-y.

Gulati S, Yadav A, Kumar N, et al (2018) Phenotypic and genotypic characterization of antioxidant enzyme system in human population exposed to radiation from mobile towers. *Mol Cell Biochem* 440(1–2):1–9. https://doi.org/10.1007/s11010-017-3150-6.

Gulati S, Kosik P, Durdik M, et al (2020) Effects of different mobile phone UMTS signals on DNA, apoptosis and oxidative stress in human lymphocytes. *Environ Pollut* 267:115632. https://doi.org/10.1016/j.envpol.2020.115632.

Guler G, Tomruk A, Ozgur E, Seyhan N (2010) The effect of radiofrequency radiation on DNA and lipid damage in non-pregnant and pregnant rabbits and their newborns. *Gen Physiol Biophys* 29(1):59–66. https://doi.org/10.4149/gpb_2010_01_59.

Guler G, Ozgur E, Keles H, et al (2011) Apoptosis resulted from radiofrequency radiation exposure of pregnant rabbits and their infants. *Bull Vet Inst Pulawy* 55:127–134.

Güler G, Tomruk A, Ozgur E, et al (2012) The effect of radiofrequency radiation on DNA and lipid damage in female and male infant rabbits. *Int J Radiat Biol* 88(4):367–373. doi.org/10.3109/09553002.2012.646349.

Guo L, Lin J, Xue Y, et al (2019) Effects of 220 MHz pulsed modulated radiofrequency field on the sperm quality in rats. *Int J Environ Res Public Health* 16(7):1286. https://doi.org/10.3390/IJERPH16071286.

Gurbuz N, Sirav B, Yuvaci H, et al (2010) Is there any possible genotoxic effect in exfoliated bladder cells of rat under the exposure of 1800 MHz GSM-like modulated radio frequency radiation (RFR)? *Electromagn Biol Med* 29(3):98–104. https://doi.org/10.3109/15368378.2010.482498.

Gurbuz N, Sirav B, Colbay M, Yetkin I, Seyhan N (2014) No genotoxic effect in exfoliated bladder cells of rat under the exposure of 1800 and 2100 MHz radio frequency radiation. *Electromagn Biol Med* 33(4):296–301.

Gurbuz N, Sirav B, Kuzay D, Ozer C, Seyhan N, et al (2015) Does radio frequency radiation induce micronuclei frequency in exfoliated bladder cells of diabetic rats? *Endocr Regul* 49:126–130. https://doi.org/10.4149/ENDO_2015_03_126.

Gürler H, Bilgici B, Akar AK, Tomak L, Bedir A (2014) Increased DNA oxidation (8-OHdG) and protein oxidation (AOPP) by low level electromagnetic field (2.45 GHz) in rat brain and protective effect of garlic. *Int J Radiat Biol* 90(10):892–896. https://doi.org/10.3109/09553002.2014.922717.

Hanci H, Odaci E, Kaya H, et al (2013) The effect of prenatal exposure to 900-MHz electromagnetic field on the 21-old-day rat testicle. *Reprod Toxicol* 42:203–209. https://doi.org/10.1016/j.reprotox.2013.09.006.

Hardell L, Hallquist A, Mild KH, et al (2002) Cellular and cordless telephones and the risk for brain tumours. *Eur J Cancer Prev* 11(4):377–386. https://doi.org/10.1097/00008469-200208000-00010.

Hardell L, Carlberg M, Söderqvist F, Mild KH, Morgan LL (2007) Long-term use of cellular phones and brain tumours: Increased risk associated with use for ≥10 years. *Occup Environ Med* 64(9):626–632. https://doi.org/10.1136/oem.2006.029751.

Hardell L, Carlberg M (2009) Mobile phones, cordless phones and the risk for brain tumours. *Int J Oncol* 35(1):5–17. https://doi.org/10.3892/ijo_00000307.

Hardell L, Carlberg M, Söderqvist F, Mild KH (2013) Case-control study of the association between malignant brain tumours diagnosed between 2007 and 2009 and mobile and cordless phone use. *Int J Oncol* 43(6):1833–1845. https://doi.org/10.3892/ijo.2013.2111.

Hardell L, Carlberg M (2020) Health risks from radiofrequency radiation, including 5G, should be assessed by experts with no conflicts of interest. *Oncol Lett* 20(4):15. https://doi.org/10. 3892/ol.2020.11876.

Hardell L, Carlberg M (2021) Lost opportunities for cancer prevention: Historical evidence on early warnings with emphasis on radiofrequency radiation. *Rev Environ Health* 36(4):585–597. https://doi.org/10.1515/reveh-2020-0168.

Harremoes P, Gee D, MacGarvin M, et al. (Eds.). (2013). *The precautionary principle in the 20th century: Late lessons from early warnings.* London: Routledge.

Harris A, Cooper M (2019) Mobile phones: Impacts, challenges, and predictions. *Hum Behav Emerg Technol* 1(1):15–17. https://doi.org/10.1002/hbe2.112.

Hatch E, Willis S, Wesselink A, et al (2021) Male cellular telephone exposure, fecundability, and semen quality: Results from two preconception cohort studies. *Hum Reprod* 36(5):1395–1404. https://doi.org/10.1093/HUMREP/DEAB001.

Haumann T, Münzenberg UWE, Maes W, Sierck P (2002) HF-radiation levels of GSM cellular phone towers in residential areas. *Proc 2nd Int Work Biol Eff EMFS* 1:327–333.

Herrala M, Mustafa E, Naarala J, Juutilainen J (2018) Assessment of genotoxicity and genomic instability in rat primary astrocytes exposed to 872 MHz radiofrequency radiation and chemicals. *Int J Radiat Biol* 94(10):883–889. https://doi.org/10.1080/09553002.2018.1450534.

Hook GJ, Zhang P, Lagroye I, et al (2004) Measurement of DNA damage and apoptosis in Molt-4 cells after in vitro exposure to radiofrequency radiation. *Radiat Res* 161(2):193–200. https://doi.org/10.1667/RR3127.

Houston BJ, Nixon B, King BV, Aitken RJ, De Iuliis GN (2018) Probing the origins of 1,800 MHz radio frequency electromagnetic radiation induced damage in mouse immortalized germ cells and spermatozoa in vitro. *Front Public Health* 6:270. https://doi.org/10.3389/fpubh.2018.00270.

Houston BJ, Nixon B, McEwan KE, et al (2019) Whole-body exposures to radiofrequency-electromagnetic energy can cause DNA damage in mouse spermatozoa via an oxidative mechanism. *Sci Rep* 9(1):17478. https://doi.org/10.1038/s41598-019-53983-9.

Huang C-Y, Chang C-W, Chen C-R, Chuang C-Y, Chiang C-S, et al (2014) Extremely low-frequency electromagnetic fields cause G1 phase arrest through the activation of the ATM-Chk2-p21 pathway. PLoS ONE 9(8): e104732. https://doi.org/ 10.1371/journal.pone.0104732.

Hui W, Xu YS, Miao Lin W, et al (2017) Protective effect of naringin against the LPS-induced apoptosis of PC12 cells: Implications for the treatment of neurodegenerative disorders. *Int J Mol Med* 39(4):819–830. https://doi.org/10.3892/ijmm.2017.2904.

Hutter HP, Moshammer H, Wallner P, Kundi M (2006) Subjective symptoms, sleeping problems, and cognitive performance in subjects living near mobile phone base stations. *Occup Environ Med* 63(5):307–313. https://doi.org/10.1136/oem.2005.020784.

Hyland GJ (2000) Physics and biology of mobile telephony. *Lancet* 356(9244):1833–1836. https://doi.org/10.1016/s0140-6736(00)03243-8.

Hyland GJ (2008) Physical basis of adverse and therapeutic effects of low intensity microwave radiation. *Indian J Exp Biol* 46(5):403–419.

ICNIRP. (1998) Guidelines for limiting exposure to time-varying electric, magnetic, and electromagnetic fields (up to 300 GHz). *Health Phys* 74(4):494–521.

ICNIRP. (2010) Guidelines for limiting exposure to time-varying electric and magnetic fields (1 Hz to 100 kHz). *Health Phys* 99(6):818–836. https://doi.org/10.1097/HP.0b013e3181f06c86.

ICNIRP. (2020) Guidelines for limiting exposure to electromagnetic fields (100 kHz to 300 GHz). *Health Phys* 118(5):483–524. https://doi.org/10.1097/HP.0000000000001210.

Ivancsits S, Diem E, Pilger A, Rüdiger HW, Jahn O (2002) Induction of DNA strand breaks by intermittent exposure to extremely-low-frequency electromagnetic fields in human diploid fibroblasts. *Mutat Res - Genet Toxicol Environ Mutagen* 519(1–2):1–13. https://doi.org/10.1016/S1383-5718(02)00109-2.

Ivancsits S, Diem E, Jahn O, Rüdiger HW (2003) Intermittent extremely low frequency electromagnetic fields cause DNA damage in a dose-dependent way. *Int Arch Occup Environ Health* 76(6):431–436. https://doi.org/10.1007/s00420-003-0446-5.

Ivancsits S, Pilger A, Diem E, Jahn O, Rüdiger HW (2005) Cell type-specific genotoxic effects of intermittent extremely low-frequency electromagnetic fields. *Mutat Res - Genet Toxicol Environ Mutagen* 583(2):184–188. https://doi.org/10.1016/j.mrgentox.2005.03.011.

Jeong YJ, Son Y, Han NK, et al (2018) Impact of long-term RF-EMF on oxidative stress and neuroinflammation in aging brains of C57BL/6 mice. *Int J Mol Sci* 19(7):2103. https://doi.org/10.3390/ijms19072103.

Juutilainen J, Heikkinen P, Soikkeli H, Mäki-Paakkanen J (2007) Micronucleus frequency in erythrocytes of mice after long-term exposure to radiofrequency radiation. *Int J Radiat Biol* 83(4):213–220. https://doi.org/10.1080/09553000601169800.

Karaca E, Durmaz B, Altug H, et al (2012) The genotoxic effect of radiofrequency waves on mouse brain. *J Neurooncol* 106(1):53–58. https://doi.org/10.1007/s11060-011-0644-z.

Kerbacher JJ, Meltz ML, Erwin DN (1990) Influence of radiofrequency radiation on chromosome aberrations in CHO cells and its interaction with DNA-damaging agents. *Radiat Res* 123(3):311–319. https://doi.org/10.2307/3577738.

Kesari KK, Luukkonen J, Juutilainen J, Naarala J (2015) Genomic instability induced by 50Hz magnetic fields is a dynamically evolving process not blocked by antioxidant treatment. *Mutat Res - Genet Toxicol Environ Mutagen* 794:46–51. https://doi.org/10.1016/j.mrgentox.2015.10.004.

Khalil AM, Gagaa MH, Alshamali AM (2012) 8-Oxo-7, 8-dihydro-2′-deoxyguanosine as a biomarker of DNA damage by mobile phone radiation. *Hum Exp Toxicol* 31(7):734–740. https://doi.org/10.1177/0960327111433184.

Khalil AM, Alemam IF, Al-Qaoud KM (2020) Association between mobile phone using and DNA damage of epithelial cells of the oral mucosa. *J Biotechnol Biomed* 3(2):50–66. https://doi.org/10.26502/jbb.2642-91280027.

Kim J, Ha CS, Lee HJ, Song K (2010) Repetitive exposure to a 60-Hz time-varying magnetic field induces DNA double-strand breaks and apoptosis in human cells. *Biochem Biophys Res Commun* 400(4):739–744. https://doi.org/10.1016/j.bbrc.2010.08.140.

Kim J, Yoon Y, Yun S, et al (2012) Time-varying magnetic fields of 60Hz at 7mT induce DNA double-strand breaks and activate DNA damage checkpoints without apoptosis. *Bioelectromagnetics* 33(5):383–393. https://doi.org/10.1002/bem.21697.

Komatsubara Y, Hirose H, Sakurai T, et al (2005) Effect of high-frequency electromagnetic fields with a wide range of SARs on chromosomal aberrations in murine m5S cells. *Mutat Res* 587(1–2):114–119. https://doi.org/10.1016/J.MRGENTOX.2005.08.010.

Koyama S, Nakahara T, Wake K, et al (2003) Effects of high frequency electromagnetic fields on micronucleus formation in CHO-K1 cells. *Mutat Res - Genet Toxicol Environ Mutagen* 541(1–2):81–89. https://doi.org/10.1016/j.mrgentox.2003.07.009.

Koyama S, Isozumi Y, Suzuki Y, Taki M, Miyakoshi J (2004) Effects of 2.45-GHz electromagnetic fields with a wide range of SARs on micronucleus formation in CHO-K1 cells. *Sci World J* 4(Suppl 2):29–40. https://doi.org/10.1100/tsw.2004.176.

Koyama S, Narita E, Shimizu Y, et al (2016) Effects of long-term exposure to 60 GHz millimeter-wavelength radiation on the genotoxicity and heat shock protein (HSP) expression of cells derived from human eye. *Int J Environ Res Public Health* 13(8):802. https://doi.org/10.3390/ijerph13080802.

Koyama S, Narita E, Suzuki Y, et al (2019) Long-term exposure to a 40-GHz electromagnetic field does not affect genotoxicity or heat shock protein expression in HCE-T or SRA01/04 cells. *J Radiat Res* 60(4):417–423. https://doi.org/10.1093/jrr/rrz017.

Kumar G, McIntosh RL, Anderson V, McKenzie RJ, Wood AW (2015) A genotoxic analysis of the hematopoietic system after mobile phone type radiation exposure in rats. *Int J Radiat Biol* 91(8):664–672. https://doi.org/10.3109/09553002.2015.1047988.

Kumar S, Kesari KK, Behari J (2010) Evaluation of genotoxic effects in male Wistar rats following microwave exposure. *Indian J Exp Biol* 48(6):586–592. https://pubmed.ncbi.nlm.nih.gov/20882761/.

Lagroye I, Anane R, Wettring BA, et al (2004a) Measurement of DNA damage after acute exposure to pulsed-wave 2450 MHz microwaves in rat brain cells by two alkaline comet assay methods. *Int J Radiat Biol* 80(1):11–20. https://doi.org/10.1080/09553000310001642911.

Lagroye I, Hook G, Wettring B, et al (2004b) Measurements of alkali-labile DNA damage and protein-DNA crosslinks after 2450 MHz microwave and low-dose gamma irradiation in vitro. *Radiat Res* 161(2):201–214. https://doi.org/10.1667/RR3122.

Lai H, Singh NP (1995) Acute low-intensity microwave exposure increases DNA single-strand breaks in rat brain cells. *Bioelectromagnetics* 16(3):207–210. https://doi.org/10.1002/bem.2250160309.

Lai H, Singh NP (1996) Single- and double-strand DNA breaks in rat brain cells after acute exposure to radiofrequency electromagnetic radiation. *Int J Radiat Biol* 69(4):513–521. https://doi.org/10.1080/095530096145814.

Lai H, Singh NP (1997) Acute exposure to a 60 Hz magnetic field increases DNA strand breaks in rat brain cells. *Bioelectromagnetics* 18(2):156–165. https://doi.org/10.1002/(sici)1521-186x(1997)18:2<156::aid-bem8>3.0.co;2-1.

Lai H, Singh NP (2004) Magnetic field-induced DNA strand breaks in brain cells of the rat. *Environ Health Perspect* 112(6):687–694. https://doi.org/10.1289/ehp.6355.

Leach V, Weller S, Redmayne M (2018) A novel database of bio-effects from non-ionizing radiation. *Rev Environ Health* 33(3):273–280. https://doi.org/10.1515/reveh-2018-0017.

Lee JW, Kim MS, Kim YJ, et al (2011) Genotoxic effects of 3T magnetic resonance imaging in cultured human lymphocytes. *Bioelectromagnetics* 32(7):535–542. https://doi.org/10.1002/bem.20664.

Li L, Bisht K, LaGroye I, et al (2001) Measurement of DNA damage in mammalian cells exposed in vitro to radiofrequency fields at SARs of 3–5 W/kg. *Radiat Res* 156(3):328–332. https://doi.org/10.1667/0033-7587(2001)156[0328:moddim]2.0.co;2.

Li R, Ma M, Li L, et al (2018) The protective effect of autophagy on DNA damage in mouse spermatocyte-derived cells exposed to 1800 MHz radiofrequency electromagnetic fields. *Cell Physiol Biochem* 48(1):29–41. https://doi.org/10.1159/000491660.

Liu C, Duan W, Xu S, et al (2013a) Exposure to 1800 MHz radiofrequency electromagnetic radiation induces oxidative DNA base damage in a mouse spermatocyte-derived cell line. *Toxicol Lett* 218(1):2–9. https://doi.org/10.1016/j.toxlet.2013.01.003.

Liu C, Gao P, Xu SC, et al (2013b) Mobile phone radiation induces mode-dependent DNA damage in a mouse spermatocyte-derived cell line: A protective role of melatonin. *Int J Radiat Biol* 89(11):993–1001. https://doi.org/10.3109/09553002.2013.811309.

Liu C, Guo H, Chen J, Guo H, Chen J (2019) Research on pulse response characteristics of wireless ultraviolet communication in mobile scene. *Opt Express* 27(8):10670–10683. https://doi.org/10.1364/OE.27.010670.

Liu Y, Bin WB, Liu KJ, et al (2015) Effect of 50 Hz extremely low-frequency electromagnetic fields on the DNA methylation and DNA methyltransferases in mouse spermatocyte-derived cell line GC-2. *BioMed Res Int* 2015:237183. https://doi.org/10.1155/2015/237183.

Lloyd DC, Saunders RD, Finnon P, Kowalczuk CI (1984) No clastogenic effect from in vitro microwave irradiation of G0 human lymphocytes. *Int J Radiat Biol* 46(2):135–141. https://doi.org/10.1080/09553008414551211.

López-Furelos A, Salas-Sánchez AA, Ares-Pena FJ, Leiro-Vidal JM, López-Martín E (2018) Exposure to radiation from single or combined radio frequencies provokes macrophage dysfunction in the RAW 264.7 cell line. *Int J Radiat Biol* 94(6):607–618. https://doi.org/10.1080/09553002.2018.1465610.

López I, Félix N, Rivera M, Alonso A, Maestú C (2021) What is the radiation before 5G? A correlation study between measurements in situ and in real time and epidemiological indicators in Vallecas, Madrid. *Environ Res* 194:110734. https://doi.org/10.1016/J.ENVRES.2021.110734.

Lu X, Oda M, Ohba T, et al (2017) Association of excessive mobile phone use during pregnancy with birth weight: An adjunct study in Kumamoto of Japan environment and children's study. *Environ Health Prev Med* 22(1):52. https://doi.org/10.1186/S12199-017-0656-1.

Luukkonen J, Liimatainen A, Juutilainen J, Naarala J (2014) Induction of genomic instability, oxidative processes, and mitochondrial activity by 50 Hz magnetic fields in human SH-SY5Y neuroblastoma cells. *Mutat Res - Fundam Mol Mech Mutagen* 760:33–41. https://doi.org/10.1016/j.mrfmmm.2013.12.002.

Lv Y, Chen S, Zhu B, et al (2021) Exposure to 50 Hz extremely-low-frequency magnetic fields induces no DNA damage in cells by gamma H2AX technology. *BioMed Res Int* 2021:8510315. https://doi.org/10.1155/2021/8510315.

Maes A, Verschaeve L, Arroyo A, De Wagter C, Vercruyssen L (1993) In vitro cytogenetic effects of 2450 MHz waves on human peripheral blood lymphocytes. *Bioelectromagnetics* 14(6):495–501. https://doi.org/10.1002/BEM.2250140602

Maes A, Collier M, Van Gorp U, Vandoninck S, Verschaeve L (1997) Cytogenetic effects of 935.2-MHz (GSM) microwaves alone and in combination with Mitomycin C. *Mutat Res - Genet Toxicol Environ Mutagen* 393(1–2):151–156. https://doi.org/10.1016/S1383-5718(97)00100-9.

Maes A, Collier M, Slaets D, Verschaeve L (2000a) Cytogenetic investigations on microwaves emitted by a 455.7 MHz car phone. *Folia Biol* 46(5):175–180.

Maes A, Collier M, Vandoninck S, Scarpa P, Verschaeve L (2000b) Cytogenetic effects of 50 Hz magnetic fields of different magnetic flux densities. *Bioelectromagnetics* 21(8):589–596.

Maes A, Collier M, Verschaeve L (2001) Cytogenetic effects of 900 MHz (GSM) microwaves on human lymphocytes. *Bioelectromagnetics* 22(2):91–96.

Maes A, Van Gorp U, Verschaeve L (2006) Cytogenetic investigation of subjects professionally exposed to radiofrequency radiation. *Mutagenesis* 21(2):139–142. https://doi.org/10.1093/MUTAGE/GEL008.

Mahmoudabadi F, Ziaei S, Firoozabadi M, Kazemnejad A (2015) Use of mobile phone during pregnancy and the risk of spontaneous abortion. *J Environ Heal Sci Eng* 13:34. https://doi.org/10.1186/S40201-015-0193-Z.

Mailankot M, Kunnath AP, Jayalekshmi H, Koduru B, Valsalan R (2009) Radio frequency electromagnetic radiation (RF-EMR) from GSM (0.9/1.8GHz) mobile phones induces oxidative stress and reduces sperm motility in rats. *Clin (Sao Paulo)* 64(6):561–565. https://doi.org/10.1590/s1807-59322009000600011.

Malyapa RS, Ahern EW, Straube WL, et al (1997a) Measurement of DNA damage after exposure to 2450 MHz electromagnetic radiation. *Radiat Res* 148(6):608–617. https://doi.org/10.2307/3579737.

Malyapa RS, Ahern EW, Straube WL, et al (1997b) Measurement of DNA damage after exposure to electromagnetic radiation in the cellular phone communication frequency band (835.62 and 847.74 MHz). *Radiat Res* 148(6):618–627. https://doi.org/10.2307/3579738.

Manikowska-Czerska E, Czerski P, Leach WM (1985) Effects of 2.45 GHz microwaves on meiotic chromosomes of male CBA/CAYmice. *J Hered* 76(1):71–73. https://doi.org/10.1093/oxfordjournals.jhered.a110027.

Manna D, Ghosh R (2016) Effect of radiofrequency radiation in cultured mammalian cells: A review. *Electromagn Biol Med* 35(3):265–301. https://doi.org/10.3109/15368378.2015.1092158.

Manti L, Braselmann H, Calabrese ML, et al (2008) Effects of modulated microwave radiation at cellular telephone frequency (1.95 GHz) on X-ray-induced chromosome aberrations in human lymphocytes in vitro. *Radiat Res* 169(5):575–583. https://doi.org/10.1667/RR1044.1.

Mariucci G, Villarini M, Moretti M, et al (2010) Brain DNA damage and 70-kDa heat shock protein expression in CD1 mice exposed to extremely low fürequency magnetic fields. *Int J Radiat Biol* 86(8):701–710. https://doi.org/10.3109/09553001003789588.

Marková E, Hillert L, Malmgren L, Persson BR, Belyaev IY (2005) Microwaves from GSM mobile telephones affect 53BP1 and γ-H2AX foci in human lymphocytes from hypersensitive and healthy persons. *Environ Health Perspect* 113(9):1172–1177. https://doi.org/10.1289/ehp.7561.

Martynyuk V, Melnyk M, Artemenko A (2016) Comparison of biological effects of electromagnetic fields with pulse frequencies of 8 and 50 Hz on gastric smooth muscles. *Electromagn Biol Med* 35(2):143–151. https://doi.org/10.3109/15368378.2015.1028072.

Mashevich M, Folkman D, Kesar A, et al (2003) Exposure of human peripheral blood lymphocytes to electromagnetic fields associated with cellular phones leads to chromosomal instability. *Bioelectromagnetics* 24(2):82–90. https://doi.org/10.1002/bem.10086.

Mazor R, Korenstein-Ilan A, Barbul A, et al (2008) Increased levels of numerical chromosome aberrations after in vitro exposure of human peripheral blood lymphocytes to radiofrequency electromagnetic fields for 72 hours. *Radiat Res* 169(1):28–37. https://doi.org/10.1667/RR0872.1.

McNamee J, Bellier P, Gajda GB, et al (2002a) DNA damage in human leukocytes after acute in vitro exposure to a 1.9 GHz pulse-modulated radiofrequency field. *Radiat Res* 158(4):534–537. https://doi.org/10.1667/0033-7587(2002)158[0534:DDIHLA]2.0.CO;2.

McNamee JP, Bellier PV, Gajda GB, et al (2002b) DNA damage and micronucleus induction in human leukocytes after acute in vitro exposure to a 1.9 GHz continuous-wave radiofrequency field. *Radiat Res* 158(4):523–533. https://doi.org/10.1667/0033-7587(2002)158[0523:DDAMII]2.0.CO;2.

McNamee JP, Bellier PV, Gajda GB, et al (2003) No evidence for genotoxic effects from 24 h exposure of human leukocytes to 1.9 GHz radiofrequency fields. *Radiat Res* 159(5):693–697. https://doi.org/10.1667/0033-7587(2003)159[0693:NEFGEF]2.0.CO;2.

Meena R, Kumari K, Kumar J, et al (2014) Therapeutic approaches of melatonin in microwave radiations-induced oxidative stress-mediated toxicity on male fertility pattern of Wistar rats. *Electromagn Biol Med* 33(2):81–91. https://doi.org/10.3109/15368378.2013.781035.

Meral O, Ozgur E, Kismali G, et al (2016) GSM-like radiofrequency exposure induces apoptosis via caspase-dependent pathway in infant rabbits. *Bratisl Med J* 117(11):672–676. https://doi.org/10.4149/BLL_2016_129.

Miller AB, Morgan LL, Udasin I, Davis DL (2018) Cancer epidemiology update, following the 2011 IARC evaluation of radiofrequency electromagnetic fields (Monograph 102). *Environ Res* 167:673–683. https://doi.org/10.1016/j.envres.2018.06.043.

Miller A, Sears M, Morgan L, et al (2019) Risks to health and well-being from radio-frequency radiation emitted by cell phones and other wireless devices. *Front Public Health* 7:223. https://doi.org/10.3389/FPUBH.2019.00223.

Mizuno K, Shinohara N, Miyakoshi J (2015) In vitro evaluation of genotoxic effects under magnetic resonant coupling wireless power transfer. *Int J Environ Res Public Health* 12(4):3853–3863. https://doi.org/10.3390/ijerph120403853.

Moon J (2020) Health effects of electromagnetic fields on children. *Clin Exp Pediatr* 63(11):422–428. https://doi.org/10.3345/CEP.2019.01494.

Nakayama M, Nakamura A, Hondou T, Miyata H (2016) Evaluation of cell viability, DNA single-strand breaks, and nitric oxide production in LPS-stimulated macrophage RAW264 exposed to a 50-Hz magnetic field. *Int J Radiat Biol* 92(10):583–589. https://doi.org/10.1080/09553002.2016.1206224.

Navarro EA, Segura J, Portolés M, Gómez-Perretta C (2003) The microwave syndrome: A preliminary study in Spain. *Electromagn Biol Med* 22(2–3):161–169. https://doi.org/10.1081/JBC-120024625.

Nordenson I, Mild KH, Andersson G, Sandström M (1994) Chromosomal aberrations in human amniotic cells after intermittent exposure to fifty hertz magnetic fields. *Bioelectromagnetics* 15(4):293–301. https://doi.org/10.1002/bem.2250150404.

Okechukwu C (2020) Does the use of mobile phone affect male fertility? A mini-review. *J Hum Reprod Sci* 13(3):174–183. https://doi.org/10.4103/JHRS.JHRS_126_19.

Ono T, Saito Y, Komura JI, et al (2004) Absence of mutagenic effects of 2.45 GHz radiofrequency exposure in spleen, liver, brain, and testis of lacZ-transgenic mouse exposed in utero. *Tohoku J Exp Med* 202(2):93–103. https://doi.org/10.1620/tjem.202.93.

Othman OE, Aly MS, El Nahas SM, Mohamed HM (2003) Mutagenic potential of radio-frequency electromagnetic fields. *Cytol (Tokyo)* 68(1):35–43. https://doi.org/10.1508/cytologia.68.35.

Pall ML (2013) Electromagnetic fields act via activation of voltage-gated calcium channels to produce beneficial or adverse effects. *J Cell Mol Med* 17(8):958–965. https://doi.org/10.1111/jcmm.12088.

Panagopoulos DJ, Messini N, Karabarbounis A, Philippetis AL, Margaritis LH (2000) A mechanism for action of oscillating electric fields on cells. *Biochem Biophys Res Commun* 272(3):634–640. https://doi.org/10.1006/bbrc.2000.2746.

Panagopoulos DJ, Karabarbounis A, Margaritis LH (2002) Mechanism for action of electromagnetic fields on cells. *Biochem Biophys Res Commun* 298(1):95–102. https://doi.org/10.1016/S0006-291X(02)02393-8.

Panagopoulos DJ, Chavdoula ED, Nezis IP, Margaritis LH (2007) Cell death induced by GSM 900-MHz and DCS 1800-MHz mobile telephony radiation. *Mutat Res - Genet Toxicol Environ Mutagen* 626(1–2):69–78. https://doi.org/10.1016/j.mrgentox.2006.08.008.

Panagopoulos DJ, Chavdoula ED, Margaritis LH (2010) Bioeffects of mobile telephony radiation in relation to its intensity or distance from the antenna. *Int J Radiat Biol* 86(5):345–357.

Panagopoulos DJ (2012) Effect of microwave exposure on the ovarian development of Drosophila melanogaster. *Cell Biochem Biophys* 63(2):121–132. https://doi.org/10.1007/s12013-012-9347-0.

Panagopoulos DJ, Johansson O, Carlo GL (2013a) Evaluation of specific absorption rate as a dosimetric quantity for electromagnetic fields bioeffects. *PLOS ONE* 8(6):e62663. https://doi.org/10.1371/journal.pone.0062663.

Panagopoulos DJ, Karabarbounis A, Lioliousis C (2013b) ELF alternating magnetic field decreases reproduction by DNA damage induction. *Cell Biochem Biophys* 67(2):703–716. https://doi.org/10.1007/s12013-013-9560-5.

Panagopoulos DJ, Johansson O, Carlo GL (2015) Polarization: A key difference between man-made and natural electromagnetic fields, in regard to biological activity. *Sci Rep* 5:14914. https://doi.org/10.1038/srep14914.

Panagopoulos D (2019a) Comparing DNA damage induced by mobile telephony and other types of man-made electromagnetic fields. *Mutat Res* 781:53–62. https://doi.org/10.1016/J.MRREV.2019.03.003.

Panagopoulos DJ (2019b) Chromosome damage in human cells induced by UMTS mobile telephony radiation. *Gen Physiol Biophys* 38(5):445–454. https://doi.org/10.4149/gpb_2019032.

Panagopoulos DJ (2020) Comparing chromosome damage induced by mobile telephony radiation and a high caffeine dose: Effect of combination and exposure duration. *Gen Physiol Biophys* 39(6):531–544. https://doi.org/10.4149/gpb_2020036.

Panagopoulos D, Karabarbounis A, Yakymenko I, Chrousos G (2021) Human-made electromagnetic fields: Ion forced oscillation and voltage-gated ion channel dysfunction, oxidative stress and DNA damage (Review). *Int J Oncol* 59(5):92. https://doi.org/10.3892/ijo.2021.5272.

Pandey N, Giri S (2018) Melatonin attenuates radiofrequency radiation (900 MHz)-induced oxidative stress, DNA damage and cell cycle arrest in germ cells of male Swiss albino mice. *Toxicol Ind Health* 34(5):315–327. https://doi.org/10.1177/0748233718758092.

Paolucci T, Pezzi L, Centra A, et al (2020) Electromagnetic field therapy: A rehabilitative perspective in the management of musculoskeletal pain - A systematic review. *J Pain Res* 13:1385–1400. https://doi.org/10.2147/JPR.S231778.

Parasuraman S, Sam A, Yee SK, Chuon BLC, Ren LY (2017) Smartphone usage and increased risk of mobile phone addiction: A concurrent study. *Int J Pharm Investig* 7(3):125–131. https://doi.org/10.4103/jphi.jphi_56_17.

Pedersen GF (1997) Amplitude modulated RF fields stemming from a GSM/DCS-1800 phone. *Wirel Netw* 3(6):489–498. https://doi.org/10.1023/A:1019158712657.

Phillips JL, Ivaschuk O, Ishida-Jones T, et al (1998) DNA damage in molt-4 T-lymphoblastoid cells exposed to cellular telephone radiofrequency fields in vitro. *Bioelectrochem Bioenerg* 45(1):103–110. https://doi.org/10.1016/S0302-4598(98)00074-9.

Popov V, Shevchenko A (2019) Analysis of standards and norms of electromagnetic irradiation levels in wireless communication systems on railway transport. *Proced Comp Sci* 149:239–245. https://doi.org/10.1016/j.procs.2019.01.129.

Port M, Abend M, Römer B, Van Beuningen D (2003) Influence of high-frequency electromagnetic fields on different modes of cell death and gene expression. *Int J Radiat Biol* 79(9):701–708. https://doi.org/10.1080/09553000310001606803.

Rageh MM, El-Gebaly RH, El-Bialy NS (2012) Assessment of genotoxic and cytotoxic hazards in brain and bone marrow cells of newborn rats exposed to extremely low-frequency magnetic field. *J Biomed Biotechnol* 2012:716023. https://doi.org/10.1155/2012/716023.

Rashmi B, Chinna S, Rodrigues C, et al (2020) Occurrence of micronuclei in exfoliated buccal mucosal cells in mobile phone users: A case-control study. *Indian J Dent Res* 31(5):734–737. https://doi.org/10.4103/ijdr.IJDR_634_18.

Reddy SB, Weller J, Desjardins-Holmes D, et al (2010) Micronuclei in the blood and bone marrow cells of mice exposed to specific complex time-varying pulsed magnetic fields. *Bioelectromagnetics* 31(6):445–453. https://doi.org/10.1002/bem.20576.

Regalbuto E, Anselmo A, De Sanctis S, et al (2020) Human fibroblasts in vitro exposed to 2.45 GHz continuous and pulsed wave signals: Evaluation of biological effects with a multimethodological approach. *Int J Mol Sci* 21(19):1–24. https://doi.org/10.3390/ijms21197069.

Rezaie-Tavirani M, Hasanzadeh H, Seyyedi S, Zali H (2017) Proteomic analysis of the effect of extremely low-frequency electromagnetic fields (ELF-EMF) with different intensities in SH-SY5Y Neuroblastoma cell line. *J Lasers Med Sci* 8(2):79–83. https://doi.org/10.15171/jlms.2017.14.

Ros-Llor I, Sanchez-Siles M, Camacho-Alonso F, Lopez-Jornet P (2012) Effect of mobile phones on micronucleus frequency in human exfoliated oral mucosal cells. *Oral Dis* 18(8):786–792. https://doi.org/10.1111/j.1601-0825.2012.01946.x.

Ross CL, Pettenati MJ, Procita J, et al (2018) Evaluation of cytotoxic and genotoxic effects of extremely low-frequency electromagnetic field on mesenchymal stromal cells. *Glob Adv Heal Med* 7:216495611877747. https://doi.org/10.1177/2164956118777472.

Russell CL (2018) 5 G wireless telecommunications expansion: Public health and environmental implications. *Environ Res* 165:484–495. https://doi.org/10.1016/j.envres.2018.01.016.

Sakuma N, Komatsubara Y, Takeda H, et al (2006) DNA strand breaks are not induced in human cells exposed to 2.1425 GHz band CW and W-CDMA modulated radiofrequency fields allocated to mobile radio base stations. *Bioelectromagnetics* 27(1):51–57. https://doi.org/10.1002/BEM.20179.

Salford LG, Brun AE, Eberhardt JL, Malmgren L, Persson BR (2003) Nerve cell damage in mammalian brain after exposure to microwaves from GSM mobile phones. *Environ Health Perspect* 111(7):881–883. https://doi.org/10.1289/ehp.6039.

Sanie-Jahromi F, Saadat I, Saadat M (2016) Effects of extremely low frequency electromagnetic field and cisplatin on mRNA levels of some DNA repair genes. *Life Sci* 166:41–45. https://doi.org/10.1016/j.lfs.2016.10.006.

Sannino A, Calabrese ML, D'Ambrosio G, Massa R, Petraglia G, Mita P, et al (2006). Evaluation of cytotoxic and genotoxic effects in human peripheral blood leukocytes following exposure to 1950-MHz modulated signal. *IEEE Trans Plasma Sci*.34:1441–1448. https://doi.org/10.1109/TPS.2006.878379.

Santini R, Santini P, Danze JM, Le Ruz P, Seigne M (2002) Enquête sur la santé de riverains de stations relais de téléphonie mobile: I/Incidences de la distance et du sexe. *Pathol Biol* 50(6):369–373. https://doi.org/10.1016/S0369-8114(02)00311-5.

Sarimov R, Malmgren LOG, Markovà E, Persson B, Belyaev IY (2004) Nonthermal GSM microwaves affect chromatin conformation in human lymphocytes similar to heat shock. *IEEE Trans Plasma Sci* 32(4):1600–1608. https://doi.org/10.1109/TPS.2004.832613.

Sarkar S, Ali S, Behari J (1994) Effect of low power microwave on the mouse genome: A direct DNA analysis. *Mutat Res Toxicol* 320(1–2):141–147. https://doi.org/10.1016/0165-1218(94)90066-3.

Scarfì MR, Fresegna AM, Villani P, Pinto R, Marino C, Sarti M, et al (2006) Exposure to radiofrequency radiation (900 MHz, GSM signal) does not affect micronucleus frequency and cell proliferation in human peripheral blood lymphocytes: An interlaboratory study. Radiat Res. 165:655–663; https://doi.org/10.1667/RR3570.1.

Schmitz C, Keller E, Freuding T, Silny J, Korr H (2004) 50-Hz magnetic field exposure influences DNA repair and mitochondrial DNA synthesis of distinct cell types in brain and kidney of adult mice. *Acta Neuropathol* 107(3):257–264. https://doi.org/10.1007/s00401-003-0799-6.

Şekeroğlu V, Akar A, Şekeroğlu ZA (2012) Cytotoxic and genotoxic effects of high-frequency electromagnetic fields (GSM 1800 MHz) on immature and mature rats. *Ecotoxicol Environ Saf* 80:140–144. https://doi.org/10.1016/j.ecoenv.2012.02.028.

Şekeroğlu ZA, Akar A, Şekeroğlu V (2013) Evaluation of the cytogenotoxic damage in immature and mature rats exposed to 900 MHz radiofrequency electromagnetic fields. *Int J Radiat Biol* 89(11):985–992. https://doi.org/10.3109/09553002.2013.809170.

Sesia S, Toufik I, Baker M (2011). *LTE-the UMTS long term evolution: From theory to practice*. John Wiley & Sons.

Shah C, Anu N, Mehul N, Sonal B (2015) Cell phone radiation and genomic damage: In vitro exposure and assessment. *Int J Innov Res Sci Eng Technol* 4:401–405. https://doi.org/10.15680/ijirset.2015.0402025.

Sharma A, Shrivastava S, Shukla S (2020) Exposure of radiofrequency electromagnetic radiation on biochemical and pathological alterations. *Neurol India* 68(5):1092–1100. https://doi.org/10.4103/0028-3886.294554.

Shen Y, Xia R, Jiang H, et al (2016) Exposure to 50 Hz-sinusoidal electromagnetic field induces DNA damage-independent autophagy. *Int J Biochem Cell Biol* 77(A):72–79. https://doi.org/10.1016/j.biocel.2016.05.009.

Singh K, Prakash C, Nirala J, et al (2020) Acute radiofrequency electromagnetic radiation exposures cause neuronal DNA damage and impair neurogenesis in the young adolescent rat brain. *bioRxiv*. https://doi.org/10.1101/2020.11.07.370627.

Singh N, Lai H (1998) 60 Hz magnetic field exposure induces DNA crosslinks in rat brain cells. *Mutat Res - Fundam Mol Mech Mutagen* 400(1–2):313–320. https://doi.org/10.1016/S0027-5107(98)00017-7.

Skamrova GB, Lantushenko AO, Shckorbatov YG, Evstigneev MP (2013) Influence of mobile phone radiation on membrane permeability and chromatin state of human buccal epithelium cells. *Biochem Biophys* 1:22–28.

Smith-Roe SL, Wyde ME, Stout MD, et al (2020) Evaluation of the genotoxicity of cell phone radiofrequency radiation in male and female rats and mice following subchronic exposure. *Environ Mol Mutagen* 61(2):276–290. https://doi.org/10.1002/em.22343.

Solek P, Majchrowicz L, Bloniarz D, Krotoszynska E, Koziorowski M (2017) Pulsed or continuous electromagnetic field induce p53/p21-mediated apoptotic signaling pathway in mouse spermatogenic cells in vitro and thus may affect male fertility. *Toxicology* 382:84–92. https://doi.org/10.1016/j.tox.2017.03.015.

Speit G, Schütz P, Hoffmann H (2007) Genotoxic effects of exposure to radiofrequency electromagnetic fields (RF-EMF) in cultured mammalian cells are not independently reproducible. *Mutat Res* 626(1–2):42–47. https://doi.org/10.1016/J.MRGENTOX.2006.08.003.

Stronati L, Testa A, Villani P, et al (2004) Absence of genotoxicity in human blood cells exposed to 50 Hz magnetic fields as assessed by comet assay, chromosome aberration, micronucleus, and sister chromatid exchange analyses. *Bioelectromagnetics* 25(1):41–48. https://doi.org/10.1002/BEM.10141.

Stronati L, Testa A, Moquet J, et al (2006) 935 MHz cellular phone radiation. An in vitro study of genotoxicity in human lymphocytes. *Int J Radiat Biol* 82(5):339–346. https://doi.org/10.1080/09553000600739173.

Su L, Yimaer A, Wei X, Xu Z, Chen G (2017) The effects of 50 Hz magnetic field exposure on DNA damage and cellular functions in various neurogenic cells. *J Radiat Res* 58(4):488–500. https://doi.org/10.1093/jrr/rrx012.

Su L, Yimaer A, Xu Z, Chen G (2018) Effects of 1800 MHz RF-EMF exposure on DNA damage and cellular functions in primary cultured neurogenic cells. *Int J Radiat Biol* 94(3):295–305. https://doi.org/10.1080/09553002.2018.1432913.

Sun C, Wei X, Yimaer A, Xu Z, Chen G (2018) Ataxia telangiectasia mutated deficiency does not result in genetic susceptibility to 50 Hz magnetic fields exposure in mouse embryonic fibroblasts. *Bioelectromagnetics* 39(6):476–484. https://doi.org/10.1002/bem.22140.

Sun Y, Zong L, Gao Z, et al (2017) Mitochondrial DNA damage and oxidative damage in HL-60 cells exposed to 900 MHz radiofrequency fields. *Mutat Res - Fundam Mol Mech Mutagen* 797–799:7–14. https://doi.org/10.1016/j.mrfmmm.2017.03.001.

Svedenstål BM, Johanson KJ (1998) Leukocytes and micronucleated erythrocytes in peripheral blood from mice exposed to 50-Hz or 20-kHz magnetic fields. *Electro- and Magnetobiol* 17(2):127–143. https://doi.org/10.3109/15368379809022558.

Svedenstål BM, Johanson KJ, Mattsson MO, Paulsson LE (1999) DNA damage, cell kinetics and ODC activities studied in CBA mice exposed to electromagnetic fields generated by transmission lines. *In Vivo (Brooklyn)* 13(6):507–513. https://pubmed.ncbi.nlm.nih.gov/10757046/.

Sykes PJ, McCallum BD, Bangay MJ, Hooker AM, Morley AA (2001) Effect of exposure to 900 MHz radiofrequency radiation on intrachromosomal recombination in pKZ1 mice. *Radiat Res* 156:495–502. https://doi.org/10.1667/0033-7587(2001)156[0495:EOETMR]2.0.CO;2.

Szmigielski S (2013) Reaction of the immune system to low-level RF/MW exposures. *Sci Total Environ* 454–455:393–400. https://doi.org/10.1016/j.scitotenv.2013.03.034.

Testa A, Cordelli E, Stronati L, et al (2004) Evaluation of genotoxic effect of low level 50 Hz magnetic fields on human blood cells using different cytogenetic assays. *Bioelectromagnetics* 25(8):613–619. https://doi.org/10.1002/bem.20048.

Thamilselvan S, Behera A, Nair SK, et al (2021) Micronuclei analysis in people residing within 25 m of radiation-exposed areas around mobile towers in Chennai, India: An observational study. *J Int Oral Heal* 13:350–355. https://doi.org/10.4103/JIOH.JIOH-358-20.

Tice R, Hook G, Donner M, McRee DI, Guy AW (2002) Genotoxicity of radiofrequency signals. I. Investigation of DNA damage and micronuclei induction in cultured human blood cells. *Bioelectromagnetics* 23(2):113–126. https://doi.org/10.1002/BEM.104.

Trosic I, Busljeta I, Kasuba V, Rozgaj R (2002) Micronucleus induction after whole-body microwave irradiation of rats. *Mutat Res - Genet Toxicol Environ Mutagen* 521(1–2):73–79. https://doi.org/10.1016/S1383-5718(02)00214-0.

Trosic I, Busljeta I, Modlic B (2004) Investigation of the genotoxic effect of microwave irradiation in rat bone marrow cells: In vivo exposure. *Mutagenesis* 19(5):361–364. https://doi.org/10.1093/mutage/geh042.

Tsarna E, Reedijk M, Birks L, et al (2019) Associations of maternal cell-phone use during pregnancy with pregnancy duration and fetal growth in 4 birth cohorts. *Am J Epidemiol* 188(7):1270–1280. https://doi.org/10.1093/AJE/KWZ092.

Udroiu I, Antoccia A, Tanzarella C, et al (2015) Genotoxicity induced by foetal and infant exposure to magnetic fields and modulation of ionising radiation effects. *PLOS ONE* 10(11):e0142259. https://doi.org/10.1371/journal.pone.0142259.

Udroiu I, Cristaldi M, Ieradi LA, et al (2008) Genotoxic and haematotoxic damage induced by ELF magnetic fields. *Eur J Oncol* 13:239–244.

Ullrich V, Apell H-J (2021) Electromagnetic fields and calcium signaling by the voltage dependent anion channel. *Open J Vet Med* 11(1):57–86. https://doi.org/10.4236/ojvm.2021.111004.

Uslu N, Demirhan O, Emre M, Seydaoğlu G (2019) The chromosomal effects of GSM-like electromagnetic radiation exposure on human fetal cells. *Biomed Res Clin Pract* 4(4). https://doi.org/10.15761/brcp.1000192.

Vijayalaxmi FMR, Dusch SJ, Dusch SJ, et al (1997) Frequency of micronuclei in the peripheral blood and bone marrow of cancer-prone mice chronically exposed to 2450 MHz radiofrequency radiation. *Radiat Res* 147(4):495–500.

Vijayalaxmi, Seaman RL, Belt ML, Doyle JM, Mathur SP, Prihoda TJ, et al (1999) Frequency of micronuclei in the blood and bone marrow cells of mice exposed to ultra-wideband electromagnetic radiation. *Int J Radiat Biol* 75(1):115–120. https://doi.org/10.1080/095530099140870.

Vijayalaxmi, Pickard W, Bisht K, Bisht KS, et al (2001) Micronuclei in the peripheral blood and bone marrow cells of rats exposed to 2450 MHz radiofrequency radiation. *Int J Radiat Biol* 77(11):1109–1115. https://doi.org/10.1080/09553000110069100.

Vijayalaxmi, Logani MK, Bhanushali A, Bhanushali A, Ziskin MC, Prihoda TJ (2004) Micronuclei in peripheral blood and bone marrow cells of mice exposed to 42 GHz electromagnetic millimeter waves. *Radiat Res* 161(3):341–345. https://doi.org/10.1667/RR3121.

Villarini M, Ambrosini MV, Moretti M, et al (2013) Brain hsp70 expression and DNA damage in mice exposed to extremely low frequency magnetic fields: A dose-response study. *Int J Radiat Biol* 89(7):562–570. https://doi.org/10.3109/09553002.2013.782449.

Veerachari SB, Vasan SS (2012) Mobile phone electromagnetic waves and its effect on human ejaculated semen: An in vitro study. *Int J Infertil Fetal Med* 3(1):15–21.

Waldmann P, Bohnenberger S, Greinert R, et al (2013) Influence of GSM signals on human peripheral lymphocytes: Study of genotoxicity. *Radiat Res* 179(2):243–253. https://doi.org/10.1667/RR2914.1.

Wang X, Liu C, Ma Q, et al (2015) 8-oxoG DNA glycosylase-1 inhibition sensitizes Neuro-2a cells to oxidative DNA base damage induced by 900 MHz radiofrequency electromagnetic radiation. *Cell Physiol Biochem* 37(3):1075–1088. https://doi.org/10.1159/000430233.

Wang Y, Liu X, Zhang Y, et al (2019) Exposure to a 50 Hz magnetic field at 100 µT exerts no DNA damage in cardiomyocytes. *Biol Open* 8(8):bio041293. https://doi.org/10.1242/bio.041293.

Winker R, Ivancsits S, Pilger A, Adlkofer F, Rüdiger HW (2005) Chromosomal damage in human diploid fibroblasts by intermittent exposure to extremely low-frequency electromagnetic fields. *Mutat Res - Genet Toxicol Environ Mutagen* 585(1–2):43–49. https://doi.org/10.1016/j.mrgentox.2005.04.013.

Wolf FI, Torsello A, Tedesco B, et al (2005) 50-Hz extremely low frequency electromagnetic fields enhance cell proliferation and DNA damage: Possible involvement of a redox mechanism. *Biochim Biophys Acta - Mol Cell Res* 1743(1–2):120–129. https://doi.org/10.1016/j.bbamcr.2004.09.005.

Xu S, Chen G, Chen C, et al (2013) Cell type-dependent induction of DNA damage by 1800 MHz radiofrequency electromagnetic fields does not result in significant cellular dysfunctions. *PLOS ONE* 8(1):e54906. https://doi.org/10.1371/journal.pone.0054906.

Xu S, Zhou Z, Zhang L, et al (2010) Exposure to 1800 MHz radiofrequency radiation induces oxidative damage to mitochondrial DNA in primary cultured neurons. *Brain Res* 1311:189–196. https://doi.org/10.1016/j.brainres.2009.10.062.

Yadav AS, Sharma MK (2008) Increased frequency of micronucleated exfoliated cells among humans exposed in vivo to mobile telephone radiations. *Mutat Res - Genet Toxicol Environ Mutagen* 650(2):175–180. https://doi.org/10.1016/j.mrgentox.2007.11.005.

Yaguchi H, Yoshida M, Ejima Y, Miyakoshi J (1999) Effect of high-density extremely low frequency magnetic field on sister chromatid exchanges in mouse m5S cells. *Mutat Res - Genet Toxicol Environ Mutagen* 440(2):189–194. https://doi.org/10.1016/S1383-5718(99)00027-3.

Yakymenko I, Tsybulin O, Sidorik E, et al (2016) Oxidative mechanisms of biological activity of low-intensity radiofrequency radiation. *Electromagn Biol Med* 35(2):186–202. https://doi.org/10.3109/15368378.2015.1043557.

Yakymenko I, Burlaka A, Tsybulin O, et al (2018) Oxidative and mutagenic effects of low intensity GSM 1800 MHz microwave radiation. *Exp Oncol* 40(4):282–287. https://doi.org/10.31768/2312-8852.2018.40(4):282-287.

Yan J, Agresti M, Bruce T, et al (2007) Effects of cellular phone emissions on sperm motility in rats. *Fertil Steril* 88(4):957–964. https://doi.org/10.1016/J.FERTNSTERT.2006.12.022.

Yao K, Wu W, Wang KJ, et al (2008) Electromagnetic noise inhibits radiofrequency radiation-induced DNA damage and reactive oxygen species increase in human lens epithelial cells. *Mol Vis* 14:964–969.

Yao KTS (1982) Cytogenetic consequences of microwave irradiation on mammalian cells incubated in vitro. *J Hered* 73(2):133–138. https://doi.org/10.1093/oxfordjournals.jhered.a109596.

Yeung AWK (2019) The 'as low as reasonably achievable' (ALARA) principle: A brief historical overview and a bibliometric analysis of the most cited publications. *Radioprotection* 54(2):103–109. https://doi.org/10.1051/radiopro/2019016.

Yokus B, Cakir DU, Akdag MZ, Sert C, Mete N (2005) Oxidative DNA damage in rats exposed to extremely low frequency electro magnetic fields. *Free Radic Res* 39(3):317–323. https://doi.org/10.1080/10715760500043603.

Yuan LQ, Wang C, Lu DF, et al (2020) Induction of apoptosis and ferroptosis by a tumor suppressing magnetic field through ROS-mediated DNA damage. *Aging (Albany NY)* 12(4):3662–3681. https://doi.org/10.18632/aging.102836.

Zendehdel R, Yu IJ, Hajipour-Verdom B, Panjali Z (2019) DNA effects of low level occupational exposure to extremely low frequency electromagnetic fields (50/60 Hz). *Toxicol Ind Health* 35(6):424–430. https://doi.org/10.1177/0748233719851697.

Zeni O, Chiavoni A, Sannino A, et al (2003) Lack of genotoxic effects (micronucleus induction) in human lymphocytes exposed in vitro to 900 MHz electromagnetic fields. *Radiat Res* 160(2):152–158. https://doi.org/10.1667/RR3014.

Zeni O, Romanò M, Perrotta A, et al (2005) Evaluation of genotoxic effects in human peripheral blood leukocytes following an acute in vitro exposure to 900 MHz radiofrequency fields. *Bioelectromagnetics* 26(4):258–265. https://doi.org/10.1002/BEM.20078.

Zeni O, Gallerano G, Romanò M, et al (2007) Cytogenetic observations in human peripheral blood leukocytes following in vitro exposure to THz radiation: A pilot study. *Health Phys* 92(4):349–357. https://doi.org/10.1097/01.HP.0000251248.23991.35.

Zeni O, Schiavoni A, Perrotta A, et al (2008) Evaluation of genotoxic effects in human leukocytes after in vitro exposure to 1950 MHz UMTS radiofrequency field. *Bioelectromagnetics* 29(3):177–184. https://doi.org/10.1002/BEM.20378.

Zhu K, Lv Y, Cheng Q, Hua J, Zeng Q (2016) Extremely low frequency magnetic fields do not induce DNA damage in human lens epithelial cells in vitro. *Anat Rec (Hoboken)* 299(5):688–697. https://doi.org/10.1002/ar.23312.

Ziemann C, Brockmeyer H, Reddy SB, et al (2009) Absence of genotoxic potential of 902 MHz (GSM) and 1747MHz (DCS) wireless communication signals: In vivo two-year bioassay in B6C3F1 mice. *Int J Radiat Biol* 85(5):454–464. https://doi.org/10.1080/09553000902818907.

Zothansiama, ZM, Lalramdinpuii M, Jagetia GC, Jagetia GC (2017) Impact of radiofrequency radiation on DNA damage and antioxidants in peripheral blood lymphocytes of humans residing in the vicinity of mobile phone base stations. *Electromagn Biol Med* 36(3):295–305. https://doi.org/10.1080/15368378.2017.1350584.

Zotti-Martelli L, Peccatori M, Scarpato R, Migliore L (2000) Induction of micronuclei in human lymphocytes exposed in vitro to microwave radiation. *Mutat Res - Genet Toxicol Environ Mutagen* 472(1–2):51–58. https://doi.org/10.1016/S1383-5718(00)00112-1.

Zotti-Martelli L, Peccatori M, Maggini V, Ballardin M, Barale R (2005) Individual responsiveness to induction of micronuclei in human lymphocytes after exposure in vitro to 1800-MHz microwave radiation. *Mutat Res - Genet Toxicol Environ Mutagen* 582(1–2):42–52. https://doi.org/10.1016/j.mrgentox.2004.12.014.

Zuo H, Lin T, Wang D, et al (2014) Neural cell apoptosis induced by microwave exposure through mitochondria-dependent caspase-3 pathway. *Int J Med Sci* 11(5):426–435. https://doi.org/10.7150/ijms.6540.

Zwamborn A, Vossen S, van Leersum B, et al (2003) Effects of global communication system radio-frequency fields on well being and cognitive functions of human subjects with and without subjective complaints. *TNO Rep* FEL-03-C148:1–89.

5 DNA and Chromosome Damage in Human and Animal Cells Induced by Mobile Telephony Electromagnetic Fields and Other Stressors

Dimitris J. Panagopoulos[1,2,3]

[1]National Center for Scientific Research "Demokritos", Athens, Greece
[2]Choremeion Research Laboratory, Medical School, National and Kapodistrian University of Athens, Athens, Greece
[3]Electromagnetic Field-Biophysics Research Laboratory, Athens, Greece

CONTENTS

Abstract 190
5.1 Introduction 190
    5.1.1 Man-made EMFs, Wireless Communications, and Genotoxic Effects 190
    5.1.2 Fruit Fly Ovarian Cells and Oogenesis 193
    5.1.3 Human Peripheral Blood Lymphocytes 194
5.2 Materials and Methods 195
    5.2.1 Fruit Fly In Vivo Experiments. Detection of Ovarian DNA Fragmentation ... 195
    5.2.2 HPBL In Vitro Experiments. Detection of Chromatid-type Aberrations 197
5.3 Results 198
    5.3.1 DNA Fragmentation in Fruit Fly Ovarian Cells 198
    5.3.2 Chromatid-type Aberrations in HPBLs 202
5.4 Discussion 205
    5.4.1 MT EMFs Are More Genotoxic than Other Types of Man-made EMFs and Other Stressors 205
    5.4.2 Safety Limits Implied by the Experimental Findings 207
    5.4.3 Physical Parameters of EMFs Responsible for Bioactivity 208
    5.4.4 Conclusions 210
References 211

Keywords: Electromagnetic fields; mobile phone radiation; wireless communications; caffeine; DNA damage; DNA fragmentation; cell death; chromatid aberrations; human lymphocytes; fruit fly; oogenesis

Abbreviations: B-field: magnetic field. DECT: Digitally Enhanced Cordless Telecommunications. E-field: electric field. ELF: Extremely Low Frequency. EMF: electromagnetic field. EMR:

electromagnetic radiation. FC: follicle cell. G: germarium. GSM: Global System for Mobile Telecommunications. HPBL: human peripheral blood lymphocytes. LTE: Long Term Evolution. M: mitosis. MF: magnetic field. MT: mobile telephony. NC: nurse cell. NR: New Radio. OC: oocyte. PCD: programmed cell death. PEF: pulsed electric field. PHA: phytohemagglutinin. RF: Radio Frequency. S: stages of oogenesis, SAR: Specific Absorption Rate. SD: Standard Deviation. SICD: stress-induced cell death. TUNEL: terminal deoxynucleotide transferase dUTP nick end labeling. ULF: Ultra Low Frequency. UMTS: Universal Mobile Telecommunications System. VGCC: voltage-gated calcium channel. VGIC: voltage-gated ion channel. VoLTE: Voice over LTE. Wi-Fi: Wireless Fidelity. 1G, 2G, 3G, 4G, 5G: first, second, third, fourth, fifth generation of MT/WC.

ABSTRACT

Induction of DNA fragmentation in fruit fly ovarian cells after *in vivo* exposure and chromatid-type aberrations in human peripheral blood lymphocytes (HPBLs) after *in vitro* exposure to mobile telephony (MT) electromagnetic fields (EMFs) from mobile phones are presented. In both cases, the biological samples were exposed in close distance to a commercially available second or third/fourth generation (2G or 3G/4G) mobile phone handset during an active phone call in "talk" mode. The DNA fragmentation in fruit fly ovarian cells induced by 2G MT EMFs was compared with that induced by 50 Hz magnetic fields (MFs) similar to or much stronger than those of high-voltage power lines or a pulsed electric field (PEF) of similar characteristics with EMFs of atmospheric discharges (lightning) under identical conditions and experimental procedures. Respectively, the degree of chromosomal damage induced by *in vitro* exposure of HPBLs to 3G/4G MT EMF was compared to that induced by a high caffeine dose (~ 290 times above the permissible single dose for an adult human) administered to blood samples of the same subjects under identical conditions and experimental procedures. In the first case, it was shown that MT EMFs are much more damaging than high-voltage power line MFs or the PEF and more damaging than previous other stressors tested on the same biological system, such as certain cytotoxic chemicals, starvation, and dehydration. In the second case, it was shown that MT EMFs are similarly and even more damaging than the above extreme caffeine dose. The combination of this caffeine dose and the 3G/4G MT EMF exposure increased dramatically the number of aberrations in the blood samples of all subjects, suggesting that MT EMF exposure may be significantly more dangerous when combined with other stressors. The above findings allow useful conclusions regarding EMF bioactivity, cell sensitivity, and relevant EMF exposure limits.

5.1 INTRODUCTION

5.1.1 Man-made EMFs, Wireless Communications, and Genotoxic Effects

There is increasing concern in the scientific community and the general public about the adverse biological and health effects of man-made electromagnetic fields (EMFs) and corresponding electromagnetic radiation (EMR), especially those employed in modern wireless communications (WC), including mobile telephony (MT), cordless domestic phones (DECT: Digitally Enhanced Cordless Telecommunications), wireless internet (Wi-Fi: Wireless Fidelity), local connections among electronic devices (Bluetooth), etc.

All WC EMFs always combine Radio Frequency (RF)/microwave (300 kHz–300 GHz) carrier waves (of the order of GHz) with Extremely Low Frequency (ELF) (3–3000 Hz) pulsing and modulation in order to increase the rate of data transfer (speech, text, images, video, Internet, etc.) and the number of subscribers using WC networks simultaneously (multiplexing) (Pedersen 1997; Holma and Toskala 2004; Sauter 2011; Panagopoulos 2011; 2017; 2019a; 2019b; 2020). Due to unpredictably variable physical parameters each moment (connectivity, position, number of subscribers using the network, atmospheric conditions, etc.) the final signal of all digital WC EMFs is unpredictably varying in time, presenting intense random variability mainly in the Ultra Low Frequency (ULF) (0–3 Hz) band, which is an additional important bioactive component (Panagopoulos 2019a; Pirard and Vatovez).

The International Agency for Research on Cancer (IARC), which is a branch of the World Health Organization (WHO), under the weight of the accumulating scientific evidence, has classified both RF (actually WC) EMFs and ELF EMFs as possibly carcinogenic to humans (Group 2B) (IARC 2002; 2013; Baan et al. 2011). Based on additional scientific evidence after the 2011 IARC classification for RF EMFs, several studies have suggested that RF/WC EMFs should be re-evaluated and classified as probably carcinogenic (Group 2A) or carcinogenic (Group 1) to humans (Hardell et al. 2013; Hardell 2017; 2019; Hardell and Nyberg 2020; Carlberg and Hardell 2017; Hardell and Carlberg 2020; Miller et al. 2018; 2019; Panagopoulos 2019b; 2020; Panagopoulos et al. 2021).

In opposition to IARC's classifications for ELF and RF man-made EMFs, the International Commission on Non-Ionizing Radiation Protection (ICNIRP), a private organization which suggests EMF exposure limits for anthropogenic EMFs to be adopted by governments, claims that there are no established biological or health effects, apart from heating, and, thus, has suggested limits significantly higher than most technical EMF exposures in the everyday environment (ICNIRP 1998; 2020).

It has been revealed, and it is extremely important, that studies performed with simulated/invariable WC EMFs emitted by generators (following IARC's recommendations) grossly underestimate the biological activity of the real-life highly variable EMFs emitted by commercially available devices (Panagopoulos et al. 2015a; Panagopoulos 2017; 2019a; 2020; Leach et al. 2018; Pall 2018; Kostoff et al. 2020). As a result, while about 50% of the studies employing simulated exposures do not find any effects, studies employing real-life WC exposures from mobile phones, Wi-Fi routers, DECT phones, MT base antennas, etc., display more than 95% consistency in showing adverse effects (Panagopoulos et al. 2015a; Panagopoulos 2017; 2019a).

The second-generation (2G) MT system is called Global System for Mobile Telecommunications (GSM) and the third/fourth generation (3G/4G) is called Universal Mobile Telecommunications System (UMTS). The 4G, which is perhaps the most common system today, is a combination of UMTS or GSM for telephony, and a different system for broadband Internet access called "UMTS Long Term Evolution" (LTE). The pure 4G MT/WC system already in operation is called Voice over LTE (VoLTE) and employs the LTE technology also for telephony. Moreover, the telecommunications industry has already begun the installation of the 5G New Radio (NR) MT/WC system. The 5G system involves even higher carrier frequencies (up to 100 GHz) in order to be able to transmit higher amounts of data per second, and a much denser network of base antennas of potentially increased power and directional beams in order to compensate for the energy scattering loss due to the higher carrier frequency (Sauter 2011; Sesia et al. 2011; Neufeld and Kuster 2018; Agiwal and Jin 2018; Dahlman et al. 2018). A part of the scientific community, including most of those who are experts in the biological and health effects of WC EMFs, has expressed strong objections to 5G installation with concerns of highly increased health risk (McClelland and Jaboin 2018; Miller et al. 2018; 2019; Panagopoulos 2019a; 2019b; Hardell and Nyberg 2020; Hardell and Carlberg 2020; Kostoff et al. 2020).

Many published peer-reviewed studies have already reported genotoxic effects (DNA or chromosome damage and related effects) of WC EMFs (or of separate RF and ELF EMFs) on a variety of organisms and cell/tissue types. The number of such studies has increased considerably in recent years, in spite of attempts to dispute some of the early ones (see reviews Phillips et al. 2009; Panagopoulos 2019a; Panagopoulos et al. 2021; Lai 2021). A recent study of the US National Toxicology Program (NTP) found that rats exposed for 9 hours (h) per day for 2 years to a simulated near field of a MT antenna emitting 2G or 3G MT EMFs developed brain and heart cancer. Moreover, the study found significantly increased DNA damage (strand breaks) in the brains of exposed animals (NTP 2018; Melnick 2019; Smith-Roe et al. 2020), confirming the fact that DNA damage is intimately related with cancer initiation. In a similar Italian life-span study, rats exposed to a simulated GSM 1800 far-field were also found to develop heart schwannomas and brain glial tumors in agreement with the NTP study (Falcioni et al. 2018). Two studies that compared the bioactivity between 2G and 3G MT EMFs emitted by an active 2G/3G mobile phone found that both types of MT EMFs induced DNA damage and histological changes on the developing liver and brain of chick embryos, with 3G (UMTS) being even more genotoxic/bioactive than 2G (GSM) (D'Silva et al. 2017; 2021).

Apart from the experimental studies which directly show genotoxic effects, there is a great number of studies showing induction of oxidative stress (OS) after even very short exposures to man-made and especially WC EMFs due to overproduction of reactive oxygen species (ROS) in cells (Yakymenko et al. 2016). Since it has become increasingly evident that the genetic damage is executed by ROS released irregularly in the cells after impairment of the intra-cellular ionic concentrations by man-made EMFs (Panagopoulos et al. 2021), the OS findings are closely related to the genetic damage findings, and several studies have already recorded both effects to be triggered concomitantly (De Iuliis et al. 2009; Burlaka et al. 2013; Yakymenko et al. 2018).

In this chapter, I describe experiments with two different biological models exposed to MT EMFs. The effects were compared with those of exposure to other types of man-made EMFs, previously tested other stressors, and an extreme caffeine dose. The first model is the oogenesis of the fruit fly *Drosophila melanogaster* (strain Oregon R, wild-type), and the second is the human peripheral blood lymphocytes (HPBLs) during mitotic cell division and especially during the most sensitive G_2-early M (mitosis) phase of their cell division cycle.

I have argued that cells and cellular functions are essentially the same in all animals/living organisms, and since all biological effects initiate at the cellular level, effects recorded in animals (and even plants) are also expected to present in humans (Panagopoulos 2017; 2019a). My prediction was absolutely verified when I found genotoxic effects to be induced by MT EMFs in HPBLs in the blood samples of healthy volunteers (Panagopoulos 2019b; 2020) similarly as in animal cells. Moreover, my prediction is verified by the plethora of studies showing similar genotoxic effects induced by man-made EMFs in different animals, cell types, and humans described in the chapters of this book and in several review studies (Phillips et al. 2009; Yakymenko et al. 2016; Panagopoulos et al. 2021; Lai 2021).

Apart from remarkable gene similarities among humans and animals (including rodents, fruit flies, etc.), the basic cellular processes are identical in all animal (and even plant) cells. All cells in humans and animals have identically constructed cell membranes and are full of billions of identical mobile ions. such as calcium (Ca^{+2}), potassium (K^+), sodium (Na^+), etc., dissolved in the cytoplasm and the extracellular aqueous solutions flowing inward and outward in the cells through membrane ion channels maintaining specific intra-cellular concentrations, the changes of which initiate and accompany all cellular events. Moreover, all cells in all living organisms have the same intracellular organelles, such as mitochondria, ribosomes, endoplasmic reticulum, nucleus containing the cell's genomic DNA with the same basic structure, chemical elements and bonds in all organisms, etc. (Alberts et al. 1994; Stryer 1995; Panagopoulos 2012a; 2013; 2019a). Finally, all cells in humans and animals (and even plants) have the most sensitive electromagnetic sensors in their cell membranes, which are none other than the voltage-gated-ion channels (VGICs), the most abundant class of ion channels (Alberts et al. 1994; Stryer 1995; Panagopoulos et al. 2021; Bertagna et al. 2021). These similarities at the cellular level between all animals are more fundamental than differences in volume, mass, shape, macroscopic functions, intelligence, etc. Thus, any cellular effect induced by EMFs in animal cells (e.g., DNA damage) is expected to be induced also in the human cells and vice versa.

A mechanism by which polarized and coherent EMFs of low frequencies (ELF/ULF) cause dysfunction of the VGICs in cell membranes leading to disruption of the cell electrochemical and redox balance and homeostasis is described in Panagopoulos et al. (2000; 2002; 2015b; 2021). Dysfunction of the VGICs in cells can lead to DNA damage by overproduction of ROS (Panagopoulos et al. 2021). This is in line with the fact that man-made EMFs induce OS in living cells and tissues

(Yakymenko et al. 2016). What has been referred to by Pall (2018) as the voltage-gated calcium channel mechanism ("VGCC mechanism") is the same mechanism. While Pall correctly noted that VGCCs (actually VGICs) play a key role in man-made EMF bioeffects (Pall 2013; 2018), he failed to refer to the already existing publications (Panagopoulos et al. 2000; 2002; 2015b), presenting this mechanism as his own discovery (!) (see Panagopoulos 2021).

5.1.2 Fruit Fly Ovarian Cells and Oogenesis

The oogenesis of the fruit fly *Drosophila melanogaster* is a model biological system, very well studied, with a good timing of developmental stages under controlled conditions, and sensitive to record effects from various environmental stressors. This is combined with the short and well-defined life cycle of the animal due to which even systemic (whole organism) effects can be observed within a few hours or days (Panagopoulos 2012a; 2012b).

Each of the two ovaries of an adult female fruit fly consists of 16 to 20 ovarioles. Each ovariole is an individual egg assembly line, with new egg chambers produced in the most anterior cyst, called germarium (G). During oogenesis, new egg chambers produced by specific stem cells bud off the germarium in each ovariole and develop through 14 successive developmental stages (S1–S14), moving toward the posterior end to be fertilized and laid through the oviduct. Stages S1–S7 constitute the early oogenesis, S8–S10 the mid-oogenesis (also called vitellogenesis), and S11–S14 the late oogenesis (also called choriogenesis). Each egg chamber consists of a cluster of 16 germ cells, surrounded by an epithelial monolayer of somatic follicle cells (FCs) responsible for building the eggshell. In the germarium, the germline cyst originates from a single cell (cystoblast) which undergoes four mitotic divisions to form the 16-cell cluster. Among the 16 germ cells, one differentiates as the oocyte (OC) – the single cell which, after fertilization, will give the embryo – and the rest become nurse cells (NCs) which will serve as nutrients for the OC. Therefore, each egg chamber in the ovaries of female *Drosophila* consists of three different cell types; a single OC, 15 NCs, and up to ~ 1200 FCs (King 1970; Margaritis et al. 1980; Nezis et al. 2000; Drummond-Barbosa and Spradling 2001; McCall 2004; Baum et al. 2005; Pritchett et al. 2009; Panagopoulos 2012a).

The NCs and the FCs, undergo programmed cell death (PCD) during the late oogenesis stages S11–S14 after they have completed their role and are no longer needed, exhibiting DNA fragmentation, actin cytoskeleton disorganization, chromatin condensation, and phagocytosis of the cellular remnants by the adjacent follicle and epithelial cells (Nezis et al. 2000; Drummond-Barbosa and Spradling 2001; McCall 2004).

Apart from the PCD, which occurs naturally in NCs and FCs during late oogenesis, stress-induced cell death (SICD) may take place when the biological system is under stress of various types, such as starvation, dehydration, heat, cold, pollution, etc. In such cases, certain egg chambers may not develop normally and then their cells commit suicide, leading the whole egg chambers to degeneration in order for the system to prevent the waste of precious nutrients, (Nezis et al. 2000; Drummond-Barbosa and Spradling 2001; McCall 2004; Panagopoulos 2012a). Both PCD and SICD occur after DNA fragmentation. The most sensitive developmental stages during oogenesis for SICD are the G referred to as the "germarium checkpoint" or "early oogenesis checkpoint", and stages S7–S8 just before the onset of vitellogenesis (S8–S10), referred to as the "mid-oogenesis checkpoint" (Drummond-Barbosa and Spradling 2001; McCall 2004). Both checkpoints have been found to be very sensitive to stress factors, such as poor nutrition (Drummond-Barbosa and Spradling 2001) or exposure to cytotoxic chemicals such as etoposide or staurosporine (Nezis et al. 2000). In all cases, the stress-induced DNA fragmentation was observed at either one of the two checkpoints and only in the NCs and FCs, not in the OC. Apart from the two checkpoints, egg chambers were not observed before our experiments (Panagopoulos et al. 2007a) to degenerate during other stages of early or mid-oogenesis (Nezis et al. 2000; Drummond-Barbosa and Spradling 2001; McCall 2004; Baum et al. 2005; Pritchett et al. 2009; Panagopoulos 2012a). Later, it was discovered that dehydration can cause SICD at all stages of early and mid-oogenesis and not only at the two checkpoints, but again not in the OC (Zhukova and Kiseleva 2011).

Many studies have been carried out with fruit flies during the past 60 years and beyond, investigating RF/WC or ELF EMF effects on various endpoints. Most of these studies have found effects on reproduction, embryogenesis, development, OS, genetic damage, and mutations (Pay et al. 1978; Ramirez et al. 1983; Delgado 1985; Ma and Chu 1993; Kikuchi et al. 1998; Koana et al. 2001; Mirabolghasemi and Azarnia 2002; Panagopoulos et al. 2004; 2007a; 2007b; 2010; 2013; Panagopoulos 2012b; 2016; Atli and Unlu 2006; Manta et al. 2014; Margaritis et al. 2014; Sagioglou et al. 2016; Yanagawa et al. 2020; Gunes et al. 2021). Reviews on other fruit fly studies with exposures to RF and ELF EMFs can be found in Panagopoulos et al. (2004; 2013), and analysis of certain important methodological and other issues can be found in Panagopoulos and Karabarbounis (2020).

In our fruit fly experiments, we studied DNA fragmentation and consequent cell death induced by different types of man-made EMFs. In other words, we studied SICD not PCD. For this reason, late oogenesis egg chambers (S11–S14) were excluded during examination, and we only examined egg chambers from G up to S10.

Here, I have compared the DNA fragmentation in the ovarian cells induced by six different man-made EMFs under identical experimental conditions and procedures. These EMFs are GSM 900 MHz; GSM 1800 MHz; alternating sinusoidal 50 Hz magnetic field (MF) at three different intensities, similar or much stronger than those of high-voltage power lines; and a pulsed electric field (PEF) similar to those of atmospheric discharges (lightning). Apart from the DNA fragmentation that all six EMFs were found to induce in fruit fly ovarian cells, the MT (GSM) EMFs and the MFs were also found to decrease fecundity, while the PEF was found to increase fecundity in fruit flies (Panagopoulos et al. 2007a; 2010; 2013; Panagopoulos 2016; 2019a). The comparison in DNA fragmentation induction among the six EMFs is made in order to see which EMF is more genotoxic on the same biological system under identical conditions and draw conclusions on which physical parameter(s) of the various EMFs might be more crucial for the bioactivity.

5.1.3 Human Peripheral Blood Lymphocytes

HPBLs – naturally arrested in the G_0 phase of the cell division cycle and usually stimulated for mitosis (M) – are a well-known model for the assessment of genotoxicity of various environmental agents, such as ionizing radiation, chemicals, smoking, pharmaceuticals, man-made EMFs, etc. (IAEA 2011). Several studies have been conducted so far to investigate the effects of MT/WC EMFs on HPBLs *in vitro* or *in vivo* (Zeni et al. 2003; 2012; Ji et al. 2004; Belyaev et al. 2005; 2009; Markova et al. 2005; Baohong et al. 2005; 2007; Stronati et al. 2006; Schwarz et al. 2008; Manti et al. 2008; El-Abd and Eltoweissy 2012; Gulati et al. 2016; Danese et al. 2017; Zothansiama et al. 2017; Panagopoulos 2019b; 2020). The majority of these studies have used simulated MT EMFs produced by generators. Regardless of simulated or real-life exposures, most of the studies have found genotoxic effects induced by the MT/WC EMFs alone or in combination with other genotoxic agents (see also Panagopoulos 2019b; 2020).

One of these studies found DNA strand breaks and chromosomal aberrations to be induced by a UMTS-like (simulated) MT EMF *in vitro* at degrees increasing with increasing exposure duration. The effects were attributed to OS caused by the EMF exposure (El-Abd and Eltoweissy 2012). In another study which employed exposures to real-life MT EMFs (Ji et al. 2004), volunteers were exposed *in vivo* by talking on their GSM (2G) mobile phones for 4 h. After the exposure, DNA damage in their blood samples was significantly increased compared to their blood samples before the exposure. Stronati et al. (2006) did not find any DNA or chromosomal damage induced *in vitro* by a GSM-like exposure during the G_0 phase. In the Danese et al. (2017) study, blood samples were exposed *in vitro* for 30 min to a real GSM (2G) signal emitted by an activated mobile phone and no significant effect on DNA double-strand breaks was reported. While the authors of this study used a 3G mobile phone for the exposures, they did not test the 3G signal but the much older (and less bioactive) 2G signal. The authors did not report the mode of mobile phone operation during the exposures ("talk", "listening", or other), and they applied an assay (the foci method), which does not detect other types of DNA damage apart from double-strand breaks, without showing any foci pictures from exposed and control samples. Moreover, the blood samples were exposed during their resting G_0 phase (like in Stronati et al. 2006), while it is known that proliferating cells are much more vulnerable than resting cells of the same kind with most sensitive phases of the cell division cycle being M and G_2 (Nias 1998; Terzoudi et al. 2011). Certainly, *in vivo* studies inevitably employ exposures during G_0 since lymphocytes are normally in this phase. In such a case, the duration of exposure has to be longer, like in the Ji et al. (2004) study. Two studies examined effects on HPBLs from people residing in the vicinity of MT base stations (within 150 m or 80 m) and, thus, exposed *in vivo* to real-life MT EMFs/EMR emitted by the base antennas. Both studies Gulati et al. (2016) and Zothansiama et al. (2017) found significantly increased genetic damage compared to control groups residing more than 800 or 300 m, respectively, away from the antennas/cell towers.

A most sensitive assay to study the sensitivity of HPBLs to various environmental stressors is the G_2 assay. This allows observation at metaphase of unrepaired DNA damage induced during G_2 or early M phase and converted into chromatid-type aberrations in cells activated for mitosis (Terzoudi and Pantelias 2006; Pantelias and Terzoudi 2010; 2011; Terzoudi et al. 2011; Panagopoulos 2019a; 2020). The reason for the increased sensitivity of the G_2 phase of the mitotic cycle is related to the existence of a checkpoint. Checkpoints in biological systems (like those in fruit fly oogenesis) exist at the most sensitive stages of metabolic procedures and are cellular damage detectors and cell-repair or cell death activators. This checkpoint during the G_2 phase identifies cellular damage and either activates repair mechanisms or drives the cell to apoptosis when the damage is not reparable in order to prevent cells with genetic damage to enter the M phase and transfer their genetic modifications to subsequent cell generations (Mendelson et al. 1981; Pantelias and Terzoudi 2011).

Caffeine in high doses is considered a blocker/abrogator of the G_2 checkpoint. Moreover, it is well-known that the addition of caffeine in cell cultures results in DNA damage and chromatid aberrations (Kuhlmann et al. 1968; Pincheira and Lopez-Saez 1991). The caffeine dose used in my HPBL experiments is the dose considered to abrogate the G_2 checkpoint, and it is frequently used in laboratories working with HPBLs to study effects of various other stressors, such as ionizing radiation (Pantelias and Terzoudi 2011; Hatzi et al. 2015). Caffeine was previously classified by IARC as possibly carcinogenic to humans (IARC 1991) but was more recently re-evaluated as non-carcinogenic (IARC 2016).

In the HPBL experiments, I have compared genotoxicity (in terms of chromatid-type aberrations) between UMTS (3G/4G) MT EMFs, a high caffeine dose, and the combination of the two stressors in order to draw conclusions on which stressor is more genotoxic, genotoxicity of combined stress conditions, and comparison of existing safety limits among WC EMFs and caffeine.

5.2 MATERIALS AND METHODS

5.2.1 Fruit Fly In Vivo Experiments. Detection of Ovarian DNA Fragmentation

In each experiment with the six different EMFs, newly emerged adult *Drosophila melanogaster* insects (fruit flies) from the stock were collected, anesthetized very lightly with diethyl ether, and separated by sex (males from females). The collected flies were placed in groups of ten males and ten females in standard laboratory glass vials, with standard food forming a smooth plane surface 1 cm thick at the bottom of the vials. The glass vials were closed with cotton plugs. Detailed descriptions were given before (Panagopoulos et al. 2004; 2007a; 2007b; 2010; 2013; Panagopoulos 2016).

The exposures to the EMFs started on the first day of each experiment (day of eclosion), 1 h after all flies were fully awakened from the anesthesia, and lasted for a total of 120 h (5 days). The net duration of exposure/sham-exposure to each individual EMF, and the field/radiation intensities ± Standard Deviation (SD) were as follows:

a) Exposure/sham-exposure to the GSM 900 or 1800 EMFs for 6 minutes (min) every 24 h (36 min total) with the mobile phone during an active phone call (or turned off during sham-exposure) in "talk" mode and in contact with the vials (RF radiation intensity averaged over 6 min ~ 0.378 ± 0.059 mW/cm^2, ELF electric (E) field ~ 19 ± 2.5 V/m, ELF magnetic (B) field ~ 0.9 ± 0.15 mG for GSM 900, and ~ 30% lower corresponding values for GSM 1800, highest SAR of the handset for human head given by the manufacturer 0.89 W/kg) (for details see Panagopoulos et al. 2010).

b) Exposure/sham-exposure to the 50 Hz alternating MF 0.1, 1.1, or 2.1 mT (1, 11, or 21 G) (with co-existing E-field intensities due to magnetic induction 0.13, 1.43, or 2.72 V/m, respectively) continuously for the 5 days (120 h total) within specially designed and constructed coils. The vials with the fruit flies were suspended by use of non-conductive materials in the center of the coils. One coil produced MF and a corresponding electric field due to magnetic induction, and the other coil was constructed with paired antiparallel turns (spirals) producing zero magnetic and electric field but maintaining identical other conditions (temperature, light, etc.) with the exposure coil (for details see Panagopoulos et al. 2013).

c) Exposure/sham-exposure to the 8 kHz (44.4 Hz pulse repetition rate), 400 kV/m PEF for 30 min every 2 h during the 5 days (30 h total) in specially designed and constructed capacitors (one connected to the PEF generator, and the other for sham-exposure not connected with its plates short-circuited to ensure zero field) (for details see Panagopoulos 2016). [This PEF roughly resembles the atmospheric EMFs produced by lightning discharges during thunderstorms. These usually have a ~ 10 kHz carrier frequency (instead of 8 kHz) with a ~ 20 Hz pulse repetition (instead of 44.4 Hz). The shape of the pulses is in both EMFs bipolar damping) (Panagopoulos and Balmori 2017)].

Then, 120 h after the beginning of exposure/sham-exposure (in the morning of the sixth day and 1 h after the final 6-min exposure to the GSM EMFs), the flies were removed from the glass vials, the females were collected, anesthetized, and dissected. Egg chambers from germarium up to S10 were collected from both ovaries of each female fly and fixed for the TUNEL (terminal deoxynucleotide transferase dUTP nick end labeling) assay, as described before (Panagopoulos et al. 2007a; 2010; 2013; Panagopoulos 2016). The glass vials with the embryos (fertilized eggs) were kept in the culture room for assessing effects on reproduction (see Panagopoulos et al. 2004; 2007b; 2010; 2013; Panagopoulos 2016).

The TUNEL assay is a known marker for DNA fragmentation (severe DNA damage including both single- and double-strand breaks). According to this assay, fluorescein dUTP (a fluorescent substance) binds through the action of terminal transferase (an enzyme that catalyzes the specific biochemical reaction) onto fragmented genomic DNA, which then becomes labeled by characteristic fluorescence. The label incorporated at the damaged sites of DNA is visualized by fluorescence microscopy (Gavrieli et al. 1992; Nezis 2000; Panagopoulos et al. 2007a).

The fruit fly groups exposed to the six different EMFs were compared with their corresponding sham-exposed groups for the percentages of egg chambers with DNA fragmentation (TUNEL-positive signal). Comparisons between groups were statistically analyzed by the single-factor Analysis of Variance test (Microsoft Excel program). The P-values provided by the statistical test represent the probabilities that differences between exposed and corresponding sham-exposed groups are due to random variation. P-values < 0.05 are generally accepted as representing statistically significant differences among compared groups (Maber 1999).

5.2.2 HPBL IN VITRO EXPERIMENTS. DETECTION OF CHROMATID-TYPE ABERRATIONS

Blood Culture and Separation into Individual Groups: Blood samples were collected from six healthy non-smoker adult volunteers (one sample from one subject in each experiment) in heparinized glass tubes for analysis of chromosomal sensitivity to mobile phone exposure, to a high caffeine dose, and to the combination of the two stressors. The subjects were both males and females, 28–42 years old, non-smokers, with "moderate" mobile phone use (no more than ~ 30 min total daily conversation on their mobile phones) and no reported history of major illnesses or any regular medication. Apart from this, no specific differences between the subjects were searched, as each subject had its own control (sham-exposed) sample.

In each experiment (one for each subject), a single culture was prepared in a 200 ml flask to ensure identical culture conditions and treatment for all individual samples. The initial whole blood sample of each subject was cultured in RPMI medium, with the addition of phytohemagglutinin (PHA) (2% of the final medium volume) to stimulate the lymphocytes (normally arrested in the G_0 phase) to enter the mitotic cycle. The single culture was incubated for 72 h, at 37°C and then subdivided into individual samples in identical 30 ml rectangular plastic flasks (for details see Panagopoulos 2019b; 2020).

Each individual sample in each experiment contained 0.5 ml blood, 5 ml culture medium, and 100 μl PHA. One individual sample was exposed to the UMTS EMF for 15 min, and another one sham-exposed. Two additional individual samples/groups were treated with caffeine. One of them was exposed for 15 min to the UMTS EMF, and the other sham-exposed. Therefore, the number of individual samples in this set of experiments were 4 (one sham-exposed/control, one UMTS 15 min-exposed, one caffeine-treated alone and sham-exposed to the EMF, and one treated with caffeine and UMTS EMF for 15 min). Right after the separation of the initial culture into individual samples and the addition of caffeine in the specific individual samples as described above, the blood samples were either exposed to the UMTS MT EMF in another room of the laboratory (called "exposure room") or sham-exposed (simply transferred for 15 min to the exposure room at the same location where the exposures took place).

Exposures were performed by a UMTS (3G) commercially available "smart" mobile phone handset. SAR value of the handset for the human head provided by the manufacturer was 0.66 W/kg. Any additional EMF emissions (such as Wi-Fi and Bluetooth) of the "smart" phone were disabled.

EMF and Caffeine Dosimetry: The RF radiation intensity and the ELF E- and B-fields (ELF-E, ELF-B) emitted by the handset during the exposures were measured at 1 cm distance from the handset. Averaged radiation intensity (power density) over 6 min was 29 ± 14 μW/cm^2, which is ~ 34 times below the corresponding ICNIRP (1998) general public limit of 1000 μW/cm^2 for the frequency of 2–6 GHz, and ~ 136 times below the latest ICNIRP (2020) corresponding limit which was raised (!) to 4000 μW/cm^2. The carrier frequency was variable at ~ 1920–1960 MHz during the exposures. Representative average ELF-E and ELF-B (from five representative instant measurements excluding background) ± SD at 100 Hz was 12 ± 4.2 V/m and 0.9 ± 0.4 mG, respectively. Corresponding average ELF-E and ELF-B (from five instant measurements excluding background) ± SD at 1500 Hz was 8 ± 4.6 V/m and 0.6 ± 0.2 mG, respectively (for details see Panagopoulos 2020).

In each individual sample treated with caffeine, 200 μl caffeine solution was added (containing 0.2 g caffeine per 10 ml PBS). Therefore, each individual caffeine-treated sample with a total volume of 5.8 ml (5 ml culture medium, 0.5 ml blood, 100 μl PHA, and 200 μl caffeine solution) contained 4 mg caffeine, and the final caffeine concentration was ~ 3.4 mM. Thus, to 0.5 ml blood of each sample (8.6% of the sample volume) corresponded ~ 0.34 mg caffeine (8.6% of the caffeine in the sample), and to the ~ 5 l whole blood contained in a 70 kg adult human body correspond ~ 3.4 g caffeine. In the case of drinking coffee, caffeine does not go directly into the blood, as it first passes through the peptic system and is distributed to all tissues. It is estimated that, from an initial amount, A g of caffeine administered by coffee drinking, about ~ A/17, is finally dissolved into the blood (Cook et al. 1996; Higdon and Frei 2006). Therefore, the amount of caffeine administered directly into each caffeine-treated blood sample corresponds to approximately 3.4 × 17 = 57.8 g caffeine taken by coffee by an adult 70 kg individual. The permitted single caffeine dose for a 70 kg adult is 0.2 g, which is ~ 290 times less than the amount of 57.8 g which corresponds to the dose used in the experiments (EFSA 2015). Thus, the caffeine dose used in the experiments was ~ 290 times higher than the permitted single dose for an adult individual of 70 kg body weight.

EMF Exposure Procedure: Two individual blood samples from each subject (one with caffeine and one without) were simultaneously exposed within the 30 ml flasks by the UMTS (3G) mobile phone handset during an active phone call ("talk" signal) for 15 min at 1 cm distance from the proximal flask wall. After the exposed samples were back in the culture room, the corresponding control (sham-exposed) samples were also transferred to the exposure room for 15 min at the same location without being exposed to the MT EMF. This was done because the background ELF-E and ELF-B and the light conditions in the two rooms were different. The temperature in both rooms was kept at 22 ± 1°C. Temperature increases within the blood samples during the 15 min exposures did not exceed 0.1°C (for details see Panagopoulos 2019b; 2020).

Metaphase Arrest, Fixation, and Observation: After exposures/sham-exposures were completed (~ 30–60 min after the beginning of the exposure procedure) and all exposed and sham-exposed samples were returned back to the culture room, all individual samples were treated with colcemid (50 μl added to each individual sample) for 60 min to arrest dividing cells at metaphase. Colcemid prohibits the formation of the attractus and, thus, prohibits metaphase cells from progressing to anaphase. The duration of colcemid treatment (60 min) right after the termination of exposure/sham-exposure, plus the exposure/sham-exposure time (1.5–2 h in total), determines the phase(s) of the cell division cycle, during which the collected metaphases were exposed. In this case, the 1.5–2 h-period determines that the cells were at the mid-late G_2 or early M (prophase) stages during the exposure/sham exposure. It also determines the duration of caffeine treatment of the specific samples.

Cells were then collected by centrifugation, treated for 10 min with hypotonic KCl solution 75 mM (Sigma-Aldrich, USA), fixed in methanol: glacial acetic acid (3:1 v/v), and stained for 10 min with 5% Giemsa solution (Merck, Germany) to be observed by light microscopy coupled with an image analysis system (Ikaros MetaSystems, Germany) to facilitate scoring.

Chromosomal damage was evaluated by the number of chromatid gaps (achromatic lesions) and chromatid breaks (terminal deletions) in cells at metaphase. For each of the four individual samples of each subject (described above), 400 metaphases identically processed from four different slides (100 cells from each slide), were blindly scored for gaps and breaks. Mean values of the total number of aberrations (gaps and breaks) per cell and SD in all samples were calculated for each individual. Gaps were scored only when extended across the full chromatid width. An aberration was considered as a "break" when the gap width was equal to or greater than the chromatid width.

Differences in the mean number of chromatid-type aberrations per cell among individual samples of each subject were statistically analyzed by application of the Student's *t*-test for unequal variances (Microsoft Excel program). The *P*-values < 0.05 for the probability that differences between individual samples are due to random variations were accepted as statistically significant.

5.3 RESULTS

5.3.1 DNA Fragmentation in Fruit Fly Ovarian Cells

The comparison of induced DNA fragmentation in ovarian cells (sum ratio of egg chambers with a TUNEL-positive signal to total number of examined egg chambers) between female fruit flies exposed to GSM 900, GSM 1800, 50 Hz MF 0.1 mT (MF1), 1.1 mT (MF2), 2.1 mT (MF3), and PEF is presented in Table 5.1, and Figure 5.1. MT EMFs (GSM 900 and 1800 MHz) were found to be significantly more detrimental than the 50 Hz MF similar to or much stronger than that of high-voltage power lines and significantly more detrimental than the PEF, inducing DNA fragmentation in a much higher degree, even though the durations of exposure to the 50 Hz MFs or to the PEF were much longer (120 h and 30 h, respectively) than the exposure to the MT EMFs (36 min). The PEF was found to induce the smallest effect despite its high intensity (400 kV/m).

More specifically, GSM 900 or GSM 1800 mobile phone radiation with a total exposure duration **36 min** increased DNA fragmentation in the egg chambers of the exposed females by up to **670%** with regard to the sham-exposed females (with the GSM 900 being more bioactive than GSM 1800, basically because of its higher intensity) (Panagopoulos et al. 2007b; 2010) (Table 5.1, Figure 5.1).* The corresponding increases in DNA fragmentation for 0.1, 1.1, and 2.1 mT 50 Hz MF exposure

TABLE 5.1
Effect of Various EMFs on Ovarian DNA Fragmentation

EMF	Exposed Groups Ratio of TUNEL-Positive to Total Number of Egg Chambers ± SD	Sham-Exp. Groups Ratio of TUNEL-Positive to Total Number of Egg Chambers ± SD	Deviation between Exposed and Sham-Exp. Groups	P-Value between Exposed and Sham-Exp. Groups
GSM 900	0.5772 ± 0.083	0.075 ± 0.038	+670%	< 0.0002
GSM 1800	0.4339 ± 0.087	0.062 ± 0.034	+600%	< 0.0005
MF 1	0.1243 ± 0.019	0.0671 ± 0.014	+85%	< 0.001
MF 2	0.1367 ± 0.02	0.0696 ± 0.018	+96%	< 0.001
MF 3	0.1407 ± 0.021	0.0655 ± 0.019	+115%	< 0.001
PEF	0.0848 ± 0.012	0.0574 ± 0.012	+48%	< 0.05

Sham-Exp: Sham-Exposed.

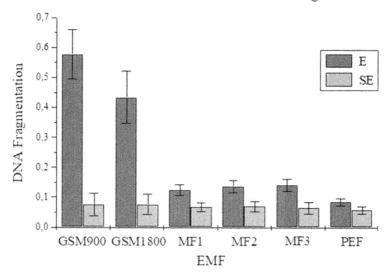

FIGURE 5.1 Ovarian DNA fragmentation (ratio of TUNEL-positive egg chambers to total number of examined egg chambers), induced by six different EMFs (GSM 900, GSM 1800, 0.1 mT MF (MF1), 1.1 mT MF (MF2), 2.1 mT MF (MF3), and 400 kV/m PEF) under identical conditions and procedures. E: exposed groups, SE: sham-exposed groups.

---

* The increases in DNA fragmentation refer to the % deviation in the ratio of egg chambers with fragmented DNA between exposed and sham-exposed females. If we referred to the difference in the percentages of DNA fragmentation between exposed and sham-exposed, the corresponding numbers would be much smaller, like in the original publications of the experiments (Panagopoulos et al. 2010; 2013; Panagopoulos 2016), and in Panagopoulos (2019a). Here, the results have been expressed in % deviation from the sham-exposed groups as in the HPBL experiments (see 5.3.2).

with regard to their corresponding sham-exposures were up to **115%** with a total exposure duration **120 h** (Table 5.1, Figure 5.1). Interestingly, the effect of the 50 Hz MF on DNA fragmentation was better correlated with the induced electric field than with the MF itself (Panagopoulos et al. 2013). Finally, the corresponding increase in DNA fragmentation due to the PEF exposure was **48%** with a total exposure duration **30 h** (Panagopoulos 2016; 2019a), (Table 5.1, Figure 5.1).

It is important to note that the mobile phone EMFs/EMR exposed the animals at the very same intensity levels as users are daily exposed by mobile phones, while the intensities of the other EMFs were significantly higher than those existing in the daily human environment: 1) The strongest MF intensity measured at the closest proximity to the most powerful high-voltage power lines is usually significantly less than 1 G or 0.1 mT (Panagopoulos et al. 2013). In our experiments exposure to 0.1 mT (1 G) caused 85% increase in ovarian DNA fragmentation (up to 8 times less than the 670% increase caused by the MT EMFs) (Table 5.1, Figure 5.1, and Panagopoulos 2019a). 2) The PEF, similar to those of atmospheric discharges (sferics), exposed the animals at 400 kV/m, while sferics are sensed by sensitive individuals at (totally polarized) intensities down to ~ 0.35 V/m (approximately ~ 1000 km from a thunderstorm). Even the closest proximity of humans to lightning discharges during a thunderstorm, normally corresponds to significantly lower PEF intensity than 400 kV/m (Panagopoulos and Balmori 2017). The exposure to the 400 kV/m PEF induced a 48% increase in DNA fragmentation with regard to the sham-exposed group (up to 14 times less than the 670% increase caused by the MT EMFs) (Table 5.1, Figure 5.1, and Panagopoulos 2019a).

From the above comparison, it follows that real-life MT EMFs are much more genotoxic than the other types of man-made EMFs and, most importantly, much more genotoxic than the 50 Hz alternating EMF, which was (and is still) accused for carcinogenicity long before the MT EMFs (Wertheimer and Leeper 1979; 1982; Miller et al. 1996; IARC 2002; Draper et al. 2005).

The recorded effect of DNA fragmentation in fruit fly ovarian cells by all six examined man-made EMFs was not accompanied by any observable heating as measured by a sensitive thermometer with 0.05°C accuracy within the mass of the food in the exposed vials during exposure to the various EMFs (Panagopoulos et al. 2004; 2007a; 2010; 2013; Panagopoulos 2016).

Previously examined stressors used in the laboratory to study SICD in *Drosophila* oogenesis, such as cytotoxic chemicals (etoposide or staurosporine) or poor nutrition were only observed to induce DNA fragmentation, exclusively in the NCs and the FCs, and exclusively at either one of the two checkpoints (G and S7–8) during early and mid-oogenesis (Nezis et al. 2000; Drummond-Barbosa and Spradling 2001; McCall 2004; Pritchett et al. 2009; Panagopoulos 2012a). Thus, they were not found to induce DNA fragmentation in the OC, neither at developmental stages other than the two checkpoints. Later, it was found that the absence of water (dehydration) can induce DNA fragmentation at more developmental stages in addition to the two checkpoints, but again, not in the OC (Zhukova and Kiseleva 2011).

Figure 5.2a shows an ovariole of a sham-exposed female fruit fly with TUNEL-negative egg chambers at all stages. Figure 5.2b shows an ovariole of a female fruit fly exposed to MF with TUNEL-positive signal (fragmented DNA) only at the germarium, and Figure 5.2c shows an ovariole of a female fruit fly exposed to the PEF with fragmented DNA only at the mid-oogenesis checkpoint (S7). The degree of damage induced by the PEF or the MF was more or less comparable to that of other cytotoxic agents examined before in the laboratory on this biological system, except for dehydration (Panagopoulos et al. 2013; Panagopoulos 2016; 2019a; Nezis et al. 2000; Drummond-Barbosa and Spradling 2001) and smaller than the damage caused by dehydration (Zhukova and Kiseleva 2011). Only in a few cases, exposure to the strongest MF (2.1 mT) caused DNA fragmentation also in the OC[*], something that was not observed with any other examined cytotoxic agent apart from man-made EMF exposure (Panagopoulos et al. 2013; Nezis et al. 2000; Drummond-Barbosa and Spradling 2001; Zhukova and Kiseleva 2011).

The mobile phone EMF exposure during normal "talk" mode was found to induce DNA fragmentation, not only at the two checkpoints but also at all developmental stages during early and

---

[*] The nucleus of the OC is distinct, as it is significantly smaller than the nuclei of the NCs (see Figure 5.2e).

mid-oogenesis (from G up to S10) and, moreover, in all three types of egg chamber cells, i.e., NCs, FCs, and the OC (Panagopoulos et al. 2007a; 2010). This had not been observed with any other EMF or previously tested stressor on this biological system. Thus, MT EMFs were found to be significantly more genotoxic than all other previously examined stress factors (etoposide, staurosporine, starvation, dehydration) at doses tested in the laboratory. Figure 5.2d shows an ovariole of a female fruit fly exposed to MT EMF with TUNEL-positive signal (fragmented DNA) at both checkpoints. Figure 5.2e shows an ovariole of an exposed fruit fly to MT EMF with fragmented DNA at all developmental stages from G to S7 and in all three types of egg chamber cells (NCs, FCs, and

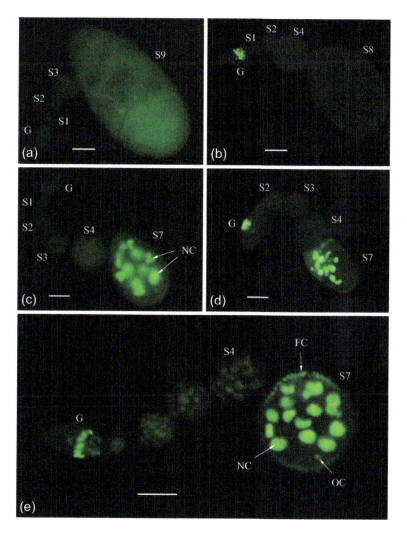

FIGURE 5.2 a) Typical TUNEL-negative fluorescence picture of an ovariole of a sham-exposed female fruit fly with egg chambers from germarium up to S9. b) Ovariole of an exposed female fruit fly (MF1) with egg chambers from germarium up to S8, with fragmented DNA only at the germarium (G) (early oogenesis checkpoint) in the NC. Egg chambers at all other developmental stages are TUNEL-negative. c) Ovariole of an exposed female fruit fly (PEF) containing egg chambers from germarium up to S7, with fragmented DNA only at S7 (mid-oogenesis checkpoint) in the NC, and TUNEL-negative at all other developmental stages. d) Ovariole of an exposed female fruit fly (GSM 900), with TUNEL-positive signal only at the two check points, germarium plus S7 egg chamber and TUNEL-negative signal at all intermediate stages. e) Ovariole (shown in greater magnification) of an exposed female fruit fly (GSM 1800), with fragmented DNA at all developmental stages from germarium (G) to S7 and in all kinds of egg chamber cells (NC, FC, OC). All Bars: 10 μm.

the OC). This and other similar photographs taken through the microscope (see Panagopoulos et al. 2007a; 2010; Panagopoulos 2011; 2012b; 2017; 2019a) are characteristic of the impressive extent of genetic damage caused by real-life MT EMFs emitted by common mobile phones.

5.3.2 Chromatid-type Aberrations in HPBLs

The results of the experiments with the HPBLs from six healthy subjects (No. 1–6) with 1600 metaphases scored from each one (400 exposed to UMTS alone, 400 to caffeine alone, 400 to a combination of UMTS and caffeine, and 400 from sham-exposed blood samples) are listed in Table 5.2 and represented in Figure 5.3. A single 15 min exposure to the UMTS MT EMF during a phone call in "talk" mode at 1 cm distance from the mobile phone increased the total number of chromosomal aberrations (chromatid gaps and breaks) by 100%–275% as compared to the sham-exposed samples, while caffeine alone (and sham-exposure to the EMF) increased the same number by 89%–250%. In four out of the six subjects, the number of aberrations induced by UMTS exposure was higher than the number of aberrations induced by caffeine. The combination of caffeine and the 15 min UMTS exposure dramatically increased the corresponding number of induced aberrations by 245%–925% compared to the sham-exposed samples (Table 5.2, Figure 5.3). In all subjects, all UMTS-exposed samples (with or without caffeine) and caffeine-treated samples differed significantly from the corresponding sham-exposed samples ($p < 0.04$) (Table 5.2). In contrast, in all subjects, the UMTS alone-treated samples did not differ significantly from the caffeine alone-treated samples ($p > 0.05$) (Table 5.2). Thus, while the effect of each stressor was very intense on the human lymphocytes, the effects of the UMTS 15 min exposure alone and the high caffeine dose alone were comparable.

The effect of UMTS MT EMF exposure on HPBLs was non-thermal. The 0.1°C temperature increase during the 15-min exposures is totally insignificant compared to the temperature difference between incubation (37°C) and room temperature in the laboratory (22°C) to which all samples were subjected during the experiments.

In Figure 5.4a, a metaphase of a sham-exposed blood sample is shown from subject No. 5 (female). This is a representative picture of a metaphase from a sham-exposed sample with all 46 chromosomes intact. Figure 5.4b shows a metaphase of a blood sample of the same subject, exposed to caffeine only (and sham-exposed to the UMTS EMF), with one chromatid achromatic lesion – gap (g). Figure 5.4c shows a metaphase of a blood sample of the same subject exposed to UMTS MT EMF (15 min) with one chromatid terminal deletion – break (b) with displaced fragment (f). Figure 5.4d shows a metaphase of a blood sample of the same subject exposed to the combination of caffeine and 15 min UMTS MT EMF with two chromatid breaks (b), and one chromatid gap (g). The chromosomal damage shown in Figures 5.4b–4d is representative of the degree of damage induced by the different exposures.

Each subject exhibited a different sensitivity to each stressor (caffeine, MT EMF exposure, or the combination of the two stressors). The differential sensitivity was also recorded in the control (sham-exposed) blood samples due to genetic and environmental factors. The mean number of total aberrations per cell between the six different healthy individuals varied in the control samples from 0.04 to 0.14, in the UMTS-exposed samples from 0.15 to 0.32, in caffeine alone-treated samples from 0.14 to 0.39, and in the samples exposed to the combination of UMTS and caffeine from 0.38 to 0.80. In some cases, subjects with fewer aberrations in their control (sham-exposed) samples exhibited higher sensitivity to the MT EMF exposure, while this was not observed with caffeine (Table 5.2).

The MT EMF exposure or the high caffeine dose alone induced mainly gaps. They also induced breaks but in significantly smaller percentages. The number of gaps induced by the UMTS exposure was 4–7 times greater than the corresponding number of induced breaks in all subjects, while the number of gaps induced by caffeine alone was 2–6 times greater than the corresponding number of induced breaks in all subjects. Both gaps and breaks and the total number of aberrations induced either by UMTS alone or by caffeine alone were significantly increased with regard to

TABLE 5.2
Chromatid-type Aberrations in HPBLs Induced by UMTS MT EMF and/or Caffeine

Subject No (Age, Sex)	Samples	Gaps in 400 Cells	Breaks in 400 Cells	Total Aberr. in 400 Cells	Mean Total Aberr. per Cell ± SD	Deviation from Sham-Exp	P-Value	Deviation from Caffeine	P-Value
1 (42, Male)	Sham-Exp	30	5	35	0.09 ± 0.03				
	UMTS	84	17	101	0.25 ± 0.08	+178%	<0.02	+39%	>0.05
	Caffeine	50	23	73	0.18 ± 0.06	+100%	<0.04		
	Caff+UMTS	110	51	161	0.40 ± 0.07	+344%	<0.01	+122%	<0.01
2 (33, Female)	Sham-Exp	37	7	44	0.11 ± 0.04				
	UMTS	70	19	89	0.22 ± 0.06	+100%	<0.03	+5%	>0.05
	Caffeine	63	22	85	0.21 ± 0.04	+91%	<0.02		
	Caff+UMTS	106	46	152	0.38 ± 0.08	+245%	<0.01	+81%	<0.03
3 (28, Male)	Sham-Exp	28	9	37	0.09 ± 0.03				
	UMTS	63	15	78	0.19 ± 0.04	+111%	<0.02	+12%	>0.05
	Caffeine	50	18	68	0.17 ± 0.04	+89%	<0.03		
	Caff+UMTS	160	162	322	0.80 ± 0.07	+789%	<0.01	+371%	<0.01
4 (40, Male)	Sham-Exp	43	15	58	0.14 ± 0.04				
	UMTS	102	26	128	0.32 ± 0.09	+129%	<0.03	-18%	>0.05
	Caffeine	97	61	158	0.39 ± 0.07	+179%	<0.01		
	Caff+UMTS	190	106	296	0.74 ± 0.10	+429%	<0.01	+90%	<0.01
5 (35, Female)	Sham-Exp	42	2	44	0.11 ± 0.01				
	UMTS	82	12	94	0.23 ± 0.03	+109%	<0.01	-12%	>0.05
	Caffeine	70	34	104	0.26 ± 0.05	+136%	<0.02		
	Caff+UMTS	217	59	276	0.69 ± 0.09	+527%	<0.01	+165%	<0.01
6 (30, Male)	Sham-Exp	15	2	17	0.04 ± 0.01				
	UMTS	56	5	61	0.15 ± 0.04	+275%	<0.01	+7%	>0.05
	Caffeine	47	8	55	0.14 ± 0.05	+250%	<0.03		
	Caff+UMTS	124	39	163	0.41 ± 0.07	+925%	<0.01	+193%	<0.01

Aberr: Aberrations, Sham-Exp: Sham-Exposed, Caff: Caffeine.

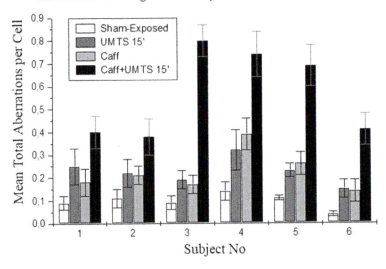

FIGURE 5.3 Mean total number of chromatid-type aberrations (gaps and breaks) per cell ± SD, in 400 cells (HPBLs) of each sample (sham-exposed, exposed to UMTS 15 min, caffeine treated, and exposed to combination of caffeine and UMTS 15 min) for each one of the six subjects (No. 1–6).

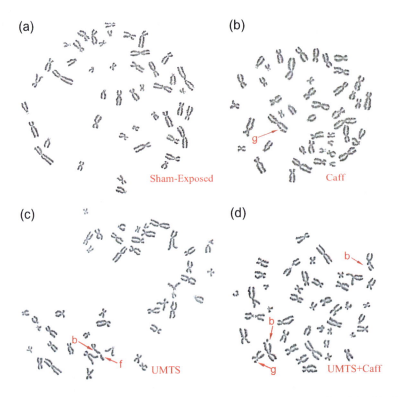

FIGURE 5.4 a) Metaphase of sham-exposed blood sample from subject No. 5 (female) with all 46 chromosomes intact. b) Metaphase of caffeine-treated blood sample from subject No. 5 with one achromatic lesion - gap (g). c) Metaphase of blood sample from subject No. 5 exposed to UMTS MT EMF with one terminal deletion – break (b) with displaced fragment (f). d) Metaphase of a blood sample from subject No. 5 exposed to a combination of caffeine and UMTS MT EMF, with two terminal deletions – breaks (b), and one achromatic lesion – gap (g).

the sham-exposed sample of each subject (Table 5.2, Figure 5.3). The combination of UMTS EMF exposure and caffeine induced mainly gaps as well, apart from one case (subject No. 3) in which the number of breaks was slightly greater than the number of gaps (Table 5.2). In this subject, the combined effect was greatest than in all other subjects, even though the individual effects of UMTS alone or of caffeine alone were smaller than in most other subjects, suggesting that the sensitivity of each subject to the combination of stressors may greatly exceed the sensitivity of this subject to each stressor alone.

5.4 DISCUSSION

5.4.1 MT EMFs Are More Genotoxic than Other Types of Man-made EMFs and Other Stressors

In this chapter, I compared results from previous studies of my group in which the fruit fly oogenesis was employed as a sensitive biological system, and the TUNEL assay was applied to record DNA fragmentation in the ovarian cells induced by six different man-made EMFs under identical conditions and procedures. The six different EMFs: were 1) GSM 900 mobile phone radiation; 2) GSM 1800 mobile phone radiation (Panagopoulos et al. 2007a; 2010); 3) 0.1 mT, 50 Hz alternating MF (MF1); 4) 1.1 mT, 50 Hz alternating MF (MF2); 5) 2.1 mT, 50 Hz alternating MF (MF3) (Panagopoulos et al. 2013); and 6) PEF (8 kHz, 44.4 Hz, 400 kV/m) (Panagopoulos 2016), similar to EMFs of atmospheric discharges (Panagopoulos and Balmori 2017).

As it becomes obvious from the comparison, the real-life MT EMFs (GSM 900, GSM 1800) are far more damaging than the 50 Hz MFs or the PEF (Table 5.1, Figure 5.1). Moreover, MT EMF exposure was found to induce DNA fragmentation in fruit fly ovarian cells more than other types of stressors examined before in the laboratory by other experimenters on this biological system, such as certain cytotoxic chemicals (etoposide, staurosporine), starvation, or dehydration (Nezis et al. 2000; Drummond-Barbosa and Spradling 2001; Zhukova and Kiseleva 2011). The real-life MT EMFs were found to induce DNA fragmentation not only at the two most sensitive developmental stages (checkpoints) but also at all developmental stages during early and mid-oogenesis (from germarium up to S10) and in all three types of egg chamber cells (i.e., not only in the NCs and FCs but also in the OC). DNA fragmentation in the OC may result, if not in cell death, in heritable mutations transferred to the next generations. Such a possibility is by far more worrying than a reduction in the offspring, as it may lead to cancerous or mutated organisms with various health problems. The 50 Hz alternating MF or the PEF were found to induce DNA fragmentation at more or less comparable degrees with the non-electromagnetic cytotoxic agents examined before in fruit fly oogenesis by other experimenters.

Previous studies that had compared simulated MT EMFs (emitted by generators or test phones) with 50 Hz MFs had found these two types of EMFs to induce effects at similar degrees on other biological models (Belyaev et al. 2005; Duan et al. 2015). In our experiments, the effect of the real-life MT EMFs was up to 8 times greater than the effect of the 50 Hz MFs, even though we used stronger MFs than the other experimenters. Thus, it is not only the conflict between results of studies with real-life and simulated MT EMFs (with real-life exposure studies finding effects in more than 95% of the cases, and simulated exposure studies finding effects in about 50% of the cases), but there is also a conflict between studies that compared effects of real-life or simulated MT EMFs with the same 50 Hz power line MF on the same biological model. It is, therefore, obvious that the significantly more pronounced effects found in our experiments with real-life MT EMFs than by other experiments with simulated MT EMFs (in both cases compared with power line MFs) are due to the significantly increased bioactivity of the real-life MT EMFs than that of the simulated MT EMFs employed in the other studies (Belyaev et al. 2005; Duan et al. 2015) (more detailed analysis in Panagopoulos 2019a).

The PEF was found to induce the smallest effect despite its very high intensity (400 kV/m). The same PEF was found before to increase fertility in fruit flies in spite of the small degree of induced

DNA fragmentation (Panagopoulos 2016). The explanation I had provided then was that the small degree of damage triggered the reaction of the reproductive system, which increased the rate of oogenesis in order to compensate the loss of a certain number of egg chambers.[*] Another explanation may be related to the fact that this EMF resembles the natural EMFs of lightning discharges. Living organisms are adapted throughout biological evolution to the natural atmospheric EMFs, and these EMFs may exert a beneficial action because they create the so-called Schumann resonances. These are stranding electromagnetic waves in the Earth's atmosphere which actually attune the brain activity of all animals. By coincidence, the pulse repetition frequency of the PEF (44.4 Hz) is very close to the sixth Schumann harmonic (Schumann 1952; Persinger 2014; Panagopoulos and Chrousos 2019).

In this chapter, I have also presented experiments with HPBLs from six healthy subjects, comparing effects of UMTS (3G/4G) MT EMF with those of a high caffeine dose (~ 290 times higher than the permitted single dose for an adult individual) in the induction of chromatid-type aberrations (gaps and breaks). It was shown that, while a single 15 min exposure of human blood samples to an active 3G (UMTS) mobile phone in "talk" mode at 1 cm distance from the handset increased chromatid-type aberrations from 100% up to 275%, the corresponding increase induced by the high caffeine dose ranged from 89% up to 250% (Table 5.2). Thus, real-life MT EMFs were found to be even more genotoxic than the extreme caffeine dose tested in these experiments, as they were similarly found to be more genotoxic than the other non-electromagnetic stressors tested before by other experimenters in fruit fly oogenesis.

The main type of aberrations induced by either MT EMF exposure or caffeine, or by the combination of the two stressors, was chromatid gaps (achromatic lesions). While chromatid breaks are more intense damages and easier to be recognized (Conger 1967), both gaps and breaks are damages of the same nature, and gaps are actually incomplete breaks (Brecher 1977). Ignoring the gaps and counting only the breaks may be another reason why certain previous studies, e.g., Stronati et al. (2006), did not find effects in HPBLs, in addition to the use of simulated MT EMFs, and to exposing during more resistant cell conditions (during the G_0 phase instead of during the cell division cycle and especially its most sensitive phases M, G_2).

Both recorded effects (chromatid-type aberrations in HPBLs, and DNA fragmentation in fruit fly ovarian cells) induced by MT and other man-made EMFs, were non-thermal, as they were not accompanied by any significant heating. As explained, in the fruit fly experiments, no temperature increase was detected within the vials with the fruit flies, and the 0.1°C highest temperature increase during the 15 min UMTS exposures in the HPBL experiments is well tolerated by the blood cells (see also Panagopoulos 2019a; 2020). Yet, it should be noted that this insignificant temperature increase, which was not recorded with GSM (2G) MT exposures, is probably due to the higher carrier frequency (1950 MHz) of UMTS (3G/4G) than the corresponding frequencies of GSM (900 and 1800 MHz) because microwave heating in all materials and living tissue increases with increasing microwave frequency (Clark et al. 2000). The upcoming 5G technology, with significantly higher carrier frequencies up to 100 GHz, much denser antenna networks, and more intense and collimated radiation beams, is expected to induce significant thermal effects in addition to the non-thermal ones which may not be tolerated by the human/animal body (Neufeld and Kuster 2018; Thielens et al. 2018; 2020; Panagopoulos 2019b; 2020; Hardell and Carlberg 2020). This may represent a great additional danger for public health, which the health authorities should have carefully investigated before allowing 5G deployment, but they did not.

The HPBL experiments showed that when MT EMF exposure is combined with a high caffeine dose, the number of induced chromatid aberrations is dramatically increased in all subjects compared to the effect of each stressor alone (Table 5.2, Figure 5.3) (see also Panagopoulos 2019b; 2020). Even though the caffeine dose used in the experiments was ~ 290 times higher than the permissible single dose for an adult individual, this result possibly implies that heavy coffee consumption

---

[*] Such a compensatory/"beneficial" effect may be related to a stimulating effect called hormesis, which is observed occasionally in biological systems under mild stress (Calabrese 2008).

combined with careless mobile phone use (attached to the head/body) may significantly increase health risks. In addition, it shows that the combination of two (or more) separate stressors (called co-stress condition) may have even greater biological effect than the sum of the individual effects of each stressor alone. Since man-made EMF exposure at different bands of the spectrum (RF, ELF, etc.) constitutes a new reality in daily life for everyone, its combination with a variety of other existing stressors (such as development, aging, sickness, infections, ionizing radiation, chemicals, pharmaceuticals, smoking, psychological stress, etc.) should be examined as a priority in future studies.

Moreover, the experiments with the HPBLs showed that the effects of MT EMF exposure on humans are dose-dependent and increase almost linearly with exposure duration (data not shown, see Panagopoulos 2020). This is in agreement with El-Abd and Eltoweissy (2012) as well as with previous results of my group regarding 2G MT EMF exposures on fruit fly reproduction (Panagopoulos and Margaritis 2010; Panagopoulos 2011; 2017).

As demonstrated in the present chapter, exposure of fruit flies to 2G (GSM) MT EMFs induces extensive DNA damage (fragmentation) in the gametes, that most usually leads to cell death and reproductive declines, as found in previous studies (Panagopoulos et al. 2004; 2007a; 2007b; 2010; Chavdoula et al. 2010; Panagopoulos 2012b). Since DNA damage is converted into chromosomal damage during the early M phase of the cell division cycle (Terzoudi and Pantelias 2006; Pantelias and Terzoudi 2010; Terzoudi et al. 2011; Tian et al. 2018), the recorded chromosomal damage induced by the UMTS (3G/4G) MT EMF and/or caffeine is apparently due to DNA damage caused by these stressors.

It should be noted that, in the HPBL experiments, the subjects probably made use of their mobile phones during the hours before blood samples were collected (as they did during the previous days and years). If their HPBLs were totally unexposed to MT EMFs (as were the fruit fly ovarian cells), especially during the 24 h prior to providing their blood samples, the number of aberrations in the control (sham-exposed) samples would, expectedly, be even smaller and the recorded effect perhaps even greater. Perhaps this is the reason why the degree of damage in the sham-exposed fruit fly cells was smaller than the degree of damage in the sham-exposed HPBLs of most subjects (Tables 5.1 and 5.2).

It has been shown that real-life MT EMFs emitted by commercially available mobile phone devices or base antennas/cell towers are, by far, more bioactive than simulated corresponding signals with invariable parameters emitted by generators (Panagopoulos et al. 2015a; Panagopoulos 2017; 2019a). This is certainly one reason, in combination with other shortcomings discussed above and in the Introduction of this chapter, why in some of the previous studies no effects of simulated MT EMFs on HPBLs were reported (Zeni et al. 2003; 2012; Stronati et al. 2006; Schwarz et al. 2008), while in the HPBL experiments described here in which a real UMTS exposure was employed, a very intense effect was found (up to 275% increase in chromatid aberrations with respect to the sham-exposed samples).

The studies that found real-life UMTS (3G) exposure to be even more genotoxic than real-life GSM (2G) (D'Silva et al. 2017; 2021) (see 5.1.1) are in line with the fact that newer types of MT EMFs (3G, 4G, 5G) transmit an increasingly higher amount/density of variable information (speech, text, images, video, Internet), making the signal increasingly complicated, unpredictably varying each moment, and increasingly more bioactive because of the inability of the living organisms to adapt. Thus, the non-thermal effects of the 5G MT EMF system being deployed are expected to be even more intense than those of 2G, 3G, 4G, and in addition, thermal effects can be expected (Neufeld and Kuster 2018; Thielens et al. 2018; 2020). This should have been seriously considered by the responsible public health authorities as emphasized above and in my recent studies (Panagopoulos 2019a; 2019b; 2020), but unfortunately it was not considered.

5.4.2 Safety Limits Implied by the Experimental Findings

In the HPBL experiments, a single MT EMF exposure at power density ~ 136 times lower than the newest 2020 ICNIRP exposure limit (or ~ 34 times lower than the corresponding 1998 ICNIRP limit) induced chromosomal aberrations to a slightly higher degree than a caffeine dose of ~ 290 times greater than the permissible single caffeine dose for an adult human. This comparison shows that the exposure limits set for WC/microwave EMFs by ICNIRP (1998; 2020) are enormously less stringent (~ 34 × 290 times or ~ 136 × 290 times respectively) than those for caffeine. Assuming linearity between caffeine dose and MT radiation intensity ($\mu W/cm^2$) regarding their abilities to induce chromatid-type chromosomal aberrations and assuming the caffeine single dose limit to be correct (since the effects of caffeine on the human organism are fast and evident in contrast to EMF health effects, which may become evident after years), the ICNIRP (2020) microwave EMF exposure limits should be lowered by ~ 40000 (= 4 × 10^4) times (136 × 290) (or the corresponding 1998 ICNIRP limits by ~ 10^4 times). By lowering the 2020 ICNIRP limit of 4000 $\mu W/cm^2$ (for 2–6 GHz averaged for 6 min) by 4 × 10^4 times (or the 1998 ICNIRP limit by 10^4 times) the limit would become 0.1 $\mu W/cm^2$ for the short-term exposures applied in the described experiments, and by lowering this by at least 100 for long-term exposures, it would become 0.001 $\mu W/cm^2$. These limits for short and long exposures respectively would be compatible with the caffeine consumption limit and realistic according to the EMF bioeffects literature (Goodman et al. 1995; Panagopoulos et al. 2002; 2010; 2015a; 2021; Panagopoulos 2011; 2017; 2019a; Burlaka et al. 2013; Yakymenko et al. 2016; 2018; Lai 2021).

The DNA damage recorded in the fruit fly experiments was induced by only a few min daily exposures for a few days for radiation intensities down to even less than 1 $\mu W/cm^2$ (see Panagopoulos et al. 2010). Setting a short-term exposure limit based on this result, this would reasonably be 0.1 $\mu W/cm^2$, which is the same as the corresponding limit deduced from the results of the HPBL experiments. As already explained, the corresponding limit for long-term (continuous) exposure in daily environments (as e.g., from various MT base antennas, devices, Wi-Fi routers, DECT phones, etc.) should be at least 100 times lower, resulting in a reasonable limit of 0.001 $\mu W/cm^2$. Such a limit for WC EMFs would provide a reasonable degree of safety, allowing for the operation of WC systems. Thus, both the fruit fly and the HPBL experiments imply the same power density safety limits for RF/WC EMFs/EMR.

5.4.3 Physical Parameters of EMFs Responsible for Bioactivity

It is evident that real-life MT/WC EMFs are significantly more bioactive than other types of manmade EMFs. The question that arises is which specific physical parameter(s) of the WC EMFs is mainly responsible for such intense bioactivity. Plea of experimental data in combination with theoretical calculations (Panagopoulos et al. 2002; 2010; 2015a; 2015b; Panagopoulos and Margaritis 2010; Panagopoulos 2019a) point that the most important physical parameters of EMF exposure in terms of bioactivity are: 1) polarization (in combination with coherence), 2) existence of ELF components (pulsing, modulation, etc.), 3) field/radiation intensity, 4) exposure duration, and 5) field variability.

Let us now examine the various physical parameters of the specific EMF exposures in the fruit fly experiments described in this chapter in order to find which parameter may be more crucial for the intense bioactivity of the MT EMFs compared to the other EMFs. 1) All six EMFs were totally (linearly) polarized (and coherent), therefore, we must exclude polarization/coherence alone as the critical parameter. 2) All six of them contained ELFs, and three of them (GSM 900, GSM 1800, PEF) were pulsed on ELF, and still the PEF caused the smallest effect on DNA fragmentation; therefore, we must also exclude ELF and pulsing. 3) Although a direct comparison in intensity is not possible because of the different frequencies and waveforms among the MT EMFs, the MF, and the PEF, the MT EMFs were at the very same intensities with those of daily use, while the other EMFs were at significantly higher intensities than environmentally existing ones, and still the effect induced by the MT EMFs was much stronger. Therefore, we must also exclude field/radiation intensity. 4) The MT EMFs were the most bioactive despite the shortest exposure duration; therefore, we must also exclude exposure duration.

The remaining parameter is the *variability* of the exposure. The parameters of the real-life MT fields (and especially intensity and waveform) vary greatly and unpredictably each moment during the exposure (even though average intensity values over a few min or more may not change very much), while the parameters of the MFs and the PEF were invariable (apart from the alternation or the pulsing of the waveforms, which were constant and predictable). But why is exposure variability so important for bioactivity?

Living organisms have been constantly exposed throughout biological evolution to terrestrial static electric and magnetic fields of average intensities of ~ 130 V/m and ~0.5 G, respectively. While no adverse health effects are connected with normal exposure to these natural ambient fields, variations in their intensities of the order of ~ 20% during "magnetic storms" or "geomagnetic pulsations" due to changes in solar activity with an average periodicity of about 11 years are connected with increased rates of animal/human health incidents, including nervous and psychic diseases, hypertensive crises, heart attacks, cerebral accidents, and mortality (Presman 1977; Dubrov 1978; Panagopoulos 2013; 2019a).

The most abundant class of ion channels in all animal cell membranes, the VGICs, switch between open and closed state whenever a change exceeding ~ 30% in the membrane voltage takes place (Liman et al. 1991; Alberts et al. 1994; Stryer 1996), and all physiological cellular effects are initiated by changes in ionic concentrations mediated by ion channel gating (Alberts et al. 1994; Panagopoulos 2013). Specifically, it is known that changes $\geq$ ~ 30 mV in the normal ~ 100 mV transmembrane voltage trigger the gating (opening or closing) of the VGICs in cell membranes (Liman et al. 1991; Panagopoulos et al. 2021).

Living organisms perceive EMFs as environmental stressors (Presman 1977; Goodman et al. 1995; Panagopoulos 2013). It is reasonably expected that cells/organisms adapt more easily when EMFs are not significantly and unexpectedly varying, in other words when their parameters are kept constant or vary only slightly, or when the variation is predictable (as e.g., with the alternating 50 Hz MFs, the PEF in the described fruit fly experiments, or the simulated MT EMFs employed in many other studies, such as Belyaev et al. 2005; Stronati et al. 2006; Schwarz et al. 2008; or Duan et al. 2015). Since living organisms do not have defenses against variations of the order of ~ 20% of natural EMFs as reported, it is reasonable to expect that they do not have defenses against EMFs, which vary unpredictably exceeding by ~ 100% or even more their average intensity (and in addition are totally polarized, coherent, pulsed, and modulated, including simultaneously several different frequencies, etc., as are the MT/WC EMFs). Similarly, since VGICs in cells respond to changes of the order of ~ 30% of the physiological membrane fields, it is realistic to expect that they will, irregularly, respond to externally applied polarized EMFs of adequate intensity, which can exert similar forces with those arising from such membrane voltage changes.

What is common between the natural EMFs in the terrestrial environment, the physiological EMFs of the cell membranes, and the man-made EMFs employed in the studies? Terrestrial and cell membrane fields are static and significantly (almost totally) polarized. They normally do not vary considerably in their intensities, but variations of the order of 20%–30% induce cellular/health effects. Man-made EMFs used in the studies are totally polarized and, at the same time (especially the MT/WC EMFs), highly variable (alternating, pulsed) with unexpected changes exceeding 100% of their normal average intensities. Thus, the common parameters during initiation of biological effects are polarization combined with variability that exceeds by 20%–30% the average intensity. It comes that variability in the EMF exposure (especially in intensity) is a crucialy important factor in order for the specific type of polarized EMF to be able to induce biological/health effects. As explained, the bioactive parameters of EMFs are: 1) polarization (combined with coherence), 2) ELFs, 3) intensity (and duration), 4) variability (unexpected changes exceeding 20%–30% of average/normal intensity). Thus, once an EMF is polarized, includes ELFs, and has adequate intensity (and duration), the parameter that makes it bioactive is variability (especially in intensity).

The extreme and unpredictable variability of the real-life WC signals concerns all existing types of WC EMFs (MT, DECT phones, Wi-Fi routers, etc.), as they all operate under the same principles combining microwave carrier signals with ELF pulsing and modulation of similar frequency bands, emitting variable information each moment, which, in turn, makes the emission variable in intensity, frequency, waveform, etc. With every new generation of WC, the amount of information transmitted each moment (speech, text, images, video, Internet, etc.) by both devices and base antennas is increased, resulting in higher variability and complexity of the signals with the living cells/organisms even more unable to adapt. The results of the recent studies that found a real 3G (UMTS) MT EMF to be more bioactive than a real 2G (GSM) MT EMF emitted by the same device on chick liver and brain (D'Silva et al. 2017; 2021) are in line with this fact.

5.4.4 Conclusions

The experiments show that real-life MT/WC EMFs are significantly more genotoxic than other types of man-made EMFs, such as high-voltage power line MFs, PEF similar to those of atmospheric discharges, and more genotoxic than previously tested non-electromagnetic stressors, such as certain cytotoxic chemicals, starvation, dehydration (Nezis et al. 2000; Drummond-Barbosa and Spradling 2001; Zhukova and Kiseleva 2011), or an extreme caffeine dose. The combination of MT EMF exposure and caffeine increased the combined effect more than the sum of the individual effects in most cases. Even though the caffeine dose used was 290 times higher than the permitted dose and, thus, much higher than the amounts of caffeine received daily by even the heaviest coffee drinkers, the magnitude of the combined effect probably suggests that mobile phone use and heavy coffee drinking is not a good combination for health.

Moreover, the present chapter shows that both human and animal cells are similarly vulnerable to MT/WC EMFs and claims that human cells are allegedly more resistant than animal cells, justifying unrestricted human exposure to WC EMFs, should only be considered irresponsible. The described effects on fruit fly ovarian cells and HPBLs are in complete agreement among them demonstrating that MT/WC EMFs are very genotoxic/bioactive, able to induce DNA damage, and consequent chromosomal damage in both human and animal cells.

Both the effects of *in vitro* exposure of HPBLs (and comparison with caffeine), and the effects of *in vivo* exposure of fruit fly ovarian cells, imply the same power density limits, 0.1 µW/cm^2 for short-term exposures and 0.001 µW/cm^2 for long-term exposures to MT/WC EMFs. These evidence-based exposure limits for acute and long-term exposure to RF/WC EMFs are 4×10^4 times and 4×10^6 times, respectively, lower than the ICNIRP (2020) limit for WC frequency bands of 2–6 GHz (4000 µW/cm^2). Thus, it is evident that ICNIRP limits provide no protection against WC EMFs.

The chapter highlights the significant finding (see also Panagopoulos 2019a) that once a specific EMF is polarized (and coherent), includes ELFs, and has adequate intensity (and duration), then variability in its parameters (especially in its intensity) is the key- parameter for its bioactivity (see also Panagopoulos et al. 2015b; Panagopoulos 2017; 2019a). The importance of field variability implies the need to define EMF exposures not only by frequency components and average intensity values but also by reporting intensity and frequency variations, pulsation, modulation, exposure duration, and – of course – polarization/coherence (which is a property of all man-made EMFs). Moreover, in published reviews of experimental studies employing MT and other types of WC EMFs, such as DECT phones, Wi-Fi, etc., it must be explicitly reported whether the described exposures are real from commercially available devices or simulated from generators, test phones, etc.

Since the health effects of all WC EMFs (including MT, Wi-Fi, DECT phones, etc.) are of utmost importance in our days, studies should be conducted to test the most sensitive biological conditions with real-life exposures and in combination with other environmental stressors, otherwise the results may be misleading for public health protection. Exposures by any type of simulated signals and within any type of exposure chambers used to produce "uniform" exposures such as "reverberation chambers" or "TEM chambers" (Ardoino et al. 2005; Wu et al. 2009) do not represent real-life exposure conditions and may produce misleading outcomes toward "no effect" findings (Panagopoulos 2017; 2019a; Pall 2018).

People should minimize exposures to man-made (and especially WC/MT) EMFs, first of all from their own WC devices close to the sources (distances smaller than 1 m) since exposure in the near field is significantly stronger than in the far field (Panagopoulos 2017; Panagopoulos and

Chrousos 2019). This can be done by using wire connections for internet and domestic telephones and keeping mobile phones at the greatest possible distance from the body during calls by use of the loudspeaker or a wired, air-tube headset. Mobile phones should be used only for short and necessary calls, with Wi-Fi and Bluetooth functions inactivated and used only occasionally and in emergencies when there is no wired Internet access. The phones should be switched off or put in airplane mode when carried on the body, kept at the greatest possible distance (at least several meters) during the day, and switched off during sleep (Panagopoulos 2011; Panagopoulos and Chrousos 2019). In addition, people should avoid living close to any antennas or high-voltage power lines, where studies find that health problems arise (Santini et al. 2002; Hallberg and Johansson 2002; Draper et al. 2005; Gulati et al. 2016; Zothansiama et al. 2017; Lopez et a 2021).

Thus, the most important conclusions in a few lines are: 1) Real-life MT/WC EMFs are much more bioactive/genotoxic than other types of man-made EMFs and more genotoxic than several other tested agents (such as certain cytotoxic chemicals, starvation, dehydration, and the extreme dose of caffeine), at least for the doses tested in the laboratory. 2) Simulated MT/WC EMFs with invariable parameters are significantly less bioactive than real-life MT/WC EMFs, and thus, experiments must be performed by commercially available mobile phones and other WC devices. 3) Both humans and animals (and probably plants as well) are similarly vulnerable to man-made EMFs. 4) The combined exposure to man-made EMFs and various other stressors in everyday life may intensify the effects even more than the sum of the separate effects induced by the individual stressors. 5) Variability of EMF exposure (especially in intensity) combined with polarization/coherence is the key for EMF bioactivity. 6) Relevant power density safety limits for short- and long-term WC EMF exposure, based on the experimental evidence, are 0.1 $\mu W/cm^2$ and 0.001 $\mu W/cm^2$, respectively, instead of the ICNIRP (2020) limit, which is 4000 $\mu W/cm^2$. 7) Minimized and cautious use of mobile phones and other WC devices, as well as avoidance of exposure from MT/WC EMFs from other devices and antennas, is required.

REFERENCES

Agiwal M, Jin H, (2018): Directional paging for 5G communications based on partitioned user ID. *Sensors*, 18(6), 1845. https://doi.org/10.3390/s18061845.

Alberts B, Bray D, Lewis J, Raff M, Roberts K, Watson JD, (1994): *Molecular biology of the cell*. Garland Publishing, Inc., New York.

Ardoino L, Lopresto V, Mancini S, Marino C, Pinto R, Lovisolo GA, (2005): A radiofrequency system for in vivo pilot experiments aimed at the studies on biological effects of electromagnetic fields. *Physics in Medicine and Biology*, 50(15), 3643–3654.

Atli E, Unlü H, (2006): The effects of microwave frequency electromagnetic fields on the development of Drosophila melanogaster. *International Journal of Radiation Biology*, 82(6), 435–441.

Baan R, Grosse Y, LaubySecretan B, El Ghissassi F, Bouvard V, Benbrahim-Tallaa L, Guha N, Islami F, Galichet L, Straif K; WHO International Agency for Research on Cancer Monograph Working Group, (2011): Carcinogenicity of radiofrequency electromagnetic fields. *Lancet Oncology*, 12(7), 624–626.

Baohong W, Jiliang H, Lifen J, Deqiang L, Wei Z, Jianlin L, Hongping D, (2005): Studying the synergistic damage effects induced by 1.8 GHz radiofrequency field radiation (RFR) with four chemical mutagens on human lymphocyte DNA using comet assay in vitro. *Mutation Research*, 578(1–2), 149–157.

Baohong W, Lifen J, Lanjuan L, Jianlin L, Deqiang L, Wei Z, Jiliang H, (2007): Evaluating the combinative effects on human lymphocyte DNA damage induced by ultraviolet ray C plus 1.8 GHz microwaves using comet assay in vitro. *Toxicology*, 232(3), 311–316.

Baum JS, St George JP, McCall K, (2005): Programmed cell death in the germline. *Seminars in Cell and Developmental Biology*, 16(2), 245–259.

Belyaev IY, Hillert L, Protopopova M, Tamm C, Malmgren LO, Persson BR, Selivanova G, Harms-Ringdahl M, (2005): 915 MHz microwaves and 50 Hz magnetic field affect chromatin conformation and 53BP1 foci in human lymphocytes from hypersensitive and healthy persons. *Bioelectromagnetics*, 26(3), 173–184.

Belyaev IY, Markovà E, Hillert L, Malmgren LOG, Persson BRR, (2009): Microwaves from UMTS/GSM mobile phones induce long-lasting inhibition of 53BP1/γ -H2AX DNA repair foci in human lymphocytes. *Bioelectromagnetics*, 30(2), 129–141.

Bertagna F, Lewis R, Silva SRP, McFadden J, Jeevaratnam K, (2021): Effects of electromagnetic fields on neuronal ion channels: A systematic review. *Annals of the New York Academy of Sciences*, 1499(1), 82–103.

Brecher S, (1977): Ultra-structural observations of x-ray induced chromatid gaps. *Mutation Research*, 42(2), 249–268.

Burlaka A, Tsybulin O, Sidorik E, Lukin S, Polishuk V, Tsehmistrenko S, Yakymenko I, (2013): Overproduction of free radical species in embryonic cells exposed to low intensity radiofrequency radiation. *Experimental Oncology*, 35(3), 219–225.

Calabrese EJ, (2008): Hormesis: Why it is important to toxicology and toxicologists. *Environmental Toxicology and Chemistry*, 27(7), 1451–1474.

Carlberg M, Hardell L, (2017): Evaluation of mobile phone and cordless phone use and glioma risk using the Bradford Hill viewpoints form 1965 on association or causation. *BioMed Research International*, 2017, 9218486.

Chavdoula ED, Panagopoulos DJ, Margaritis LH, (2010): Comparison of biological effects between continuous and intermittent exposure to GSM-900 MHz mobile phone radiation: Detection of apoptotic cell death features. *Mutation Research*, 700(1–2), 51–61.

Clark DE, Folz DC, West JK, (2000): Processing materials with microwave energy. *Materials Science and Engineering, Part A*, 287(2), 153–158.

Conger AD, (1967): Real chromatid deletions versus gaps. *Mutation Research*, 4(4), 449–459.

Cook DG, Peacock JL, Feyerabend C, Carey IM, Jarvis MJ, Anderson HR, Bland JM, (1996): Relation of caffeine intake and blood caffeine concentrations during pregnancy to fetal growth: Prospective population based study. *BMJ*, 313(7069), 1358–1362.

Dahlman E, Parkvall S, Skoeld J, (2018): *5G NR: The next generation wireless access technology*. Academic Press, Elsevier, London.

Danese E, Lippi G, Buonocore R, Benati M, Bovo C, Bonaguri C, Salvagno GL, Brocco G, Roggenbuck D, Montagnana M, (2017): Mobile phone radiofrequency exposure has no effect on DNA double strand breaks (DSB) in human lymphocytes. *Annals of Translational Medicine*, 5(13), 272.

De Iuliis GN, Newey RJ, King BV, Aitken RJ, (2009): Mobile phone radiation induces reactive oxygen species production and DNA damage in human spermatozoa in vitro. *PLOS ONE*, 4(7), e6446.

Delgado JMR, (1985): Biological effects of extremely low frequency electromagnetic fields. *Journal of Bioelectricity*, 4(1), 75–91.

Draper G, Vincent T, Kroll ME, Swanson J, (2005): Childhood cancer in relation to distance from high voltage power lines in England and Wales: A case-control study. *BMJ*, 330(7503), 1290.

Drummond-Barbosa D, Spradling AC, (2001): Stem cells and their progeny respond to nutritional changes during Drosophila oogenesis. *Developments in Biologicals*, 231(1), 265–278.

D'Silva MH, Swer RT, Anbalagan J, Rajesh B, (2017): Effect of radiofrequency radiation emitted from 2G and 3G cell phone on developing liver of chick embryo - A comparative study. *Journal of Clinical and Diagnostic Research*, 11(7), 5–9.

D'Silva MH, Swer RT, Anbalagan J, Bhargavan R, (2021): Assessment of DNA damage in chick embryo brains exposed to 2G and 3G cell phone radiation using alkaline comet assay technique. *Journal of Clinical and Diagnostic Research*, 15:AC01–AC04. https://doi.org/10.7860/jcdr/2021/47115.14441.

Duan W, Liu C, Zhang L, He M, Xu S, Chen C, Pi H, Gao P, Zhang Y, Zhong M, Yu Z, Zhou Z, (2015): Comparison of the genotoxic effects induced by 50 Hz extremely low-frequency electromagnetic fields and 1800 MHz radiofrequency electromagnetic fields in GC-2 cells. *Radiation Research*, 183(3), 305–314.

Dubrov AP, (1978): *The geomagnetic field and life*. Plenum Press, New York.

EFSA (European Food Safety Authority), (2015): Scientific opinion on the safety of caffeine. *EFSA Journal*, 13(5), 4102.

El-Abd SF, Eltoweissy MY, (2012): Cytogenetic alterations in human lymphocyte culture following exposure to radiofrequency field of mobile phone. *Journal of Applied Pharmaceutical Science*, 2(2), 16–20.

Falcioni L, Bua L, Tibaldi E, Lauriola M, De Angelis L, Gnudi F, Mandrioli D, Manservigi M, Manservisi F, Manzoli I, Menghetti I, Montella R, Panzacchi S, Sgargi D, Strollo V, Vornoli A, Belpoggi F, (2018): Report of final results regarding brain and heart tumors in Sprague-Dawley rats exposed from prenatal life until natural death to mobile phone radiofrequency field representative of a 1.8 GHz GSM base station environmental emission. *Environmental Research*, 165, 496–503.

Gavrieli Y, Sherman Y, Ben-Sasson SA, (1992): Identification of programmed cell death in situ via specific labeling of nuclear DNA fragmentation. *Journal of Cell Biology*, 119(3), 493–501.

Goodman EM, Greenebaum B, Marron MT, (1995): Effects of electro-magnetic fields on molecules and cells. *International Review of Cytology*, 158, 279–338.

Gulati S, Yadav A, Kumar N, Kanupriya, ANK, Kumar R, Gupta R, Gupta R, (2016): Effect of GSTM1 and GSTT1 polymorphisms on genetic damage in humans populations exposed to radiation from mobile towers. *Archives of Environmental Contamination and Toxicology*, 70(3), 615–625.

Gunes M, Ates K, Yalcin B, Akkurt S, Ozen S, Kaya B, (2021): An evaluation of the genotoxic effects of electromagnetic radiation at 900 MHz, 1800 MHz, and 2100 MHz frequencies with a SMART assay in Drosophila melanogaster. *Electromagnetic Biology and Medicine*, 40(2), 254–263.

Hallberg O, Johansson O, (2002): Melanoma incidence and frequency modulation (FM) broadcasting. *Archives of Environmental Health*, 57(1), 32–40.

Hardell L, Carlberg M, Hansson Mild K, (2013): Use of mobile phones and cordless phones is associated with increased risk for glioma and acoustic neuroma. *Pathophysiology*, 20(2), 85–110.

Hardell L, (2017): World Health Organization, radiofrequency radiation and health - A hard nut to crack (Review). *International Journal of Oncology*, 51(2), 405–413.

Hardell L, (2019): Notes on parliament hearing in Tallinn, Estonia June 4, 2019 as regards the deployment of the fifth generation, 5G, of wireless communication. *World Academy of Sciences Journal*, 1, 275–282.

Hardell L, Carlberg M, (2020): Health risks from radiofrequency radiation, including 5G, should be assessed by experts with no conflicts of interest. *Oncology Letters*, 20(4), 15.

Hardell L, Nyberg R, (2020): Appeals that matter or not on a moratorium on the deployment of the fifth generation, 5G, for microwave radiation. *Molecular and Clinical Oncology*. https://doi.org/10.3892/mco.2020.1984.

Hatzi VI, Karakosta M, Barszczewska K, Karachristou I, Pantelias G, Terzoudi GI, (2015): Low concentrations of caffeine induce asymmetric cell division as observed in vitro by means of the CBMN-assay and iFISH. *Mutation Research/Genetic Toxicology and Environmental Mutagenesis*, 793, 71–78.

Higdon JV, Frei B, (2006): Coffee and health: A review of recent human research. *Critical Reviews in Food Science and Nutrition*, 46(2), 101–123.

Holma H, Toskala A, (2004): *WCDMA for UMTS, radio access for third generation mobile communications*. John Wiley & Sons Ltd., Chichester, England

IAEA, (2011): *Cytogenetic dosimetry: Applications in preparedness for and response to radiation emergencies*. International Atomic Energy Agency, Vienna.

IARC, (1991): *Coffee, tea, mate, methylxanthines, and methylglyoxal*, Vol. 51. International Agency for Research on Cancer, World Health Organization, Lyon.

IARC, (2002): *Non-ionizing radiation, part 1: Static and extremely low-frequency (ELF) electric and magnetic fields*, Vol. 80. International Agency for Research on Cancer, Lyon.

IARC, (2013): *Non-Ionizing radiation, part 2: Radiofrequency electromagnetic fields*, Vol. 102. International Agency for Research on Cancer, Lyon.

IARC, (2016): *Coffee, mate and very hot beverages*, Vol. 116. International Agency for Research on Cancer, Lyon.

ICNIRP, (1998): Guidelines for limiting exposure to time-varying electric, magnetic and electromagnetic fields (up to 300 GHz). *Health Physics*, 74, 494–522.

ICNIRP, (2020): Guidelines for limiting exposure to electromagnetic fields (100 kHz to 300 GHz). *Health Physics*, 118(5), 483–524.

Ji S, Oh E, Sul D, Choi JW, Park H, Lee E, (2004): DNA Damage of lymphocytes in volunteers after 4 hours use of mobile phone. *Journal of Preventive Medicine and Public Health*, 37(4), 373–380.

Kikuchi T, Ogawa M, Otaka Y, Furuta M, (1998): Multigeneration exposure test of Drosophila melanogaster to ELF magnetic fields. *Bioelectromagnetics*, 19(6), 335–340.

King RC, (1970): *Ovarian development in Drosophila melanogaster*. Academic Press, New York.

Koana T, Okada MO, Takashima Y, Ikehata M, Miyakoshi J, (2001): Involvement of eddy currents in the mutagenicity of ELF magnetic fields. *Mutation Research*, 476(1–2), 55–62.

Kostoff RN, Heroux P, Aschner M, Tsatsakis A, (2020): Adverse health effects of 5G mobile networking technology under real-life conditions. *Toxicology Letters*, 323, 35–40.

Kuhlmann W, Fromme HG, Heege EM, Ostertag W, (1968): The mutagenic action of caffeine in higher organisms. *Cancer Research*, 28(11), 2375–2389.

Lai H, (2021): Genetic effects of non-ionizing electromagnetic fields. *Electromagnetic Biology and Medicine*, 40(2), 264–273.

Leach V, Weller S, Redmayne M, (2018): A novel database of bio-effects from non-ionizing radiation. *Reviews on Environmental Health*, 33(3), 1–8.

Liman ER, Hess P, Weaver F, Koren G, (1991): Voltage-sensing residues in the S4 region of a mammalian K+ channel. *Nature*, 353(6346), 752–756.

López I, Félix N, Rivera M, Alonso A, Maestú C, (2021): What is the radiation before 5G? A correlation study between measurements in situ and in real time and epidemiological indicators in Vallecas, Madrid. *Environmental Research*, 194, 110734.

Ma TH, Chu KC, (1993): Effect of the extremely low frequency (ELF) electromagnetic field (EMF) on developing embryos of the fruit fly (Drosophila melanogaster L). *Mutation Research*, 303(1), 35–39.

Maber J, (1999): *Data analysis for biomolecular sciences*. Longman, Harlow, England.

Manta AK, Stravopodis DJ, Papassideri IS, Margaritis LH, (2014): Reactive oxygen species elevation and recovery in Drosophila bodies and ovaries following short-term and long-term exposure to DECT base EMF. *Electromagnetic Biology and Medicine*, 33(2), 118–131.

Manti L, Braselmann H, Calabrese ML, Massa R, Pugliese M, Scampoli P, Sicignano G, Grossi G, (2008): Effects of modulated microwave radiation at cellular telephone frequency (1.95 GHz) on X-ray-induced chromosome aberrations in human lymphocytes in vitro. *Radiation Research*, 169(5), 575–583.

Margaritis LH, Kafatos FC, Petri WH, (1980): The eggshell of Drosophila melanogaster; fine structure of the layers and regions of the wild-type eggshell. *Journal of Cell Science*, 43, 1–35.

Margaritis LH, Manta AK, Kokkaliaris KD, Schiza D, Alimisis K, Barkas G, Georgiou E, Giannakopoulou O, Kollia I, Kontogianni G, Kourouzidou A, Myari A, Roumelioti F, Skouroliakou A, Sykioti V, Varda G, Xenos K, Ziomas K, (2014): Drosophila oogenesis as a bio-marker responding to EMF sources. *Electromagnetic Biology and Medicine*, 33(3), 165–189.

Markova E, Hillert L, Malmgren L, Persson BR, Belyaev IY, (2005): Microwaves from GSM mobile telephones affect 53BP1 and gamma-H2AX foci in human lymphocytes from hypersensitive and healthy persons. *Environmental Health Perspectives*, 113(9), 1172–1177.

McCall K, (2004): Eggs over easy: Cell death in the Drosophila ovary. *Developmental Biology*, 274(1), 3–14.

McClelland S, Jaboin JJ, (2018): The radiation safety of 5G wi-fi: Reassuring or Russian roulette? *International Journal of Radiation Oncology Biology Physics*, 101(5), 1274–1275.

Melnick RL, (2019): Commentary on the utility of the national toxicology program study on cell phone radiofrequency radiation data for assessing human health risks despite unfounded criticisms aimed at minimizing the findings of adverse health effects. *Environmental Research*, 168, 1–6.

Mendelsohn J, Hudic D, Castagnola J, (1981): DNA synthesis and proliferation of human lymphocytes in vitro: Ill. Fate of cycling cells in aging cultures of phytohemagglutinin stimulated human lymphocytes. *Journal of Cellular Physiology*, 106(1), 13–22.

Miller AB, To T, Agnew DA, Wall C, Green LM, (1996): Leukemia following occupational exposure to 60-Hz electric and magnetic fields among Ontario electric utility workers. *American Journal of Epidemiology*, 144(2), 150–160.

Miller AB, Morgan LL, Udasin I, Davis DL, (2018): Cancer epidemiology update, following the 2011 IARC evaluation of radiofrequency electromagnetic fields (Monograph 102). *Environmental Research*, 167, 673–683.

Miller AB, Sears ME, Morgan LL, Davis DL, Hardell L, Oremus M, Soskolne CL, (2019): Risks to health and well-being From radio-frequency radiation emitted by cell phones and other wireless devices. *Frontiers in Public Health*, 7, 223. https://doi.org/10.3389/fpubh.2019.00223.

Mirabolghasemi G, Azarnia M, (2002): Developmental changes in Drosophila melanogaster following exposure to alternating electromagnetic fields. *Bioelectromagnetics*, 23(6), 416–420.

Neufeld E, Kuster N, (2018): Systematic derivation of safety limits for time-varying 5G radiofrequency exposure based on analytical models and thermal dose. *Health Physics*, 115(6), 705–711.

Nezis IP, Stravopodis DJ, Papassideri I, Robert-Nicoud M, Margaritis LH, (2000): Stage-specific apoptotic patterns during Drosophila oogenesis. *European Journal of Cell Biology*, 79(9), 610–620.

Nias AHW, (1998): *An introduction to radiobiology*. J.Wiley & Sons, Chichester, England.

NTP (National Toxicology Program), (2018): Toxicology and carcinogenesis studies in Hsd: Sprague Dawley SD rats exposed to whole-body radio frequency radiation at a frequency (900 MHz) and modulations (GSM and CDMA) used by cell phones, NTP TR 595, Department of Health and Human Services, USA.

Pall ML, (2013): Electromagnetic fields act via activation of voltage-gated calcium channels to produce beneficial or adverse effects. *Journal of Cellular and Molecular Medicine*, 17(8), 958–965.

Pall ML, (2018): Wi-fi is an important threat to human health. *Environmental Research*, 164, 405–416.

Panagopoulos DJ, Messini N, Karabarbounis A, Filippetis AL, Margaritis LH, (2000): A mechanism for action of oscillating electric fields on cells. *Biochemical and Biophysical Research Communications*, 272(3), 634–640.

Panagopoulos DJ, Karabarbounis A, Margaritis LH, (2002): Mechanism for action of electromagnetic fields on cells. *Biochemical and Biophysical Research Communications*, 298(1), 95–102.

Panagopoulos DJ, Karabarbounis A, Margaritis LH, (2004): Effect of GSM 900-MHz mobile phone radiation on the reproductive capacity of Drosophila melanogaster. *Electromagnetic Biology and Medicine*, 23(1), 29–43.

Panagopoulos DJ, Chavdoula ED, Nezis IP, Margaritis LH, (2007a): Cell death induced by GSM 900 MHz and DCS 1800 MHz mobile telephony radiation. *Mutation Research*, 626(1–2), 69–78.

Panagopoulos DJ, Chavdoula ED, Karabarbounis A, Margaritis LH, (2007b): Comparison of bioactivity between GSM 900 MHz and DCS 1800 MHz mobile telephony radiation. *Electromagnetic Biology and Medicine*, 26(1), 33–44.

Panagopoulos DJ, Chavdoula ED, Margaritis LH, (2010): Bioeffects of mobile telephony radiation in relation to its intensity or distance from the antenna. *International Journal of Radiation Biology*, 86(5), 345–357.

Panagopoulos DJ, Margaritis LH, (2010): The effect of exposure duration on the biological activity of mobile telephony radiation. *Mutation Research*, 699(1/2), 17–22.

Panagopoulos DJ, (2011): Analyzing the health impacts of modern telecommunications microwaves. In: LV Berhardt (Ed.), *Advances in medicine and biology*, Vol. 17. Nova Science Publishers, Inc., New York, 1–55.

Panagopoulos DJ, (2012a): Gametogenesis, embryonic and post-embryonic development of Drosophila melanogaster, as a model system for the assessment of radiation and environmental genotoxicity. In: M Spindler-Barth (Ed.), *Drosophila melanogaster: Life cycle, genetics and development*. Nova Science Publishers, New York, 1–38.

Panagopoulos DJ, (2012b): Effect of microwave exposure on the ovarian development of Drosophila melanogaster. *Cell Biochemistry and Biophysics*, 63(2), 121–132.

Panagopoulos DJ, (2013): Electromagnetic interaction between environmental fields and living systems determines health and well-being. In: MH Kwang and SO Yoon (Eds), *Electromagnetic fields: Principles, engineering applications and biophysical effects*. Nova Science Publishers, New York, 87–130.

Panagopoulos DJ, Karabarbounis A, Lioliousis C, (2013): ELF alternating magnetic field decreases reproduction by DNA damage induction. *Cell Biochemistry and Biophysics*, 67(2), 703–716.

Panagopoulos DJ, Johansson O, Carlo GL, (2015a): Real versus simulated mobile phone exposures in experimental studies. *BioMed Research International*, 2015, 607053.

Panagopoulos DJ, Johansson O, Carlo GL, (2015b): Polarization: A key difference between man-made and natural electromagnetic fields, in regard to biological activity. *Scientific Reports*, 5, 14914. https://doi.org/10.1038/srep14914.

Panagopoulos DJ, (2016): Pulsed electric field increases reproduction. *International Journal of Radiation Biology*, 92(2), 94–106.

Panagopoulos DJ, (2017): Mobile telephony radiation effects on insect ovarian cells: The necessity for real exposures bioactivity assessment. The key role of polarization, and the "ion forced-oscillation mechanism". In: CD Geddes (Ed.), *Microwave effects on DNA and proteins*. Springer, Cham, Switzerland, 1–48.

Panagopoulos DJ, Balmori A, (2017): On the biophysical mechanism of sensing atmospheric discharges by living organisms. *Science of the Total Environment*, 599–600, 2026–2034.

Panagopoulos DJ, (2019a): Comparing DNA damage induced by mobile telephony and other types of man-made electromagnetic fields. *Mutation Research Reviews*, 781, 53–62.

Panagopoulos DJ, (2019b): Chromosome damage in human cells induced by UMTS mobile telephony radiation. *General Physiology and Biophysics*, 38(5), 445–454.

Panagopoulos DJ, Chrousos GP, (2019): Shielding methods and products against man-made electromagnetic fields: Protection versus risk. *Science of the Total Environment*, 667C, 255–262.

Panagopoulos DJ, (2020): Comparing chromosome damage induced by mobile telephony radiation and a high caffeine dose: Effect of combination and exposure duration. *General Physiology and Biophysics*, 39(6), 531–544.

Panagopoulos DJ, Karabarbounis A, (2020): Comments on "diverse radiofrequency sensitivity and radiofrequency effects of mobile or cordless phone near fields exposure in Drosophila melanogaster". *Advances in Environmental Studies*, 4(1), 271–276.

Panagopoulos DJ, Karabarbounis A, Yakymenko I, Chrousos GP, (2021): Mechanism of DNA damage induced by human-made electromagnetic fields. *International Journal of Oncology*, 59(5), 92.

Panagopoulos DJ, (2021): Comments on Pall's "Millimeter (MM) wave and microwave frequency radiation produce deeply penetrating effects: The biology and the physics". *Reviews on Environmental Health*. https://doi.org/10.1515/REVEH-2021-0090.

Pantelias GE, Terzoudi GI, (2010): Functional cell-cycle chromatin conformation changes in the presence of DNA damage result into chromatid breaks: A new insight in the formation of radiation-induced chromosomal aberrations based on the direct observation of interphase chromatin. *Mutation Research*, 701(1), 27–37.

Pantelias GE, Terzoudi GI, (2011): A standardized G2-assay for the prediction of individual radiosensitivity. *Radiotherapy and Oncology*, 101(1), 28–34.

Pay TL, Andersen FA, Jessup GL, (1978): A comparative study of the effects of microwave radiation and conventional heating on the reproductive capacity of *Drosophila melanogaster*. *Radiation Research*, 76(2), 271–282.

Pedersen GF, (1997): Amplitude modulated RF fields stemming from a GSM/DCS-1800 phone. *Wireless Networks*, 3(6), 489–498.

Persinger MA, (2014): Schumann resonance frequencies found within quantitative electroencephalographic activity: Implications for earth-brain interactions. *International Letters of Chemistry, Physics and Astronomy*, 11(1), 24–32.

Phillips JL, Singh NP, Lai H, (2009): Electromagnetic fields and DNA damage. *Pathophysiology*, 16(2–3), 79–88.

Pincheira J, López-Sáez JF, (1991): Effects of caffeine and cycloheximide during G_2 prophase in control and X-ray-irradiated human lymphocytes. *Mutation Research*, 251(1), 71–77.

Pirard W, Vatovez B: *Study of pulsed character of radiation emitted by wireless telecommunication systems*. Institut scientifique de service public, Liège, Belgium. https://www.issep.be/wp-content/uploads/7IWSBEEMF_B-Vatovez_W-Pirard.pdf.

Presman AS, (1977): *Electromagnetic fields and life*. Plenum Press, New York.

Pritchett TL, Tanner EA, McCall K, (2009): Cracking open cell death in the Drosophila ovary. *Apoptosis*, 14(8), 969–979.

Ramirez E, Monteagudo JL, Garcia-Gracia M, Delgado JMR, (1983): Oviposition and development of Drosophila modified by magnetic fields. *Bioelectromagnetics*, 4(4), 315–326.

Sagioglou NE, Manta AK, Giannarakis IK, Skouroliakou AS, Margaritis LH, (2016): Apoptotic cell death during Drosophila oogenesis is differentially increased by electromagnetic radiation depending on modulation, intensity and duration of exposure. *Electromagnetic Biology and Medicine*, 35(1), 40–53.

Santini R, Santini P, Danze JM, Le Ruz P, Seigne M, (2002): Study of the health of people living in the vicinity of mobile phone base stations: I. Influences of distance and sex. *Pathologie biologie*, 50(6), 369–373.

Sauter M, (2011): *From GSM to LTE: An introduction to mobile networks and mobile broadband*, John Wiley & Sons, Chichester, UK.

Schumann WO, (1952): Uber die strahlunglosen eigenschwingungen einer leitenden Kugel, die von einer Luftschicht und einer Ionospharenh ulle umgeben ist (On the characteristic oscillations of a conducting sphere which is surrounded by an air layer and an ionospheric shell). *Zeitschrift für Naturforschung*, 7A, 149–154.

Schwarz C, Kratochvil E, Pilger A, Kuster N, Adlkofer F, Ruediger HW, (2008): Radiofrequency electromagnetic fields (UMTS, 1,950 MHz) induce genotoxic effects in vitro in human fibroblasts but not in lymphocytes. *International Archives of Occupational and Environmental Health*, 81(6), 755–767.

Sesia S, Toufik I, Baker M (Eds.), (2011): *LTE – The UMTS long term evolution*. John Wiley & Sons Ltd., West Sussex.

Smith-Roe SL, Wyde ME, Stout MD, Winters JW, Hobbs CA, Shepard KG, Green AS, Kissling GE, Shockley KR, Tice RR, Bucher JR, Witt KL, (2020): Evaluation of the genotoxicity of cell phone radiofrequency radiation in male and female rats and mice following subchronic exposure. *Environmental and Molecular Mutagenesis*, 1(2), 276–290.

Stronati L, Testa A, Moquet J, Edwards A, Cordelli E, Villani P, Marino C, Fresegna AM, Appolloni M, Lloyd D, (2006): 935 MHz cellular phone radiation. An in vitro study of genotoxicity in human lymphocytes. *International Journal of Radiation Biology*, 82(5), 339–346.

Stryer L, (1996): *Biochemistry*, 4th ed. W.H. Freeman and Co, New York.

Terzoudi GI, Pantelias GE, (2006): Cytogenetic methods for biodosimetry and risk individualization after exposure to ionizing radiation. *Radiation Protection Dosimetry*, 122(1–4), 513–520.

Terzoudi GI, Hatzi VI, Donta-Bakoyianni C, Pantelias GE, (2011): Chromatin dynamics during cell cycle mediate conversion of DNA damage into chromatid breaks and affect formation of chromosomal aberrations: Biological and clinical significance. *Mutation Research*, 711(1–2), 174–186.

Thielens A, Bell D, Mortimore DB, Greco MK, Martens L, Joseph W, (2018): Exposure of insects to radiofrequency electromagnetic fields from 2 to 120 GHz. *Scientific Reports*, 8(1), 3924. https://doi.org/10.1038/s41598-018-22271-3.

Thielens A, Greco MK, Verloock L, Martens L, Joseph W, (2020): Radio-frequency electromagnetic field exposure of western honey bees. *Scientific Reports*, 10(1), 461. https://doi.org/10.1038/s41598-019-56948-0.

Tian XL, Lu X, Feng JB, Cai TJ, Li S, Tian M, Liu QJ, (2018): Alterations in histone acetylation following exposure to ^{60}Co γ-rays and their relationship with chromosome damage in human lymphoblastoid cells. *Radiation and Environmental Biophysics*, 57(3), 215–222.

Wertheimer N, Leeper E, (1979): Electrical wiring configurations and childhood cancer. *American Journal of Epidemiology*, 109(3), 273–284.

Wertheimer N, Leeper E, (1982): Adult cancer related to electrical wires near the home. *International Journal of Epidemiology*, 11(4), 345–355.

Wu GW, Liu XX, Wu MX, Zhao JY, Chen WL, Lin RH, Lin JM, (2009): Experimental study of millimeter wave-induced differentiation of bone marrow mesenchymal stem cells into chondrocytes. *International Journal of Molecular Medicine*, 23(4), 461–467.

Yakymenko I, Tsybulin O, Sidorik E, Henshel D, Kyrylenko O, Kyrylenko S, (2016): Oxidative mechanisms of biological activity of low-intensity radiofrequency radiation. *Electromagnetic Biology and Medicine*, 35(2), 186–202.

Yakymenko I, Burlaka A, Tsybulin I, Brieieva I, Buchynska L, Tsehmistrenko I, Chekhun F, (2018): Oxidative and mutagenic effects of low intensity GSM 1800 MHz microwave radiation. *Experimental Oncology*, 40(4), 282–287.

Yanagawa A, Tomaru M, Kajiwara A, Nakajima, Quemener ED, Steyer JP, Mitani T, (2020): Impact of 2.45 GHz microwave irradiation on the fruit fly, Drosophila melanogaster. *Insects*, 11(9), 598.

Zeni O, Chiavoni AS, Sannino A, Antolini A, Forigo D, Bersani F, Scarfì MR, (2003): Lack of genotoxic effects (micronucleus induction) in human lymphocytes exposed *in vitro* to 900 MHz electromagnetic fields. *Radiation Research*, 160(2), 152–158.

Zeni O, Sannino A, Romeo S, Massa R, Sarti M, Reddy AB, Prihoda TJ, Vijayalaxmi, Scarfì MR, (2012): Induction of an adaptive response in human blood lymphocytes exposed to radiofrequency fields: Influence of the universal mobile telecommunication system (UMTS) signal and the specific absorption rate. *Mutation Research*, 747(1), 29–35.

Zhukova MV, Kiseleva EV, (2011): Effects of starvation on the lifespan and apoptosis in the ovarian cells of Drosophila melanogaster. *Russian Journal of Genetics: Applied Research*, 1(4), 315–320.

Zothansiama, Zosangzuali M, Lalramdinpuii M, Jagetia GC, (2017): Impact of radiofrequency radiation on DNA damage and antioxidants in peripheral blood lymphocytes of humans residing in the vicinity of mobile phone base stations. *Electromagnetic Biology and Medicine*, 36(3), 295–305.

6 The Impacts of Wireless Communication Electromagnetic Fields on Human Reproductive Biology

Kasey Miller[1], Kiara Harrison[1], Jacinta H. Martin[1], Brett Nixon[1], and Geoffry N. De Iuliis[1]

[1]Reproductive Science Group, School of Environmental and Life Sciences, College of Engineering, Science and Environment, University of Newcastle, Callaghan, Australia

CONTENTS

Abstract 221
6.1 Introduction 221
    6.1.1 Scope and Intent of This Chapter 221
    6.1.2 Environmental Factors Affect Reproductive Potential in Both Humans and Animals 222
    6.1.3 The Role of Reproductive Sciences in Understanding the Health Impacts of WC EMFs 222
    6.1.4 ROS Are Key Mediators of Gamete Quality and Pathways Leading toward Cellular Damage 223
    6.1.5 Germ Cells and EMF Research 223
6.2 The Proximity of WCDs/EMF-Sources to Our Reproductive Systems 224
    6.2.1 The Simulation of WC EMFs for Reproduction Research 225
6.3 Limitations in the Field of Research 227
    6.3.1 Exposure Limits for WC EMFs 228
6.4 A Synopsis on the Current State of Human Infertility 228
    6.4.1 Female Infertility 229
    6.4.2 Male Infertility 230
    6.4.3 ART and Associated Risks 231
6.5 The Impacts of WC EMFs on Fertility 232
    6.5.1 Impacts on Embryo Development and Progeny 232
    6.5.2 Impacts on the Female 233
    6.5.3 Impacts on the Male 238

	6.5.4 Testicular Cancer	239
6.6	Impact on Sperm Function and DNA Integrity	239
	6.6.1 Understanding Impacts on Sperm Function from Murine Models	240
	6.6.2 Human Cohort and In Vitro Studies	245
6.7	Mechanism of Action for Man-made EMFs on Reproduction	246
	6.7.1 Thermal and Non-thermal Effects	247
	6.7.2 Effects Governed by Intensity and Frequency Including Tissue Penetration Depth	248
	6.7.3 The Impact of Man-made EMFs on Reproductive Hormone Profiles	248
	6.7.4 Proposed Mechanism of Man-made EMF Action (Oxidative Mediated Mechanism)	248
6.8	Man-made EMFs and OS	249
	6.8.1 OS in Reproductive Systems	249
	6.8.2 Antioxidants and Their Roles in Advancing the Field	251
	6.8.3 The Electron Transport Chain	252
	6.8.4 Evidence of Mitochondrial Disturbance in the Female Reproductive Tract	252
6.9	Discussion and Conclusions	255
	6.9.1 Mitochondrial ROS and DNA Damage	256
	6.9.2 Spermatozoa and the Male Reproductive System as a Study Model for the Effects of WC EMFs	257
	6.9.3 Where to Next and Recommendations	258
Funding		259
Acknowledgement		259
References		259

Keywords: electromagnetic fields; electromagnetic radiation; extremely low frequency; radio frequency; wireless communications; fertility; reproduction

Abbreviations: ART: assisted reproductive technology. ASBS: ATP synthase beta subunit. BBT: basal body temperature. CAT: catalase. DCFH-DA: 2'-7' dichlorofluorescin diacetate. ELF: Extremely Low Frequency. EMF: electromagnetic field. EMR: electromagnetic radiation. ETC: electron transport chain. FSH: follicle-stimulating hormone. GSM: Global System for Mobile Telecommunications. GPx: glutathione peroxidase. GSH: reduced glutathione. hCG: human chorionic gonadotropin. HSG: hysterosalpingogram. HSP: heat shock protein. HYOUI: hypoxia up-regulated protein 1 precursor. ICSI: intra-cytoplasmic sperm injection. IP: intraperitoneal. IVF: *in vitro* fertilization. LDH-X: lactate dehydrogenase-X. LH: luteinizing hormone. LPO: lipid peroxidation. MDA: malondialdehyde. mmW: millimeter waves. MT: mobile telephony. NOS: nitric oxide synthases. OS: oxidative stress. OSI: oxidative stress index. PUFA: polyunsaturated fatty acid. RF: Radio Frequency. ROS: reactive oxygen species. sncRNA: small non-coding RNA. SAR: Specific Absorption Rate. SOD: superoxide dismutase. TAS: total antioxidant status. TOS: total oxidant status. TRPV1: transient receptor potential subfamily V member 1 cation channel. TUNEL: terminal deoxynucleotide transferase dUTP nick end labeling. VDAC: voltage-dependent anion channel. VGIC: voltage-gated ion channel. VLF: Very Low Frequency. WC: wireless communications. WCDs: wireless communication devices. 4HNE: 4- hydroxynonenal. 8-OH-dG: 8-hydroxy-2'-deoxyguanosine.

ABSTRACT

The domain of reproductive biology underpins our understanding of human fertility and forms an important part of the debate on the safety of wireless communication (WC) electromagnetic fields (EMFs). While studies on the effects of anthropogenic EMFs on reproduction are of clear importance, recent evidence suggests that such studies are well placed to provide much-anticipated mechanistic insights on the health impacts of EMFs. Resolution of the biophysical mechanism(s) of action is one of the most important keys required to unlock scientific progression and enable accurate assessment of health risk. Growing recourse to assisted reproductive technologies (ART) across developed nations has justifiably given rise to concern about our decreasing collective fertility as a species. While this issue is certainly multi-factorial, the rise of anthropogenic EMF exposures and especially those of WC technology has aligned with a simultaneous global decline in male semen quality parameters. This well recognized link to reproductive health clearly underlines the unique sensitivity of our reproductive systems to environmental change and has prompted investigation of the impact of novel environmental insults such as WC EMFs. The current picture of how WC EMFs impact reproduction is not yet completely clear, but the field offers strong evidence of negative impacts on the cells, tissues, and processes that influence fertility. Accordingly, here we summarize the highest quality evidence outlining effects of WC EMFs on reproductive tissues and germ cells, and based on this, we propose a plausible mechanism for the molecular nature of the interaction of WC EMF with our biology. We also highlight some of the controversies in this field, including those pertaining to policy. Against this background, we contend that, in parallel with our advancing research, revising the safety limits of anthropogenic EMF exposures to our population is warranted.

6.1 INTRODUCTION

6.1.1 Scope and Intent of This Chapter

The use of wireless communication devices (WCDs), such as mobile phones and corresponding mobile telephony (MT) base station antennas, cordless domestic phones, etc., is now an integral part of our lives with the uptake of the ubiquitous technology still rapidly accelerating. These devices and other appliances, such as Wi-Fi (Wireless Fidelity) routers/emitters for connection to the internet, are almost universally incorporated into our living environment, emitting polarized electromagnetic fields (EMFs) and corresponding electromagnetic radiation (EMR) with Radio Frequency (RF) (300 kHz–300 GHz) and more specifically microwave (300 MHz–300 GHz) carrier signals, combined with Extremely Low Frequency (ELF) (3–3000 Hz) modulation and pulsation and displaying random variability of the final signal in the Ultra Low Frequency (ULF) (0–3 Hz) band (Pedersen 1997; Zwamborn et al. 2003; Hyland 2008; ARPANSA 2014; Panagopoulos 2019). This has generated a persistent and novel environmental stressor for living organisms, with the effects of the EMFs emitted by these devices and corresponding antennas on biological systems under active debate. With the inherent importance of human reproduction, the quality of which underpins our future population, both in terms of numbers and health status, the effects of these EMFs on reproductive biology are forming a key part of the "EMF safety" discussion. The potential that our current exposures could be influencing the health trajectory of future populations sets apart the reproductive branch of the larger field. Despite the public demand for answers and the quality of science conducted to date, progress toward ruling out significant general health risks, has been slow. Consequently, the public demand for a definitive health risk assessment of WCDs continues to go unmet. This chapter will provide an update on recent advances made in research addressing the effects of wireless communications (WC) EMF emissions on reproductive biology and reproductive health. While we strongly note that this field is still controversial with many conflicting conclusions published, our intent is to highlight research that is driving progression of the field. The male germline has provided key insights to date; however, we will discuss the benefits of both *in vitro* and *in vivo* based studies for both male and female reproductive systems and also the power of human studies complemented by whole body animal model exposure studies. Indeed, by harnessing the sensitivity of reproductive systems, we have learned that germ cells, spermatozoa in particular, are impacted by man-made EMFs, with these impacts manifesting via clear observations and defined cellular pathways. These effects have implications for risks associated with our fertility, but importantly, this information will also be vital if we are to prove a putative mechanism of action toward defining any risks of this novel and pervasive man-made environmental factor on public health.

6.1.2 Environmental Factors Affect Reproductive Potential in Both Humans and Animals

Human fertility is under pressure. While there is recent evidence to suggest waning human fecundity, this measure of fertility over the whole population is complex. Despite these complexities, we are observing a clear global decline in the quality of semen parameters and a loss of female fertility (Faddy et al. 1992; Spandorfer et al. 1998; Maher et al. 2003; Aitken and De Iuliis 2007; Wang and Sauer 2006; George and Kamath 2010; Agarwal et al. 2014a; Simon et al. 2017; Chambers 2017; Barratt et al. 2018), with the latter being largely associated with the advancing age of first-time mothers. The root causes of our fertility decline, including those associated with age, are still unknown; however, novel environmental insults are coming to the forefront in the pursuit to explore this decline in fertility and eliminate the potential of a cascading health burden. Amongst the urgent need to investigate impacts of environmental influences on reproduction, the importance of novel factors, such as WC EMFs, is firmly highlighted. Indeed, there is now a considerable body of evidence that associates WC EMF exposures (mainly from mobile phones) with poor fertility metrics (Gajda et al. 2002; Fejes et al. 2005; Wdowiak et al. 2007; Agarwal et al. 2009; De Iuliis et al. 2009a; Mailankot et al. 2009; Al-Damegh 2012; Ghanbari et al. 2013; Gorpinchenko et al. 2014; Furtado-Filho et al. 2014; Zalata et al. 2015; Zhang et al. 2016; Houston et al. 2018a; 2019). In this vein, anthropogenic WC EMFs have been reported to induce significant reproductive defects in animal models, such as rodents (Magras and Xenos 1997; Yan et al. 2007; Gul et al. 2009), insects (Panagopoulos et al. 2004; 2007a; 2007b; 2010; 2013a; Panagopoulos and Margaritis 2010a; 2010b; Sharma and Kumar 2010), amphibians (Balmori 2006), and birds (Balmori 2005; Batellier et al. 2008; Burlaka et al. 2013). Collectively, this accentuates the importance of continued scientific research in reproductive science and medicine.

6.1.3 The Role of Reproductive Sciences in Understanding the Health Impacts of WC EMFs

Despite the significant advancements made in reproductive science and medicine, the origins of numerous forms of infertility remain unknown, and this is particularly true for the approximated 30%–40% of men (Jungwirth et al. 2015; Agarwal et al. 2021) diagnosed with idiopathic infertility. Reproductive research has historically provided significant insight into conditions causing a patient's infertility, particularly in regard to those driven by disrupted hormone signaling, including those instigated by environmental factors. As such, examination of an individual's hormone profile, together with anatomical assessment of the reproductive tract routinely underpin a patient's clinical assessment. So successful are these criteria that hormonal disruption, along with physical malformation of the reproductive tract, account for about 70% of the individuals presenting to assisted reproductive technology (ART) clinics (Sadeghi 2015), with clearly defined diagnoses for infertility. These criteria are not, however, all encompassing, leaving substantial gaps in our understanding of infertility and a large cohort (~ 30%) of individuals devoid of rational therapy options. In parallel with declining fertility in our species, mounting evidence indicates that the ARTs that provide the only recourse for infertility, may themselves, not be risk free, with the use of such technologies linked to elevated risks of adverse health in the offspring (Sánchez-Calabuig et al. 2014; Chen and Heilbronn 2017; Novakovic et al. 2019; Bergh and Wennerholm 2020). Intriguingly, this risk appears to originate and is perpetuated by the paternal germline (Aitken et al. 2020; Aitken and Baker 2020; Aitken 2018). In support of this theory, our own research group has contributed significantly to the understanding of how such germline perturbations arise and subsequently impact embryo development and offspring health (Aitken et al. 1998; 2012a; 2020; Aitken and De Iuliis 2007; Lee et al. 2009; Aitken 2020; Aitken and Baker 2020).

Given this and the knowledge that a major proportion of WC EMF studies highlight effects on the human reproductive system, we contend that analysis of this system provides a critical avenue for the necessary advancements in our understanding of the health implications of WC technologies. This is made possible by virtue of the male germline's innate and measurable susceptibly to environmental alterations, especially when compared to most somatic cells (Koppers et al. 2008; Lee et al. 2009; De Iuliis et al. 2009b; Aitken et al. 2012a; 2012b; 2013). The anatomy of the human male reproductive system also places it in a uniquely vulnerable position for chronic EMF exposure from devices stored, or used, in the pants pockets or the vicinity of the testes, including Wi-Fi enabled laptop computer use. With the rise of mobile data technology, the types of exposures are also no longer limited to scenarios such as "standby mode", call initiation, or hands-free mode, as was the focus in recent past years. Instead, the constant data (internet) connection afforded by current WC technologies, elevates the risk of chronic exposure and justifies a more detailed risk assessment of these devices. We note that new mobile "smart phone" devices emit separate signals (from separate antennas) for telephony and for Internet connection, exposing the users to different pulsing WC EMFs simultaneously. Finally, because the recognition that fertility issues in men are often an early indicator of poor overall health, the reproductive system has particular utility in examining the effects of a novel exposure on an individual's overall wellbeing (Aitken et al. 1998; 2020).

6.1.4 ROS ARE KEY MEDIATORS OF GAMETE QUALITY AND PATHWAYS LEADING TOWARD CELLULAR DAMAGE

Of the numerous discoveries that have had significant impact on our understanding of reproductive biology and fertility treatment, the knowledge that reactive oxygen species (ROS) are both physiologically critical signaling molecules for animal biology (Shapiro 1991; Wong and Wessel 2005; Kazama and Hino 2012), and common mediators of infertility, has been paramount. This is true for both the male and female germline (Mihalas et al. 2017a; 2017b; 2018). In this vein, we and others have demonstrated that excess levels of ROS create a state of oxidative stress (OS), which arises from the aberrant regulation of ROS and antioxidant molecules. A small level of ROS can then precipitate further ROS generation, lipid peroxidation (LPO) and electrophilic aldehyde formation, mitochondrial disruption, and antioxidant exhaustion (Figure 6.1). This state can then lead to mitochondrial collapse, DNA damage, amyloid plaque formation, dysregulated meiosis (Lord and Aitken 2013; Mihalas et al. 2017a; 2017b; 2018; Cafe et al. 2021), losses of sperm and oocyte function (Vernet et al. 2004; De Iuliis et al. 2009b; Noblanc et al. 2012; Aitken et al. 2020), and cell death.

Unsurprisingly, there is now a wealth of data to show that cellular OS may be induced by environmentally relevant levels of WC EMFs in various cell types/tissues (Yakymenko et al. 2016), including those that make up the male reproductive tract and germline (Agarwal et al. 2009; De Iuliis et al. 2009a; Al-Damegh 2012; Ghanbari et al. 2013; Liu et al. 2013b; Houston et al. 2019). The latter is likely attributable to the inherent vulnerability and unique cell biology of the male germline as discussed below (Lenzi et al. 2000; Wdowiak et al. 2007; Aitken et al. 2012b; 2015).

6.1.5 GERM CELLS AND EMF RESEARCH

In addition to the biological suitability of the germline for studying the effects of WC EMFs on reproduction, these unique cells provide enhanced scope for experimental design, including the non-invasive assessment of human cells after "real world" exposures and via controlled *in vitro*

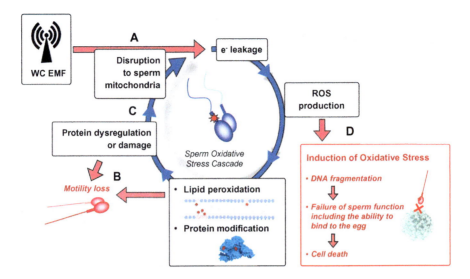

FIGURE 6.1 Oxidative stress cascade in spermatozoa: (A) Initial disruption may occur by interaction with the electron transport chain (ETC) or mitochondrial outer membrane voltage-dependent anion channel (VDAC) proteins (sperm mitochondria are all located in the midpiece (neck) of the mature gamete), leading to electron leakage. Subsequent lipid peroxidation (LPO) leads to motility loss (B) as well as mitochondrial protein damage (C) causing a positive feedback loop of ROS production, and a state of oxidative stress (OS) (D).

studies on ejaculated spermatozoa. In addition to providing important background context in reproductive biology in this chapter, we shall explore the documented effects of WC EMFs through a cross section of reproductive studies with a particular focus on the germlines and that of the male reproductive system. A key narrative is the critical importance that reproductive biology is playing in unravelling the science in this field, ultimately, providing rational information on the health risks posed by WC EMFs and informing the policy that regulates its use and continued development.

6.2 THE PROXIMITY OF WCDs/EMF-SOURCES TO OUR REPRODUCTIVE SYSTEMS

The exponential increase in WCD use over the past 20 years (Gorpinchenko et al. 2014) has driven the rapid evolution of digital cellular WC technologies. The deployment of each technological generation has overcome its predecessor's limitations, advancing the WC network to meet this growing demand for high data transfer speed, reliability, and low latency (Ohmori et al. 2001; Authority 2015). Unsurprisingly, WC EMFs are now ubiquitous in our modern environment, prompting many to raise concerns and investigate relationships between the use of WCDs held at the ear (i.e., mobile or cordless phones) and brain malignancies (Prasad et al. 2017; Leng 2016; Karipidis et al. 2018; Miller et al. 2018; 2019). However, with the common practice of storing mobile phones in the pants pockets or trouser belt, the reproductive system is also highly exposed to WC EMFs and, thus, any associated forms of cellular perturbation.

Indeed, the proximity of mobile phones and other WCDs to our reproductive system, and in particular the testes, is driving increasing concern with nearly a thousand publications having reported on this subject since 1990. These articles, although often conflicting, do infer consequences to the reproductive system following WCD exposure. Therefore, continued investigation into testicular function and WCD exposure is of paramount importance, and perhaps might uncover an association between this technology and the ever-increasing rise in human infertility. For this reason, a more careful use of WCDs is advocated.

As communication networks evolve, increasingly higher carrier frequencies are required to facilitate greater data bandwidth demands, and consequently, the amount of energy they can transmit is simultaneously increased, along with increasing variability of the signals, which is considered an important bioactive factor (Panagopoulos 2019; Kostoff et al. 2020). This relationship has formed much of the concern around fifth generation (5G) of MT and future generations of the WC technology; however, the WC industry reassures their consumers that the emissions of these new technologies are still well within accepted guidelines.

The rate of energy absorbed from RF EMF sources per unit mass of localized tissue is generally defined as the Specific Absorption Rate (SAR). The majority of carrier frequencies still used are between 850 and 2600 MHz, with SAR values between 0.12 to 1.6 W/kg of body mass. While these energy levels may impart only subtle, if any, thermal effect on human tissues, there is compelling evidence to suggest that these ranges can disrupt human fertility (La Vignera et al. 2012). Relevant literature exploring the effects of man-made EMFs on the male reproductive system indicates that those cell populations most vulnerable to this insult are the Leydig cells, seminiferous tubules, and spermatozoa (La Vignera et al. 2012). These effects are obviously exerted via non-ionizing energies with non-thermal properties, which are well established to cause spermatogenesis disruption leading to a decrease in sperm quantity and quality (Agarwal et al. 2008). Increasing experimental evidence advocates that OS and DNA damage are key mechanisms by which mobile phone and other WCDs use causes damage (Deepinder et al. 2007).

Although we and others use the SAR as an exposure metric, we should also acknowledge that it has been criticized as being prone to providing non-objective measurements and failing to take into account microscopic absorption that might be of importance, especially in the case of non-thermal effects. It has been proposed that the power density of the incident EMR (in mW/cm^2) should be used as a direct and more accurate metric (Baker et al. 2004; Panagopoulos et al. 2013b). WC EMFs at environmentally relevant intensities (e.g., mobile phones and other WCDs) are not found to induce any considerable heating in exposed tissues beyond 0.1°C–0.3°C, which is not considered significant for inducing biological effects (Belyaev 2005; Dawe et al. 2006; Panagopoulos et al. 2013b). Yet, this is still untested for the 5G systems under deployment, which may be able to induce more significant heating due to the higher carrier frequencies and beam forming at increased intensities (Neufeld and Kuster 2018).

Given the unique exteriorization from the human body, the temperature of the testis is 2°C–3°C lower than rectal temperature, with 35°C considered optimal for spermatogenesis (Saikhun et al. 1998). A review of 5G WC-related EMF studies and expected health effects has highlighted that there is an essential need for more research into local heat impacts on body surfaces, such as the skin and eyes, with improvement to study design necessary for safety assessment (Simkó and Mattsson 2019). The attention to exposed surface area (and not volume) requires further consideration because of the very shallow penetration depth of 5G and millimeter wave* (mmW) carrier frequencies. Although such frequencies have a very shallow penetration depth, contrastingly, associated ELFs (due to pulsation and modulation of the WC EMFs) have considerable tissue penetration depth, calling attention to the accountability of both surface area and volume. Investigation into whether there are any plausible health-related effects associated with the skin is under way (Karipidis et al. 2021) and, while requiring careful corroboration in the scientific community, pronounces the further significance for the potentially vulnerable exteriorized human testes (Miller and Torday 2019). Discussion concerning non-thermal modes of action in human reproductive systems, which constitute the vast majority of recorded effects, is the feature in this chapter and is addressed below; however, possible micro-thermal impacts are clearly not to be neglected, as they form a key part of the debate and must be explicitly considered in any research design in this field.

6.2.1 The Simulation of WC EMFs for Reproduction Research

As in the larger field, simulating environmentally relevant exposures for reproductive systems comes with complex challenges. Temperature rises in cells and tissue can be suitably assessed

---

* Millimeter waves are called microwaves with wavelengths of the order of mm, or frequencies 30–300 GHz.

using highly controlled, estimated SARs (Starkey 2016). Environmentally relevant (real) signals with complex field patterns and variable peak field strengths and intervals between transmissions can be more difficult in estimating their average energies; however, they may influence biological and reproductive systems in ways that controlled simulated patterns may not. Another important variable to consider, which is difficult to emulate in controlled simulations, is that biological responses can be greater during intermittent exposures rather than continuous (Diem et al. 2005; Remondini et al. 2006) and, furthermore, are contingent on the pulse characteristics for the same average power (Oscar and Hawkins 1977). Additionally, real-life WC EMFs are intrinsically variable, making their bioactivity greater than those WC EMFs that are simulated (Panagopoulos 2019; Kostoff et al. 2020).

The vulnerability of the germline to WC EMFs throughout testicular developmental stages is governed by a myriad of factors, making WC EMF research designs highly complex to accurately simulate. The biophysical attributes that must be considered by researchers when investigating the relationship between mobile phone use and the reproductive system include radiation intensity and included frequencies, exposure time and duration, modulation, signal intensity windows, polarization, free space propagation, proportion of developing germ cells exposed, and the water content of the tissue (Belyaev 2005; 2010; Blackman 2009; Agarwal et al. 2011; Stefanovic et al. 2017; Kesari et al. 2018; Panagopoulos 2019). Figure 6.2 shows a system for exposing cells *in vitro*, with a waveguide apparatus housed within a Faraday cage placed in an incubator with a central antenna.

Many animal studies using rats, mice, and rabbits have shown that WC EMFs can negatively interfere with the male reproductive system (Kesari et al. 2018). The effects of mobile phone use on the human male reproductive system have been mainly investigated by *in vitro* exposure systems using isolated spermatozoa, yet these are complemented by a handful of clinical studies (La Vignera et al. 2012) and *in vivo* exposure models featuring men keeping their phones within their pants pockets (Fejes et al. 2005). Predominantly, the abnormalities detected during these clinical studies appear to be directly related to the duration of mobile phone use and the distance of the device from the body (intensity of radiation/field). Potential confounding factors that should be accounted for in clinical studies are other lifestyle and environmental

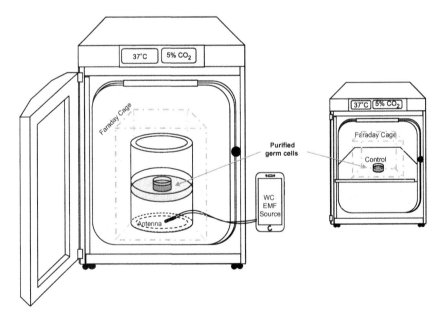

FIGURE 6.2 Waveguide apparatus for gamete irradiation *in vitro*: Representation of a typical exposure system for purified cells to simulated or phone generated (real-life) WC EMFs.

factors associated with semen quality such as age, drug and alcohol use, body mass index, and appropriate nutritional intake (Zhang et al. 2016; Sajeda and Al-Watter 2011). Clinical study constraints involving patients presenting to fertility clinics heavily relies on surveying the habits of mobile phone use, including duration of possession, daily standby position, and daily use times. These self-reported exposure habits can be difficult to define criterion parameters and accurately and objectively report device usage. Furthermore, these studies require large cohorts of population participation to accurately interpret findings (Fejes et al. 2005; Kilgallon and Simmons 2005; Wdowiak et al. 2007). Another challenge to consider during reported mobile phone device habits and emulated exposures is the presence of common metal objects, including coins, rings, and zippers, that are frequently kept in the same trouser pockets as mobile phones, which can possibly exacerbate the effects of WC EMFs by locally increasing the EMF intensities (Lewis et al. 2017).

Challenges and limitations that need consideration within animal research include the requirement of containment. Animal housing during RF exposures is diverse within the research field, spanning from restrained containment to unrestrained containment, with each presenting limitations for the experimental design. Animals allowed to freely move around their housing environment during exposures pose experimental limitations, as varied exposure levels to EMFs may occur as animal proximity to the source deviates, especially given the small, targeted area of the reproductive system. Accordingly, in an attempt to circumvent these animal digressions, increased scaling of animal sample size must be considered in conjunction with appropriate statistical analysis. Contrastingly, animals that are restrained throughout exposures, although ensuring a targeted and direct exposure, present additional limitations pertaining to stress-related outcomes associated with the restraint. Furthermore, murine models used to emulate environmental EMF exposures may pose some confounding limitations to clinical translation due to the non-pendulous structure of their scrotum (Cairnie and Harding 1981).

Due to the demand for higher data rates, 5G and future generation WC systems rely on highly directional radiation by beam-forming antennas instead of traditional antennas with omnidirectional broadcasting radiation. Therefore, utilization of directional antennas alongside dosimetry relevance in extant exposure systems is useful in understanding and accepting a mechanism of action in which WC EMFs interact with the reproductive system. Urgent clarification around the detailed mechanism of impacts of non-thermal effects from WC EMF exposure is required.

6.3 LIMITATIONS IN THE FIELD OF RESEARCH

Past research regarding the impact of ELF and RF EMFs on human reproduction specifically reflects a raft of RF EMF studies focusing on general biology and health, with several inconsistencies and contradictory results. While the quality of experimental design is now much improved, dosimetry continues to be a complex issue. The heterogeneity of data and lack of confidence in the conclusions of some early studies underpins the limitations to our current ability to drive advancements and provide definitive outcomes. An example of the limitations faced are those parameters such as exposure duration, which varies widely from study to study, with few looking at the effect of chronic timelines. The inherent difficulties of comparing published studies and modeling chronic exposures, in turn, limits a clear consensus.

Today, investigators conclude that there are significant negative impacts on reproductive systems (La Vignera et al. 2012; Singh et al. 2018; Negi and Singh 2021), and well-designed studies are now available showing that man-made EMFs induce a suite of deleterious effects on a variety of cells and tissue types, including OS and micronuclei formation, which implicates a genotoxic assault and DNA fragmentation (Cairnie and Harding 1981; Dawson et al. 1998; Cecconi et al. 2000; Zmyślony et al. 2000; Rodriguez et al. 2003; Lai and Singh 2004; Aitken et al. 2005; Fejes et al. 2005; Delimaris et al. 2006; Lixia et al. 2006; Erogul et al. 2006; Oral et al. 2006; Depinder et al. 2007; Jung et al. 2007; Yan et al. 2007;

Panagopoulos et al. 2007a; 2010; 2013a; Panagopoulos and Margaritis 2010a; 2010b; Panagopoulos 2019; Agarwal et al. 2008; 2009; 2011; Falzone et al. 2008; De Iuliis et al. 2009a; 2009b; Mailankot et al. 2009; Singh et al. 2009; Lee et al. 2010; Sarookhani et al. 2011; Saygin et al. 2011; Falzone et al. 2011; Sajeda and Al-Watter 2011; Gutschi et al. 2011; Imai et al. 2011; Kesari et al. 2011; 2018; Al-Damegh 2012; La Vignera et al. 2012; Rago et al. 2013; Burlaka et al. 2013; Kumar et al. 2013; 2014; Ghanbari et al. 2013; Mortazavi et al. 2013a; Shahin et al. 2013; 2017; Liu et al. 2013a; 2013b; 2015; Meena et al. 2014; Gorpinchenko et al. 2014; Mihai et al. 2014; Adams et al. 2014; Bakacak et al. 2015; Sangun et al. 2015; Zalata et al. 2015; Zhang et al. 2016; Alchalabi et al. 2016a; 2016b; Yakymenko et al. 2016; Tumkaya et al. 2016; Sepehrimanesh et al. 2017; Lewis et al. 2017; Lin et al. 2017; Roozbeh et al. 2018; Singh et al. 2018; Houston et al. 2018a; 2019; Okechukwu 2020; Negi and Singh 2021)

6.3.1 Exposure Limits for WC EMFs

International public health guidelines for RF EMFs emitted by mobile phones were constrained to a general public SAR limit of 2.0 W/kg for head and trunk by the International Commission on Non-Ionizing Radiation Protection (ICNIRP 1998; 2020). Variation in this threshold is, however, observed between countries. Australia, India, the United States (US), and Canada, for example, have governed a SAR limit of 1.6 W/kg, whereas Europe enforces a 2.0 W/kg threshold (Dahal 2013; ARPANSA 2014). Power density limit for the general public for mobile phone frequencies (around 2 GHz) in most of these countries is 1 mW/cm^2. Certain countries have set tighter limits. For example, Russia, Poland, and Italy have adopted a corresponding limit of 10 µW/cm^2 (Madjar 2016). Due to the significant technological development behind this infrastructure and drastic increases in global accessibility, the resultant increase in WC EMF exposure is huge. In addition to new scientific data in dosimetry and health risks (Baan et al. 2011), the ICNIRP guidelines are widely considered to be outdated (Hardell 2017; Bortkiewicz 2019). ICNIRP (2009) described human male infertility research as "limited available data on other non-cancer outcomes show no effect of RF field exposure", and that "no convincing evidence that low level [acute] exposure results in any adverse outcomes on testicular function" for humans. However, animal-based studies, which primarily utilized rodents and fruit flies, report clear negative effects on fertility (Magras and Xenos 1997; Aitken et al. 2005; Yan et al. 2007; Panagopoulos et al. 2007a; 2007b; 2010; Panagopoulos and Margaritis 2010a; 2010b) at levels well below the ICNIRP limits. Advancements in research now indicate a more robust link between exposures and negative influences on human fertility, predominantly on the male germline, with additional effects on reproductive organs and changes in testosterone levels also of note (Naeem 2014; Lin et al. 2017; Roozbeh et al. 2018; Kesari et al. 2018; Negi and Singh 2021). Given further scientific evidence and consensus from continued investigations, reassessment of the current ICNIRP guidelines is required. However, despite the evidence, these limits have been further increased in the latest ICNIRP guidelines, which raised the average localized power density value, over 6 min of local heating, from 1 mW/cm^2 (ICNIRP 1998) to 4 mW/cm^2 (ICNIRP 2020).

The Scientific Committee on Emerging and Newly Identified Health Risks (SCENIHR) has published a multitude of opinions on health issues related to consumer products (Ahlbom et al. 2008). The purpose of the SCENIHR's most recently published review (Samaras et al. 2015) was to update opinions and acknowledge significant gaps in understanding. The review reported no health effects from exposure or co-exposure from available WC technology. Published comments on the SCENIHR (Samaras et al. 2015) review emphasize the lack of acknowledgement of well accepted health effects and underestimation of the importance of mechanistic evidence associated with induced OS (Henshaw and O'Carroll 2009; Redmayne 2016).

6.4 A SYNOPSIS ON THE CURRENT STATE OF HUMAN INFERTILITY

Infertility refers to a biological inability for a couple of reproductive age to successfully conceive after a full year of actively trying, including failure to sustain a pregnancy to term (Vander Borght and Wyns 2018). Moreover, it can be further defined into primary and secondary; primary infertility refers to non-pregnancy after 12 months, whereas secondary infertility refers to loss of fertility after being able to achieve pregnancy at least once (Tabong and Adongo 2013). Although age-related fertility declines have been documented to afflict both men and women, the etiology is disproportionately skewed toward females. Altogether, infertility is experienced by approximately 12%–15% of couples globally (Sharlip et al. 2002). However, males and females within a couple often present with co-existing factors, and as such, it is recommended that partners be investigated both individually and together (Agarwal et al. 2015; Vander Borght and Wyns 2018).

6.4.1 Female Infertility

Approximately 9% of women in developed countries and 13% of women worldwide are reported to experience fertility problems, which can be caused by a variety of medical conditions such as polycystic ovary syndrome, uterine fibroids, endometriosis, endometrial fibroids, pelvic inflammatory disease, and premature ovarian insufficiency (Vander Borght and Wyns 2018; Barbieri 2019). While most female fertility is assessed after 1 year of unprotected intercourse in which no conception is achieved, earlier assessment is often warranted in cases of oligomenorrhea (infrequent menstrual cycles), amenorrhea (menstrual absence), patient age 35 years or older, or in suspected and diagnosed cases of aforementioned pelvic pathologies (Smith et al. 2003). As age also plays a significant role in female fertility, with a sharp decline generally occurring around 35 years of age, the reported median age for final birth is 40–41 years in most studied populations (Spira 1988; te Velde and Pearson 2002). Such age-related attenuation of female fertility is broadly attributed to two main factors: a numerical reduction in the ovarian pool of oocytes owing to periodic loss of oocytes associated with the monthly menstrual cycle, and commensurate loss of quality and, hence, the developmental potential of the remaining oocyte reserve. Both scenarios reflect the fact that women are born with a finite number of oocytes, which are gradually depleted over their reproductive lifespan. The attendant decline in oocyte quality is evidenced by an increased prevalence of oocytes harboring abnormal chromosome numbers (i.e., aneuploidy), which escalates to the point where ~ 50% of the oocytes in women aged > 40 are considered non-viable (Carolan and Frankowska 2011; Fragouli et al. 2011; Hassold and Chiu 1985; Hsueh et al. 2015; Kuliev et al. 2011; Chambers et al. 2017). These defects manifest an elevated risk of miscarriage and conceiving a child with chromosomal abnormalities. Overall, these data demonstrate the irreparable nature of age-associated female infertility. Although complex etiologies undoubtedly contribute to the deterioration of oocyte quality, increasing attention has focused on the pervasive impact of environmental factors. Indeed, the prolonged lifespan of the meiotically arrested oocyte places this cell at elevated risk of environmentally induced lesions, many of which commonly result in dysregulation of protein homeostasis. Whilst oocytes can reduce this threat via activation of a network of surveillance, repair, and proteolytic pathways, these defenses are themselves prone to age-related defects, thus, reducing their ability to remove damaged proteins (Mihalas et al. 2017a; 2017b; 2018).

Contemporary clinical assessment of female reproductive defects is often aimed at ruling out anovulation and tubal obstruction, generally involving analysis of ovulatory function concurrently with assessment for male subfertility (Collins et al. 1993; Hunt and Vollenhoven 2020). Methodology ranges from determining luteinizing hormone (LH) levels from urine samples and progesterone concentration from mid-luteal phase serum samples, sonohysterogram and/or hysterosalpingogram (HSG) examination of uterine cavity, and HSG or laparoscopic assessment of tubal patency (Smith et al. 2003). Meanwhile, methods such as postcoital assessment, biopsy of endometrial tissue, and recording of basal body temperature (BBT) are regularly considered as antiquated forms of fertility assessment and, as such, are often discarded in favor of less invasive commercialized tests with higher predictive values, such as cervical mucus observation in conjunction with urinary ovulation prediction kits that detect LH (Leiva et al. 2017; Su et al. 2017).

6.4.2 Male Infertility

Half of all human infertility issues experienced by couples are attributed to male-assigned factors, with male subfertility reported to affect an estimated 7% of all men (Lotti and Maggi 2015). Male fertility is adversely impacted by a variety of genetic disorders and medical conditions, such as varicocele, retrograde ejaculation, erectile dysfunction, undescended testicles, and insufficient testosterone levels (Krausz 2011). Moreover, significant changes in lifestyle and environmental exposures over the previous 50 years, primarily experienced in westernized countries and including the rise of WC technologies and handheld WCDs, have increasingly been linked with reductions in fertility (Adams et al. 2014; Okechukwu 2020). Illustrative of this phenomenon, the period spanning between 1973 and 2011 was accompanied by a decline in average human sperm concentrations within an ejaculate (Figure 6.3), equating to 1.4% year on year and amounting to an overall reduction of 50%–60% (Levine et al. 2017). These declining sperm numbers are reportedly accompanied by reduced sperm quality as assessed by readouts of motility and vitality (Carlsen et al. 1992).

Despite intensifying research, there remains no clear explanation for the underlying causes of the decreasing trends in semen quality. In recent years, there has been growing appreciation that a man's general and reproductive health are closely intertwined, with strong associations established between male infertility and future health (Eisenberg et al. 2014; 2015a; 2015b; Glazer et al. 2017). It has been hypothesized that shared genetic pathways, environment, and lifestyle factors, possibly acting *in utero*, could play a leading role (Bonde et al. 2016). Linked to this work, mounting evidence indicates that semen quality can serve as a biological marker for future male health, as several large epidemiological studies of > 50,000 men from Europe and the US report consistent association between poor semen quality and mortality. Indeed, decreased sperm parameters (concentration, count, motility, morphology) are linked to a 2.3-fold increased risk of death in the following 8 years, with a > 90% predictability (equivalent to the risk due to smoking or diabetes) (Jensen et al. 2009; Eisenberg et al. 2014; Eisenberg et al. 2015a).

Current epidemiological evidence is compatible with an association between male infertility and paternal environmental exposures, as is the link between male infertility and risk of chronic disease and mortality. However, the relatively small number of related studies and insufficient adjustment of confounders preclude definitive statements about the causality of these associations, and as yet, the pathogenic pathways connecting these conditions are uncertain (Kirk et al. 2018). Important

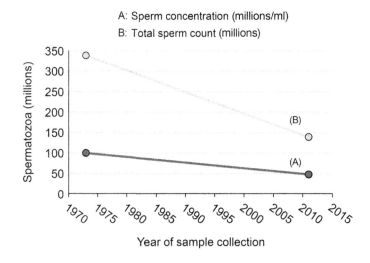

FIGURE 6.3 Decline of semen parameters 1970–2010. The graphs are based on data by Levine et al. (2017), and there is building evidence that semen quality continues to fall.

insights have been gained from preclinical animal studies, which have revealed that a range of paternal stressors linked to reduced fertilization potential also induce genetic and epigenetic alterations within the paternal germline (Lotti and Maggi 2015; Ilacqua et al. 2018; Sengupta et al. 2017). Moreover, in addition to defects in testicular development, such studies have increasingly implicated downstream passage through the epididymis as a key site of vulnerability for sperm maturation, during which the transcriptionally and translationally active spermatozoa can sustain pronounced proteomic and epigenomic changes without the facility to repair such damage (Katen et al. 2017; Houston et al. 2019; Trigg et al. 2021). Collectively, these data highlight that male fertility and, more specifically, the clinical assessment of basic sperm parameters, represent readily accessible biomarkers and a potentially powerful untapped resource with which to diagnose diseases early and predict the long-term health of an individual.

6.4.3 ART AND ASSOCIATED RISKS

ART, such as *in vitro* fertilization (IVF) and intra-cytoplasmic sperm injection (ICSI), have seen a boom in technological advancement and research focus in recent years (Wang and Sauer 2006). Use of ART services is often linked to male infertility, wherein sperm cells are unable to reach the egg within the female reproductive tract due to a combination of low sperm count, insufficient motility, and/or loss of vitality (Tournaye 2012). The most common ART process involves several steps, beginning with stimulation of female hormone production to increase the number of eggs recovered during ovulation, followed by either retrieval of matured eggs under light anesthesia or intra-uterine insemination during which a sperm sample is deposited within the uterus. External to the body, sperm cells are introduced to eggs by one of two means: IVF, in which sperm from a male donor or partner and eggs are combined until fertilization occurs, or ICSI, which involves a single sperm being injected into each extracted egg (Wang and Sauer 2006; Zegers-Hochschild et al. 2009). Fertilized eggs are then given time (2–5 days, depending on clinical practice) for embryo development before one or more embryos are reintroduced to the woman's uterus (Farquhar and Marjoribanks 2018; Brown et al. 2016; Van Voorhis 2006). In occasions that several embryos develop, they can be frozen and stored for later procedures (Konc et al. 2014; Simopoulou et al. 2018; Nagy et al. 2014). Two weeks after embryo transfer, clinical testing for pregnancy by assessing for human chorionic gonadotropin (hCG) levels in blood takes place. If non-pregnancy occurs, the woman will need to experience a menstrual cycle before deciding to try again (Van Voorhis 2006; Wang and Sauer 2006). If embryos are stored from previous attempts, the woman does not need to undergo the egg retrieval processes again (Nagy et al. 2014). However, should a positive test result occur, an ultrasound inspection is scheduled for 2 weeks later to assess development (Iyoke et al. 2013). It should be noted that even in the occurrence of a clinical pregnancy, miscarriage can still occur (Fragouli et al. 2011).

Aligning with the multi-step nature of ART procedures, these services are often associated with considerable out-of-pocket expenses, and much like any other medical procedure, there are associated risks, such as ovarian hyperstimulation syndrome in which an excessive response to fertility drugs occurs, multiple births (twins and triplets), premature labor, caesarean delivery, and low birth weight (Simopoulou et al. 2018; Kalra et al. 2011). In Australia and New Zealand during 2019, 88,929 IVF cycles were initiated, representing a substantial annual increase of 5.6% compared to the use of this technology in 2014 (Newman et al. 2021).

Although ARTs can improve fertilization rates, such as when sperm motility prevents gamete interaction, pre-existing damage to the paternal genome or epigenome cannot be alleviated, therefore, carrying an associated risk of adversative clinical outcomes for conceived progeny (Simon et al. 2017). Research has further found compelling evidence that links a diverse array of environmental stressors to modifications of small non-coding RNA (sncRNA) profiles in the paternal germline, which, in turn, significantly implicates the possibility of phenotypes being transmitted transgenerationally, ultimately affecting the health and wellbeing of the offspring (Rodgers et al. 2015; Hur et al. 2017; Trigg et al. 2019; Nixon et al. 2019).

6.5 THE IMPACTS OF WC EMFs ON FERTILITY

The reproductive system and, in particular, the male germline is providing new insights into how the non-ionizing anthropogenic EMFs of WC networks can affect biology and human health. Several observations in other systems and cell types overlap those discovered in reproduction. However, through studies utilizing spermatozoa, OS is now a widely accepted impact of these EMFs. In fact, regardless of sex, ELF, and RF/WC EMFs have been consistently linked with induced OS via elevating levels of ROS leading to impacted fertility (Kesari et al. 2018). Compiled in Table 6.1 is a summary of relevant literature pertaining to man-made EMF exposure in the RF/WC, ELF, and Very Low Frequency (VLF) (3–30 kHz) bands and its impacts on reproduction and development. Given that WC EMFs always combine ELF and RF (and even VLF to a lesser extent) EMFs, we examine, here, effects from all these frequency bands, mainly ELF and RF/WC EMFs.

6.5.1 Impacts on Embryo Development and Progeny

The effect of man-made EMF exposure on implantation and embryo development has been investigated in several animal models; however, the reproductive risk of exposure in non-pregnant women has not been extensively researched. In combination with numerous EMF studies, several studies utilizing ELF oscillating EMFs modeling those produced by power lines indicate adverse effects on development in these model systems. Superovulated mice that were mated following exposures to ELF EMFs (50 Hz, 0.5 mT) 4 h/day, 6 days a week for 2 weeks (Pourlis 2009) experienced a significant decline in total blastocyst number together with a complementary increase in DNA fragmentation, thus, highlighting the detrimental effects that ELF EMF exposure poses during the preimplantation stage of development (Borhani et al. 2011). In a porcine model, ELF EMFs have similarly been shown to affect early embryo development by delaying cleavage of fertilized eggs in the oviduct of swine following 50 Hz, 0.75 mT EMF exposure 4 hours (h) prior to ovulation (Bernabò et al. 2010). In *Drosophila*, exposure to sinusoidal ELFs at 50 Hz reflective of intensities usually found close to high-voltage power lines or higher, led to a decrease in reproduction, due to cellular DNA fragmentation, and death (Panagopoulos et al. 2013a). These same consequences, with greater severity, have also been identified following 2G Global System for Mobile Telecommunications (GSM) mobile phone exposures for as little as a few minutes (min) daily for a few days (Panagopoulos et al. 2004; 2007a; 2007b; 2010; Panagopoulos and Margaritis 2010a; 2010b; Panagopoulos 2019). In humans, epidemiological studies conducted on pregnant women working in office spaces have revealed that EMFs generated from a computer monitor may lead to birth defects and spontaneous abortion (Bergqvist 1984; Goldhaber et al. 1988; Bryant and Love 1989). Studies of this nature, however, are still limited and, while intriguing, further investigation and independent verification is required.

Following implantation, fetal organogenesis and tissue development is generally sensitive to environmental/external factors as shown in established adult murine systems (Alchalabi et al. 2016a; 2016b). On the background of the proven developmental impacts of these external influences, potential sensitivity to physical factors such as man-made EMFs is also of interest here. Indeed, this paradigm has launched many investigations, and despite some inconsistencies, the majority of studies suggest that WC EMFs do have developmental toxicity. In one such study, the biological effects of 1800 MHz GSM (2G) EMFs were examined throughout pregnancy and extended to monitoring of the offspring after birth (Alchalabi et al. 2016a; 2016b). These data indicate that exposure during a murine pregnancy leads to asymmetrical distribution of implantation sites in uterine horns, reductions in fetal weight, and an increase in fetal anomalies. After birth, there was a sustained decrease in weekly weight gain and reduced functional and physiological development in the pups and evidence for the induction of OS in the mothers. Similarly, rabbits exposed to GSM-like EMF (15 min/day for 7 days) during mid-pregnancy, exhibited oxidative destruction in hepatic tissue as a result of ROS production (Tomruk et al. 2010). Importantly, this finding has since been confirmed by an independent research group indicating that the 1800 MHz GSM-like exposures of non-pregnant and pregnant rabbits (15 min/day for 7 days) resulted in detectable ROS production leading to LPO and oxidative DNA damage (Güler et al. 2012).

Intriguingly studies have now linked EMF-induced developmental disruptions during pregnancy to changes in the behavior of the exposed offspring (Aldad et al. 2012). These include impaired memory, hyperactivity, and decreased anxiety, all of which persist into adulthood. Indeed, *in utero* exposure to a Wi-Fi 2.45 GHz EMF for 2 h/day for 30 days recapitulated the previously observed phenomenon of asymmetrical implantation sites but also induced DNA damage in the brain cells of the pups (Singh et al. 2009), which would be expected to impact their behavior. *In utero* exposure to a continuous-wave field at the same carrier frequency (2 h/day for 45 days, SAR 0.023 W/kg), conducted by another research group, induced OS in the liver, kidney, and ovary tissue, which was linked to a disruption of implantation and embryonic morphologies and was concomitant with DNA damage in the developing brain cells (Shahin et al. 2013). In line with these developmental issues, postnatal growth restriction and delayed puberty was also found in female rats, likely as a consequence of the OS encountered in the brain and ovarian tissues. In terms of the reproductive consequences, the authors suggest that the induction of OS likely arose as a consequence of alterations in the biosynthesis of the heat shock proteins (HSPs) HSP70 and HSP105 (Alchalabi et al. 2016a; 2016b), both of which are well known to play critical roles in fertilization and early embryonic development (Matwee et al. 2001). While we will not cover this specific aspect in detail here, such findings are in agreement with the knowledge that a host of developmental processes are underpinned by ROS and impinged by OS (Agarwal et al. 2005; Dundar et al. 2009; Sangun et al. 2015).

Building on these animal studies, human studies also allude to adverse effects of man-made and especially WC EMF exposure during pregnancy. One such study involving a review of the health effects on teachers, the unborn offspring of teaching staff, and school-aged children following the installation of WC EMF emitting technologies both within and surrounding schools (such as Wi-Fi and MT antennae), found that children were among the population groups who were most sensitive to WC EMF exposure (Redmayne and Johansson 2015). As such, a cohort of related studies have since emerged aimed at determining the specific molecular and neural effects of EMFs on the health and wellbeing of fetuses and young people. It has been reasoned that the relatively thin and small size of young skulls may increase the penetrative capacity of mobile phone EMR (Gandhi and Kang 2001; Gandhi et al. 2012). Additionally, stem cells have been shown to be more susceptible to the effects of MT/WC EMFs than differentiated cells (Markovà et al. 2010). This cell type decreases in density throughout life, but is at its highest density during embryonic development, suggesting that fetuses and infants could be more vulnerable than adults (Ahmed et al. 2017). In support of this conclusion, French researchers have recently pointed to a potential correlation between chronic mobile phone use by pregnant women and reduced fetal growth (Boileau et al. 2020). Their study analyzed medical charts from 1,378 women from the central southwest of France, of which 99.3% reported an average daily use of mobile phones for 29.8 min. Women who indicated an average daily usage period greater than 30 min were significantly more likely to have children with lowest fetal growth. Previous studies had found no association between maternal use of mobile phones and reduction in fetal growth (Mortazavi et al. 2013b; Tsarna et al. 2019). Yet the finding of Boileau et al. (2020) is supported by another study, which reported that prenatal exposure to mobile phone EMFs is associated with behavioral problems in children (Divan et al. 2008). Regardless of this, these data indicate that this area deserves considerably more attention to bear out the effects and mechanisms by which man-made EMFs may influence pregnancy, fetal growth, and offspring outcomes.

6.5.2 Impacts on the Female

Female reproductive health is considered by some to be less at-risk from man-made EMF exposure than male reproduction because of the substantial potential for cellular repair during female

TABLE 6.1
Summary of Effects of WC EMFs on Reproduction

EMF	Frequency and Wavelength	Sources	Male Reproduction	
RF	300 kHz– 300 GHZ (1 mm–1 km)	Mobile phones, mobile telephony base antennas, Wi-Fi, Bluetooth, smart meters	(+) Acrosome reaction, (N) sperm capacitation, (−) sperm vitality, (−) sperm motility & kinetics, (−) sperm count, (+) sperm lipid peroxidation, (−) sperm binding to hemizona, (−) sperm head morphometry, (+) sperm pathological morphology, (+) sperm and testicular mtROS/ROS, (+) sperm DNA damage and fragmentation, (−) sperm ZP adhesion, (+) *Clusterin (CLU)* expression, (−) acrosin activity, (−) total antioxidant capacity, (+) testosterone levels, (−) decreased luteinizing hormone levels, (−) seminiferous tubule and testicular weight, (+) mouse spermatocyte (GC-2) DNA damage, (+) apoptotic spermatozoa, (−) blc-2 protein and mRNA expression, (+) bax, cytochrome C and caspase 3and 8 expression in spermatozoa, (+) increase in xanthine oxidase activity in testes, (−) Leydig cell number	Houston et al. 2018a; 2019; Gorpinchenko et al. 2014; Zalata et al. 2015; Agarwal et al. 2009; Fejes et al. 2005; Ghanbari 2013; De Iuliis et al. 2009; Yan et al. 2007a; Erogul et al. 2006; Falzone et al. 2011; Falzone et al. 2008; Gutschi et al. 2011; Kumar et al. 2013; 2014; Liu, et al. 2013; 2015; Meena et al. 2014b; Mortazavi et al. 2013a; Rago et al. 2013; Saygin et al. 2011; Zhang et al. 2016
			(N) Interstitial tissue %, (N/−) testes weight/size, (N) testicular biopsy score count, (−/+) seminiferous tubule diameter, (+) disorganized seminiferous tubule sperm cycle, (−) germinal epithelium height, (−) serum testosterone, (+) testicular catalase activity, (−) testicular and epididymal serum and tissue glutathione, glutathione peroxidase, and superoxide dismutase levels (+) testicular and epididymal lipid peroxidation, (+) malondialdehyde, (+) free radicals, (−) histone kinase, (−) size of seminiferous tubule lumen, (+) apoptotic bodies (caspase 3), (−) mouse Leydig (TM3) cell proliferation and secretion of testosterone, (−) levels of testicular LDH-X, (+) levels of serum FSH, (+) upregulation of proteins (ATP synthase beta subunit (ASBS) and hypoxia regulated protein 1 precursor (HYOU1) involved in testicular and spermatogenic signaling pathways	Gye and Park 2012; Al-Damegh 2012b; Mailankot et al. 2009; Kesari et al 2011; Kumar et al 2013; Lin et al. 2017; Meena et al. 2014b; Sarookhani et al. 2011; Sepehrimanesh et al. 2017
			(N) Testes, epididymal, seminal vesicle or prostate weight change, (N) sperm count, (N) sperm vitality, (N) sperm motility, (N) sperm concentration, (N) semen volume, (N) change in testicular histology	Lee et al. 2010; Imai et al. 2011; Lewis et al. 2017; Tumkaya et al. 2016
VLF	3–30 kHz (10–100 km)	CRT monitors	(N) Spermatogenesis	Dawson et al. 1998
ELF	3–3000 Hz (100–100000 km)	Military communications systems	(N) Testes weight, (−) sperm vitality, (−) seminiferous tubule morphology	Lee et al. 2004

((N): No effect. (−): Decrease in the measured parameter after exposure. (+): Increase in the measured parameter after exposure.)

Female Reproduction		Embryo Development		EMF
(−) Primordial and ovarian follicle count, (+) endometrial apoptosis, (+) OS, (+) endometrial, uterine, and ovarian histopathological impairment and apoptosis, (N) oestrodial levels, (N) uterine wet mass, (+) endometrial lipid peroxidation and apoptosis, (+) total antioxidant capacity, (+) serum LH, (+) ovarian and uterine levels of ROS, NO, lipid peroxidation, total carbonyl content, (−) number of developing corpus lutea, (−) levels of serum LH, FSH, (−) levels of sex steroids E2 and P4, (−) expression steroidogenic factor-1 (SF-1/Ad4BP), steroidogenic acute regulatory protein (StAR), 3β-hydroxysteroid dehydrogenase (3β-HSD) and cytochrome P450 aromatase (P450arom/CYP 19 A1) in ovary and, estrogen receptor-α (ER-α) and estrogen receptor-β (ER-β) in ovary, (−) antioxidant enzyme activity (superoxide dismutase, catalase, glutathione peroxidase) in ovary	Gye and Park 2012; Gul et al. 2009; Alchalabi et al. 2016; Bakacak et al. 2015; Oral et al. 2006; Sangun et al. 2015; Shahin et al. 2013; 2017.	(+/N) Blastocyst development, (N) fertilization	Houston et al. 2018a; 2019	RF
		(+) dsDNA breaks, (N) mitochondrial function, (N) chromosomal modifications, (N) nuclear apoptosis, (+) abnormality of implantation sites in the uterus	Nikolova et al. 2005; Shahin et al. 2013; Singh et al. 2009	
(−) Formation of 53BP1 foci in mesenchymal stem cells	Markovà et al 2010	(−) Fetal and postnatal growth	Boileau et al. 2020; Sangun et al. 2015	
(+) Macrophages in corpus luteum and growing follicles	Roozbeh et al. 2018	(−) Newborn litter size	Magras and Xenos 1997	
(−) Reproductive capacity, (−) oviposition, (−) egg chambers, (+) cell death, (+) DNA fragmentation	Panagopoulos et al. 2004; 2007a; 2007b; 2010; Panagopoulos and Margaritis 2010a; 2010b	(+) Chicken embryo mortality	Batellier et al. 2008	
(−) Egg laying rate in honeybees	Sharma and Kumar 2010	(+) Superoxide and nitrogen oxide in quail embryos, (+) oxidative stress (8OHdG), (−) levels of superoxide dismutase and catalase activity	Burlaka et al. 2013	
(+) Estrus cycle length	Jung et al. 2007	(N) Embryogenesis or fetus maturation	Dawson et al. 1998	VLF
(N) Ovum maturation and ovulation, fertilization, and implantation	Dawson et al. 1998			
(−) Reproductive capacity, (+) oocyte cell death, (+) DNA fragmentation	Panagopoulos et al 2013	(N) Anogenital distance, (N) preputial separation, (N) testes weight, (N) testes development, (N) male F1 reproductive capacity	Chung et al. 2005	ELF
(−) Antrum follicle formation, (−) estradiol release and granulosa cell DNA synthesis	Cecconi et al. 2000	(+) DNA fragmentation, (−) blastocyst development	Borhani et al. 2011	
(+) Estrus cycle length	Rodriguez et al. 2003			

germline development. However, due to the elongated lifespan of female germ cells, in contrast to their relatively short-lived male counterparts, the risk from chronic exposures deserves consideration. As such, current female-focused research tends to utilize pulsed and modulated RF (WC) EMF exposures, as their ELF pulsing/modulation components are known to be able to penetrate deep enough into the tissue (Cecconi et al. 2000; Rodriguez et al. 2003; Jung et al. 2007; Panagopoulos et al. 2015), and the current scientific landscape has a less than complete view of potential risks to human female fertility. Based on the nature of the research exposure model and limited *in vitro* cell availability, whole animal exposures are heavily exploited by female-focused *in vivo* studies (Jung et al. 2007; Rodriguez et al. 2003; Cecconi et al. 2000; Bakacak et al. 2015). As such, a significant body of evidence has been generated using a *Drosophila melanogaster* model to investigate the effects of GSM 900 MHz/1800 MHz mobile phone radiation on female reproductive capacity and, in particular, the impact experienced during early- and mid-oogenesis (Panagopoulos et al. 2007a; 2007b; 2010; Panagopoulos and Margaritis 2010a; 2010b; Chavdoula et al. 2010).

During early- and mid-oogenesis, exposure to GSM 900 MHz or 1800 MHz radiation from mobile phones has been thoroughly investigated in *Drosophila* (Panagopoulos et al. 2007a). These exposures were conducted during the first 6 days of a fly's adult life and consisted of a single daily dose of radiation delivered for 6 min/day. One h after the last exposure on the sixth day, the ovaries were dissected into individual ovarioles and prepared for DNA fragmentation analysis (using the terminal deoxynucleotide transferase dUTP nick end labeling assay (TUNEL)). Regardless of which type of exposure (GSM 900 or 1800 MHz) was used, these flies exhibited a non-thermal decrease in the number of laid eggs (oviposition), which correlated to the significant elimination of large numbers of early and mid-oogenesis staged egg chambers. This loss was determined to occur as a consequence of DNA fragmentation and affected the oocyte, follicle cells, and nurse cells within the egg chambers. Interestingly, although both types of radiation affected fruit fly reproduction considerably, the effect of GSM 1800 MHz appeared to be more modest than the effect of GSM 900 MHz, mainly due its slightly lower intensity, and secondarily due to its increased frequency (Panagopoulos et al. 2007b). In another study by the same researchers, GSM 900/1800 EMF exposure was able to decrease the reproductive capacity of *Drosophila* even when the active mobile phone was positioned at a distance of 100 cm away (Panagopoulos et al. 2010), with diminishing effects observed with increasing distance/decreasing intensities. This distance/dose exposure revealed the highest effects at distances of 0 cm (252–378 $\mu W/cm^2$) and 20–30 cm (10 $\mu W/cm^2$ radiation intensity) with effects not evident at distances further than 1 m (<1 $\mu W/cm^2$ radiation intensity). In conjunction with proximity of the fruit fly to the WC EMF source, these studies also indicated that the exposure duration of both GSM 900 MHz and 1800 MHz EMFs affects reproduction at degrees increasing linearly with increasing exposure duration, suggesting that short-term exposures might also have cumulative effects (Panagopoulos and Margaritis 2010a). Expanding the investigation into these possible cumulative effects, the authors performed intermittent exposures with 10-min intervals between successive exposures (Chavdoula et al. 2010) using this fruit fly model. The resulting data indicate that intermittent exposure was able to decrease the fly's reproductive capacity by altering the actin-cytoskeleton network of the egg chambers and inducing DNA fragmentation to levels equivalent of those observed in animals that were continuously exposed for the same total duration. Interestingly, longer intervals between the exposures allowed the fruit fly to recover and partly overcome the aforementioned reproductive consequences, possibly reflecting a putative compensatory mechanism that involves DNA repair or replacement of the defective egg chambers with new ones.

In mice, females subjected to a 20 kHz VLF EMF for 6 weeks displayed an extension of the estrous cycle (Jung et al. 2007) and fewer cycles overall taking place in exposed female mice, compared to their sham-exposed controls. In female rats, this trend was not recapitulated following exposure to 10 kHz, 0.2 mT sine wave EMF (Dawson et al. 1998). In fact, continuous exposure of adult Wistar female rats to 50 Hz ELF EMF for 3 months also failed to result in either physiological or hormone perturbations, with no alterations detected in the weights of the uteri and ovaries, progesterone or estrogen levels, relative to the period of the estrous cycle (Aydin et al. 2009). However, in bovine, a 60 Hz EMF applied for 16 h per day did instigate an extended estrous cycle in the cows examined (Rodriguez et al. 2003). To further evaluate these exposures *in vitro*, mouse follicles, were subjected to a 33 Hz ELF EMF at 5-day intervals. This resulted in defects in follicle growth, but in contrast, follicles exposed to a 50 Hz EMF did not see these same effects. Interestingly, exposure for only 3 days at both frequencies (33 Hz and 50 Hz ELF EMF) led to significant inhibition of antrum formation in follicles cultured *in vitro* (Cecconi et al. 2000).

Additionally, in rats, examination of ovarian histology following RF exposures at similar carrier frequencies of MT (Gye and Park 2012; Bakacak et al. 2015) indicated that a 900 MHz EMF in close proximity to the ovaries for 15 min a day for a total of 15 days, at 0.18 W/kg and 4 W/kg SARs, was able to significantly reduce the number of primordial follicles in the ovary when compared to the control (Gye and Park 2012; Bakacak et al. 2015). This phenomenon of reduced ovarian follicle counts following exposure has been further demonstrated in a study using Swiss Albino rats exposed to mobile phone EMF *in utero* (Gul et al. 2009). Building on this study, female mice were irradiated with GSM 1.8 GHz mobile phone EMF emitting in various modes, such as during stand-by, receiving, or dialing (Shahin et al. 2017), similarly leading to a reduction in the number of corpus luteum (physiological cysts in the ovary that release progesterone to prepare the uterus for pregnancy), primary, preantral, antral, and pre-ovulatory follicles, coinciding with an elevation in atretic follicles (follicles with vacuolization due to oocyte degeneration and apoptotic bodies), nitrosative stress, and OS in the ovary (Figure 6.4). Upon histological examination, the ovary, uterus, and hypothalamus of exposed specimens exhibited markers of increased levels of ROS 2'-7' dichlorofluorescin diacetate (DCFH-DA) and dysfunctional antioxidant enzyme activity, such as reduced superoxide dismutase (SOD), catalase (CAT), glutathione peroxidase (GPx), and ovarian steroidogenic enzyme activity (reduced 3β-hydroxysteroid dehydrogenase (3β-HSD), 17β-hydroxysteroid, steroidogenic factor 1 (SF-1), steroidogenic acute regulatory protein (StAR), aromatase (P450arom), and estrogen receptor alpha (ERα) and beta (ERβ)). Decreases in uterine diameter and thickness of the endometrium and myometrium were also observed together with evidence of endometrial and

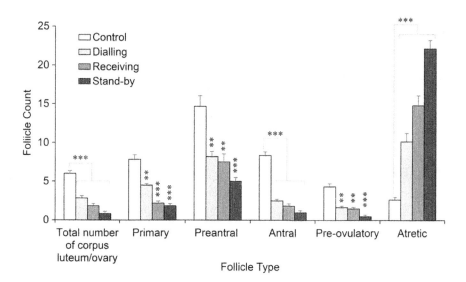

FIGURE 6.4 Ovarian follicle counts. Effect of mobile phone EMF exposure during various modes of operation (dialing, receiving, or stand-by) on the reproductive system of female mice: (i) Total number of corpus luteum, (ii) total number of different types of healthy follicles (primary, preantral, antral, pre-ovulatory) excluding the primordial and atretic follicles, and (iii) total number of atretic (degenerating) follicles. Significant reduction in total number of healthy follicles and corpus luteum was observed in all the exposed groups of mice compared to the control, and significant increase in the number of atretic follicles. Values are expressed as means ± SEM ($N = 12$). Significance of difference from respective control, **$p < 0.01$, ***$p < 0.001$. [SEM: Standard Error of the Mean]

myometrial degeneration. Significant increases in the levels of ROS, total nitrite and nitrate concentration, malondialdehyde (MDA) levels, and carbonyl content were similarly observed in the hypothalamus and uterus across all exposed groups when compared to the control group. Subsequent antioxidant enzyme activity levels (SOD, CAT, and GPx) were significantly reduced in the hypothalamus, ovary, and uterus tissues. In addition to reduced ovarian steroidogenic enzymes (3β-HSD, SF-1, StAR, P450arom, ERα and ERβ) downregulation of cytochrome P-450scc and reduced levels of serum LH, follicle-stimulating hormone (FSH), estrogen, and progesterone were also recorded within the ovaries of mobile phone irradiated mice. Together, such data indicate that long-term exposures of this type could reduce folliculogenesis and steroidogenesis, impairing fertility in mice. Conversely in rats, estrogenic activity was not affected by RF EMF exposure (1.439 GHz for 4 h a day for 3 days) (Yamashita et al. 2010).

While some of these data are difficult to fully rationalize, taken together, they show that the estrous cycle regulated by ovarian steroids may be sensitive to WC EMF exposure. A study focusing on the potential effects to ovarian histology explored the effects of RF exposures at similar carrier frequencies of mobile telephony in female rats (Gye and Park 2012; Bakacak et al. 2015). A 900 MHz EMF exposure in close proximity to the ovaries for 15 min a day for a total of 15 days, at 0.18 W/kg and 4 W/kg SARs, resulted in significantly fewer primordial follicles compared to the control (Bakacak et al. 2015; Gye and Park 2012). Despite their advantage in allowing researchers to design short, intermediate, and chronic exposure studies through which adverse effects of EMFs can be characterized under a spectrum of doses, a notable issue of utilizing animal models, such as those discussed above, is that the exposures used in these studies may not directly reflect the environmental exposures on humans because of the differences in organ surface and depth. These limitations call for more detailed studies on humans to fully characterize how WC EMFs affect our reproductive potential.

6.5.3 Impacts on the Male

Concern over WC EMFs and their impact on male fertility is steadily growing (Ozguner et al. 2005a; Yan et al. 2007; Imai et al. 2011; Al-Damegh 2012; Narayanan et al. 2015), with exposure to WC signals having been linked to clear negative impacts on semen quality (De Iuliis et al. 2009a; Agarwal et al. 2009; Ghanbari et al. 2013; Tumkaya et al. 2016). Notwithstanding some controversy regarding the experimental design of such exposures (Marchionni et al. 2006; Wang and Sauer 2006; Kesari et al. 2011), the principle that these EMFs can elicit a detrimental impact on sperm function is now well supported (Marchionni et al. 2006; Yao et al. 2008; Mailankot et al. 2009; Furtado-Filho et al. 2014; Kahya et al. 2014). The loss of sperm function observed includes reduced overall motility associated with diminished motility parameters. These outcomes are extended to further functional losses, including reduced ability to engage in interactions with the outer shell of the oocyte (a structure known as the zona pellucida) and overall reduced fertilization potential (Falzone et al. 2011). Many studies have also concluded that this reduced function appears in concert with elevated ROS and/or a state of OS. We and others have clarified an important biological link between OS and sperm function and highlighted that, along with the damage to lipids and proteins, sperm DNA also suffers damage via an oxidative mechanism. This possibility is a most critical aspect of studying the effects of WC EMFs on reproduction, as even for the lay person, it is abundantly clear that sperm DNA integrity is a key driver of not only fertilization success but also the health of the offspring.

Significant impairment of sperm motility and vitality, as well as the integrity of the paternal genome these cells carry, have been widely observed endpoints among recent studies testing the effects of WC EMFs, thus, supporting a direct effect of WC EMFs on mature spermatozoa. Notably, these effects parallel significant increases in mitochondrial and cytosolic superoxide free radical generation (Agarwal et al. 2009; De Iuliis et al. 2009a), underpinning the importance of ROS as a causative factor in EMF-induced sperm dysfunction. Interestingly, equivalent defects are very commonly observed in about 50% of men who present with fertility issues (Brugh and Lipshultz

2004; Hirsh 2003; Kumar and Singh 2015; Tremellen 2008; Iwasaki and Gagnon 1992; Zini et al. 1993; Ochsendorf et al. 1994; Shekarriz et al. 1995). While we do not suggest that such effects are uniquely driven by man-made EMF exposures, the pathways of OS induction and their clinical consequences closely mirror the results observed under experimental exposure to WC EMFs. It is now well accepted in the field of male infertility research that factors causing OS, such as diet, chemical exposures, and even advancing age, drive poor sperm function and DNA damage. The stand-alone evidence that supports the principle that WC EMFs can induce sperm damage through an OS mechanism is further bolstered by the robust clinical data advancing male infertility research. Of particular concern is the damage to the paternal genome in the male germline incurred by OS insult. With the recognized impact of sperm DNA damage on both fertility and health risk of the progeny, this alone provides a clear imperative to resolve the mechanism(s) by which WC EMFs affect this important biological system. Given the breath of research in this area, we will address the impacts of WC EMFs on sperm function in more detail in Section 6.6.

6.5.4 Testicular Cancer

Earlier data analysis studies (Skakkebaek et al. 2001; Somer et al. 2012) had examined the correlation between testicular cancer rates and mobile phone subscriptions between 1991 and 1998 and discerned no statistical evidence between rates of incidence of testicular tumors and the exponential increase in mobile subscription rates; however, the researchers conceded that because of the long cancer latency times, there is a need for continued monitoring. Although no further evidence is currently available on this subject, we contend that it should form a priority for future research given that testicular cancer is the second most common cancer in reproductively aged men (those aged 20 to 39), and considering that the rate of men diagnosed with testicular cancer has grown by more than 50% over the past 30 years (AIHW 2020).

6.6 IMPACT ON SPERM FUNCTION AND DNA INTEGRITY

Given the anatomical position of the male reproductive system and the thin outer layer of the scrotum that houses the testes, these primary male reproductive organs are more exposed to external factors, including EMFs, than organs housed deeper inside the body, such as the liver or kidneys. The vulnerability of the male reproductive tract was first established by Wdowiak et al. (2007) who demonstrated that males who use mobile phones exhibit increased rates of abnormal sperm morphology and decreased sperm motility compared to those who did not use these devices. Furthermore, these effects were exacerbated with longer exposure to mobile phone EMFs. More recent reports on larger cohorts have substantiated these initial findings and reiterated the outcome of sperm morphological damage (Zhang et al. 2016) and losses in motility (Agarwal et al. 2008; De Iuliis et al. 2009a; Mortazavi et al. 2013a) associated with mobile phone use and have, thus, bolstered the importance of this type of research.

Importantly, data produced by these studies also indicate that man-made EMFs can induce sperm DNA damage through non-thermal mechanisms (Houston et al. 2018a). As an example, a 900 MHz or 1.7 GHz RF EMF is capable of inducing DNA breakage in cauda epididymal spermatozoa and embryonic stem cells in mice (Aitken et al. 2005; Nikolova et al. 2005), while an 1800 MHz EMF can induce DNA breaks in human fibroblasts and rat granulosa cells as assessed via the comet assay (Kesari et al. 2018). These observations are especially concerning because human spermatozoa are capable of harboring considerable DNA damage that is independent of any pronounced effects on the cell's motility (Aitken et al. 1998). Damaged spermatozoa, therefore, can still have fertilization potential, whereupon the egg would then bear responsibility for repairing the paternal DNA prior to the initiation of S-phase of the first mitotic division (Aitken et al. 2014). The reason that the egg would bear the responsibility of repairing the sperm DNA is because a mature spermatozoon is transcriptionally and translationally silent. This unique silencing occurs largely because of the almost crystalline compaction of the paternal DNA within the minute sperm head. Further elevating the potential threat of sperm DNA damage for fertilization and embryo development (Lord and Aitken 2015) is that the egg's reparative capabilities are still controversial. In this context, eggs have been shown to possess limited repair options for the forms of DNA damage (Martin et al. 2019), including the oxidative DNA damage that are a hallmark of man-made EMF exposure (Martin et al. 2018). Indeed, eggs are deficient in the recognition and subsequent repair mechanisms needed to resolve oxidative DNA damage (Aitken et al. 2014). This knowledge, taken together with the diminishing quality of eggs and general repair machinery that accompanies advancing female age, has obvious implications for the mutational load passed onto the offspring. Indeed, we may already be observing the impacts of this male borne damage manifesting as further fertility losses in men, who were conceived through ART, in which rigorous natural sperm selection barriers are bypassed (Merchant et al. 2011; Sánchez-Calabuig et al. 2014; Sakkas et al. 2015; Belva et al. 2016; Nasr-Esfahani et al. 2012).

Utilizing *in vitro* methods to investigate sperm provides many benefits; however, extrapolation of the conclusions drawn to clinical relevance must be made with caution. The absence of natural "barriers" between the spermatozoa and the EMF source is not taken into consideration, and/or it is not possible to be accurately modeled. Such "barriers" include the scrotum, the seminiferous tubules, and the epididymis, which house spermatozoa at various points during their lifetime, and the blood–testis barrier and general biokinetics present within any biological system. Devices are capable of emitting simple RF EMF modes, approximating the emissions of WCDs and their networks (Gajda et al. 2002), but many of these simulated systems do not contain ELF pulsations, modulations, and variability components that are present within all MT/WC signals (Panagopoulos 2019). This, therefore, also decreases the direct translatability of these findings to human health.

Despite this, while *in vitro* models alone may not be entirely capable of predicting the clinical consequences of these exposures, they are excellent at predicting effect types, defining mechanisms and magnitudes of interactions. In this vein, with mounting evidence that WC EMFs drive damage in gametes, including a loss of DNA integrity, the spermatozoon is also playing a key role in deciphering the nature of these effects and how DNA damage arises despite the non-ionizing energies. To date, work on the sperm cell has confirmed that the mitochondrion is a key organelle in this respect, which might underpin the mechanism of action for these exposures and deserves considerably more attention.

6.6.1 Understanding Impacts on Sperm Function from Murine Models

Before we delve into the current literature surrounding the mechanism by which man-made EMF exposures induce damage in spermatozoa, we first need to address the breadth of effects reported. To do so, the following sections survey the current literature regarding murine and human exposures and list the sometimes-conflicting effects demonstrated in each study.

Whole body exposures to 900 MHz EMF in Sprague Dawley rats showed no significant effect on testicular or epididymis weights, percentage of interstitial tissue, sperm count, plasma LH and FSH levels, or germ cell apoptosis (Ozguner et al. 2005a; Lee et al. 2010). However, structures in the testis, such as the diameter of the seminiferous tubule and the mean height of the germinal epithelium as well as serum total testosterone levels, were all significantly decreased (Ozguner et al. 2005a). Similarly, carrier frequencies of 905 MHz without pulsation and modulation in C57BL/6 mice showed no evidence of gross histological change in the testes, or any changes in the early embryonic development of embryos sired/conceived by their sperm (Houston et al. 2019). These observations align with clinical data that do not support that a decline in human fertility is directly linked to WCD usage over the past 35 years. However, this may be an oversimplistic view.

Looking in more detail at murine model studies, exposures to WC EMFs do precipitate adverse impacts, specifically declines in cell vitality and motility of the mature spermatozoa retrieved from the epididymis of C57BL/6 mice (Aitken et al. 2005; Houston et al. 2019). Exposure of mouse caudal epididymal sperm and embryonic stem cell lines *in vitro* corroborate the above findings and, in addition, exhibit elevated levels of mitochondrial ROS, DNA oxidation, and DNA fragmentation (Sarkar et al. 1994; Aitken et al. 2005; Nikolova et al. 2005; Houston et al. 2019). Proteomic analysis of the testicular tissue of the same species also exposed to a 900 MHz carrier frequency for 30 consecutive days exhibited over expression of ATP synthase beta subunit (ASBS) and hypoxia up-regulated protein 1 precursor (HYOUI), proteins associated with carcinogenic risk and reproductive damage (Sepehrimanesh et al. 2017). Indeed, these proteins play critical roles in spermatogenesis and maintenance of testicular function (Sepehrimanesh et al. 2017). Exposure of rats at higher frequencies, specifically 1.95 GHz and 2.45 GHz in the absence of modulation and pulsations, portraited no effect on the weights of testes, epididymis, seminal vesicles, and prostate, and neither sperm counts within the testis or epididymis was impacted (Imai et al. 2011). However, exposure of Wistar rats to a 2.45 GHz EMF resulted in the decline of Leydig cell quantity with an attendant increase in apoptotic cells present within the seminiferous tubules (Saygin et al. 2011).

Mice exposed to a 900 MHz carrier frequency at 90 mW/kg SAR for 12 h/day for 7 days revealed significant damage to DNA integrity of mature spermatozoa, specifically the mitochondrial genome and the nuclear β-globin locus (ELF measurements not provided) (Aitken et al. 2005). All other parameters, such as the number of sperm produced and functional competence of these cells, remained unperturbed. These findings highlight that this EMF regimen does not impact testicular function or the developing germ cells themselves given microscopic sperm parameters were unaffected; however, on this time scale, there was a significant genotoxic effect on the more vulnerable epididymal sperm that have reached maturity, and, as previously described (Sections 6.2 and 6.5), have been shown to have differing sensitivities to exogenous factors (Figure 6.5).

Investigations into the role of ROS in the effects of WC EMFs using male Wistar rats, under a 2 h/day for 35 days exposure at a SAR of 0.9W/kg at 900 MHz (Kesari et al. 2011) demonstrated a highly significant increase in CAT activity in the sperm of the exposed animals. Similarly, other antioxidant enzymes, such as GPx and SOD, exhibited a considerable reduction in activity. In these samples, the total ROS levels were determined to be elevated in the exposed cohort. Supporting this finding, was the elevated level of MDA, an electrophilic aldehyde produced through the common OS pathways of LPO. In addition, examination of histone kinase, a proxy for the cell cycle function, displayed a complementary decrease within the exposed group (Kumar et al. 2014), suggesting that EMF exposure can perturb mitosis.

Since OS and ROS appear to be common themes within the man-made EMF bioeffects story, considerable effort has been placed on examining the effects of antioxidants alongside many exposure regimens. In 2010, the protective effects of the powerful antioxidant hormone melatonin against oxidatively driven testicular impairment by EMFs was demonstrated in a Wistar rat model (Meena et al. 2014). The experimental regimen used in this study featured a longer exposure period at 2 h/day for 45 days (2.45 GHz carrier frequency with 50 Hz modulation frequency, at SAR 0.14 W/kg or power density 0.21 mW/cm^2) together with the intra-peritoneal (IP) administration of melatonin (2 mg/kg). Under these conditions, melatonin proved effective in elevating the activity of testicular lactate dehydrogenase isoenzyme and in reducing ROS and MDA levels when compared to exposed, but melatonin-free controls. Furthermore, the treatment groups that received the melatonin supplementation during WC EMF exposures had reduced levels of protein carbonyl content, xanthine oxidase, DNA fragmentation (assessed via the Comet assay), and increased testosterone levels within the testes (Meena et al. 2014).

In our own study, spermatozoa isolated from the mouse caudal epididymis did not show active ROS production after 6 h exposure to 1.8 GHz carrier at 0.15 W/kg (without pulsation/modulation) *in vitro*. However, after 3 h, 20% of these exposed cells expressed elevated DNA damage as detected by the alkaline comet assay (Houston et al. 2018). When the comet assay is performed under alkaline conditions, it is capable of detecting DNA double-strand breaks, single-strand breaks, alkali-labile sites, DNA-DNA/DNA-protein cross-linking, and incomplete excision repair sites (Pu et al. 2015) as opposed to oxidative DNA damage. These data could, therefore, suggest that other pathways may

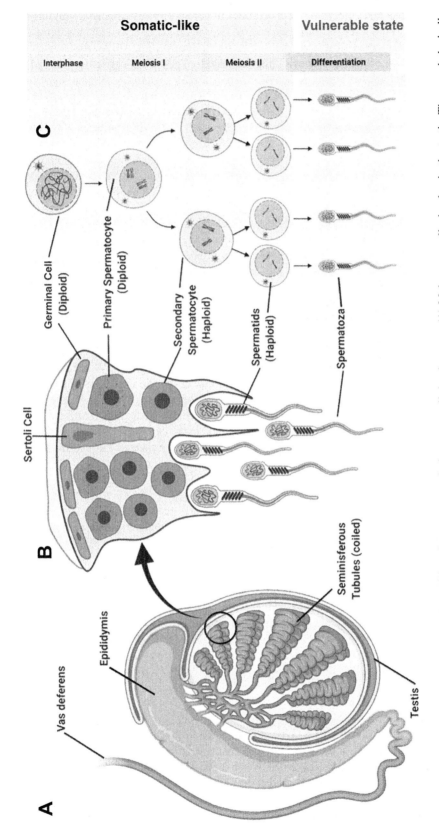

FIGURE 6.5 Testis cross section and vulnerability of spermatozoa during germ cell development. (A) Male germ cells develop in the testis. The morphologically mature spermatozoa then enter the epididymis for further maturation. In this tissue, limited protection from external stressors is available, and spermatozoa are at their most vulnerable phase. (B) Germ cell development in the testis utilizes Sertoli cells to nourish and afford protections from stressors. (C) In the early stages of germ cell development, their morphology is somatic-like. This cytoplasmic volume, the potential to translate new proteins, and the association with Sertoli cells offers protection from stress, whereas morphologically mature cells have limited capacity for such protection.

also have important roles in the effects seen following irradiation of this type. It might also explain some of the inconsistencies mentioned above, if different exposure regimens were to be shown to trigger different pathways that may culminate in similar endpoints (apoptosis, motility loss, and DNA damage).

In our study, we failed to detect ROS in appreciable levels in exposed cells. However, we did discover hallmarks of an OS state, including elevated expression of oxidative DNA damage (8-hydroxy-2′-deoxyguanosine: 8-OH-dG) following 4 h of RF EMF exposure, which might be indicative of a short-lived ROS "burst" rather than a sustained ROS state. In congruence with other studies, our observation of significant RF EMF-induced DNA fragmentation and oxidative DNA damage in exposed mouse spermatozoa was accompanied by a clear decrease in sperm motility parameters, including progressive, rapid motility, and straight-line velocity, but this was not associated with any compromise of the sperm cell's ability for capacitation as assessed by protein phosphotyrosine expression, which was unchanged between groups.

Studies focused on cell lines, including those from mice and humans, have critical roles to play in unraveling any mysteries that may underlie a definitive understanding of health risk from WC EMFs. Our research group has previously shown the importance of the sperm mitochondria in the onset of WC EMF-induced OS that then leads to impaired cell function (Aitken and De Iuliis 2007; De Iuliis et al. 2009b; Aitken et al. 2010; 2012a; 2012b; 2014; 2015) together with the oxidative pathways that are uniquely amplified in spermatozoa (Lenzi et al. 2000; Aitken et al. 2012b; 2013; 2015). This strengthens our hypothesis that the unique properties of the male germline render these reproductive cells extremely sensitive to the physical interaction of man-made EMFs. Accordingly, we have investigated the RF EMF-induced perturbation of mitochondrial function in early-stage mouse germ cells and compared this response to somatic cell lines (Houston et al. 2018). In one such study, we highlighted the increased vulnerability of male germ cells, mouse primary spermatogonia, spermatogonia-like (GC1), and spermatocyte-like (GC2) cell lines compared to somatic cell types such as granulosa (COV434), mouse fibroblasts (McCoy), and human embryonic kidney (HEK293) cell lines. During the exposure of these cells to a 1.8 GHz carrier frequency at 0.15 W/kg, the generation of mitochondrial ROS was clearly observed by the marker MitoSOX-red in GC1 cells after 2 h of exposure ($p < 0.001$), GC2s after 4 h of exposure ($p < 0.05$), and isolated primary spermatogonia ($p < 0.05$) after 2 h of exposure. It was fascinating to then observe no mitochondrial ROS response in the somatic cell lines, even after 6 h of exposure. Interestingly, this exposure did not precipitate any cell death across the various cell types tested.

Probing further into the mechanism behind the mitochondrial ROS production, we discovered that an electron transport chain (ETC) inhibitor, antimycin A, that specifically targets electron flow to complex III, amplified mitochondrial ROS generation in GC1 cells, but only while an EMF was applied (Houston et al. 2018a). Antimycin A, in the absence of the EMF, did elevate levels of mitochondrial ROS compared to the untreated control; however, the additional rise in mitochondrial ROS during the applied EMF suggested a synergistic interaction, implicating complex III as a potential molecular target of man-made EMFs. Although preliminary, these findings yield a testable hypothesis that man-made EMFs target complex III of the ETC, a unifying mechanism that could potentially account for many observations made in the larger field of studying the impacts of man-made EMFs. The plausibility of this mechanism is enhanced when considering the voltage associated with the inner mitochondrial membrane wherein the ETC resides and its discrete role in shuttling protons to maintain the voltage. Man-made EMFs have, indeed, been linked to biological effects originating from systems associated with membrane voltage (De Iuliis et al. 2009a; Murphy 2009; Perry et al. 2011; Lu et al. 2012; Panagopoulos et al. 2015; Liu et al. 2015).

In our corresponding study (Houston et al. 2019) using a whole-body exposure, we demonstrated that the ROS pathways induced at the mitochondria may also be responsible for the measured effects we observe after exposures *in vivo*. Using a carrier frequency of 905 MHz at 2 W/kg for 12 h/day, clear signs of cellular disruption were observed, with significant declines in sperm vitality and total, progressive, and rapid motility following 5 weeks of exposure. Again, these functional impacts were also associated with increased generation of mitochondrial ROS, which was detected following 1 and 3 weeks of exposure. Utilizing both the halo* and comet assays, DNA fragmentation was identified in the exposed spermatozoa after 3 and 5 weeks of exposure. As predicted, a significant increase in DNA oxidative damage (8-OH-dG) was detected; however, this was first identified as early as 1 week of exposure and held true for all subsequent time points examined. Despite these findings, but in line with similar published studies (Ozguner et al. 2005a; Tumkaya et al. 2016), no measurable effects on testis morphology were observed, and further, the cauda epididymal sperm isolated from irradiated mice maintained the ability to progress through fertilization and support blastocyst development after IVF. While such results could again signal the subtleties involved in the biology of man-made EMF effects, the ability of these cells harboring proven DNA damage to participate in embryo development, provides even more impetus to fully characterize any incurred paternal genomic damage by man-made EMFs and define their mechanism of action.

In an earlier study, while also appreciating the strengths of an *in vitro* waveguide system, we focused our probing on the pathways in spermatozoa that could be perturbed or that interact with man-made EMF exposures. The robust nature of dose dependent studies for determining drug or inhibitor activities and actions was applied here. Purified human spermatozoa were exposed to a 1.8 GHz carrier frequency for 16 h (De Iuliis et al. 2009a) across the SAR range of 0.4–27.5 W/kg. The cells exhibited a clear power dose-dependent decline in both motility and vitality, and importantly, this decline occurred in parallel with increased production of ROS. Of note, was that a significant reduction of both motility and vitality was observed at a power level as low as 1.0 W/kg (within ICNIRP limits). Total and mitochondrial ROS generation were independently assessed, and both were significantly elevated from above that of the control at an SAR of 1.0 W/kg and 2.8 W/kg, respectively. Additionally, highly significant relationships between SAR, oxidative DNA damage, and DNA fragmentation (TUNEL assay) were also observed under these exposure regimes. The assessment of oxidative sperm DNA damage marker, 8-OH-dG, revealed a strong positive correlation (coefficient of determination $R^2=0.727$) with mitochondrial ROS generation, suggesting that, as SAR (intensity of EMR) increases, mitochondrial ROS generation is elevated, leading to more extensive oxidative DNA damage. The motility and viability parameters, as well as the evidence of DNA fragmentation that tracks the EMF power doses, can all be precipitated by OS events arising from the mitochondria. Indeed, the relationship of DNA damage with mitochondrial ROS and 8-OH-dG formation exhibited strong positive correlations ($R^2=0.861$ and $R^2=0.725$, respectively).

While we have now discussed the idea that OS and DNA damage induced by man-made EMFs arise from interaction with sperm mitochondria, a growing body of data also demonstrates that sperm function, including motility, maturation, and the ability to bind to the oocyte, can also be affected by EMFs. In spermatozoa, as in essentially all other cell types, HSPs provide quality control in a process known as proteostasis. This process in sperm is arguably more important, as it is responsible for stabilizing the limited complement of proteins in a transcriptionally and translationally silent cell. During post-testicular maturation within the epididymis and female reproductive tract, proteins within these morphologically mature cells rely heavily on post-translational modifications (Gervasi and Visconti 2017; Björkgren and Sipilä 2019) and transit to a discrete subcellular location, influenced by HSPs. Disruption of HSP functionality, unsurprisingly, causes spermatogenesis to arrest, maturation to become compromised, and fertilization to be inhibited (Dun et al. 2012). As such, HSPs are likely an important aspect of protecting or buffering these germ cells from externally induced stresses. While we have studied this area in the context of general sperm cell biology, other studies have investigated changes to HSPs in metabolically active cells under the effect of man-made EMFs. Short-term exposures, perhaps as expected, were found to cause upregulation of HSPs, including HSP60, HSP70, HSP70/1, and HSP90 (Weisbrot et al. 2003; Gerner et al. 2010).

---

* An assay for the assessment of genomic DNA strand breakage in single cells.

Altogether, there is compelling evidence that WC EMFs negatively impact the reproductive systems of animals, inducing a reduction of gamete quality, with this finding confirmed by separate, publications from several independent research groups.

6.6.2 Human Cohort and In Vitro Studies

In a study of a large cohort, mobile phone use among 2,110 men who had attended an Austrian fertility clinic between 1993 and 2007 was assessed in combination with semen analysis (Gutschi et al. 2011). This study revealed a significant difference in sperm morphology, with 68% pathological morphology in those patients using mobile phones, compared to 58.1% with no mobile phone use. This observation indicates perturbation of the testicular process of spermatogenesis. Additionally, patients who used mobile phones showed a hormone shift with significantly higher testosterone and LH levels. In an analogous study, 63 subjects were evaluated and partitioned into four groups based on active mobile phone use (Rago et al. 2013). Semen parameter analysis revealed no overt impact of mobile phone use; however, the authors reported a significantly higher percentage of sperm DNA fragmentation (recorded by the TUNEL assay) in the cohort that had more than 4 h of mobile phone usage a day and carried the phone in their pants. A study utilizing 371 clinical patients, also correlating semen parameters with mobile phone use habits, reported that higher daily mobile phone usage may negatively impact sperm. Indeed, mobile phone possession duration and the daily duration of use were negatively correlated, albeit modestly, with the percent of rapid progressive motile sperm (Pearson's correlation coefficient $r = -0.12$ and $r = -0.19$, respectively) and positively correlated with the percentage of slow progressive motile sperm ($r = 0.12$ and $r = 0.28$, respectively) (Fejes et al. 2005).

In vitro sperm exposure studies provide, perhaps, a little more consistency than the human studies mentioned above. While the exposure levels in the *in vitro* studies may be elevated beyond what these cells might experience in the body, they, nevertheless, provide more robust and convincing data on the processes that precipitate the observed EMF effects.

Indeed, the common carrier frequency of 900 MHz employed in 2G (GSM) MT is heavily utilized for *in vitro* studies, which has meant greater comparability between studies. As a result, some common observations arising from *in vitro* exposure models include decreased motility and alterations in sperm kinematic parameters (Erogul et al. 2006; De Iuliis et al. 2009a; Agarwal et al. 2008; 2009; Mortazavi et al. 2013; Houston et al. 2018). Despite the bulk of studies using the common carrier frequency, there are a few that have also utilized this GSM band but employ authentic mobile phones as the source of EMFs. One such study applied a peak power of 2 W (average power density 0.02 mW/cm^2) for a total time of only 5 min. In this study, the authors recorded a highly significant loss in rapid and slow progressive motility and an increase in non-motile sperm compared to non-exposed cells (Erogul et al. 2006). Another study using an experimental design with a carrier frequency of 870 MHz at 1.46 W/kg SAR for 60 min observed significant declines in the motility ($p = 0.003$) and vitality ($p < 0.01$) across 32 individual semen samples (Agarwal et al. 2009). In another study using simulated MT/WC EMFs (such as those utilizing generators and waveguide systems), motility parameters associated with hyperactivation, such as straight-line velocity and beat cross frequency, were also negatively affected after direct *in vitro* exposure. These observations were made after exposure to 900 MHz at 5.7 W/kg in a waveguide system that simulated MT/WC EMFs produced by mobile phones (Falzone et al. 2008). A summary of the impact of WC EMFs on the reproductive and endocrine system can be found in Figure 6.6.

Although *in vitro* sperm exposure studies appear more consistent in their findings than the aforementioned human studies, it is difficult to understand how potential temperature rises were monitored and controlled for, especially in earlier studies. This is, in fact, a critical consideration when interpreting this body of data, given that similar sperm defects can be elicited by heat stress alone (Houston et al. 2018b). Potential bulk thermal artifacts are practically eliminated through use of a water temperature buffering system built into the waveguide and, in these systems, sperm

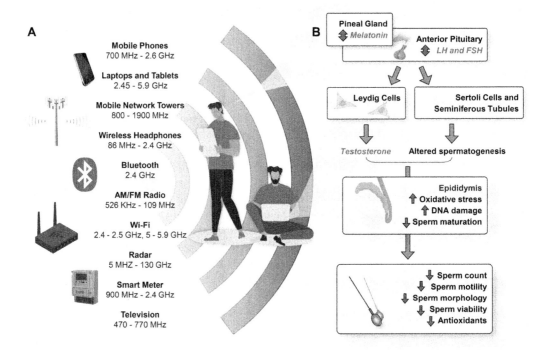

FIGURE 6.6 Impact of WC EMFs on hormone disruption, sperm function, and fertility: (A) WC EMFs in the environment and their carrier frequencies. (B) WC EMFs can impact the male reproductive tract, particularly where spermatozoa are vulnerable in the epididymis, but also act via disruption of the pineal and pituitary glands in the brain. The luteinizing hormone (LH) and follicle-stimulating hormone (FSH) signaling molecules from the brain are intimately entangled in the regulation of reproductive hormones and function, with disruptions leading to negative reproductive outcomes.

functional assessments following 1-h exposures to 900 MHz 2 W/kg in a waveguide apparatus (Falzone et al. 2011) have shown that the sperm acrosome reaction is unaffected by this exposure regimen; however, the ability of these cells to engage with the outer vestiges of the oocyte, the zona pellucida, was reduced markedly using the hemizona assay,* which could account for the decreases in fertilization potential and embryo development.

One observation that might explain how the human spermatozoa exposed to a pulsed 900 MHz EMF at 2 W/kg were unable to engage with the zona pellucida is the gross morphological alterations in sperm head area and acrosome cap size, with this reduction in cellular dimension taking place over the brief 1-h exposure period (Falzone et al. 2011). While a potential mechanism to account for these apparent morphological changes is not immediately obvious, these findings provide an interesting avenue for further advancements in the field and speak to yet another intriguing avenue of research.

6.7 MECHANISM OF ACTION FOR MAN-MADE EMFs ON REPRODUCTION

On the backdrop of a publicly debated field, the pursuit of a testable physico-biological mechanism of action for man-made EMFs is of critical importance. The carrier frequencies utilized by wireless devices and their networks would have the potential to heat materials if the devices were to utilize higher power densities. Any bulk thermal heating that may take place experimentally can be controlled and somewhat managed by design, however, accounting for potential localized or "micro-heating" events remains elusive. In the early years of bioelectromagnetic research the

---

* A diagnostic test for the tight binding of human spermatozoa to the human zona pellucida to predict fertilization potential.

thermal interaction with biology was considered the only mechanism to be plausible in accounting for EMF effects (Goodman et al. 1995; Creasey and Goldberg 2001). While there are still comments in the field claiming that non-ionizing fields at non-thermal intensities cannot cause any damage (ICNIRP 1998; 2020), it is now very well accepted and demonstrated that man-made EMFs can indeed affect living tissue through non-thermal pathways. Some of the key pathways implicated in this response involve the mitochondria, the voltage-gated ion channels (VGICs), and the induction of ROS. While it is true that these EMFs do not have the energy profiles needed to directly damage biomolecules by breaking chemical bonds, it is now also clear that cellular and tissue damage can result, experimentally, through indirect mechanisms. However, whether the mechanisms are direct or not, in our view, does not lessen the urgency to advance this field.

6.7.1 Thermal and Non-thermal Effects

The physical mechanism behind the thermal effects of high amplitude RF and mmW EMFs are well captured in a number of articles, with an explanation provided by Challis (2005). In this theory, water is likely to be the key biological molecule responsible for transmitting the energy of these EMFs in the form of temperature rises in tissues, similar to a microwave oven heating food. However, it should be noted that, while this phenomenon occurs in a microwave oven at 1000 W of power, mobile phones operate at a maximum of 1–2 W. Non-thermal effects of EMFs on biology are associated with the vast majority of experimental studies (Adey 1981; 1993; Goodman et al. 1995; Velizarov et al. 1999; Belyaev 2005; Panagopoulos et al. 2004; 2010; 2013b) and even with high energy or ionizing photons, e.g., x-rays or gamma radiation. The probability that non-ionizing EMFs can directly induce biological changes, for example, create DNA damage, is exceedingly low, therefore, understanding the potential for indirect non-thermal mechanisms is justified.

The testes are highly sensitive to thermal fluctuations, as spermatogenesis in humans may be interrupted at temperatures above 35°C (Saikhun et al. 1998; Mieusset et al. 1985; Mieusset and Bujan 1994). To account for this, the testes reside external to the rest of the body in the scrotum and, together with the rich circulatory system of the male reproductive system, allows for an efficient maintenance of temperature. This phenomenon is clearly evident in the semen parameters of a man who has recently suffered a fever event (Sergerie et al. 2007). Every interaction between a tissue and RF EMFs of considerable intensity has the capacity to induce a transfer of energy that may contribute to a rise in temperature (Foster and Glaser 2007). However, it seems that these energy transfers into skin and other superficial tissues, can be absorbed and, thus, disperse any micro-thermal gains rapidly, buffering the body from bulk or sustained heating, at least for RF/WC EMFs at environmentally relevant intensities. Despite their unique proximity to modern day WC EMF sources, this appears to be true also for the testes. Accordingly, non-thermal mechanisms appear to drive the majority of modifications seen in reproductive tissues. Indeed, there are now well-established non-thermal links to increased free radical production in these systems.

The perturbation of calcium flux and calcium channels by man-made EMFs (as well as of other cations such as sodium and potassium) also has a strong backing in the literature. This non-thermal mode of action may also have key roles in disturbing both the male and female germ cells, as they progress through their developmental and functional stages. Calcium, sodium, potassium, and other critical ions play key roles in all cellular function and specifically in the gain of functional competence in spermatozoa, whereas calcium signaling and oscillations are a hallmark of fertilization events in the oocyte and in early embryo development (Wang and Machaty 2013; Stewart and Davis 2019). Interestingly, EMF-induced impairment of VGICs have clear links to an onset of ROS production in a range of cell types (Panagopoulos et al. 2000; 2002; 2015; 2021; Creasey and Goldberg 2001; Pall 2013; 2015; 2018; Bertagna et al. 2021), placing these channels as a prominent target of interest.

6.7.2 Effects Governed by Intensity and Frequency Including Tissue Penetration Depth

As already discussed, the strength of an EMF exposure in the RF band is measured by the incident radiation intensity (power density) and/or the SAR, the rate at which energy is absorbed by unit mass of living tissue. The SAR values are also dependent on the density and conductivity of target tissues but are also influenced by factors such as the frequency of electromagnetic waves. As the EMF frequency increases, the wave penetrative capacity in materials, including biological tissue, decreases (Gabriel et al. 1996). However, EMFs across a large range of tested WC frequencies have been reported to increase free radical production in clinical settings (Kıvrak et al. 2017). Studies have reported effects across a large range of SARs from ~ 30 W/kg, down to 0.1 W/kg, or power densities down to ~ 0.2 µW/cm^2 (Magras and Xenos 1997; Burlaka et al. 2013). Our previously mentioned study (De Iuliis et al. 2009a) has shown a dose-dependent effect on spermatozoa (*in vitro*) across these SAR magnitudes, and another study has shown an intensity-dependent effect on ovarian DNA fragmentation for intensities ranging from 378 µW/cm^2 down to < 1 µW/cm^2 (Panagopoulos et al. 2010).

6.7.3 The Impact of Man-made EMFs on Reproductive Hormone Profiles

Dysfunction of key hormonal systems can have significant consequences for the fertility capacity of an individual. These perturbations may take place in any of the systems/organs that regulate hormonal control for reproductive function, including the hypothalamus, pituitary, the testes, and ovaries. Not surprisingly, man-made EMFs have been shown to disrupt brain function, with several studies highlighting disturbances that subsequently impact the reproductive endocrine system (Gye and Park 2012). *In vivo* studies utilizing rat models, have suggested that WC EMF exposure, even short-term with SARs as low as 0.016 W/kg or power densities less than 1 µW/cm^2, may alter the permeability of the blood-brain barrier, disrupt ion and protein transportation across cellular membranes, and cause infertility (Salford et al. 1994; Magras and Xenos 1997; Pall 2013). Consequently, exposure has been linked with disrupted secretion of a wide variety of gonadotropins and androgens, including antigonadotrophic, growth, thyroid stimulating, and FSH in the pituitary gland, as well as melatonin in the pineal gland – all of which have downstream consequences on the fertility of both sexes (Kesari et al. 2018). As an example, melatonin is an essential reproductive hormone wielding antigonadotrophic control on testosterone and progesterone secretion through the direct regulation of upstream signaling hormones such as the secretion of LH in the pituitary. Worryingly, man-made EMFs have been repeatedly linked with reduced melatonin secretion (Burch et al. 2002; El-Helaly and Abu-Hashem 2010).

As another example, mobile phone exposures have also been linked with substantially decreased pituitary gonadotropin production, in both male rabbit and female mouse *in vivo* studies (Sarookhani et al. 2011; Shahin et al. 2017). In the male, the Leydig cells, which reside in the interstitial tissue between seminiferous tubule structures within the testes, are the primary producers of male androgens. Leydig cells are stimulated by LH to synthesize and secrete testosterone, which in turn, provides feedback to the brain, ultimately regulating testosterone levels. In male rats, EMF exposures with frequencies up to 10 GHz have been linked with the disruption of testosterone synthesis and secretion by reducing Leydig cell density and disturbing seminiferous tubule activity (Kumar et al. 2013). Altogether, these data unequivocally speak to an interaction between WC EMFs and hormone regulation (Figure 6.6). Moreover, these effects could underpin many of the male fertility-related consequences discussed in previous sections (see Section 6.4).

6.7.4 Proposed Mechanism of Man-made EMF Action (Oxidative Mediated Mechanism)

In view of a weight of evidence supporting the ability of WC EMFs to specifically interact with mitochondrial function, instigating ROS production, increasing the expression of mitochondrial apoptotic markers (Liu et al. 2015), and decreasing mitochondrial membrane electrical potential (Lu et al. 2012), a compelling hypothesis about this voltage-reliant organelle may serve as the key gateway for the detrimental effects of man-made EMFs. Indeed, we have proposed that these EMFs potentiate the leakage of electrons from complex III of the ETC. Such electron leakage is suspected to occur via interference with proton transport through the transmembrane subunits of the inner mitochondrial membrane; however, this idea is not completely characterized.

We and others, believe that the observed effects, and particularly those on reproduction, are caused by the ability of polarized and coherent ELF EMFs (all manmade EMFs and those emitted from WCDs/antennas are totally polarized and coherent and contain ELFs) to force the oscillation of ions in phase and in parallel with the applied oscillating EMF as suggested by Panagopoulos et al. (2000; 2002; 2015; 2021). This may then induce dysfunction of VGICs potentially perturbing the strict membrane potentials enforced in the intermembrane compartments of the mitochondria, which otherwise stabilize proton flow. As a consequence, this destabilization or reduced proton emigration, ultimately yields a reduction in ATP production (Perry et al. 2011).

Under these conditions, when the NADH/NAD+ ratio is high and associated with compromised mitochondrial respiration, superoxide anion free radical ($O_2\bullet^-$) is rapidly formed within the ETC (Murphy 2009), thereby driving a state of OS. Moreover, since mitochondria are responsible for the majority of ROS production normally found within spermatozoa (Koppers et al. 2008), it is indeed conceivable that disrupting the function of these organelles may account for the elevated ROS production observed with WC EMF exposure in many studies, including our own (De Iuliis et al. 2009a; Yakymenko et al. 2016; Panagopoulos et al. 2021).

An important feature of this plausible mechanism is that it accounts for the subtle or variable species-specific alterations that modulated and pulsed RF EMFs have been recorded to induce in terms of sperm motility. This is due to the different pathways used to provide energy for the spermatozoa of species such as humans, mice, and rats which are not exclusively dependent on oxidative phosphorylation (Storey 2008). While changes due to WC EMF exposures may lead to no overt visible impacts, such as morphological defects in the mature gamete, subtle perturbations of both genetic and epigenetic factors arising from OS may have significant implications for embryogenesis and, therefore, the health of future generations.

6.8 MAN-MADE EMFs AND OS

6.8.1 OS in Reproductive Systems

OS refers to the rate of ROS production exceeding a biological system's antioxidant capacity or, in other words, when redox homeostasis is lost in favor of unregulated oxidative reactions and cascades. While the impact of ROS is now an established research subfield advancing germ cell biology and ART, the physiological roles of ROS were first described in the sperm of aquatic species. Standing on the seminal works in this field (Jones et al. 1979; Aitken and Clarkson 1987; Aitken et al. 1998; 2012; 2014; 2015; Aitken and De Iuliis 2007), we now have a good understanding of ROS production and ensuing OS pathways in human spermatozoa. Indeed, at least half of all male infertility can now be attributed to OS (Kesari et al. 2018), resulting either from excessive production of ROS, such as oxygen-containing, free radicals, peroxides, reactive nitrogen species, and/or a lack of antioxidant capacity (Bansal and Bilaspuri 2011; Elmussareh et al. 2015; Wagner et al. 2018).

ROS can be potent oxidants, but as a group, their reaction kinetics are diverse. More reactive species look to acquire electrons from a wide variety of nearby stable molecules, including proteins, lipids, and nucleic acids (Kothari et al. 2010). Direct damage to the paternal genome by ROS induced by man-made EMFs has been shown to induce single- and double-strand DNA breaks (Houston et al. 2018a). While damaging important germ cell components directly, ROS also elicit a range of secondary insults, such as those arising from LPO (Figure 6.7) and their resulting reactive aldehyde products (Aitken et al. 2012a). Both direct and indirect modes ultimately lead to a decline

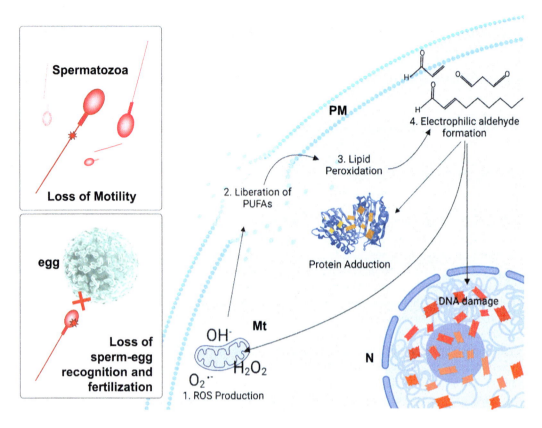

FIGURE 6.7 Lipid peroxidation cascades facilitate cellular damage and the induction of oxidative stress: Polyunsaturated fatty acids (PUFAs) in the sperm plasma membrane (PM) provide substrates for lipid peroxidation. Reactive products (such as electrophilic aldehydes) can then react with proteins and DNA, inducing further cellular damage, including the compounding disruption of the mitochondria (Mt). Ultimately, this leads to loss of sperm function. This mechanism has also been observed in oocytes (Mihalas et al. 2018). [N: nucleus].

of semen parameters (Agarwal et al. 2014; Elmussareh et al. 2015). In oocytes, ROS have also been implicated in effecting critical meiotic processes leading to aneuploidy. This observation, interestingly, recapitulates the mechanisms observed in the *in vitro* and *in vivo* aging oocyte reserves in mouse models (Lord et al. 2015; Lord et al. 2013), fitting the "free-radical theory of aging" paradigm (Mihalas et al. 2017a; 2017b; 2018). The effects in oocytes seem to progress through an LPO mechanism that leads to the dysfunction of tubulin, intimately disrupting the meiotic process (Mihalas et al. 2017a). Within sperm, mitochondria that are subjected to LPO can manifest with a decrease of mitochondrial membrane potential, electron leakage, further ROS production, and reduced capacity for ATP synthesis (Xiao et al. 2017). In the male, LPO of the sperm plasma membrane is enhanced, given the membrane's unusually rich composition of polyunsaturated fatty acids that, otherwise, provide sperm with the required membrane properties for efficient motility and ability to interact with the oocyte (Aitken et al. 2012b). LPO of the sperm membrane alters the characteristics of the specialized lipid bilayer, resulting in loss of cell function, such as immobility (Nowicka-Bauer and Nixon 2020). Conversely, intercellular physiological levels of ROS are tightly regulated in high-quality spermatozoa, with these low concentrations playing key roles in physiological maturation processes, including hyperactivation, capacitation, acrosome reactions, and signaling required for fertilization (Iommiello et al. 2015). The principle means by which biological systems (including germ cells) regulate ROS levels is through the expression of antioxidant enzymes and biomolecules (Drevet 2006; Drevet and Aitken 2020; O'Flaherty 2019; 2020; Nowicka-Bauer and Nixon 2020).

These antioxidant systems have been widely studied in both male (Drevet 2006; Drevet and Aitken 2020; O'Flaherty 2019; 2020) and female germ cells (Agarwal et al. 2005; Ruder et al. 2008; Wang et al. 2017; Aitken 2020). Dysregulation of the antioxidant system is, thus, another avenue driving OS in biological systems.

6.8.2 Antioxidants and Their Roles in Advancing the Field

Given the prevalence of oxidative processes that lead to germ cell damage, the approach to intervene via antioxidants supplementation is warranted. However, the application of antioxidant formulations, particularly for male infertility has met with only limited, if any, success (Ahmadi et al. 2016; Ali et al. 2021). While the reasons behind the deficiencies are beyond the scope of this chapter, they perhaps highlight the need for a deeper understanding of the detailed redox chemistry that underpins their potential therapeutic benefits. Despite the current clinical limitations, appropriate pharmacological intervention has a clear role in restricting OS, thereby protecting germlines from acquiring and passing on damage sustained as a consequence of environmental exposures. The few studies that have sought to investigate the utility of antioxidants have begun to yield some critical mechanistic insights into the emerging complexities around the nature of EMF biophysical interactions.

In six highlighted studies that have explored the action of three common antioxidants – melatonin, ascorbic acid (vitamin C), and α-tocopherol (vitamin E) – it has been shown that they have the ability to inhibit ROS processes following exposure to WC EMFs and restore fertility potential which is otherwise impacted by oxidative mediated damage in several murine models and cell lines (Oral et al. 2006; Al-Damegh 2012; Liu et al. 2013a; Meena et al. 2014; Solek et al. 2018; Pandey and Giri 2018).

Melatonin is a hormone critical for temporal signaling in animals but is also a potent endogenous antioxidant (Tan et al. 2015). Accordingly, this molecule has been extensively tested as an inhibitor of OS. In reproductive circles, melatonin supplementation has been successful in subduing damage induced by OS and continues to be an important tool for elucidating the role of ROS in WC EMF-induced damage. Studies (Meena et al. 2014; Pandey and Giri 2018) have shown clear benefits of melatonin in combating the adverse impact of man-made EMFs in the spermatozoa of male Wistar rats and albino Swiss mice, respectively. Indeed, both studies find that melatonin supplementation to the animal while exposed reduced ROS abundance and evidence of oxidative damage (MDA, protein carbonyls, and DNA damage) and restored appropriate levels of testosterone and testis specific lactate dehydrogenase-X (LDH-X).

Vitamins C and E are also popular choices for antioxidant investigations and can work synergistically when supplemented together. Vitamin C, being highly polar, can differentiate in its effects from vitamin E, which gets distributed into hydrophobic tissue and cellular membranes. These vitamins were tested in male Wistar albino rats (Al-Damegh 2012), female Wistar albino rats (Oral et al. 2006), and in mouse spermatocyte-derived cells (GC-2) *in vitro* (Liu et al. 2013a). In both the male and female rats exposed to WC EMFs (carrier frequencies: 900, 1800, 1900 MHz), the combination of vitamins C and E resulted in marked reductions in EMF-induced pathologies, including reduced MDA levels in the endometrium of female rats and the reduction of two additional LPO markers, conjugated dienes and hydroperoxides, in the testis of male rats, while also moderating levels of endogenous antioxidant molecules. The exposure to WC EMF (1800 MHz at SAR 4 W/kg) *in vitro* initiated oxidative DNA damage in the germ cell lines which was eliminated in cells supplemented with vitamin E. A subset of cells in the Liu et al. (2013a) study was also exposed solely to hydrogen peroxide* (H_2O_2) at a level that mimicked the impacts of EMF exposure alone. Importantly, the same protective effects of vitamin supplementation were observed in these cells verifying the role of ROS in these effects. In female rats, WC EMF exposure initiated a range of apoptotic markers

---

* Hydrogen peroxide converts to hydroxyl radical (OH•), which is the most potent ROS.

which were similarly able to be modulated by vitamins C and E (Guney et al. 2007). WC EMF-induced changes in the testes structure of male rats, such as an increase in seminiferous tubule diameter, were also ameliorated by antioxidant supplementation (Al-Damegh 2012).

The supplementation of antioxidant molecules for both whole body and *in vitro* exposure studies is a key aspect to advance our knowledge on the link between WC EMFs and ROS production in the reproductive systems. In other animal systems, there are reports that claim both "protective" effects and no effect of antioxidants under WC EMF exposures. In the limited pool of data approaching the effects of antioxidants on challenged reproductive systems, the evidence that antioxidants, such as vitamins C and E and melatonin, suppress oxidative insults that are otherwise elevated under WC EMF exposure is exciting and provides further support for the theory that ROS is the primary mechanism behind negative effects observed in reproductive systems following WC EMF exposures.

6.8.3 THE ELECTRON TRANSPORT CHAIN

The physiological concentrations of ROS in biological systems, including the germlines, largely originate from the normal process of oxidative phosphorylation that utilizes the mitochondrial ETC (Figure 6.8). A small number of electrons that "leak" from the mitochondrial system are primarily intercepted by O_2, leading to the formation of superoxide anion free radical ($O_2\cdot^-$) and, subsequently, other ROS (Turrens 2003). The magnitude of ROS leakage varies between the ETC complexes, with complex I (NADH oxidase) responsible for a bulk of the superoxide generation (Quinlan et al. 2013), alongside complex III, which also has a prominent role in general ROS production. The mode of emigration of ROS from complex I to the matrix has been suggested to be more damaging in spermatozoa, driving significant peroxidative damage (Koppers et al. 2008). ROS generated at complex III escape to the intermembrane space, where they encounter a pool, albeit finite, of mitochondrial antioxidant protections.

The movement of electrons through the ETC is logically a highly regulated process; however, perturbations of the electron flow elevate ROS production. Such perturbations can be driven pharmacologically, however, evidence is also mounting that EMFs emitted by WCDs and antennas can also drive ROS generation (De Iuliis et al. 2009a, Burlaka et al. 2013) through disruption of the electron stream (Martino and Castello 2011) and/or interference with proton transport (Panagopoulos et al. 2000; 2002; 2015; 2021) through the transmembrane complexes of the inner mitochondrial membrane. Disruption of the mitochondria in this manner is also associated with increased expression of mitochondrial apoptotic markers (Liu et al. 2015) and decreased mitochondrial membrane electrical potential (Lu et al. 2012). There is also evidence suggesting that EMFs may impair the conformation of key antioxidant proteins (Lu et al. 2012), preventing them from participating in the elimination of free radicals generated during normal respiration (De Iuliis et al. 2009a; Burlaka et al. 2013).

6.8.4 EVIDENCE OF MITOCHONDRIAL DISTURBANCE IN THE FEMALE REPRODUCTIVE TRACT

The impact of WC EMF-associated OS in the female reproductive system has been investigated in the ovaries of exposed Wistar rats to EMF (2.45 MHz, 0.1 W/kg, 1 h/day) during prenatal (embryos exposed *in utero*) and postnatal (animals exposed to EMF from 21 days of age until puberty) development (Sangun et al. 2015). Total measurement of oxidative and antioxidant capacities of ovarian tissue was completed by determination of the total oxidant status (TOS) and total antioxidant status (TAS). These measurements reflect an equivalent tissue or intra-cellular concentration of H_2O_2, and TOS in exposed animals at both exposure periods was found to be significantly higher than in the control group ($p < 0.05$). Furthermore, the oxidative stress index (OSI) of ovarian tissue in the prenatal and postnatal group, calculated from the TOS and TAS assessments was significantly higher than those in the control group ($p < 0.05$). The study interprets the increases in TOS and OSI within the ovary indicative of EMF-induced chronic OS, and these results are interestingly in

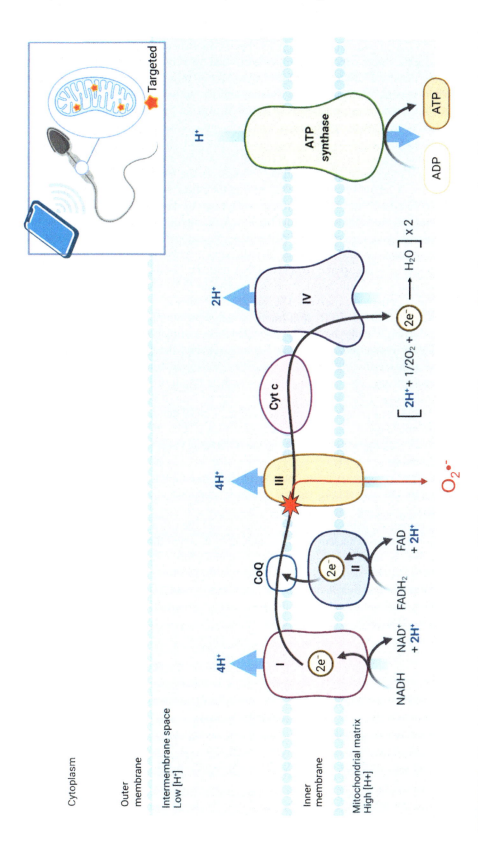

FIGURE 6.8 The mitochondrial electron transport chain (ETC) is the main source of ROS in spermatozoa: Disruption or damage of any of the main complexes in the system can result in electron leakage, ultimately contributing to elevation of $O_2^{\bullet-}$ and other ROS concentrations in sperm. Some evidence suggests that complex III may be implicated in leaking electrons under WC EMF exposures (Houston et al. 2018a).

accordance with other non-mammalian studies (Burlaka et al. 2013; Manta et al. 2017). The impact of WC EMF-induced OS on ovarian physiology has been established in *Drosophila melanogaster*, in which adult female flies were exposed to radiation emitted by a commercial mobile phone (GSM 1800 MHz) at power density 27 µW/cm^2 (SAR 0.15 W/kg) for 30 min (Manta et al. 2017). Following exposure, a 1.6-fold increase in ovarian ROS content and the activation of follicle apoptosis was observed. Furthermore, micro-array analysis of exposed ovarian follicles revealed 168 differentially expressed genes associated with multiple metabolic and cellular processes related to stress and apoptosis. The authors suggested that WC EMF-dependent increase in ROS likely induces a non-targeted, genome-wide transcriptional perturbation of gene expression profiling, leading to vulnerable ovarian cell subpopulations and follicle apoptosis in the fruit fly. An elegant *in ovo* study in which the exposure of Japanese quail embryos during the initial stages of development to GSM 900 MHz mobile phone EMF for 38 h (1.6 days), 5 days, or 10 days at very low average power density (0.25 µW/cm^2) (about 1000 times below the power density level of the mobile phone to the head) demonstrated a persistent overproduction of superoxide anion ($O_2\cdot^-$) and nitrogen oxide (NO•) free radicals in all exposure periods and found an associated increase in LPO (specifically in thiobarbituric acid reactive substances) and oxidative DNA damage (8-OH-dG) ($p < 0.001$) (Burlaka et al. 2013). Consequently, a significant decline in antioxidant SOD and CAT enzyme activity in the exposed embryonic cells was observed. The specific detection of superoxide generation using electron paramagnetic resonance (EPR) spin-trapping technique supports that WC EMF-induced superoxide overproduction in embryonic cells originates from a mitochondrial pathway.

However, the site of interaction, such as ETC complexes, and the exact mechanism by which WC EMF exposure induces free radical ($O_2\cdot^-$ and NO•) overproduction within the mitochondria are not yet fully recognized, and more focused research is required (Friedman et al. 2007; De Iuliis et al. 2009a). Increased concentrations of NO• may be caused by an additional expression of NO• synthases (NOS) or through the direct activation of NOS molecules present at the time of EMF exposure. This stable generation of NO• may perturbate ETC complexes, triggering increased generation of superoxide and potentially peroxynitrite (ONOO$^-$) (Valko et al. 2006) and subsequent downstream DNA damage.

Female Sprague Dawley rats exposed to 1800 MHz GSM radiation for 2 h per day for 30 and 60 days were examined for OS-induced ovarian and uterine biochemical and histopathological changes (Alchalabi et al. 2016b). Ovarian histopathological changes in exposed groups included vacuolation of interstitial cells, granulosa cells, luteal cells, ooplasm, and the disruption and thinning of zona pellucida. Furthermore, cellular nucleus changes indicating degeneration at Graafian follicles[*] and micro-nuclei formation in oocytes and luteal cells[†] were observed in the exposed groups. Uterine histopathological analysis revealed glandular and luminal epithelium cell apoptosis coupled with infiltration of eosinophils, polymorphonucleocyte lymphocyte, and macrophage activation in the myometrium and endometrium layers which activated excessive production of reactive oxygen and nitrogen species. Biochemical analysis of ovarian tissue revealed a significant reduction in GPx activity alongside increases in MDA levels across both exposure groups. Similarly, uterine tissue within the 60-day exposure group had significantly reduced GPx activity and increased MDA levels compared to those in the 30-day exposure group. Furthermore, the GSM EMF exposure in both exposure groups resulted in a significant reduction in serum melatonin.

The increase in MDA levels accompanied with a reduction of melatonin and decreased activity of GPx due to the over consumption of reduced glutathione (GSH) are likely responsible for the development of LPO presenting as an end product of oxidative damage in ovarian tissue (Chernoff et al. 1992; Tilly and Tilly 1995) as seen in other studies (Koyu et al. 2009; Avci et al. 2012; Tkalec et al. 2013). These findings suggest that WC EMF-induced alterations of antioxidant capacity associated

---

[*] Liquid-filled cavities in the mammalian ovaries within which oocytes develop. During ovulation the follicle ruptures to release the enclosed oocyte.
[†] Cells developing from the follicular cells surrounding the ovarian follicle.

with excessive production of free radicals may contribute to ovarian and endometrial oxidative damage, which could be related to the pathogenesis and progression of endometritis.

Further, oxidatively induced impairment and apoptosis of rat endometrium has been recorded following GSM 900 MHz simulated mobile phone exposures (Oral et al. 2006). Biochemical analysis revealed that MDA levels in endometrial tissue of exposed rats were higher than in the control groups ($p < 0.01$); however, when administered with antioxidants (vitamin E and vitamin C), MDA levels were significantly reduced ($p < 0.01$). Immunohistochemical analysis of apoptotic marker caspase-3 in exposed rat endometrium showed mild to moderate cytoplasmic reactivity and moderate to strong immunostaining of stromal cells and capillary endothelial cells of stroma. However, intensities of caspase-3 staining in epithelial and stromal cells of the endometrium were significantly decreased in the EMF + vitamin C and E group ($p < 0.01$). Caspase-8 immunoreactivity in the EMF-exposed group displayed moderate cytoplasmic reactivity in endometrial surface epithelial cells and strong staining in stromal and capillary endothelial cells compared to the EMF + vitamin group ($p < 0.01$), which displayed very weak cytoplasmic staining. Significantly increased immunolabeling of Bax was detected in surface and glandular epithelial cells and stromal cells of the endometrium within the EMF treatment group when compared to control and EMF + vitamin group ($p < 0.01$). Conversely, Bcl-2 immunolabeling in epithelial, glandular, and stromal cells in the EMF group were significantly weaker than the control group ($p < 0.01$). In a similar study, increased endometrial levels of NO• were recorded in exposed rats alongside decreased endometrial SOD, CAT, and GPx activities, compared to control populations (Guney et al. 2007). Likewise, co-administration of vitamin C and vitamin E displayed ameliorative properties causing NO• levels to decrease and antioxidant enzyme activity to increase, respectively.

Effects of mobile phone (900 and 1800 MHz) and Wi-Fi (2450 MHz) pulsing EMF exposures on uterine OS and hormone levels in pregnant rats and their offspring have also been assessed by other research groups (Yüksel et al. 2016; Altun et al. 2018). Following 365-day exposures, maternal rat plasma TOS concentrations were significantly higher in all three exposed groups compared to the control population ($p < 0.05$). Uterine LPO levels within 4-, 5-, and 6-week old offspring were significantly higher across all three exposure groups compared to the control group ($p < 0.05$). These findings were coupled with decreased levels of GPx activity in the 4- and 5-week old offspring across all three exposure conditions ($p < 0.05$). Furthermore, offspring within these exposure groups displayed significant reductions in prolactin (6-weeks old), progesterone and estrogen levels (4-, 5-, and 6-weeks old) compared to the control group. The authors proposed that the transient receptor potential subfamily V member 1 cation channel (TRPV1) may be a possible molecular pathway linking the observed alterations in hormone and OS levels in the uterus of maternal rats following the year-long exposure to WC EMFs. TRPV1 likely mediates entry of Ca^{2+} into the uterus, leading to accumulative OS and opening of mitochondrial membrane pores, leading to downstream mitochondrial dysfunction, including the swelling and rupture of the outer membrane, releasing superoxide and hydrogen peroxide.

6.9 DISCUSSION AND CONCLUSIONS

Here we have provided an overview of the state of research seeking to elucidate the impacts of WC EMFs on reproductive health and function. Despite limitations, we are now seeing a strong body of evidence emerging that suggests WC EMFs are capable of damaging reproductive tissue and the germ cells themselves. The data collated in the field to date support the ability of WC EMFs to impair biological systems and, indeed, the reproductive cells and tissues. Undoubtedly, the complexities of EMF interactions with biology can hinder rapid progress in the field, and while some of the data presented to date may not prove negative health impacts, based on the reviewed data, we conclude that there is a clear potential for this. The involvement of ROS or an OS state and the disturbances of VGICs are two of the leading mediators of the man-made EMF-induced effects observed in reproductive systems.

Animal models have indicated that alterations driven by WC EMFs can reduce fertility. The investigation on reproductive systems is making important impacts for the general field, with the male reproductive system in particular offering unique and sensitive molecular insights, over decades of study, into how WC EMFs may interact with living organisms. On the back of the biological evidence, the potential for these EMFs to impact our reproductive capacity and general health is clear.

WC EMF-induced dysfunction of VGICs in the cytoplasm and mitochondrial membranes leads to the overproduction of ROS as described by Panagopoulos et al. (2021). Thus, man-made and especially WC EMFs have the potential to promote an imbalance of ROS promoting a state of OS. Importantly, for the male germline, we and others have identified key vulnerabilities to ROS, including a positive feedback loop mediated by LPO and the formation of reactive aldehydes, to disrupt aldehyde dehydrogenase (Aitken et al. 2012a) in complex II of the ETC, further disrupting electron flow.

6.9.1 Mitochondrial ROS and DNA Damage

A solid body of evidence has now shown that, in a range of biological models, sperm cells generate ROS in response to man-made EMF exposures, including the direct exposure of WC EMFs on cells and within murine models. Several studies, including our own (Houston et al. 2019), provide data directly linking the mitochondria to the EMF-induced OS. Exposing unrestrained male mice to 905 MHz EMF, for 12 h per day up to 5 weeks revealed the role of sperm mitochondria. The testes of the exposed mice exhibited no evidence of gross histological change or elevated OS, but by contrast, after 5 weeks, the exposure adversely impacted the vitality and motility profiles of mature epididymal spermatozoa, with elevated DNA oxidation and fragmentation also observed (Figure 6.9). These spermatozoa exhibited increased production of mitochondrial ROS after 1 week of exposure. All periods of exposure did not impair the fertilization competence of spermatozoa nor their ability to support early embryonic development (Houston et al. 2019) in line with other studies (Ozguner et al. 2005a; Tumkaya et al. 2016). However, the ability of damaged cells to participate in embryo development and pass on or propagate potentially transgenerational impacts provides even more

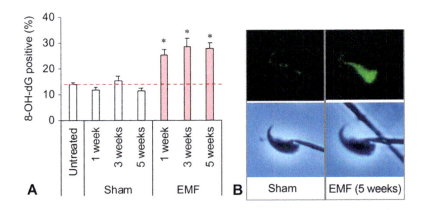

FIGURE 6.9 WC EMFs can induce oxidative damage in the spermatozoa of exposed mice: (A) The sperm from mice exposed for 12 h/day for up to 5 weeks exhibit a key oxidative stress biomarker 8-hydroxy-2′-deoxyguanosine (8-OH-dG). Data presented as mean + SEM. *$p < 0.05$. (B) Representative images of spermatozoa stained with the 8-OH-dG antibody (up) and not stained (down). The localization of the oxidative DNA damage biomarker 8-OH-dG in the sperm nucleus is evident only in the EMF-exposed spermatozoa (Houston et al. 2019) [SEM: Standard Error of the Mean. Sham: Sham-exposed. EMF: EMF-exposed].

impetus to fully characterize any paternal genomic damage caused by WC EMFs and determine in more detail the underlying mechanism of action. Other relevant studies detailing mitochondrial ROS-mediated DNA damage induced by external factors have also been described in Sections 6.5, 6.6, and 6.7 of this chapter.

Sperm mitochondrial ROS leading to DNA damage of the paternal genome is now well established in male germ cell biology (Aitken and De Iuliis 2009; Aitken and Baker 2020). With growing evidence implicating sperm mitochondria as a key target for EMF-induced perturbations, concerns about the effects of WC EMFs on male fertility are justified. These concerns resonate because of the potential for damaged spermatozoa to not only negatively influence the fertility of the man but also the health of the offspring. Conceivably, a compounding of effects through the male germline over generations could lead to significant health impacts in the future. While the paternal genome has been the focus in transmission of adverse health risk to the offspring, both genetic and epigenetic factors can be influenced by OS events, and thus, the importance of paternal epigenetic dysregulation is now gaining significant interest (Reilly et al. 2016; Sharma et al. 2016).

A possible explanation for these observations is that WC EMFs can affect mitochondrial membranes to produce large amounts of endogenous oxygen radicals leading to the excessive consumption and depletion of SOD, GPx, and CAT enzyme activities, leading to a systemic biological redox imbalance. The change in MDA levels in exposed rat endometrium coupled with the ameliorative effects of antioxidants, vitamin C and E (Oral et al. 2006), supports the hypothesis that LPO may be one of the molecular mechanisms involved with EMF-induced damage. The antiapoptotic activity of these antioxidants has been demonstrated by several studies already reported in this chapter, in which it was shown that they directly inhibit free radical-mediated apoptosis through their scavenging properties and ability to maintain the Bcl-2 protein in its functional form via membrane stabilization action, therefore inhibiting the release of mitochondrial cytochrome C (Serbecic and Beutelspacher 2005).

6.9.2 SPERMATOZOA AND THE MALE REPRODUCTIVE SYSTEM AS A STUDY MODEL FOR THE EFFECTS OF WC EMFs

Given that about 50% of all infertility involves a male factor (Agarwal et al. 2014), and with sperm quality determining the health trajectory of future generations (Liu et al. 2012; Aitken et al. 2020; Aitken and Baker 2020; Lee et al. 2009), detailed investigations into the impact of environmental factors on this system is well warranted. Besides the utility for reproductive research, the attributes of the male reproductive system and the specialized nature of spermatozoa make this system a sensitive and non-invasive model to investigate potentially wider ranging environmentally induced health impacts (Aitken, et al. 2004; 2020).

We have discussed that the cellular architecture and properties of the spermatozoon render this cell type uniquely sensitive to external factors that may initiate an OS response. Similarly, the vulnerabilities toward reaching an OS state also impact sperm DNA integrity. Altogether, this provides a model to monitor the effects of such external factors whose effects may be more difficult to measure in somatic cells. Nevertheless, the phenomena of OS induced by WC EMFs, characterized by excessive generation of ROS, is now well supported (Houston et al. 2018a) across several systems including *Drosophila* (Manta et al. 2017), mouse fibroblasts (Hou et al. 2015), cultured breast cancer cells (Kahya et al. 2014), rat heart tissue (Ozguner et al. 2005b), human lens epithelial cells (Yao et al. 2008), and studies on mammalian spermatozoa (Agarwal et al. 2009; Kesari et al. 2011). Our previously published work on human spermatozoa (De Iuliis et al. 2009a) is further strengthened by our studies on mouse spermatogonial-like and spermatocyte-like germ cell lines GC1 and GC2, respectively, and primary spermatogonial cells. These data identify that both male and female germ cells are more sensitive to WC EMFs than somatic cell types, including granulosa, mouse McCoy, and human embryonic kidney cell lines (Houston et al. 2018a).

The proximity of WCDs to our reproductive system, in particular the testicles, and how they influence testicular development and function is driving increasing concern and is of paramount relevance to this globally established industry. Spermatozoa are manufactured within the testis and, in order to achieve fertilization, these cells must then undergo maturation transformations during the transit through the epididymis process of spermatogenesis in the testes. These processes that culminate in the development and maturation of a single sperm cell and its ability to bind to an oocyte, dictate the trajectory of the developing embryo and health of the offspring. Previously published work has highlighted the susceptibility of male germ cells within the testes as being particularly more sensitive to WC EMFs compared to other somatic cell types (Gutschi et al. 2011; Sepehrimanesh et al. 2017; Houston et al. 2018a; 2019); however, some further studies have implicated downstream passage through the epididymis as a key site of vulnerability for sperm maturation (Reilly et al. 2016; Hutcheon et al. 2017) in which the cell sheds most of its cytoplasm and many defense capabilities. The anatomy of the male reproductive system specifically, with only a thin outer layer of the scrotum for protection against penetrating EMFs, places it in a unique vulnerable position to receive chronic EMF exposure from devices stored, or used, in the vicinity of the testes.

6.9.3 WHERE TO NEXT AND RECOMMENDATIONS

A predominant phenomenon we and others have observed after WC EMF exposure is the onset of an OS state in tissues, cells, and systems. This clearly outlines antioxidant-based factors as a potential intervention option. *In vivo* application of antioxidants to mice has been shown to improve sperm concentration, motility, morphology, and DNA integrity during WC EMF exposure. However, any protective effects of antioxidants on exposed human reproductive systems have yet to be confirmed.

Dosimetry and well-defined exposure conditions are of fundamental importance for research that attempts to link experimental observations to clinical relevance. Awareness of the importance of this aspect of experimental design for the field is now accelerating. In order to significantly advance the field and moving forward as a research community, we must determine and adopt standard guidelines for experimental design.

The location in which we largely transport mobile devices and its proximity to our reproductive systems reinforces the importance of the work in this field. In addition to this, the sensitivity of germ cells and their fundamental role in determining the trajectory of future populations highlights those reproductive systems, in particular spermatozoa, as a critical focus for the field. Whereas subtle effects of WC EMFs on tissue and somatic cells may be well buffered against prolonged health effects, damage in germ cells may include genetic and epigenetic modifications that remain uncorrected and could conceivably be compounded over generations to precipitate severe negative health outcomes only after decades. The paradigm that sperm cell damage leads to transgenerational effects is well established (Aitken and Baker 2020; Aitken et al. 2020; Lee et al. 2009; Liu et al. 2012). Public health legislation for reproductive health should urgently be updated to align with the latest scientific developments.

From the very first studies that investigated the biological and health effects of man-made EMFs, it was perhaps alarming to learn that these EMFs may have any impact on biology (Adey 1981; 1993; Goodman et al. 1995). While there has been resistance to this idea from certain industries and parts of the community, along with necessary scientific skepticism, our current period of discovery can now confirm these effects experimentally. This general position is also reflected in the contributions made in the reproduction-specific research subfield. What is clear from the combined work to date, including important work on reproductive tissues, is that the significant adverse effects observed warrant the urgent research advancements that are required to elucidate more details. The need for broader research focus, such as the potential for accumulative damage via chronic OS, provide plausible mechanisms for health impacts in future generations, precipitated by the recorded biological changes. The urgency to pursue this research question is, therefore, linked to avoiding a potential compounding future health burden due to chronic exposures of our population to WC EMFs. Until then, reassessing the appropriate limits of man-made EMF and especially WC EMF exposures is necessary (Harremoes et al. 2013).

FUNDING

KM and KH are supported by a NHMRC Project Grant (APP1156997) awarded to GND and BN.

ACKNOWLEDGEMENT

Some content contained in Figures 6.5 and 6.7 was created with Biorender.com

REFERENCES

Adams JA, Galloway TS, Mondal D, Esteves SC, Mathews F (2014) "Effect of mobile telephones on sperm quality: A systematic review and meta-analysis." *Environ Int* 70:106–112. https://doi.org/10.1016/j.envint.2014.04.015.

Adey WR (1981) "Tissue interactions with nonionizing electromagnetic fields." *Physiol Rev* 61(2):435–514. https://doi.org/10.1152/physrev.1981.61.2.435.

Adey WR (1993) "Biological effects of electromagnetic fields." *J Cell Biochem* 51(4):410–416. https://doi.org/10.1002/jcb.2400510405.

Agarwal A, Gupta S, Sharma RK (2005) "Role of oxidative stress in female reproduction." *Reprod Biol Endocrinol* 3(1):28. https://doi.org/10.1186/1477-7827-3-28.

Agarwal A, Deepinder F, Sharma RK, Ranga G, Li J (2008) "Effect of cell phone usage on semen analysis in men attending infertility clinic: An observational study." *Fertil Steril* 89(1):124–128. https://doi.org/10.1016/j.fertnstert.2007.01.166.

Agarwal A, Desai NR, Makker K, Varghese A, Mouradi R, *et al* (2009) "Effects of radiofrequency electromagnetic waves (RF-EMW) from cellular phones on human ejaculated semen: An in vitro pilot study." *Fertil Steril* 92(4):1318–1325. https://doi.org/10.1016/j.fertnstert.2008.08.022.

Agarwal A, Singh A, Hamada A, Kesari K (2011) "Cell phones and male infertility: A review of recent innovations in technology and consequences." *Int Braz J Urol* 37(4):432–454. https://doi.org/10.1590/s1677-55382011000400002.

Agarwal A, Virk G, Ong C, du Plessis SS (2014) "Effect of oxidative stress on male reproduction." *World J Mens Health* 32(1):1–17. https://doi.org/10.5534/wjmh.2014.32.1.1.

Agarwal A, Mulgund A, Hamada A, Chyatte MR (2015) "A unique view on male infertility around the globe." *Reprod Biol Endocrinol* 13(1):37. https://doi.org/10.1186/s12958-015-0032-1.

Agarwal A, Baskaran S, Parekh N, Cho CL, Henkel R, *et al* (2021) "Male infertility." *Lancet* 397(10271):319–333. https://doi.org/10.1016/s0140-6736(20):32667-2.

Ahlbom A, Bridges J, de Seze R, Hillert L, Juutilainen J, *et al* (2008) "Possible effects of electromagnetic fields (EMF) on human health–Opinion of the scientific committee on emerging and newly identified health risks (SCENIHR)." *Toxicology* 246(2–3):248–250. https://doi.org/10.1016/j.tox.2008.02.004.

Ahmadi S, Bashiri R, Ghadiri-Anari A, Nadjarzadeh A (2016) "Antioxidant supplements and semen parameters: An evidence based review." *Int J Reprod Biomed* 14(12):729–736.

Ahmed AS, Sheng MH, Wasnik S, Baylink DJ, Lau KW (2017) "Effect of aging on stem cells." *World J Exp Med* 7(1):1–10. https://doi.org/10.5493/wjem.v7.i1.1.

AIHW (2020): Australian Institute of Health and Welfare, Cancer data in Australia,Web report 2020. Edited by Cancer series no. 122. Canberra, Austaralia.

Aitken RJ, Clarkson JS (1987) "Cellular basis of defective sperm function and its association with the genesis of reactive oxygen species by human spermatozoa." *J Reprod Fertil* 81(2):459–469. https://doi.org/10.1530/jrf.0.0810459.

Aitken RJ, Gordon E, Harkiss D, Twigg JP, Milne P, *et al* (1998) "Relative impact of oxidative stress on the functional competence and genomic integrity of human Spermatozoa1." *Biol Reprod* 59(5):1037–1046. https://doi.org/10.1095/biolreprod59.5.1037.

Aitken RJ, Koopman P, Lewis SE (2004) "Seeds of concern." *Nature* 432(7013):48–52. https://doi.org/10.1038/432048a.

Aitken RJ, Bennetts LE, Sawyer D, Wiklendt AM, King BV (2005) "Impact of radio frequency electromagnetic radiation on DNA integrity in the male germline." *Int J Androl* 28(3):171–179. https://doi.org/10.1111/j.1365-2605.2005.00531.x.

Aitken RJ, De Iuliis GN (2007) "Origins and consequences of DNA damage in male germ cells." *Reprod Biomed Online* 14(6):727–733. https://doi.org/10.1016/S1472-6483(10):60676-1.

Aitken RJ, De Iuliis GN (2009) "On the possible origins of DNA damage in human spermatozoa." *Mol Hum Reprod* 16(1):3–13. https://doi.org/10.1093/molehr/gap059.

Aitken RJ, De Iuliis GN, Finnie JM, Hedges A, McLachlan RI (2010) "Analysis of the relationships between oxidative stress, DNA damage and sperm vitality in a patient population: Development of diagnostic criteria." *Hum Reprod* 25(10):2415–2426. https://doi.org/10.1093/humrep/deq214.

Aitken RJ, De Iuliis GN, Gibb Z, Baker MA (2012a) "The Simmet lecture: New horizons on an old landscape–Oxidative stress, DNA damage and apoptosis in the male germ line." *Reprod Domest Anim* 47(Suppl 4):7–14. https://doi.org/10.1111/j.1439-0531.2012.02049.x.

Aitken RJ, Whiting S, De Iuliis GN, McClymont S, Mitchell LA, Baker MA (2012b) "Electrophilic aldehydes generated by sperm metabolism activate mitochondrial reactive oxygen species generation and apoptosis by targeting succinate dehydrogenase." *J Biol Chem* 287(39):33048–33060. https://doi.org/10.1074/jbc.M112.366690.

Aitken RJ, Smith TB, Lord T, Kuczera L, Koppers AJ, et al (2013) "On methods for the detection of reactive oxygen species generation by human spermatozoa: Analysis of the cellular responses to catechol oestrogen, lipid aldehyde, menadione and arachidonic acid." *Andrology* 1(2):192–205. https://doi.org/10.1111/j.2047-2927.2012.00056.x.

Aitken RJ, Smith TB, Jobling MS, Baker MA, De Iuliis GN (2014) "Oxidative stress and male reproductive health." *Asian J Androl* 16(1):31–38. https://doi.org/10.4103/1008-682x.122203.

Aitken RJ, Gibb Z, Baker MA, Drevet J, Gharagozloo P (2015) "Causes and consequences of oxidative stress in spermatozoa." *Reprod Fertil Dev* 28(1–2):1–10. https://doi.org/10.1071/rd15325.

Aitken RJ (2018) "Not every sperm is sacred; a perspective on male infertility." *Mol Hum Reprod* 24(6):287–298. https://doi.org/10.1093/molehr/gay010.

Aitken RJ (2020) "Impact of oxidative stress on male and female germ cells: Implications for fertility." *Reproduction* 159(4):R189–R201. https://doi.org/10.1530/rep-19-0452.

Aitken RJ, Baker MA (2020) "The role of genetics and oxidative stress in the etiology of male infertility—A unifying hypothesis?" *Front Endocrinol* 11. https://doi.org/10.3389/fendo.2020.581838.

Aitken RJ, De Iuliis GN, Nixon B (2020) "The sins of our forefathers: Paternal impacts on de novo mutation rate and development." *Annu Rev Genet* 54(1):1–24. https://doi.org/10.1146/annurev-genet-112618-043617.

Alchalabi ASH, Aklilu E, Aziz AR, Malek F, Ronald SH, Khan MA (2016a) "Different periods of intrauterine exposure to electromagnetic field: Influence on female rats' fertility, prenatal and postnatal development." *Asian Pac J Reprod* 5(1):14–23. https://doi.org/10.1016/j.apjr.2015.12.003.

Alchalabi ASH, Rahim H, Aklilu E, Al-Sultan II, Aziz AR, et al (2016b) "Histopathological changes associated with oxidative stress induced by electromagnetic waves in rats' ovarian and uterine tissues." *Asian Pac J Reprod* 5(4):301–310. https://doi.org/10.1016/j.apjr.2016.06.008.

Aldad TS, Gan G, Gao X-B, Taylor HS (2012) "Fetal radiofrequency radiation exposure from 800–1900 MHz-rated cellular telephones affects neurodevelopment and behavior in mice." *Sci Rep* 2(1):312. https://doi.org/10.1038/srep00312.

Al-Damegh MA (2012) "Rat testicular impairment induced by electromagnetic radiation from a conventional cellular telephone and the protective effects of the antioxidants vitamins C and E." *Clin (Sao Paulo)* 67(7):785–792. https://doi.org/10.6061/clinics/2012(07):14.

Ali M, Martinez M, Parekh N (2021) "Are antioxidants a viable treatment option for male infertility?" *Andrologia* 53(1):e13644. https://doi.org/10.1111/and.13644.

Altun G, Deniz ÖG, Yurt KK, Davis D, Kaplan S (2018) "Effects of mobile phone exposure on metabolomics in the male and female reproductive systems." *Environ Res* 167:700–707. https://doi.org/10.1016/j.envres.2018.02.031.

ARPANSA (2014) "Report by the ARPANSA radiofrequency expert panel on review of radiofrequency health effects research – Scientific literature 2000–2012." Edited by ARPANSA, Technical Report No. 164.

Authority (2015) *Communications report: 2013–14 series. Australian communications and media authority.* Australian Government, Canberra.

Avci B, Akar A, Bilgici B, Tunçel ÖK (2012) "Oxidative stress induced by 1.8 GHz radio frequency electromagnetic radiation and effects of garlic extract in rats." *Int J Radiat Biol* 88(11):799–805. https://doi.org/10.3109/09553002.2012.711504.

Aydin M, Cevik A, Kandemir FM, Yuksel M, Apaydin AM (2009) "Evaluation of hormonal change, biochemical parameters, and histopathological status of uterus in rats exposed to 50-Hz electromagnetic field." *Toxicol Ind Health* 25(3):153–158. https://doi.org/10.1177/0748233709102717.

Baan R, Grosse Y, Lauby-Secretan B, El Ghissassi F, Bouvard V, et al (2011) "Carcinogenicity of radiofrequency electromagnetic fields." *Lancet Oncol* 12(7):624–626. https://doi.org/10.1016/s1470-2045(11)70147-4.

Bakacak M, Bostancı MS, Attar R, Yıldırım ÖK, Yıldırım G, et al (2015) "The effects of electromagnetic fields on the number of ovarian primordial follicles: An experimental study." *Kaohsiung J Med Sci* 31(6):287–292. https://doi.org/10.1016/j.kjms.2015.03.004.

Baker KB, Tkach JA, Nyenhuis JA, Phillips M, Shellock FG, et al (2004) "Evaluation of specific absorption rate as a dosimeter of MRI-related implant heating." *J Magn Reson Imaging* 20(2):315–320. https://doi.org/10.1002/jmri.20103.

Balmori A (2005) "Possible effects of electromagnetic fields from phone masts on a population of white stork (Ciconia ciconia)." *Electromagn Biol Med* 24(2):109–119. https://doi.org/10.1080/15368370500205472.

Balmori A (2006) "The incidence of electromagnetic pollution on the amphibian decline: Is this an important piece of the puzzle?" *Toxicol Environ Chem* 88(2):287–299. https://doi.org/10.1080/02772240600687200.

Bansal AK, Bilaspuri GS (2011) "Impacts of oxidative stress and antioxidants on semen functions." *Vet Med Int* 2011:686137. https://doi.org/10.4061/2011/686137.

Barbieri RL (2019) "Chapter 22 - Female infertility." In: JF Strauss, RL Barbieri (Eds.), *Yen and Jaffe's reproductive endocrinology (eighth edition)*. Philadelphia: Elsevier, 556–581.e7.

Barratt CLR, De Jonge CJ, Sharpe RM (2018) "'Man up': The importance and strategy for placing male reproductive health centre stage in the political and research agenda." *Hum Reprod* 33(4):541–545. https://doi.org/10.1093/humrep/dey020.

Batellier F, Couty I, Picard D, Brillard JP (2008) "Effects of exposing chicken eggs to a cell phone in 'call' position over the entire incubation period." *Theriogenology* 69(6):737–745. https://doi.org/10.1016/j.theriogenology.2007.12.006.

Belva F, Bonduelle M, Roelants M, Michielsen D, Van Steirteghem A, et al (2016) "Semen quality of young adult ICSI offspring: The first results." *Hum Reprod* 31(12):2811–2820. https://doi.org/10.1093/humrep/dew245.

Belyaev I (2005) "Non-thermal biological effects of microwaves." *Microw Rev* 11, 13–29.

Belyaev I (2010) "Dependence of non-thermal biological effects of microwaves on physical and biological variables: Implications for reproducibility and safety standards." *Eur J Oncol (Library)* 5:187–218.

Bergh C, Wennerholm UB (2020) "Long-term health of children conceived after assisted reproductive technology." *Ups J Med Sci* 125(2):152–157. https://doi.org/10.1080/03009734.2020.1729904.

Bergqvist UO (1984) "Video display terminals and health: A technical and medical appraisal of the state of the art." *Scand J Work Environ Health* 10(Suppl 2):1–87.

Bernabò N, Tettamanti E, Russo V, Martelli A, Turriani M, et al (2010) "Extremely low frequency electromagnetic field exposure affects fertilization outcome in swine animal model." *Theriogenology* 73(9):1293–1305. https://doi.org/10.1016/j.theriogenology.2009.12.010.

Bertagna F, Lewis R, Silva SRP, McFadden J, Jeevaratnam K (2021) "Effects of electromagnetic fields on neuronal ion channels: A systematic review." *Ann N Y Acad Sci* 1499(1):82–103. https://doi.org/10.1111/nyas.14597.

Björkgren I, Sipilä P (2019) "The impact of epididymal proteins on sperm function." *Reproduction* 158(5):R155–R167. https://doi.org/10.1530/rep-18-0589.

Blackman C (2009) "Cell phone radiation: Evidence from ELF and RF studies supporting more inclusive risk identification and assessment." *Pathophysiology* 16(2–3):205–216. https://doi.org/10.1016/j.pathophys.2009.02.001.

Boileau N, Margueritte F, Gauthier T, Boukeffa N, Preux P-M, et al (2020) "Mobile phone use during pregnancy: Which association with fetal growth?" *J Gynecol Obstet Hum Reprod* 49(8):101852. https://doi.org/10.1016/j.jogoh.2020.101852.

Bonde JP, Flachs EM, Rimborg S, Glazer CH, Giwercman A, et al (2016) "The epidemiologic evidence linking prenatal and postnatal exposure to endocrine disrupting chemicals with male reproductive disorders: A systematic review and meta-analysis." *Hum Reprod Update* 23(1):104–125. https://doi.org/10.1093/humupd/dmw036.

Borhani N, Rajaei F, Salehi Z, Javadi A (2011) "Analysis of DNA fragmentation in mouse embryos exposed to an extremely low-frequency electromagnetic field." *Electromagn Biol Med* 30(4):246–252. https://doi.org/10.3109/15368378.2011.589556.

Bortkiewicz A (2019) "Health effects of radiofrequency electromagnetic fields (RF EMF)." *Ind Health* 57(4):403–405. https://doi.org/10.2486/indhealth.57_400.

Brown J, Daya S, Matson P (2016) "Day three versus day two embryo transfer following in vitro fertilization or intracytoplasmic sperm injection." *Cochrane Database Syst Rev* 12. https://doi.org/10.1002/14651858.CD004378.pub3.

Brugh VM, 3rd, Lipshultz LI (2004) "Male factor infertility: Evaluation and management." *Med Clin North Am* 88(2):367–385. https://doi.org/10.1016/s0025-7125(03)00150-0.

Bryant HE, Love EJ (1989) "Video display terminal use and spontaneous abortion risk." *Int J Epidemiol* 18(1):132–138. https://doi.org/10.1093/ije/18.1.132.

Burch JB, Reif JS, Noonan CW, Ichinose T, Bachand AM, et al (2002) "Melatonin metabolite excretion among cellular telephone users." *Int J Radiat Biol* 78(11):1029–1036. https://doi.org/10.1080/09553000210166561.

Burlaka A, Tsybulin O, Sidorik E, Lukin S, Polishuk V, et al (2013) "Overproduction of free radical species in embryonal cells exposed to low intensity radiofrequency radiation." *Exp Oncol* 35(3):219–225.

Cafe SL, Nixon B, Ecroyd H, Martin JH, Skerrett-Byrne DA, Bromfield EG (2021) "Proteostasis in the male and female germline: A new outlook on the maintenance of reproductive health." *Front Cell Dev Biol* 9. https://doi.org/10.3389/fcell.2021.660626.

Cairnie AB, Harding RK (1981) "Cytological studies in mouse testis irradiated with 2.45-GHz continuous-wave microwaves." *Radiat Res* 87(1):100–108. https://doi.org/10.2307/3575544.

Carlsen E, Giwercman A, Keiding N, Skakkebaek NE (1992) "Evidence for decreasing quality of semen during past 50 years." *Br Med J* 305(6854):609–613. https://doi.org/10.1136/bmj.305.6854.609.

Carolan M, Frankowska D (2011) "Advanced maternal age and adverse perinatal outcome: A review of the evidence." *Midwifery* 27(6):793–801. https://doi.org/10.1016/j.midw.2010.07.006.

Cecconi S, Gualtieri G, Di Bartolomeo A, Troiani G, Cifone MG, Canipari R (2000) "Evaluation of the effects of extremely low frequency electromagnetic fields on mammalian follicle development." *Hum Reprod* 15(11):2319–2325. https://doi.org/10.1093/humrep/15.11.2319.

Challis LJ (2005) "Mechanisms for interaction between RF fields and biological tissue." *Bioelectromagnetics* 26(S7):S98–S106. https://doi.org/10.1002/bem.20119.

Chambers GM, Paul RC, Harris K, Fitzgerald O, Boothroyd CV, et al (2017) "Assisted reproductive technology in Australia and New Zealand: Cumulative live birth rates as measures of success." *Med J Aust* 207(3):114–118. https://doi.org/10.5694/mja16.01435.

Chavdoula ED, Panagopoulos DJ, Margaritis LH (2010) "Comparison of biological effects between continuous and intermittent exposure to GSM-900-MHz mobile phone radiation: Detection of apoptotic cell-death features." *Mutat Res Genet Toxicol Environ Mutagen* 700(1):51–61. https://doi.org/10.1016/j.mrgentox.2010.05.008.

Chen M, Heilbronn LK (2017) "The health outcomes of human offspring conceived by assisted reproductive technologies (ART)." *J Dev Orig Health Dis* 8(4):388–402. https://doi.org/10.1017/s2040174417000228.

Chernoff N, Rogers JM, Kavet R (1992) "A review of the literature on potential reproductive and developmental toxicity of electric and magnetic fields." *Toxicology* 74(2):91–126. https://doi.org/10.1016/0300-483X(92)90132-X.

Collins J, Burrows E, Willan A (1993) "Infertile couples and their treatment in Canadian academic infertility clinics." *Royal Commission on New Reproductive Technologies. Treatment of Infertility: Current Practices and Psychosocial Implications* 10:233–329.

Creasey WA, Goldberg RB (2001) "A new twist on an old mechanism for EMF bioeffects?" *EMF Health Report* 9(2):1–11. https://www.emfsa.co.za/research-and-studies/creasey-wa-goldberg-rb-2001-a-new-twist-on-an-old-mechanism-for-emf/.

Dahal KP (2013) "Mobile communication and its adverse effects." *Himalayan Phys* 4(0):51–59. https://doi.org/10.3126/hj.v4i0.9429.

Dawe AS, Smith B, Thomas DW, Greedy S, Vasic N, et al (2006) "A small temperature rise may contribute towards the apparent induction by microwaves of heat-shock gene expression in the nematode Caenorhabditis elegans." *Bioelectromagnetics* 27(2):88–97. https://doi.org/10.1002/bem.20192.

Dawson BV, Robertson IG, Wilson WR, Zwi LJ, Boys JT, et al (1998) "Evaluation of potential health effects of 10 kHz magnetic fields: A rodent reproductive study." *Bioelectromagnetics* 19(3):162–171. https://doi.org/10.1002/(sici)1521-186x(1998):19:3<162::aid-bem4>3.0.co;2-#.

De Iuliis GN, Newey RJ, King BV, Aitken RJ (2009a) "Mobile phone radiation induces reactive oxygen species production and DNA damage in human spermatozoa in vitro." *PLOS ONE* 4(7):e6446. https://doi.org/10.1371/journal.pone.0006446.

De Iuliis GN, Thomson LK, Mitchell LA, Finnie JM, Koppers AJ, et al (2009b) "DNA damage in human spermatozoa is highly correlated with the efficiency of chromatin remodeling and the formation of 8-hydroxy-2'-deoxyguanosine, a marker of oxidative stress." *Biol Reprod* 81(3):517–524. https://doi.org/10.1095/biolreprod.109.076836.

Deepinder F, Makker K, Agarwal A (2007) "Cell phones and male infertility: Dissecting the relationship." *Reprod Biomed Online* 15(3):266–270. https://doi.org/10.1016/s1472-6483(10)60338-0.

Delimaris J, Tsilimigaki S, Messini-Nicolaki N, Ziros E, Piperakis SM (2006) "Effects of pulsed electric fields on DNA of human lymphocytes." *Cell Biol Toxicol* 22(6):409–415. https://doi.org/10.1007/s10565-006-0105-1.

Diem E, Schwarz C, Adlkofer F, Jahn O, Rüdiger H (2005) "Non-thermal DNA breakage by mobile-phone radiation (1800 MHz) in human fibroblasts and in transformed GFSH-R17 rat granulosa cells in vitro." *Mutat Res* 583(2):178–183. https://doi.org/10.1016/j.mrgentox.2005.03.006.

Divan HA, Kheifets L, Obel C, Olsen J (2008) "Prenatal and postnatal exposure to cell phone use and behavioral problems in children." *Epidemiology* 19(4):523–529. https://doi.org/10.1097/EDE.0b013e318175dd47.

Drevet JR (2006) "The antioxidant glutathione peroxidase family and spermatozoa: A complex story." *Mol Cell Endocrinol* 250(1–2):70–79. https://doi.org/10.1016/j.mce.2005.12.027.

Drevet JR, Aitken RJ (2020) "Oxidation of sperm nucleus in mammals: A physiological necessity to some extent with adverse impacts on oocyte and offspring." *Antioxidants (Basel, Switzerland)* 9(2):95. https://doi.org/10.3390/antiox9020095.

Dun M, Aitken R, Nixon B (2012) "The role of molecular chaperones in spermatogenesis and the post-testicular maturation of mammalian spermatozoa." *Hum Reprod Update* 18(4):420–435. https://doi.org/10.1093/humupd/dms009.

Dundar B, Cesur G, Comlekci S, Songur A, Gokcimen A, et al (2009) "The effect of the prenatal and postnatal long-term exposure to 50 Hz electric field on growth, pubertal development and IGF-1 levels in female Wistar rats." *Toxicol Ind Health* 25(7):479–487. https://doi.org/10.1177/0748233709345942.

Eisenberg ML, Li S, Behr B, Cullen MR, Galusha D, et al (2014) "Semen quality, infertility and mortality in the USA." *Hum Reprod* 29(7):1567–1574. https://doi.org/10.1093/humrep/deu106.

Eisenberg ML, Li S, Behr B, Pera RR, Cullen MR (2015a) "Relationship between semen production and medical comorbidity." *Fertil Steril* 103(1):66–71. https://doi.org/10.1016/j.fertnstert.2014.10.017.

Eisenberg ML, Li S, Brooks JD, Cullen MR, Baker LC (2015b) "Increased risk of cancer in infertile men: Analysis of U.S. claims data." *J Urol* 193(5):1596–1601. https://doi.org/10.1016/j.juro.2014.11.080.

El-Helaly M, Abu-Hashem E (2010) "Oxidative stress, melatonin level, and sleep insufficiency among electronic equipment repairers." *Indian J Occup Environ Med* 14(3):66–70. https://doi.org/10.4103/0019-5278.75692.

Elmussareh M, Mahrous A, Kayes O (2015) "Antioxidant therapy for male subfertility: Myth or evidence-based?" *Trends Urol Men's Health* 6(1). https://doi.org/10.1002/tre.439.

Erogul O, Oztas E, Yildirim I, Kir T, Aydur E, et al (2006) "Effects of electromagnetic radiation from a cellular phone on human sperm motility: An in vitro study." *Arch Med Res* 37(7):840–843. https://doi.org/10.1016/j.arcmed.2006.05.003.

Faddy MJ, Gosden RG, Gougeon A, Richardson SJ, Nelson JF (1992) "Accelerated disappearance of ovarian follicles in mid-life: Implications for forecasting menopause." *Hum Reprod* 7(10):1342–1346. https://doi.org/10.1093/oxfordjournals.humrep.a137570.

Falzone N, Huyser C, Fourie F, Toivo T, Leszczynski D, Franken D (2008) "In vitro effect of pulsed 900 MHz GSM radiation on mitochondrial membrane potential and motility of human spermatozoa." *Bioelectromagnetics* 29(4):268–276. https://doi.org/10.1002/bem.20390.

Falzone N, Huyser C, Becker P, Leszczynski D, Franken DR (2011) "The effect of pulsed 900-MHz GSM mobile phone radiation on the acrosome reaction, head morphometry and zona binding of human spermatozoa." *Int J Androl* 34(1):20–26. https://doi.org/10.1111/j.1365-2605.2010.01054.x.

Farquhar C, Marjoribanks J (2018) "Assisted reproductive technology: An overview of cochrane reviews." *Cochrane Database Syst Rev* 8(8):Cd010537. https://doi.org/10.1002/14651858.CD010537.pub5.

Fejes I, Závaczki Z, Szöllosi J, Koloszár S, Daru J, et al (2005) "Is there a relationship between cell phone use and semen quality?" *Arch Androl* 51(5):385–393. https://doi.org/10.1080/014850190924520.

Foster KR, Glaser R (2007) "Thermal mechanisms of interaction of radiofrequency energy with biological systems with relevance to exposure guidelines." *Health Phys* 92(6):609–620. https://doi.org/10.1097/01.Hp.0000262572.64418.38.

Fragouli E, Alfarawati S, Goodall N, Sánchez-García J, Colls P, Wells D (2011) "The cytogenetics of polar bodies: Insights into female meiosis and the diagnosis of aneuploidy." *Mol Hum Reprod* 17(5):286–295.

Friedman J, Kraus S, Hauptman Y, Schiff Y, Seger R (2007) "Mechanism of short-term ERK activation by electromagnetic fields at mobile phone frequencies." *Biochem J* 405(3):559–568. https://doi.org/10.1042/bj20061653.

Furtado-Filho OV, Borba JB, Dallegrave A, Pizzolato TM, Henriques JA, et al (2014) "Effect of 950 MHz UHF electromagnetic radiation on biomarkers of oxidative damage, metabolism of UFA and antioxidants in the livers of young rats of different ages." *Int J Radiat Biol* 90(2):159–168. https://doi.org/10.3109/09553002.2013.817697.

Gabriel S, Lau RW, Gabriel C (1996) "The dielectric properties of biological tissues: III. Parametric models for the dielectric spectrum of tissues." *Phys Med Biol* 41(11):2271–2293. https://doi.org/10.1088/0031-9155/41/11/003.

Gajda GB, McNamee JP, Thansandote A, Boonpanyarak S, Lemay E, Bellier PV (2002) "Cylindrical waveguide applicator for in vitro exposure of cell culture samples to 1.9-GHz radiofrequency fields." *Bioelectromagnetics* 23(8):592–598. https://doi.org/10.1002/bem.10055.

Gandhi OP, Kang G (2001) "Calculation of induced current densities for humans by magnetic fields from electronic article surveillance devices." *Phys Med Biol* 46(11):2759–2771. https://doi.org/10.1088/0031-9155/46/11/301.

Gandhi OP, Morgan LL, de Salles AA, Han YY, Herberman RB, Davis DL (2012) "Exposure limits: The underestimation of absorbed cell phone radiation, especially in children." *Electromagn Biol Med* 31(1):34–51. https://doi.org/10.3109/15368378.2011.622827.

George K, Kamath MS (2010) "Fertility and age." *J Hum Reprod Sci* 3(3):121–123. https://doi.org/10.4103/0974-1208.74152.

Gerner C, Haudek V, Schandl U, Bayer E, Gundacker N, et al (2010) "Increased protein synthesis by cells exposed to a 1800-MHz radio-frequency mobile phone electromagnetic field, detected by proteome profiling." *Int Arch Occup Environ Health* 83(6):691–702. https://doi.org/10.1007/s00420-010-0513-7.

Gervasi MG, Visconti PE (2017) "Molecular changes and signaling events occurring in spermatozoa during epididymal maturation." *Andrology* 5(2):204–218. https://doi.org/10.1111/andr.12320.

Ghanbari M, Mortazavi SB, Khavanin A, Khazaei M (2013) "The effects of cell phone waves (900 MHz-GSM band) on sperm parameters and total antioxidant capacity in rats." *Int J Fertil Steril* 7(1):21–28.

Glazer CH, Bonde JP, Eisenberg ML, Giwercman A, Hærvig KK, et al (2017) "Male infertility and risk of nonmalignant chronic diseases: A systematic review of the epidemiological evidence." *Semin Reprod Med* 35(3):282–290. https://doi.org/10.1055/s-0037-1603568.

Goldhaber MK, Polen MR, Hiatt RA (1988) "The risk of miscarriage and birth defects among women who use visual display terminals during pregnancy." *Am J Ind Med* 13(6):695–706. https://doi.org/10.1002/ajim.4700130608.

Goodman E, Greenebaum B, Marron M (1995) "Effects of electromagnetic fields on molecules and cells." *Int Rev Cytol* 158:279–338. https://doi.org/10.1016/S0074-7696(08):62489-4.

Gorpinchenko I, Nikitin O, Banyra O, Shulyak A (2014) "The influence of direct mobile phone radiation on sperm quality." *Cent Eur J Urol* 67(1):65–71. https://doi.org/10.5173/ceju.2014.01.art14.

Gul A, Celebi H, Uğraş S (2009) "The effects of microwave emitted by cellular phones on ovarian follicles in rats." *Arch Gynecol Obstet* 280(5):729–733. https://doi.org/10.1007/s00404-009-0972-9.

Güler G, Tomruk A, Ozgur E, Sahin D, Sepici A, et al (2012) "The effect of radiofrequency radiation on DNA and lipid damage in female and male infant rabbits." *Int J Radiat Biol* 88(4):367–373. https://doi.org/10.3109/09553002.2012.646349.

Guney M, Ozguner F, Oral B, Karahan N, Mungan T (2007) "900 MHz radiofrequency-induced histopathologic changes and oxidative stress in rat endometrium: Protection by vitamins E and C." *Toxicol Ind Health* 23(7):411–420. https://doi.org/10.1177/0748233707080906.

Gutschi T, Mohamad Al-Ali B, Shamloul R, Pummer K, Trummer H (2011) "Impact of cell phone use on men's semen parameters." *Andrologia* 43(5):312–316. https://doi.org/10.1111/j.1439-0272.2011.01075.x.

Gye MC, Park CJ (2012) "Effect of electromagnetic field exposure on the reproductive system." *Clin Exp Reprod Med* 39(1):1–9. https://doi.org/10.5653/cerm.2012.39.1.1.

Hardell L (2017) "World Health Organization, radiofrequency radiation and health - A hard nut to crack (review)." *Int J Oncol* 51(2):405–413. https://doi.org/10.3892/ijo.2017.4046.

Harremoes P, Gee D, MacGarvin M, Stirling A, Keys J, et al (Eds.). (2013) *The precautionary principle in the 20th century: Late lessons from early warnings*. London: Routledge.

Hassold T, Chiu D (1985) "Maternal age-specific rates of numerical chromosome abnormalities with special reference to trisomy." *Hum Genet* 70(1):11–17. https://doi.org/10.1007/bf00389450.

Henshaw DL, O'Carroll MJ (2009) *Scientific committee on emerging and newly identified health risks (SCENIHR)*. Brussels: European Commission.

Hirsh A (2003) "Male subfertility." *BMJ* 327(7416):669–672. https://doi.org/10.1136/bmj.327.7416.669.

Hou Q, Wang M, Wu S, Ma X, An G, et al (2015) "Oxidative changes and apoptosis induced by 1800-MHz electromagnetic radiation in NIH/3T3 cells." *Electromagn Biol Med* 34(1):85–92. https://doi.org/10.3109/15368378.2014.900507.

Houston BJ, Nixon B, King BV, Aitken RJ, De Iuliis GN (2018a) "Probing the origins of 1,800 MHz radio frequency electromagnetic radiation induced damage in mouse immortalized germ cells and spermatozoa in vitro." *Front Public Health* 6. https://doi.org/10.3389/fpubh.2018.00270.

Houston BJ, Nixon B, Martin JH, De Iuliis GN, Trigg NA, et al (2018b) "Heat exposure induces oxidative stress and DNA damage in the male germ line." *Biol Reprod* 98(4):593–606. https://doi.org/10.1093/biolre/ioy009.

Houston BJ, Nixon B, McEwan KE, Martin JH, King BV, et al (2019) "Whole-body exposures to radiofrequency-electromagnetic energy can cause DNA damage in mouse spermatozoa via an oxidative mechanism." *Sci Rep* 9(1):17478. https://doi.org/10.1038/s41598-019-53983-9.

Hsueh AJ, Kawamura K, Cheng Y, Fauser BC (2015) "Intraovarian control of early folliculogenesis." *Endocr Rev* 36(1):1–24. https://doi.org/10.1210/er.2014-1020.

Hunt S, Vollenhoven B (2020) "Assessment of female fertility in the general practice setting." *Aust J Gen Pract* 49(6):304–308.

Hur SSJ, Cropley JE, Suter CM (2017) "Paternal epigenetic programming: Evolving metabolic disease risk." *J Mol Endocrinol* 58(3):R159–R168. https://doi.org/10.1530/jme-16-0236.

Hutcheon K, McLaughlin EA, Stanger SJ, Bernstein IR, Dun MD, et al (2017) "Analysis of the small non-protein-coding RNA profile of mouse spermatozoa reveals specific enrichment of piRNAs within mature spermatozoa." *RNA Biol* 14(12):1776–1790. https://doi.org/10.1080/15476286.2017.1356569.

Hyland GJ (2008) "Physical basis of adverse and therapeutic effects of low intensity microwave radiation." *Indian J Exp Biol* 46(5):403–419.

ICNIRP (1998) "Guidelines for limiting exposure to time-varying electric, magnetic, and electromagnetic fields (up to 300 GHz)." *Health Phys* 74(4):494–521.

ICNIRP (2009) "Statement on the 'Guidelines for limiting exposure to time-varying electric, magnetic, and electromagnetic fields (up to 300 GHz)'." *Health Phys* 97(3):257–258.

ICNIRP (2020) "Guidelines for limiting exposure to electromagnetic fields (100 kHz to 300 GHz)." *Health Phys* 118(5):483–524.

Ilacqua A, Izzo G, Emerenziani GP, Baldari C, Aversa A (2018) "Lifestyle and fertility: The influence of stress and quality of life on male fertility." *Reprod Biol Endocrinol* 16(1):115. https://doi.org/10.1186/s12958-018-0436-9.

Imai N, Kawabe M, Hikage T, Nojima T, Takahashi S, Shirai T (2011) "Effects on rat testis of 1.95-GHz W-CDMA for IMT-2000 cellular phones." *Syst Biol Reprod Med* 57(4):204–209. https://doi.org/10.3109/19396368.2010.544839.

Iommiello VM, Albani E, Di Rosa A, Marras A, Menduni F, et al (2015) "Ejaculate oxidative stress is related with sperm DNA fragmentation and round cells." *Int J Endocrinol* 2015:321901. https://doi.org/10.1155/2015/321901.

Iwasaki A, Gagnon C (1992) "Formation of reactive oxygen species in spermatozoa of infertile patients." *Fertil Steril* 57(2):409–416. https://doi.org/10.1016/s0015-0282(16):54855-9.

Iyoke CA, Ugwu GO, Ezugwu FO, Ajah LO, Mba SG (2013) "The role of ultrasonography in in-vitro fertilization and embryo transfer (IVF-ET)." *Niger J Med* 22(3):162–170.

Jensen TK, Jacobsen R, Christensen K, Nielsen NC, Bostofte E (2009) "Good semen quality and life expectancy: A cohort study of 43,277 men." *Am J Epidemiol* 170(5):559–565. https://doi.org/10.1093/aje/kwp168.

Jones R, Mann T, Sherins R (1979) "Peroxidative breakdown of phospholipids in human spermatozoa, spermicidal properties of fatty acid peroxides, and protective action of seminal plasma*." *Fertil Steril* 31(5):531–537. https://doi.org/10.1016/S0015-0282(16):43999-3.

Jung KA, Ahn HS, Lee YS, Gye MC (2007) "Effect of a 20 kHz sawtooth magnetic field exposure on the estrous cycle in mice." *J Microbiol Biotechnol* 17(3):398–402.

Jungwirth A, Diemer T, Kopa Z, Krausz C, Tournaye H (2015) *European association of urology (EAU) guidelines on male infertility*. Arnhem: European Association of Urology.

Kahya M, Nazıroğlu M, Çiğ B (2014) "Selenium reduces mobile phone (900 MHz)-induced oxidative stress, mitochondrial function, and apoptosis in breast cancer cells." *Biol Trace Elem Res* 160(2). https://doi.org/10.1007/s12011-014-0032-6.

Kalra SK, Ratcliffe SJ, Coutifaris C, Molinaro T, Barnhart KT (2011) "Ovarian stimulation and low birth weight in newborns conceived through in vitro fertilization." *Obstet Gynecol* 118(4):863–871. https://doi.org/10.1097/AOG.0b013e31822be65f.

Karipidis K, Elwood M, Benke G, Sanagou M, Tjong L, Croft RJ (2018) "Mobile phone use and incidence of brain tumour histological types, grading or anatomical location: A population-based ecological study." *BMJ, (Open)* 8(12):e024489. https://doi.org/10.1136/bmjopen-2018-024489.

Karipidis K, Mate R, Urban D, Tinker R, Wood A (2021) "5G mobile networks and health-a state-of-the-science review of the research into low-level RF fields above 6 GHz." *J Expo Sci Environ Epidemiol* 31(4):585–605. https://doi.org/10.1038/s41370-021-00297-6.

Katen AL, Sipilä P, Mitchell LA, Stanger SJ, Nixon B, Roman SD (2017) "Epididymal CYP2E1 plays a critical role in acrylamide-induced DNA damage in spermatozoa and paternally mediated embryonic resorptions†." *Biol Reprod* 96(4):921–935. https://doi.org/10.1093/biolre/iox021.

Kazama M, Hino A (2012) "Sea urchin spermatozoa generate at least two reactive oxygen species; the type of reactive oxygen species changes under different conditions." *Mol Reprod Dev* 79(4):283–295. https://doi.org/10.1002/mrd.22025.

Kesari K, Kumar S, Behari J (2011) "Effects of radiofrequency electromagnetic wave exposure from cellular phones on the reproductive pattern in male Wistar rats." *Appl Biochem Biotechnol* 164(4):546–559. https://doi.org/10.1007/s12010-010-9156-0.

Kesari KK, Agarwal A, Henkel R (2018) "Radiations and male fertility." *Reprod Biol Endocrinol* 16(1):118. https://doi.org/10.1186/s12958-018-0431-1.

Kilgallon SJ, Simmons LW (2005) "Image content influences men's semen quality." *Biol Lett* 1(3):253–255. https://doi.org/10.1098/rsbl.2005.0324.

Kirk M, Smurthwaite K, Braunig J, Trevenar S, D'Este C, et al (2018) "The PFAS health study: Systematic literature review", The Australian National University, Canberra.

Kıvrak EG, Yurt KK, Kaplan AA, Alkan I, Altun G (2017) "Effects of electromagnetic fields exposure on the antioxidant defense system." *J Microsc Ultrastruct* 5(4):167–176. https://doi.org/10.1016/j.jmau.2017.07.003.

Konc J, Kanyó K, Kriston R, Somoskői B, Cseh S (2014) "Cryopreservation of embryos and oocytes in human assisted reproduction." *BioMed Res Int* 2014:307268. https://doi.org/10.1155/2014/307268.

Koppers AJ, De Iuliis GN, Finnie JM, McLaughlin EA, Aitken RJ (2008) "Significance of mitochondrial reactive oxygen species in the generation of oxidative stress in spermatozoa." *J Clin Endocrinol Metab* 93(8):3199–3207. https://doi.org/10.1210/jc.2007-2616.

Kostoff RN, Heroux P, Aschner M, Tsatsakis A (2020) "Adverse health effects of 5G mobile networking technology under real-life conditions." *Toxicol Lett* 323:35–40. https://doi.org/10.1016/j.toxlet.2020.01.020.

Kothari S, Thompson A, Agarwal A, du Plessis SS (2010) "Free radicals: Their beneficial and detrimental effects on sperm function." *Indian J Exp Biol* 48(5):425–435.

Koyu A, Ozguner F, Yilmaz H, Uz E, Cesur G, Ozcelik N (2009) "The protective effect of caffeic acid phenethyl ester (CAPE) on oxidative stress in rat liver exposed to the 900 MHz electromagnetic field." *Toxicol Ind Health* 25(6):429–434. https://doi.org/10.1177/0748233709106821.

Krausz C (2011) "Male infertility: Pathogenesis and clinical diagnosis." *Best Pract Res Clin Endocrinol Metab* 25(2):271–285. https://doi.org/10.1016/j.beem.2010.08.006.

Kuliev A, Zlatopolsky Z, Kirillova I, Spivakova J, Cieslak Janzen J (2011) "Meiosis errors in over 20,000 oocytes studied in the practice of preimplantation aneuploidy testing." *Reprod Biomed Online* 22(1):2–8. https://doi.org/10.1016/j.rbmo.2010.08.014.

Kumar N, Singh AK (2015) "Trends of male factor infertility, an important cause of infertility: A review of literature." *J Hum Reprod Sci* 8(4):191–196. https://doi.org/10.4103/0974-1208.170370.

Kumar S, Behari J, Sisodia R (2013) "Influence of electromagnetic fields on reproductive system of male rats." *Int J Radiat Biol* 89(3):147–154. https://doi.org/10.3109/09553002.2013.741282.

Kumar S, Nirala JP, Behari J, Paulraj R (2014) "Effect of electromagnetic irradiation produced by 3G mobile phone on male rat reproductive system in a simulated scenario." *Indian J Exp Biol* 52(9):890–897.

La Vignera S, Condorelli RA, Vicari E, D'Agata R, Calogero AE (2012) "Effects of the exposure to mobile phones on male reproduction: A review of the literature." *J Androl* 33(3):350–356. https://doi.org/10.2164/jandrol.111.014373.

Lai H, Singh NP (2004) "Magnetic-field-induced DNA strand breaks in brain cells of the rat." *Environ Health Perspect* 112(6):687–694. https://doi.org/10.1289/ehp.6355.

Lee HJ, Pack JK, Kim TH, Kim N, Choi SY, et al (2010) "The lack of histological changes of CDMA cellular phone-based radio frequency on rat testis." *Bioelectromagnetics* 31(7):528–534. https://doi.org/10.1002/bem.20589.

Lee KM, Ward MH, Han S, Ahn HS, Kang HJ, et al (2009) "Paternal smoking, genetic polymorphisms in CYP1A1 and childhood leukemia risk." *Leuk Res* 33(2):250–258. https://doi.org/10.1016/j.leukres.2008.06.031.

Leiva RA, Bouchard TP, Abdullah SH, Ecochard R (2017) "Urinary luteinizing hormone tests: Which concentration threshold best predicts ovulation?" *Front Public Health* 5. https://doi.org/10.3389/fpubh.2017.00320.

Leng L (2016) "The relationship between mobile phone use and risk of brain tumor: A systematic review and meta-analysis of trails in the last decade." *Chin Neurosurg Jl* 2(1):38. https://doi.org/10.1186/s41016-016-0059-y.

Lenzi A, Gandini L, Maresca V, Rago R, Sgrò P, et al (2000) "Fatty acid composition of spermatozoa and immature germ cells." *Mol Hum Reprod* 6(3):226–231. https://doi.org/10.1093/molehr/6.3.226.

Levine H, Jørgensen N, Martino-Andrade A, Mendiola J, Weksler-Derri D, et al (2017) "Temporal trends in sperm count: A systematic review and meta-regression analysis." *Hum Reprod Update* 23(6):646–659. https://doi.org/10.1093/humupd/dmx022.

Lewis RC, Mínguez-Alarcón L, Meeker JD, Williams PL, Mezei G, et al (2017) "Self-reported mobile phone use and semen parameters among men from a fertility clinic." *Reprod Toxicol (Elmsford, N.Y.)* 67:42–47. https://doi.org/10.1016/j.reprotox.2016.11.008.

Lin YY, Wu T, Liu JY, Gao P, Li KC, et al (2017) "1950 MHz radio frequency electromagnetic radiation inhibits testosterone secretion of mouse Leydig cells." *Int J Environ Res Public Health* 15(1). https://doi.org/10.3390/ijerph15010017.

Liu C, Duan W, Xu S, Chen C, He M, et al (2013a) "Exposure to 1800 MHz radiofrequency electromagnetic radiation induces oxidative DNA base damage in a mouse spermatocyte-derived cell line." *Toxicol Lett* 218(1):2–9. https://doi.org/10.1016/j.toxlet.2013.01.003.

Liu C, Gao P, Xu SC, Wang Y, Chen CH, et al (2013b) "Mobile phone radiation induces mode-dependent DNA damage in a mouse spermatocyte-derived cell line: A protective role of melatonin." *Int J Radiat Biol* 89(11):993–1001. https://doi.org/10.3109/09553002.2013.811309.

Liu Q, Si T, Xu X, Liang F, Wang L, Pan S (2015) "Electromagnetic radiation at 900 MHz induces sperm apoptosis through bcl-2, bax and caspase-3 signaling pathways in rats." *Reprod Health* 12(1):65. https://doi.org/10.1186/s12978-015-0062-3.

Liu WM, Pang RT, Chiu PC, Wong BP, Lao K, et al (2012) "Sperm-borne microRNA-34c is required for the first cleavage division in mouse." *Proc Natl Acad Sci U S A* 109(2):490–494. https://doi.org/10.1073/pnas.1110368109.

Lixia S, Yao K, Kaijun W, Deqiang L, Huajun H, et al (2006) "Effects of 1.8 GHz radiofrequency field on DNA damage and expression of heat shock protein 70 in human lens epithelial cells." *Mutat Res* 602(1–2):135–142. https://doi.org/10.1016/j.mrfmmm.2006.08.010.

Lord T, Aitken RJ (2013) "Oxidative stress and ageing of the post-ovulatory oocyte." *Reproduction* 146(6):R217–R227. https://doi.org/10.1530/rep-13-0111.

Lord T, Nixon B, Jones KT, Aitken RJ (2013) "Melatonin prevents postovulatory oocyte aging in the mouse and extends the window for optimal fertilization in Vitro1." *Biol Reprod* 88(3). https://doi.org/10.1095/biolreprod.112.106450.

Lord T, Aitken RJ (2015) "Fertilization stimulates 8-hydroxy-2'-deoxyguanosine repair and antioxidant activity to prevent mutagenesis in the embryo." *Dev Biol* 406(1):1–13. https://doi.org/10.1016/j.ydbio.2015.07.024.

Lord T, Martin JH, Aitken RJ (2015) "Accumulation of electrophilic aldehydes during postovulatory aging of mouse oocytes causes reduced fertility, oxidative stress, and apoptosis." *Biol Reprod* 92(2):33. https://doi.org/10.1095/biolreprod.114.122820.

Lotti F, Maggi M (2015) "Ultrasound of the male genital tract in relation to male reproductive health." *Hum Reprod Update* 21(1):56–83. https://doi.org/10.1093/humupd/dmu042.

Lu YS, Huang BT, Huang YX (2012) "Reactive oxygen species formation and apoptosis in human peripheral blood mononuclear cell induced by 900 MHz mobile phone radiation." *Oxid Med Cell Longev* 2012:740280. https://doi.org/10.1155/2012/740280.

Madjar HM (2016) "Human radio frequency exposure limits: An update of reference levels in Europe, USA, Canada, China, Japan and Korea." *2016 International Symposium on Electromagnetic Compatibility - EMC EUROPE* 2016:467–473.

Magras IN, Xenos TD (1997) "RF radiation-induced changes in the prenatal development of mice." *Bioelectromagnetics* 18(6):455–461. https://doi.org/10.1002/(sici)1521-186x(1997):18:6<455::aid-bem8>3.0.co;2-1.

Maher ER, Brueton LA, Bowdin SC, Luharia A, Cooper W, et al (2003) "Beckwith-Wiedemann syndrome and assisted reproduction technology (ART)." *J Med Genet* 40(1):62–64. https://doi.org/10.1136/jmg.40.1.62.

Mailankot M, Kunnath AP, Jayalekshmi H, Koduru B, Valsalan R (2009) "Radio frequency electromagnetic radiation (RF-EMR) from GSM (0.9/1.8 GHz) mobile phones induces oxidative stress and reduces sperm motility in rats." *Clin (Sao Paulo, Brazil)* 64(6):561–565. https://doi.org/10.1590/s1807-59322009000600011.

Manta AK, Papadopoulou D, Polyzos AP, Fragopoulou AF, Skouroliakou AS, et al (2017) "Mobile-phone radiation-induced perturbation of gene-expression profiling, redox equilibrium and sporadic-apoptosis control in the ovary of Drosophila melanogaster." *Fly (Austin)* 11(2):75–95. https://doi.org/10.1080/19336934.2016.1270487.

Marchionni I, Paffi A, Pellegrino M, Liberti M, Apollonio F, et al (2006) "Comparison between low-level 50 Hz and 900 MHz electromagnetic stimulation on single channel ionic currents and on firing frequency in dorsal root ganglion isolated neurons." *Biochim Biophys Acta* 1758(5):597–605. https://doi.org/10.1016/j.bbamem.2006.03.014.

Markovà E, Malmgren LOG, Belyaev IY (2010) "Microwaves from mobile phones inhibit 53BP1 focus formation in human stem cells more strongly Than in differentiated cells: Possible mechanistic link to cancer risk." *Environ Health Perspect* 118(3):394–399. https://doi.org/10.1289/ehp.0900781.

Martin JH, Bromfield EG, Aitken RJ, Lord T, Nixon B (2018) "Double strand break DNA repair occurs via non-homologous end-joining in mouse MII oocytes." *Sci Rep* 8(1):9685. https://doi.org/10.1038/s41598-018-27892-2.

Martin JH, Aitken RJ, Bromfield EG, Nixon B (2019) "DNA damage and repair in the female germline: Contributions to ART." *Hum Reprod Update* 25(2):180–201. https://doi.org/10.1093/humupd/dmy040.

Martino CF, Castello PR (2011) "Modulation of hydrogen peroxide production in cellular systems by low level magnetic fields." *PLOS ONE* 6(8):e22753. https://doi.org/10.1371/journal.pone.0022753.

Matwee C, Kamaruddin M, Betts DH, Basrur PK, King WA (2001) "The effects of antibodies to heat shock protein 70 in fertilization and embryo development." *Mol Hum Reprod* 7(9):829–837. https://doi.org/10.1093/molehr/7.9.829.

Meena R, Kumari K, Kumar J, Rajamani P, Verma HN, Kesari KK (2014) "Therapeutic approaches of melatonin in microwave radiations-induced oxidative stress-mediated toxicity on male fertility pattern of Wistar rats." *Electromagn Biol Med* 33(2):81–91. https://doi.org/10.3109/15368378.2013.781035.

Merchant R, Gandhi G, Allahbadia GN (2011) "In vitro fertilization/intracytoplasmic sperm injection for male infertility." *Indian J Urol* 27(1):121–132. https://doi.org/10.4103/0970-1591.78430.

Mieusset R, Grandjean H, Mansat A, Pontonnier F (1985) "Inhibiting effect of artificial cryptorchidism on spermatogenesis." *Fertil Steril* 43(4):589–594. https://doi.org/10.1016/s0015-0282(16):48502-x.

Mieusset R, Bujan L (1994) "The potential of mild testicular heating as a safe, effective and reversible contraceptive method for men." *Int J Androl* 17(4):186–191. https://doi.org/10.1111/j.1365-2605.1994.tb01241.x.

Mihai CT, Rotinberg P, Brinza F, Vochita G (2014) "Extremely low-frequency electromagnetic fields cause DNA strand breaks in normal cells." *J Environ Health Sci Eng* 12(1):15. https://doi.org/10.1186/2052-336x-12-15.

Mihalas BP, De Iuliis GN, Redgrove KA, McLaughlin EA, Nixon B (2017a) "The lipid peroxidation product 4-hydroxynonenal contributes to oxidative stress-mediated deterioration of the ageing oocyte." *Sci Rep* 7(1):6247. https://doi.org/10.1038/s41598-017-06372-z.

Mihalas BP, Redgrove KA, McLaughlin EA, Nixon B (2017b) "Molecular mechanisms responsible for increased vulnerability of the ageing oocyte to oxidative damage." *Oxid Med Cell Longev* 2017:4015874. https://doi.org/10.1155/2017/4015874.

Mihalas BP, Bromfield EG, Sutherland JM, De Iuliis GN, McLaughlin EA, et al (2018) "Oxidative damage in naturally aged mouse oocytes is exacerbated by dysregulation of proteasomal activity." *J Biol Chem* 293(49):18944–18964. https://doi.org/10.1074/jbc.RA118.005751.

Miller AB, Morgan LL, Udasin I, Davis DL (2018) "Cancer epidemiology update, following the 2011 IARC evaluation of radiofrequency electromagnetic fields (Monograph 102)." *Environ Res* 167:673–683. https://doi.org/10.1016/j.envres.2018.06.043.

Miller AB, Sears ME, Morgan LL, Davis DL, Hardell L, et al (2019) "Risks to health and well-being from radio-frequency radiation emitted by cell phones and other wireless devices." *Front Public Health* 7:223–223. https://doi.org/10.3389/fpubh.2019.00223.

Miller JWB, Torday JS (2019) "Reappraising the exteriorization of the mammalian testes through evolutionary physiology." *Communicat Integr Biol* 12(1):38–54. https://doi.org/10.1080/19420889.2019.1586047.

Mortazavi S, Parsanezhad M, Kazempour M, Ghahramani P, Mortazavi A, Davari M (2013a) "Male reproductive health under threat: Short term exposure to radiofrequency radiations emitted by common mobile jammers." *J Hum Reprod Sci* 6(2):124–128. https://doi.org/10.4103/0974-1208.117178.

Mortazavi SM, Shirazi KR, Mortazavi G (2013b) "The study of the effects of ionizing and non-ionizing radiations on birth weight of newborns to exposed mothers." *J Nat Sci Biol Med* 4(1):213–217. https://doi.org/10.4103/0976-9668.107293.

Murphy MP (2009) "How mitochondria produce reactive oxygen species." *Biochem J* 417(1):1–13. https://doi.org/10.1042/bj20081386.

Naeem Z (2014) "Health risks associated with mobile phones use." *Int J Health Sci (Qassim)* 8(4):V–VI.

Nagy ZP, Nel-Themaat L, Chang CC, Shapiro DB, Berna DP (2014) "Cryopreservation of eggs." *Methods Mol Biol* 1154:439–454. https://doi.org/10.1007/978-1-4939-0659-8_20.

Narayanan SN, Kumar RS, Karun KM, Nayak SB, Bhat PG (2015) "Possible cause for altered spatial cognition of prepubescent rats exposed to chronic radiofrequency electromagnetic radiation." *Metab Brain Dis* 30(5):1193–1206. https://doi.org/10.1007/s11011-015-9689-6.

Nasr-Esfahani MH, Deemeh MR, Tavalaee M (2012) "New era in sperm selection for ICSI." *Int J Androl* 35(4):475–484. https://doi.org/10.1111/j.1365-2605.2011.01227.x.

Negi P, Singh R (2021) "Association between reproductive health and nonionizing radiation exposure." *Electromagn Biol Med* 40(1):92–102. https://doi.org/10.1080/15368378.2021.1874973.

Neufeld E, Kuster N (2018) "Systematic derivation of safety limits for time-varying 5G radiofrequency exposure based on analytical models and thermal dose." *Health Phys* 115(6):705–711. https://doi.org/10.1097/hp.0000000000000930.

Newman JE, Paul RC, Chambers GM (2021) *Assisted reproductive technology in Australia and New Zealand 2019*. Sydney: National Perinatal Epidemiology and Statistics Unit, the University of New South Wales, Sydney.

Nikolova T, Czyz J, Rolletschek A, Blyszczuk P, Fuchs J, et al (2005) "Electromagnetic fields affect transcript levels of apoptosis-related genes in embryonic stem cell-derived neural progenitor cells." *FASEB J* 19(12):1686–1688. https://doi.org/10.1096/fj.04-3549fje.

Nixon B, De Iuliis GN, Dun MD, Zhou W, Trigg NA, Eamens AL (2019) "Profiling of epididymal small non-protein-coding RNAs." *Andrology* 7(5):669–680. https://doi.org/10.1111/andr.12640.

Noblanc A, Peltier M, Damon-Soubeyrand C, Kerchkove N, Chabory E, et al (2012) "Epididymis response partly compensates for spermatozoa oxidative defects in snGPx4 and GPx5 double mutant mice." *PLOS ONE* 7(6):e38565. https://doi.org/10.1371/journal.pone.0038565.

Novakovic B, Lewis S, Halliday J, Kennedy J, Burgner DP, et al (2019) "Assisted reproductive technologies are associated with limited epigenetic variation at birth that largely resolves by adulthood." *Nat Commun* 10(1):3922. https://doi.org/10.1038/s41467-019-11929-9.

Nowicka-Bauer K, Nixon B (2020) "Molecular changes induced by oxidative stress that impair human sperm motility." *Antioxidants (Basel, Switzerland)* 9(2):134. https://doi.org/10.3390/antiox9020134.

Ochsendorf FR, Thiele J, Fuchs J, Schüttau H, Freisleben HJ, et al (1994) "Chemiluminescence in semen of infertile men." *Andrologia* 26(5):289–293. https://doi.org/10.1111/j.1439-0272.1994.tb00804.x.

O'Flaherty C (2019) "Orchestrating the antioxidant defenses in the epididymis." *Andrology* 7(5):662–668. https://doi.org/10.1111/andr.12630.

O'Flaherty C (2020) "Reactive oxygen species and male fertility." *Antioxidants* 9(4):287.

Ohmori S, Yamao Y, Nakajima N (2001) "The future generations of mobile communications based on broadband access methods." *Wirel Pers Commun* 17(2–3):175–190. https://doi.org/10.1023/a:1011248901400.

Okechukwu CE (2020) "Does the use of mobile phone affect male fertility? A mini-review." *J Hum Reprod Sci* 13(3):174–183. https://doi.org/10.4103/jhrs.JHRS_126_19.

Oral B, Guney M, Ozguner F, Karahan N, Mungan T, et al (2006) "Endometrial apoptosis induced by a 900-MHz mobile phone: Preventive effects of vitamins E and C." *Adv Ther* 23(6):957–973. https://doi.org/10.1007/BF02850217.

Oscar KJ, Hawkins TD (1977) "Microwave alteration of the blood-brain barrier system of rats." *Brain Res* 126(2):281–293. https://doi.org/10.1016/0006-8993(77):90726-0.

Ozguner F, Altinbas A, Ozaydin M, Dogan A, Vural H, et al (2005) "Mobile phone-induced myocardial oxidative stress: Protection by a novel antioxidant agent caffeic acid phenethyl ester." *Toxicol Ind Health* 21(9):223–230. https://doi.org/10.1191/0748233705th228oa.

Ozguner M, Koyu A, Cesur G, Ural M, Ozguner F, et al (2005) "Biological and morphological effects on the reproductive organ of rats after exposure to electromagnetic field." *Saudi Med J* 26(3):405–410.

Pall ML (2013) "Electromagnetic fields act via activation of voltage-gated calcium channels to produce beneficial or adverse effects." *J Cell Mol Med* 17(8):958–965. https://doi.org/10.1111/jcmm.12088.

Pall ML (2015) "Scientific evidence contradicts findings and assumptions of Canadian safety panel 6: Microwaves act through voltage-gated calcium channel activation to induce biological impacts at non-thermal levels, supporting a paradigm shift for microwave/lower frequency electromagnetic field action." *Rev Environ Health* 30(2):99–116. https://doi.org/10.1515/reveh-2015-0001.

Pall ML (2018) "Wi-fi is an important threat to human health." *Environ Res* 164:405–416. https://doi.org/10.1016/j.envres.2018.01.035.

Panagopoulos DJ, Messini N, Karabarbounis A, Philippetis AL, Margaritis LH (2000) "A mechanism for action of oscillating electric fields on cells." *Biochem Biophys Res Commun* 272(3):634–640. https://doi.org/10.1006/bbrc.2000.2746.

Panagopoulos DJ, Karabarbounis A, Margaritis LH (2002) "Mechanism for action of electromagnetic fields on cells." *Biochem Biophys Res Commun* 298(1):95–102. https://doi.org/10.1016/s0006-291x(02):02393-8.

Panagopoulos DJ, Karabarbounis A, Margaritis LH (2004) "Effect of GSM 900-MHz mobile phone radiation on the reproductive capacity of Drosophila melanogaster." *Electromagn Biol Med* 23(1):29–43. https://doi.org/10.1081/JBC-120039350.

Panagopoulos DJ, Chavdoula ED, Nezis IP, Margaritis LH (2007a) "Cell death induced by GSM 900-MHz and DCS 1800-MHz mobile telephony radiation." *Mutat Res* 626(1–2):69–78. https://doi.org/10.1016/j.mrgentox.2006.08.008.

Panagopoulos DJ, Chavdoula ED, Karabarbounis A, Margaritis LH (2007b) "Comparison of bioactivity between GSM 900 MHz and DCS 1800 MHz mobile telephony radiation." *Electromagn Biol Med* 26(1):33–44. https://doi.org/10.1080/15368370701205644.

Panagopoulos DJ, Chavdoula ED, Margaritis LH (2010) "Bioeffects of mobile telephony radiation in relation to its intensity or distance from the antenna." *Int J Radiat Biol* 86(5):345–357. https://doi.org/10.3109/09553000903567961.

Panagopoulos D, Margaritis L (2010a) "The identification of an intensity "window" on the bioeffects of mobile telephony radiation." *Int J Radiat Biol* 86(5):358–366. https://doi.org/10.3109/09553000903567979.

Panagopoulos DJ, Margaritis LH (2010b) "The effect of exposure duration on the biological activity of mobile telephony radiation." *Mutat Res Genet Toxicol Environ Mutagen* 699(1–2):17–22. https://doi.org/10.1016/j.mrgentox.2010.04.010.

Panagopoulos DJ, Karabarbounis A, Lioliousis C (2013a) "ELF alternating magnetic field decreases reproduction by DNA damage induction." *Cell Biochem Biophys* 67(2):703–716. https://doi.org/10.1007/s12013-013-9560-5.

Panagopoulos DJ, Johansson O, Carlo GL (2013b) "Evaluation of specific absorption rate as a dosimetric quantity for electromagnetic fields bioeffects." *PLOS ONE* 8(6):e62663. https://doi.org/10.1371/journal.pone.0062663.

Panagopoulos DJ, Johansson O, Carlo GL (2015) "Polarization: A key difference between man-made and natural electromagnetic fields, in regard to biological activity." *Sci Rep* 5(1):14914. https://doi.org/10.1038/srep14914.

Panagopoulos DJ (2019) "Comparing DNA damage induced by mobile telephony and other types of man-made electromagnetic fields." *Mutat Res Rev Mutat Res* 781:53–62.

Panagopoulos DJ, Karabarbounis A, Yakymenko I, Chrousos GP (2021) "Human-made electromagnetic fields: Ion forced-oscillation and voltage-gated ion channel dysfunction, oxidative stress and DNA damage" (review). *Int J Oncol* 59(5):92. https://doi.org/10.3892/ijo.2021.5272.

Pandey N, Giri S (2018) "Melatonin attenuates radiofrequency radiation (900 MHz)-induced oxidative stress, DNA damage and cell cycle arrest in germ cells of male swiss albino mice." *Toxicol Ind Health* 34(5):315–327. https://doi.org/10.1177/0748233718758092.

Pedersen GF (1997) "Amplitude modulated RF fields stemming from a GSM/DCS-1800 phone." *Wirel Netw* 3(6):489–498. https://doi.org/10.1023/A:1019158712657.

Perry SW, Norman JP, Barbieri J, Brown EB, Gelbard HA (2011) "Mitochondrial membrane potential probes and the proton gradient: A practical usage guide." *BioTechniques* 50(2):98–115. https://doi.org/10.2144/000113610.

Pourlis AF (2009) "Reproductive and developmental effects of EMF in vertebrate animal models." *Pathophysiology* 16(2–3):179–189. https://doi.org/10.1016/j.pathophys.2009.01.010.

Prasad M, Kathuria P, Nair P, Kumar A, Prasad K (2017) "Mobile phone use and risk of brain tumours: A systematic review of association between study quality, source of funding, and research outcomes." *Neurol Sci* 38(5):797–810. https://doi.org/10.1007/s10072-017-2850-8.

Pu X, Wang Z, Klaunig JE (2015) "Alkaline comet assay for assessing DNA damage in individual cells." *Curr Protoc Toxicol* 65(1):3.12.1–3.12.11. https://doi.org/10.1002/0471140856.tx0312s65.

Quinlan CL, Perevoshchikova IV, Hey-Mogensen M, Orr AL, Brand MD (2013) "Sites of reactive oxygen species generation by mitochondria oxidizing different substrates." *Redox Biol* 1(1):304–312. https://doi.org/10.1016/j.redox.2013.04.005.

Rago R, Salacone P, Caponecchia L, Sebastianelli A, Marcucci I, et al (2013) "The semen quality of the mobile phone users." *J Endocrinol Invest* 36(11):970–974. https://doi.org/10.3275/8996.

Redmayne M, Johansson O (2015) "Radiofrequency exposure in young and old: Different sensitivities in light of age-relevant natural differences." *Rev Environ Health* 30(4):323–335. https://doi.org/10.1515/reveh-2015-0030.

Redmayne M (2016) "International policy and advisory response regarding children's exposure to radio frequency electromagnetic fields (RF-EMF)." *Electromagn Biol Med* 35(2):176–185. https://doi.org/10.3109/15368378.2015.1038832.

Reilly JN, McLaughlin EA, Stanger SJ, Anderson AL, Hutcheon K, *et al* (2016) "Characterisation of mouse epididymosomes reveals a complex profile of microRNAs and a potential mechanism for modification of the sperm epigenome." *Sci Rep* 6:31794. https://doi.org/10.1038/srep31794.

Remondini D, Nylund R, Reivinen J, Poulletier de Gannes F, Veyret B, *et al* (2006) "Gene expression changes in human cells after exposure to mobile phone microwaves." *Proteomics* 6(17):4745–4754. https://doi.org/10.1002/pmic.200500896.

Rodgers AB, Morgan CP, Leu NA, Bale TL (2015) "Transgenerational epigenetic programming via sperm microRNA recapitulates effects of paternal stress." *Proc Natl Acad Sci U S A* 112(44):13699–13704. https://doi.org/10.1073/pnas.1508347112.

Rodriguez M, Petitclerc D, Burchard JF, Nguyen DH, Block E, Downey BR (2003) "Responses of the estrous cycle in dairy cows exposed to electric and magnetic fields (60 Hz) during 8-h photoperiods." *Anim Reprod Sci* 77(1–2):11–20. https://doi.org/10.1016/s0378-4320(02):00273-7.

Roozbeh N, Abdi F, Amraee A, Atarodi Kashani Z, Darvish L (2018) "Influence of radiofrequency electromagnetic fields on the fertility system: Protocol for a systematic review and meta-analysis." *JMIR Res Protoc* 7(2):e33. https://doi.org/10.2196/resprot.9102.

Ruder EH, Hartman TJ, Blumberg J, Goldman MB (2008) "Oxidative stress and antioxidants: Exposure and impact on female fertility." *Hum Reprod Update* 14(4):345–357. https://doi.org/10.1093/humupd/dmn011.

Sadeghi MR (2015) "Unexplained infertility, the controversial matter in management of infertile couples." *J Reprod Infertil* 16(1):1–2.

Saikhun J, Kitiyanant Y, Vanadurongwan V, Pavasuthipaisit K (1998) "Effects of sauna on sperm movement characteristics of normal men measured by computer-assisted sperm analysis." *Int J Androl* 21(6):358–363. https://doi.org/10.1046/j.1365-2605.1998.00138.x.

Sajeda S, Al-Watter Y (2011) "Effect of mobile phone usage on semen analysis in infertile men." *Tikrit J Pharm Sci* 7(1):77–82.

Sakkas D, Ramalingam M, Garrido N, Barratt CL (2015) "Sperm selection in natural conception: What can we learn from mother nature to improve assisted reproduction outcomes?" *Hum Reprod Update* 21(6):711–726. https://doi.org/10.1093/humupd/dmv042.

Salford LG, Brun A, Sturesson K, Eberhardt JL, Persson BRR (1994) "Permeability of the blood-brain barrier induced by 915 MHz electromagnetic radiation, continuous wave and modulated at 8, 16, 50, and 200 Hz." *Microsc Res Tech* 27(6):535–542. https://doi.org/10.1002/jemt.1070270608.

Samaras T, Leitgeb N, Auvinen A, Danker-Hopfe H, Hansson Mild K, *et al* (2015) *SCENIHR (scientific committee on emerging and newly identified health risks), potential health effects of exposure to electromagnetic fields (EMF), 27 January, 2015.*

Sánchez-Calabuig MJ, López-Cardona AP, Fernández-González R, Ramos-Ibeas P, Fonseca Balvís N, *et al* (2014) "Potential health risks associated to ICSI: Insights from animal models and strategies for a safe procedure." *Front Public Health* 2. https://doi.org/10.3389/fpubh.2014.00241.

Sangun O, Dundar B, Darici H, Comlekci S, Doguc DK, Celik S (2015) "The effects of long-term exposure to a 2450 MHz electromagnetic field on growth and pubertal development in female Wistar rats." *Electromagn Biol Med* 34(1):63–71. https://doi.org/10.3109/15368378.2013.871619.

Sarkar S, Ali S, Behari J (1994) "Effect of low power microwave on the mouse genome: A direct DNA analysis." *Mutat Res* 320(1–2):141–147. https://doi.org/10.1016/0165-1218(94):90066-3.

Sarookhani M, Asiabanha M, Safari A, Zaroushani V, Ziaeiha M (2011) "The influence of 950 MHz magnetic field (mobile phone radiation) on sex organ and adrenal functions of male rabbits." *Afr J Biochem Res* 5:65–68.

Saygin M, Caliskan S, Karahan N, Koyu A, Gumral N, Uguz A (2011) "Testicular apoptosis and histopathological changes induced by a 2.45 GHz electromagnetic field." *Toxicol Ind Health* 27(5):455–463. https://doi.org/10.1177/0748233710389851.

Scientific Committee on Emerging Newly Identified Health Risks (2015) "Opinion on potential health effects of exposure to electromagnetic fields." *Bioelectromagnetics* 36(6):480–484. https://doi.org/10.1002/bem.21930.

Sengupta P, Dutta S, Krajewska-Kulak E (2017) "The disappearing sperms: Analysis of reports published between 1980 and 2015." *Am J Mens Health* 11(4):1279–1304. https://doi.org/10.1177/1557988316643383.

Sepehrimanesh M, Kazemipour N, Saeb M, Nazifi S, Davis DL (2017) "Proteomic analysis of continuous 900-MHz radiofrequency electromagnetic field exposure in testicular tissue: A rat model of human cell phone exposure." *Environ Sci Pollut Res Int* 24(15):13666–13673. https://doi.org/10.1007/s11356-017-8882-z.

Serbecic N, Beutelspacher SC (2005) "Anti-oxidative vitamins prevent lipid-peroxidation and apoptosis in corneal endothelial cells." *Cell Tissue Res* 320(3):465–475. https://doi.org/10.1007/s00441-004-1030-3.

Sergerie M, Mieusset R, Croute F, Daudin M, Bujan L (2007) "High risk of temporary alteration of semen parameters after recent acute febrile illness." *Fertil Steril* 88(4):970.e1–970.e7. https://doi.org/10.1016/j.fertnstert.2006.12.045.

Shahin S, Singh VP, Shukla RK, Dhawan A, Gangwar RK, et al (2013) "2.45 GHz microwave irradiation-induced oxidative stress affects implantation or pregnancy in mice, Mus musculus." *Appl Biochem Biotechnol* 169(5):1727–1751. https://doi.org/10.1007/s12010-012-0079-9.

Shahin S, Singh SP, Chaturvedi CM (2017) "Mobile phone (1800 MHz) radiation impairs female reproduction in mice, Mus musculus, through stress induced inhibition of ovarian and uterine activity." *Reprod Toxicol* 73:41–60. https://doi.org/10.1016/j.reprotox.2017.08.001.

Shapiro BM (1991) "The control of oxidant stress at fertilization." *Science* 252(5005):533–536. https://doi.org/10.1126/science.1850548.

Sharlip ID, Jarow JP, Belker AM, Lipshultz LI, Sigman M, et al (2002) "Best practice policies for male infertility." *Fertil Steril* 77(5):873–882. https://doi.org/10.1016/s0015-0282(02):03105-9.

Sharma U, Conine CC, Shea JM, Boskovic A, Derr AG, et al (2016) "Biogenesis and function of tRNA fragments during sperm maturation and fertilization in mammals." *Science* 351(6271):391–396. https://doi.org/10.1126/science.aad6780.

Sharma VP, Kumar NR (2010) "Changes in honeybee behaviour and biology under the influence of cellphone radiations." *Curr Sci* 98(10):1376–1378.

Shekarriz M, Thomas AJ, Jr., Agarwal A (1995) "Incidence and level of seminal reactive oxygen species in normal men." *Urology* 45(1):103–107. https://doi.org/10.1016/s0090-4295(95):97088-6.

Simkó M, Mattsson M-O (2019) "5G wireless communication and health effects-A pragmatic review based on available studies Regarding 6 to 100 GHz." *Int J Environ Res Public Health* 16(18):3406. https://doi.org/10.3390/ijerph16183406.

Simon L, Zini A, Dyachenko A, Ciampi A, Carrell DT (2017) "A systematic review and meta-analysis to determine the effect of sperm DNA damage on in vitro fertilization and intracytoplasmic sperm injection outcome." *Asian J Androl* 19(1):80–90. https://doi.org/10.4103/1008-682x.182822.

Simopoulou M, Sfakianoudis K, Bakas P, Giannelou P, Papapetrou C, et al (2018) "Postponing pregnancy through oocyte cryopreservation for social reasons: Considerations Regarding clinical practice and the socio-psychological and bioethical issues involved." *Med (Kaunas, Lithuania)* 54(5):76. https://doi.org/10.3390/medicina54050076.

Singh R, Nath R, Mathur AK, Sharma RS (2018) "Effect of radiofrequency radiation on reproductive health." *Indian J Med Res* 148(Suppl):S92–s99. https://doi.org/10.4103/ijmr.IJMR_1056_18.

Singh VP, Singh P, Chaturvedi CM, Shukla RK, Dhawan A, et al (2009) "2.45 GHz low level CW microwave radiation affects embryo implantation sites and single strand DNA damage in brain cells of mice, mus musculus." *2009 international conference on emerging trends in electronic and photonic devices & systems*, 22–24 December 2009.

Skakkebaek NE, Rajpert-De Meyts E, Main KM (2001) "Testicular dysgenesis syndrome: An increasingly common developmental disorder with environmental aspects." *Hum Reprod* 16(5):972–978. https://doi.org/10.1093/humrep/16.5.972.

Smith S, Pfeifer SM, Collins JA (2003) "Diagnosis and management of female infertility." *JAMA* 290(13):1767–1770. https://doi.org/10.1001/jama.290.13.1767.

Solek P, Majchrowicz L, Koziorowski M (2018) "Aloe arborescens juice prevents EMF-induced oxidative stress and thus protects from pathophysiology in the male reproductive system in vitro." *Environ Res* 166:141–149. https://doi.org/10.1016/j.envres.2018.05.035.

Somer RA, Surapaneni R, Beach DF (2012) "Relationship of testicular cancer incidence and cellular phone use." *J Clin Oncol* 30(15):e12008–e12008. https://doi.org/10.1200/jco.2012.30.15_suppl.e12008.

Spandorfer SD, Avrech OM, Colombero LT, Palermo GD, Rosenwaks Z (1998) "Effect of parental age on fertilization and pregnancy characteristics in couples treated by intracytoplasmic sperm injection." *Hum Reprod* 13(2):334–338. https://doi.org/10.1093/humrep/13.2.334.

Spira A (1988) "The decline of fecundity with age." *Maturitas Suppl* 1(1):15–22. https://doi.org/10.1016/0378-5122(88):90004-7.

Starkey SJ (2016) "Inaccurate official assessment of radiofrequency safety by the advisory group on non-ionising radiation." *Rev Environ Health* 31(4):493–503. https://doi.org/10.1515/reveh-2016-0060.

Stefanovic M, Panic SR, de Souza RAA, Reig J (2017) "Recent advances in RF propagation modeling for 5G systems." *Int J Antennas Propag* 2017:4701208. https://doi.org/10.1155/2017/4701208.

Stewart TA, Davis FM (2019) "An element for development: Calcium signaling in mammalian reproduction and development." *Biochim Biophys Acta (BBA) - Mol Cell Res* 1866(7):1230–1238. https://doi.org/10.1016/j.bbamcr.2019.02.016.

Storey BT (2008) "Mammalian sperm metabolism: Oxygen and sugar, friend and foe." *Int J Dev Biol* 52(5–6):427–437. https://doi.org/10.1387/ijdb.072522bs.

Su HW, Yi YC, Wei TY, Chang TC, Cheng CM (2017) "Detection of ovulation, a review of currently available methods." *Bioeng Transl Med* 2(3):238–246. https://doi.org/10.1002/btm2.10058.

Tabong PT-N, Adongo PB (2013) "Infertility and childlessness: A qualitative study of the experiences of infertile couples in Northern Ghana." *BMC Preg Childbirth* 13(1):72. https://doi.org/10.1186/1471-2393-13-72.

Tan D-X, Manchester LC, Esteban-Zubero E, Zhou Z, Reiter RJ (2015) "Melatonin as a potent and inducible endogenous antioxidant: Synthesis and metabolism." *Molecules (Basel, Switzerland)* 20(10):18886–18906. https://doi.org/10.3390/molecules201018886.

te Velde ER, Pearson PL (2002) "The variability of female reproductive ageing." *Hum Reprod Update* 8(2):141–154. https://doi.org/10.1093/humupd/8.2.141.

Tilly JL, Tilly KI (1995) "Inhibitors of oxidative stress mimic the ability of follicle-stimulating hormone to suppress apoptosis in cultured rat ovarian follicles." *Endocrinology* 136(1):242–252. https://doi.org/10.1210/endo.136.1.7828537.

Tkalec M, Stambuk A, Srut M, Malarić K, Klobučar GI (2013) "Oxidative and genotoxic effects of 900 MHz electromagnetic fields in the earthworm Eisenia fetida." *Ecotoxicol Environ Saf* 90:7–12. https://doi.org/10.1016/j.ecoenv.2012.12.005.

Tomruk A, Guler G, Dincel AS (2010) "The influence of 1800 MHz GSM-like signals on hepatic oxidative DNA and lipid damage in nonpregnant, pregnant, and newly born rabbits." *Cell Biochem Biophys* 56(1):39–47. https://doi.org/10.1007/s12013-009-9068-1.

Tournaye H (2012) "Male factor infertility and ART." *Asian J Androl* 14(1):103–108. https://doi.org/10.1038/aja.2011.65.

Tremellen K (2008) "Oxidative stress and male infertility—A clinical perspective." *Hum Reprod Update* 14(3):243–258. https://doi.org/10.1093/humupd/dmn004.

Trigg NA, Eamens AL, Nixon B (2019) "The contribution of epididymosomes to the sperm small RNA profile." *Reproduction* 157(6):R209–R223. https://doi.org/10.1530/rep-18-0480.

Trigg NA, Skerrett-Byrne DA, Xavier MJ, Zhou W, Anderson AL, et al (2021) "Acrylamide modulates the mouse epididymal proteome to drive alterations in the sperm small non-coding RNA profile and dysregulate embryo development." *Cell Rep* 37(1):109787. https://doi.org/10.1016/j.celrep.2021.109787.

Tsarna E, Reedijk M, Birks LE, Guxens M, Ballester F, et al (2019) "Associations of maternal cell-phone use during pregnancy with pregnancy duration and fetal growth in 4 birth cohorts." *Am J Epidemiol* 188(7):1270–1280. https://doi.org/10.1093/aje/kwz092.

Tumkaya L, Kalkan Y, Bas O, Yilmaz A (2016) "Mobile phone radiation during pubertal development has no effect on testicular histology in rats." *Toxicol Ind Health* 32(2):328–336. https://doi.org/10.1177/0748233713500820.

Turrens JF (2003) "Mitochondrial formation of reactive oxygen species." *J Physiol* 552(2):335–344. https://doi.org/10.1113/jphysiol.2003.049478.

Valko M, Rhodes CJ, Moncol J, Izakovic M, Mazur M (2006) "Free radicals, metals and antioxidants in oxidative stress-induced cancer." *Chem Biol Interact* 160(1):1–40. https://doi.org/10.1016/j.cbi.2005.12.009.

Van Voorhis BJ (2006) "Outcomes from assisted reproductive technology." *Obstet Gynecol* 107(1):183–200. https://doi.org/10.1097/01.AOG.0000194207.06554.5b.

Vander Borght M, Wyns C (2018) "Fertility and infertility: Definition and epidemiology." *Clin Biochem* 62:2–10. https://doi.org/10.1016/j.clinbiochem.2018.03.012.

Velizarov S, Raskmark P, Kwee S, (1999): "The effects of radiofrequency fields on cell proliferation are non-thermal", *Bioelectrochemistry and Bioenergetics*, 48, 177–180.

Vernet P, Aitken RJ, Drevet JR (2004) "Antioxidant strategies in the epididymis." *Mol Cell Endocrinol* 216(1–2):31–39. https://doi.org/10.1016/j.mce.2003.10.069.

Wagner H, Cheng JW, Ko EY (2018) "Role of reactive oxygen species in male infertility: An updated review of literature." *Arab J Urol* 16(1):35–43. https://doi.org/10.1016/j.aju.2017.11.001.

Wang C, Machaty Z (2013) "Calcium influx in mammalian eggs." *Reproduction* 145(4):R97–R105. https://doi.org/10.1530/rep-12-0496.

Wang J, Sauer MV (2006) "In vitro fertilization (IVF) a review of 3 decades of clinical innovation and technological advancement." *Ther Clin Risk Manag* 2(4):355–364. https://doi.org/10.2147/tcrm.2006.2.4.355.

Wang S, He G, Chen M, Zuo T, Xu W, Liu X (2017) "The role of antioxidant enzymes in the ovaries." *Oxid Med Cell Longev* 2017:4371714. https://doi.org/10.1155/2017/4371714.

Wdowiak A, Wdowiak L, Wiktor H (2007) "Evaluation of the effect of using mobile phones on male fertility." *Ann Agric Environ Med* 14(1):169–172.

Weisbrot D, Lin H, Ye L, Blank M, Goodman R (2003) "Effects of mobile phone radiation on reproduction and development in Drosophila melanogaster." *J Cell Biochem* 89(1):48–55.

Wong JL, Wessel GM (2005) "Reactive oxygen species and Udx1 during early sea urchin development." *Dev Biol* 288(2):317–333. https://doi.org/10.1016/j.ydbio.2005.07.004.

Xiao M, Zhong H, Xia L, Tao Y, Yin H (2017) "Pathophysiology of mitochondrial lipid oxidation: Role of 4-hydroxynonenal (4-HNE) and other bioactive lipids in mitochondria." *Free Radic Biol Med* 111:316–327. https://doi.org/10.1016/j.freeradbiomed.2017.04.363.

Yakymenko I, Tsybulin O, Sidorik E, Henshel D, Kyrylenko O, Kyrylenko S (2016) "Oxidative mechanisms of biological activity of low-intensity radiofrequency radiation." *Electromagn Biol Med* 35(2):186–202. https://doi.org/10.3109/15368378.2015.1043557.

Yamashita H, Hata K, Yamaguchi H, Tsurita G, Wake K, et al (2010) "Short-term exposure to a 1439-MHz TDMA signal exerts no estrogenic effect in rats." *Bioelectromagnetics* 31(7):573–575. https://doi.org/10.1002/bem.20593.

Yan JG, Agresti M, Bruce T, Yan YH, Granlund A, Matloub HS (2007) "Effects of cellular phone emissions on sperm motility in rats." *Fertil Steril* 88(4):957–964. https://doi.org/10.1016/j.fertnstert.2006.12.022.

Yao K, Wu W, Wang K, Ni S, Ye P, et al (2008) "Electromagnetic noise inhibits radiofrequency radiation-induced DNA damage and reactive oxygen species increase in human lens epithelial cells." *Mol Vis* 14:964–969.

Yüksel M, Nazıroğlu M, Özkaya MO (2016) "Long-term exposure to electromagnetic radiation from mobile phones and wi-fi devices decreases plasma prolactin, progesterone, and estrogen levels but increases uterine oxidative stress in pregnant rats and their offspring." *Endocrine* 52(2):352–362. https://doi.org/10.1007/s12020-015-0795-3.

Zalata A, El-Samanoudy AZ, Shaalan D, El-Baiomy Y, Mostafa T (2015) "In vitro effect of cell phone radiation on motility, DNA fragmentation and clusterin gene expression in human sperm." *Int J Fertil Steril* 9(1):129–136. https://doi.org/10.22074/ijfs.2015.4217.

Zegers-Hochschild F, Adamson GD, de Mouzon J, Ishihara O, Mansour R, et al (2009) "International committee for monitoring assisted reproductive technology (ICMART) and the World Health Organization (WHO) revised glossary of ART terminology, 2009." *Fertil Steril* 92(5):1520–1524. https://doi.org/10.1016/j.fertnstert.2009.09.009.

Zhang G, Yan H, Chen Q, Liu K, Ling X, et al (2016) "Effects of cell phone use on semen parameters: Results from the MARHCS cohort study in Chongqing, China." *Environ Int* 91:116–121. https://doi.org/10.1016/j.envint.2016.02.028.

Zini A, de Lamirande E, Gagnon C (1993) "Reactive oxygen species in semen of infertile patients: Levels of superoxide dismutase- and catalase-like activities in seminal plasma and spermatozoa." *Int J Androl* 16(3):183–188.

Zmyslony M, Palus J, Jajte J, Dziubaltowska E, Rajkowska E (2000) "DNA damage in rat lymphocytes treated in vitro with iron cations and exposed to 7 mT magnetic fields (static or 50 Hz)." *Mutat Res Fundam Mol Mech Mutagen* 453(1):89–96. https://doi.org/10.1016/S0027-5107(00):00094-4.

Zwamborn APM, Vossen SHJ, van Leersum BJA, Ouwens MA, Makel WN (2003) *Effects of global communication system radio-frequency fields on well being and cognitive functions of human subjects with and without subjective complaints*. FEL-03-C148. The Hague, the Netherlands: TNO Physics and Electronics Laboratory. (Available in http: //home.tiscali.be/milieugezondheid/ dossiers/gsm/TNO_rapport_Nederland_sept_2003.pdf.)

7 Effects of Wireless Communication Electromagnetic Fields on Human and Animal Brain Activity

Haitham S. Mohammed

Biophysics Department, Faculty of Science, Cairo University, Giza, Egypt

CONTENTS

Abstract	276
7.1 Introduction	276
7.1.1 Electroencephalography (EEG) Signals	276
7.1.2 Clinical Significance of EEG	279
7.1.3 Wireless Communication Technology, Electromagnetic Fields, and the Brain	280
7.2 WC EMFs and Human Brain Activity	283
7.2.1 EEG Changes in Healthy Subjects Induced by WC EMF Exposure	283
7.2.2 EEG Changes in Diseased Subjects Induced by WC EMF Exposure	286
7.3 WC EMFs and Animal Brain Activity	287
7.4 Discussion	288
References	291

Keywords: electromagnetic fields, electromagnetic radiation, wireless technology, wireless communication, brain activity, electroencephalography, radio frequency, extremely low frequency, sleep.

Abbreviations: BCI: brain–computer interface. CNS: central nervous system. CT: computed tomography. DECT: Digitally Enhanced Cordless Telecommunications. EEG: electroencephalography. ELF: Extremely Low Frequency. EMF: electromagnetic field. EMR: electromagnetic radiation. ERD: event-related desynchronization. ERS: event-related synchronization. GABA: γ-aminobutyric acid. GSM: Global System for Mobile Telecommunications. LTE: Long-Term Evolution. MRI: magnetic resonance imaging. MT: mobile telephony. NREM sleep: non-rapid eye movement sleep. PET: positron emission tomography. SWS: slow-wave sleep. UMTS: Universal Mobile Telecommunications System. REM sleep: rapid eye movement sleep. RF: Radio Frequency. ULF: Ultra Low Frequency. VGCC: voltage-gated calcium channel. VGIC: voltage-gated ion channel. WC: wireless communications. 1G, 2G, 3G, 4G, 5G: first, second, third, fourth, fifth generation of MT/WC systems.

DOI: 10.1201/9781003201052-10

ABSTRACT

The wide and increasing use of telecommunication equipment has necessitated the study of its effects on biological systems and, in particular, on brain activity. Due to the electrical nature of communication between neuronal cells in the brain, the effects of anthropogenic electromagnetic fields (EMFs) and corresponding electromagnetic radiation (EMR) on the human and animal brain have become the focus of many studies. Electroencephalography (EEG) as a direct and sensitive tool for monitoring brain functional changes can be implemented to decipher these effects. Pulsation and modulation of the wireless communication (WC) electromagnetic signals at low frequencies produce complex radiation patterns with components in the Radio Frequency (RF)/microwave and the Extremely Low Frequency (ELF) bands. This mixed type of EMFs/EMR we call wireless communication EMFs/EMR (WC EMFs/EMR). Increasing experimental and theoretical evidence emphasizes the crucial role of the ELF signal pulsation/modulation in the effects of WC EMFs/EMR on human and animal EEG, even at intensities well below the officially accepted limits for human exposure. The duration of exposure is an additional important parameter for the induced effects. The vast majority of recorded effects of WC EMFs/EMR on the human/animal brain are not accompanied by any significant heating, and thus, they are categorized as non-thermal effects. This chapter highlights the concepts related to the human and animal EEG and its alterations induced by anthropogenic EMFs and especially WC EMFs/EMR. Effects on wake and sleep human and animal EEG are described. The importance of animal studies is discussed, and the need for methodological standardization in experimental studies is emphasized. Proposed mechanisms for the action of anthropogenic EMFs on brain activity are reviewed. More studies investigating the non-thermal effects of WC EMFs/EMR on the human and animal brain are needed in order to further explore the effects, the interaction mechanisms, and the consequences of anthropogenic EMFs on health and wellbeing.

7.1 INTRODUCTION

7.1.1 Electroencephalography (EEG) Signals

During neuronal activity, the movement of mobile ions (basically Na^+ and K^+) across the neuronal membrane generates an electric current. This current creates an electric field in the active cell's extracellular space. Neurons (neural cells), as information transmission and processing units, use electrical impulses and chemical signals to transmit information and communicate with each other. Typically, neurons have two types of extensions called axons and dendrites (Figure 7.1)

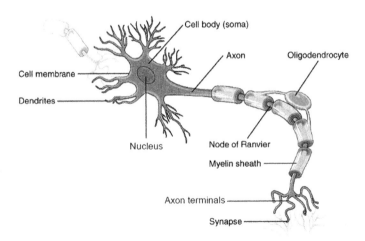

FIGURE 7.1 The basic components of a neural cell (neuron).

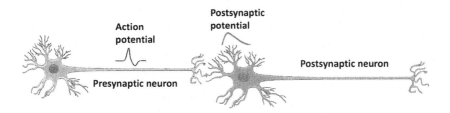

FIGURE 7.2 Action potential and post-synaptic potential in a neuronal synapse.

that originate from their cell bodies. Dendrites, with their multiple branches, act as receiving inputs to the neuron from the other connected neurons, whereas the axon transmits the electrical impulse to the neuronal terminals to release neurotransmitters. Different neurons possess different neurotransmitters that, upon release in the synapse, cause either excitation or inhibition to the post-synaptic neuron.

During impulse propagation along the neuronal axon, the voltage-gated ion channels (VGICs) are recruited; however, after the release of neurotransmitters in the synaptic cleft (i.e., space between neurons), the ligand-gated ion channels are involved in initiating the corresponding electrical signals in the post-synaptic cell. In neuronal tissue, there are two types of electrical potentials (Figure 7.2): the action potential, which is generated within the cell and lasts only a few milliseconds (usually 1–2 ms), and the post-synaptic potential, which is generated between neuronal cells as a result of neurotransmitter release in the synaptic cleft and lasts for a longer time (>15–20 ms).

Because of the relatively long duration of the post-synaptic potentials (both excitatory and inhibitory), they can be spatially and temporally summed, becoming the primary contributors to the final electroencephalography (EEG) signal that can be recorded from the human/animal scalp.

EEG signals recorded by metal electrodes attached at specific positions on the scalp according to the International 10–20 system (Figure 7.3) originate primarily in the cortex's pyramidal neuronal cells. These cells are vertically oriented to the surface of the skull, and their activities are summed providing the final EEG signals (Figure 7.4). The polarity of the surface potential is determined by the location of the synaptic activity within these cortical neurons. When the afferent excitatory

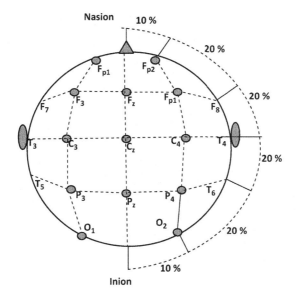

FIGURE 7.3 Locations for electrode attachment on the human scalp according to the International 10–20 system for EEG recordings.

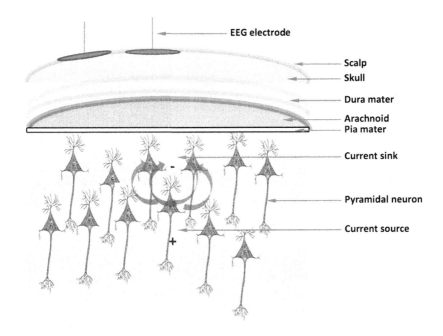

FIGURE 7.4 Diagrammatic drawing that explains the origin of EEG signals. Pyramidal neurons are the main contributors to EEG. The synchronized post-synaptic potentials of these vertically aligned neurons are the main sources of EEG signals recorded at the scalp.

synaptic activity occurs at the apical part of the cortical neuron's dendrites (closer to the scalp), the intracellular flow of ionic current due to the neuronal membrane depolarization will be downward, forming a current sink in the extracellular space (Figure 7.4) (due to the flow of Na^+/K^+ ions from the extracellular space through the ion channels on the neuronal membrane), while at the distal part of the dendrites, the ionic current will flow out of the neuron (current source) and move extracellularly in the opposite direction (upward). Then the deflection of the electrical potential recorded by the EEG on the scalp will be upward (positive). However, when the afferent synaptic activity occurs at the distal part of the neuron's dendrites (in regard to the scalp), it will induce the reverse process (downward extracellular current), causing a negative deflection in the electrical potential recorded by the scalp EEG (Figure 7.4).

EEG signals, in addition to being modulated by various evoking stimuli from the sensory system are of a spontaneous nature. The intrinsic oscillatory electrical activity of the neuron involved in the genesis of EEG rhythms, as well as the continuous subcortical activity directed toward the cortex, account for these spontaneous rhythms.

In general, the amplitude and frequency of EEG signals are affected by the mental state and activity in both healthy and diseased conditions. In comparison to other diagnostic methods [e.g., magnetic resonance imaging (MRI)], quantitative analysis of EEG signals provides a diagnostic tool that is unique in its temporal resolution. Normal brain waves are categorized by increasing frequency into five frequency bands (delta, theta, alpha, beta, and gamma), with frequencies ranging from 0.1 to 100 Hz and amplitude in the micro-Volt (μV) range due to the great attenuation that takes place in the tissues between the signal origin and the recording electrodes (Figure 7.5).

Apart from the fact that the frequency bands are conserved in both human and animal brain activity, similarities and differences are found between the human and animal EEG. The slow delta waves (0.1–4 Hz) that characterize deep or non-rapid eye movement (NREM) sleep generated mainly by the oscillations in thalamocortical neurons are similar in both humans and animals (Sullivan et al. 2014). However, the theta rhythm which originates in the hippocampus and is associated with rapid eye movement (REM) sleep and locomotion in rodents (6–10 Hz) is observed at a

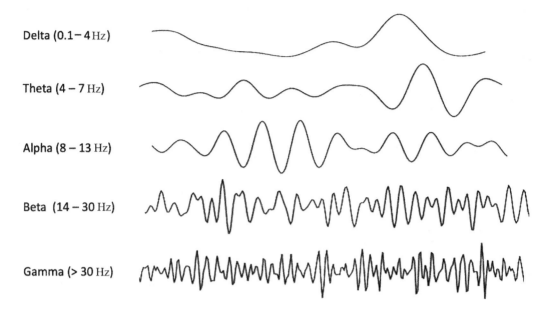

FIGURE 7.5 Normal human EEG rhythms (amplitude versus time) classified into different frequency bands.

lower frequency (4–7 Hz) in humans (Mitchell et al. 2008). Zhang et al. (2019), in a cross-species investigation on the resting state EEG, reported certain EEG differences between humans and rats attributed to the different sensory sensitivities among the species.

EEG, as a diagnostic tool, has several advantages that other diagnostic methods do not have. Such advantages are non-invasiveness, safety, and low cost compared to other medical imaging techniques such as positron emission tomography (PET), computed tomography (CT), and MRI. Moreover, EEG has additional advantages, such as real-time monitoring of brain activity with high temporal resolution (milliseconds). As a result, EEG recording and analysis are used in a wide range of applications, including brain disease diagnosis and prognosis as well as the brain-computer interface (BCI) in which the recorded EEG signals are decoded and analyzed through a computerized system in an attempt to facilitate/restore the movement of paralyzed limbs (Carmena et al. 2003).

7.1.2 Clinical Significance of EEG

The lower frequency brain activities recorded by EEG in the range of 0.1–4 Hz are called delta waves (Figure 7.5). Studies have revealed the electrophysiological background of several distinct activities in this frequency range during sleep as well as in the waking state (Harmony 2013; Knyazev 2012). In the context of the neuro-cognitive processes, these waves have been correlated with a specific cognitive function: the memory (Hickey and Race 2021). Delta waves are the dominant frequencies characterizing slow-wave sleep (SWS), a state of deep sleep.

The hallmark of the other type of sleep, called REM sleep, is the theta oscillations (4–7 Hz) (Figure 7.5), which originate in the hippocampus and are very well characterized in rodents. These waves have been recently associated with learning and memory in humans (Herweg et al. 2020).

The alpha waves are in the frequency range of 8–13 Hz (Figure 7.5). They correspond to cognitive signals and are related to attention, memory, and cognitive processing (Minarik et al. 2018). Their highest amplitude/power is recorded in the occipital, temporal, and parietal areas of the brain during the eye-closed waking state (Niedermeyer and Da Silva 2005). However, they are greatly attenuated when eyes are open. The alpha rhythm has been suggested to be involved in the active inhibition of task-irrelevant processes (when irrelevant stimuli are presented during the active processing of a targeted task) and in the binding processes of working memory and perception, in which differentiation/categorization between stimuli is taking place. The alpha band is considered the most sensitive in showing effects induced by EMF exposure (Klimesch et al. 2007; Palva and Palva 2007).

It is worth mentioning that the sensitivity of the EEG and especially of the alpha waves to EMFs/EMR could be linked to the atmospheric electromagnetic oscillations called Schumann resonances (Schuman and Koenig 1954) with a fundamental frequency at 7.83 Hz which overlap with the electrical potentials produced by the human and animal brain, especially in the alpha band. Saroka and Persinger (2014) provided quantitative evidence through mathematical solutions and measured data for a direct relationship between Schuman resonances and human cerebral cortical activity.

With regard to higher frequencies, beta and gamma rhythms (Figure 7.5) are recorded in cortical areas. The beta oscillations are associated with the cortical output from pyramidal neurons of deeper layers and can be distinguished into beta-1 waves (14–20 Hz) and beta-2 waves (20–30 Hz). The beta frequencies in animals and humans are related mainly to motor functions and alertness. The gamma frequency band refers to frequency components greater than 30 Hz and has been linked to the so-called "feature binding hypothesis", which is related to perception (Roelfsema et al. 1997).

7.1.3 Wireless Communication Technology, Electromagnetic Fields, and the Brain

Due to technological advancements, especially in wireless communications (WC) during the past few decades, humans are exposed to anthropogenic electromagnetic fields (EMFs) and corresponding electromagnetic radiation (EMR) constantly during their lives unlike previous generations. Our forefathers did not voluntarily expose their heads to the radiation emitted by mobile phones, or cordless domestic phones [called Digitally Enhanced Cordless Telecommunications (DECT) phones] and did not live in the presence of these and other WC devices such as Wi-Fi (Wireless Fidelity) for wireless connection to the Internet or Bluetooth for wireless connection among electronic devices. All these types of WC EMFs/EMR combine Radio Frequency (RF) carrier waves (300 kHz–300 GHz) which are modulated and pulsed by Extremely Low Frequency (ELF) signals (3–3000 Hz) and on/off pulsations (Pedersen 1997; Hyland 2000; 2008; Huber et al. 2000; 2002; 2005; Panagopoulos 2019). Actually, most, if not all, types of anthropogenic EMFs/EMR, that are simply called "RF", even those that do not carry information in the form of modulation (e.g., from radars), in reality, contain pulsations in the ELF band and some form of random variability, mainly in the Ultra Low Frequency (ULF) band (0–3 Hz) (Puranen and Jokela 1996; Pirard and Vatovez 2014). Moreover, all anthropogenic EMFs/EMR are totally polarized and coherent and, for this reason, significantly more bioactive than natural EMFs/EMR (Panagopoulos et al. 2015). The complex high and low frequency polarized EMFs/EMR emitted by WC devices/antennas, we shall call WC EMFs/EMR in the present chapter. Since all oscillating EMFs emit EMR, and all types of polarized and coherent (man-made) EMR possess electric and magnetic fields, the terms WC EMFs and WC EMR are used equally in the chapter.

The widespread use of WC devices emitting polarized EMFs/EMR has raised global concerns about the potential adverse effects of these fields/radiations on human health and wellbeing (Carpenter 2013; Duhaini 2016). Several international organizations, such as the private non-governmental organization (NGO) called the International Commission on Non-Ionizing Radiation Protection (ICNIRP), have established ELF and RF exposure limits which, in fact, should be significantly reduced based on the published scientific literature on the biological and health effects of the man-made EMFs. Although the non-thermal biological and health effects (not accompanied by any significant tissue heating) of man-made EMFs at environmentally encountered intensities have been documented for a long time in the scientific literature, the exposure limits set by the ICNIRP continue to rely exclusively on the thermal (heating) effects of EMFs/EMR (ICNIRP 1998; 2009; 2020). This is attributed by the ICNIRP to its own claim that there is no biophysical mechanism accounting for the non-thermal effects of man-made EMFs in living tissue and to the discrepancy in the findings among experimental studies. However, ample evidence has long since highlighted a great number of non-thermal adverse biological and health effects of both ELF and RF EMFs (Adey 1981; 1993; Goodman et al. 1995; Magras and Xenos 1997; Ivancsits et al. 2003; Diem et al. 2005; Hardell et al. 2007; 2009; 2013; Blackman 2009; Phillips et al. 2009; Mohammed et al. 2013; Hinrikus et al. 2015; Yakymenko et al. 2016; 2018; Panagopoulos 2017; 2019; 2020; Miller et al. 2018; 2019), and a plausible biophysical/biochemical non-thermal mechanism has also been published and verified experimentally (Panagopoulos et al. 2000; 2002; 2015; 2021).

The brain's electrical function/activity, as well as its proximity to EMF/EMR sources such as mobile or cordless phones, makes it one of the most affected organs, especially by WC EMF exposures. Therefore, many studies have focused on exploring the effects of WC EMFs in the central nervous system (CNS) and particularly in the brain. It has been demonstrated that exposure to various types of anthropogenic/technical EMFs causes excitation, suppression, or alteration of brain electrical signals (Hossmann and Hermann 2003; Zhang et al. 2017; Ohayon et al. 2019).

EEG is one of the most sensitive techniques that can reflect the status/activity of the brain with a high temporal resolution. Human EEG recordings during WC EMF exposure, in either eye-open or eye-closed waking conditions, have been extensively employed to evaluate the effects of WC EMFs/EMR on brain activity. Other studies have employed sleep EEG recordings to investigate the effects of WC EMF/EMR on the structure and quality of sleep. Due to longer exposure durations in the sleep studies, almost all studies carried out with the aim of searching for changes in sleep parameters due to WC EMF/EMR exposure have found such effects (Huber et al. 2002; Schreier et al. 2006; Schmid et al. 2012; Danker-Hopfe et al. 2016; Eggert et al. 2020; Liu et al. 2021)

Studies investigating the effects of WC EMFs on the human brain activity are many in bioelectromagnetic research during the past 20 years and beyond, and usually employ either commercially available mobile phone devices or generators that emit radiation in the RF range plus ELF pulsations mimicking the signals of the mobile phones. The main reason that many studies investigate the effects of mobile phone emissions on brain activity is that mobile phones are the most common WC devices used by billions of users worldwide and have become necessary tools in everyday life. At the same time, their adverse effects on health are a subject of intense scientific research and concern among scientists and among the general public (Hardell et al. 2007; 2009; 2013; Blackman 2009; Yakymenko et al. 2016; 2018; Panagopoulos 2017; 2019; 2020; Miller et al. 2018; 2019).

The first-generation (1G) mobile phones appeared on the market during the 1980s and were analog devices operating at 450 or 800 MHz, emitting and receiving modulated continuous signals. They were followed by the 2G Global System for Mobile Telecommunications (GSM) digital system in the 1990s, which is still in use and works at carrier frequencies 900, 1800, and 1900 MHz. The carrier signal is phase- and amplitude-modulated and is emitted in pulses ("frames") repeated at a rate of 217 Hz, with additional pulsations at 2 and 8 Hz (Pedersen 1997; Curwen and Whalley 2008). 2G mobile phones and generators simulating their signals have been extensively employed in studies researching the effects of mobile telephony (MT) EMFs on human and animal EEG (Dogan et al. 2012). The Universal Mobile Telecommunications System (UMTS) employed in 3G and 4G MT, which started operating in the late 2000s with 3G, uses carrier frequencies 1900–2200 MHz and pulsations at 100 and 1500 Hz (Hyland 2008). The Long Term Evolution (LTE) system (operated in 4G), which started around 2015, and the 5G WC system that is currently being deployed employ higher carrier frequencies, up to 2.6 GHz (4G), and up to 100 GHz (5G), respectively (Yang et al. 2016; Neufeld and Kuster 2018; Panagopoulos 2020).

Most published studies investigating the effects of WC EMFs/EMR on the human/animal EEG have focused on the effects of 2G (GSM) mobile phone EMFs. A smaller number of studies have focused on the effects of 3G (UMTS) and 4G (UMTS/LTE) mobile phone EMFs (Huber et al. 2000; 2002; 2005; Lv et al. 2014; Roggeveen et al. 2015a; 2015b; Yang et al. 2017; Vecsei et al. 2018a; 2018b). The majority of studies have been carried out using generators emitting simulated mobile phone signals of lower power density than the commercially available mobile phones.

These generators emit pulsing RF EMR at power densities that are incapable of raising the tissue temperature at any noticeable level. For example, in Mohammed et al. (2013), we applied a simulated GSM signal produced by a generator with a power density (0.025 mW/cm^2 or 25 µW/cm^2) about 10 times lower than the power density of a mobile phone EMR on a user's head when the phone is in touch with the ear, which is also not found to induce any significant heating (Panagopoulos 2017; 2019). Thus, any changes in the brain physiology reported in the literature, resulting from such lower-power WC EMF/EMR-like exposures by use of such generators are definitely attributed to non-thermal mechanisms. Only one study has reported that exposure to pulsing RF EMFs/EMR at a Specific Absorption Rate (SAR) of 1–2 W/kg, similar to that of a mobile phone touching the head (power density values were not provided in this study), could induce a (micro)thermal effect in the brain (Loughran et al. 2019).

Apart from the non-thermal nature of the vast majority of reported effects, another noteworthy aspect is that the majority of the devices employed to mimic emissions from mobile phones and other WC devices/antennas, emit EMFs that are pulsed by ELF pulses like the real-life WC devices/ antennas (Figure 7.6). More importantly, they can be programmed to emit the carrier (RF) signal with or without the ELF pulsation and study the corresponding effects separately, while in real mobile phone emissions the pulsations are inseparable from the overall signal. The necessary combination of RF and ELF in all WC EMF/EMR types implies that studies investigating the effects of WC EMFs/EMR on living organisms/tissues are just as important as studies investigating the effects of purely ELF exposures, and both types of studies are complementary to each other.

Several studies have found that the ELF EMFs which accompany WC EMFs/EMR in the form of modulation and pulsation constitute the most significant parameter that induces the changes in the human/animal brain activity (Huber et al. 2002; 2005; Mohammed et al. 2013). This finding has been specifically documented by studies that compared the effect of continuous-wave (non-pulsed) EMR (the RF carrier alone) with pulsed or modulated EMR of identical other parameters and reported a difference in the induced effects between these two types of exposures. Important modulation types and differences between continuous and pulsed RF signals are shown in Figure 7.6.

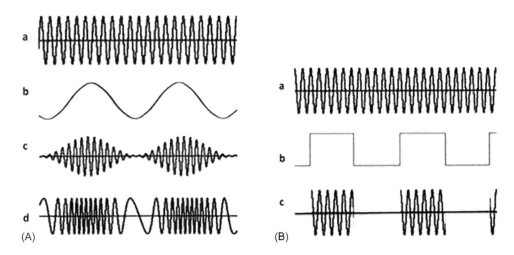

FIGURE 7.6 (A) Analog modulation of carrier RF signal. a) A continuous-wave carrier RF signal, b) ELF modulating signal, c) Amplitude modulation of the carrier signal, and d) Frequency modulation of the carrier signal. (B) Pulsing or digital modulation of a carrier RF signal. a) Carrier RF signal, b) Digital rectangular modulating pulses (on/off pulsation), and c) Pulsing (digital modulation) of the carrier signal.

7.2 WC EMFs AND HUMAN BRAIN ACTIVITY

7.2.1 EEG Changes in Healthy Subjects Induced by WC EMF Exposure

Significant evidence has been accumulated regarding changes in the EEG rhythms of healthy subjects induced by WC EMFs/EMR exposure. Alpha rhythm changes have been the focus of several studies that investigated EEG alterations taking place during or after exposure of humans to WC EMFs/EMR. Despite the fact that different studies have used different methodologies and EMF/EMR parameters, such as intensity, frequency, duration, and pulsation, a general conclusion regarding the induced changes in brain rhythms, specifically in the alpha rhythm, can be drawn. A basic finding is an increase or decrease in the EEG power* of the alpha rhythm. The number of studies that have reported an increase in the recorded power (Reiser et al. 1995; Von Klitzing 1995; Lebedeva et al. 2000; Croft et al. 2002; Huber et al. 2000; 2002; Curcio et al. 2005; Regel et al. 2007; Hinrikus et al. 2008; 2017; Croft et al. 2008; Croft et al. 2010; Suhhova et al. 2013) is greater than the number of studies that found a corresponding decrease (D'Costa et al. 2003; Maby et al. 2006; Vecchio et al. 2012; Perentos et al. 2013; Ghosn et al. 2015) or the number of studies that found no effect (Röschke and Mann 1997; Hietanen et al. 2000; Perentos et al. 2007).

Other studies have reported either an increase or decrease in the recorded EEG power of the other frequency bands; delta, theta, beta, and gamma (Delta: Croft et al. 2002; Maby et al. 2006; Theta: Reiser et al. 1995; Maby et al. 2006; Beta: Reiser et al. 1995; D'Costa et al. 2003; Maby et al. 2006; Hinrikus et al. 2008; 2017; Suhhova et al. 2013; Gamma: Curcio et al. 2015).

The ELF modulation/pulsation of the WC EMFs/EMR has been found to play a most significant role in the effects of this type of radiation in the brain. As reported already, several studies have found an increase in the alpha band power as a result of WC EMF/EMR exposure (Figure 7.7). Most of these studies found that the ELF pulsation/modulation of the RF EMR is a key factor for the induced changes in the EEG signals and reported that the EEG power changes in the various frequency bands of the EEG are related to the specific ELF pulsing/modulation frequency that was applied during the WC EMF/EMR exposure (Von Klitzing 1995; Huber et al. 2002; Regel et al. 2007; Hinrikus et al. 2008; 2017; Suhhova et al. 2013).

For example, Huber et al. (2002), through a double-blind experimental procedure, investigated the effects of a handset-like signal, simulating the radiation emitted from GSM mobile phones, with

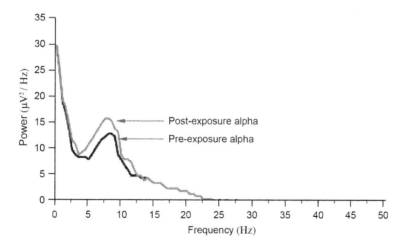

FIGURE 7.7 Increase in the alpha band power of human EEG after exposure to WC EMFs/EMR. The unit ($\mu V^2/Hz$) refers to the spectral power density of the recorded signal in the different EEG frequency bands.

---

* The term EEG power refers to the absolute magnitude of the total activity in the specific frequency band of the EEG.

a carrier frequency at 900 MHz and ELF pulsations at 2, 8, 217, and 1736 Hz, plus the corresponding harmonics, on the waking and sleep EEG of 32 healthy young male volunteers. The authors combined the EEG recordings with PET imaging to correlate the EEG changes due to EMF/EMR exposure, with the changes in the regional cerebral blood flow observed in the PET scans. Exposure to the pulsed EMF for 30 minutes (min) at one side of the head increased both the relative cerebral blood flow at this side and the power of the alpha waves during the waking state. The results were in agreement with previous findings of the same authors (Huber et al. 2000). In addition, the recording of EEG signals during sleep after the EMF exposure revealed that the pulsed EMF was capable of causing modification in the EEG signal power in the frequency range of 12.25–13.5 Hz, which corresponds to sleep spindle oscillatory frequencies (characteristic pattern of EEG signals in the range of 11–16 Hz occurring during NREM sleep or deep sleep), while the corresponding continuous-wave EMF/EMR (without ELF pulsations) was unable to induce such a change in the EEG signals. Based on this observation, the researchers suggested that the ELF pulsation of the WC EMFs/EMR is necessary to induce waking and sleep EEG changes. Thus, the authors demonstrated that: a) the WC EMF exposure during the waking state can modify EEG during both waking and subsequent sleep states, b) the EEG changes are accompanied by changes in the cerebral blood flow at the exposure side of the head, c) the ELF pulsation of the WC EMF is the key parameter producing the effects, and d) the induced changes in brain activity outlast the exposure period (Huber et al. 2000; 2002).

Hinrikus et al. (2008) extended the findings regarding the EMF signal pulsation/modulation effects on the EEG of healthy subjects, by investigating the effect of different signal pulsation frequencies on the EEG rhythms. The authors exposed 30 healthy subjects to 450 MHz EMR emitted by a generator at a power density of 0.16 mW/cm^2 pulsed at 7, 14, and 21 Hz. Like Huber et al. they found that the pulsing frequency plays a key role in the induced changes in the EEG power. While the EMF pulsations at 14 and 21 Hz increased the EEG power in the alpha and beta frequency bands, no increase was recorded to be induced by the EMF pulsations at 7 Hz.

Other studies did not find a significant difference between the effects of continuous and pulsed RF EMF/EMR (Perentos et al. 2007) or found a decrease in the spectral power of the alpha band (D'Costa et al. 2003; Perentos et al. 2013; Ghosn et al. 2015) by either pulsed or continuous-wave EMR exposure. The main concern with those studies that found negative or contradicting results is that they used short exposure durations.

Some authors have argued that the negative results reported in the literature may be due to short exposure durations and that a minimum exposure duration should be applied in order to record an effect. For example, a study carried out by Roscke and Mann (1997) found no effects after 3.5 min MT EMF exposure (power density 0.05 mW/cm^2). Another study (Hietanen et al. 2000) also reported non-significant effects of EMF exposure on EEG activity after 20 min of exposure (peak output power of 1–2 W).

In fact, it is not the exposure duration alone but in combination with the applied field/radiation intensity. Hinrikus et al. (2008) (described above) found significant effects by only a few min exposure but at a relatively high power density (0.16 mW/cm^2). In our experiments with rats (Mohammed et al. 2013), we found effects with a lower power density (0.025 mW/cm^2) but long-term exposures (1 h per day for 1 month). Therefore, both parameters (intensity/power density and duration of exposure) are important in inducing EEG effects.

Another important factor that may contribute to different outcomes in the various studies is the source of the EMF used in the study. Some studies used simulated WC EMF signals in order to be able to adjust the different parameters that characterize the EMF signals such as the pulsation and the power density (Huber et al. 2002; 2005; Perentos et al. 2007; Hinrikus et al. 2008). Other studies have utilized commercially available mobile phone devices which emit real EMF signals which can only slightly be manipulated by the experimenter but represent the actual conditions of real-life mobile phone exposures (Krause et al. 2000a; 2000b; Vecchio et al. 2007; 2012; Roggeveen et al. 2015a; 2015b). Real EMF/EMR exposures by mobile phones and other WC devices are found to be much more bioactive than simulated exposures. The reason is attributed to the variable nature of the real signals in comparison to the invariable signals emitted by generators and the organism's greater difficulty to adapt to the variable signals (Panagopoulos 2017; 2019).

Recording of resting human EEG signals during the waking state may be carried out in eye-open or eye-closed conditions. Assessment of the effects of WC EMFs on the EEG of an awake human under these conditions is found to be different in different studies. Some studies have found effects during the eye-closed condition but were unable to record any effect during the eye-open condition (Ghosn et al. 2015). This finding has been attributed to the lower amplitude of alpha waves when the subjects have their eyes open, resulting in a higher difficulty to detect changes. Under the closed-eye condition, the amplitude of alpha waves is higher, facilitating the detection of effects (Barry et al. 2007; Chen et al. 2008). Other studies, however, reported effects only during the eye-open condition (Croft et al. 2002) and argued that closing the eyes makes the person drowsy, which in turn affects the alpha rhythm and acts as noise that reduces the probability of detecting WC EMF/EMR-related changes on the alpha rhythm.

Experimental protocols to assess the effects of WC EMFs/EMR on task-related waking EEG rhythms and cognitive-motor performance have been designed. In this kind of experiments, phasic relative amplitude* decreases [called event-related desynchronization (ERD)] and increases [called event-related synchronization (ERS)] in the EEG bands are recorded. A Finnish group of researchers investigated the effects of EMF exposure by a commercially available GSM mobile phone on the ERD/ERS of different frequency bands in the EEG of normal subjects while they were performing auditory memory tasks (Krause et al. 2000a) or visual working memory tasks (Krause et al. 2000b) and found that the WC EMF exposure induced an alteration in the amplitude of the alpha band during the cognitive processes ("phases"). They suggested that the effects of WC EMF exposure on the brain oscillatory activity are not task-specific because they found similar effects during auditory and visual working memory tasks.

Vecchio et al. (2012) reported that GSM radiation can have a reducing effect on the resting EEG, reducing the power of the alpha band in a way connected with the task-related cortical activity. Reduction of the alpha band power and shifting of the EEG power toward higher frequencies has been connected with increasing attention and decreased reaction time to various stimuli (Kirschfeld 2008). Thus, the change in the alpha band power of the subjects after exposure to the mobile phone EMF is connected with a change in attention and reaction time induced by the EMF exposure. Vecchio et al. (2012) hypothesized that the EMF exposure affects the oscillations of the cortical activities engaged in the processing of sensorimotor information during the cognitive task. They also suggested that the mechanism underlying these changes may be related to the effects of EMF exposure on the ionic channels (Na^+, K^+, Ca^{2+}, etc.) of the cell membrane and the modulation in Ca^{2+} cellular homeostasis.

Indeed, such a mechanism for the action of man-made EMFs on cells was already described by Panagopoulos et al. (2000; 2002). Today, this mechanism has been verified by many experimental studies which combined EMF exposure with ion channel blockers (in most cases calcium channel blockers), as reviewed by Pall (2013), or with-patch clamp ion current recordings, as recently reviewed by Bertagna et al. (2021).

Compared with the results obtained from waking EEG studies, sleep EEG results are more consistent among them in showing the effects of WC EMF/EMR exposure on brain activity. One effect of WC EMFs/EMR on sleep brain activity observed in several studies is the increase in the EEG power in the frequency range of 12–14 Hz (sleep spindle frequencies) during NREM sleep (Huber et al. 2000; 2002; Loughran et al. 2005; Regel et al. 2007; Schmid et al. 2012). Most of the published literature demonstrates a consensus regarding the key role of the ELF pulsations of the WC EMFs in generating the observed effects. However, the specific ELF pulsation frequency does not seem to be

---

* Phasic relative amplitude is called the amplitude/power during a specific phase/state in relation to the mean amplitude/power during the experiment.

crucial for inducing the sleep EEG effects. Other frequencies in the delta and theta bands have also been found to be affected by the ELF pulsations of WC EMFs/EMR (Schmid et al. 2012). Despite these well-documented and repeatedly reported findings concerning the effects of WC EMFs on brain activity during sleep, there is no solid conclusion about the biological role of these changes on the subject's health and wellbeing (Mann and Roschke, 2004).

EEG signals have been used to test the individual sensitivity to WC EMFs. A study carried out with this aim (Hinrikus et al. 2008) exposed four groups of healthy humans to microwave radiation (450 MHz) pulsed at different low frequencies (7, 14, 21, 40, 70, 217, and 1000 Hz). In all cases, the pulsing EMF/EMR exposure increased the EEG power of the exposed subjects except for the case of the 1000 Hz pulsations. Moreover, different subjects were affected by different pulsing frequencies (apart from the 1000 Hz that did not affect anyone). The differential sensitivity was attributed to the variability of the physiological state of the brain among individuals. Another study found differences in the effects of WC EMR (900 MHz pulsed at 2 Hz) on the sleep EEG of each subject (intra-individual) induced by two identical exposures with 2 weeks intermittence between them (each exposure session had a different effect) and between different exposed subjects (inter-individual) (Lustenberger et al. 2015). Therefore, it could be argued that studies employing EEG as a tool to investigate the effects of WC EMFs/EMR on the human brain should have a large sample size (many individuals) in order to overcome individual variations and detect changes that would otherwise go undetected with smaller sample sizes.

7.2.2 EEG Changes in Diseased Subjects Induced by WC EMF Exposure

Aside from studying EEG changes induced by WC EMF exposure in healthy human subjects, EEG signal analysis has been used to evaluate changes in brain activity in epileptic patients. The activity of epileptic seizures is characterized by high amplitude and sharp frequency spiking in the beta and gamma bands (Curcio et al. 2015). A slight reduction in the frequency (number of spikes per unit time) of this spiking activity, an increase in the power of the gamma frequency band, and an increase in the inter-hemispheric coherence (the synchronization between brain areas located in different sites on the two brain hemispheres) have been reported in 12 patients with focal epilepsy exposed for 45 min to WC EMR (902.4 MHz, pulsed at 8.33 and 217 Hz, with peak power 2 W) emitted from a GSM mobile phone device (Curcio et al. 2015). The authors carried out their experiments in a cross-over and double-blind way. EEG signals were recorded from each patient in exposed (mobile phone on) and sham-exposed (mobile phone off) conditions, and the recordings were separated by one week. Each exposure/sham-exposure session lasted for 45 min, and the patients were exposed in a counterbalanced way (i.e., half of the patients were exposed in the first session and sham-exposed in the second session, while the other half of the patients received the same procedure in reversed order). Spiking activity was automatically determined by using a special computer program, according to which the spikes were detected when the amplitude of the EEG signal exceeded the background activity by one-third and had a frequency of more than 13 Hz (beta and gamma bands). Furthermore, in order to evaluate the differences between cortical areas, the authors calculated the EEG coherence for each frequency band between different brain sites located in the two brain hemispheres. They hypothesized that the reduction in the spiking activity resulted from the activation of the inhibitory neurons surrounding the epileptogenic focus. They argued that the changes recorded in the fast (gamma) band as opposed to changes in the alpha band in healthy subjects, are specific to the epileptic patients because of their abnormal cortical excitability.

Recently, Azmy et al. (2020), in a cross-sectional study (i.e., collecting data at a certain time period), found that EEG recordings after exposure to WC EMF from a mobile phone (GSM 900 MHz, SAR 1.03 W/kg in the head) for 30 min are different in the normal control subjects than in persons with epilepsy. Epileptic patients under the influence of the WC EMF exposure displayed a significant increase in the rate of their interictal EEG spiking activity (between epileptic seizures), and one patient with a normal basal record developed an epileptic seizure during the exposure. The study suggested that epileptic patients might be at high risk from the effects of acute WC EMF exposure and recommended further research by using chronic or repeated exposure to be carried out with epileptic patients.

Due to the scarcity of reported data concerning the effects of WC EMF exposure on non-healthy persons with abnormal EEG activities, no firm conclusions can be drawn, and more research on groups of people under pathological conditions is required to assess the effects of WC EMF exposure on them.

7.3 WC EMFs AND ANIMAL BRAIN ACTIVITY

Animals, unlike humans, have less variation in brain activity between individuals and can be better controlled by the experimenter. Animal studies that used EEG recordings to investigate the effects of WC EMFs on brain activity have demonstrated both similar and contradicting results to those of human studies. In agreement with the results of human studies, pulsed/modulated EMFs have been found to be more bioactive than the corresponding continuous-wave EMFs on animal EEG (Mohammed et al. 2013; Liu et al. 2021).

Long-term exposure of rats (1 month, 1 h/day) to non-thermal (25 $\mu W/cm^2$, SAR averaged over 1 g in the head 0.245 W/kg) pulsed 900 MHz EMF/EMR with ELF pulsations at 8 or 16 Hz (close to the frequencies of the EEG waves) has been found in our experiments to significantly alter their sleep EEG pattern (Mohammed et al. 2013). Moreover, we found that the REM sleep state was more susceptible to changes induced by the WC EMF exposure than the NREM sleep state. Theta and beta frequency bands during REM sleep were significantly decreased in amplitude/power compared to their control values. Interestingly, these affected frequency bands are close to the applied EMF pulsation frequencies (8 and 16 Hz). Furthermore, we found an increase in the REM sleep latency that indicated a change in the sleep structure of the EMF-exposed animals with respect to the unexposed. These findings could be related to memory impairments and changes in the corresponding sleep parameters.

Changes in the typical waveform of sleep EEG have also been reported in mice exposed to 2.4 GHz EMF/EMR pulsed at 100 Hz, whereas the non-pulsed (continuous wave) corresponding EMF/EMR had little effect (Liu et al. 2021). These results, along with Mohammed et al. (2013), in rodents are in agreement with the results of Huber et al. (2002) in humans regarding the crucial role of the ELF pulsations of the WC EMFs in inducing biological effects.

Moreover, animal EEG studies have revealed that repeated exposure to ELF- modulated EMR induces an increase in beta-2 (17.8–30.5 Hz) waves in the cortex and hypothalamus (Vorobyov et al. 2010). This additional effect produced by the ELF pulsation/modulation of the EMR further emphasizes the findings of the human EEG studies that the ELF pulsation/modulation of the WC EMFs/EMR is the key factor that induces the changes in the EEG signals. The study pointed also to an important effect that is rarely investigated in EMF studies, namely the cumulative effect of the WC EMF/EMR exposures. Because, today, humans are constantly exposed to man-made EMF sources in their daily lives, the cumulative effect of man-made EMF exposure should be at the forefront of current research.

EEG signal variations are intimately linked to the alterations in the level of neurotransmitters in different brain areas. For example, arousal (increase in the EEG frequency) can be induced by the activation of cholinergic receptors, and conversely, the attenuation in the cholinergic activity is related to SWS (Detari, 2000). On the other hand, abnormalities detected in the analysis of the EEG signals are considered as a direct reflection of imbalances in neurotransmitter systems. For example, EEG seizure activity (in either normal or epileptic persons) is triggered because of a failure in cholinergic (excitatory) or γ-aminobutyric acid (GABA)-ergic (inhibitory) control, as the balance between excitatory and inhibitory neurotransmitter systems is important for normal brain functioning (Friedman et al. 2007). It was suggested that anthropogenic EMF/EMR exposure represents a stress-producing stimulus that is responsible for increasing the levels of serotonin and catecholamines (neurotransmitters sensitive to stress, as their levels are elevated in a stressful situation) (Lowry et al. 2003). Prolonged exposure to WC EMFs/EMR (900 MHz, 0.02 mW/cm^2, 4 months) has been reported to cause variations in the inhibitory and excitatory neurotransmitter levels in different brain areas of the exposed rats, and these variations have been linked to the mechanisms that underlie the adverse effects of mobile phone radiation (Noor et al. 2011).

Another significant aspect reported in animal studies is the long-term and delayed effects of WC EMF exposure. This concerns the effects that appear not only during the exposure but also after the cessation of the exposure, and they are attributed to the exposure. These long-lasting effects have been reported in several studies that investigated WC EMF/EMR effects on animals (Petrova et al. 2005; Li et al. 2008; Vorobyov et al. 2010; Mohammed et al. 2011). Petrova et al. (2005) reported a delayed effect of pulsed EMF/EMR exposure with a pulse repetition frequency of 6 Hz. They reported a suppression of EEG frequencies in rats close to the pulsing frequency and a decrease in the inter-hemispheric synchronization. Although the exposure duration was 1 h, the effect persisted in the exposed animals for 22 h after the end of the exposure. They suggested that the exposure affected the animals' circadian system. Such effects found on animals in which the changes in brain function induced by pulsed EMFs (WC EMFs/EMR) outlast the exposure period have also been found with humans (Huber et al. 2000).

The chronic effects of exposure of Wistar rats to pulsed 2.45 GHz microwaves at an average power density of 1 mW/cm^2 for 3 h daily for 30 days were investigated by Li et al. (2008). They found that the exposure induced significant deficits in spatial learning and memory performance and an increase in plasma corticosterone levels compared to control animals. In another study, Vorobyov et al. (2010) found that the repeated exposure to WC EMFs/EMR (915 MHz, pulsed at 4 Hz, 0.3 mW/cm^2) several times per day for 5 days affects brain activity at the 3.2–6 Hz and 17.8–30.5 Hz frequency bands in the cortex and at the 6–17.8 Hz band in the hypothalamus. They argued that the effects of the repeated exposure were cumulative.

More evidence on the cumulative effects of WC EMF/EMR exposure on the nervous system was provided by another long-term exposure study of ours (Mohammed et al. 2011). In these experiments, we recorded EEG signals from animals exposed to WC EMFs/EMR (900 MHz, 0.02 mW/cm^2, SAR 1.165 W/kg, pulsed at 217 Hz) after 1, 2, and 4 months of constant exposure and after 1 month from the end of the exposures to determine delayed effects. Quantitative analysis of EEG signals revealed a shift in the EEG power from beta to delta frequency bands. Specifically, a decrease in the power of the beta frequency band occurred concomitantly with an increase in the power of the delta frequency band. This change persisted for more than a month after the termination of the exposure. The prolonged response to the WC EMF/EMR exposure seems to be associated with the cumulative nature of the effects and should be further investigated by future studies.

7.4 DISCUSSION

Although we can safely conclude that WC EMF/EMR exposure affects the human and animal brain activity with the ELF pulsations playing a crucial role in the effects, methodological differences between different studies make the drawing of a detailed picture of these effects difficult (Danker-Hopfe et al. 2019; Dalecki, et al. 2021).

Although several studies have confirmed the presence of EEG effects due to anthropogenic and, specifically, WC EMF exposure, few studies have discussed the underlying mechanisms related to the changes that take place in the EEG signals as a result of the interaction between man-made EMF exposure and the brain. Furthermore, even fewer studies have explored the consequences of these changes on a subject's health and wellbeing. It has been found that stress, anxiety, and depression increased linearly with increasing exposure time to man-made EMFs. Szemerszky et al. (2010) reported that the long-term (4–6 weeks) and short-term (5 days) exposure to ELF EMFs (50 Hz,

0.5 mT) could induce depression and metabolic disturbances in rats. Van Wijngaarden et al. (2000) indicated that exposure of electric utility workers to ELF EMFs could lead to depression and suicidal behavior. They hypothesized that chronic man-made EMF exposure may affect pineal gland function, disrupting normal circadian rhythmicity and melatonin synthesis, as originally suggested by Wilson (1988).

A significant finding is the correlation between regional cerebral blood flow and changes in sleep and waking EEG due to exposure to pulsed WC EMFs/EMR (900 MHz) for 30 min as reported by Huber et al. (2002). This study combined data from PET and EEG during waking and sleep states and revealed an increase in relative regional cerebral blood flow and enhancement in the alpha frequency power in the waking state and during Stage 2 sleep. The study suggested that the change in cerebral blood flow resulted in an alteration in the cortical neuronal activity that modified neuronal loops involved in sleep spindle generation and waking alpha enhancement. The study findings, along with other studies by the same and other authors, also provide strong evidence for the association between EEG changes and the ELF pulsing/modulation of the WC EMFs/EMR (Huber et al. 2000; 2002; 2005; Hinrikus et al. 2008; Vorobyov et al. 2010; Mohammed et al. 2013; Liu et al. 2021).

There are also contradictory findings reported by different studies, with most of the studies reporting effects, and some others reporting no effects on the same endpoints. While many studies have found similar results corroborating each other, the inability to replicate results in certain cases has also been reported (Wagner et al. 1998). We have argued (Mohammed et al. 2013) that the use of short duration rather than chronic exposures, in addition to relatively low intensity, may account for the fact that several studies found no effects of WC EMFs on the brain activity (Sienkiewicz et al. 2000; Cassel et al. 2004; Cosquer et al. 2005). In our experiments (Mohammed et al. 2011; 2013), the animals were exposed to the WC EMF for 30 consecutive days. This relatively long period of exposure allowed the radiation effects to be accumulated and ended up with effects that might not have appeared with short-duration exposures. This possibly explains the absence of effects in some studies that applied short duration and relatively low intensity exposures.

Despite the fact that the non-thermal effects of WC EMFs emitted by wireless devices on the human and animal brain have been documented in the EEG literature, the ICNIRP (2020) still sets its limits based exclusively on the thermal effects of RF radiation. For example, in our studies (Mohammed et al. 2011; 2013), we found significant adverse effects on rat brain activity to be induced by a power density of 20-25 $\mu W/cm^2$ at 900 MHz (meaning that any realistic safety limit has to be significantly lower), while the corresponding ICNIRP (2020) limit for the general public (averaged over 30 min) is 450 $\mu W/cm^2$. Therefore, there is a demand that the limits should be urgently revised and become compatible with the findings of the studies. Moreover, conflicts of interest and industry funding in the scientific community and the health organizations have also been reported to significantly affect study outcomes and exposure regulations (Leach et al. 2018; Hardell and Carlberg 2020; 2021).

One proposed solution for resolving inconsistencies among experimental findings is to establish a standardized protocol for exposure and recording conditions so that the results obtained are comparable among different studies and unified firm conclusions can be reached. Several criteria should be considered in such a standardized protocol. The study design should be specified, and the examination of subjects in a study should be randomized and double-blinded. The determination of the study design could make studies comparable with each other and would direct the experimenters toward standard methods that avoid bias or misinterpretation of the results. The conditions of the EEG signal recording (eye-opened/eye-closed) and the conditions of the recording environment (light, temperature, humidity, background EMFs in the laboratory, etc.) should be determined. The session of EEG recording should be carried out at a fixed time of the day to avoid the EEG diurnal variation and last for a minimum period of time to overcome EEG fluctuations. The analysis of the EEG signals should follow the same procedures, analyzed signals should come from the same type of scalp electrodes, and the limits of the EEG frequency bands should be standardized. Similar parameters of the EMF source should be used (frequencies included in the final signal, output power, distance from source, exposure duration). The health status, age, gender, and the intake of nervous system-modulating substances (alcohol, caffeine, nicotine, etc.) should be controlled and taken into consideration when selecting the subjects for the study.

Moreover, there is a need for further research on the mechanisms underlying the non-thermal effects of WC EMFs on brain activity and function. A biophysical mechanism has been introduced by Panagopoulos et al. (2000; 2002; 2015; 2021) showing that the forced oscillation of mobile ions (K^+, Na^+, Cl^-, Ca^{2+}) in all cells caused by polarized anthropogenic EMFs can irregularly gate the VGICs on cell membranes, disrupting the cell electrochemical balance and initiating the effects described in the literature. This mechanism has predicted that the lower (ELF) frequencies are the most bioactive and that the pulsed EMR can have stronger effects than the continuous-wave EMR. These predictions have already been verified by several EEG studies. Impaired function of VGICs has been confirmed for calcium, sodium, and potassium channels after ELF or WC EMF exposure (Lisi et al. 2006; Pall 2013; Li et al. 2014; Cecchetto et al. 2020; Zheng et al. 2021; Bertagna et al. 2021). The implication of this mechanism in human/animal brain physiology should be further explored experimentally.

In close connection with the aforementioned mechanism, Pall (2016) has argued that the voltage-gated calcium channels (VGCCs) are the main locations where the action of the EMF on the nervous system occurs. Indeed, the principal role played by the VGCCs in the release of the neurotransmitters makes the impairment of these channels due to EMF/EMR exposure directly linked to the variations displayed in the EEG signals.

Hinrikus et al. (2017) proposed a different model to explain the low-level modulated microwave effects on the brain activity based on the effects of microwave radiation on the water molecules that result in diffusion changes, which in turn, may affect the dynamics of the neurotransmitters and the neuronal resting membrane potential.

Loughran et al. (2019) suggested that WC EMR induces changes in the EEG signals through some thermal mechanism. They stabilized skin temperature via a specialized water perfusion suit that was worn by the study-participants during WC EMR exposure. By recording EEG and skin temperature during a 30 min exposure to a simulated GSM signal they found a 0.18°C elevation in finger temperature by the highest applied intensity (SAR = 2 W/Kg). The authors deduced that the effects on the EEG are associated with thermoregulatory alterations and are mediated by some thermal mechanism. However, such small temperature elevations after 30 min of intense exposure (close to a mobile phone) not exceeding 0.1–0.3°C have been reported by other experimenters as well (Dawe et al. 2006; Panagopoulos 2020), and it is unlikely that such a slight heating of living tissue can be the cause of the recorded EEG changes or other biological effects. As underlined in the present chapter, in the vast majority of the EEG studies the recorded effects were not accompanied by any significant heating in the exposed tissues, not even at the level of 0.1–0.3°C.

In conclusion, the present chapter has highlighted the following points:

1. The significance of EEG recordings in assessing the variations induced in the nervous system by WC EMF exposure.
2. Most of the examined data reveal that WC EMF exposure causes changes in the EEG signals.
3. The presented data emphasize the non-thermal action of WC EMFs/EMR at environmentally encountered intensities/levels on brain physiology, as recorded by human and animal EEG studies.
4. The ELF pulsation/modulation of the WC EMFs/EMR plays a major role in the induced effects.
5. A standardized protocol of exposure parameters, procedure, and EEG signal analysis should be developed and implemented by laboratories in order to be able to compare their results and draw firm conclusions about the effects of different parameters of the EMF exposure.
6. There are only a few studies that investigate the effects of WC EMF exposure in diseased or abnormal persons, and more research is needed in this field.

7. Studies investigating the mechanisms of action and biological consequences of brain physiology changes induced by anthropogenic EMF exposures are necessary for future research.

EEG recordings and analysis, as well as their correlation with other measured parameters, such as neurotransmitters, may provide a great opportunity to decipher the detailed mechanism of the effects occurring during human and animal brain exposure to this new ambient environmental stressor, which is the anthropogenic EMFs and, more particularly, the WC EMFs/EMR.

REFERENCES

Adey WR, (1981): Tissue interactions with non-ionizing electromagnetic fields. *Physiological Reviews* 61(2), 435–514.
Adey WR, (1993): Biological effects of electromagnetic fields. *Journal of Cellular Biochemistry* 51(4), 410–416.
Azmy R, Shamloul R, Elsawy NA, Elkholy S, Maher E, (2020): Effects of mobile phones electromagnetic radiation on patients with epilepsy: An EEG study. *Egyptian Journal of Neurology, Psychiatry and Neurosurgery* 56(1), 36. https://doi.org/10.1186/s41983-020-00167-2.
Barry RJ, Clarke AR, Johnstone SJ, Magee CA, Rushby JA, (2007): EEG differences between eyes-closed and eyes-open resting conditions. *Clinical Neurophysiology* 118(12), 2765–2773.
Bertagna F, Lewis R, Silva SRP, McFadden J, Jeevaratnam K, (2021): Effects of electromagnetic fields on neuronal ion channels: A systematic review. *Annals of the New York Academy of Sciences* 1499(1), 82–103.
Blackman C, (2009): Cell phone radiation: Evidence from ELF and RF studies supporting more inclusive risk identification and assessment. *Pathophysiology* 16(2–3), 205–216.
Carmena JM, Lebedev MA, Crist RE, O'Doherty JE, Santucci DM, Dimitrov DF, Patil PG, Henriquez CS, Nicolelis MA, (2003): Learning to control a brain-machine interface for reaching and grasping by primates. *PLOS Biology* 1(2), E42. https://doi.org/10.1371/ journal.pbio.0000042.
Carpenter DO, (2013): Human disease resulting from exposure to electromagnetic fields. *Reviews on Environmental Health* 28(4), 159–172. https://doi.org/10.1515/ reveh-2013-0016.
Cassel JC, Cosquer B, Galani R, Kuster N, (2004): Whole-body exposure to 2.45 GHz electromagnetic fields does not alter radial-maze performance in rats. *Behavioural Brain Research* 155(1), 37–43.
Cecchetto C, Maschietto M, Boccaccio P, Vassanelli S, (2020): Electromagnetic field affects the voltage-dependent potassium channel Kv1.3. *Electromagnetic Biology and Medicine* 39(4), 316–322.
Chen AC, Feng W, Zhao H, Yin Y, Wang P, (2008): EEG default mode network in the human brain: Spectral regional field powers. *Neuroimage* 41(2), 561–574.
Cosquer B, Kuster N, Cassel JC, (2005): Whole-body exposure to 2.45 GHz electromagnetic fields does not alter 12-arm radial maze with reduced access to spatial cues in rats. *Behavioural Brain Research* 161(2), 331–334.
Croft RJ, Chandler JS, Burgess AP, Barry RJ, Williams JD, Clarke AR, (2002): Acute mobile phone operation affects neural function in humans. *Clinical Neurophysiology* 113(10), 1623–1632.
Croft RJ, Hamblin D, Spong J, Wood A, McKenzie R, Stough C, (2008): The effect of mobile phone electromagnetic fields on the alpha rhythm of human electroencephalogram. *Bioelectromagnetics* 29(1), 1–10.
Croft RJ, Leung S, McKenzie RJ, Loughran SP, Iskra S, Hamblin DL, Cooper NR, (2010): Effects of 2G and 3G mobile phones on human alpha rhythms: Resting EEG in adolescents, young adults, and the elderly. *Bioelectromagnetics* 31(6), 434–444.
Curcio G, Ferrara M, Moroni F, D'Inzeo G, Bertini M, De Gennaro L, (2005): Is the brain influenced by a phone call? An EEG study of resting wakefulness. *Neuroscience Research* 53(3), 265–270.
Curcio G, Mazzucchi E, Della Marca G, Vollono C, Rossini PM, (2015): Electromagnetic fields and EEG spiking rate in patients with focal epilepsy. *Clinical Neurophysiology* 126(4), 659–666.
Curwen P, Whalley J, (2008): Mobile communications in the 21st century. In: AC Harper's, RV Buress (Eds.), *Mobile telephones: Networks, applications and performance*. Nova Science Publishers, New York, 29–75.
Dalecki A, Verrender A, Loughran SP, Croft RJ, (2021): The effect of GSM electromagnetic field exposure on the waking electroencephalogram: Methodological influences. *Bioelectromagnetics* 42(4), 317–328. https://doi.org/10.1002/bem.22338.

Danker-Hopfe H, Dorn H, Bolz T, Peter A, Hansen M-L, Eggert T, Sauter C, (2016): Effects of mobile phone exposure (GSM 900 and WCDMA/UMTS) on polysomnography based sleep quality: An intra- and inter-individual perspective. *Environmental Research* 145, 50–60.

Danker-Hopfe H, Eggert T, Dorn H, Sauter C, (2019): Effects of RF-EMF on the human resting-state EEG-the inconsistencies in the consistency. Part 1: Non-exposure-related limitations of comparability between studies. *Bioelectromagnetics* 40(5), 291–318. https://doi.org/10.1002/bem.22194.

Dawe AS, Smith B, Thomas DW, Greedy S, Vasic N, Gregory A, Loader B, de Pomerai DI, (2006): A small temperature rise may contribute towards the apparent induction by microwaves of heat-shock gene expression in the nematode Caenorhabditis elegans. *Bioelectromagnetics* 27(2), 88–97.

D'Costa H, Trueman G, Tang L, Abdel-Rahman U, Abdel-Rahman W, Ong K, Cosic I, (2003): Human brain wave activity during exposure to radiofrequency field emissions from mobile phones. *Australasian Physical and Engineering Sciences in Medicine* 26(4), 162–167.

Detari L, (2000): Tonic and phasic influence of basal forebrain unit activity on the cortical EEG. *Behavioural Brain Research* 115(2), 159–170.

Diem E, Schwarz C, Adlkofer F, Jahn O, Rudiger H, (2005): Non-thermal DNA breakage by mobile-phone radiation (1800 MHz) in human fibroblasts and in transformed GFSH-R17 rat granulosa cells in vitro. *Mutation Research* 583(2), 178–183.

Dogan M, Turtay MG, Oguzturk H, Samdanci E, Turkoz Y, Tasdemir S, Alkan A, Bakir S, (2012): Effects of electromagnetic radiation produced by 3G mobile phones on rat brains: Magnetic resonance spectroscopy, biochemical, and histopathological evaluation. *Human and Experimental Toxicology*, 31(6), 557–564. https://doi.org/10.1177/ 0960327111412092.

Duhaini I, (2016): The effects of electromagnetic fields on human health. *Physica Medica* 32(3), 213.

Eggert T, Dorn H, Sauter C, Schmid G, Danker-Hopfe H, (2020): RF-EMF exposure effects on sleep – Age doesn't matter in men! *Environmental Research* 191, 110173.

Friedman A, Behrens CJ, Heinemann U, (2007): Cholinergic dysfunction in temporal lobe epilepsy. *Epilepsia* 48(Suppl 5), 126–130.

Ghosn R, Yahia-Cherif L, Hugueville L, Ducorps A, Lemarechal JD, Thuroczy G, de Seze R, Selmaoui B, (2015): Radiofrequency signal affects alpha band in resting electroencephalogram. *Journal of Neurophysiology* 113(7), 2753–2759.

Goodman EM, Greenebaum B, Marron MT, (1995): Effects of electromagnetic fields on molecules and cells. *International Review of Cytology* 158, 279–338.

Hardell L, Carlberg M, Söderqvist F, Mild KH, Morgan LL, (2007): Long-term use of cellular phones and brain tumours: Increased risk associated with use for > or =10 years. *Occupational and Environmental Medicine* 64(9), 626–632.

Hardell L, Carlberg M, Hansson Mild K, (2009): Epidemiological evidence for an association between use of wireless phones and tumor diseases. *Pathophysiology* 16(2–3), 113–122.

Hardell L, Carlberg M, Hansson Mild K, (2013): Use of mobile phones and cordless phones is associated with increased risk for glioma and acoustic neuroma. *Pathophysiology* 20(2), 85–110.

Hardell L, Carlberg M, (2020): Health risks from radiofrequency radiation, including 5G, should be assessed by experts with no conflicts of interest. *Oncology Letters* 20(4), 15.

Hardell L, Carlberg M, (2021): Lost opportunities for cancer prevention: Historical evidence on early warnings with emphasis on radiofrequency radiation. *Reviews on Environmental Health*. https://doi.org/10.1515/reveh-2020-0168.

Harmony T, (2013): The functional significance of delta oscillations in cognitive processing. *Frontiers in Integrative Neuroscience* 7, 83. https://doi.org/10.3389/ fnint.2013.00083.

Herweg NA, Solomon EA, Kahana MJ, (2020): Theta oscillations in human memory. *Trends in Cognitive Sciences* 24(3), 208–227. https://doi.org/10.1016/j.tics.2019.12.006.

Hickey P, Race E, (2021): Riding the slow wave: Exploring the role of entrained low-frequency oscillations in memory formation. *Neuropsychologia*, 160, 107962. https://doi.org/10.1016/j.2021.107962.

Hietanen M, Kovala T, Hämäläinen A-M, (2000): Human brain activity during exposure to radiofrequency fields emitted by cellular phones. *Scandinavian Journal of Work, Environment and Health*, 26 (2), 87–92.

Hinrikus H, Bachmann M, Lass J, Karai D, Tuulik V, (2008): Effect of low frequency modulated microwave exposure on human EEG: Individual sensitivity. *Bioelectromagnetics* 29(7), 527–538.

Hinrikus H, Lass J, Karai D, Pilt K, Bachmann M, (2015): Microwave effect on diffusion: A possible mechanism for non-thermal effect. *Electromagnetic Biology and Medicine* 34(4), 327–333.

Hinrikus H, Bachmann M, Karai D, Lass J, (2017): Mechanism of low-level microwave radiation effect on nervous system. *Electromagnetic Biology and Medicine* 36(2), 202–212. https://doi.org/10.1080/15368378.2016.1251451.

Hossmann KA, Hermann DM, (2003): Effects of electromagnetic radiation of mobile phones on the central nervous system. *Bioelectromagnetics* 24(1), 49–62. https://doi.org/ 10.1002/bem.10068.

Huber R, Graf T, Cote KA, Wittmann L, Gallmann E, Matter D, Schuderer J, Kuster N, Borbely AA, Achermann P, (2000): Exposure to pulsed high-frequency electromagnetic field during waking affects human sleep EEG. *NeuroReport* 11(15), 3321–3325.

Huber R, Treyer V, Borbely AA, Schuderer J, Gottselig JM, Landolt HP, Werth E, Berthold T, Kuster N, Buck A, Achermann P, (2002): Electromagnetic fields, such as those from mobile phones, alter regional cerebral blood flow and sleep and waking EEG. *Journal of Sleep Research* 11(4), 289–295.

Huber R, Treyer V, Schuderer J, Berthold T, Buck A, Kuster N, Landolt H-P, Achermann P, (2005): Exposure to pulse-modulated radio frequency electromagnetic fields affects regional cerebral blood flow. *European Journal of Neuroscience* 21(4), 1000–1006.

Hyland GJ, (2000): Physics and biology of mobile telephony. *Lancet* 356(9244), 1833–1836.

Hyland GJ, (2008): Physical basis of adverse and therapeutic effects of low intensity microwave radiation. *Indian Journal of Experimental Biology* 46(5), 403–419.

ICNIRP, (1998): Guidelines for limiting exposure to time-varying electric, magnetic and electromagnetic fields (up to 300 GHz). *Health Physics* 74, 494–522.

ICNIRP, (2009): ICNIRP statement on the "Guidelines for limiting exposure to time-varying electric, magnetic, and electromagnetic fields (up to 300 GHz)". *Health Physics* 97(3), 257–258. https://doi.org/10.1097/HP.0b013e3181aff9db.

ICNIRP, (2020): ICNIRP guidelines for limiting exposure to electromagnetic fields (100 kHz to 300 GHz. *Health Physics* 118(5), 483–524.

Ivancsits S, Diem E, Jahn O, Rüdiger HW, (2003): Intermittent extremely low frequency electromagnetic fields cause DNA damage in a dose-dependent way. *International Archives of Occupational and Environmental Health* 76(6), 431–436.

Kirschfeld K, (2008): Relationship between the amplitude of alpha waves and reaction time. *NeuroReport*, 19(9), 907–910.

Klimesch W, Sauseng P, Hanslmayr S, (2007): EEG alpha oscillations: The inhibition-timing hypothesis. *Brain Research Reviews* 53(1), 63–88.

Knyazev GG, (2012): EEG delta oscillations as a correlate of basic homeostatic and motivational processes. *Neuroscience and Biobehavioral Reviews* 36(1), 677–695. https://doi.org/10.1016/j.neubiorev.2011.10.002.

Krause CM, Sillanmäki L, Koivisto M, Häggqvist A, Saarela C, Revonsuo A, Laine M, Hämäläinen H, (2000a): Effects of electromagnetic field emitted by cellular phones on the EEG during a memory task. *NeuroReport* 11(4), 761–764. https://doi.org/10.1097/00001756-200003200-00021.

Krause CM, Sillanmäki L, Koivisto M, Häggqvist A, Saarela C, Revonsuo A, Laine M, Hämäläinen H, (2000b): Effects of electromagnetic fields emitted by cellular phones on the electroencephalogram during a visual working memory task. *International Journal of Radiation Biology* 76(12), 1659–1667. https://doi.org/10.1080/09553000050201154.

Leach V, Weller S, Redmayne M, (2018): A novel database of bio-effects from non-ionizing radiation. *Reviews on Environmental Health* 33(3):273–280.

Lebedeva N, Sulimov A, Sulimova O, Kotrovskaya T, Gailus T, (2000): Cellular phone electromagnetic field effects on bioelectric activity of human brain. *Critical Reviews in Biomedical Engineering* 28(1–2), 323–338.

Li M, Wang Y, Zhang Y, Zhou Z, Yu Z, (2008): Elevation of plasma corticosterone levels and hippocampal glucocorticoid receptor translocation in rats: A potential mechanism for cognition impairment following chronic low-power-density microwave exposure. *Journal of Radiation Research (Tokyo)* 49(2), 163–170.

Li Y, Yan X, Liu J, Li L, Hu X, Sun H, Tian J, (2014): Pulsed electromagnetic field enhances brain-derived neurotrophic factor expression through L-type voltage-gated calcium channel- and Erk-dependent signaling pathways in neonatal rat dorsal root ganglion neurons. *Neurochemistry International* 75, 96–104.

Lisi A, Ledda M, Rosola E, Pozzi D, D'Emilia E, Giuliani L, Foletti A, Modesti A, Morris SJ, Grimaldi S, (2006): Extremely low frequency electromagnetic field exposure promotes differentiation of pituitary corticotrope-derived AtT20 D16V cells. *Bioelectromagnetics* 27, 641–651.

Liu L, Deng H, Tang X, Lu Y, Zhou J, Wang X, Zhao Y, Huang B, Shi Y, (2021): Specific electromagnetic radiation in the wireless signal range increases wakefulness in mice. *Proceedings of the National Academy of Sciences of the USA*, 118(31), e2105838118. https://doi.org/10.1073/pnas.2105838118.

Loughran SP, Wood AW, Barton JM, Croft RJ, Thompson B, Stough C, (2005): The effect of electromagnetic fields emitted by mobile phones on human sleep. *NeuroReport* 16(17), 1973–1976.

Loughran SP, Verrender A, Dalecki A, Burdon CA, Tagami K, Park J, Taylor N, Croft RJ, (2019): Radiofrequency electromagnetic field exposure and the resting EEG: Exploring the thermal mechanism hypothesis. *International Journal of Environmental Research and Public Health*, 16(9), 1505. https://doi.org/10.3390/ijerph16091505.

Lowry CA, Plant A, Shanks N, Ingram CD, Lightman SL, (2003): Anatomical and functional evidence for a stress-responsive, monoamine-accumulating area in the dorsomedial hypothalamus of adult rat brain. *Hormones and Behavior* 43(1), 254–262.

Lustenberger C, Murbach M, Tüshaus L, Wehrle F, Kuster N, Achermann P, Huber R, (2015): Inter-individual and intra-individual variation of the effects of pulsed RF EMF exposure on the human sleep EEG. *Bioelectromagnetics* 36(3), 169–177. https://doi.org/10.1002/bem.21893.

Lv B, Su C, Yang L, Xie Y, Wu T, (2014): Whole brain EEG synchronization likelihood modulated by long term evolution electromagnetic fields exposure. Annual International Conference of the IEEE Engineering in Medicine and Biology Society. *IEEE Engineering in Medicine and Biology Society. Annual International Conference* 2014, 986–989.

Maby E, Le Bouquin Jeannes R, Faucon G, (2006): Short-term effects of GSM mobiles phones on spectral components of the human electroencephalogram. *Annual International Conference of the IEEE Engineering in Medicine and Biology Society IEEE* 1, 3751–3754.

Magras IN, Xenos TD, (1997): RF radiation-induced changes in the prenatal development of mice. *Bioelectromagnetics* 18(6), 455–461.

Mann K, Röschke J, (2004): Sleep under exposure to high-frequency electromagnetic fields. *Sleep Medicine Reviews* 8(2), 95–107. https://doi.org/10.1016/S1087-0792(03)00004-2.

Miller AB, Morgan LL, Udasin I, Davis DL, (2018): Cancer epidemiology update, following the 2011 IARC evaluation of radiofrequency electromagnetic fields (Monograph 102). *Environmental Research* 167, 673–683.

Miller AB, Sears ME, Morgan LL, Davis DL, Hardell L, Oremus M, Soskolne CL, (2019): Risks to health and well-being From radio-frequency radiation emitted by cell phones and other wireless devices. *Frontiers in Public Health* 7, 223. https://doi.org/10.3389/fpubh.2019.00223.

Minarik T, Berger B, Sauseng P, (2018): The involvement of alpha oscillations in voluntary attention directed towards encoding episodic memories. *Neuroimage* 166, 307–316.

Mitchell DJ, McNaughton N, Flanagan D, Kirk IJ, (2008): Frontal-midline theta from the perspective of hippocampal 'theta'. *ProgNeurobiol* 86(3), 156–185.

Mohammed HS, Radwan NM, Nawal AA, (2011): Long-term low-level electromagnetic radiation causes changes in the EEG of freely-moving rats. *Romanian Journal of Biophysics* 21, 43–51.

Mohammed HS, Fahmy HM, Radwan NM, Elsayed AA, (2013): Non-thermal continuous and modulated electromagnetic radiation fields effects on sleep EEG of rats. *Journal of Advanced Research* 4(2), 181–187. https://doi.org/10.1016/j.jare.2012.05.005.

Neufeld E, Kuster N, (2018): Systematic derivation of safety limits for time-varying 5G radiofrequency exposure based on analytical models and thermal dose. *Health Physics* 115(6), 705–711.

Niedermeyer E, Da Silva FL, (2005): *Electroencephalography: Basic principles, clinical applications, and related fields*. Lippincott Williams & Wilkins, Philadelphia, USA.

Noor NA, Mohammed HS, Ahmed NA, Radwan NM, (2011): Variations in amino acid neurotransmitters in some brain areas of adult and young male albino rats due to exposure to mobile phone radiation. *European Review for Medical and Pharmacological Sciences* 15(7), 729–742.

Ohayon MM, Stolc V, Freund FT, Milesi C, Sullivan SS, (2019): The potential for impact of man-made super low and extremely low frequency electromagnetic fields on sleep. *Sleep Medicine Reviews* 47, 28–38. https://doi.org/10.1016/j.smrv.2019.06.001.

Pall ML, (2013): Electromagnetic fields act via activation of voltage-gated calcium channels to produce beneficial or adverse effects. *Journal of Cellular and Molecular Medicine* 17(8), 958–965. https://doi.org/10.1111/jcmm.12088.

Pall ML, (2016): Microwave frequency electromagnetic fields (EMFs) produce widespread neuropsychiatric effects including depression. *Journal of Chemical Neuroanatomy* 75(Pt B), 43–51. https://doi.org/10.1016/j.jchemneu.2015.08.001.

Palva S, Palva JM, (2007): New vistas for a-frequency band oscillations. *Trends in Neurosciences* 30(4), 150–157.

Panagopoulos DJ, Messini N, Karabarbounis A, Philippetis AL, Margaritis LH, (2000): A mechanism for action of oscillating electric fields on cells. *Biochemical and Biophysical Research Communications* 272(3), 634–640.

Panagopoulos DJ, Karabarbounis A, Margaritis LH, (2002): Mechanism for action of electromagnetic fields on cells. *Biochemical and Biophysical Research Communications* 298(1), 95–102. https://doi.org/10.1016/s0006-291x(02)02393-8.

Panagopoulos DJ, Johansson O, Carlo GL, (2015): Polarization: A key difference between man-made and natural electromagnetic fields, in regard to biological activity. *Scientific Reports* 5, 14914.

Panagopoulos DJ, (2017): Mobile telephony radiation effects on insect ovarian cells. The necessity for real exposures bioactivity assessment. The key role of polarization, and the ion forced-oscillation mechanism. In: CD Geddes (Ed.), *Microwave effects on DNA and proteins*, 1–48, Springer, Cham, Switzerland.

Panagopoulos DJ, (2019): Comparing DNA damage induced by mobile telephony and other types of man-made electromagnetic fields. *Mutation Research Reviews* 781, 53–62.

Panagopoulos DJ, (2020): Comparing chromosome damage induced by mobile telephony radiation and a high caffeine dose. Effect of combination and exposure duration. *General Physiology and Biophysics* 39(6), 531–544.

Panagopoulos DJ, Karabarbounis A, Yakymenko I, Chrousos GP, (2021): Mechanism of DNA damage induced by human-made electromagnetic fields. *International Journal of Oncology* 59(5), 92.

Pedersen GF, (1997): Amplitude modulated RF fields stemming from a GSM/DCS-1800 phone. *Wireless Networks* 3(6), 489–498.

Perentos N, Croft RJ, McKenzie RJ, Cvetkovic D, Cosic I, (2007): Comparison of the effects of continuous and pulsed mobile phone like RF exposure on the human EEG. *Australasian Physics & Engineering Sciences in Medicine* 30(4), 274–280.

Perentos N, Croft RJ, McKenzie RJ, Cosic I, (2013): The alpha band of the resting electroencephalogram under pulsed and continuous radio frequency exposures. *IEEE Transactions on Bio-Medical Engineering* 60(6), 1702–1710.

Petrova EV, Gulyaeva NV, Titarov SI, Rozhnov YV, Koval'zon VM, (2005): Actions of pulsed ultra-broadband electromagnetic irradiation on the EEG and sleep in laboratory animals. *Neuroscience and Behavioral Physiology* 35(2), 165–170. https://doi.org/10.1007/s11055-005-0062-9.

Phillips JL, Singh NP, Lai H, (2009): Electromagnetic fields and DNA damage. *Pathophysiology* 16(2–3), 79–88.

Pirard W, Vatovez B, (2014): *Study of pulsed character of radiationemitted by wireless telecommunication systems*. Institut scientifique de service public, Liège, Belgium. https://www.issep.be/wp-content/uploads/7IWSBEEMF_B-Vatovez_W-Pirard.pdf.

Puranen L, Jokela K, (1996): Radiation hazard assessment of pulsed microwave radars. *Journal of Microwave Power and Electromagnetic Energy* 31(3), 165–177.

Regel SJ, Gottselig JM, Schuderer J, Tinguely G, Rétey JV, Kuster N, Landolt H-P, Achermann P, (2007): Pulsed radio-frequency radiation affects cognitive performance and the waking electroencephalogram. *NeuroReport* 18(8), 803–807.

Reiser H, Dimpfel W, Schober F, (1995): The influence of electromagnetic fields on human brain activity. *European Journal of Medical Research* 1(1), 27–32.

Roelfsema PR, Engel AK, König P, Singer W, (1997): Visuomotor integration is associated with zero time-lag synchronization among cortical areas. *Nature* 385(6612), 157–161. https://doi.org/10.1038/385157a0.

Roggeveen S, van Os J, Viechtbauer W, Lousberg R, (2015a): EEG changes due to experimentally induced 3G mobile phone radiation. *PLOS ONE* 10(6), e0129496. https://doi.org/10.1371/journal.pone.0129496.

Roggeveen S, van Os J, Lousberg R, (2015b): Does the brain detect 3G mobile phone radiation peaks? An explorative in-depth analysis of an experimental study. *PLOS ONE* 10(5), e0125390. https://doi.org/10.1371/journal.pone.0125390.

Röschke J, Mann K, (1997): No short-term effects of digital mobile radio telephone on the awake human electroencephalogram. *Bioelectromagnetics* 18(2), 172–176.

Saroka KS, Persinger MA, (2014): Quantitative evidence for direct effects between ionosphere Schumann resonances and human cerebral cortical activity. *International Letters of Chemistry, Physics and Astronomy* 39, 166–194.

Schmid MR, Loughran SP, Regel SJ, Murbach M, Bratic Grunauer A, Rusterholz T, Bersagliere A, Kuster N, Achermann P, (2012): Sleep EEG alterations: Effects of different pulse-modulated radio frequency electromagnetic fields. *Journal of Sleep Research* 21(1), 50–58. https://doi.org/10.1111/j.1365-2869.2011.00918.x.

Schreier N, Huss A, Roosli M, (2006): The prevalence of symptoms attributed to electromagnetic field exposure: A cross-sectional representative survey in Switzerland. *Sozial- und Praventivmedizin* 51(4), 202–209.

Schumann WO, Koenig H, (1954): Uber die Beobachtung von "atmospherics" bei geringsten Frequenzen. *Naturwissenschaften* 8, 183–184.

Sienkiewicz ZJ, Blackwell RP, Haylock RG, Saunders RD, Cobb BL, (2000): Low-level exposure to pulsed 900 MHz microwave radiation does not cause deficits in the performance of a spatial learning task in mice. *Bioelectromagnetics* 21(3), 151–158.

Suhhova A, Bachmann M, Karai D, Lass J, Hinrikus H, (2013): Effect of microwave radiation on human EEG at two different levels of exposure. *Bioelectromagnetics* 34(4), 264–274.

Sullivan D, Mizuseki K, Sorgi A, Buzsáki G, (2014): Comparison of sleep spindles and theta oscillations in the hippocampus. *Journal of Neuroscience* 34(2), 662–674.

Szemerszky R, Zelena D, Barna I, Bárdos G, (2010): Stress-related endocrinological and psychopathological effects of short-and long-term 50 Hz electromagnetic field exposure in rats. *Brain Research Bulletin* 81(1), 92–99. https://doi.org/10.1016/j.brainresbull.2009.10.015.

Vecchio F, Babiloni C, Ferreri F, Curcio G, Fini R, Del Percio C, Rossini PM, (2007): Mobile phone emission modulates interhemispheric functional coupling of EEG alpha rhythms. *European Journal of Neuroscience* 25(6), 1908–1913.

Vecchio F, Buffo P, Sergio S, Iacoviello D, Rossini PM, Babiloni C, (2012): Mobile phone emission modulates event-related desynchronization of alpha rhythms and cognitive-motor performance in healthy humans. *Clinical Neurophysiology* 123(1), 121–128.

Vecsei Z, Knakker B, Juhász P, Thuróczy G, Trunk A, Hernádi I, (2018a): Short-term radiofrequency exposure from new generation mobile phones reduces EEG alpha power with no effects on cognitive performance. *Scientific Reports* 8(1), 18010. https://doi.org/10.1038/s41598-018-36353-9.

Vecsei Z, Thuróczy G, Hernádi I, (2018b): The effect of a single 30-min long term evolution mobile phone-like exposure on thermal pain threshold of young healthy volunteers. *International Journal of Environmental Research and Public Health* 15(9), 1849–1860.

Van Wijngaarden E, Savitz DA, Kleckner RC, Cai J, Loomis D, (2000): Exposure to electromagnetic fields and suicide among electric utility workers: A nested case-control study. *Occupational and Environmental Medicine* 57(4), 258–263.

Von Klitzing L, (1995): Medical/biological study (experimental study) low-frequency pulsed electromagnetic fields influence EEG of man. *Physica medica* 11, 77–80.

Vorobyov V, Janać B, Pesić V, Prolić Z, (2010): Repeated exposure to low-level extremely low frequency-modulated microwaves affects cortex-hypothalamus interplay in freely moving rats: EEG study. *International Journal of Radiation Biology* 86(5), 376–383. https://doi.org/10.3109/09553000903567938.

Wagner P, Röschke J, Mann K, Hiller W, Frank C, (1998): Human sleep under the influence of pulsed radiofrequency electromagnetic fields: A polysomnographic study using standardized conditions. *Bioelectromagnetics* 19(3), 199–202.

Wilson BW, (1988): Chronic exposure to ELF fields may induce depression. *Bioelectromagnetics* 9(2), 195–205.

Yakymenko I, Tsybulin O, Sidorik E, Henshel D, Kyrylenko O, Kyrylenko S, (2016): Oxidative mechanisms of biological activity of low-intensity radiofrequency radiation. *Electromagnetic Biology and Medicine* 35(2), 186–202.

Yakymenko I, Burlaka A, Tsybulin I, Brieieva I, Buchynska L, Tsehmistrenko I, Chekhun F, (2018): Oxidative and mutagenic effects of low intensity GSM 1800 MHz microwave radiation. *Experimental Oncology* 40(4), 282–287.

Yang L, Chen Q, Lv B, Wu T, (2017): Long-term evolution electromagnetic fields exposure modulates the resting state EEG on alpha and beta bands.*Clinical EEG and Neuroscience* 48(3), 168–175.

Zhang J, Sumich A, Wang GY, (2017): Acute effects of radiofrequency electromagnetic field emitted by mobile phone on brain function. *Bioelectromagnetics* 38(5), 329–338. https://doi.org/10.1002/bem.22052.

Zhang F, Wang F, Yue L, Zhang H, Peng W, Hu L, (2019): Cross-species investigation on resting state electroencephalogram. *Brain Topography* 32(5), 808–824. https://doi.org/10.1007/s10548-019-00723-x.

Zheng Y, Xia P, Dong L, Tian L, Xiong C, (2021): Effects of modulation on sodium and potassium channel currents by extremely low frequency electromagnetic fields stimulation on hippocampal CA1 pyramidal cells. *Electromagnetic Biology and Medicine* 17, 1–12.

8 Electro-hypersensitivity as a Worldwide, Man-made Electromagnetic Pathology: *A Review of the Medical Evidence*

Dominique Belpomme[1,2] and Philippe Irigaray[2]

[1]Medical Oncology Department, Paris University, Paris, France
[2]European Cancer and Environment Research Institute, Brussels, Belgium

CONTENTS

Abstract		298
8.1	Introduction	299
8.2	Review of the Bioclinical Evidence	301
	8.2.1 Historical, Scientific, and Institutional Background	301
	8.2.2 How to Define EHS	302
	8.2.3 Association with MCS	303
	8.2.4 Physical Symptoms: Toward the Individualization of a Specific Neurologic Syndrome	304
	8.2.5 First Complete Clinical Description of Symptoms in EHS Self-reported Patients	306
	8.2.6 Search for Reliable Disease Biomarkers	311
	8.2.7 The Use of Biomarkers for Objective Diagnosis	313
	8.2.7.1 Low-grade Inflammation	313
	8.2.7.2 BBB Disruption/Opening	315
	8.2.7.3 Autoimmune Response and the Open Question of EHS-associated Impaired Immunity	316
	8.2.7.4 6-Hydroxymelatonin Sulfate in Urine	316
	8.2.7.5 Nitrosative-Oxidative Stress	318
	8.2.7.6 Disease Presentation without Biomarker Detection	322
	8.2.8 The Critical Role of Neurotransmitters. EHS as a Proven Brain Pathological Disorder	323
	8.2.9 Genotyping of Drug Metabolism-related Enzymes and Possible Screening Intervention	324
	8.2.10 Cerebral Imaging	325
	8.2.11 Diagnostic Criteria, Sensitivity, Specificity, and Reproducibility of the Biological Methods	328
8.3	Pathogenesis and Etiology	329

	8.3.1	Pathophysiological Significance of Biomarkers	329
	8.3.2	Search for Bioclinical Triggers, Pathogenesis, and Etiology	331
		8.3.2.1 Symptomatic and Biological Triggers in EHS Patients. The Importance of Good Clinical Practice in Carrying out Well-designed Provocation Studies	332
		8.3.2.2 Search for Hypersensitivity Characterization	335
		8.3.2.3 Search for Etiology	341
8.4	Therapeutics and Methods of Prevention		343
8.5	Social Consequences of EHS and How to Counteract the Increasing Problem		345
8.6	General Discussion and Conclusion		345
References			348

Keywords: biomarkers, EHS, electro-hypersensitivity, electromagnetic field, extremely low frequency, MCS, multiple chemical sensitivity, non-ionizing radiation, oxidative stress, provocation studies, radio frequency.

Abbreviations: 5-HT: 5-hydroxytryptamine, i.e., serotonin. 6-OHMS: 6-hydroxymelatonin sulfate. AD: adrenaline. BBB: blood–brain barrier. BBF: brain blood flow. CAT: catalase. CNS: central nervous system. CoQ10: coenzyme Q10. CRP: C reactive protein. CT: cerebral tomosphygmography. DA: dopamine. DCS: Digital Cellular System. DECT: Digitally Enhanced Cordless Telecommunications. DMN: default mode network. DOPAC: dihydroxyphenyl-acetic acid. ECG: electrocardiogram. EEG: electroencephalogram. EHS: electro-hypersensitivity. ELF: Extremely Low Frequency. EMF: electromagnetic field. EMG: electromyogram. EMR: electromagnetic radiation. ESP: electric skin potential. FDA: Food and Drug Administration. fMRI: functional magnetic resonance imaging. FPP: fermented papaya preparation. GPx: glutathione peroxidase. GR: glutathione reductase. GSH: reduced glutathione. GluT: total glutathione. GSM: Global System for Mobile Telecommunications. GSSG: oxidized glutathione. GST: glutathione S-transferase. Hb: hemoglobin. HRV: heart rate variability. hs-CRP: high-sensitivity C reactive protein. HSP: heat shock protein. ICD: International Classification of Diseases. IEI: idiopathic environmental intolerance. IEI-EMF: idiopathic environmental intolerance attributed to EMF. IgE: immunoglobulin E. IgG: immunoglobulin G. IPCS: International Program on Chemical Safety. MCS: multiple chemical sensitivity. MDA: malondialdehyde. MRI: magnetic resonance imaging. MT: mobile telephony. NIEHS: US National Institute of Environmental Health Sciences. NIH: US National Institute of Health. NMRI: Naval Medical Research Institute. norAD: noradrenaline. NOS: nitric oxide synthase. NTT: nitrotyrosine. OS: oxidative stress. P_0: myelin protein zero. PET: positron emission tomography. PNS: peripheral nervous system. RBC: red blood cells. RF: Radio Frequency. ROS: reactive oxygen species. SCBF: skin capillary blood flow. SD: Standard Deviation. SE: Standard Error. SNP: single nucleotide polymorphic. SOD: superoxide dismutase. TBARs: thiobarbituric acid-reactive substances. TDU: transcranial doppler ultrasound. TILT: toxicant-induced loss of tolerance. UCTS: ultrasonic cerebral tomosphygmography. VDT: visual display terminal. VGIC: voltage-gated ion channel. VLF: Very Low Frequency. WC: wireless communication. WHO: World Health Organization. Wi-Fi: Wireless Fidelity. WLAN: wireless local area network.

ABSTRACT

Much of the controversy over the causes of electro-hypersensitivity (EHS) and multiple chemical sensitivity (MCS) lies in the absence of both recognized clinical criteria and objective biomarkers for widely accepted diagnosis. However, there are, presently, sufficient clinical, biological, and radiological data for EHS to be acknowledged as a distinctly well-defined, objectively identified, and characterized neurologic pathological disorder. Therefore, patients who self-report suffering from EHS should be diagnosed and treated on the basis of currently available biological tests and the use of suitable cerebral imaging. Because we have shown that EHS is frequently associated with MCS in EHS patients and that both those individualized clinical entities share a common pathophysiological mechanism for symptom occurrence, it appears that EHS and MCS can be identified as a unique neurologic pathological syndrome, whatever their precise causal origin is. In this review, we distinguish the etiology of EHS itself from the environmental causes that trigger symptoms and subsequent pathophysiological changes after EHS occurrence. Contrary to present scientifically unfounded claims, we indubitably refute the hypothesis of a nocebo effect to account for the genesis of EHS and its presentation in EHS self-reported patients. We also refute the erroneous concept that EHS could be reduced to a vague "functional impairment". The hypersensitivity that characterizes EHS appears to be a persistent and most often irreversible pathological state, as is also the case for sensitivity to chemicals in MCS-bearing patients. Taken into consideration the WHO-proposed causality criteria, we argue that EHS may, in fact, be causally related to increased exposure to manmade electromagnetic fields (EMFs) and, in a limited number of cases, to marketed environmental chemicals. We, therefore, appeal to all governments and international health institutions and, more particularly, the WHO to urgently consider this growing EHS-associated pandemic plague and to acknowledge EHS as a new real disorder.

8.1 INTRODUCTION

It has been shown that man-made (artificial) electromagnetic fields (EMFs) and corresponding electromagnetic radiation (EMR) differ physically from those of natural origin (Panagopoulos et al. 2015) in that they are totally polarized and coherent and that their specific polarized physical properties may account for their toxicological effects. EMF emissions associated with wireless communications (WC) involve microwave/Radio Frequency (RF) carrier signals (300 kHz–300 GHz) which are modulated and pulsed by Extremely Low Frequencies (ELF) (3–3000 Hz) in order to transmit increasing amounts of information and simultaneously accommodate increasing numbers of users. Thus, WC EMFs/EMR are a combination of ELF and RF EMFs (Panagopoulos 2019). Both ELF and RF EMFs have been categorized as possibly carcinogenic to humans by the International Agency for Research on Cancer (IARC 2002; 2013). Toxicological effects induced by artificial EMFs have been shown at the whole-organism level in wild animals (Balmori 2005; 2006; Sharma and Kumar 2010; Cucurachi et al. 2013; Cammaerts and Johansson 2014; Engels et al. 2014; Redlarski et al. 2015; Odemer and Odemer 2019), particularly in insects, by using simulation models (Thielens et al. 2018) as well as in laboratory animals (Panagopoulos et al. 2007; 2010; Soffritti et al. 2016; Falcioni et al. 2018; National Toxicology Program 2018a; 2018b) and more than suspected in humans (Jutilainen et al. 1993; Ahlbom et al. 2000; IARC 2002; 2013; Draper et al. 2005; Garcia et al. 2008; Myung et al. 2009; De Iuliis et al. 2009; Davanipour et al. 2009; Hardell et al. 2013; Mahmoudabadi et al. 2015; Terzi et al. 2016; Belpomme et al. 2018; Irigaray et al. 2018a; Miller et al. 2019; Panagopoulos 2020; Carles et al. 2020; Belpomme and Irigaray 2020). Moreover, pathophysiological effects have been elucidated at tissue and cell levels in humans using *in vitro* (Belyaev et al. 2009; Ma et al. 2014; Chen et al. 2014; Eghlidospour et al. 2017) and *in vivo* research methods (Sonmez et al. 2010; Aldad et al. 2012; Guxens et al. 2013; Balassa et al. 2013; Roshangar et al. 2014; Zhang et al. 2015; Odaci et al. 2016) and at the molecular level (Mathie et al. 2003; Blank and Goodman 2009; Fragopoulou et al. 2012; Beyer et al. 2013; Zuo et al. 2014; Xiong et al. 2015; Zalata et al. 2015), including genetic (Nordenseon et al. 1994; Lai and Singh 1995; Ivancsits et al. 2002; 2003; Diem et al. 2005; Panagopoulos et al. 2007; 2010; Lai 2021), and/or epigenetic molecular processes (Blank and Goodman 1999; 2009; 2011; Belyaev et al. 2006; Rodríguez-De la Fuente et al. 2010; Yang et al. 2012; Leon et al. 2014; Dasdag et al. 2015a).

Although this scientific panorama incriminating the toxic effects of man-made EMFs in all living organisms appears sufficiently clear for most scientists, a few of them – and most of the national or international health bodies with apparent close links to the industry – still dispute the existence of adverse non-thermal or micro-thermal EMF-related effects (Belayev 2005). This is the case with the German non-governmental organization (NGO), called the International Commission on Non-Ionizing Radiation Protection (ICNIRP) (Hardell 2017; Hardell and Carlberg 2020). ICNIRP (1998; 2010; 2020) promotes the scientifically obsolete opinion that the only tissue and health effects induced by man-made EMFs are of thermal nature, which, unfortunately, ICNIRP shares with the World Health Organization (WHO). Particularly, some of these mainstream bodies deny the potential role of man-made EMFs in electro-hypersensitivity (EHS) symptomatic occurrence and, contrary to the present WHO official position of recognizing EHS as a morbid condition, even deny the existence of EHS itself (Belpomme and Irigaray 2021).

We have previously published evidence that (a) EHS is a distinct, newly identified, and objectively characterized neurologic pathological disorder which can be clinically diagnosed, treated, and prevented using peripheral blood and urine molecular biomarkers, and cerebral imaging (Belpomme and Irigaray 2020); (b) EHS and multiple chemical sensitivity (MCS) are associated in some EHS patients, both displaying similar clinical presentation and abnormal biological and radiological changes; thus, EHS and MCS could, in fact, be two etiopathogenic aspects of a unique common pathological syndrome (Belpomme et al. 2015; 2016); (c) EHS, like many other chronic pathological disorders or diseases, may be associated with detectable low-grade inflammation and nitroso-oxidative stress with consequent blood–brain barrier (BBB) disruption/opening (Belpomme et al. 2015; Belpomme and Irigaray 2020), as in Alzheimer's disease (Heneka and O'Banion 2007; Bell and Zlokovic 2009; Erickson and Banks 2013) and in other psychoneurologic pathologies (Patel and Frey 2015) for which such BBB disruption/opening has been shown; and (d) the mechanism through which EHS and related symptoms occur is now understood (Belpomme et al. 2015; Belpomme and Irigaray 2020).

In a recent international scientific consensus report, many scientists have recognized molecular biomarkers and imaging to be of critical value to the study of EHS (Belpomme et al. 2021). However, a pending important question is the role of EMF exposure in triggering clinical symptoms and biological changes and in causing EHS itself. In fact, a great difficulty comes from the confusion existing in the scientific literature in addressing the problem of man-made EMF toxicity and EHS causality. As emphasized in this recent consensus report, a clear-cut distinction must be made between the daily environmental causes that trigger pathophysiological changes and clinical symptoms after EHS has occurred (pathogenesis) and the causal origin of EHS itself (etiology). At present, the lack of reliable EMF-exposure provocation studies may explain why most mainstream medical, sanitary, and societal bodies still believe 1) that there is no sufficient scientific proof to assert that clinical symptoms experienced by EHS self-reported patients are really caused by EMF exposure and 2) that EHS genesis could not be the consequence of excessive man-made EMF exposure in genetically susceptible people. Additionally, since the WHO officially stated in 2005 (WHO 2005), and more recently (WHO 2014), that EHS is a "disabling condition" associated with "non-specific symptoms that lack apparent toxicological or physiological basis or independent verification" and with "no clear diagnosis criteria", it is erroneously thought that EHS cannot be diagnosed medically and is not causally related with EMF exposure.

To the contrary, we show, here, that EHS is, in fact, a new and worldwide man-made EMF-related environmental pathology that can be medically diagnosed, treated, and prevented. To this end, we (a) recall the historical background of EHS; (b) specify how it can be defined; (c) show that it may be associated with MCS; (d) describe the clinical symptomatic picture occurring in the so-called EHS or EHS/MCS self-reported patients; (e) show how EHS can, in fact, be objectively identified and characterized using biomarkers and cerebral imaging; (f) describe the clinical and biological diagnostic criteria which can be used in medical practice and research and discuss the sensitivity, specificity, and the reproducibility of the method used; (g) report that an etiopathological model, based on the underlying molecular mechanisms resulting from biomarkers and imaging findings, is now possible; (h) discuss the problem of EMF-related bioclinical triggers, hypersensitivity pathogenesis and etiology (providing arguments that EHS-associated bioclinical changes and EHS itself can, in fact, be causally related with man-made EMFs and, in some cases, with marketed environmental chemicals); (i) overview the differing presently available empirical therapeutic and preventive methods; (j) analyze the societal consequences of EHS on the patient's daily life and the public health measures that should urgently be taken by governments and health institutions to counteract the present worldwide emerging intolerance to man-made EMFs and the present growing EHS-associated pandemic plague. Most specifically, we appeal to the WHO to promptly rectify its present official negative EHS position because much medical progress in understanding EHS and taking care of patients has been made since its 2005 and 2014 official statements.

8.2 REVIEW OF THE BIOCLINICAL EVIDENCE

8.2.1 Historical, Scientific, and Institutional Background

The term electromagnetic hypersensitivity, which was then commonly named EHS, was first proposed in 1991 by William Rea to identify the pathological condition of patients reporting health effects while being experimentally exposed to EMF versus sham-exposed and being compared to healthy controls in a controlled area (Rea et al. 1991). This term was then re-used in 1997 in a report provided by a European group of experts for the European Commission to clinically describe this unusual pathological condition, which may implicate EMF exposure as symptomatic trigger (Bergqvist and Vogel 1997). In 2004, because of the worldwide increase of EHS, the WHO organized an international scientific workshop in Prague to define and characterize EHS. Although not acknowledging EHS as being caused by EMF exposure, because of a lack of available correlation studies, the Prague working group clearly defined EHS as "a phenomenon where individuals experience adverse health effects while using or being in the vicinity of devices emanating electric, magnetic, or electromagnetic fields" (Mild et al. 2006). According to a previous 1996 WHO-sponsored International Program on Chemical Safety (IPCS) conference in Berlin on MCS (Report of the Workshop on Multiple Chemical Sensitivities 1996), it was recommended to qualify such unknown new environmental pathological conditions under the term of "idiopathic environmental intolerance (IEI)". Thus, following the 2004 Prague workshop, instead of using the term EHS, it was recommended to use the term IEI attributed to EMF (IEI-EMF) to name this particular pathological condition because of the lack of a proven causal link between EHS and EMF exposure and no proven pathophysiological mechanism linking EMF exposure with clinical symptoms. However, because the term EHS was in common use, the WHO officially acknowledged EHS as an adverse health condition in its 2005 fact sheet No. 296 (WHO 2005) and further reported on public health and mobile phone use in its 2014 fact sheet No. 193, claiming a lack of proven causal link between EMF emissions from mobile phones and health effects and no proven underlying pathophysiological mechanism accounting for such effects (WHO 2014). But it was already shown that mobile phone and WC EMFs cause oxidative stress (OS) by generating reactive oxygen species (ROS) in the cells, and DNA damage (for example Lai and Singh 1995; Ivancsits et al. 2002; 2003; Diem et al. 2005; Panagopoulos et al. 2007; 2010; 2021; De Iuliis et al. 2009; Phillips et al. 2009; Yakymenko et al. 2016; and many more studies), and the biophysical mechanism of action was also known already (Panagopoulos et al. 2000; 2002). Thus, the WHO ignored a great number of significant scientific findings.

Indeed, since the 2005 and 2014 WHO official positions and according to expressed requirements (Repacholi 1998), much progress has been made to understand the biophysical and biochemical mechanisms leading to the health effects of EMFs (Panagopoulos et al. 2015; 2021; Yakymenko et al. 2016) and their pathophysiological effects (Belpomme et al. 2015; 2018; Irigaray et al. 2018a) and, more particularly, to identify and characterize EHS as a new pathological disorder (Belpomme and Irigaray 2020). Such progress on EMF effects and EHS was summarized in an international consensus meeting held in 2015 at the Royal Belgium Academy of Medicine in Brussels (Carpenter and Belpomme 2015) and, since then (as reported here), further new progress has been made. Table 8.1 summarizes the historical scientific steps and WHO statements concerning MCS and EHS.

TABLE 8.1
Historical Steps Identifying EHS and MCS

Year	Development	Reference
1962	First identification and description of MCS	(Randolph 1962)
1991	First identification and description of EHS	(Rea et al. 1991)
1996	WHO-sponsored workshop: MCS classified as idiopathic environmental intolerance (IEI) (Berlin)	(Report of the Workshop on Multiple Chemical Sensitivities 1996)
1997	Possible health implication of EMF exposure: a report prepared by a European group of experts for the European Commission (Stockholm)	(Bergqvist and Vogel 1997)
1999	Definition of MCS:1999 consensus meeting (Atlanta USA)	(Bartha et al. 1999)
2004	WHO-sponsored workshop: identification of idiopathic environmental intolerance attributed to EMF (Prague)	(Mild et al. 2006)
2005	WHO fact sheet No. 292, aiming at defining EHS	(WHO 2005)
2014	WHO fact sheet No. 193: EMF and public health; mobile phone	(WHO 2014)
2015	Fourth Paris Appeal Colloquium; a focus on EMF and EHS (Brussels)	(Carpenter and Belpomme 2015)
2021	The critical importance of molecular biomarkers and imaging in the study of EHS. An international scientific consensus report	(Belpomme et al. 2021)

8.2.2 How to Define EHS

There remains persisting confusion between EHS, which was acknowledged by the WHO (2005), and IEI-EMF, which was proposed one year before during the 2004 WHO-sponsored Prague meeting (Mild et al. 2006). Although the clinical identification of EHS was proposed in 1991, a major pending question is how to define EHS and how to distinguish it from IEI-EMF, as both individualized pathological entities have been symptomatically described and recognized by the WHO. According to present developments in the scientific literature, there is, indeed, a lack of identifying criteria allowing a clear distinction of EHS from IEI-EMF (Köteles et al. 2013; Belpomme and Irigaray 2021). EHS, as indicated above, is presently considered by the WHO as a disability condition not proven to be causally related to EMFs and, not specifically subject to medical diagnosis and treatment, while IEI-EMF is defined as an idiopathic environmental intolerance condition possibly attributed to EMF.

There are several reasons accounting for the difficulty in distinguishing EHS from IEI-EMF and defining EHS more properly. 1. Most scientists did not focus their research on the precise clinical description of symptoms in EHS self-reported patients, as it was usually considered that these symptoms are all subjective and non-specific. They mainly focused on causality, following the Prague IEI-EMF recommendation, by using so-called provocation studies in EHS self-reported patients in an attempt to prove intolerance to EMFs. Unfortunately, this was done before EHS was clearly biologically and clinically defined, so many of these studies used incorrect methodology (see below and Section 8.3.2.1) (Belpomme et al. 2021). 2. As already reported in 1997, clinical symptom occurrence in EHS patients may depend on the type of electromagnetic devices used by the patients and/or the type of electromagnetic sources situated in their close environment. For example cutaneous lesions are associated with the use of visual display terminals (VDT) (Lindén and Rolfsen 1981; Nilsen 1982; Knave et al. 1985; Berg 1988; Bergqvist and Wahlberg 1994); headache and ear problems are associated with excessive use of mobile phones; headache and other general symptoms with Wireless Fidelity (Wi-Fi) (wireless internet connection, also called WLAN: wireless local area network) exposure, exposure to various antennas, or exposure to EMFs from electric power transmission lines (Bergqvist 1984). All these different types of EMF exposure encouraged the misleading assertion that symptoms are so diverse and non-specific that they could not be considered as part of one general characteristic symptomatic picture. 3. Provocation studies – aimed at testing the causal role of EMF exposure versus sham-exposure in triggering EMF-related symptoms – instead of testing EHS self-reported patients, have most frequently investigated healthy volunteers (see Section 8.3.2). Moreover, due to a lack of an established and carefully designed objective methodology (Belpomme et al. 2021), most provocation tests applied in EHS self-reported patients (contrary to many of those applied in healthy volunteers) have provided negative results, leading to improper assessment of EHS in most studies, while the EMF-induced clinical and biological toxic effects in normal people have been clearly established (see Section 8.3.2). This lack of a rigorous clinical method, particularly the lack of objective inclusion and endpoint assessment criteria used to investigate EHS patients (Belpomme et al. 2021), may explain why the definition and cause of EHS are still controversial, to such a point that – contrary to the WHO official reports – some scientists still question the existence of EHS itself (Genuis and Lipp 2012; Belpomme and Irigaray 2021). Moreover, this lack of identifying criteria may explain why EHS description/symptomatology overlaps various types of environmental intolerance illnesses such as MCS, fibromyalgia, chronic fatigue syndrome, the building sickness syndrome, the "Persian Gulf War syndrome", and other diversely designated environmental illnesses (Palmquist et al. 2014). 4. Research did not focus on biomarkers and radiological investigation analysis to objectively identify and characterize EHS as a new pathological disorder, independently of its putative EMF-related symptomatic trigger and causal origin (Belpomme et al. 2021).

Because the term "EHS" is much more prevalent in scientific literature than "IEI-EMF", and because according to Baliatsas (2012), EHS is "the only parameter that clearly distinguishes sensitive from control individuals using the causal attribution of symptomatology to EMF exposure", we shall consider EHS reported by EHS patients as the clinical cornerstone of the environmental pathology we are dealing with and shall describe its symptomatic and biological characterization on evidence provided by face-to-face medical questioning and physical examination of so-called EHS self-reported patients. However, we will propose below a more restrictive definition of EHS and a clear-cut distinction between EHS and IEI-EMF on the basis of their different clinical and pathophysiological properties. In this way, we hope to provide a more precise definition of these two clinically individualized and recognized disorders and to promote more efficient future research. Thus, we propose that designation of EHS be restricted to the presumable pathological intra-corporal acquisition of hypersensitivity to EMFs (as is the case for sensitivity to chemicals in MCS patients), while IEI-EMF will be defined as presumable EMF-related environmental intolerance (see Section 8.3.2 on bioclinical triggers and on hypersensitivity pathogenesis and etiology). We note that such a proposed pathophysiological definition, based on a decrease in the environmental tolerance threshold to better define EHS and MCS, is similar to that of toxicant-induced loss of tolerance (TILT) proposed by Miller (1999) and may, in fact, introduce a better definition for environmental sensitivity-related diseases.

However, regardless of this distinction, it clearly appears that, in the absence of any recognized biomarkers and radiological criteria, the attribution of deleterious health outcomes to man-made EMF exposure, as reported by EHS self-reported patients, constitutes an inevitable first necessary research step, considering EHS in these patients as a new working framework entity to investigate (Baliatsas et al. 2012). In any case, the need of reliable provocation EMF exposure studies to prove the connection of the symptoms and biomarkers with man-made EMFs remains.

8.2.3 Association with MCS

In 1962, using a controlled environment, the American allergist T.G. Randolph first clearly described, clinically, what is today commonly called MCS (Randolph 1962). However, according to Rea (2016), several precursors, including Hippocrates had already mentioned people who could not tolerate odor-associated food and drink.

Since Randolph's seminal description, MCS has been acknowledged as idiopathic environmental intolerance during the 1996 Berlin WHO-sponsored workshop and has been more precisely identified clinically during a 1999 consensus meeting in Atlanta (Bartha et al. 1999) by characteristic criteria as: "(a) a chronic condition, (b) with reproducibility, (c) in response to low levels of exposure, (d) to multiple unrelated chemicals and, (e) improves when incitants are removed". To these, a sixth diagnostic criterion has been added, specifying that "symptoms occur in multiple organ systems" (Lacour et al. 2005).

Contrary to some early belief, including that of Randolph (1962) and more recently that of the WHO in the International Classification of Diseases (ICD)10,* MCS should not be considered as an allergic disorder but as an environmental intolerance-related disorder with neurogenic inflammation manifested by sensitivity to environmental chemicals (mostly man-made chemicals) leading to the increased sensation of odor, burning and irritation of the airway, involving the nose, the oral cavity and the pharyngo-laryngeal tract, and sometimes the bronchus (Bell et al. 1992; Meggs 1993; 1997; Hillert et al. 2007).

Using the six aforementioned internationally recognized MCS clinical criteria, we have shown for the first time that MCS is associated with EHS in about 30% of EHS self-reported cases, meaning that the two different etiopathogenic processes may, in fact, be part of a unique common pathological syndrome (Belpomme et al. 2015). This hypothesis is reinforced by the fact that many EHS patients, in addition to reporting hypersensitivity to EMFs, also report intolerance to many chemicals, without, however, reaching the MCS diagnostic standards (based on the six aforementioned international criteria). This could explain why, in addition to EHS, MCS should also be studied clinically and pathophysiologically, if we want to clearly understand what EHS really is and correctly enlighten its pathophysiology (see Section 8.3.1 on pathophysiology).

8.2.4 Physical Symptoms: Toward the Individualization of a Specific Neurologic Syndrome

Clinical symptoms presumably related to microwave exposure were initially reported by Soviet scientists (Dodge 1969; Carpenter 2015). They consisted of headaches, fatigue, loss of appetite, insomnia, loss of concentration and immediate memory, transient cardiovascular dysfunction, and labile emotional behavior. Some or all of these symptoms were described particularly in people exposed to microwave radar EMR. During the Soviet period, this symptomatologic description was not acknowledged by Western scientists. However, in 1971 and then in 1972 in a revised form, the US Naval Medical Research Institute (NMRI) was able to count more than 2,500 references on the biological and clinical response to RF/microwave EMR published up to April 1972 in the global scientific literature (Glaser 1972).

In 1979, the clinical symptoms reported to be caused by microwaves were referred to as a new clinical syndrome named "microwave syndrome" (Pollack 1979). This particular clinical syndrome was considered to involve the nervous system and to be characterized by symptoms such as fatigue, headaches, dysesthesia, and various central nervous system (CNS) autonomic effects occurring in RF/microwave equipment workers. This microwave syndrome is symptomatically tantamount to the experimentally identified pathological disorder termed hypersensitivity to EMFs (i.e., EHS) by Rea (1991). Although the microwave syndrome and the EHS are very similar in their symptomatic picture, they differ in the acknowledgment of their etiology, as in the microwave syndrome, the etiology is ascertained (microwave), while in EHS, the causal role of EMF exposure remains a pending question for many scientists. Therefore, we believe that the microwave syndrome must not be confused with EHS, although they are similar symptomatically.

A first approach in describing the adverse health effects associated with exposure to man-made EMFs was made in Sweden by Bergqvist (1984), who reported in a well-documented overview article the clinical symptoms occurring in people using VDT. Recorded symptoms included eye problems, ocular disturbance with change in visual performance, musculoskeletal discomfort,

---

* In the ICD-10 code T78.4, MCS is registered under the category "other and unspecified allergy".

facial skin rashes, stress and psychological distress particularly involving mood disturbance, and adverse pregnancy outcomes. Although it was shown that there was an increased number and mobilization of mast cells in the skin of normal volunteers using VDT or television (TV) (Johansson et al. 2001), suggesting this phenomenon could be EMF-related, no clear causal relationship could be established between symptom occurrence and VDT- or TV-related EMF exposure, neither between symptoms and VDT use, except for eye problems and musculoskeletal discomfort. Thus, this observational study could not relate specifically any symptom occurrence to EMF exposure.

Following this VDT study, Bergqvist and Vogel (1997), with other European experts working for the European Commission conducted a multinational questionnaire-based survey and reported in 1997 that patients who claim to be electro-hypersensitive have frequently "neurasthenia" symptoms, headache and skin symptoms and, less frequently, sleep disturbance and anxiety. However, again, these symptoms were considered non-specific and not causally related to EMF exposure. In fact, this large multinational questionnaire-based survey was unable to clinically define the real symptomatic picture presented by so-called EHS patients and its possible connection with EMF exposure.

However, in 1998, it was reported by the US National Institute of Environmental Health Sciences (NIEHS) that health effects could be caused by exposure to power-line frequency (50–60 Hz) electric and magnetic fields (NIEHS 1998), while in 2000, an increased prevalence of headache among mobile phone users was observed in Singapore (Chia et al. 2000). Then in 2002, a study in France described the clinical symptoms ascribed to mobile phone use in a French engineering school (Santini et al. 2002), and a year later, similar symptoms were ascribed to mobile telephony (MT) base station proximity (Santini et al. 2003). In both epidemiological studies, the most common symptoms were headaches, loss of concentration and immediate memory, sleep disturbance, and tiredness. In addition, burning sensation to the face, pricking and warmth in the ear were reported among mobile phone users, while skin problems, visual and hearing disturbances, dizziness, irritability, and depressive tendency were reported in people living in less than 300 meters from MT base stations. These latter symptoms were also described by another study in Australia (McLean 2008).

In fact, many studies focused on the symptomatic risk of EMF-exposed people from the general population but not specifically on EHS self-reported patients. All these general population-based studies were based on telephone survey or mailed or web-based questionnaires and not on face-to-face questioning and physical examination of the individuals. Moreover, most of these studies in the general population investigated one or few self-reported symptoms such as headache (Chia et al. 2000; Milde-Busch et al. 2010; Sudan et al. 2012; Auvinen et al. 2019), tinnitus (Frei et al. 2012; Medeiros and Sanchez 2016; Auvinen et al. 2019), sleep disturbance (Hutter et al. 2006; Mohler et al. 2012; Monazzam et al. 2014; Huss et al. 2015; Eyvazlou et al. 2016; Tettamanti et al. 2020), cognitive deficiency (Hutter et al. 2006), psychiatric symptoms (Silva et al. 2015), and eye cataracts (Zaret 1973). Thus, they did not report a detailed description of the complete symptomatic picture associated with EMF exposure.

Other studies were based on the various healthy population categories exposed to EMFs, such as EMF-exposed workers (Broadbent et al. 1985; Liu et al. 2014); mobile phone and/or computer users (Chia et al. 2000; Santini et al. 2002; Korpinen and Pääkkönen 2009; Augner et al. 2012; Sudan et al. 2012; Küçer and Pamukçu 2014; Szyjkowska et al. 2014; Cho et al. 2016a; 2016b; Medeiros and Sanchez 2016; Stalin et al. 2016; Kumar et al. 2018; Auvinen et al. 2019); people exposed to ELF EMFs, mainly 50–60 Hz from power lines and transformers (McMahan and Meyer 1995; NIEHS 1998; Baliatsas et al. 2011; Barsam et al. 2012; Bolte et al. 2015; Monazzam et al. 2014; Bonhomme-Faivre et al. 1998; Barsam et al. 2012; Bagheri Hosseinabad et al. 2019); or people exposed to EMFs from MT base station antennas at relatively close distances (Santini et al. 2003; Hutter et al. 2006; McLean 2008; Augner and Hacker 2009; Röösli et al. 2010a; Baliatsas et al. 2011; Bortkiewicz et al. 2012; Monazzam et al. 2014; Malek et al. 2015; Martens et al. 2017; 2018).

In addition, there are many studies in infants, children, and teenagers (Mortazavi et al. 2007; Söderqvist et al. 2008; Milde-Busch et al. 2010; Sudan et al. 2012; Byun et al. 2013; Zheng et al.

2014; 2015; Chiu et al. 2015; Huss et al. 2015; Schoeni et al. 2016; 2017; Roser et al. 2016; Calvente et al. 2016; Durusov et al. 2017; Liu et al. 2019; Cabré-Riera et al. 2019) showing higher susceptibility and sensitivity of these groups to man-made EMFs than adults (Kheifets et al. 2005; McInerny 2012). But these epidemiological studies (although most symptoms they recorded were very similar to those described in the microwave syndrome and to those reported by the EHS self-reported patients) did not record all symptoms and did not find a statistically significant association of symptoms with EMF exposure, with the exception of headache, cognitive deficiency, and sleep disturbance, which appeared most often to be correlated with EMF intensity and/or duration of exposure (see Section 8.3.2).

Surprisingly, only a few studies have focused specifically on the description of the health symptoms of EHS self-reported patients. Such a clinical observational investigation prevents erroneous conclusions, such as that symptoms are subjective, non-specific, and not causally related to EMF exposure (Levallois 2002; Röösli 2008; Röösli et al. 2010b; Baliatsas et al. 2014). More recently, studies allowing a more precise description of symptoms in such patients were conducted in Finland (Hagström et al. 2013) and in Netherlands (van Dongen et al. 2014). In both studies, the percentage of women was higher in the EHS group than in the general population, suggesting some gender-related genetic susceptibility, as reported in our studies. In the Dutch study, the number of symptoms was higher among people recruited by NGOs than in the general population (van Dongen et al. 2014), while in the Finnish study, it was shown that the number of symptoms during the acute phase of EHS is higher than before (Hagström et al. 2013). Table 8.2 summarizes all known major original published studies reporting the symptomatic picture in EHS patients.

All these studies provide a similar but not exhaustive symptomatic picture in EHS self-reported patients, although the selection procedures were different among studies, and the EMF sources were often not well characterized.

In fact, as emphasized by Carpenter (2015), the strongest evidence that EHS is a real syndrome, similar to the microwave syndrome (which was first described), comes from initial cases reported from 1980 to 2000 of acute high intensity exposure of healthy people to microwaves, which resulted in "prolonged illness" (Williams and Webb 1980; Forman et al. 1982; Schilling 1997; 2000; Reeves 2000). Such clinical abnormalities had also been described in 1998 in mobile phone users by Hocking (1998). The same researchers documented neurological changes in C-fiber nerves induced by mobile phone exposure (Hocking and Westerman 2002). Furthermore, in 2011, in a single EHS case, it was demonstrated that such neurological abnormalities were part of a novel neurological syndrome (Mc Carty et al. 2011). The study employed a double-blind EMF exposure (60 Hz, 300 V/m electric field, continuous-wave or pulsed by 10 Hz on/off pulsations) versus sham-exposure provocation test, specifically designed to minimize unintentional sensory reactions. In this exemplary experimental study, the symptoms were causally attributed primarily to the pulsed EMF exposure in comparison with sham-exposure, rather than to the continuous-wave EMF exposure (see Section 8.3.2, specificaly 8.3.2.1).

8.2.5 First Complete Clinical Description of Symptoms in EHS Self-reported Patients

Since 2009, we have constituted and prospectively maintained a database which presently includes more than 3,000 EHS and/or MCS self-reported cases. It appears, so far, to be the most important series of such patients worldwide, and it was registered by the French Committee for the Protection of Persons (CPP), with registration number 2017-A02706-47, as well as in the European Clinical Trials Database (EudraCT), with registration number 2018-001056-36. Moreover, all included patients gave their informed consent for clinical and biological research investigation and were anonymously registered.

By examining this database, we confirmed that EHS in EHS self-reported patients is a newly identified and well-characterized neurological syndrome which can be diagnosed, treated, and prevented using objective biomarkers and cerebral imaging (Belpomme and Irigaray 2020). We have

TABLE 8.2
Original Published Studies Describing the Symptomatic Picture of EHS Self-reported Patients

Authors (Country)	Study Type	EMF Source/Exposure	Effective/Evaluable Cases
Dodge 1969 (USA)	Observational study	Microwaves	391 cases vs 100 controls
Rea et al. 1991 (USA)	Provocation study	0.1 Hz–5 MHz magnetic field	25 patients vs sham, and vs 25 healthy controls
Bergqvist and Vogel 1997 (International)	Nation-wide questionnaire-based survey	General EMF exposure	72 EHS patients
Hillert et al. 2002 (Sweden)	Population-based questionnaires	EMF all types	15.000 participants (general population), including 1.5% EHS patients
Navarro et al. 2003 (Spain)	Questionnaire-based survey and EMF power density measurements	WC EMFs	101 persons close to MT base station
Oberfeld et al. 2004 (Spain)	Questionnaire-based survey and EMF measurement	WC EMFs	201 persons close to two GSM 900–1800 base stations
Schreier et al. 2006 (Switzerland)	Telephone interviews, cross-sectional study	50/60 Hz EMF residential/personal exposures	2,048 participants, including 5% (107) EHS patients
Schüz et al. 2006 (Germany)	Questionnaire-based survey via internet	EMF all types, including mobile phone use and MT base stations	192 persons with health complaints, including 107 EHS patients
Röösli et al. 2010 (Switzerland)	Population-based questionnaire and weekly measurements	EMF all types, including MT base station proximity, mobile phone and cordless phone use, and WLAN (Wi-Fi).	1,375 participants (general population), including 8% (130) EHS patients
Johansson et al. 2010 (Sweden)	Questionnaire-based survey	EMF all types, including domestic appliance and computer, and mobile phone use	45 cases with mobile phone use and 71 EHS patients compared with a population-based sample of 106 subjects and 43 controls
Kato and Johansson 2012 (Japan)	Questionnaire-based survey	EMF all types, including medical device use, mobile and cordless phone use, and proximity to MT base stations	75 EHS patients
Hagström et al. 2013 (Finland)	Questionnaire-based survey via internet	EMF all types, (selection of 50 electrical devices)	194 EHS patients
van Dongen et al. 2014 (Netherlands)	Questionnaire-based survey via internet	EMF all types	188 people sensitive to EMF versus 937 people non-sensitive to EMF
Nordin et al. 2014 (Sweden)	Questionnaire-based survey	EMF all types	113 EHS patients versus 48 controls
Baliatsas et al. 2014 (Netherlands)	Questionnaire-based survey and electronic medical records	EMF all types, including proximity to MT base stations, mobile phone use, domestic appliance, and WLAN (Wi-Fi).	5,789 respondents, including 514 (8.8%) cases with general environmental sensitivity and 202 cases (3.5%) with IEI-EMF (EHS), and the rest of respondents (5,073 cases) used as controls
Belpomme and Irigaray 2020 (France)	Clinical examinations	EMF all types	50 EHS patients in comparison with 50 normal subjects and 50 MCS patients

TABLE 8.3
Summary of Demographic Data Based on the First 727 Serially Individualized Cases of EHS and/or MCS Issued from the Database

Demographic Data	EHS	MCS	EHS/MCS
Number of cases (%)	521 (71.7%)	52 (7.1%)	154 (21.2%)
Age (mean ± SD)	48.2 ± 12.9	48.5 ± 10.3	46.7 ± 11.2
Age (median, range)	48 (16–83)	47 (31–70)	46 (22–76)
Sex ratio (women/men)	344/177	34/18	117/37
Female percentage	66%	65%	76%

Source: Belpomme et al. 2015; Belpomme and Irigaray 2020
[SD: Standard Deviation]

previously reported the detailed demographic data and symptomatic descriptions that were provided by the analysis of EHS and/or MCS self-reported cases. We did not use telephone interviews or internet-based questionnaire surveys, but face-to-face interviews and physical examinations, a method which minimizes patient-dependent subjective or imprecise analysis.

Table 8.3 depicts the demographic data characterizing the population sample of the first 727 already evaluable cases in the database so far being serially investigated. A noteworthy finding, also observed in other studies, is that women appear to be much more susceptible to EHS and/or MCS than men. Indeed, in our series of EHS and/or MCS self-reported patients two thirds are female with no difference between the EHS and MCS groups of patients. The female predominance appears to be more pronounced among patients with both EHS and MCS, where three out of four cases are females. In this series of patients, median age is about 47 years and does not differ between the EHS, MCS, and EHS/MCS individualized groups.

As indicated in **Figure 8.1**, all age categories are represented and mainly include young and old adults and even adolescents with EHS. This may be due to their excessive use of wireless technology (mostly mobile phones, Wi-Fi connected computers, and other devices) and their increased

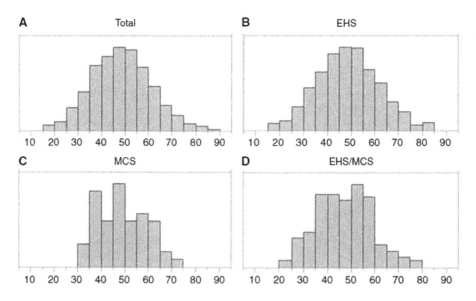

FIGURE 8.1 Relative number of patients in different age categories in the total 727 evaluable cases (A) and in the three groups of EHS (B), MCS (C) and EHS/MCS (D) patients, issued from the database (Belpomme et al. 2015).

sensitivity to EMFs at an early age (Kheifets et al. 2005). In fact, outside the present series of patients, we have observed that infants and children may also be suffering from intolerance to EMF and possibly from EHS. We have also observed that they may display clinical symptoms similar to those of the adults but with increased attention- and concentration-deficit hyperactivity symptoms.

We also compared, retrospectively, the symptomatic picture in EHS self-reported cases to that of apparently healthy people and to patients with MCS or with both EHS and MCS (Belpomme et al. 2018; Belpomme and Irigaray 2020). As depicted in Table 8.4, clinical symptoms in EHS-bearing patients were found to be significantly more frequent than in apparently healthy controls.

TABLE 8.4
Clinical Symptoms in EHS Patients in Comparison with Those of Seemingly Healthy Controls, MCS, and EHS/MCS patients*

Clinical Symptoms	EHS (%) $n = 50$	Controls (%) $n = 50$	p^{**}	MCS (%) $n = 50$	p^{***}	EHS/MCS (%) $n = 50$	p^{****}
Headache	88	0	< 0.0001	80	0.122	96	0.065
Fatigue	88	12	< 0.0001	72	0.0047	94	0.216
Dysesthesia	82	0	< 0.0001	67	0.0149	96	0.002
Concentration/attention deficiency	76	0	< 0.0001	67	0.21	88	0.041
Sleep disturbance	74	6	< 0.0001	47	< 0.0001	92	0.001
Ear heat/otalgia	70	0	< 0.0001	16	< 0.0001	90	< 0.001
Dizziness	70	0	< 0.0001	52	0.0137	68	0.878
Loss of immediate memory	70	6	< 0.0001	56	0.04	84	0.028
Tinnitus	60	6	< 0.0001	35	< 0.001	88	< 0.0001
Depression tendency	60	0	< 0.0001	29	< 0.0001	76	0.022
Transitory cardiovascular abnormalities	50	0	< 0.0001	36	0.046	56	0.479
Myalgia	48	6	< 0.0001	48	1	76	< 0.0001
Ocular deficiency	48	0	< 0.0001	43	0.478	56	0.322
Balance disorder	42	0	< 0.0001	40	0.885	52	0.202
Hyperacusis	40	6	< 0.0001	20	< 0.001	52	0.118
Anxiety/panic	38	0	< 0.0001	19	0.003	28	0.176
Arthralgia	30	18	0.067	24	0.611	56	< 0.001
Irritability	24	6	< 0.001	14	0.071	24	1
Suicidal ideation	20	0	< 0.0001	9	0.027	40	0.003
Emotional behavior	20	12	0.176	16	0.461	20	1
Skin lesions	16	0	< 0.0001	14	0.692	45	< 0.0001
Impaired thermoregulation	14	0	< 0.0001	6	0.236	8	0.258
Confusion	8	0	0.007	0	0.0038	20	0.023

Source: Belpomme et al. 2018; Belpomme and Irigaray 2020

[n: number of cases. p: probability that difference is due to random variation. *These data result from the clinical analysis of 150 consecutive clinically evaluable cases in the database. Symptoms in EHS patients were compared retrospectively with symptoms obtained from a series of 50 apparently normal subjects used as controls. These symptoms were also compared to those occurring in MCS and EHS/MCS patients. Percentages of patients with symptoms were compared by using the chi-square independence test. **p-value between EHS patients and normal controls. ***p-value between EHS patients and MCS patients. ****p-value between EHS patients and EHS/MCS patients.]

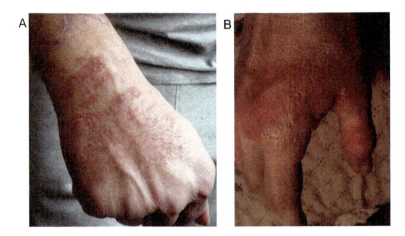

FIGURE 8.2 Examples of skin lesions observed on the hand of an EHS patient (A), and of an EHS/MCS patient (B). (Photographs of subjects registered in the database).

Some of these symptoms were found to be not subjective in contrast to results published in the scientific literature obtained from questionnaire-based surveys of patients with the microwave syndrome (Dodge 1969; Navarro et al. 2003) or in EHS self-reported patients (Bergqvist and Vogel 1997; Hillert et al. 2002; Mild et al. 2006; Schreier et al. 2006; Schüz et al. 2006; Johansson et al. 2010; Röösli et al. 2010b; Kato and Johansson 2012; Baliatsas et al. 2014; Nordin et al. 2014; van Dongen et al. 2014) and statements by the WHO (2005). In many cases, they were confirmed by family members. Moreover, we were able to detect by neurological examination a Romberg sign[*] in about 2%–5% of the cases and observe cutaneous lesions in the face, forearms, or hands (**Figure 8.2**) in 16% and 45% of the cases in EHS and EHS/MCS patients, respectively. Although many symptoms were considered as non-specific, the overall clinical picture, resulting from their prevalence and association, strongly suggests that EHS can be recognized and identified as a typical neurologic disorder regardless of its causal origin. This is also the case for MCS and EHS/MCS combined syndrome (Belpomme et al. 2015; Belpomme and Irigaray 2020). Moreover, we should emphasize that the existence of subjective and non-specific symptoms is not an argument against the medical reality of EHS in suffering patients. In medicine, except for rare pathognomonic symptoms, most symptoms are commonly subjective and not specific to diseases. Rather it is their association within the framework of a clinical picture which contributes to disease identification and characterization.

Table 8.4 also reveals that between EHS patients and MCS patients there is no statistically significant difference for headache, myalgia, arthralgia, balance disorder, concentration/attention deficiency, emotivity, irritability, skin lesions, and impaired thermoregulation. By contrast, dysesthesia, ear heat/otalgia, tinnitus, hyperacusis, dizziness, loss of immediate memory, sleep disturbance, fatigue, depression tendency, and suicidal ideation, appear to be statistically more frequent in EHS than in MCS. Moreover, in the cases of EHS/MCS combined syndrome, most of the symptoms such as headache, dysesthesia, myalgia, arthralgia, tinnitus, and cognitive capability (including loss of immediate memory, concentration/attention deficiency, and tempo-spatial confusion), were found to be significantly more frequent than in EHS alone. This suggests that the presence of an additional chemical sensitivity component to EHS is associated with a more severe pathology. This was especially the case for skin lesions in EHS/MCS patients, which, as previously mentioned, were detected in 45% of the cases, and for physical and psychological suffering and depressive tendency with underlying suicidal ideation in 40% of the cases.

The aforementioned clinical observations strongly suggest that EHS and EHS/MCS are objective somatic disorders which cannot be claimed to originate from non-EMF-related psychologic or

---

[*] Objective neurological imbalance when closing the eyes.

psychiatric causes (Frick et al. 2002) or from simple undefined functional impairment (Belpomme and Irigaray 2020, 2021; Belpomme et al. 2021).

8.2.6 Search for Reliable Disease Biomarkers

The identification and measurement of reliable biomarkers is a critical clinical research step for identifying and characterizing diseases. This is definitely the case for any new pathological disorder or clinical syndrome, such as EHS, MCS, or other environmental intolerance-related illnesses. However, to our knowledge, such bioclinical objective approaches had still not been considered as identified criteria in EHS and MCS studies (Baliatsas et al. 2012). Usually, blood laboratory findings in EHS patients are normal, with the exception, in several cases, of non-specific biological abnormalities of thyroid and liver dysfunctions (Dahmen et al. 2009). We, thus, searched for characteristic biomarkers from previous experimental data obtained from *in vitro* and *in vivo* animal studies of intolerance to man-made EMFs and chose biological tests which could be used clinically for the diagnosis and care of EHS and/or MCS self-reported patients.

Biomarkers used in clinical practice and research have been defined by the Biomarker Definition Working Group of the US National Institute of Health (NIH) (Strimbu and Tavel 2010), and more recently by the Food and Drug Administration (FDA)-NIH Biomarker Working Group (FDA-NIH Biomarker Working Group 2016), as measurable and objectively evaluable indicators that can be routinely and repeatedly used to objectively characterize disease presentation and evolution. Biomarkers should, therefore, be distinguished from laboratory findings, as they must be routinely used and repeatedly measurable by biological tests, allowing them to be associated with sufficient sensitivity, specificity, and reproducibility (see Section 8.2.11 on diagnostic criteria).

Accordingly, we routinely measured the inflammation-associated high-sensitivity C reactive protein (hs-CRP) in the peripheral blood of EHS and/or MCS patients, and the levels of vitamin D2-D3, as it has been suggested that low levels of its metabolite, the secosteroid 25 hydroxy-vitamin D (25-D), could be a consequence of inflammatory and/or autoimmune processes (Albert et al. 2009). It is also reported that vitamin D deficiency is associated with abnormal development and functioning of the CNS (Tuohimaa et al. 2009; Eyles et al. 2009). Moreover, it has been shown that, upon brain injury, degeneration, or infection, the inflammatory response may trigger degranulation of mast cells, leading to a massive release of histamine in the blood (Marquardt 1983; Rocha et al. 2014). For this reason, we systematically measured the levels of histamine in the peripheral blood of the patients. In addition, as the best-known mast cell degranulation mechanism involves the crosslinking of immunoglobulin E (IgE) to its high affinity specific cell surface receptor (Greaves and Sabroe 1996), we also measured total IgE levels in the peripheral blood. It is well known that histamine is a potent mediator of inflammation, which is able to increase BBB permeability through oxidative/nitrosative stress (Mayhan 1996; Abbott 2000).

Therefore, we looked for possible oxidative/nitrosative stress-related biomarkers that could be involved in BBB disruption and identified nitrotyrosine (NTT) because it results from the toxic effects of peroxynitrite (ONOO⁻) (an important non-radical ROS often detected after EMF exposure) on proteins and is considered as a biomarker of BBB disruption (Tan et al. 2004; Phares et al. 2007; Pacher et al. 2007; Yang et al. 2013). During the inflammatory process, it is well known that cells produce excessive amounts of superoxide anion ($O_2\cdot^-$) and nitric oxide (NO•) free radicals and that, although NO• is a relatively weak free radical ROS resulting from the activation of nitric oxide synthase (NOS), its excessive intracellular production is associated with cytotoxic properties because of the formation of the extremely reactive oxygen/nitrogen species peroxynitrite (ONOO⁻) according to the reaction:

$$NO\cdot + O_2\cdot^- \rightarrow ONOO^-$$

TABLE 8.5
Disease Biomarkers Investigated Routinely in Self-reported EHS and/or MCS Patients, and Their Normal Levels

Biomarker	Author, Year	Sample Type	Normal Range
Low-grade Inflammation			
High-sensitivity C reactive protein (hs-CRP)	Pearson et al. 2003	Serum	≤ 3 mg/l
Histamine	Lebel et al. 1996	Plasma	≤ 10 nM
IgE	Dessaint et al. 1975	Serum	≤ 100 U/ml
Heat shock protein 27 (HSP27)	De and Roach 2004	Serum	≤ 5 ng/ml
Heat shock protein 70 (HSP70)	Pockley et al. 1998	Serum	≤ 5 ng/ml
Vitamin D2-D3	Belsey et al. 1971	Serum	≥ 30 ng/ml
Blood–Brain Barrier Disruption/Opening			
Nitrotyrosine (NTT)	Ischiropoulos et al. 1992	Serum	≥ 0.6 µg/l and ≤ 0.9 µg/ml
Protein S100B	Smit et al. 2005	Serum	≤ 0.105 µg/l
Autoimmune Response			
Anti-myelin P_0 protein autoantibodies	Arnold et al. 1985	Plasma	Negative
Melatonin Metabolites			
6-Hydroxymelatonin sulfate (6-OHMS)	Schumacher et al. 1989	Urine	≥ 5 ng/l and ≤ 40 ng/l

Source: Belpomme et al. 2015; Belpomme and Irigaray 2020

(Mayhan 1996; Pacher et al. 2007; Gunaydin and Houk 2009). In addition, the calcium-binding protein S100B produced and released predominantly by perivascular astrocytes has also been shown to be such a biomarker of BBB disruption/opening (Kapural et al. 2002; Marchi et al. 2004; Koh and Lee 2014), although it is not specific and, to our knowledge, not dependent on OS.

As indicated in Table 8.5 free NTT, the protein-combined NTT, and the S100B protein were important biomarkers that we routinely measured in the peripheral blood of EHS and/or MCS patients to evidence BBB disruption/opening. These two BBB indicators (NTT and S100B) are, thus, important to be assessed in EHS and/or MCS patients in an attempt to prove that man-made EMFs as well as chemicals could be causally related to EHS and/or MCS respectively (see Sections 8.3.2 and, particularly, 8.3.2.3 on etiology).

We also considered that non-thermal WC EMF exposure could induce a repetitive cellular stress, leading to continuous heat shock protein (HSP) over-expression and release in exposed tissues, particularly in the brain (de Pomerai et al. 2000; French et al. 2001; Yang et al. 2012; Kesari et al. 2014; Ikwegbue et al. 2017). HSPs are a family of highly conserved proteins with chaperone functions acting to maintain the structural conformation of cellular proteins. HSP over-expression under inflammatory-related cellular stress conditions promotes an anti-inflammatory response (Berberian 1990; Georgopoulos and Welch 1993; Hartl 1996). We, thus, speculated that the major inducible stress protein HSP70, which was shown to oppose neuronal apoptosis (Yenari et al. 2005; Sabirzhanov et al. 2012) and BBB disruption (Kelly and Yenari 2002; Yenari et al. 2005) could be involved. This could also be the case for HSP27 (Leszczynski et al. 2002; Leak et al. 2013). However, under chronic man-made EMF exposure, it has been reported that intra-cellular HSP70 levels may decline and, consequently, decrease cell protection (Di Carlo et al. 2002). We, thus, systematically measured HSP70 and HSP27 levels in the peripheral blood of EHS and/or MCS patients in order to determine whether these chaperone proteins are a marker of man-made EMF and/or chemical chronic exposure, as it has been shown in experimental studies using non-thermal EMF exposure (de Pomerai et al. 2000; French et al. 2001; Blank and Goodman 2009; Yang et al. 2012; Kesari et al. 2014).

Moreover, during oxidative/nitrosative stress, proteins may be extensively modified and denatured, acquiring new epitopes,* i.e., new antigenic determinants, which can result in their loss of specificity and biological activity, hence in the synthesis of autoantibodies (Ohmori and Kanayama 2003; Profumo et al. 2011). This is the case for EMF exposure which has been shown to alter DNA replication and mitosis, form abnormal proteins (Lin et al. 1997; Tsurita et al. 1999; de Pomerai et al. 2000), and produce oxidation-related immunoglobulin G (IgG) autoantibodies (Bozic et al. 2007). Consequently, we hypothesized that, under exposure to environmental levels of man-made EMFs and/or chemicals, proteins such as myelin protein zero (myelin P_0) – the most abundant protein in the peripheral nervous system (PNS) – may be denatured to such a degree that it acquires autoantigenic properties. Note that the Schwann cells† and the myelinating glia,‡ i.e., the non-neuronal tissue components of the PNS, express myelin P_0 throughout their development until the formation of mature myelin. Therefore, we systematically searched for the presence of autoantibodies against myelin P_0 in the blood of EHS and/or MCS patients.

Finally, given the fact that in our series of investigated cases, many patients had sleep disturbances (see Table 8.4) and that some effects of EMF exposure have been reported to be mediated by the pineal hormone melatonin (Burch et al. 1999), we also systematically measured melatonin metabolism in these patients. However, measurement of endogenous melatonin in urine is not useful because of its very low unmetabolized levels (Kovács et al. 2000), and thus, we measured the levels of its metabolite 6-hydroxymelatonin sulfate (6-OHMS) and calculated the 6-OHMS/creatinine ratio in urine to reduce the variability of 6-OHMS measurement attributed to urine dilution because creatinine is excreted in a relatively constant amount.

8.2.7 The Use of Biomarkers for Objective Diagnosis

Much of the controversy over the acknowledgment of EHS as a new emerging pathology lies in the absence of reliable disease biomarkers for its diagnosis. Although the EHS-associated symptomatic picture can contribute to the diagnosis of EHS, the use of peripheral blood and urine biomarkers is needed to obtain objective assessment.

The biomarkers we use routinely for identifying and characterizing EHS and MCS are summarized in Table 8.5. The use of additional biomarkers such as nitro-oxidative stress biomarkers or cerebral neurotransmitters and genotyping of drug metabolism-related enzymes will be described separately (see Sections 8.2.7.6, 8.2.8, and 8.2.9).

8.2.7.1 Low-grade Inflammation

We have shown that using different peripheral blood biomarkers, low-grade inflammation, BBB disruption/opening, and anti-myelin P_0 autoimmune response could be detected in many EHS patients, suggesting that oxidative/nitrosative stress (see Section 8.2.7.5) may cause these different types of biological alterations. This is similarly the case in MCS and EHS/MCS patients in whom such alterations were also found.

The results provided by the biomarker tests we have routinely used, and their biological significance, have been previously discussed (Belpomme et al. 2015). An increase in hs-CRP levels was found in 14.7% of the overall cases, more precisely in 15%, 13.5%, and 14.3%, of the cases in the EHS, MCS, and EHS/MCS groups, respectively (Table 8.6), suggesting that, in such pathological disorders, patients display some type of systemic (whole body) inflammation. We, thus, systematically looked for causes of inflammation/infection in these patients, but with the exception of three cases, we did not find any. Furthermore, because hs-CRP is considered to be a non-specific

---

* An epitope, also known as antigenic determinant, is the part of an antigen that is recognized by the immune system, specifically by antibodies, B cells, or T cells.
† Cells in the peripheral nervous system that produce the myelin sheath around neuronal axons.
‡ The connective tissue of the nervous system is composed of several different cell types, such as astrocytes and micro- and macro-glial cells.

TABLE 8.6
Biomarker Levels in the Peripheral Blood of Patients with EHS and/or MCS

Biomarker Normal Values	EHS (Mean ± SE)	Above Normal*		MCS (Mean ± SE)	Above Normal*			EHS/MCS (Mean ± SE)	Above Normal*		
		Ratio	%		Ratio	%	p^{**}		Ratio	%	p^{***}
hs-CRP < 3 mg/l	10.3 ± 1.9	78/521	15	5.3 ± 1.7	3/52	6	0.5	6.9 ± 1.7	22/154	14.3	0.36
Histamine < 10 nmol/l	13.6 ± 0.2	182/491	37	23.5 ± 4.5	18/44	40.9	0.91	13.6 ± 0.4	59/142	41.5	0.52
IgE < 100 U/ml	329.5 ± 43.9	115/521	22	150.9 ± 18.3	8/52	15.4	0.23	385 ± 70	38/154	24.7	0.53
HSP 70 < 5 ng/ml	8.2 ± 0.2	91/486	18.7	5.9 ± 0.5	4/52	12	0.03	8 ± 0.3	36/142	25.4	0.72
HSP 27 < 5 ng/ml	7.3 ± 0.2	123/476	25.8	6.8 ± 0.1	6/52	8****	0.59	7.2 ± 0.3	42/132	31.8	0.56

Source: Belpomme et al. 2015; Belpomme and Irigaray 2020

[SE: standard error. *p*: probability that difference is due to random variation. *Evaluable cases. **Comparison between the EHS and MCS groups for marker mean level values by the two-tailed t-test. ***Comparison between the EHS and EHS/MCS groups by the two-tailed t-test. ****With the exception of MCS, for which there is a significantly lower percentage above normal for HSP 27, the percentages above normal in EHS and EHS/MCS for all the investigated biomarkers did not differ significantly among the three groups (as found by the chi-square independence test).]

biomarker of age-related cognitive decline or of dementia and, more particularly, of Alzheimer's disease (Schmidt et al. 2002; Dik et al. 2005), we systematically searched for Alzheimer's disease in these patients. In two cases, Alzheimer's disease was discovered among patients in our database. It was possibly due to previous excessive EMF exposure (Sobel et al. 1995; 1996; Qiu et al. 2004; Garcia et al. 2008; Davanipour and Sobel 2009; Söderqvist et al. 2010), but it was not proven as specifically related to EHS. We, thus, considered these as two non-evaluable cases in the analysis.

An important finding was that histamine in the peripheral blood was increased in nearly 40% of all patients and that this increase did not differ between the three investigated groups (Table 8.6). This finding suggests that histamine is not only a clinical biomarker of EHS and MCS but also plays a crucial role in the pathogenesis of both disorders. This molecule has been shown to be not only a neurotransmitter produced by and released from the CNS but also an inflammatory mediator produced by and released from mast cells in many allergic inflammatory (Marquardt 1983) and neuro-inflammatory processes (Greaves and Sabroe 1996) (see Section 8.3.1 on pathophysiology).

Increased levels of circulating IgE were also found in 22%, 15.4%, and 24.7% of the EHS, MCS, and MCS/EHS groups, respectively, with no statistically significant difference between groups. This was not associated with the co-existence of a possible unrelated allergy in the majority of patients. Since histamine release from mast cells involves the high affinity of IgE with its membrane receptor situated on the surface of mast cells (Greaves and Sabroe 1996; Gazerani et al. 2003), we searched for a correlation between histamine and IgE levels in the peripheral blood of the patients. However, such a correlation was not found.

As indicated in Table 8.6, depending on the group considered, increased levels of the HSP70 and HSP27 chaperone proteins were detected in the peripheral blood in about 7%–19% and about 11%–26% of the patients, respectively. Collectively, 25%–40% of the patients had increased levels of HSP70 and/or HSP27, without statistically significant difference between the three groups. That means that HSP70 and HSP27 are circulating biomarkers associated not only with EMF exposure, as it is the case in animal experimental studies (de Pomerai et al. 2000; French et al. 2001; Blank and Goodman 2009; Yang et al. 2012; Kesari et al. 2014) but also with chemical exposure, as shown in our clinical study (Belpomme et al. 2015). However, in this study, HSP70 and HSP27 increased levels were found to be more frequent in EHS than in MCS patients. It should be noted that we also found a pronounced decrease in the vitamin D levels in all three groups of patients.

8.2.7.2 BBB Disruption/Opening

The BBB has been known since the seminal work of Ehrlich (1885). Its major role is to protect the brain from the circulating toxic substances in the peripheral blood which could reach the brain tissues. Mammalian BBB is composed of brain capillary endothelial cells with the adjoining pericytes and extracellular matrix (Rubin and Staddon 1999; Abbott 2000; Saunders et al. 2014; Moretti et al. 2015). In the fetal and early postnatal periods, endothelial cells acquire a set of characteristics, including high electrical resistance of intercellular tight junctions (Rubin and Staddon 1999). There is evidence that the BBB can be affected by environmental stressors, including man-made EMFs which can disrupt/open the BBB and, subsequently, render the brain unprotected from blood toxic substances (Salford et al. 1994, 2003; Nordal and Wong 2005; Nittby et al. 2008; 2009; Stam 2010). This has, particularly, been demonstrated experimentally in rats (Oscar and Hawkins 1977; Merritt et al. 1978; Salford et al. 1994; 2003; Eberhardt et al. 2008; Ding et al. 2009). However, the role of man-made EMF exposure at non-thermal levels in opening the BBB is still controversial (Nittby et al. 2008; Stam 2010), although it has been shown to be possible at micro-thermal EMF levels. Studies suggest that EMF-related OS can induce endothelial cell apoptosis, gene expression changes, and alteration of the brain micro-environment, initiating and worsening brain injury. During this process, as we will describe below, the mast cells and the angiogenic mediators they can release seem to be involved (Ribatti 2015) (see Section 8.3.1 on pathophysiology). As discussed above, proteins S100B and NTT have been shown to be BBB disruption biomarkers (Kapural et al. 2002; Marchi et al. 2004; Kho and Lee 2014; Tan et al. 2004; Phares et al. 2007; Pacher et al. 2007; Yang et al. 2013). We have, thus, looked for these two biomarkers in the peripheral blood of patients with EHS, MCS, or EHS/MCS. As indicated in Table 8.7, the levels of circulating S100B protein have been found to be increased in overall 15% of the patients, with no statistically significant differences between the three groups. This finding confirms previously reported data showing that glia-derived S100B protein is a biomarker of hypoperfusion-associated brain damage or dysfunction (Donato 2001; Marchi et al. 2004; Michetti et al. 2012; Stamataki et al. 2013; Kho and Lee 2014) and, more particularly, of neuro-degenerative diseases such as Alzheimer's disease (Sheng et al. 1997) and amyotrophic lateral sclerosis (Migheli et al. 1999). This finding also differs from the results in non-EHS healthy subjects, for whom S100B protein levels have been found to be normal within 2 hours (h) following GSM (Global System for Mobile Telecommunications) mobile phone use (Söderqvist et al. 2009a; 2009b; 2015).

Similarly, increased NTT blood levels have been detected in 20%–30% of the overall cases, with no statistical difference between the three groups. Finally, as indicated in Table 8.7, it appears that increased levels of S100B protein and/or NTT can be detected in approximately 50%–60% of the cases when both biomarkers are measured simultaneously. Since as previously indicated, proteins S100B and NTT are potential markers of BBB disruption, we consider that such disruption may be detected clinically on the basis of the positive results obtained in more than 50% of the cases by using both biomarkers, regardless of their EHS and/or MCS etiopathogenic presentation.

8.2.7.3 Autoimmune Response and the Open Question of EHS-associated Impaired Immunity

Although it has been shown that chronic exposure to man-made EMFs may impair the immune defense system (Liburdy 1979, 1980; Liburdy and Wyant 1984; Dasdag et al. 2002; Johansson 2009; Szmigielski 2013; Rosado et al. 2018), making the organism more vulnerable to opportunistic infections (Shanson and FRCPath 1989; Bunch and Crook 1998), to our knowledge, there has been no specific published immunological investigation so far conducted in EHS self-reported patients.

We should note that, as far as the present COVID-19 viral infection is concerned, we have not practically observed any excess of COVID-19 Delta variant cases in the EHS patients we have examined clinically, although chronic man-made EMF-related immune deficiency in the general population might have furthered such a COVID-19-related viral infection. Extensive research in this field should be initiated to answer this still open important question.

As indicated in Table 8.8, we have, however, detected autoantibodies against myelin P_0 in 17% to nearly 29% of the overall patients, with no significant difference between the three investigated groups, suggesting that, in these patients, EHS and MCS are associated with some type of autoimmune response. Additionally, it has been shown that such autoimmune process may result from long-term exposure to man-made EMFs (Grigoriev et al. 2010).

8.2.7.4 6-Hydroxymelatonin Sulfate in Urine

6-hydroxymelatonin sulfate (6-OHMS) and creatinine were measured in the 24-h urine samples of a number of patients from the three (EHS, MCS, and EHS/MCS) groups. Many had a decrease in the 6-OHMS and in the 6-OHMS/creatinine ratio, suggesting that these patients have decreased antioxidant defenses (Brzezinski 1997; Baydas et al. 2006) and, thus, may be at higher risk of chronic diseases. Since decreased melatonin production has been associated with sleep disturbances (see Section 8.2.4), this might explain why such patients present sleep deficiency. However, by contrast, a limited number of patients (about 8%–12%) in all groups had an increase in 6-OHMS level and, consequently, an increase in the 6-OHMS/creatinine ratio, for which there is presently no clear explanation. This increase could be due to an unexplained EMF-related melatonin overproduction by the pineal gland or to a decreased OS-related melatonin consumption, or to both alterations (see Section 8.3.1).

TABLE 8.7
Mean (± SE) Levels of S100B and NTT in EHS and/or MCS Patients Peripheral Blood

Normal Values	EHS Mean ± SE	Above Normal Ratio***	%	MCS Mean ± SE	Above Normal (%) Ratio***	%	p*	EHS/MCS Mean ± SE	Above Normal (%) Ratio***	%	p**
S100B < 0.105 µg/l	0.20 ± 0.03	73/495	14.7	0.25 ± 0.05	6/51	11.18	0.56	0.17 ± 0.03	28/142	19.7	0.69
NTT < 0.9 µg/ml	1.36 ± 0.12	77/259	29.7	1.26 ± 0.13	6/29	20.7	0.85	1.40 ± 0.12	22/76	28.9	0.86
Increased NTT and/or S100B	–	133/250	53.2	–	12/22	54.5	–	–	46/73	63	–

Source: Belpomme et al. 2015; Belpomme and Irigaray 2020

[SE: standard error. *p*: probability that difference is due to random variation. *Comparison between the EHS and MCS groups of patients by the two-tailed t-test. **Comparison between the EHS and EHS/MCS groups by the two-tailed t-test. ***In 495 EHS patients, the S100B biomarker was used alone, while it was used alone in 51 and 142 patients with MCS and EHS/MCS, respectively. Likewise, in 259 EHS patients, the NTT biomarker was used alone, while it was used alone in 29 and 76 patients with MCS and EHS/MCS, respectively. For both tests performed at the same time, there were 250 EHS patients and only 22 and 73 patients with MCS and EHS/MCS, respectively.]

TABLE 8.8
Percentages of EHS and/or MCS Patients with Positive Test for Myelin P_0 Protein Autoantibodies

Marker	EHS		MCS		EHS/MCS	
	Positive Ratio	Positive %	Positive Ratio	Positive %	Positive Ratio	Positive %
Autoantibodies against myelin P_0 protein (qualitative test)	109/477	22.9	8/47	17	30/140	23.6

Source: Belpomme et al. 2015; Belpomme and Irigaray 2020

8.2.7.5 Nitrosative-Oxidative Stress

The evidence of OS is a very important research step for the objective diagnosis of EHS and/or MCS and their acknowledgment as a common syndrome.

A first scientific attempt showing OS in EHS patients was made by De Luca et al. (2014), who compared the level of OS enzymes in peripheral red blood cells (RBC) of the patients with that of control healthy subjects. They showed, for the first time, that glutathione S-transferase (GST), reduced glutathione (GSH), and catalase (CAT) activities in RBC were significantly decreased in EHS patients in comparison to those in healthy subjects, while glutathione peroxidase (GPx) and superoxide dismutase (SOD) activities were significantly increased in these patients. This finding suggests that EHS can be identified and characterized by the detection of OS. A similar biological signature was found in MCS patients, but the increase in SOD activity was found to be significantly more elevated than in EHS patients, and the decreases in CAT and GSH activities were found to be significantly more profound. This suggests that OS in MCS patients appears to be more evident than in EHS patients. The detection of increased OS in EHS and MCS patients was further confirmed by the demonstration of a significant decrease in co-enzyme Q10 (CoQ10) level in the plasma of both types of patients compared to that of normal healthy individuals and, consequently, by the existence of a significant increase in the oxidized CoQ10 versus total CoQ10 ratio in these patients in comparison with normal controls (De Luca et al. 2010; 2014).

As indicated in Table 8.9, we also prospectively studied a series of patients from our database for oxidative/nitrosative stress by measuring in their plasma thiobarbituric acid-reactive substances (TBARs), including malondialdehyde (MDA) for lipid peroxidation, thiol group molecules, GSH and oxidized glutathione (GSSG) for OS assessment, and NTT for peroxynitrite-induced oxidative/nitrosative stress assessment. In addition, we measured SOD activity in RBC and glutathione reductase (GR) and GPx in RBC and plasma (Irigaray et al. 2018a).

In Tables 8.10, 8.11, and 8.12 and in **Figures 8.3** and **8.4**, we show the results obtained with the different biomarkers we used to detect nitroso-oxidative stress in EHS patients, confirming the results obtained by De Luca et al. (2014). As indicated in Tables 8.10 and 8.11, 30%–50% of EHS patients were found to have a statistically significant increase in TBARs, MDA, GSSG, and NTT biomarker levels in their blood plasma, while as indicated in Table 8.12, SOD and GPx activities in RBC were significantly increased in 60% and 19% of the patients, respectively. Detailed results of this analysis are also presented in Figures 8.3 and 8.4.

Finally, when considering either one of the three main nitroso-oxidative stress biomarkers, TBARs, GSSG, and NTT, we found that nearly 70%–80% of EHS patients displayed at least one increased biomarker. More precisely, 43% of these patients displayed one increased biomarker, 21% two, and 15% three increased biomarkers. We have, thus, concluded that the routine clinical use of oxidative/nitrosative stress biomarkers can help diagnose EHS in patients presenting clinically EHS-associated symptoms described in Table 8.4.

TABLE 8.9
OS-associated Biomarkers, Antioxidative Non-enzymatic Proteins, and Antioxidative Enzymes in the Plasma and/or Red Blood Cells in EHS Self-reported Patients

Biomarker	Sample Type	Introduced By
Oxidative Stress		
MDA	Plasma	Londero and Lo Greco 1996
TBARS	Plasma	Okhawa et al. 1979
GSSG	Plasma	Akerboom and Sies 1981
NTT	Plasma	Ischiropoulos et al. 1992
Antioxidative non-enzymatic proteins		
Total thiol	Plasma	Jocelyn 1987
GSH	Plasma	Akerboom and Sies 1981
GluT	Plasma	Akerboom and Sies 1981
Antioxidative enzymes		
SOD	RBC	Marklund and Marklund 1974
GR	Plasma/RBC	Mannervik 2001
GPx	Plasma/RBC	Günzler et al. 1974

[GluT: total glutathione (including both GSH and GSSG).]

TABLE 8.10
OS Biomarkers in the Peripheral Blood of EHS Self-reported Patients

		EHS Patients Mean ± SD	p-Value*	EHS Patients with Values above Upper Normal Limits			
OS Biomarkers	Normal Values (Range)			Ratio of Cases	% Cases	Mean ± SD	p-Value**
TBARS (μM)	2.5 ± 0.18 (2.13–2.86)	2.85 ± 0.06	0.013	15/32	48.88	3.14 ± 0.17	< 0.0001
MDA (μM)	1.46 ± 0.17 (1.12–1.81)	1.76 ± 0.06	0.053	14/32	43.75	2.10 ± 0.19	< 0.0001
GSSG (μM)	12.4 ± 3.4 (5.5–19.3)	20.74 ± 1.74	0.051	13/32	40.63	29.46 ± 9.95	< 0.0001
NTT (μg/ml)	0.75 ± 0.08 (0.6–0.9)	0.78 ± 0.35	0.79	20/60	33.33	1.19 ± 0.21	< 0.0001

Source: Irigaray et al. 2018a

[*p-values obtained for comparisons between biomarker values in EHS patients and the normal range values. **p-values obtained for comparison between EHS patients with values above the upper normal limits and the normal values. Comparisons were made by the two-tailed Student's *t*-test.]

TABLE 8.11
Non-enzymatic Protein Biomarkers Measured in the Peripheral Blood of EHS Self-reported Patients

				EHS Patients with Values above Upper Normal Limits			
OS Biomarkers	Normal Mean Values (± SD) and Normal Range	EHS Patients Mean ± SD	p-Value*	Ratio of Cases	% Cases	Mean ± SD	p-Value**
GSH (μM)	965 ± 118 (729–1203)	794.62 ± 34.74	0.012	6/32	18.75	639.47 ± 69.27	< 0.0001
GSH/GSSG ratio	84.15 ± 29.35 (40.1–155)	46.92 ± 3.68	<0.0001	13/32	40.63	29.77 ± 4.72	< 0.0001
GluT (μM)	989 ± 120 (749–1228)	873.47 ± 27.85	0.041	6/32	18.75	669.83 ± 9.67	< 0.0001
GSH/GluT ratio	99 ± 0.19 (94.1–99.9)	95.25 ± 0.33	0.0009	9/32	29.13	92.86 ± 1.29	< 0.0001

Source: Irigaray et al. 2018a

[*p-values obtained for comparisons between biomarker values in EHS patients and the normal range values. **p-values obtained for comparison between EHS patients with values above the upper normal limits and the normal control values. Comparisons were made by the two-tailed Student's *t*-test.]

TABLE 8.12
Antioxidant/Detoxifying Enzymatic Activity Measured in Red Blood Cells and the Plasma of EHS Self-reported Patients

				EHS Patients with Values above Upper Normal Limits			
Anti-OS Enzymes	Normal Values (Range)	EHS Patients Mean ± SD	p-Value*	Ratio of Cases	% Cases	Mean ± SD	p-Value**
SOD (RBC) (U/mg Hb)	1.34 ± 0.06 (1.22–1.46)	1.50 ± 0.02	0.002	19/32	59.38	1.57 ± 0.08	< 0.0001
GPx (RBC) (U/g Hb)	44.1 ± 8.2 (27.8–60.5)	51.92 ± 1.62	0.044	6/32	18.75	66.70 ± 4.76	< 0.0001
GPx (plasma) (U/l)	375 ± 37.5 (300–450)	379.28 ± 9.30	0.83	3/32	9.38	469.67 ± 26.31	< 0.0001
GR (RBC) (U/g Hb)	8.9 ± 2.1 (4.7–13.2)	9.42 ± 0.34	0.56	2/32	6.25	14.15 ± 0.35	< 0.0001
GR (plasma) (U/l)	54 ± 9 (33–75)	61.69 ± 9.17	0.16	0	0	–	–

Source: Irigaray et al. 2018a

[*p-values obtained for comparisons between biomarker values in EHS patients and the normal range values. **p-values obtained for comparison between EHS patients with values above the upper normal limits and the normal control values. Comparisons were made by the two-tailed Student's *t*-test. RBC: red blood cells. Hb: hemoglobin.]

Electro-hypersensitivity as a Man-made EMF Pathology

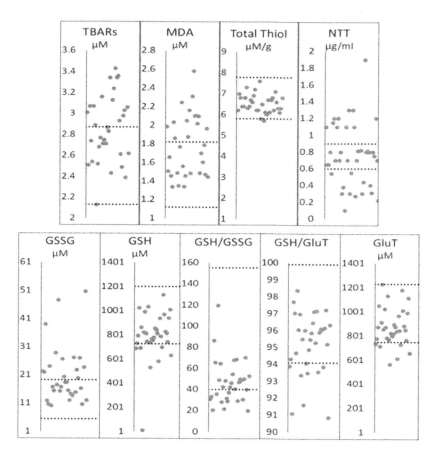

FIGURE 8.3 Values of OS biomarkers in the plasma of EHS self-reported patients in comparison with normal values. Normal values are within the dashed lines. Data obtained from 60 EHS self-reported patients. Total thiol-free sulfhydryl group including GSH.

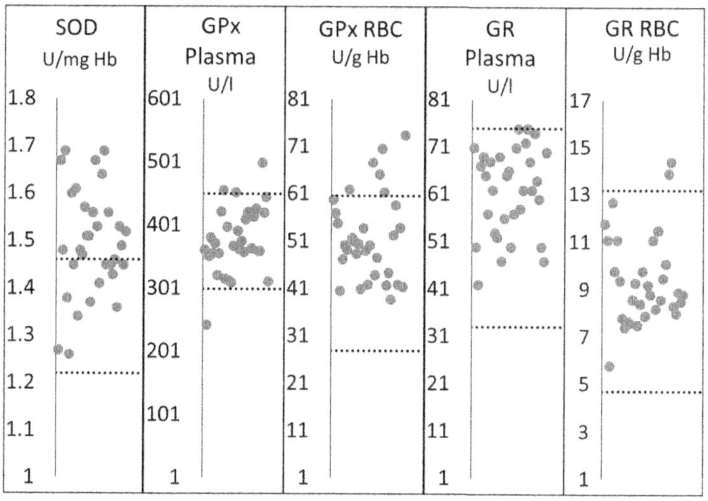

FIGURE 8.4 Specific activity of antioxidant/detoxifying enzymes measured in the plasma and red blood cells of EHS self-reported patients in comparison with normal range values (see Irigaray et al. 2018a). Normal values are within the dashed lines. The patients with elevated plasma levels for GPx and GR were different from those with elevated RBC levels. Hb: hemoglobin; RBC: red blood cell.

8.2.7.6 Disease Presentation without Biomarker Detection

The described, biomarkers have been found to characterize EHS in about 70%–80% of the cases, but as indicated in Table 8.13 and **Figure 8.5**, in 20%–30% of the EHS self-reported evaluable cases so far investigated, no biomarker could be detected. However, no particular/different symptoms could be identified to distinguish these non-contributing cases from the other diagnosed EHS cases. Thus, in such cases other biomarkers and cerebral imaging should be used to objectively diagnose EHS (see below).

TABLE 8.13
Summary of Results Expressed in Percentages of EHS Patients, According to Main Biological Effects Assessed by the Use of Classical Biomarkers*

Biological Effect	Percentage of EHS Patients (%)
Oxidative-nitrosative stress	~ 70–80
Low-grade inflammation	~ 70–80
BBB opening	~ 50–60
Undetermined	~ 20–30

[*We note that the overall percentages found to be positive to each main biological effect are approximate because results obtained by the different tests overlap each other, and the different biological effects also overlap each other.]

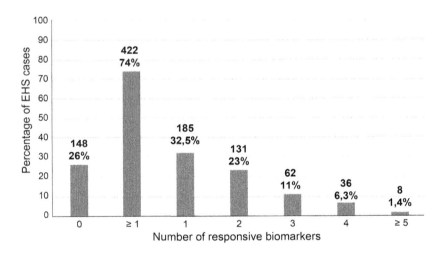

FIGURE 8.5 Percentages and numbers of EHS patients (total number 570), according to the number of biomarkers that were positive in each one of them. [Included biomarkers: histamine, 6-OHMS, HSP, autoantibodies to myelin P_0, S100B, and NTT].

Presently, we have no explanation for these atypical presentations that could not be diagnosed as EHS by using the above routine biomarkers. Possibly these cases could be characterized by using other inflammation-related biomarkers, such as substance P*, other not known released inflammatory mediators, or other OS-associated molecules which contribute to BBB disruption/opening. The lack of positive results obtained from these routinely used biomarkers could be due to the possibility that they were not released in the blood and/or urine from the brain or from other tissues where

---

* Neuro-peptide recognized as an inflammatory mediator released by mastocytes.

they are produced. It could also be due to their lability, which possibly made them not expressed at the time of testing. This latter hypothesis seems however to be unlikely, since we repeatedly tested these biomarkers in such non-contributing cases and confirmed that the above biomarkers were not detectable at any time. However, in such non-contributing cases, the diagnosis of EHS could be accomplished by the use of neurotransmitters and imaging techniques, although these methods are also not specific (see Sections 8.2.8 and 8.2.10).

8.2.8 THE CRITICAL ROLE OF NEUROTRANSMITTERS. EHS AS A PROVEN BRAIN PATHOLOGICAL DISORDER

As reported in Table 8.4, it is clearly established clinically that symptoms occurring in EHS patients mainly consist of headaches, sleep disturbances, dizziness, and cognitive deficiency, including loss of immediate memory and attention/concentration deficiency, depression tendency with suicidal ideation, and emotional behavior. All these symptoms suggest a neurological disorder of the brain. In recent years, many studies have shown that the CNS is an important target for man-made EMF exposures (Bas et al. 2009; Sonmez et al. 2010; Saikhedkar et al. 2014; Megha et al. 2015a; Gökçek-Saraç et al. 2021). Several studies have focused on the metabolism and transport of neurotransmitters in the brain (Inaba et al. 1992; Noor et al. 2011; Saikhedkar et al. 2014; Gökçek-Saraç et al. 2021). Neurotransmitters* are specific molecules acting as messengers of synaptic transmission of the nerve impulses (Sheffler et al. 2021). Importantly, man-made EMFs have been demonstrated to affect the metabolism and transport of neurotransmitters in various regions of the brain and, thus, significantly affect brain function (Noor et al. 2011; Aboul Ezz et al. 2013; Hu et al. 2021).

For example, it has been shown that exposure to mobile phone EMFs can reduce dopamine (DA) production in the hippocampus of rats and decrease learning and memory capacity (Aboul Ezz et al. 2013). But it has also been shown that DA content in the brain of mouse embryos increases after lower intensity mobile phone EMF exposure, whereas it decreases after higher intensity exposure (Ji et al. 2012), a finding suggesting that EMF-induced metabolic abnormalities of monoamine neurotransmitters are dose dependent. A similar neuro-biological change has been evidenced for adrenaline (AD) and noradrenaline (norAD) neurotransmitters synthetized and secreted by postganglionic sympathetic neurons and adrenergic nerve endings in the brain (Silverberg et al. 1978). Here too, long-term exposure to mobile phone EMF may lead to abnormal levels of these neurotransmitters in the brain, depending on the intensity of EMF combined with the duration of exposure. At higher intensity/longer exposure duration, the mobile phone EMF could decrease these two monoamine neurotransmitters in the brain in rats (Megha et al. 2015b), whereas at lower intensity, it could increase norAD (Cao et al. 2000). It is well known that 5-hydoxytryptamine (5-HT), known as serotonin, is massively produced in the gastrointestinal tract, with only a small amount synthetized in the brain. This neurotransmitter is mainly localized in the pineal gland, the hypothalamus, and the cerebral cortex, more particularly, in the neuronal synapses (Charnay and Léger 2010). Since 1986, many physiological functions have been discovered to be associated with 5-HT. They include mood, cognition, memory, pain, sleep, and body temperature (Petkov and Konstantinova 1986), all of which have been found to be disturbed in EHS patients (see Table 8.4). Disruption of any of these physiological functions has been reported as indicator of EMF-induced brain injury (Lai et al. 2021). It has been shown in rats that after long-term intense microwave EMF exposure (2.856 GHz with average power density 5, 10, 20, or 30 mW/cm^2) the content of 5-HT in the hippocampus and the cerebrospinal fluid is increased and may be associated with decrease in learning and memory performance (Li et al. 2015). Finally, all these animal studies suggest that long-term exposure to WC EMFs can change the levels of neurotransmitters in the brain and induce brain-related

---

* Most neurotransmitters are synthesized from common cellular metabolites by enzymes expressed specifically in neurons. Neurotransmitters are loaded into synaptic vesicles by vesicular transporters using a proton gradient as the source of energy. Distinct transporters are responsible for loading different transmitters.

TABLE 8.14
Ratio-Percentage of EHS Patients with Altered Levels of Various Neurotransmitters in Their Urine

Neurotransmitter	Patients	Percentage (%)
Dopamine increase	17/42	41
3-4 DOPAC decrease	18/42	43
Noradrenaline increase	11/42	26
Adrenaline increase	8/42	19
Adrenaline decrease	12/42	22
Serotonin increase	4/42	9.5
Serotonin decrease	5/42	12

Source: Belpomme and Irigaray 2020
[3-4 DOPAC: 3,4-Dihydroxyphenylacetic acid.]

neurologic symptomatic alterations, such as cognitive and emotional defects, memory loss, and sleep disturbances.

Considering these experimental data, we investigated whether the levels of the three previous brain categories of biogenic amines (DA, AD/norAD, 5-HT) could be altered in the urine of EHS patients. Table 8.14 summarizes our data. We have demonstrated for the first time that EHS in EHS self-reported patients is associated with changes in the normal levels of these cerebral neurotransmitters and that various abnormal neurotransmitter profiles exist among patients, meaning that the levels of neurotransmitters differ from one patient to the other. Using such additional biomarkers, we have, therefore, confirmed that EHS is a well-established new brain-related neurologic disorder and that, according to EMF-related experimental data on neurotransmitters in animals, it may be causally related to anthropogenic EMF exposures (see Section 8.3.2).

Although measurement of these neurotransmitters in the urine of EHS patients may help diagnose EHS objectively, particularly in non-contributing cases of EHS (see Section 8.2.7.6), we have presently no clear physiopathological interpretation for the different profiles we observed in these patients. We have still not investigated the other categories of neurotransmitters, such as the amino acid/peptide categories, the nitric oxide or the neuro-peptide substance P. The effects of EMF exposure on neurotransmitters, as well as the possible mechanism underlying EMF-induced neurotransmitter changes, have been discussed recently in a well-documented review article on the biological effects of microwave EMF exposures on neurotransmitters in the brain of experimental animals (Hu et al. 2021). New research on neurotransmitters should expand our understanding of EHS as a brain disorder.

8.2.9 Genotyping of Drug Metabolism-related Enzymes and Possible Screening Intervention

As shown above, EHS and MCS could be characterized by their OS status and its association with low-grade inflammation, which may include cytokine profiling. But as reported initially by De Luca et al. (2010), genotyping of drug metabolism-related enzymes[*] could not be used for diagnosing MCS and screening intervention[†] because genetic susceptibility detected on the basis of genetic polymorphisms of xenobiotic-metabolizing enzymes did not appear to differ from that of

---

[*] The method that detects small genetic inter-individual differences that can lead to variability in the pharmacokinetic activity of enzymes that metabolize xenobiotics or drugs.

[†] Screening intervention is a method designed to identify pathological conditions which pre-exist to disease occurrence in asymptomatic patients, thus, enabling earlier diagnosis and preventive management.

the normal population. However, in a more recent study, the same research group has shown that genotyping of drug metabolism-related enzymes, although it is not specific, could in fact be used as a feasible tool not only for objectively diagnosing EHS and/or MCS, but also for screening intervention i.e., for the detection of people who may be at risk of EHS and/or MCS in the general population (Caccamo et al. 2013). Such data showing the existence of a genetic predisposition in MCS patients, involving genetic variations in drug metabolism enzymes had previously been reported (McKeown-Eyssen et al. 2004), but such genetic susceptibility in these patients was thought to need scientific confirmation (Berg et al. 2010). Such a genetic (and/or epigenetic) approach might lead to a more precise determination of the significance of genetic predisposition in EHS and/or MCS. But such an approach should not be confused with the pathophysiological characterization of hypersensitivity to EMFs and/or chemicals (Belpomme and Irigaray 2021) (see Sections 8.3.2 and 8.3.2.2). In fact, from genetic polymorphism studies, it was found that xenobiotic sensor- or metabolism-related gene variants associated with MCS and with other sensitivity-related illnesses concern the cytochrome P450 isoenzymes CYP2C9, CYP2C19, and CYP2D6, which may be considered as risk factors for MCS but also as potential discriminants of other environmental sensitivity illnesses, depending on specific combinations of their mutated alleles (Caccamo et al. 2013). A similar alteration of specific genotypic pattern was identified in EHS patients with altered distribution (compared to normal control) of the CYP2C19*1/*2* single nucleotide polymorphic variant and a 9.7-fold increased risk of developing EHS for the haplotype (null) $GSTT_1$ + (null) $GSTM_1$ variants[†] (De Luca et al. 2014). Finally, this approach may open the way of promising future screening intervention to detect as early as possible people at risk of EHS and/or MCS in the general population.

8.2.10 Cerebral Imaging

Classical brain imaging, including brain computerized tomography (CT), brain magnetic resonance imaging (MRI), and brain angioscans usually display normal patterns in EHS patients and in MCS or EHS/MCS patients. The normality in the results of these methods is not an argument against the diagnosis of these pathological disorders. We have shown that the use of other imaging techniques could be helpful in the diagnosis of EHS and MCS. In fact, as indicated in Table 8.15, by using transcranial Doppler ultrasound (TDU) (Purkayastha and Sorond 2012) in EHS patients, we showed a decrease in the mean pulsatility index[‡] in one or both middle cerebral arteries, (i.e., for one artery in 25% and 31% of the cases for the right and left artery, respectively, and for both arteries in 50% of the cases). Moreover, for the EHS/MCS group of patients, there was a decrease in the mean pulsatility index for one artery in 20% of the cases and for both arteries in 50% of the cases.

Regarding resistance in the brain blood flow (BBF), we found that, in EHS patients, BBF resistance was increased in one artery in 6.25% of the cases and in both arteries in 18.75% of the cases, while in EHS/MCS patients, BBF resistance was increased in one artery in 5%–10% of the cases and in both arteries in 25% of the cases. Also, the mean BBF velocity was below normal in 9.75%-40% and above normal in 5%–18.75% of the cases, depending on the group considered (EHS or EHS/MCS) (see Table 8.15). This suggests that, in EHS and/or MCS patients, BBF may be abnormal in one or both of the middle cerebral arteries (Belpomme and Irigaray 2020). Using positron emission tomography (PET), an increase in regional BBF has been shown to be induced by pulsed MT EMFs in certain areas of the brain of healthy volunteers (Huber et al. 2002; 2005; Aalto et al. 2006), whereas BBF was found to be decreased in other areas (Aalto et al. 2006). Currently, we

---

[*] Reference laboratories usually report a diplotype genotype, which is composed of inherited maternal and paternal star-alleles (e.g., CYP2C19*1/*2, in which an individual inherited a *1 allele and a *2 allele).

[†] The glutathione S-transferase theta 1 ($GSTT_1$) and glutathione S-transferase mu 1 ($GSTM_1$) are key molecules in cellular detoxification. Their null alleles cannot encode active enzymes because large segments of the gene are deleted.

[‡] Pulsatility index is calculated from flow velocity to assess pulsatility. It is defined as the difference between maximum and minimum blood flow velocity normalized to the average velocity: $PI = (V_{max} - V_{min})/V_{mean}$ (PI is the pulsatility index and V is the velocity).

TABLE 8.15
Resistance Index, Pulsatility Index, and Mean BBF Velocity in the Right and Left Middle Cerebral Arteries in 32 EHS and 20 EHS/MCS Patients Using TDU

EHS	Normal Values			Percentage of Patients					
		(Mean ± SE)		Below Normal (%)			Above Normal (%)		
Index	Right and Left	Right	Left	Right Only	Left Only	Both	Right Only	Left Only	Both
Resistance index	< 0.75	0.62 ± 0.03	0.65 ± 0.04	–	–	–	6.25	6.25	18.75
Pulsatility index	> 0.60	0.55 ± 0.02	0.55 ± 0.03	25	31.25	50	–	–	–
Mean BBF velocity	62 ± 12	59.56 ± 5.98	61.35 ± 5.27	9.75	9.75	31.25	–	9.25	18.75

EHS/MCS	Normal Values			Percentage of Patients					
		(Mean ± SE)		Below Normal (%)			Above Normal (%)		
Index	Right and Left	Right	Left	Right Only	Left Only	Both	Right Only	Left Only	Both
Resistance index	< 0.75	0.79 ± 0.09	0.64 ± 0.04	–	–	–	5	10	25
Pulsatility index	> 0.60	0.48 ± 0.03	0.61 ± 0.02	20	0	65	–	–	–
Mean BBF velocity	62 ± 12	53.03 ± 9.09	51.77 ± 7.63	20	20	40	10	10	5

Source: Belpomme and Irigaray 2020

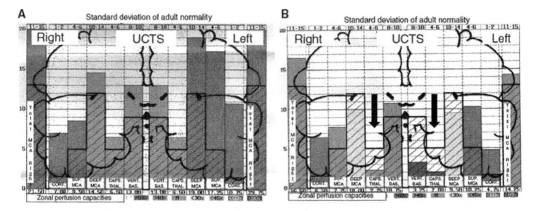

FIGURE 8.6 Examples of UCTS diagrams obtained from the database exploring the overall centimetric ultrasound tissue pulsatility in the two temporal lobes of a normal subject (A) and of an EHS patient (B) (according to Irigaray et al. 2018c; Belpomme and Irigaray 2020). The height of the columns expresses the pulsometric index (PI). The bold horizontal lines in each column represent the median normal PI. Mean values of PI in each explored area are recorded from the cortex to the internal part of each temporal lobe (i.e., from left to right for the right lobe, and from right to left for the left lobe). In A (normal subject), all values are over the median normal PI values, whereas in B (EHS self-reported patient), values in the capsulothalamic areas (as indicated by the two bold arrows, the fifth column from the surface to the internal part of the right temporal lobe and the second column from the internal part to the surface of the left temporal lobe respectively) are significantly below median normal values, suggesting that the limbic system and the thalamus in each temporal lobe may be involved in EHS, as exemplified in this patient.

have no explanation of our findings with TDU analysis in comparison to the findings with PET in healthy subjects.

In addition to our aforementioned data with TDU, by using ultrasonic cerebral tomosphygmography (UCTS) applied to the temporal lobes, we showed that there is a significant decrease in mean pulsometric index[*] in the middle cerebral artery-dependent tissue areas of these lobes, especially in the capsulothalamic area, which corresponds to the limbic system and the thalamus (Irigaray et al. 2018c). As shown in **Figure 8.6**, this tissue hypopulsation – mainly detected in the capsulothalamic area of these lobes – suggests that EHS and/or MCS are associated with a capillary BBF decrease localized in these two brain structures (the limbic system and the thalamus), meaning that they may be associated with some vascular and/or neuronal dysfunction (Irigaray et al. 2018c; Belpomme and Irigaray 2020). Although these abnormalities are not specific (as they may be similar to those found in Alzheimer's disease and other neuro-degenerative disorders) we recently showed that UCTS could be, so far, one of the most accurate imaging techniques to contribute to the diagnosis of EHS and/or MCS and to follow treated patients (Irigaray et al. 2018c). This finding was confirmed by another independent study using cases registered in our database (Greco 2020).

These brain abnormalities, however, do not seem to be restricted to the limbic system and the thalamus because, using TDU as indicated above, we showed that, in EHS and/or MCS patients, BBF of the middle cerebral arteries could be abnormal and may persist for a long time after diagnosis (see below). Furthermore, using functional MRI (fMRI) in EHS patients who are chronically exposed to ELF EMFs, regional BBF changes were also reported mainly in the frontal lobes as an abnormal default mode network[†] (DMN). The changes were manifested as hyperconnectivity of this DMN in association with a decrease in cerebral BBF and metabolic processes in the two hyperconnected components (Heuser and Heuser 2017). In **Figure 8.7**, abnormal DMN is represented

---

[*] The mean pulsometric index is similar with the pulsatility index in TDU but determined here differently.

[†] The DMN is an inter-connected and anatomically defined set of brain regions.

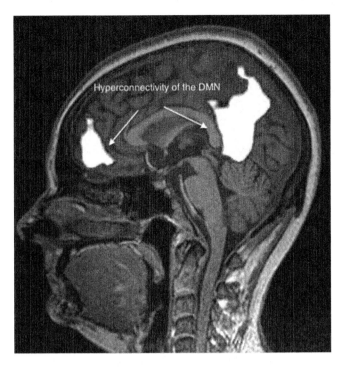

FIGURE 8.7 Abnormal functional MRI brain scan showing hyperconnectivity of DMN in patients complaining of EHS after long-term exposure to man-made EMFs (according to Heuser and Heuser 2017).

with fragmented hyperconnectivity of the anterior and posterior components, which may lead to decreased BBF and/or metabolism in the bi-frontal lobes.

8.2.11 Diagnostic Criteria, Sensitivity, Specificity, and Reproducibility of the Biological Methods

To our knowledge, there are presently no pathognomonic clinical symptoms which allow a clear diagnosis of EHS and distinction from MCS. However, although many symptoms (but not all) are subjective and not specific, their association in a thorough overall clinical picture may allow such a diagnosis. As shown in Table 8.4, the six most characteristic common symptoms of EHS and MCS are headache, dysesthesia, tinnitus, dizziness, fatigue, and cognitive deficiency. To this symptomatic core, myalgia and muscular spasm (which may correlate with fibromyalgia), transitory cardiovascular symptoms, balance disorder, skin lesions, and depressive tendency (the latter not existing before EHS occurrence) could be added.

To this relatively stereotypical clinical picture, we should add the six following clinical diagnostic criteria that we have used to include EHS self-reported patients in our database and in previous studies and that we recommend for use in medical practice: (a) absence of known pathology accounting for the observed clinical symptoms; (b) reproducibility of symptom occurrence under man-made EMF exposure from suspected environmental sources; (c) regression or disappearance of symptoms in the case of presumed environmental source avoidance; (d) chronic evolution of symptoms occurrence; (e) symptomatic picture similar to that described above and published in scientific literature; (f) no pre-existing co-existing pathology such as atherosclerosis, diabetes, cancer, and/or neuro-degenerative or psychiatric diseases that would render the interpretation of clinical and biological data difficult.

Most of these criteria were close to those used in defining MCS in the 1999 Atlanta consensus meeting (Bartha et al. 1999) and in its subsequent proposed revision (Lacour et al. 2005) and were adapted to EHS using similar basic criteria, in agreement with those reported by the WHO (2006). We agree that criteria (b) and (c) are dependent on patient's subjective feelings. However, as documented in previous studies (Belpomme et al. 2015, 2018; Irigaray et al. 2018a), instead of considering the self-reported diagnosis of EHS claimed by the patients alone, we recommend that all available anamneses and clinical examination-related pre-inclusion data be analyzed as additional comprehensive information before considering that the case fulfills all the six preceding clinical diagnostic criteria. Similar clinical criteria (although they were apparently less constraining) were also proposed in 2012 by Baliatsas et al. (2012). They included: (a) "self-report of being (hyper)sensitive to EMF", (b) "attribution of symptoms to at least one EMF source", (c) "absence of medical or psychiatric/psychological disorder capable of accounting for these symptoms", and (d) "symptoms should occur soon (up to 24 hours) after the individual perceives an exposure source or exposed area".

Nevertheless, as previously emphasized, such a clinical approach is quite insufficient to make an objective diagnosis of EHS and/or MCS. Hence, there was a need for introducing suitable biomarkers and imaging technique analysis. This is what was confirmed recently in the abovementioned multi-national consensus report (Belpomme et al. 2021). In patients for whom they can be detected, these biomarkers can help not only establish the diagnosis of EHS and/or MCS objectively but also provide assessment of therapeutic intervention during patient follow-up (see Section 8.4). We should, however, emphasize that, in EHS patients, not all symptoms are acute and reversible, as is usually the case with headache, dysesthesia, dizziness, or loss of cognitive capability, but they may be chronic, such as sleep disturbance, fatigue, and depressive tendency.

In addition to the clinical criteria, the use of biological tests and radiological investigation is necessary. However, to get objective results, there is a need to ensure sensitivity, specificity, and reproducibility of the biological tests. The general methodology we have used to identify and characterize EHS and/or MCS includes clinical examination, use of routine biomarkers, and imaging techniques (Califf 2018; Reeves 2000; Belpomme et al. 2015; 2021).

Regarding the specificity of our general methodology, a major consideration is that all biomarkers proposed to objectively diagnose EHS are, in fact, not specific since OS, low-grade inflammation, BBB disruption/opening, and anti-myelin P_0 autoimmune response, may also be found in many other chronic diseases and pathologic disorders (Chrousos and Gold 1992; Holmstrom and Finkel 2014).

Regarding sensitivity, the general methodology was not able to identify and characterize all EHS and/or MCS cases since as described in Section 8.2.7.6, there were 20%–30% of cases that could not be identified as EHS. These cases, however, could be diagnosed on the basis of the recorded levels of brain neurotransmitters, genetically related drug metabolism enzymes, and the characteristic abnormalities detected by imaging techniques.

Finally, the reproducibility of our proposed general methodology must be tested in different laboratories because most of the biological results we have provided were obtained from a unique research laboratory in Paris (France). We should emphasize that the sensitivity and specificity of this methodology need to be improved. Moreover, the results obtained by the clinical use of these biomarkers may correlate with symptoms and disease type, but not with initial causes. Consequently, these tests cannot be used to identify the causal origin of EHS and/or MCS nor their hypersensitivity-associated properties. They can be used only to objectively diagnose these disorders and establish their nosological identification and classification as being part of a common new brain neurologic syndrome (see Sections 8.3.1 and 8.3.2).

8.3 PATHOGENESIS AND ETIOLOGY

8.3.1 Pathophysiological Significance of Biomarkers

As reported above, an important finding of our studies is that histamine could play a critical role in detecting EMF-related and/or chemical-related pathophysiological responses. The fact that histamine levels were not found to be increased in all patients does not exclude the possibility that these patients may have local histamine production and release in tissues exposed to man-made EMFs and/or chemical stressors. This consideration may also apply to other biomarkers.

Histamine is not just a neuro-inflammation mediator, as it also plays a critical pathophysiological role as a neurotransmitter in the brain. For example, neuronal histamine has been shown to be involved in the sleep cycle, motor activity, synaptic plasticity, and memory (Marquardt 1983; Wada et al. 1991; Onodera et al. 1994; Haas et al. 2008; Panula and Nuutinen 2013). All these functions have been found to be affected in EHS and/or MCS patients. Histamine release from sympathetic nerves can be experimentally induced by nerve stimulation (Chen et al. 2010), and it has been shown that histamine H1 receptor may play a major role in the regulation of sympathetic nerve activity (Murakami et al. 2015). This may explain why EHS and/or MCS patients may manifest transitory sympathetic-related digestive symptoms such as nausea and abdominal pain or cardiovascular symptoms such as heart rate variability (HRV), arrhythmia, and/or arterial pressure instability when exposed to man-made EMF and/or to chemical stressors (Havas 2013). Moreover, following ischemic-hypoxic damage, histamine release from nerve endings has been found to be enhanced, possibly contributing to some neuro-protection (Adachi 2005).

However, histamine is also a unique molecule which fulfills all criteria that have been historically considered for defining an inflammatory mediator (Dale 1906). Histamine is mainly produced and stored in perivascular tissue resident mast cells and circulating basophils and is released in inflammatory tissues through established mechanisms, predominantly involving cell surface receptors. Regarding histamine release from skin mast cells, the best-known degranulation mechanism involves circulating IgE and the high affinity IgE cell surface receptor on mast cells (Greaves and Sabroe 1996).

In addition, mast cells are predominantly found at host/environment interfaces such as the skin, and also in the respiratory and gastrointestinal tracts (Padawer 1963), and are closely associated with blood vessels playing a crucial sentinel role in host defense (Marshall 2004). Consequently, more precise investigations remain to be done in EHS and/or MCS patients to determine which mast cell-associated tissues are the sources of histamine release. Since brain mast cells have been shown to be critical regulators in the pathogenesis of CNS diseases, including stroke, traumatic injury, and neuro-degenerative diseases (Rinne et al. 2002; Adachi 2005), we systematically looked for brain pathological alterations in EHS and/or MCS self-reported patients. As indicated above (Section 8.2.8), the results obtained from cerebral imaging and from cerebral neurotransmitters and their metabolites argue for a predominant pathological role of mast cells in the brain of EHS patients.

On the basis of clinical symptoms, cerebral biomarkers, and cerebral imaging analysis, there are strong arguments to consider EHS and MCS as parts of a common new brain neurological syndrome. This, however, does not exclude the possibility of a direct effect of man-made EMFs and/or chemical exposure on the skin, the bronchus, and the gastrointestinal tract, suggesting that other mast cells than those of the brain may also contribute to the release of histamine in the peripheral blood. Whichever tissue is the source of histamine release, we must recall that mast cells are critical regulators in the pathogenesis of CNS diseases and that brain mast cells are involved in BBB disruption/opening by releasing locally the angiogenic mediators stored in their granules, leading to vaso-proliferative reactions occurring in such pathological conditions. Regarding EHS and MCS, this is what our data strongly suggest.

We have, thus, summarized in **Figure 8.8** a hypothetical model for the effects of environmental stressors on the brain, in which we specify two main underlying steps: a) a primary nitroso-OS-related local inflammatory response with histamine release from mastocytes and reactive proliferating glia cells (gliosis) in addition to the parallel role of cerebral vascular hypoperfusion due to a decrease in BBF (as shown in Section 8.2.10), leading to brain hypoxia and b) a nitroso-OS-related inflammatory amplification, with BBB disruption/opening and transmigration of circulating

Electro-hypersensitivity as a Man-made EMF Pathology 331

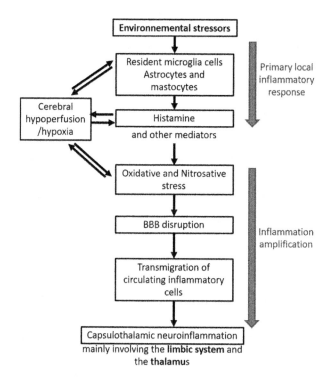

FIGURE 8.8 Hypothetical model for the effects of anthropogenic EMFs and/or chemicals in the brain (Belpomme et al. 2015; Belpomme and Irigaray 2020).

inflammatory cells into the brain leading to brain neuro-inflammation, particularly in the capsulo-thalamic area.

8.3.2 SEARCH FOR BIOCLINICAL TRIGGERS, PATHOGENESIS, AND ETIOLOGY

The cause of symptomatic occurrence in EHS patients and the causal origin of EHS itself are still debated among scientists, some of them refuting the causal effect of man-made EMFs in triggering symptoms not only in EHS patients (Levallois 2002; Röösli 2008; Röösli et al. 2010b) but also in normal people (Baliatsas et al. 2015) to such a point that governmental and official institutional bodies claim that there is no proof that anthropogenic EMF exposure is toxic for the general population and causally related to EHS genesis and EHS-associated symptom occurrence. Thus, these institutions ignore the plethora of studies showing adverse biological and health effects of anthropogenic EMF exposures and do not really worry about the suffering of EHS patients. In contrast, they approve the growing and ubiquitous development of WC technologies and associated EMF emissions accused of a variety of toxico-biological effects, without considering any public health consequences. A major difficulty here is that EHS patients are not only associated with hypersensitivity to anthropogenic EMFs but may also be sensitive to multiple chemicals (see Sections 8.2.2, 8.2.3, and 8.3.1).

There are three distinct scientific questions to address: a) what are the environmental triggers of clinical symptoms and biological changes in EHS patients, b) how can we better define EHS, and c) what is the etiology of EHS in genetically susceptible subjects, and how is it generated.

We agree with the WHO that any causality research must satisfy the four following causality criteria: a) "The existence of biological effects and health hazards can only be established when research results are replicated in independent laboratories or supported by related studies", b) "There is agreement with accepted scientific principles", c) "The underlying mechanism is understood", and d) "A dose-response can be established" (WHO 2006).

Taking into account these four criteria, we describe and discuss here the present medical state-of-the-art regarding the three above distinct scientific questions. First, we argue that man-made EMFs can really trigger pathophysiological changes and clinical symptoms in EHS patients, confirming the claim of EHS self-reported patients. Second, we discuss how to better define EHS and propose a hypothetical underlying process accounting for its genesis. Third, we discuss the etiology of EHS and argue that excessive EMF exposure can be the main cause of EHS genesis in genetically susceptible people.

8.3.2.1 Symptomatic and Biological Triggers in EHS Patients. The Importance of Good Clinical Practice in Carrying out Well-designed Provocation Studies

The purpose of provocation studies is to prove that EHS patients display symptoms described in Table 8.4 during or after exposure to man-made ELF and/or WC EMFs, i.e., to anthropogenic electromagnetic sources. As indicated in the present chapter, the EHS-associated neurological symptoms are similar to those described in the microwave syndrome, which was considered at that time as *evidently* caused by microwave EMFs in exposed workers. Against all standard medical practice, the clinical symptoms reported by the EHS patients have not been considered as medically assessed and recognized, but simply considered as "self-reported symptoms", meaning they were not considered "functional symptoms", as is commonly the case in medicine since Hippocrates. Hence, they are not accepted as a valuable descriptive clinical tool to identify and diagnose EHS due to their reported subjectivity and non-specificity. In fact, as can be soundly deduced from any face-to-face questioning and physical examination of EHS patients, there is *a priori* no medical reason to dismiss the patients' feelings and to believe they make up or mistake each time they attribute their symptoms to anthropogenic EMF exposure.

In a previous survey based on the subjective description of EHS by EHS patients, the putative causal relationship between experienced symptoms and EMF exposure has been categorized in three stages according to the patient's own feeling, depending on the intensity, duration, and number of symptoms. Stage 1: Temporary symptoms leading patients to seek their causal origin. Stage 2: Persisting symptoms of increased intensity, duration, and number, leading patients to suspect that anthropogenic EMF sources could be causally involved. Stage 3: Intolerable symptoms (including the neurovegetative ones), leading patients to acquire the certitude that anthropogenic EMF exposure is causally involved (Bergqvist and Vogel 1997). Such a characterization makes EHS patients a heterogeneous group, which may explain why many experimentally designed provocation studies using EMF exposure versus sham-exposure have provided inconsistent results, i.e., EHS patients could not correctly discriminate EMF exposure from sham-exposure (Rubin et al. 2010; 2011). Thus, in combination with the negative results of certain epidemiological studies, it was considered that the causative role of EMFs could not be proved (Levallois 2002; Röösli et al. 2010b; Rubin et al. 2011).

However, as indicated in Table 8.16, several of these provocation studies performed in EHS patients faced methodological shortcomings. A major criticism, as emphasized in the 2021 consensus report, is that provocation studies were carried out before EHS had been objectively diagnosed using biomarkers and imaging techniques. This fact, in combination with methodological issues (see Table 8.16), has resulted in negative findings (Belpomme et al. 2021). We, thus, consider it scientifically unjustified to speculate that the claims of patients reporting EHS are unfounded and that their subjective symptomatic feeling could be due to psychosomatic or nocebo effects (Belpomme et al. 2021; Belpomme and Irigaray 2021).

An additional important reason for negative results in provocation studies is the fact that, in cases of chronic inflammation and chronic suffering, the subjects' responses to EMF stimuli may be confused without clearly discriminating exposure/sham-exposure changes, especially when changes occur in a high rate with short durations (this is related with issues No. 5–8 in Table 8.16). In other words, the responses of chronically affected patients to short-time stimuli may not be those reasonably expected. This led us to consider that the modalities of the provocation studies should

TABLE 8.16
Methodological Shortcomings in Provocation Studies with Negative Results

No	Methodological Problem	Studies
1	Lack of precise inclusion criteria. No objective criteria based on molecular biomarkers and imaging techniques.	(Röösli 2008; Röösli et al. 2010b; Baliatsas et al. 2012; Schmiedchen et al. 2019)
2	No clear consideration of medical anamnesis and degree of EHS severity.	(Baliatsas et al. 2012; Schmiedchen et al. 2019)
3	No consideration of an association with MCS.	(Belpomme et al. 2015)
4	No consideration that EHS patients are intolerant to specific types of man-made EMFs.	(Röösli 2008; Röösli et al. 2010b; Baliatsas et al. 2012; Schmiedchen et al. 2019)
5	Too short exposure duration.	(Baliatsas et al. 2012; Eltiti et al. 2015)
6	Symptom recording performed too early.	(Baliatsas et al. 2012; Schmiedchen et al. 2019)
7	Endpoint criteria depending on subjective statements.	(Röösli 2008; Rubin et al. 2010; 2011; Baliatsas et al. 2012; Eltiti et al. 2015; Schmiedchen et al. 2019)
8	Possible EHS-associated psychological conditioning due to past suffering.	(Dieudonné 2020)
9	Possible significant EMF levels during sham-exposure.	(Phillips 2002)

Source: Belpomme et al. 2021

be adapted to the biological condition of the organism and that subjective endpoints should be dismissed.

In fact, not all provocation studies have provided negative results. Indeed, in several single- or double-blind provocation studies, technical ELF and/or WC EMFs have been shown to trigger clinical and biological health effects. As indicated in Table 8.17, various objective clinical and pathophysiological changes have been evidenced in these patients. Clinical effects include HRV and/or blood pressure variability (Havas et al. 2010; Havas 2013; Koppel et al. 2018), altered pupillary light reflex (Rea et al. 1991), reduced visual perception (Trimmel and Schweiger 1998), and abnormal movement during sleep (Mueller and Schierz 2004), which have all been established by objective clinical evaluation. Pathophysiological effects include altered electroencephalogram (EEG) during sleep (Arnetz et al. 2007; Lustenberger et al. 2013), altered electromyogram (EMG) (Tuengler and von Klitzing 2013; von Klitzing 2021), altered skin capillary blood flow (SCBF) (Tuengler and von Klitzing 2013; Loos et al. 2013), electric skin potential (ESP), and skin conductivity changes (Tuengler and von Klitzing 2013). All the aforementioned recorded effects by provocation studies during or after man-made EMF exposure from various sources were objectively measured.

Moreover, as already mentioned, in a single EHS case double-blind experiment (see Section 8.2.4), EMF exposure-related intolerance in comparison with sham-exposure has also been reported to be induced by pulsing 60 Hz (with 10 Hz on/off pulsations) rather than continuous-wave 60 Hz ELF electric field exposure. As stated by the authors, this means that "the statistically reliable somatic reactions to subliminal EMF exposure were obtained under conditions that reasonably excluded the causative effect of any psychological process" (McCarty et al. 2011).

Such positive effects recorded by provocation studies have also been independently shown in two earlier EHS case reports (Hocking and Westerman 2002, 2003) and in more recent studies showing effects of pulsed microwave EMF exposure on HRV and the autonomic nervous system (Havas et al. 2010; Havas 2013). Provocation studies using HRV, SCBF, ESP, and EMG recordings in EHS patients have also provided positive findings (Tuengler and von Klitzing 2013; von Klitzing 2021). The authors proposed to combine continuous measurements of HRV, SCBP, and ESP before, during, and after EMF exposure versus sham-exposure to assess distinction of EHS patients from individuals suffering from other pathological conditions (Tuengler and von Klitzing 2013).

TABLE 8.17
Provocation Studies in EHS Self-reported Patients Using EMF Exposure Versus Sham-exposure and/or Comparison with Healthy Controls, with Positive Results

Authors	Endpoints	Type of Study	Subjects	Results
Rea et al. 1991 (USA)	Pupillary light reflex	Double-blind EMF versus sham provocation study	25 EHS patients versus 25 healthy controls	16/25 EHS patients consistently reported symptoms, compared with 0/25 healthy controls
Trimmel and Schweiger 1998 (Austria)	Attention, perception, and memory tests	Double-blind provocation study	36 EHS versus 30 healthy controls	Reduced performance of visual attention and perception by combining a 50 Hz magnetic field with acoustic noise exposure, compared to the effects of noise only
Mueller and Schierz 2004 (Switzerland)	Sleep disturbance	Double-blind cross-over provocation study	54 EHS cases	Cases moved away from area with maximum EMF levels
Arnetz et al. 2007 (USA)	Sleep EEG	Double-blind case–control study compared to sham	38 IEI-EMF and 31 healthy controls	Exposure caused longer latency to deep sleep from sleep onset and reduced cerebral slow wave
Mc Carty et al. 2011 (USA)	Symptomatic responses and EMF field perception	Single-blind provocation study, EMF versus sham-exposure	A single female EHS case	In the first experiment, the EHS person reported somatic reactions with ELF EMF exposure than with sham-exposure. In the second experiment, she reported significantly more intense symptoms with pulsed EMF exposure than with continuous-wave EMF exposure. In the third experiment, she was not able to perceive EMF consciously
Havas et al. 2010 (Canada)	HRV, RBC clumping	Single-blind provocation study EMF versus sham-exposure	25 EHS self-reported patient	40% of EHS patients experienced some changes in their HRV during pulsed microwave exposure
Tuengler and von Klitzing 2013 (Germany)	HRV, capillary blood flow and SEP	Single-blind provocation l study	Several types of EHS patients	Modifications of biological parameters caused by EMF exposure
Koppel et al. 2018 (Estonia)	HRV	Single-blind provocation study	108 EHS patients	HRV significantly lower during EMF exposure than non-exposure
von Klitzing 2021 (Germany)	ECG and EMG	Single-blind provocation study	5 EHS patients	Modification of EMG caused by WC EMF exposure.

We, thus, agree that provocation studies are important when they use suitable methodology, and we are presently seeking the precise explanations for the negative results obtained in previous provocation studies (Rubin et al. 2010, 2011). The already available scientific data strongly suggest that man-made EMF exposures are causally involved in triggering adverse clinical symptoms and pathophysiological effects in EHS patients. Moreover, the available data dismiss the hypothesis of a causal psycho-pathological effect to account for the EHS-associated symptom occurrence (see Section 8.3.2.3), even if we accept that, in addition to the somatic disease, some patients may present personality traits that make them psychologically more vulnerable to environmental stressors (Hillert et al. 1999; Frick et al. 2002).

Many, if not all, of the clinical and pathophysiological changes that have been shown to be triggered by man-made EMF exposure in positive provocation studies are compatible with our hypothetical brain effect model summarized in Figure 8.8. As shown in Table 8.17, the recorded changes include the EHS-associated acute and reversible sympathetic and parasympathetic symptoms, such as HRV, pupillary light reflex, and other acute neurological symptoms, such as attention/memory loss and sleep disturbance, and, above all, objective biophysical cerebral and transient skin parameters changes. There is, so far, no available specific blood or urine biomarker of EMF exposure. To increase the objectivity and effectiveness of provocation studies, one could use some of the above-reported disease biomarkers and imaging techniques.

As already reported (see Section 8.2.11), among the symptoms presented in EHS patients, not all are acute and reversible. In the case of no treatment and no protective measures, chronic symptoms such as loss of immediate and retrospective memory, mental confusion, insomnia, chronic fatigue, depressive tendency with possible suicidal ideation may persist and even become irreversible due to the progressive occurrence of some neuronal lesional state, eventually leading to cerebral atrophy. Such evolution may occur in the case of chronic brain vascular insufficiency caused by persisting high resistance of the BBF and low pulsatility in the cerebral middle arteries (see Section 8.2.8). This evolution to such an irreversible pathological state may result from the persistence of hypersensitivity to EMFs and/or chemicals during many years and from the chronic exposure to these environmental stressors. In such cases, longitudinal studies* should be carried out to understand more precisely why patients, after an initial period of symptomatic improvement due to initial treatment, may later relapse, possibly in the form of irreversible psychiatric brain disease due to the persistence of EHS with no protection against man-made EMFs and/or chemicals.

8.3.2.2 Search for Hypersensitivity Characterization

We have previously proposed to define EHS as the intra-corporal acquisition of a pathological state of hypersensitivity to man-made EMFs in genetically or epigenetically predisposed EHS patients, as is the case for marketed man-made chemicals in MCS patients. By contrast, IEI could be defined as the environmental intolerance to man-made EMFs, chemicals, or other stressors, without the necessary acquisition of a state of hypersensitivity (see Section 8.2.2). More precisely, we have proposed to define EHS clinically and biologically as a decrease in the physiological CNS-related EMF tolerance threshold, meaning that intolerance to EMF in EHS patients could occur even for weak EMF intensities, while intolerance to EMF in non-EHS people could occur for higher EMF intensities (Belpomme and Irigaray 2021). A similar pathophysiological process involving a decrease in the CNS-related chemical tolerance threshold could apply to MCS and could result, similarly, in chemical intolerance for even very low concentrations of multiple environmental chemicals. Because MCS may be associated with EHS (see Section 8.2.3), the triggering role of low concentration multiple environmental chemicals in addition to that of weak EMF exposures may explain the difficulty to design and carry out effective provocation studies in EHS patients.

While the present medical state-of-the-art must avoid any psychological interpretation for EHS occurrence and symptomatic development, a pending question remains: Could the provocation studies prove hypersensitivity to man-made EMFs (i.e., that EHS patients are more sensitive to man-made EMFs than "healthy" non-EHS subjects)? And could these patients detect the presence or absence of WC or ELF EMFs better than other persons? Relative to this important dual question is that it was initially supposed that healthy volunteers would not show any symptomatic and pathophysiological responses under exposure to EMFs when compared to no exposure (Wagner et al. 2000; Kleinlogel 2008; Valentini et al. 2010; Baliatsas et al. 2015). Likewise, it was hypothesized that healthy individuals would not show any difference in their responses to EMF exposure from

---

* In other words, studies during the whole life of patients.

EHS patients in case–control studies (Landgrebe et al. 2008) or double-blind studies (Lowden et al. 2011). To the contrary, EHS patients, depending on the endpoint, would exhibit typical responses during and/or after EMF-exposure (Tuengler and von Klitzing 2013). This hypothesis was, however, challenged by biological studies (Markovà et al. 2005), epidemiological studies (Röösli 2008), and provocation studies (Rubin et al. 2011), showing no evidence that short-term exposure to WC EMFs causes symptoms in EHS patients or that these patients were able to detect WC or ELF EMF exposure better that healthy subjects.

Furthermore, contrary to the hypothesis of no effect of man-made EMF exposure in normal healthy individuals, many provocation studies in healthy volunteers have evidenced biological effects induced by man-made EMF exposures, such as alterations of sleep EEG (Mann and Röschke 1996; Schmid et al. 2012) and resting EEG (von Klitzing 1995; Huber et al. 2002; Ghosn et al. 2015; Loughran et al. 2019), alteration of evoked electrical potentials (Carrubba et al. 2008), changes in the EEG alpha rhythm (Croft et al. 2008; Vecchio et al. 2012), and in the EEG slow beta, fast beta, and gamma rhythms (Roggeveen et al. 2015). Man-made EMFs (especially WC EMFs) have also been shown to alter the brain response during a memory task (Krause et al. 2000), affect the sleep-dependent performance improvement (Lustenberger et al. 2013), modify the human performance in attention tasks (Crasson et al. 1999), induce annoyance, modify smells (Carlsson et al. 2005), and influence cognitive performance (Verrender et al. 2016). In addition, it has been reported that man-made EMF exposure decreases slow brain potentials at central and temporo-parieto-occipital brain region (Freude et al. 1998), increases brain glucose metabolism activity (Volkow et al. 2011) and oxygen consumption at the frontal cortex (Curcio et al. 2009), alters the rate of hemoglobin deoxygenation (Mousavy et al. 2009; Muehsam et al. 2013), influences electric properties of human blood measured by impedance spectroscopy (Sosa et al. 2005), increases blood viscosity (Tao and Huang 2011), modifies brain vascularization (Huber et al. 2002; Aalto et al. 2006), alters blood pressure-associated baro-reflex activity (Braune et al. 1998), or induces vagal nerve stimulation at ECG and EEG (Burgess et al. 2016). In addition, it has been shown that some of these alterations in healthy people are dependent on the inspiration/expiration ratio (Béres et al. 2018).

All these studies in healthy people are summarized in Table 8.18. In fact, demonstrating hypersensitivity to EMFs (i.e., the specific pathophysiological identification of EHS) should be the main objective of further research in EHS patients. But to reach such an objective, EHS patients should be rigorously compared with normal healthy controls in case–control or randomized studies to demonstrate their increased intolerance to man-made EMFs after having determined suitable pre-defined objective endpoints. To our knowledge, such comparative studies have not yet been performed.

Considering the above-reported EMF-induced positive findings in normal people, the scientific demonstration of a specific decrease in the tolerance threshold of EHS patients, using such comparative methods will be extremely difficult. Therefore, research on hypersensitivity to EMFs in EHS patients may still remain an open question for a long time. The pathophysiological mechanism of EHS is still unknown. There are, however, some scientific indications supporting the hypothesis of a pathophysiological response, accounting specifically for the existence of such a hypersensitive state in EHS patients. (a) At a molecular level, the voltage-gated ion channels (VGICs) in cell membranes, the most abundant class of ion channels in cells of all organisms, are perhaps the most important EMF sensors and, thus, possibly a plausible target for polarized and coherent (man-made) EMFs (Bawin and Adey 1976; Liburdy 1992; Walleczek 1992; Balcavage et al. 1996; Panagopoulos et al. 2000; 2002; 2015; 2021; Pall 2013; Bertagna et al. 2021). According to Panagopoulos's biophysical model (Panagopoulos et al. 2000; 2002; 2015; 2021), man-made EMFs can irregularly open or close VGICs and, thus, disrupt any cell's electrochemical balance, which in turn, is a trigger for ROS production. In addition, neurons would be highly electrosensitive through their higher percentage of VGICs because they convey electric pulses, and thus, they are even more sensitive to applied EMFs (Panagopoulos et al. 2000; 2002; 2015; 2021). However, it is unknown whether such VGIC-related molecular process can explain the unique and specific pathophysiological mechanism of EHS, leading to the decrease of the brain-located EMF tolerance threshold (Salford et al. 2003; Bertagna et

TABLE 8.18
Provocation/Observational Studies with Positive Findings in Healthy Volunteers

Study, Country	Observed Endpoints	Type of Study	Evaluable Cases	Effect of EMF Exposure
von Klitzing 1995 (Germany)	Changes in resting EEG	Observational study involving low-frequency (217 Hz) exposure	17 healthy students	Alteration in the range of alpha activity during and after exposure for some hours
Mann et al. 1996 (Germany)	Changes in sleep EEG	Single-blind study involving GSM (2G) mobile phone EMF (pulsed 900 MHz) exposure	24 healthy male volunteers	Temporal pattern of cortisol secretion differs between placebo and exposure night
Braune et al. 1998 (Germany)	Blood pressure (BP), heart rate, capillary perfusion, and subjective wellbeing	Single-blind placebo-controlled study involving mobile phone EMF (pulsed 900 MHz) exposure	7 healthy volunteers	BP associated baro-reflex with activity alteration
Freude et al; 1998 (Germany)	Slow brain potentials (SBP)	Single-blind study involving mobile phone EMF (pulsed 916.2 MHz) exposure	16 healthy young peoples	Significant decrease of SBP at central and temporo-parieto-occipital brain regions
Crasson et al. 1999 (Belgium)	Changes in event-related potentials (ERP) and EEG/psycho-physiological and psychological behavior	Two double-blind experimental studies involving 50 Hz exposure and sham	21 healthy male volunteers	Low-level 50 Hz MF may have a slight influence on ERP and reaction time under circumstances of sustained attention
Krause 2000 (Finland)	Changes in EEG (during a memory task)	Single-blind study involving mobile phone EMF exposure	16 healthy volunteers	Mobile phone EMF modifies the brain responses
Croft et al. 2002 (Australia)	Effects of active mobile phone on the neurological system	Single-blind cross-over study involving RFR (900 MHz) exposure	24 healthy volunteers	Mobile phone EMF exposure affects brain functioning
Huber et al. 2002 (Switzerland)	Effect of EMF on waking regional cerebral blood flow (rCBF) and on waking and sleep EEG in humans	Double-blind study involving two types of MT EMF (a "base-station-like" and a "handset-like" signal) vs. sham control exposure	16 healthy young, right-handed, male subjects	Pulsed EMF increase waking rCBF and pulsation of EMF is necessary to induce waking and sleep EEG changes
Curcio et al. 2005 (Italia)	Effects of GSM on the nervous system	MT EMF (902.4 MHZ) exposure	20 healthy volunteers	Pulsed EMFs affect normal brain functioning

(Continued)

TABLE 8.18 (CONTINUED)
Provocation/Observational Studies with Positive Findings in Healthy Volunteers

Study, Country	Observed Endpoints	Type of Study	Evaluable Cases	Effect of EMF Exposure
Carlsson et al. 2005 (Sweden)	Annoyance related to electrical and chemical factors in a Swedish general population	Cross-sectional study involving different electrical equipment	13,604 subjects, representative of the population of Scania, Sweden	Connection between environmental annoyance, wellbeing and functional capacity
Huber et al. 2005 (Switzerland)	Effect of MT EMF on waking regional cerebral blood flow (rCBF)	Double-blind study involving two types of MT EMF (a "base-station-like" and a "handset-like" signal) vs. sham control exposure	12 healthy young males	Only "handset-like" (with stronger ELF pulsation) MT EMF exposure affected rCBF
Aalto et al. 2006 (Finland)	Effects of an active mobile phone on rCBF	Double-blind, counterbalanced study design with subjects performing a computer-controlled verbal working memory task	12 healthy volunteers	EMF emitted by a mobile phone affects rCBF in humans
Croft et al. 2008 (Australia)	Effects of GSM mobile phone EMF on the nervous resting system	Double-blind cross-over study. Mobile phone EMF (pulsed 895 MHz) exposure versus sham	120 healthy volunteers	Alpha power enhancement during MP exposure
Carrubba et al. 2008 (USA)	Evoked brain electrical potentials, EEG normal humans, and patients with epilepsy	Review on different normal human studies	Different normal human studies	Changes in brain activity
Curcio et al. 2009 (Italy)	Oxygenation of the frontal cortex by functional near-infrared spectroscopy (fNIRS)	Double-blind case–control study of GSM mobile phone signal (pulsed 902.4 MHz) compared to sham	31 healthy students	Slight influence in frontal cortex
Moussavy et al. 2009 (Iran)	Structure and function of hemoglobin	Experimental study involving simulated mobile phone EMF (pulsed 910 MHz and 940 MHz) exposure	Human adult hemoglobin prepared from human RBC of healthy donors	MT-like EMF decreases oxygen affinity and modifies tertiary structure of hemoglobin, depending on field intensity and time of exposure
Carrubba et al. 2010 (USA)	Effects of GSM MT ELF pulsation (217 Hz) on the nervous system	Double-blind study	20 healthy volunteers	MT ELF pulsation triggers evoked potentials during ordinary mobile phone use

Study, Country	Observed Endpoints	Type of Study	Evaluable Cases	Effect of EMF Exposure
Lowden et al. 2011 (Sweden)	Sleep EEG	Double-blind study with simulated GSM mobile phone EMF (pulsed 884 MHz) exposure versus sham	48 healthy volunteers	GSM MT-like exposure increases alpha range in sleep EEG
Volkow et al. 2011 (USA)	Brain glucose metabolism (PET scan)	Single-blind study involving 50-min mobile phone (pulsed 837.8 MHz) exposure	47 healthy participants	Increased brain glucose metabolism in the region closest to the device
Tao and Huang 2011 (USA)	Blood viscosity	Experimental study involving 1.3 T magnetic pulse to a small sample of blood	Human blood from healthy donors	After 1 min of exposure, blood viscosity was reduced by 33%
Vecchio et al. 2012 (Italy)	Changes in event-related desynchronization (ERD) at resting EEG	Placebo-controlled double-blind study involving GSM mobile phone EMF (pulsed 902.4 MHz) exposure	11 healthy volunteers	The peak amplitude of ERD and the reaction time to go stimuli were modulated by the effect on the cortical activity
Schmid et al. 2012 (Switzerland)	Resting EEG and polysomnography cognitive/behavioral endpoints	Double-blind cross-over study involving GSM-like EMF (pulsed 900 MHz) exposure	30 young healthy men	GSM MT (pulsed) EMF alters brain function. More significant effect with lower frequency pulsation
Muehsam et al. 2013 (USA)	Structure and function of hemoglobin	Experimental study involving a pulsed RF EMF (27.12 MHz) or a static magnetic field exposure	Human adult hemoglobin prepared from human RBC of healthy donors	Exposure for 10–30 min to either pulsed RF EMF or static magnetic field increased the rate of deoxygenation of hemoglobin occurring several min to several hours after the end of EMF exposure
Lustenberger et al. 2013 (Switzerland)	Brain activity during sleep EEG	Double-blind cross-over study involving pulsed RF (900 MHz) EMF exposure	16 healthy males	Pulsed RF EMF affects ongoing brain activity during sleep

(Continued)

TABLE 8.18 (CONTINUED)
Provocation/Observational Studies with Positive Findings in Healthy Volunteers

Study, Country	Observed Endpoints	Type of Study	Evaluable Cases	Effect of EMF Exposure
Ghosn et al. 2015 (France)	Changes in resting EEG effects of GSM MT EMF on the neurological system	Double-blind case–control study involving MT EMF (pulsed 900 MHz) exposure	26 healthy volunteers	During and after exposure, the alpha band power is significantly decreased with closed eyes compared to sham
Roggeveen et al. 2015 (Netherland)	Changes in resting EEG	Single-blind, cross-over study involving UMTS (3G) mobile phone EMF (pulsed 1.9291–1.9397 GHz) exposure	31 young female	All brain waves except delta changed significantly due to exposure at the ear, with stronger effects with ipsilateral exposure
Burgess et al. 2016 (UK)	Resting EEG and ECG (HRV)	Blinded randomized provocation study with a standardized TETRA signal versus sham	164 police officers and 60 volunteers	Vagal nerve stimulation at ECG and EEG
Verrender et al. 2016 (Australia)	Visual discrimination task and modified Sternberg working memory task	Double-blind cross-over study involving pulsed RF (920 MHz) EMF exposure	36 healthy volunteers	Cognitive performance was faster relative to sham in a working memory task during pulsed RF EMF exposure
Bères et al. 2018 (Hungary)	Heart rate asymmetry (HRA) and HRV parameters using repeated measures	Double-blind cross-over study involving GSM mobile phone EMF (pulsed 1800 MHz) exposure	20 healthy volunteers	Increase HRV under 1:1 breathing and mobile phone exposure
Loughran et al. 2019 (Australia)	Change in resting EEG	Double-blind cross-over study involving GSM-like (pulsed 920 MHz) EMF exposure versus sham	36 healthy volunteers	Alpha activity increases during high exposure condition compared to sham-exposure

al. 2021). (b) Alike all animals, humans are all sensitive to EMFs but normally not hypersensitive. Electromagnetic "receptors", whose role is, in fact, still unknown (Johnsen and Lohmann 2005; Nordmann et al. 2017), have been characterized as "cryptochroms" in retina (Gegear et al. 2010; Grehl et al. 2016) and as "magnetosomes" in the human brain (particularly in the hippocampus) and the meninges (Kirschvink et al. 1992; Dunn et al. 1995; Maher et al. 2016). Magnetosomes are supposedly located in the area corresponding to the observed EHS-associated biological abnormalities. These hypothetical receptors presumably consist of ferrous magnetite (Fe_3O_4) (graigite) or maghemite (Fe_2O_3) crystals (Kirschvink et al. 1992). Biogenic magnetite has been proposed to be a target capable of producing damaging biological effects (Kirschvink 1996; Johnsen and Lohmann 2005). (c) There are many molecular targets of EMFs in cells, including proteins (Porcelli et al. 1997; Bohr and Bohr 2000), chromatin, and DNA (Belyaev and Kravchenko 1994; Blank and Goodman 1999; Goodman and Blank 2002; Belyaev et al. 2005; Blank 2005; Lai 2021).

It has been shown in laboratory animals that neurons are particularly vulnerable to environmental stressors and that EMFs and/or chemicals can particularly damage them (Frey 1993; Redmayne and Johansson 2014; Megha et al. 2015a; 2015b). In addition, EMFs and/or chemicals can change neurotransmitter concentrations; alter synaptic-related proteins, particularly in the hippocampus (Bas et al. 2009; Leone et al. 2014; Teimori et al. 2016; Tan et al. 2019); and induce apoptosis (Joubert et al. 2008; Sonmez et al. 2010; Zuo et al. 2014; Odaci et al. 2016; Eghlidospour et al. 2017). Because, clinically, EHS appears to be an acquired, persisting, and, so far, irreversible state, one hypothesis could be that man-made EMFs and/or chemicals may have adversely affected or even destroyed specific neurons and their connections with the brain. This may occur mainly in the hippocampus where molecular interactions could lead to protein conformational abnormalities, epigenetic alterations and/or multiple DNA strand breaks, and finally, to neuronal destruction (Pihan and Doxsey 1999; Bohr and Bohr 2000; Foster 2000; Trosko 2000; Pinton et al. 2020). This is certainly a path for further neuro-pathophysiological research in order to better characterize (hyper)sensitivity in EHS and/or MCS patients via specific CNS neurological investigations.

8.3.2.3 Search for Etiology

The uncertainty in the results of several provocation studies performed in EHS self-reported patients and their misinterpretation by hypothesizing some nocebo effect explain the great confusion existing between research for a triggering effect of EMF exposure and the EHS etiology. Interpreting the negative results provided by several provocation studies in EHS patients as not arising from incorrect methodology (Blackman 2009; Schmiedchen et al. 2019) but from some nocebo effect, considering EHS as a psychological disease (Rubin et al. 2010; 2011), has been – in our view – a flawed interpretation. Indeed, the so-called nocebo effect is a hypothesis that needs to be validated by suitable experimental studies (Belpomme et al. 2021; Belpomme and Irigaray 2021). This has not occurred. To the contrary, on the basis of several interviews of a limited number of EHS patients, it has been shown that the psychological behavior associated with EHS in suffering patients is secondary to disease occurrence and, therefore, not the cause but the consequence of EHS (Dieudonné 2016). More significantly, the molecular and radiological abnormalities that have been detected in EHS patients clearly demonstrate that EHS is a neurological somatic disease not a psychological one. Similarly, MCS has not only been shown to be associated with increased sensitivity to chemicals but to be caused by acute or chronic initial toxic episodes triggered by multiple – mostly synthetic – environmental chemicals in genetically susceptible subjects (Bartha et al. 1999). EHS could also be defined not only as a specific state of intolerance to man-made EMFs but as mainly caused by excessive man-made EMF exposure. This critical interpretation has been provided by several scientists who analyzed the microwave syndrome (Carpenter 2014; 2015). The issue was more recently discussed in a review analyzing the underlying mechanisms for EHS development by Stein and Udasin (2020).

From the analysis of the scientific literature and our own findings, we now consider several important facts/arguments to establish causation by EMF exposure:

1. EHS appeared subsequently to anthropogenic EMF environmental pollution, with a highly increasing prevalence since the massive deployment of WC technology (Bandara and Carpenter 2018).
2. EHS occurrence is not restricted to regions or countries, but it is a worldwide plague with pandemic extension, as is the case for the worldwide expansion of the man-made EMF-emitting technologies (see Section 8.5).
3. Several main EHS-associated symptoms, such as sleep disturbance (Davis 1997), depressive tendency (Poole et al. 1993; Verkasalo et al. 1997), and suicide risk (Perry et al. 1981;

Johnston 2008) have been shown in independent epidemiological studies to result from dose-dependent EMF exposure, implying that excessive man-made EMF exposure (in terms of intensity and/or duration) is the cause of the EHS symptoms (Perry et al. 1981; Poole et al. 1993; Davis 1997; Verkasalo et al. 1997; Johnston 2008).

4. EHS symptoms are similar to those of the microwave syndrome, implying that EHS may be caused by microwave (WC EMF) exposure.
5. As documented in the present chapter (Sections 8.2.6–8.2.11 and 8.3.1), EHS patients are characterized particularly by low-grade inflammation, nitroso-oxidative stress, BBB disruption/opening, and brain neurotransmitter changes, which all have been shown by different independent experimental animal studies to be caused by man-made EMF exposures (Salford et al. 1994; 2003; Cao et al. 2000; Eberhardt et al. 2008; Nittby et al. 2009; Yang et al. 2012; Aboul Ezz et al. 2013; Megha et al. 2015a; 2015b; Saili et al. 2015; Hu et al. 2021).
6. As shown in Table 8.19, EHS patients are usually associated with past history of excessive exposure to WC (RF combined with ELF) EMFs, and/or ELF EMFs, suggesting that man-made EMFs are the most plausible causal factor in inducing EHS.
7. In addition to the well-documented interaction of man-made EMFs with VGICs (Panagopoulos et al. 2002; 2015; 2021), many independent *in vitro* and *in vivo* studies demonstrate that man-made EMFs can interact with endogenous physiological electric fields which control cellular biological functions (Weisenseel 1983; Nuccitelli 1988; 2003; Borgens 1988; Blanchard and Blackman 1994; Shi and Borgens 1995; McCaig and Zhao 1997; McCaig et al. 2005; Yao et al. 2009; Del Giudice et al. 2011; Funk 2015). When externally applied, man-made EMFs distort the physiological endogenous EMFs, they also distort the corresponding cellular functions, resulting in adverse biological/health effects. This is particularly the case for brain, heart, and muscles, suggesting a plausible role of man-made EMF exposure as a causing agent (Frey 1993; Vander Vorst et al. 2006).
8. It has been shown that man-made EMFs/EMR are totally polarized and coherent and, thus, differ from natural EMFs/EMR, which are only partially polarized in certain occasions, and this key difference accounts for their harmful and toxic effects on biomolecules, cells, and tissues in contrast to natural EMFs, which are vital for life (Panagopoulos et al. 2015).

TABLE 8.19
Percentage of EHS Cases Affected by Various Reported Sources

Sources	Features	EHS Cases (%)
Mobile phone	Combined RF and ELF EMFs	37
Mobile phone/DECT		8
DECT		7
Energy-saving lamps/mobile phone		1.4
Wi-Fi		16
MT base station antennas		3
Cathode-ray screen	VLF EMFs	9
High-voltage power lines	ELF EMFs	2.7
Power transformer		1.7
Railway		0.8
Chemicals	Non-EMF sources	11
Idiopathic		2.4

Source: Belpomme and Irigaray 2020
[VLF: Very Low Frequency (3–30 kHz).]

9. The neurotoxic effects of polarized and coherent (man-made) EMFs are also well-documented. There are, indeed, many *in vitro* and *in vivo* animal studies (Bas et al. 2009; Sonmez et al. 2010; Yang et al. 2012; Aldad et al. 2012; Deshmukh et al. 2013; Balassa et al. 2013; Furtado-Filho et al. 2015; Megha et al. 2015a; Zhang et al. 2015; Odaci et al. 2016; Sırav and Seyhan 2016) and human studies showing neurological and mainly brain toxic effects of man-made EMFs (Gandhi et al. 1996; Cardis et al. 2008; Dasdag et al. 2012; Belpomme et al. 2018).

10. At molecular level, it has been shown that non-thermal EMF exposures can lead to genetic (DNA/chromosome) damage in humans and animals (Lai and Singh 1995, 2004; Phillips et al. 2009; Panagopoulos 2019; 2020; Lai 2021), chromosomal instability (Semin et al. 1995; Mashevich et al. 2003), effects on mitosis (Bickmore and Carothers 1995; Pihan and Doxsey 1999), alterations in protein synthesis and structure (Lin et al. 1997; Bohr and Bohr 2000; de Pomerai et al. 2000), chromatin alterations (Belyaev and Kravchenko 1994; Belyaev et al. 2005), and inhibition of DNA repair (Belyaev 2005; Belyaev et al. 2006).

11. It has been claimed by the ICNIRP (1998; 2010; 2020) that since man-made EMFs are non-ionizing, they do not have sufficient energy to directly damage DNA. But it has been shown that man-made (ELF and WC) EMFs, even at non-thermal levels, can indirectly cause DNA damage and mutations. They can do this by generating free radicals/ROS which, in turn, damage DNA (Yakymenko et al. 2016; Panagopoulos et al. 2021). ROS/free radicals may also induce consequent changes in gene expression and regulation through epigenetic mechanisms (Blank and Goodman 1999; Foster 2000; Goodman and Blank 2002; Leone et al. 2014; Dasdag 2015a, 2015b; Belpomme and Irigaray 2022). Finally, they may possibly cause changes in charge distributions, leading to structural changes in molecules (Blank and Soo 2001; Blank 2008). It has been shown that man-made EMFs can induce rapid synthesis of heat shock proteins, which accounts for the anti-inflammatory response (Lin et al. 1999, 2001; Blank 2005; Blank and Goodman 1999; 2011) that we have found to occur in EHS patients (see Section 8.2.6). Such epigenetic effects may be involved, in addition to the above-mentioned neuronal damage.

The above facts/arguments indicate a causal role of anthropogenic EMFs in inducing EHS (Belpomme and Irigaray 2022). Although man-made EMF exposure appears to be the main cause of EHS, MCS may precede the occurrence of EHS in some EHS cases. Thus, we have hypothesized that chemicals may also be implicated as causing agents in EHS genesis in a limited number of cases (11%) (see Table 8.19). Additionally, in association with the causal role of EMFs and/or chemicals, there may be some independent risk factors associated with EHS genesis, such as pre-existing depression, psychiatric comorbidity[*] (Meg Tseng et al. 2011), brain trauma, chronic immunosuppression leading to opportunistic infections[†], or congenital brain malformation, which could further the EMF- and/or chemical-related EHS genesis in genetically and/or epigenetically predisposed individuals. Future research should focus on these risk factors with biophysical/biological and epidemiological studies.

8.4 THERAPEUTICS AND METHODS OF PREVENTION

Presently there is no standardized treatment for EHS and no internationally recognized medical needs for persons with environmental sensitivity, including EHS and/or MCS (Gibson et al. 2015). Furthermore, the medical approach proposed to counteract the suffering state of the patients has not yet been well defined and adapted to this condition. In most cases, psychological methods do not solve the medical problem and do not help patients. Definitely, the minimization/avoidance of

---

[*] The co-occurrence of more than one disorder in the same individual.
[†] Infections that occur more often (or are more severe) in people with weakened immune systems.

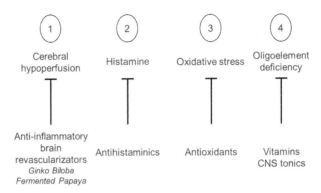

FIGURE 8.9 Proposed therapies (in addition to the avoidance of pollutants).

exposure to anthropogenic EMFs and chemical pollutants is the first and permanent necessity and the most important step. Furthermore, as summarized in **Figure. 8.9**, there are some empirical treatments that could be recommended on the basis of the results provided by biological investigations and the use of biomarkers.

For example, we have found that EHS patients frequently display a profound deficit in vitamins and trace elements, especially vitamin D and zinc, which should be corrected (Irigaray et al. 2018b; Belpomme and Irigaray 2020). Antihistamines could be used in cases of increased histamine in the blood. Furthermore, antioxidants such as glutathione and, more specifically, anti-oxidative/nitrosative stress medications could also be used in case of oxidative/nitrosative stress detection (Irigaray et al. 2018a). Moreover, we showed that natural products such as fermented papaya preparation (FPP) and ginkgo biloba can restore brain pulsatility in the various tissue areas of the temporal lobes that depend upon middle cerebral artery, thereby improving brain hemodynamics and, consequently, brain oxygenation (Irigaray et al. 2018b). Since FPP was shown to possess some antioxidant, anti-inflammatory, and immune-modulating properties (Aruoma et al. 2010; Somanah et al. 2012; Murakamai et al. 2012), we recommend the use of this widely available natural product. In our empirical experience, there was a similar beneficial effect of this therapeutic method in MCS patients. However, there remains a lack of adequate clinically controlled trials proving an overall verification for the positive effect of all these proposed empirical therapeutic and protective components in addition to exposure avoidance.

In addition, a possible protective effect of partial and/or intermittent EMF-shielding is discussed because it may not be tolerated when applied continuously to people (Wever 1970; 1973; 1979; Panagopoulos and Chrousos 2019). In experiments carried out in the Max Planck institute in Germany by Wever (1970; 1973; 1979) with the participation of many volunteers staying in a completely shielded apartment for a few weeks, it was shown that the shielding of the natural atmospheric (Schumann) electromagnetic resonances led to internal desynchronization of the participants, which is a serious medical syndrome. It should be experimentally tested in EHS patients, as to our knowledge, such a protective effect has not been proved in well-conducted controlled clinical trials or long-term experiments. Likewise, the possible beneficial effect of grounding/earthing should be tested by long-term experiments (Chevalier et al. 2012; Díaz-Del Cerro et al. 2020).

Avoidance of artificial EMF sources and of many (mostly commercial) man-made chemicals is definitely the first priority in any case. Moreover, removal of mercury-based dental amalgam fillings is recommended in EHS and/or MCS patients. A similar consideration reasonably applies for metallic plates and screws used in bone restoring surgery, apart from cases when they cannot be removed for surgical reasons. We do not recommend the systematic use of chelation to withdraw metals including mercury from the organism because of possible severe complications and no established scientific justification in controlled trials.

In the case of no treatment and no protection against anthropogenic EMFs and/or multiple chemical environmental stressors, EHS may evolve toward cerebral atrophy or a neuro-degenerative and psychiatric disorder, possibly closely related to some atypical Alzheimer's disease. Yet, while treating and protecting patients as early as possible, we never observed the occurrence of true Alzheimer's disease during the follow-up of any of the 3,000 cases recorded since 2009 in our database. By contrast, regression and even disappearance of symptoms of intolerance may occur after treatment and protection of patients. However, it should be noted that, in our experience (hyper)sensitivity to EMFs and/or MCS-related chemicals never disappears completely. As explained, it seems to be related with some CNS-associated pathological process, requiring strong and persistent anti-EMF and anti-chemical prevention. Thus, both EHS and MCS disorders cannot be merely reduced to some vague spontaneously reversible so-called "functional impairment".

8.5 SOCIAL CONSEQUENCES OF EHS AND HOW TO COUNTERACT THE INCREASING PROBLEM

Due to intolerable symptomatic feelings, the individual societal cost of EHS and/or MCS is very high. Exclusion from human society concerns many patients with loss of employment and/or change of residence in order to avoid noxious EMF and/or chemical exposures. In many cases, people with EHS are led to the transformation of their living place into a Faraday (metal) cage, and even modification of clothing for metal shielding against man-made (most usually WC) EMFs, methods which can be risky, as explained above (Wever 1970; 1973; 1979; Panagopoulos and Chrousos 2019). Additionally, without appropriate medical care and protection, patients may develop depression with suicidal ideation. To this, we should add the current lack of knowledge of medical practitioners on this issue and the lack of official acknowledgment of EHS and MCS, in spite of recent medical surveys to educate medical practitioners on the problem of EHS (Slottje et al. 2017).

Whatever the causal origin of EHS and MCS, there is compelling evidence that these two pathological disorders constitute an unsolved, large, and increasing global health problem. As emphasized (see Section 8.3.2.3), there are strong arguments incriminating man-made and especially WC EMFs as well as man-made chemicals introduced in the environment in the past few decades (Belpomme et al. 2007; Irigaray and Belpomme 2010; Pall 2018; Bandara and Carpenter 2018). Indeed, the increased numbers of EHS cases worldwide, and the fact that this plague began as soon as WC technology became widespread, argue that man-made and especially WC EMFs are the cause of EHS.

In Table 8.20, the percentages of EHS occurrence among the population are estimated to range from 0.7% to 13.3% (average 3%–5%) in most countries worldwide, meaning that millions of people may, in fact, be affected by EHS. Similar worldwide percentages may account for MCS (Genuis 2010). However, because these estimations were based on non-objective criteria for identifying EHS (Baliatsas et al. 2012), these data require confirmation by more objective investigations.

Given the eight billions of the global population – with most of them exposed daily by mobile and/or cordless phones, Wi-Fi, and other WC devices and man-made EMF sources – and the present and future development of 5G (fifth generation of MT/WC) (Hardell and Nyberg 2020; Hardell and Carlberg 2020), it is expected that these percentages will significantly increase in the next years (Hallberg and Oberfeld 2006), in as much as the manufacturers of WC and other EMF technologies and industrial chemicals will continue developing their products (Bandara and Carpenter 2018).

8.6 GENERAL DISCUSSION AND CONCLUSION

In this chapter, we have addressed all issues related to the historical background, diagnosis, treatment, and prevention of EHS. After extensive examination of this pathological condition, we define EHS as an endogenous pathological state of decreased CNS-related tolerance threshold to man-made EMFs. In addition, by applying the WHO-proposed causality criteria, we argue that EHS is really caused by anthropogenic EMFs, as it is claimed by the EHS patients.

TABLE 8.20
Percentages of People with Self-reported EHS in Different Countries

Study, Country	Year of Results	Sample Size (Age)*	% People who Contributed to the Survey	Estimated % People with EHS
Hillert et al. 2002, Sweden	1997	15,000 (19–80)	73	1.5
Palmquist et al. 2014, Sweden	2010	3,406	40	2.7
Schreier et al. 2006, Switzerland	2004	2,048 (> 14)	55.1	5
Röösli et al. 2010a, Switzerland	2008	1,122 (30–60)	37	8.6
Röösli et al. 2010b, Switzerland	2009	1,122 (30–60)	37	7.7
Blettner et al. 2009, Germany	2004	30,047	58.6	10.3
Kowall et al. 2012, Germany (phase 1)	2004	30,047	58.4	8.7
Kowall et al. 2012, Germany (phase 2)	2006	30,047	58.4	7.2
Levallois et al. 2002, USA	1998	2,072	58.3	3.2
Korpinen et al. 2009, Finland	2002	6,121	40.8	0.7
Eltiti et al. 2007, UK	2005	3,633	18.2	4
Meg Tseng et al. 2011, Taiwan	2007	1,251	11.5	13.3
Schröttner and Leitgeb 2008, Austria	2008	460	88	3.5
Furubayashi et al. 2009, Japan	2007	2,472	62.3	1.2
Batiatsas et al. 2014, Netherlands	2011	5,789	39.6	3.5
Van Dongen et al. 2014, Netherlands	Before 2013	1,009	60	7

[*Where provided, age of included patients is indicated in brackets]

The VGIC mechanism (Panagopoulos et al. 2000; 2002; 2015; 2021), which can explain the general toxic effect of man-made EMF exposure in the living cells, may not be enough alone to specifically account for EHS genesis, as it must be additionally explained how the initial cellular imbalance caused by man-made EMFs through this mechanism results finally in the decreased CNS-related tolerance threshold to man-made EMFs, which is a chronic systemic disorder. We believe our work has elucidated several possible biological processes for that.

By using a large series of patients, and by investigating several biomarkers in the peripheral blood and urine and suitable imaging techniques, we have shown that EHS is a distinct brain neurologic pathological disorder which can be objectively diagnosed and treated. Moreover, we have shown that although they differ in their etiology and pathogenesis, EHS and MCS share a similar clinical and biological signature, and, therefore, they must be considered medically as parts of a unique environmental intolerance-related neurological syndrome. This finding, mainly based on the use of biomarkers and suitable imaging techniques, must evidently be confirmed by other research groups. We hope that our findings, in combination with those of other researchers, will accelerate research progress and result in the acknowledgment of EHS and MCS as real pathological disorders. This is what many scientists have agreed recently in a scientific consensus report stipulating the critical role of biomarkers and imaging techniques in the study of EHS (Belpomme et al. 2021). Indeed, further research efforts should be made to prove the causal role of EMFs in triggering EHS-associated symptoms and EHS genesis itself.

Moreover, in this chapter, we have analyzed the medical diagnostic criteria of EHS and have emphasized that symptom occurrence should be reproducible under the influence of man-made EMF exposure, a necessary diagnostic criterion to distinguish EHS from IEI-EMF. Additionally, we have discussed the sensitivity, specificity, and reproducibility of the biomarkers we have routinely used and have reported that they are not specific and do not account for 20%–30% of the EHS cases. Thus, to allow an objective diagnosis of EHS in these cases, it is necessary to use other biomarkers (such as brain neurotransmitters) and various cerebral imaging techniques.

The various collected independent data we have provided in the chapter regarding the etiology of EHS fulfill the causation criteria proposed by the WHO (2006) (described in Section 8.3.3), as they confirm a dose-response dependency of symptom occurrence in epidemiological studies, and they show that the pathophysiological changes observed in cells and animals exposed to man-made EMFs are similar to the pathophysiological changes observed in EHS patients. Moreover, all these data provided by different independent research teams are consistent with general scientific principles. These data, therefore, strongly suggest the role of man-made EMFs as a causal agent of EHS-associated symptoms and pathophysiological changes. In addition, it is clearly demonstrated by many independent provocation studies that excessive man-made EMF exposure is noxious for healthy people.

Consequently, there is sufficient available scientific evidence to strongly recommend protective measures against the present unrestricted man-made electromagnetic pollution by applying the Precautionary Principle to protect particularly pregnant women, infants, children, and teenagers who have been proved to be more susceptible and sensitive to anthropogenic EMFs (Carpenter and Sage 2008).

As reported in this chapter, since the 2005 and 2014 WHO official statements, much progress has been made in the identification and understanding of EHS (and MCS) and the bioclinical health effects of man-made EMFs and/or chemicals in the human organism. But EHS and MCS have not yet been clearly characterized by the WHO. Research on the health effects of man-made non-thermal EMF exposures has now clearly demonstrated the occurrence of toxic effects in animals and humans. But, as noted above, although VGICs on cell membranes (particularly in neural cells) may indeed represent a major transducer for EMF-induced effects on the CNS (Panagopoulos et al. 2000; 2002; 2015; 2021; Bertagna et al. 2021), their implication in EHS genesis is still unclear, as a lot of work is still needed to fill the knowledge gap between cellular/neuronal imbalance and CNS disorder.

Contrary to the unrealistic claims by the ICNIRP still denying the existence of non-thermal biological and health effects of man-made EMFs, we strongly believe that critical research progress has recently been made, making the non-thermal man-made EMF effects today common knowledge among experts in bioelectromagnetics and related areas. Indeed, it appears that man-made EMFs, in addition to environmental chemicals, are the cause of EHS as well as the cause for certain types of cancer (IARC 2002; 2013).

Present research on EHS mainly focuses on the role of man-made EMFs as symptomatic triggers and causes of EHS. In addition, the health care needs of people with environmental sensitivities, such as EHS or MCS, should be carefully addressed in the present socioeconomic environment and medical challenges (Gibson et al. 2015) to ensure appropriate sanitary measures.

Today's level of scientific knowledge engenders a great ethical responsibility of scientists and governments and, most of all, of national and international health bodies to uncover the adverse health effects of the increasing man-made EMF exposures and to oppose the emerging and growing worldwide EHS and MCS global plagues. This means that suitable public health measures must urgently be taken to decrease man-made and especially WC EMF exposures, and recognize EHS and MCS as new pathologies.

We, therefore, strongly ask the WHO to include EHS and MCS in the future versions of the WHO ICDs on the basis of their clinical and pathophysiological identification and characterization, just as has been done for the other already included and recognized diseases.

REFERENCES

Aalto, S., Haarala, C., Brück, A., Sipilä, H., Hämäläinen, H. and Rinne, J. O., (2006): Mobile phone affects cerebral blood flow in humans. *J Cereb Blood Flow Metab* 26(7):885–90.

Abbott, N. J., (2000): Inflammatory mediators and modulation of blood-brain barrier permeability. *Cell Mol Neurobiol* 20(2):131–47.

Aboul Ezz, H. S., Khadrawy, Y. A., Ahmed, N. A., Radwan, N. M. and El Bakry, M. M., (2013): The effect of pulsed electromagnetic radiation from mobile phone on the levels of monoamine neurotransmitters in four different areas of rat brain. *Eur Rev Med Pharmacol Sci* 17(13):1782–8.

Adachi, N., (2005): Cerebral ischemia and brain histamine. *Brain Res Brain Res Rev* 50(2):275–86.

Ahlbom, A., Day, N., Feychting, M., et al., (2000): A pooled analysis of magnetic fields and childhood leukaemia. *Br J Cancer* 83(5):692–98.

Akerboom, T. P. and Sies, H., (1981): Assay of glutathione, glutathione disulfide, and glutathione mixed disulfides in biological samples. *Methods Enzymol* 77:373–82.

Albert, P. J., Proal, A. D. and Marshall, T. G., (2009): Vitamin D: The alternative hypothesis. *Autoimmun Rev* 8(8):639–44.

Aldad, T. S., Gan, G., Gao, X. B. and Taylor, H. S., (2012): Fetal radiofrequency radiation exposure from 800–1900 MHz-rated cellular telephones affects neurodevelopment and behavior in mice. *Sci Rep* 2:312.

Arnetz, B. B., Akerstedt, T., Hillert, L., Lowden, A., Kuster, N. and Wiholm, C., (2007): The effects of 884 MHz GSM wireless communication signals on self-reported symptom and sleep (EEG): An experimental provocation study. *PIERS Online* 3(7):1148–50.

Arnold, W., Pfaltz, R. and Altermatt, H. J., (1985): Evidence of serum antibodies against inner ear tissues in the blood of patients with certain sensorineural hearing disorders. *Acta Otolaryngol* 99(3–4):437–44.

Aruoma, O. I., Hayashi, Y., Marotta, F., Mantello, P., Rachmilewitz, E. and Montagnier, L., (2010): Applications and bioefficacy of the functional food supplement fermented papaya preparation. *Toxicology* 278(1):6–16.

Augner, C. and Hacker, G. W., (2009): Are people living next to mobile phone base stations more strained? Relationship of health concerns, self-estimated distance to base station, and psychological parameters. *Indian J Occup Environ Med* 13(3):141–5.

Augner, C., Gnambs, T., Winker, R. and Barth, A., (2012): Acute effects of electromagnetic fields emitted by GSM mobile phones on subjective well-being and physiological reactions: A meta-analysis. *Sci Total Environ* 424:11–5.

Auvinen, A., Feychting, M., Ahlbom, A., et al., (2019): Headache, tinnitus and hearing loss in the international cohort study of mobile phone use and health (COSMOS) in Sweden and Finland. *Int J Epidemiol* 48(5):1567–79.

Bagheri Hosseinabadi, M., Khanjani, N., Ebrahimi, M. H., Haji, B. and Abdolahfard, M., (2019): The effect of chronic exposure to extremely low-frequency electromagnetic fields on sleep quality, stress, depression and anxiety. *Electromagn Biol Med* 38(1):96–101.

Balassa, T., Varró, P., Elek, S., Drozdovszky, O., Szemerszky, R., Világi, I. and Bárdos, G., (2013): Changes in synaptic efficacy in rat brain slices following extremely low-frequency magnetic field exposure at embryonic and early postnatal age. *Int J Dev Neurosci* 31(8):724–30.

Balcavage, W. X., Alvager, T., Swez, J., Goff, C. W., Fox, M. T., Abdullyava, S. and King, M. W., (1996): A mechanism for action of extremely low frequency electromagnetic fields on biological systems. *Biochem Biophys Res Commun* 222(2):374–78.

Baliatsas, C., Bolte, J., Yzermans, J., Kelfkens, G., Hooiveld, M., Lebret, E. and van Kamp, I., (2015): Actual and perceived exposure to electromagnetic fields and non-specific physical symptoms: An epidemiological study based on self-reported data and electronic medical records. *Int J Hyg Environ Health* 218(3):331–44.

Baliatsas, C., van Kamp, I., Hooiveld, M., Yzermans, J. and Lebret, E., (2014): Comparing non-specific physical symptoms in environmentally sensitive patients: Prevalence, duration, functional status and illness behavior. *J Psychosom Res* 76(5):405–13.

Baliatsas, C., van Kamp, I., Kelfkens, G., Schipper, M., Bolte, J., Yzermans, J. and Lebret, E., (2011): Non-specific physical symptoms in relation to actual and perceived proximity to mobile phone base stations and powerlines. *BMC Public Health* 11:421.

Baliatsas, C., van Kamp, I., Lebret, E. and Rubin, G. J., (2012): Idiopathic environmental intolerance attributed to electromagnetic fields (IEI-EMF): A systematic review of identifying criteria. *BMC Public Health* 12:643.

Balmori, A., (2005): Possible effects of electromagnetic fields from phone masts on a population of White Stork (Ciconia ciconia). *Electromagn Biol Med* 24(2):109–19.

Balmori, A., (2006): The incidence of electromagnetic pollution on the amphibian decline: Is this an important piece of the puzzle? *Toxicol Environ Chem* 88(2):287–99.

Bandara, P. and Carpenter, D. O., (2018): Planetary electromagnetic pollution: It is time to assess its impact. *Lancet Planet Health* 2(12):e512–4.

Barsam, T., Monazzam, M. R., Haghdoost, A. A., Ghotbi, M. R. and Dehghan, S. F., (2012): Effect of extremely low frequency electromagnetic field exposure on sleep quality in high voltage substations. *Iran J Environ Health Sci Eng* 9(1):15.

Bartha, L., Baumzweiger, W., Buscher, D. S., et al., (1999): Multiple chemical sensitivity: A 1999 consensus. *Arch Environ Health* 54(3):147–9.

Bas, O., Odaci, E., Kaplan, S., Acer, N., Ucok, K. and Colakoglu, S., (2009): 900 MHz electromagnetic field exposure affects qualitative and quantitative features of hippocampal pyramidal cells in the adult female rat. *Brain Res* 1265:178–85.

Bawin, S. M. and Adey, W. R., (1976): Sensitivity of calcium binding in cerebral tissue to weak environmental electric fields oscillating at low frequency. *Proc Natl Acad Sci U S A* 73(6):1999–2003.

Baydas, G., Ozer, M., Yasar, A., Koz, S. T. and Tuzcu, M., (2006): Melatonin prevents oxidative stress and inhibits reactive gliosis induced by hyperhomocysteinemia in rats. *Biochem (Mosc)* 71:S91–5.

Bell, I. R., Miller, C. S. and Schwartz, G. E., (1992): An olfactory-limbic model of multiple chemical sensitivity syndrome: Possible relationships to kindling and affective spectrum disorders. *Biol Psychiatry* 32(3):218–42.

Bell, R. D. and Zlokovic, B. V., (2009): Neurovascular mechanisms and blood-brain barrier disorder in Alzheimer's disease. *Acta Neuropathol* 118(1):103–13.

Belpomme, D., Irigaray, P., Hardell, L., Clapp, R., Montagnier, L., Epstein, S. and Sasco, A. J., (2007): The multitude and diversity of environmental carcinogens. *Environ Res* 105(3):414–29.

Belpomme, D., Campagnac, C. and Irigaray, P., (2015): Reliable disease biomarkers characterizing and identifying electrohypersensitivity and multiple chemical sensitivity as two etiopathogenic aspects of a unique pathological disorder. *Rev Environ Health* 30(4):251–71.

Belpomme, D., Campagnac, C. and Irigaray, P., (2016): Corrigendum to: Reliable disease biomarkers characterizing and identifying electrohypersensitivity and multiple chemical sensitivity as two etiopathogenic aspects of a unique pathological disorder. *Rev Environ Health*. https://doi.org/10.1515/reveh-2015-8888.

Belpomme, D., Hardell, L., Belyaev, I., Burgio, E. and Carpenter, D. O., (2018): Thermal and non-thermal health effects of low intensity non-ionizing radiation: An international perspective. *Environ Pollut* 242(A):643–58.

Belpomme, D. and Irigaray, P., (2020): Electrohypersensitivity as a newly identified and characterized neurologic pathological disorder: How to diagnose, treat, and prevent it. *Int J Mol Sci* 21(6):1915.

Belpomme, D., Carlo, G. L., Irigaray, P., et al., (2021): The critical importance of molecular biomarkers and imaging in the study of electrohypersensitivity. A scientific consensus international report. *Int J Mol Sci* 22(14):7321.

Belpomme, D. and Irigaray, P., (2021): Why scientifically unfounded and misleading claim should be dismissed to make true research progress in the acknowledgment of electrohypersensibility as a new worldwide emerging pathology. *Rev Environ Health*. https://doi.org/10.1515/reveh-2021-0104.

Belpomme, D. and Irigaray, P., (2022): Why electrohypersensitivity and related symptoms are caused by non-ionizing man-made electromagnetic fields: An overview and medical assessment. *Environ Res.* 212(Pt A):113374.

Belsey, R., Deluca, H. F. and Potts, J. T. Jr., (1971): Competitive binding assay for vitamin D and 25-OH vitamin D. *J Clin Endocrinol Metab* 33(3):554–7.

Belyaev, I., (2005): Non-thermal biological effects of microwaves. *Microw Rev* 11:3–29.

Belyaev, I. Y., Hillert, L., Protopopova, M., Tamm, C., Malmgren, L. O., Persson, B. R., Selivanova, G., Harms-Ringdahl, M., (2005): 915 MHz microwaves and 50 Hz magnetic field affect chromatin conformation and 53BP1 foci in human lymphocytes from hypersensitive and healthy persons. *Bioelectromagnetics* 26(3):173–84.

Belyaev, I. Y., Koch, C. B., Terenius, O., et al., (2006): Exposure of rat brain to 915 MHz GSM microwaves induces changes in gene expression but not double stranded DNA breaks or effects on chromatin conformation. *Bioelectromagnetics* 27(4):295–306.

Belyaev, I. Y., Markovà, E., Hillert, L., Malmgren, L. O. and Persson, B. R., (2009): Microwaves from UMTS/GSM mobile phones induce long-lasting inhibition of 53BP1/gamma-H2AX DNA repair foci in human lymphocytes. *Bioelectromagnetics* 30(2):129–41.

Belyaev, S. Y. and Kravchenko, V. G., (1994): Resonance effect of low-intensity millimeter waves on the chromatin conformational state of rat thymocytes. *Z Naturforsch C J Biosci* 49(5–6):352–8.

Berberian, P. A., Myers, W., Tytell, M., Challa, V. and Bond, M. G., (1990): Imunohistochemical localization of heat shock protein-70 in normalappearing and atherosclerotic specimens of human arteries. *Am J Pathol* 136(1):71–80.

Béres, S., Németh, Á., Ajtay, Z., Kiss, I., Németh, B. and Hejjel, L., (2018): Cellular phone irradiation of the head affects heart rate variability depending on inspiration/expiration ratio. *In Vivo* 32(5):1145–53.

Berg, M., (1988): Skin problems in workers using visual display terminals: A study of 201 patients. *Contact Dermatitis* 19(5):335–41.

Berg, N. D., Rasmussen, H. B., Linneberg, A., et al., (2010): Genetic susceptibility factors for multiple chemical sensitivity revisited. *Int J Hyg Environ Health* 213(2):131–9.

Bergqvist, U. and Vogel, E., (1997): Possible health implications of subjective symptoms and electromagnetic fields. In *A report prepared by a European group of experts for the European commission, DGV*; Arbete Och Hälsa, 19. Stockholm: Swedish National Institute for Working Life. http://www2.niwl.se/forlag/en/.

Bergqvist, U. and Wahlberg, J. E., (1994): Skin symptoms and disease during work with visual display terminals. *Contact Dermatitis* 30(4):197–204.

Bergqvist, U. O., (1984): Video display terminals and health: A technical and medical appraisal of the state of the art. *Scand J Work Environ Health* 10:1–87.

Bertagna, F., Lewis, R., Silva, S. R. P., McFadden, J. and Jeevaratnam, K., (2021): Effects of electromagnetic fields on neuronal ion channels: A systematic review. *Ann N Y Acad Sci*. https://doi.org/10.1111/nyas.14597.

Beyer, C., Christen, P., Jelesarov, I. and Fröhlich, J., (2013): Experimental system for real-time assessment of potential changes in protein conformation induced by electromagnetic fields. *Bioelectromagnetics* 34(6):419–28.

Bickmore, W. A. and Carothers, A. D., (1995): Factors affecting the timing and imprinting of replication on a mammalian chromosome. *J Cell Sci* 108(8):2801–9.

Blackman, C., (2009): Cell phone radiation: Evidence from ELF and RF studies supporting more inclusive risk identification and assessment. *Pathophysiology* 16(2–3):205–16.

Blanchard, J. P. and Blackman, C. F., (1994): Clarification and application of an ion parametric resonance model for magnetic field interactions with biological systems. *Bioelectromagnetics* 15(3):217–38.

Blank, M., (2005): Do electromagnetic fields interact with electrons in the Na,K-ATPase? *Bioelectromagnetics* 26(8):677–83.

Blank, M., (2008): Protein and DNA reactions stimulated by electromagnetic fields. *Electromagn Biol Med* 27(1):3–23.

Blank, M. and Goodman, R., (1999): Electromagnetic fields may act directly on DNA. *J Cell Biochem* 75(3):369–74.

Blank, M. and Goodman, R., (2009): Electromagnetic fields stress living cells. *Pathophysiology* 16(2–3):71–8.

Blank, M. and Goodman, R., (2011): DNA is a fractal antenna in electromagnetic fields. *Int J Radiat Biol* 87(4):409–15.

Blank, M. and Soo, L., (2001): Electromagnetic acceleration of electron transfer reactions. *J Cell Biochem* 81(2):278–83.

Blettner, M., Schlehofer, B., Breckenkamp, J., et al., (2009): Mobile phone base stations and adverse health effects: Phase 1 of a population-based, cross-sectional study in Germany. *Occup Environ Med* 66(2):118–23.

Bohr, H. and Bohr, J., (2000): Microwave-enhanced folding and denaturation of globular proteins. *Phys Rev E Stat Phys Plasmas Fluids Relat Interdiscip Topics* 61(4 Pt B):4310–4.

Bolte, J. F., Baliatsas, C., Eikelboom, T. and van Kamp, I., (2015): Everyday exposure to power frequency magnetic fields and associations with non-specific physical symptoms. *Environ Pollut* 196:224–9.

Bonhomme-Faivre, L., Marion, S., Bezie, Y., Auclair, H., Fredj, G. and Hommeau, C., (1998): Study of human neurovegetative and hematologic effects of environmental low-frequency (50-Hz) electromagnetic fields produced by transformers. *Arch Environ Health* 53(2):87–92.

Borgens, R. B., (1988): Stimulation of neuronal regeneration and development by steady electrical fields. In S. G. Waxman ed., *Advances in Neurology*, 47:547–564.

Bortkiewicz, A., Gadzicka, E., Szyjkowska, A., Politański, P., Mamrot, P., Szymczak, W. and Zmyślony, M., (2012): Subjective complaints of people living near mobile phone base stations in Poland. *Int J Occup Med Environ Health* 25(1):31–40.

Bozic, B., Cucnik, S., Kveder, T. and Rozman, B., (2007): Autoimmune reactions after electro-oxidation of IgG from healthy persons: Relevance of electric current and antioxidants. *Ann N Y Acad Sci* 1109:158–66.

Braune, S., Wrocklage, C., Raczek, J., Gailus, T. and Lücking, C. H., (1998): Resting blood pressure increase during exposure to a radio-frequency electromagnetic field. *Lancet* 351(9119):1857–8.

Broadbent, D. E., Broadbent, M. H., Male, J. C. and Jones, M. R., (1985): Health of workers exposed to electric fields. *Br J Ind Med* 42(2):75–84.

Brzezinski, A., (1997): Melatonin in humans. *N Engl J Med* 336(3):186–95.

Bunch, C. and Crook, D. W. M., (1998): Opportunistic infections. In Delves P.J., Roitt I.M. (Eds): *Encyclopedia of immunology* (2nd ed.). San Diego, CA: Academic Press, pp. 1881–4, ISBN: 0-12-226765-6.

Burch, J. B., Reif, J. S., Yost, M. G., Keefe, T. J. and Pitrat, C. A., (1999): Reduced excretion of a melatonin metabolite in workers exposed to 60 Hz magnetic fields. *Am J Epidemiol* 150(1):27–36.

Burgess, A. P., Fouquet, N. C., Seri, S., et al., (2016): Acute exposure to terrestrial trunked radio (TETRA) has effects on the electroencephalogram and electrocardiogram, consistent with vagal nerve stimulation. *Environ Res* 150:461–9.

Byun, Y. H., Ha, M., Kwon, H. J., et al., (2013): Mobile phone use, blood lead levels, and attention deficit hyperactivity symptoms in children: A longitudinal study. *PLOS ONE* 8(3):e59742.

Cabré-Riera, A., Torrent, M., Donaire-Gonzalez, D., Vrijheid, M., Cardis, E. and Guxens, M., (2019): Telecommunication devices use, screen time and sleep in adolescents. *Environ Res* 171:341–7.

Caccamo, D., Cesareo, E., Mariani, S., et al., (2013): Xenobiotic sensor- and metabolism-related gene variants in environmental sensitivity-related illnesses: A survey on the Italian population. *Oxid Med Cell Longev* 2013:831969.

Califf, R. M., (2018): Biomarker definitions and their applications. *Exp Biol Med (Maywood)* 243(3):213–21.

Calvente, I., Pérez-Lobato, R., Núñez, M. I. et al., (2016): Does exposure to environmental radiofrequency electromagnetic fields cause cognitive and behavioral effects in 10-year-old boys? *Bioelectromagnetics* 37(1):25–36.

Cammaerts, M. C. and Johansson, O., (2014): Ants can be used as bio-indicators to reveal biological effects of electromagnetic waves from some wireless apparatus. *Electromagn Biol Med* 33(4):282–8.

Cao, Z., Zhang, H., Tao, Y. and Liu, J., (2000): Effects of microwave radiation on lipid peroxidation and the content of neurotransmitters in mice. *Wei Sheng Yan Jiu* 29(1):28–9.

Cardis, E., Deltour, I., Mann, S., et al., (2008): Distribution of RF energy emitted by mobile phones in anatomical structures of the brain. *Phys Med Biol* 53(11):2771–83.

Carles, C., Esquirol, Y., Turuban, M., et al., (2020): Residential proximity to power lines and risk of brain tumor in the general population. *Environ Res* 185:109473.

Carlsson, F., Karlson, B., Ørbaek, P., Osterberg, K. and Ostergren, P. O., (2005): Prevalence of annoyance attributed to electrical equipment and smells in a Swedish population, and relationship with subjective health and daily functioning. *Public Health* 119(7):568–77.

Carpenter, D. O., (2014): Excessive exposure to radiofrequency electromagnetic fields may cause the development of electrohypersensitivity. *Altern Ther Health Med* 20(6):40–2.

Carpenter, D. O., (2015): The microwave syndrome or electro-hypersensitivity: Historical background. *Rev Environ Health* 30(4):217–22.

Carpenter, D. O. and Belpomme, D., (2015): Idiopathic environmental intolerance. *Rev Environ Health* 30(4):207.

Carpenter, D. O. and Sage, C., (2008): Setting prudent public health policy for electromagnetic field exposures. *Rev Environ Health* 23(2):91–117.

Carrubba, S. and Marino, A. A., (2008): The effects of low-frequency environmental-strength electromagnetic fields on brain electrical activity: A critical review of the literature. *Electromagn Biol Med* 27(2):83–101.

Carrubba, S., Frilot, C. 2nd, Chesson, A. L. Jr. and Marino, A. A., (2010): Mobile-phone pulse triggers evoked potentials. *Neurosci Lett* 469(1):164–8.

Charnay, Y. and Léger, L., (2010): Brain serotonergic circuitries. *Dial Clin Neurosci* 12(4):471–87.

Chen, C., Ma, Q., Liu, C., et al., (2014): Exposure to 1800 MHz radiofrequency radiation impairs neurite outgrowth of embryonic neural stem cells. *Sci Rep* 4:5103.

Chen, Y. Y., Lv, J., Xue, X. Y., He, G. H., Zhou, Y., Jia, M. and Luo, X. X., (2010): Effects of sympathetic histamine on vasomotor responses of blood vessels in rabbit ear to electrical stimulation. *Neurosci Bull* 26(3):219–24.

Chevalier, G., Sinatra, S. T., Oschman, J. L., Sokal, K. and Sokal, P., (2012): Earthing: Health implications of reconnecting the human body to the earth's surface electrons. *J Environ Public Health*:291541. https://doi.org/10.1155/2012/291541.

Chia, S. E., Chia, H. P. and Tan, J. S., (2000): Prevalence of headache among handheld cellular telephone users in Singapore: A community study. *Environ Health Perspect* 108(11):1059–62.

Chiu, C. T., Chang, Y. H., Chen, C. C., Ko, M. C. and Li, C. Y., (2015): Mobile phone use and health symptoms in children. *J Formos Med Assoc* 114(7):598–604.

Cho, Y. M., Lim, H. J., Jang, H., et al., (2016a): A cross-sectional study of the association between mobile phone use and symptoms of ill health. *Environ Health Toxicol* 31:e2016022.

Cho, Y. M., Lim, H. J., Jang, H., et al., (2016b): A follow-up study of the association between mobile phone use and symptoms of ill health. *Environ Health Toxicol* 32:e2017001.

Chrousos, G. P. and Gold, P. W., (1992): The concepts of stress and stress system disorders: Overview of physical and behavioral homeostasis. *JAMA* 267(9):1244–52.

Crasson, M., Legros, J. J., Scarpa, P. and Legros, W., (1999): 50 Hz magnetic field exposure influence on human performance and psychophysiological parameters: Two double-blind experimental studies. *Bioelectromagnetics* 20(8):474–86.

Croft, R. J., Chandler, J. S., Burgess, A. P., Barry, R. J., Williams, J. D. and Clarke, A. R., (2002): Acute mobile phone operation affects neural function in humans. *Clin Neurophysiol* 113(10):1623–32.

Croft, R. J., Hamblin, D. L., Spong, J., Wood, A. W., McKenzie, R. J. and Stough, C., (2008): The effect of mobile phone electromagnetic fields on the alpha rhythm of human electroencephalogram. *Bioelectromagnetics* 29(1):1–10.

Cucurachi, S., Tamis, W. L., Vijver, M. G., Peijnenburg, W. J., Bolte, J. F. and de Snoo, G. R., (2013): A review of the ecological effects of radiofrequency electromagnetic fields (RF-EMF). *Environ Int* 51:116–40.

Curcio, G., Ferrara, M., Limongi, T., et al., (2009): Acute mobile phones exposure affects frontal cortex hemodynamics as evidenced by functional near-infrared spectroscopy. *J Cereb Blood Flow Metab* 29(5):903–10.

Curcio, G., Ferrara, M., Moroni, F., D'Inzeo, G., Bertini, M. and De Gennaro, L., (2005): Is the brain influenced by a phone call? An EEG study of resting wakefulness. *Neurosci Res* 53(3):265–70.

Dahmen, N., Ghezel-Ahmadi, D. and Engel, A., (2009): Blood laboratory findings in patients suffering from self-perceived electromagnetic hypersensitivity (EHS). *Bioelectromagnetics* 30(4):299–306.

Dale, H. H., (1906): On some physiological actions of ergot. *J Physiol* 34(3):163–206.

Dasdag, S., Akdag, M. Z., Erdal, M. E., et al., (2015a): Long term and excessive use of 900 MHz radiofrequency radiation alter microRNA expression in brain. *Int J Radiat Biol* 91(7):555–61.

Dasdag, S., Akdag, M. Z., Erdal, M. E., et al., (2015b): Effects of 2.4 GHz radiofrequency radiation emitted from wi-fi equipment on microRNA expression in brain tissue. *Int J Radiat Biol* 91(7):555–61.

Dasdag, S., Akdag, M. Z., Kizil, G., Kizil, M., Cakir, D. U. and Yokus, B., (2012): Effect of 900 MHz radio frequency radiation on beta amyloid protein, protein carbonyl, and malondialdehyde in the brain. *Electromagn Biol Med* 31(1):67–74.

Dasdag, S., Sert, C., Akdag, Z. and Batun, S., (2002): Effects of extremely low frequency electromagnetic fields on hematologic and immunologic parameters in welders. *Arch Med Res* 33(1):29–32.

Davanipour, Z. and, Sobel, E., (2009): Long-term exposure to magnetic fields and the risks of Alzheimer's disease and breast cancer: Further biological research. *Pathophysiology* 16(2–3):149–56.

Davis, S., (1997): Weak residential magnetic fields affect melatonin in humans. *Microw News* 17(6):novembre–décembre 1997. http://microwavenews.com/news/backissues/n-d97issue.pdf.

De Iuliis, G. N., Newey, R. J., King, B. V. and Aitken, R. J., (2009): Mobile phone radiation induces reactive oxygen species production and DNA damage in human spermatozoa in vitro. *PLOS ONE* 4(7):e6446.

De Luca, C., Scordo, M. G., Cesareo, E., et al., (2010): Biological definition of multiple chemical sensitivity from redox state and cytokine profiling and not from polymorphisms of xenobiotic-metabolizing enzymes. *Toxicol Appl Pharmacol* 248(3):285–92.

De Luca, C., Thai, J. C., Raskovic, D., Cesareo, E., Caccamo, D., Trukhanov, A. and Korkina, L., (2014): Metabolic and genetic screening of electromagnetic hypersensitive subjects as a feasible tool for diagnostics and intervention. *Mediators Inflamm* 2014:924184.

De Pomerai, D., Daniells, C., David, H., et al., (2000): Non-thermal heat-shock response to microwaves. *Nature* 405(6785):417–8.

De, A. K. and Roach, S. E., (2004): Detection of the soluble heat shock protein 27 (hsp27) in human serum by an ELISA. *J Immunoassay Immunochem* 25(2):159–70.

Del Giudice, E., Stefanini, P., Tedeschi, A. and Vitiello, G., (2011): The interplay of biomolecules and water at the origin of the active behavior of living organisms. *J Phys Conf S* 329, Article ID: 012001.

Deshmukh, P. S., Megha, K., Banerjee, B. D., Ahmed, R. S., Chandna, S., Abegaonkar, M. P. and Tripathi, A. K., (2013): Detection of low level microwave radiation induced deoxyribonucleic acid damage vis-à-vis genotoxicity in brain of Fischer rats. *Toxicol Int* 20(1):19–24.

Dessaint, J. P., Bout, D., Wattre, P. and Capron, A., (1975): Quantitative determination of specific IgE antibodies to Echinococcus granulosus and IgE levels in sera from patients with hydatid disease. *Immunology* 29(5):813–23.

Di Carlo, A., White, N., Guo, F., Garrett, P. and Litovitz, T., (2002): Chronic electromagnetic field exposure decreases HSP70 levels and lowers cytoprotection. *J Cell Biochem* 84(3):447–54.

Díaz-Del Cerro, E., Vida, C., Martínez de Toda, I., Félix, J. and De la Fuente, M., (2020): The use of a bed with an insulating system of electromagnetic fields improves immune function, redox and inflammatory states, and decrease the rate of aging. *Environ Health* 19(1):118.

Diem, E., Schwarz, C., Adlkofer, F., Jahn, O. and Rüdiger, H., (2005): Non-thermal DNA breakage by mobile-phone radiation (1800 MHz) in human fibroblasts and in transformed GFSH-R17 rat granulosa cells in vitro. *Mutat Res* 583(2):178–83.

Dieudonné, M., (2016): Does electromagnetic hypersensitivity originate from nocebo responses? Indications from a qualitative study. *Bioelectromagnetics* 37(1):14–24.

Dieudonné, M., (2020): Electromagnetic hypersensitivity: A critical review of explanatory hypotheses. *Environ Health* 19(1):48.

Dik, M. G., Jonker, C., Hack, C. E., Smit, J. H., Comijsn, H. C. and Eikelenboom, P., (2005): Serum inflammatory proteins and cognitive decline in older persons. *Neurology* 64(8):1371–7.

Ding, G. R., Li, K. C., Wang, X. W., et al., (2009): Effect of electromagnetic pulse exposure on brain micro vascular permeability in rats. *Biomed Environ Sci* 22(3):265–8.

Dodge, C., (1969): *Clinical and hygenic aspects of exposure to electromagnetic radiation*. Bioscience Division of US Navy, US Naval Observatory. Washington DC.

Donato, R., (2001): S100: A multigenic family of calcium-modulated proteins of the EF-hand type with intracellular and extracellular functional roles. *Int J Biochem Cell Biol* 33(7):637–68.

Draper, G., Vincent, T., Kroll, M. E. and Swanson, J., (2005): Childhood cancer in relation to distance from high voltage power lines in England and Wales: A case-control study. *BMJ* 330(7503):1290.

Dunn, J. R., Fuller, M., Zoeger, J., et al., (1995): Magnetic material in the human hippocampus. *Brain Res Bull* 36(2):149–53.

Durusoy, R., Hassoy, H., Özkurt, A. and Karababa, A. O., (2017): Mobile phone use, school electromagnetic field levels and related symptoms: A cross-sectional survey among 2150 high school students in Izmir. *Environ Health* 16(1):51.

Eberhardt, J. L., Persson, B. R., Brun, A. E., Salford, L. G. and Malmgren, L. O., (2008): Blood-brain barrier permeability and nerve cell damage in rat brain 14 and 28 days after exposure to microwaves from GSM mobile phones. *Electromagn Biol Med* 27(3):215–29.

Eghlidospour, M., Ghanbari, A., Mortazavi, S. M. J. and Azari, H., (2017): Effects of radiofrequency exposure emitted from a GSM mobile phone on proliferation, differentiation, and apoptosis of neural stem cells. *Anat Cell Biol* 50(2):115–23.

Ehrlich, P., (1885): *Das Sauerstoffbedürfnis des Organismus: Eine farbanalytische Studie*. Berlin: Hirschwald-Verlag.

Eltiti, S., Wallace, D., Zougkou, K., Russo, R., Joseph, S., Rasor, P. and Fox, E., (2007): Development and evaluation of the electromagnetic hypersensitivity questionnaire. *Bioelectromagnetics* 28(2):137–51.

Eltiti, S., Wallace, D., Russo, R. and Fox, E., (2015): Aggregated data from two double blind base station provocation studies comparing individuals with idiopathic environmental intolerance with attribution to electromagnetic fields and controls. *Bioelectromagnetics* 36(2):96–107.

Engels, S., Schneider, N. L., Lefeldt, N., et al., (2014): Anthropogenic electromagnetic noise disrupts magnetic compass orientation in a migratory bird. *Nature* 509(7500):353–6.

Erickson, M. A. and Banks, W. A., (2013): Blood-brain barrier dysfunction as a cause and consequence of Alzheimer's disease. *J Cereb Blood Flow Metab* 33(10):1500–13.

Eyles, D. W., Feron, F., Cui, X., et al., (2009): Developmental vitamin D deficiency causes abnormal brain development. *Psychoneuroendocrinology* 34:S247–57.

Eyvazlou, M., Zarei, E., Rahimi, A. and Abazari, M., (2016): Association between overuse of mobile phones on quality of sleep and general health among occupational health and safety students. *Chronobiol Int* 33(3):293–300.

Falcioni, L., Bua, L., Tibaldi, E., et al., (2018): Report of final results regarding brain and heart tumors in Sprague-Dawley rats exposed from prenatal life until natural death to mobile phone radiofrequency field representative of a 1.8 GHz GSM base station environmental emission. *Environ Res* 165:496–50.

FDA-NIH Biomarker Working Group., (2016): *BEST (biomarkers, EndpointS, and other tools) resource.* Silver Spring (MD); Bethesda (MD): Food and Drug Administration (US); National Institutes of Health (US). www.ncbi.nlm.nih.gov/books/NBK326791/.

Forman, S. A., Holmes, C. K., McManamon, T. V. and Wedding, W. R., (1982): Psychological symptoms and intermittent hypertension following acute microwave exposure. *J Occup Med* 24(11):932–4.

Foster, K. R., (2000): Thermal and nonthermal mechanisms of interaction of radio-frequency energy with biological systems. *IEEE Trans Plasma Sci* 28(1):15–23.

Fragopoulou, A. F., Samara, A., Antonelou, M. H., et al., (2012): Brain proteome response following whole body exposure of mice to mobile phone or wireless DECT base radiation. *Electromagn Biol Med* 31(4):250–74.

Frei, P., Mohler, E., Braun-Fahrländer, C., Fröhlich, J., Neubauer, G., Röösli, M. and QUALIFEX-Team., (2012): Cohort study on the effects of everyday life radio frequency electromagnetic field exposure on non-specific symptoms and tinnitus. *Environ Int* 38(1):29–36.

French, P. W., Penny, R., Laurence, J. A. and McKenzie, D. R., (2001): Mobile phones, heat shock proteins and cancer. *Differentiation* 67(4–5):93–7.

Freude, G., Ullsperger, P., Eggert, S. and Ruppe, I., (1998): Effects of microwaves emitted by cellular phones on human slow brain potentials. *Bioelectromagnetics* 19(6):384–7.

Frey, A. H., (1993): Electromagnetic field interactions with biological systems. *FASEB J* 7(2):272–81.

Frick, U., Rehm, J. and Eichhammer, P., (2002): Risk perception, somatization, and self report of complaints related to electromagnetic fields–A randomized survey study. *Int J Hyg Environ Health* 205(5):353–60.

Funk, R. H., (2015): Endogenous electric fields as guiding cue for cell migration. *Front Physiol* 6:143.

Furtado-Filho, O. V., Borba, J. B., Maraschin, T., Souza, L. M., Henriques, J. A., Moreira, J. C. and Saffi, J., (2015): Effects of chronic exposure to 950 MHz ultra-high-frequency electromagnetic radiation on reactive oxygen species metabolism in the right and left cerebral cortex of young rats of different ages. *Int J Radiat Biol* 91(11):891–7.

Furubayashi, T., Ushiyama, A., Terao, Y., et al., (2009): Effects of short-term W-CDMA mobile phone base station exposure on women with or without mobile phone related symptoms. *Bioelectromagnetics* 30(2):100–13.

Gandhi, O. P., Lazzi, G. and Furse, C. M., (1996): Electromagnetic absorption in the human head and neck for mobile telephones at 835 and 1900 MHz. *IEEE Trans Microw Theor Tech* 44(10):1884–97.

García, A. M., Sisternas, A. and Hoyos, S. P., (2008): Occupational exposure to extremely low frequency electric and magnetic fields and Alzheimer disease: A meta-analysis. *Int J Epidemiol* 37(2):329–40.

Gazerani, P., Pourpak, Z., Ahmadiani, A., Hemmati, A. and Kazemnejad, A., (2003): A correlation between migraine, histamine and immunoglobulin E. *Scand J Immunol* 57(3):286–90.

Gegear, R. J., Foley, L. E., Casselman, A. and Reppert, S. M., (2010): Animal cryptochromes mediate magnetoreception by an unconventional photochemical mechanism. *Nature* 463(7282):804–7.

Genuis, S. J. and Lipp, C. T., (2012): Electromagnetic hypersensitivity: Fact or fiction? *Sci Total Environ* 414:103–12.

Genuis, S. J., (2010): Sensitivity-related illness: The escalating pandemic of allergy, food intolerance and chemical sensitivity. *Sci Total Environ* 408(24):6047–61.

Georgopoulos, C. and Welch, W. J., (1993): Role of the major heat shock proteins as molecular chaperones. *Annu Rev Cell Biol* 9:601–34.

Ghosn, R., Yahia-Cherif, L., Hugueville, L., et al., (2015): Radiofrequency signal affects alpha band in resting electroencephalogram. *J Neurophysiol* 113(7):2753–9.

Gibson, P. R., Kovach, S. and Lupfer, A., (2015): Unmet health care needs for persons with environmental sensitivity. *J Multidisc Healthc* 8:59–66.

Glaser, Z. R., (1972): Bibliography of reported biological phenomena ('effects') and clinical manifestation attributed to microwave and radio-frequency radiation. Project MF12.524.015-00043 Report No. 2. Second Printing, with Revisions, Corrections, and Additions. Naval Medical Research Institute. National Naval Medical Center Bethesda, Maryland 20014, U.S.A.

Goodman, R. and Blank, M., (2002): Insights into electromagnetic interaction mechanisms. *J Cell Physiol* 192(1):16–22.

Gökçek-Saraç, Ç., Akçay, G., Karakurt, S., Ateş, K., Özen, Ş. and Derin, N., (2021): Possible effects of different doses of 2.1 GHz electromagnetic radiation on learning, and hippocampal levels of cholinergic biomarkers in Wistar rats. *Electromagn Biol Med* 40(1):179–90.

Greaves, M. W. and Sabroe, R. A., (1996): Histamine: The quintessential mediator. *J Dermatol* 23(11):735–40.

Greco, F., (2020): Technical assessment of ultrasonic cerebral Tomosphygmography and new scientific evaluation of its clinical interest for the diagnosis of electrohypersensitivity and multiple chemical sensitivity. *Diagnostics (Basel)* 10(6):427.

Grehl, S., Martina, D., Goyenvalle, C., Deng, Z. D., Rodger, J. and Sherrard, R. M., (2016): In vitro magnetic stimulation: A simple stimulation device to deliver defined low intensity electromagnetic fields. *Front Neural Circuits* 10:85.

Grigoriev, Y. G., Grigoriev, O. A., Ivanov, A. A., Lyaginskaya, A. M., Merkulov, A. V., Stepanov, V. S. and Shagina, N. B., (2010): Autoimmune process after long-term low-level exposure to electromagnetic field (experimental results). Part 1. Mobile communications and changes in electromagnetic conditions for the population: Need for additional substantiation of existing hygienic standards. *Biophysics* 55(6):1041–45.

Gunaydin, H. and Houk, K. N., (2009): Mechanisms of peroxynitrite-mediated nitration of tyrosine. *Chem Res Toxicol* 22(5):894–8.

Günzler, W. A., Kremers, H. and Flohé, L., (1974): An improved coupled test procedure for glutathione peroxidase (EC 1-11-1-9) in blood. *Z Klin Chem Klin Biochem* 12(10):444–8.

Guxens, M., van Eijsden, M., Vermeulen, R., et al., (2013): Maternal cell phone and cordless phone use during pregnancy and behaviour problems in 5-year-old children. *JECH* 67(5):432–8.

Haas, H. L., Sergeeva, O. A. and Selbach, O., (2008): Histamine in the nervous system. *Physiol Rev* 88(3):1183–241.

Hagström, M., Auranen, J. and Ekman, R., (2013): Electromagnetic hypersensitive Finns: Symptoms, perceived sources and treatments, a questionnaire study. *Pathophysiology* 20(2):117–22.

Hallberg, Ö. and Oberfeld, G., (2006): Letter to the editor: Will we all become electrosensitive? *Electromagn Biol Med* 25(3):189–91.

Hardell, L., (2017): World Health Organization, radiofrequency radiation and health - A hard nut to crack. *Int J Oncol* 51(2):405–13.

Hardell, L. and Carlberg, M., (2020): Health risks from radiofrequency radiation, including 5G, should be assessed by experts with no conflicts of interest. *Oncol Lett* 20(4):15.

Hardell, L., Carlberg, M. and Hansson Mild, K., (2013): Use of mobile phones and cordless phones is associated with increased risk for glioma and acoustic neuroma. *Pathophysiology* 20(2):85–110.

Hardell, L. and Nyberg, R., (2020): Appeals that matter or not on a moratorium on the deployment of the fifth generation, 5G, for microwave radiation. *Mol Clin Oncol* 12(3):247–57.

Hartl, F. U., (1996): Molecular chaperones in cellular protein folding. *Nature* 381(6583):571–9.

Havas, M., Marrongelle, J., Pollner, B., Kelley, E., Rees, C. and Tully, L., (2010): Provocation study using heart rate variability shows microwave radiation from 2.4 GHz cordless phone affects autonomic nervous system. In Giuliani L, Soffritti M (Eds): *Non-thermal effects and mechanisms of interaction between electromagnetic fields and living matter*. Fidenza: Mattioli, pp. 273–300. ISBN: 9788862611664.

Havas, M., (2013): Radiation from wireless technology affects the blood, the heart, and the autonomic nervous system. *Rev Environ Health* 28(2–3):75–84.

Heneka, M. T. and O'Banion, M. K., (2007): Inflammatory processes in Alzheimer's disease. *J Neuroimmunol* 184(1–2):69–91.

Heuser, G. and Heuser, S. A., (2017): Functional brain MRI in patients complaining of electrohypersensitivity after long term exposure to electromagnetic fields. *Rev Environ Health* 32(3):291–9.

Hillert, L., Hedman, B. K., Söderman, E. and Arnetz, B. B., (1999): Hypersensitivity to electricity: Working definition and additional characterization of the syndrome. *J Psychosom Res* 47(5):429–38.

Hillert, L., Berglind, N., Arnetz, B. B. and Bellander, T., (2002): Prevalence of self-reported hypersensitivity to electric or magnetic fields in a population-based questionnaire survey. *Scand J Work Environ Health* 28(1):33–41.

Hillert, L., Musabasic, V., Berglund, H., Ciumas, C. and Savic, I., (2007): Odor processing in multiple chemical sensitivity. *Hum Brain Mapp* 28(3):172–82.

Hocking, B., (1998): Preliminary report: Symptoms associated with mobile phone use. *Occup Med (Lond)* 48(6):357–60.

Hocking, B. and Westerman, R., (2002): Neurological changes induced by a mobile phone. *Occup Med (Lond)* 52(7):413–5.

Hocking, B. and Westerman, R., (2003): Neurological effects of radiofrequency radiation. *Occup Med (Lond)* 53(2):123–7.

Holmstrom, K. M. and Finkel, T., (2014): Cellular mechanisms and physiological consequences of redox-dependent signaling. *Nat Rev Mol Cell Biol* 15(6):411–21.

Hu, C., Zuo, H. and Li, Y., (2021): Effects of radiofrequency electromagnetic radiation on neurotransmitters in the brain. *Front Public Health* 9:691880.

Huber, R., Treyer, V., Borbély, A. A., et al., (2002): Electromagnetic fields, such as those from mobile phones, alter regional cerebral blood flow and sleep and waking EEG. *J Sleep Res* 11(4):289–95.

Huber, R., Treyer, V., Schuderer, J., et al., (2005): Exposure to pulse-modulated radio frequency electromagnetic fields affects regional cerebral blood flow. *Eur J Neurosci* 21(4):1000–6.

Huss, A., van Eijsden, M., Guxens, M., et al., (2015): Environmental radiofrequency electromagnetic fields exposure at home, mobile and cordless phone use, and sleep problems in 7-year-old children. *PLOS ONE* 10(10):e0139869.

Hutter, H. P., Moshammer, H., Wallner, P. and Kundi, M., (2006): Subjective symptoms, sleeping problems, and cognitive performance in subjects living near mobile phone base stations. *Occup Environ Med* 63(5):307–13.

IARC (International Agency for Research on Cancer)., (2002): Non-ioizing radiation, part 1: Static and extremely low-frequency (ELF) electric and magnetic fields. In *IARC monographs on the evaluation of carcinogenic risks to humans*. Lyon: IARC Press, Vol. 80, p. 341.

IARC (International Agency for Research on Cancer)., (2013): Non-ionization radiation, part 2: Radiofrequency electromagnetic fields. In *IARC monographs on the evaluation of carcinogenic risks to humans*. Lyon: IARC Press, Vol. 102, p. 406.

ICNIRP (The International Commission on Non-Ionizing Radiation Protection)., (1998): Guidelines for limiting exposure to time-varying electric, magnetic and electroomagnetic fields (up to 300 GHz). *Health Phys* 74:494–522.

ICNIRP (The International Commission on Non-Ionizing Radiation Protection)., (2010): Guidelines for limiting exposure to time-varying electric and magnetic fields (1 Hz to 100 kHz). *Health Phys* 99(6):818–36.

ICNIRP (The International Commission on Non-Ionizing Radiation Protection)., (2020): Guidelines for limiting exposure to electromagnetic fields (100 kHz to 300 GHz). *Health Phys* 118(5):483–524.

Ikwegbue, P. C., Masamba, P., Oyinloye, B. E. and Kappo, A. P., (2017): Roles of heat shock proteins in apoptosis, oxidative stress, human inflammatory diseases, and cancer. *Pharmaceuticals (Basel)* 11(1):2.

Inaba, R., Shishido, K., Okada, A. and Moroji, T., (1992): Effects of whole body microwave exposure on the rat brain contents of biogenic amines. *Eur J Appl Physiol Occup Physiol* 65(2):124–8.

Irigaray, P. and Belpomme, D., (2010): Basic properties and molecular mechanisms of exogenous chemical carcinogens. *Carcinogenesis* 31(2):135–48.

Irigaray, P., Caccamo, D. and Belpomme, D., (2018a): Oxidative stress in electrohypersensitivity self reporting patients: Results of a prospective in vivo investigation with comprehensive molecular analysis. *Int J Mol Med* 42(4):1885–98.

Irigaray, P., Garrel, C., Houssay, C., Mantello, P. and Belpomme, D., (2018b): Beneficial effects of a fermented papaya preparation for the treatment of electrohypersensitivity self-reporting patients: Results of a phase I–II clinical trial with special reference to cerebral pulsation measurement and oxidative stress analysis. *Funct Foods Health Dis* 8(2):122–44.

Irigaray, P., Lebar, P. and Belpomme, D., (2018c): How ultrasonic cerebral Tomosphygmography can contribute to the diagnosis of electrohypersensitivity. *J Clin Diagn Res* 6:143.

Ischiropoulos, H., Zhu, L., Chen, J., Tsai, M., Martin, J. C., Smith, C. D. and Beckman, J. S., (1992): Peroxynitrite-mediated tyrosine nitration catalyzed by superoxide dismutase. *Arch Biochem Biophys* 298(2):431–7.

Ivancsits, S., Diem, E., Pilger, A., Rüdiger, H. W. and Jahn, O., (2002): Induction of DNA strand breaks by intermittent exposure to extremely-low-frequency electromagnetic fields in human diploid fibroblasts. *Mutat Res* 519(1–2):1–13.

Ivancsits, S., Diem, E., Jahn, O. and Rüdiger, H. W., (2003): Age-related effects on induction of DNA strand breaks by intermittent exposure to electromagnetic fields. *Mech Ageing Dev* 124(7):847–50.

Ji, J., Zhang, Y. H., Yang, X. Q., Jiang, R. P., Guo, D. M. and Cui, X., (2012): The influence of microwave radiation from cellular phone on fetal rat brain. *Electromagn Biol Med* 31(1):57–66.

Jocelyn, P. C., (1987): Spectrophotometric assay of thiols. *Methods Enzymol* 143:44–67.

Johansson, A., Nordin, S., Heiden, M. and Sandström, M., (2010): Symptoms, personality traits, and stress in people with mobile phone-related symptoms and electromagnetic hypersensitivity. *J Psychosom Res* 68(1):37–45.

Johansson, O., (2009): Disturbance of the immune system by electromagnetic fields-A potentially underlying cause for cellular damage and tissue repair reduction which could lead to disease and impairment. *Pathophysiology* 16(2–3):157–77.

Johansson, O., Gangi, S., Liang, Y., Yoshimura, K., Jing, C. and Liu, P. Y., (2001): Cutaneous mast cells are altered in normal healthy volunteers sitting in front of ordinary TVs/PCs–Results from open-field provocation experiments. *J Cutan Pathol* 28(10):513–9.

Johnsen, S. and Lohmann, K. J., (2005): The physics and neurobiology of magnetoreception. *Nat Rev Neurosci* 6(9):703–12.

Johnston, L., (2008): Suicides "linked to phone masts". *Express.co.uk*, 22 juin 2008. www.express.co.uk/post s/view/49330/Suicides-linked-tophone-masts.

Joubert, V., Bourthoumieu, S., Leveque, P. and Yardin, C., (2008): Apoptosis is induced by radiofrequency fields through the caspase-independent mitochondrial pathway in cortical neurons. *Radiat Res* 169(1):38–45.

Juutilainen, J., Matilainen, P., Saarikoski, S., Läärä, E. and Suonio, S., (1993): Early pregnancy loss and exposure to 50-Hz magnetic fields. *Bioelectromagnetics* 14(3):229–36.

Kapural, M., Krizanac-Bengez, Lj., Barnett, G., et al., (2002): Serum S-100beta as a possible marker of blood-brain barrier disruption. *Brain Res* 940(1–2):102–4.

Kato, Y. and Johansson, O., (2012): Reported functional impairments of electrohypersensitive Japanese: A questionnaire survey. *Pathophysiology* 19(2):95–100.

Kelly, S. and Yenari, M. A., (2002): Neuroprotection: Heat shock proteins. *Curr Med Res Opin* 18:s55–60.

Kesari, K. K., Meena, R., Nirala, J., Kumar, J. and Verma, H. N., (2014): Effect of 3G cell phone exposure with computer controlled 2-D stepper motor on non-thermal activation of the hsp27/p38MAPK stress pathway in rat brain. *Cell Biochem Biophys* 68(2):347–58.

Kheifets, L., Repacholi, M., Saunders, R. and van Deventer, E., (2005): The sensitivity of children to electromagnetic fields. *Pediatrics* 116(2):e303–13.

Kirschvink, J. L., (1996): Microwave absorption by magnetite: A possible mechanism for coupling nonthermal levels of radiation to biological systems. *Bioelectromagnetics* 17(3):187–94.

Kirschvink, J. L., Kobayashi-Kirschvink, A. and Woodford, B. J., (1992): Magnetite biomineralization in the human brain. *Proc Natl Acad Sci U S A* 89(16):7683–7.

Kleinlogel, H., Dierks, T., Koenig, T., Lehmann, H., Minder, A. and Berz, R., (2008): Effects of weak mobile phone - Electromagnetic fields (GSM, UMTS) on event related potentials and cognitive functions. *Bioelectromagnetics* 29(6):488–97.

Knave, B. G., Wibom, R. I., Voss, M., Hedström, L. D. and Bergqvist, U. O., (1985): Work with video display terminals among office employees. I. Subjective symptoms and discomfort. *Scand J Work Environ Health* 11(6):457–66.

Koh, S. X. and Lee, J. K., (2014): S100B as a marker for brain damage and blood-brain barrier disruption following exercise. *Sports Med* 44(3):369–85.

Koppel, T., Vilcane, I. and Ahonen, M., (2018): 50 Hz magnetic field affects heart rate variability–An experimental study. In *Proceedings of the 2018 EMF-Med 1st world conference on biomedical applications of electromagnetic fields (EMF-Med)*, Split, Croatia, 10–13 September 2018; New York: IEEE, pp. 1–2. https://doi.org/10.23919/EMF-MED.2018.8526072.

Korpinen, L. H. and Pääkkönen, R. J., (2009): Self-report of physical symptoms associated with using mobile phones and other electrical devices. *Bioelectromagnetics* 30(6):431–7.

Köteles, F., Szemerszky, R., Gubányi, M., et al., (2013): Idiopathic environmental intolerance attributed to electromagnetic fields (IEI-EMF) and electrosensibility (ES) - Are they connected? *Int J Hyg Environ Health* 216(3):362–70.

Kovács, J., Brodner, W., Kirchlechner, V., Arif, T. and Waldhauser, F., (2000): Measurement of urinary melatonin: A useful tool for monitoring serum melatonin after its oral administration. *J Clin Endocrinol Metab* 85(2):666–70.

Kowall, B., Breckenkamp, J., Blettner, M., Schlehofer, B., Schüz, J. and Berg-Beckhoff, G., (2012): Determinants and stability over time of perception of health risks related to mobile phone base stations. *Int J Public Health* 57(4):735–43.

Krause, C. M., Sillanmäki, L., Koivisto, M., et al., (2000): Effects of electromagnetic field emitted by cellular phones on the EEG during a memory task. *NeuroReport* 11(4):761–4.

Küçer, N. and Pamukçu, T., (2014): Self-reported symptoms associated with exposure to electromagnetic fields: A questionnaire study. *Electromagn Biol Med* 33(1):15–7.

Kumar, G., Gupta, N. and Sinha, N. K., (2018): Mobile phone users and its effect of hearing in terms of distortion product otoacoustic emission (DPOAE). *J Evol Med Dent Sci* 7(52):5520–3.

Lacour, M., Zunder, T., Schmidtke, K., Vaith, P. and Scheidt, C., (2005): Multiple chemical sensitivity syndrome (MCS)—Suggestions for an extension of the U.S. MCS-case definition. *Int J Hyg Environ Health* 208(3):141–51.

Lai, H., (2021): Genetic effects of non-ionizing electromagnetic fields. *Electromagn Biol Med* 40(2):264–73.

Lai, H. and Singh, N. P., (1995): Acute low-intensity microwave exposure increases DNA single-strand breaks in rat brain cells. *Bioelectromagnetics* 16(3):207–10.

Lai, H. and Singh, N. P., (2004): Magnetic-field-induced DNA strand breaks in brain cells of the rat. *Environ Health Perspect* 112(6):687–94.

Lai, Y. F., Wang, H. Y. and Peng, R. Y., (2021): Establishment of injury models in studies of biological effects induced by microwave radiation. *Mil Med Res* 8(1):12.

Landgrebe, M., Frick, U., Hauser, S., Langguth, B., Rosner, R., Hajak, G. and Eichhammer, P., (2008): Cognitive and neurobiological alterations in electromagnetic hypersensitive patients: Results of a case-control study. *Psychol Med* 38(12):1781–91.

Leak, R. K., Zhang, L., Stetler, R. A., et al., (2013): HSP27 protects the blood-brain barrier against ischemia-induced loss of integrity. *CNS Neurol Disord Drug Targets* 12(3):325–37.

Lebel, B., Arnoux, B., Chanez, P., Bougeard, Y. H., Daures, J. P., Bousquet, J. and Campbell, A. M., (1996): Ex vivo pharmacologic modulation of basophil histamine release in asthmatic patients. *Allergy* 51(6):394–400.

Leone, L., Fusco, S., Mastrodonato, A., et al., (2014): Epigenetic modulation of adult hippocampal neurogenesis by extremely low-frequency electromagnetic fields. *Mol Neurobiol* 49(3):1472–86.

Leszczynski, D., Joenväärä, S., Reivinen, J. and Kuokka, R., (2002): Non-thermal activation of the hsp27/p38MAPK stress pathway by mobile phone radiation in human endothelial cells: Molecular mechanism for cancer- and blood-brain barrier-related effects. *Differentiation* 70(2–3):120–9.

Levallois, P., (2002): Hypersensitivity of human subjects to environmental electric and magnetic field exposure: A review of the literature. *Environ Health Perspect* 110:613–8.

Levallois, P., Neutra, R., Lee, G. and Hristova, L., (2002): Study of self-reported hypersensitivity to electromagnetic fields in California. *Environ Health Perspect* 110:619–23.

Li, H. J., Peng, R. Y., Wang, C. Z., et al., (2015): Alterations of cognitive function and 5-HT system in rats after long term microwave exposure. *Physiol Behav* 140:236–46.

Liburdy, R. P., (1979): Radiofrequency radiation alters the immune system: Modulation of T- and B-lymphocyte levels and cell-mediated immunocompetence by hyperthermic radiation. *Radiat Res* 77(1):34–46.

Liburdy, R. P., (1980): Radiofrequency radiation alters the immune system. II. Modulation in vivo lymphocyte circulation. *Radiat Res* 83(1):66–73.

Liburdy, R. P., (1992): Calcium signalling in lymphocytes and ELF fields: Evidence for an electric field metric and a site of interaction involving the calcium ion channel. *FEBS Lett* 301(1):53–9.

Liburdy, R. P. and Wyant, A., (1984): Radiofrequency radiation and the immune system. Part 3. In vitro effects on human immunoglobin and on murine T- and B-lymphocytes. *Int J Radiat Biol Relat Stud Phys Chem Med* 46(1):67–81.

Lin, H., Blank, M. and Goodman, R., (1999): A magnetic field-responsive domain in the human HSP70 promoter. *J Cell Biochem* 75(1):170–6.

Lin, H., Blank, M., Rossol-Haseroth, K. and Goodman, R., (2001): Regulating genes with electromagnetic response elements. *J Cell Biochem* 81(1):143–8.

Lin, H., Opler, M., Head, M., Blank, M. and Goodman, R., (1997): Electromagnetic field exposure induces rapid, transitory heat shock factor activation in human cells. *J Cell Biochem* 66(4):482–88.

Lindén, V. and Rolfsen, S., (1981): Video computer terminals and occupational dermatitis. *Scand J Work Environ Health* 7(1):62–4.

Liu, H., Chen, G., Pan, Y., et al., (2014): Occupational electromagnetic field exposures associated with sleep quality: A cross-sectional study. *PLOS ONE* 9(10):e110825.

Liu, S., Wing, Y. K., Hao, Y., Li, W., Zhang, J. and Zhang, B., (2019): The associations of long-time mobile phone use with sleep disturbances and mental distress in technical college students: A prospective cohort study. *Sleep* 42(2). doi: 10.1093/sleep/zsy213

Londero, D. and Lo Greco, P., (1996): Automated high-performance liquid chromatographic separation with spectrofluorometric detection of a malondialdehyde-thiobarbituric acid adduct in plasma. *J Chromatogr A* 729(1–2):207–10.

Loos, N., Thuróczy, G., Ghosn, R., et al., (2013): Is the effect of mobile phone radiofrequency waves on human skin perfusion non-thermal? *Microcirculation* 20(7):629–36.

Loughran, S. P., Verrender, A., Dalecki, A., et al., (2019): Radiofrequency electromagnetic field exposure and the resting EEG: Exploring the thermal mechanism hypothesis. *Int J Environ Res Public Health* 16(9):1505.

Lowden, A., Akerstedt, T., Ingre, M., et al., (2011): Sleep after mobile phone exposure in subjects with mobile phone-related symptoms. *Bioelectromagnetics* 32(1):4–14.

Lustenberger, C., Murbach, M., Dürr, R., Schmid, M. R., Kuster, N., Achermann, P. and Huber, R., (2013): Stimulation of the brain with radiofrequency electromagnetic field pulses affects sleep-dependent performance improvement. *Brain Stimul* 6(5):805–11.

Ma, Q., Deng, P., Zhu, G., et al., (2014): Extremely low-frequency electromagnetic fields affect transcript levels of neuronal differentiation-related genes in embryonic neural stem cells. *PLOS ONE* 9(3):e90041.

Maher, B. A., Ahmed, I. A., Karloukovski, V., et al., (2016): Magnetite pollution nanoparticles in the human brain. *Proc Natl Acad Sci U S A* 113(39):10797–801.

Mahmoudabadi, F. S., Ziaei, S., Firoozabadi, M. and Kazemnejad, A., (2015): Use of mobile phone during pregnancy and the risk of spontaneous abortion. *J Environ Health Sci Eng* 13:34.

Malek, F., Rani, K. A., Rahim, H. A. and Omar, M. H., (2015): Effect of short-term mobile phone base station exposure on cognitive performance, body temperature, heart rate and blood pressure of Malaysians. *Sci Rep* 5:13206.

Mann, K. and Röschke, J., (1996): Effects of pulsed high-frequency electromagnetic fields on human sleep. *Neuropsychobiology* 33(1):41–7.

Mannervik, B., (2001): Measurement of glutathione reductase activity. *Curr Protoc Toxicol* Chapter 7:Unit7.2. doi: 10.1002/0471140856.tx0702s00

Marchi, N., Cavaglia, M., Fazio, V., Bhudia, S., Hallene, K. and Janigro, D., (2004): Peripheral markers of blood-brain barrier damage. *Clin Chim Acta* 342(1–2):1–12.

Marklund, S. and Marklund, G., (1974): Involvement of the superoxide anion radical in the autoxidation of pyrogallol and a convenient assay for superoxide dismutase. *Eur J Biochem* 47(3):469–74.

Markovà, E., Hillert, L., Malmgren, L., Persson, B. R. and Belyaev, I. Y., (2005): Microwaves from GSM mobile telephones affect 53BP1 and gamma-H2AX foci in human lymphocytes from hypersensitive and healthy persons. *Environ Health Perspect* 113(9):1172–7.

Marquardt, D. L., (1983): Histamine. *Clin Rev Allergy* 1(3):343–51.

Marshall, J. S., (2004): Mast-cell responses to pathogens. *Nat Rev Immunol* 4(10):787–99.

Martens, A. L., Slottje, P., Smid, T., Kromhout, H., Vermeulen, R. C. H. and Timmermans, D. R. M., (2018): Longitudinal associations between risk appraisal of base stations for mobile phones, radio or television and non-specific symptoms. *J Psychosom Res* 112:81–9.

Martens, A. L., Slottje, P., Timmermans, D. R. M., Kromhout, H., Reedijk, M., Vermeulen, R. C. H. and Smid, T., (2017): Modeled and perceived exposure to radiofrequency electromagnetic fields from mobile-phone base stations and the development of symptoms over time in a general population cohort. *Am J Epidemiol* 186(2):210–9.

Mashevich, M., Folkman, D., Kesar, A., Barbul, A., Korenstein, R., Jerby, E. and Avivi, L., (2003): Exposure of human peripheral blood lymphocytes to electromagnetic fields associated with cellular phones leads to chromosomal instability. *Bioelectromagnetics* 24(2):82–90.

Mathie, A., Kennard, L. E. and Veale, E. L., (2003): Neuronal ion channels and their sensitivity to extremely low frequency weak electric field effects. *Radiat Prot Dosim* 106(4):311–6.

Mayhan, W. G., (1996): Role of nitric oxide in histamine-induced increases in permeability of the blood-brain barrier. *Brain Res* 743(1–2):70–6.

McCaig, C. D., Rajnicek, A. M., Song, B. and Zhao, M., (2005): Controlling cell behavior electrically: Current views and future potential. *Physiol Rev* 85(3):943–78.

McCaig, C. D. and Zhao, M., (1997): Physiological electric fields modify cell behaviour. *BioEssays* 19(9):819–26.

McCarty, D. E., Carrubba, S., Chesson, A. L., Frilot, C., Gonzalez-Toledo, E. and Marino, A. A., (2011): Electromagnetic hypersensitivity: Evidence for a novel neurological syndrome. *Int J Neurosci* 121(12):670–6.

McInerny, T. K., (2012): The cell phone right to know act H.R., 6358. Federal Legislation on Wireless. Letter on behalf of the American Academy of Pediatrics (AAP) to Dennis Kucinich (Washington DC). December 12. https://skyvisionsolutions.files.wordpress.com/2013/04/aap-letter.pdf.

McKeown-Eyssen, G., Baines, C., Cole, D. E., Riley, N., Tyndale, R. F., Marshall, L. and Jazmaji, V., (2004): Case-control study of genotypes in multiple chemical sensitivity: CYP2D6, NAT1, NAT2, PON1, PON2 and MTHFR. *Int J Epidemiol* 33(5):971–8.

McLean, L., (2008): The impacts of radiofrequency radiation from mobile phone antennas. EMR Australia report. http://www.champs-electromagnetiques.com/images_doc/australie_antennerelais.pdf.

McMahan, S. and Meyer, J., (1995): Symptom prevalence and worry about high voltage transmission lines. *Environ Res* 70(2):114–8.

Medeiros, L. N. and Sanchez, T. G., (2016): Tinnitus and cell phones: The role of electromagnetic radiofrequency radiation. *Braz J Otorhinolaryngol* 82(1):97–104.

Meg Tseng, M. C., Lin, Y. P. and Cheng, T. J., (2011): Prevalence and psychiatric comorbidity of self-reported electromagnetic field sensitivity in Taiwan: A population-based study. *J Formos Med Assoc* 110(10):634–41.

Meggs, W. J., (1993): Neurogenic inflammation and sensitivity to environmental chemicals. *Environ Health Perspect* 101(3):234–8.

Meggs, W. J., (1997): Hypothesis for induction and propagation of chemical sensitivity based on biopsy studies. *Environ Health Perspect* 105:473–8.

Megha, K., Deshmukh, P. S., Banerjee, B. D., Tripathi, A. K., Ahmed, R. and Abegaonkar, M. P., (2015a): Low intensity microwave radiation induced oxidative stress, inflammatory response and DNA damage in rat brain. *Neurotoxicology* 51:158–65.

Megha, K., Deshmukh, P. S., Ravi, A. K., Tripathi, A. K., Abegaonkar, M. P. and Banerjee, B. D., (2015b): Effect of low-intensity microwave radiation on monoamine neurotransmitters and their key regulating enzymes in rat brain. *Cell Biochem Biophys* 73(1):93–100.

Merritt, J. H., Chamness, A. F. and Allen, S. J., (1978): Studies on blood-brain barrier permeability after microwave-radiation. *Radiat Environ Biophys* 15(4):367–77.

Michetti, F., Corvino, V., Geloso, M. C., Lattanzi, W., Bernardini, C., Serpero, L. and Gazzolo, D., (2012): The S100B protein in biological fluids: More than a lifelong biomarker of brain distress. *J Neurochem* 120(5):644–59.

Migheli, A., Cordera, S., Bendotti, C., Atzori, C., Piva, R. and Schiffer, D., (1999): S-100beta protein is upregulated in astrocytes and motor neurons in the spinal cord of patients with amyotrophic lateral sclerosis. *Neurosci Lett* 261(1–2):25–8.

Mild, K. H., Repacholi, M., van Deventer, E. and Ravazzani, P., (2006): Electromagnetic hypersensitivity. In *Proceedings of the WHO international seminar and working group meeting on EMF hypersensitivity*, Prague, Czech Republic, 25–27 October 2004; Geneva: World Health Organization. ISBN: 92-4-159412-8.

Milde-Busch, A., von Kries, R., Thomas, S., Heinrich, S., Straube, A. and Radon, K., (2010):The association between use of electronic media and prevalence of headache in adolescents: Results from a population-based cross-sectional study. *BMC Neurol* 10:12.

Miller, A. B., Sears, M. E., Morgan, L. L., Davis, D. L., Hardell, L., Oremus, M. and Soskolne, C. L., (2019): Risks to health and well-being from radio-frequency radiation emitted by cell phones and other wireless devices. *Front Public Health* 7:223.

Miller, C. S., (1999): Are we on the threshold of a new theory of disease? Toxicant-induced loss of tolerance and its relationship to addiction and abdiction. *Toxicol Ind Health* 15(3–4):284–94.

Mohler, E., Frei, P., Fröhlich, J., Braun-Fahrländer, C., Röösli, M. and QUALIFEX-Team., (2012): Exposure to radiofrequency electromagnetic fields and sleep quality: A prospective cohort study. *PLOS ONE* 7(5):e37455.

Monazzam, M. R., Hosseini, M., Matin, L. F., Aghaei, H. A., Khosroabadi, H. and Hesami, A., (2014): Sleep quality and general health status of employees exposed to extremely low frequency magnetic fields in a petrochemical complex. *J Environ Health Sci Eng* 12:78.

Moretti, R., Pansiot, J., Bettati, D., et al., (2015): Blood-brain barrier dysfunction in disorders of the developing brain. *Front Neurosci* 9:40.

Mortazavi, S. M. J., Ahmadi, J. and Shariati, M., (2007): Prevalence of subjective poor health symptoms associated with exposure to electromagnetic fields among university students. *Bioelectromagnetics* 28(4):326–30.

Mousavy, S. J., Riazi, G. H., Kamarei, M., et al., (2009): Effects of mobile phone radiofrequency on the structure and function of the normal human hemoglobin. *Int J Biol Macromol* 44(3):278–85.

Muehsam, D., Lalezari, P., Lekhraj, R., et al., (2013): Non-thermal radio frequency and static magnetic fields increase rate of hemoglobin deoxygenation in a cell-free preparation. *PLOS ONE* 8(4):e61752.

Mueller, C. H. and Schierz, C., (2004): Project NEMESIS: Doubleblind study on effects of 50 Hz EMF on sleep quality, physiological parameters and field perception in people suffering from electrical hypersensitivity. In K. H. Mild, M. Repacholi, E. van Deventer and P. Ravazzani ed., *Electromagnetic hypersensitivity: Proceedings of the international workshop on EMF hypersensitivity*. Prague, World Health Organization, pp. 107–21.

Murakamai, S., Takayama, F., Egashira, T., Imao, M. and Morio, A., (2012): Protective effect of fermented papaya preparation on stress-induced acute gastric mucosal lesion. *J Biophys Chem* 3:311–6.

Murakami, M., Yoshikawa, T., Nakamura, T., et al., (2015): Involvement of the histamine H1 receptor in the regulation of sympathetic nerve activity. *Biochem Biophys Res Commun* 458(3):584–9.

Myung, S. K., Ju, W., McDonnell, D. D., Lee, Y. J., Kazinets, G., Cheng, C. T. and Moskowitz, J. M., (2009): Mobile phone use and risk of tumors: A meta-analysis. *J Clin Oncol* 27(33):5565–72.

National Toxicology Program., (2018a): *NTP technical report on the toxicology and carcinogenesis studies in Hsd: Sprague Dawley SD rats exposed to whole-body radio frequency radiation at a frequency (900 MHz) and modulations (GSM and CDMA) used by cell phones.* Washington, DC: NTP. https://ntp.niehs.nih.gov/ntp/about_ntp/trpanel/2018/march/tr595peerdraft.pdf.

National Toxicology Program., (2018b): *NTP technical report on the toxicology and carcinogenesis studies in B6C3F1/N mice exposed to whole-body radio frequency radiation at a frequency (1900 MHz) and modulations (GSM and CDMA) used by cell phones.* Washington, DC: NTP. https://ntp.niehs.nih.gov/ntp/about_ntp/trpanel/2018/march/tr596peerdraft.

Navarro, E. A., Segura, J., Portoles, M. and Gomez-Perretta, C., (2003): The microwave syndrome: A preliminary study in Spain. *Electromagn Biol Med* 22(2–3):161–9.

NIEHS (National Institute of Environmental Health Sciences)., (1998): Assessment of health effects from exposure to power-line frequency electric and magnetic fields. Publication No. 98-3981. Research Triangle Park, NC: NIEHS.

Nilsen, A., (1982): Facial rash in visual display unit operators. *Contact Dermatitis* 8(1):25–8.

Nittby, H., Brun, A., Eberhardt, J., Malmgren, L., Persson, B. R. and Salford, L. G., (2009): Increased blood-brain barrier permeability in mammalian brain 7 days after exposure to the radiation from a GSM-900 mobile phone. *Pathophysiology* 16(2–3):103–12.

Nittby, H., Grafström, G., Eberhardt, J. L., Malmgren, L., Brun, A., Persson, B. R. and Salford, L. G., (2008): Radiofrequency and extremely low-frequency electromagnetic field effects on the blood-brain barrier. *Electromagn Biol Med* 27(2):103–26.

Noor, N. A., Mohammed, H. S., Ahmed, N. A. and Radwan, N. M., (2011): Variations in amino acid neurotransmitters in some brain areas of adult and young male albino rats due to exposure to mobile phone radiation. *Eur Rev Med Pharmacol Sci* 15(7):729–42.

Nordal, R. A. and Wong, C. S., (2005): Molecular targets in radiation-induced blood-brain barrier disruption. *Int J Radiat Oncol Biol Phys* 62(1):279–87.

Nordenson, I., Mild, K. H., Andersson, G. and Sandström, M., (1994): Chromosomal aberrations in human amniotic cells after intermittent exposure to fifty hertz magnetic fields. *Bioelectromagnetics* 15(4):293–301.

Nordin, S., Neely, G., Olsson, D. and Sandström, M., (2014): Odor and noise intolerance in persons with self-reported electromagnetic hypersensitivity. *Int J Environ Res Public Health* 11(9):8794–805.

Nordmann, G. C., Hochstoeger, T. and Keays, D. A., (2017): Magnetoreception-A sense without a receptor. *PLOS Biol* 15(10):e2003234.

Nuccitelli, R., (1988): Ionic currents in morphogenesis. *Experientia* 44(8):657–66.

Nuccitelli, R. (2003): Endogenous electric fields during development, regeneration and wound healing. *Radiat Prot Dosimetry*. 106(4):375–383.

Oberfeld, G., Navarro, E., Portoles, M., Ceferino, M. and Gomez-Perretta, C., (2004): The microwave syndrome further aspect of a Spanish study. In *Proceedings of the international conference in Kos*. Greece, pp. 1–5.

Odaci, E., Hanci, H., İkinci, A., et al., (2016): Maternal exposure to a continuous 900-MHz electromagnetic field provokes neuronal loss and pathological changes in cerebellum of 32-day-old female rat offspring. *J Chem Neuroanat* 75(B):105–10.

Odemer, R. and Odemer, F., (2019): Effects of radiofrequency electromagnetic radiation (RF-EMF) on honey bee queen development and mating success. *Sci Total Environ* 661:553–62.

Ohkawa, H., Ohishi, N. and Yagi, K., (1979): Assay for lipid peroxides in animal tissues by thiobarbituric acid reaction. *Anal Biochem* 95(2):351–8.

Ohmori, H. and Kanayama, N., (2003): Mechanisms leading to autoantibody production: Link between inflammation and autoimmunity. *Curr Drug Targets Inflamm Allergy* 2(3):232–41.

Onodera, K., Yamatodani, A., Watanabe, T. and Wada, H., (1994): Neuropharmacology of the histaminergic neuron system in the brain and its relationship with behavioral disorders. *Prog Neurobiol* 42(6):685–702.

Oscar, K. J. and Hawkins, T. D., (1977): Microwave alteration of the blood-brain barrier system of rats. *Brain Res* 126(2):281–93.

Pacher, P., Beckman, J. S. and Liaudet, L., (2007): Nitric oxide and peroxynitrite in health and disease. *Physiol Rev* 87(1):315–424.

Padawer, J., (1963): Quantitative studies with mast cells. *Ann N Y Acad Sci* 103(1):87–138.

Pall, M. L., (2013): Electromagnetic fields act via activation of voltage-gated calcium channels to produce beneficial or adverse effects. *J Cell Mol Med* 17(8):958–65.

Pall, M. L., (2018): Wi-fi is an important threat to human health. *Environ Res* 164:405–16.

Palmquist, E., Claeson, A. S., Neely, G., Stenberg, B. and Nordin, S., (2014): Overlap in prevalence between various types of environmental intolerance. *Int J Hyg Environ Health* 217(4–5):427–34.

Panagopoulos, D. J., Messini, N., Karabarbounis, A., Filippetis, A. L. and Margaritis, L. H., (2000): A mechanism for action of oscillating electric fields on cells. *Biochem Biophys Res Commun* 272(3):634–40.

Panagopoulos, D. J., Karabarbounis, A. and Margaritis, L. H., (2002): Mechanism for action of electromagnetic fields on cells. *Biochem Biophys Res Commun* 298(1):95–102.

Panagopoulos, D. J., Chavdoula, E. D., Nezis, I. P. and Margaritis, L. H., (2007): Cell Death induced by GSM 900 MHz and DCS 1800 MHz mobile telephony radiation. *Mutat Res* 626(1–2):69–78.

Panagopoulos, D. J., Chavdoula, E. D. and Margaritis, L. H., (2010): Bioeffects of mobile telephony radiation in relation to its intensity or distance from the Antenna. *Int J Radiat Biol* 86(5):345–57.

Panagopoulos, D. J., Johansson, O. and Carlo, G. L., (2015): Polarization: A key difference between man-made and natural electromagnetic fields, in regard to biological activity. *Sci Rep* 12:14914.

Panagopoulos, D. J., (2019): Comparing DNA damage induced by mobile telephony and other types of man-made electromagnetic fields. *Mutat Res Rev Mutat Res* 781:53–62.

Panagopoulos, D. J. and Chrousos, G. P., (2019): Shielding methods and products against man-made electromagnetic fields: Protection versus risk. *Sci Total Environ* 667C:255–62.

Panagopoulos, D. J., (2020): Comparing chromosome damage induced by mobile telephony radiation and a high caffeine dose: Effect of combination and exposure duration. *Gen Physiol Biophys* 39(6):531–44.

Panagopoulos, D. J., Karabarbounis, A., Yakymenko, I. and Chrousos, G. P., (2021): Human-made electromagnetic fields: Ion forced-oscillation and voltage-gated ion channel dysfunction, oxidative stress and DNA damage (review). *Int J Oncol* 59(5):92. https://doi.org/10.3892/ijo.2021.5272.

Panula, P. and Nuutinen, S., (2013): The histaminergic network in the brain: Basic organization and role in disease. *Nat Rev Neurosci* 14(7):472–87.

Patel, J. P. and Frey, B. N., (2015): Disruption in the blood-brain barrier: The missing link between brain and body inflammation in bipolar disorder? *Neural Plast* 2015:708306.

Pearson, T. A., Mensah, G. A., Alexander, R. W., et al.; Centers for Disease Control and Prevention; American Heart Association., (2003): Markers of inflammation and cardiovascular disease: Application to clinical and public health practice: A statement for healthcare professionals from the Centers for Disease Control and Prevention and the American Heart Association. *Circulation* 107(3):499–511.

Perry, F. S., Reichmanis, M., Marino, A. A. and Becker, R. O., (1981): Environmental power-frequency magnetic fields and suicide. *Health Phys* 41(2):267–77.

Petkov, V. D. and Konstantinova, E., (1986): Effects of the ergot alkaloid elymoclavine on the level and turnover of biogenic monoamines in the rat brain. *Arch Int Pharmacodyn Ther* 281(1):22–34.

Phares, T. W., Fabis, M. J., Brimer, C. M., Kean, R. B. and Hooper, D. C., (2007): A peroxynitrite-dependent pathway is responsible for blood-brain barrier permeability changes during a central nervous system inflammatory response: TNF-alpha is neither necessary nor sufficient. *J Immunol* 178(11):7334–43.

Phillips, A., (2002): Peer review and quality of science. Comments Posted at 07/08/2006 on Powerwatch Website. https://www.powerwatch.org.uk/columns/aphilips/index.asp.

Phillips, J. L., Singh, N. P. and Lai, H., (2009): Electromagnetic fields and DNA damage. *Pathophysiology* 16(2–3):79–88.

Pihan, G. A. and Doxsey, S. J., (1999): The mitotic machinery as a source of genetic instability in cancer. *Semin Cancer Biol* 9(4):289–302.

Pinton, G., Ferraro, A., Balma, M. and Moro, L., (2020): Specific low-frequency electromagnetic fields induce expression of active KDM6B associated with functional changes in U937 cells. *Electromagn Biol Med* 39(2):139–53.

Pockley, A. G., Shepherd, J. and Corton, J. M., (1998): Detection of heat shock protein 70 (Hsp70) and anti-Hsp70 antibodies in the serum of normal individuals. *Immunol Investig* 27(6):367–77.

Pollack, H., (1979): The microwave syndrome. *Bull N Y Acad Med* 55(11)):1240–3.

Poole, C., Kavet, R., Funch, D. P., Donelan, K., Charry, J. M. and Dreyer, N. A., (1993): Depressive symptoms and headaches in relation to proximity of residence to an alternating-current transmission line right-of-way. *Am J Epidemiol* 137(3):318–30.

Porcelli, M., Cacciapuoti, G., Fusco, S., Massa, R., d'Ambrosio, G., Bertoldo, C., De Rosa, M. and Zappia, V., (1997): Non-thermal effects of microwaves on proteins: Thermophilic enzymes as model system. *FEBS Lett* 402(2–3):102–6.

Profumo, E., Buttari, B. and Riganò, R., (2011): Oxidative stress in cardiovascular inflammation: Its involvement in autoimmune responses. *Int J Inflam* 2011:295705.

Purkayastha, S. and Sorond, F., (2012): Transcranial doppler ultrasound: Technique and application. *Semin Neurol* 32(4):411–20.

Qiu, C., Fratiglioni, L., Karp, A., Winblad, B. and Bellander, T., (2004): Occupational exposure to electromagnetic fields and risk of Alzheimer's disease. *Epidemiology* 15(6):687–94.

Randolph, T. G., (1962): *Human ecology and susceptibility to the chemical environment*. Charles C Thomas publisher, Springfield, US.

Rea, W. J., (2016): History of chemical sensitivity and diagnosis. *Rev Environ Health* 31(3):353–61.

Rea, W. J., Pan, Y., Fenyves, E. F., et al., (1991): Electromagnetic field sensitivity. *J Bioelectr* 10:214–56.

Redlarski, G., Lewczuk, B., Żak, A., et al., (2015): The influence of electromagnetic pollution on living organisms: Historical trends and forecasting changes. *BioMed Res Int* 2015:234098.

Redmayne, M. and Johansson, O., (2014): Could myelin damage from radiofrequency electromagnetic field exposure help explain the functional impairment electrohypersensitivity? A review of the evidence. *J Toxicol Environ Health B Crit Rev* 17(5):247–58.

Reeves, G. I., (2000): Review of extensive workups of 34 patients over-exposed to radiofrequency radiation. *Aviat Space Environ Med* 71(3):206–15.

Repacholi, M. H., (1998): Low-level exposure to radiofrequency electromagnetic fields: Health effects and research needs. *Bioelectromagnetics* 19(1):1–19.

Report of the workshop on Multiple Chemical Sensitivities (MCS), Berlin, Germany, (21–23 February 1996). https://apps.who.int/iris/handle/10665/26723/browse?authority=Multiple+Chemical+Sensitivity&type=mesh.

Ribatti, D., (2015): The crucial role of mast cells in blood-brain barrier alterations. *Exp Cell Res* 338(1):119–25.

Rinne, J. O., Anichtchik, O. V., Eriksson, K. S., et al., (2002): Increased brain histamine levels in Parkinson's disease but not in multiple system atrophy. *J Neurochem* 81(5):954–60.

Rocha, S. M., Pires, J., Esteves, M., Graça, B. and Bernardino, L., (2014): Histamine: A new immunomodulatory player in the neuron-glia crosstalk. *Front Cell Neurosci* 8:120.

Rodríguez-De la Fuente, A. O., Alcocer-González, J. M., Heredia-Rojas, J. A., et al., (2010): Effect of 60 Hz electromagnetic fields on the activity of hsp70 promoter: An in vivo study. *Cell Biol Int Rep* 19(1):e00014.

Roggeveen, S., van Os, J., Viechtbauer, W. and Lousberg, R., (2015): EEG changes due to experimentally induced 3G mobile phone radiation. *PLOS ONE* 10(6):e0129496.

Röösli, M., (2008): Radiofrequency electromagnetic field exposure and non-specific symptoms of ill health: A systematic review. *Environ Res* 107(2):277–87.

Röösli, M., Frei, P., Mohler, E. and Hug, K., (2010a): Systematic review on the health effects of exposure to radiofrequency electromagnetic fields from mobile phone base stations. *Bull World Health Organ* 88(12):887–96.

Röösli, M., Mohler, E. and Frei, P., (2010b): Sense and sensibility in the context of radiofrequency electromagnetic field exposure. *C R Phys* 11(9–10):576–84.

Rosado, M. M., Simkó, M., Mattsson, M. O. and Pioli, C., (2018): Immune-modulating perspectives for low frequency electromagnetic fields in innate immunity. *Front Public Health* 6:85.

Roser, K., Schoeni, A. and Röösli, M., (2016): Mobile phone use, behavioural problems and concentration capacity in adolescents: A prospective study. *Int J Hyg Environ Health* 219(8):759–69.

Roshangar, L., Hamdi, B. A., Khaki, A. A., Rad, J. S. and Soleimani-Rad, S., (2014): Effect of low-frequency electromagnetic field exposure on oocyte differentiation and follicular development. *Adv Biomed Res* 3:76.

Rubin, G. J., Nieto-Hernandez, R. and Wessely, S., (2010): Idiopathic environmental intolerance attributed to electromagnetic fields (formerly 'electromagnetic hypersensitivity'): An updated systematic review of provocation studies. *Bioelectromagnetics* 31(1):1–11.

Rubin, G. J., Hillert, L., Nieto-Hernandez, R., van Rongen, E. and Oftedal, G., (2011): Do people with idiopathic environmental intolerance attributed to electromagnetic fields display physiological effects when exposed to electromagnetic fields? A systematic review of provocation studies. *Bioelectromagnetics* 32(8):593–609.

Rubin, L. L. and Staddon, J. M., (1999): The cell biology of the blood-brain barrier. *Annu Rev Neurosci* 22:11–28.

Sabirzhanov, B., Stoica, B. A., Hanscom, M., Piao, C. S. and Faden, A. I., (2012): Over-expression of HSP70 attenuates caspase-dependent and caspase-independent pathways and inhibits neuronal apoptosis. *J Neurochem* 123(4):542–54.

Saikhedkar, N., Bhatnagar, M., Jain, A., Sukhwal, P., Sharma, C. and Jaiswal, N., (2014): Effects of mobile phone radiation (900 MHz radiofrequency) on structure and functions of rat brain. *Neurol Res* 36(12):1072–9.

Saili, L., Hanini, A., Smirani, C., et al., (2015): Effects of acute exposure to WIFI signals (2.45 GHz) on heart variability and blood pressure in Albinos rabbit. *Environ Toxicol Pharmacol* 40(2):600–5.

Salford, L. G., Brun, A., Sturesson, K., Eberhardt, J. L. and Persson, B. R., (1994): Permeability of the blood-brain barrier induced by 915 MHz electromagnetic radiation, continuous wave and modulated at 8, 16, 50, and 200 Hz. *Microsc Res Tech* 27(6):535–42.

Salford, L. G., Brun, A. E., Eberhardt, J. L., Marmgren, L. and Persson, B. R., (2003): Nerve cell damage in mammalian brain after exposure to microwaves from GSM mobile phones. *Environ Health Perspect* 111(7):881–3.

Santini, R., Santini, P., LeRuz, P., Danze, J. M. and Seigne, M., (2003): Survey study of people living in the vicinity of cellular phone base stations. *Electromagn Biol Med* 22(1):41–9.

Santini, R., Seigne, M., Bonhomme-Faivre, L., Bouet, S., Defrasme, E. and Sage, M., (2002): Symptoms experienced by users of digital cellular phones: A study of a French engineering school. *Electromagn Biol Med* 21(1):81–8.

Saunders, N. R., Dreifuss, J. J., Dziegielewska, K. M., Johansson, P. A., Habgood, M. D., Møllgård, K. and Bauer, H. C., (2014): The rights and wrongs of blood-brain barrier permeability studies: A walk through 100 years of history. *Front Neurosci* 8:404.

Schilling, C. J., (1997): Effects of acute exposure to ultrahigh radiofrequency radiation on three antenna engineers. *Occup Environ Med* 54(4):281–4.

Schilling, C. J., (2000): Effects of exposure to very high frequency radiofrequency radiation on six antenna engineers in two separate incidents. *Occup Med (Lond)* 50(1):49–56.

Schmid, M. R., Loughran, S. P., Regel, S. J., et al., (2012): Sleep EEG alterations: Effects of different pulse-modulated radio frequency electromagnetic fields. *J Sleep Res* 21(1):50–8.

Schmidt, R., Schmidt, H., Curb, J. D., Masaki, K., White, L. R. and Launer, L. J., (2002): Early inflammation and dementia: A 25-year follow-up of the Honolulu-Asia aging study. *Ann Neurol* 52(2):168–74.

Schmiedchen, K., Driessen, S. and Oftedal, G., (2019): Methodological limitations in experimental studies on symptom development in individuals with idiopathic environmental intolerance attributed to electromagnetic fields (IEI-EMF) - A systematic review. *Environ Health* 18(1):88.

Schoeni, A., Roser, K., Bürgi, A. and Röösli, M., (2016): Symptoms in Swiss adolescents in relation to exposure from fixed site transmitters: A prospective cohort study. *Environ Health* 15(1):77.

Schoeni, A., Roser, K. and Röösli, M., (2017): Symptoms and the use of wireless communication devices: A prospective cohort study in Swiss adolescents. *Environ Res* 154:275–83.

Schreier, N., Huss, A. and Röösli, M., (2006): The prevalence of symptoms attributed to electromagnetic field exposure: A cross-sectional representative survey in Switzerland. *Soz Praventivmed* 51(4):202–9.

Schröttner, J. and Leitgeb, N., (2008): Sensitivity to electricity—Temporal changes in Austria. *BMC Public Health* 8:310.

Schumacher, M., Nanninga, A. and Leidenberger, F., (1989): S-35 and 1–125 radioimmunoassays for the measurement of 6-sulphatoxymelatonin in human urine. *Acta Endrocinol* 120:132.

Schüz, J., Petters, C., Egle, U. T., et al., (2006): The "Mainzer EMF-Wachhund": Results from a watchdog project on self-reported health complaints attributed to exposure to electromagnetic fields. *Bioelectromagnetics* 27(4):280–7.

Semin, I. A., Shvartsburg, L. K. and Dubovik, B. V., (1995): Changes in the secondary structure of DNA under the influence of external low-intensity electromagnetic field. *Radiats Biol Radioecol* 35(1):36–41.

Shanson, D. C. and FRCPath, M. B., (1989): *Opportunistic infections in microbiology in clinical practice* (2nd ed.), 151–67. ISBN: 9780723614036.

Sharma, V. P. and Kumar, N. R., (2010): Changes in honeybee behaviour and biology under the influence of cellphone radiations. *Curr Sci* 98:1376–8.

Sheffler, Z. M., Reddy, V. and Pillarisetty, L. S., (2021): *Physiology, Neurotransmitters*. Treasure Island, FL: StatPearls Publishing.

Sheng, J. G., Mrak, R. E. and Griffin, W. S., (1997): Glial-neuronal interactions in Alzheimer disease: Progressive association of IL-1alpha+ microglia and S100beta+ astrocytes with neurofibrillary tangle stages. *J Neuropathol Exp Neurol* 56(3):285–90.

Shi, R. and Borgens, R. B., (1995): Three-dimensional gradients of voltage during development of the nervous system as invisible coordinates for the establishment of embryonic pattern. *Dev Dyn* 202(2):101–14.

Silva, D. F., Barros, W. R., Almeida Mda, C. and Rêgo, M. A., (2015): Exposure to non-ionizing electromagnetic radiation from mobile telephony and the association with psychiatric symptoms. *Cad Saude Publ* 31(10):2110–26.

Silverberg, A. B., Shah, S. D., Haymond, M. W. and Cryer, P. E., (1978): Norepinephrine: Hormone and neurotransmitter in man. *Am J Physiol* 234(3):E252–6.

Sırav, B. and Seyhan, N., (2016): Effects of GSM modulated radio-frequency electromagnetic radiation on permeability of blood-brain barrier in male & female rats. *J Chem Neuroanat* 75(B):123–7.

Slottje, P., van Moorselaar, I., van Strien, R., Vermeulen, R., Kromhout, H. and Huss, A., (2017): Electromagnetic hypersensitivity (EHS) in occupational and primary health care: A nation-wide survey among general practitioners, occupational physicians and hygienists in the Netherlands. *Int J Hyg Environ Health* 220(2 Pt B):395–400.

Smit, L. H., Korse, C. M. and Bonfrer, J. M., (2005): Comparison of four different assays for determination of serum S-100B. *Int J Biol Markers* 20(1):34–42.

Sobel, E., Davanipour, Z., Sulkava, R., et al., (1995): Occupations with exposure to electromagnetic fields: A possible risk factor for Alzheimer's disease. *Am J Epidemiol* 142(5):515–24.

Sobel, E., Dunn, M., Davanipour, Z., Qian, Z. and Chui, H. C., (1996): Elevated risk of Alzheimer's disease among workers with likely electromagnetic field exposure. *Neurology* 47(6):1477–81.

Söderqvist, F., Carlberg, M. and Hardell, L., (2009a): Use of wireless telephones and serum S100B levels: A descriptive cross-sectional study among healthy Swedish adults aged 18–65 years. *Sci Total Environ* 407(2):798–805.

Söderqvist, F., Carlberg, M., Hansson Mild, K. and Hardell, L., (2009b): Exposure to an 890-MHz mobile phone-like signal and serum levels of S100B and transthyretin in volunteers. *Toxicol Lett* 189(1):63–6.

Söderqvist, F., Carlberg, M. and Hardell, L., (2008): Use of wireless telephones and self-reported health symptoms: A population-based study among Swedish adolescents aged 15–19 years. *Environ Health* 7:18.

Söderqvist, F., Carlberg, M. and Hardell, L., (2015): Biomarkers in volunteers exposed to mobile phone radiation. *Toxicol Lett* 235(2):140–6.

Söderqvist, F., Hardell, L., Carlberg, M. and Mild, K. H., (2010): Radiofrequency fields, transthyretin, and Alzheimer's disease. *J Alzheimers Dis* 20(2):599–606.

Soffritti, M., Tibaldi, E., Padovani, M., et al., (2016): Synergism between sinusoidal-50 Hz magnetic field and formaldehyde in triggering carcinogenic effects in male Sprague-Dawley rats. *Am J Ind Med* 59(7):509–21.

Somanah, J., Aruoma, O. I., Gunness, T. K., et al., (2012): Effects of a short-term supplementation of a fermented papaya preparation on biomarkers of diabetes mellitus in a randomized Mauritian population. *Prev Med* 54:S90–7.

Sonmez, O. F., Odaci, E., Bas, O. and Kaplan, S., (2010): Purkinje cell number decreases in the adult female rat cerebellum following exposure to 900 MHz electromagnetic field. *Brain Res* 1356:95–101.

Sosa, M., Bernal-Alvarado, J., Jiménez-Moreno, M., et al., (2005): Magnetic field influence on electrical properties of human blood measured by impedance spectroscopy. *Bioelectromagnetics* 26(7):564–70.

Stalin, P., Abraham, S. B., Kanimozhy, K., Prasad, R. V., Singh, Z. and Purty, A. J., (2016): Mobile phone usage and its health effects among adults in a semi-urban area of Southern India. *J Clin Diagn Res* 10(1):LC14–6.

Stam, R., (2010): Electromagnetic fields and the blood-brain barrier. *Brain Res Rev* 65(1):80–97.

Stamataki, E., Stathopoulos, A., Garini, E., et al., (2013): Serum S100B protein is increased and correlates with interleukin 6, hypoperfusion indices, and outcome in patients admitted for surgical control of hemorrhage. *Shock* 40(4):274–80.

Stein, Y. and Udasin, I. G., (2020): Electromagnetic hypersensitivity (EHS, microwave syndrome) - Review of mechanisms. *Environ Res* 186:109445.

Strimbu, K. and Tavel, J. A., (2010): What are biomarkers? *Curr Opin HIV AIDS* 5(6):463–6.

Sudan, M., Kheifets, L., Arah, O., Olsen, J. and Zeltzer, L., (2012): Prenatal and postnatal cell phone exposures and headaches in children. *Open Pediatr Med J* 6(2012):46–52.

Szmigielski, S., (2013): Reaction of the immune system to low-level RF/MW exposures. *Sci Total Environ* 454–455:393–400.

Szyjkowska, A., Gadzicka, E., Szymczak, W. and Bortkiewicz, A., (2014): The risk of subjective symptoms in mobile phone users in Poland–An epidemiological study. *Int J Occup Med Environ Health* 27(2):293–303.

Tan, K. H., Harrington, S., Purcell, W. M. and Hurst, R. D., (2004): Peroxynitrite mediates nitric oxide-induced blood-brain barrier damage. *Neurochem Res* 29(3):579–87.

Tan, S. Z., Tan, P. C., Luo, L. Q., et al., (2019): Exposure effects of terahertz waves on primary neurons and neuron-like cells under nonthermal conditions. *Biomed Environ Sci* 32(10):739–54.

Tao, R. and Huang, K., (2011): Reducing blood viscosity with magnetic fields. *Phys Rev E Stat Nonlin Soft Matter Phys* 84(1 Pt 1):011905.

Teimori, F., Khaki, A. A., Rajabzadeh, A. and Roshangar, L., (2016): The effects of 30 mT electromagnetic fields on hippocampus cells of rats. *Surg Neurol Int* 7:70.

Terzi, M., Ozberk, B., Deniz, O. G. and Kaplan, S., (2016): The role of electromagnetic fields in neurological disorders. *J Chem Neuroanat* 75(B):77–84.

Tettamanti, G., Auvinen, A., Åkerstedt, T., et al., (2020): Long-term effect of mobile phone use on sleep quality: Results from the cohort study of mobile phone use and health (COSMOS). *Environ Int* 140:105687.

Thielens, A., Bell, D., Mortimore, D. B., Greco, M. K., Martens, L. and Joseph, W., (2018): Exposure of insects to radio-frequency electromagnetic fields from 2 to 120 GHz. *Sci Rep* 8(1):3924.

Trimmel, M. and Schweiger, E., (1998): Effects of an ELF (50 Hz, 1 mT) electromagnetic field (EMF) on concentration in visual attention, perception and memory including effects of EMF sensitivity. *Toxicol Lett* 96–97:377–82.

Trosko, J. E., (2000): Human health consequences of environmentally-modulated gene expression: Potential roles of ELF-EMF induced epigenetic versus mutagenic mechanisms of disease. *Bioelectromagnetics* 21(5):402–6.

Tsurita, G., Ueno, S., Tsuno, N. H., Nagawa, H. and Muto, T., (1999): Effects of exposure to repetitive pulsed magnetic stimulation on cell proliferation and expression of heat shock protein 70 in normal and malignant cells. *Biochem Biophys Res Commun* 261(3):689–94.

Tuengler, A. and von Klitzing, L., (2013): Hypothesis on how to measure electromagnetic hypersensitivity. *Electromagn Biol Med* 32(3):281–90.

Tuohimaa, P., Keisala, T., Minasyan, A., Cachat, J. and Kalueff, A., (2009): Vitamin D, nervous system and aging. *Psychoneuroendocrinology* 34:S278–86.

Valentini, E., Ferrara, M., Presaghi, F., De Gennaro, L. and Curcio, G., (2010): Systematic review and meta-analysis of psychomotor effects of mobile phone electromagnetic fields. *Occup Environ Med* 67(10):708–16.

van Dongen, D., Smid, T. and Timmermans, D. R., (2014): Symptom attribution and risk perception in individuals with idiopathic environmental intolerance to electromagnetic fields and in the general population. *Perspect Public Health* 134(3):160–8.

Vander Vorst, A., Rosen, A. and Kotsuka, Y., (2006): RF-microwave interaction with biological tissues. 2006 Wiley IEE Press, Hoboken (New Jersey). ISBN:047173277X, 9780471732778.

Vecchio, F., Buffo, P., Sergio, S., Iacoviello, D., Rossini, P. M. and Babiloni, C., (2012): Mobile phone emission modulates event-related desynchronization of α rhythms and cognitive-motor performance in healthy humans. *Clin Neurophysiol* 123(1):121–8.

Verkasalo, P. K., Kaprio, J., Varjonen, J., Romanov, K., Heikkilä, K. and Koskenvuo, M., (1997): Magnetic fields of transmission lines and depression. *Am J Epidemiol* 146(12):1037–45.

Verrender, A., Loughran, S. P., Dalecki, A., McKenzie, R. and Croft, R. J., (2016): Pulse modulated radiofrequency exposure influences cognitive performance. *Int J Radiat Biol* 92(10):603–10.

Volkow, N. D., Tomasi, D., Wang, G. J., et al., (2011): Effects of cell phone radiofrequency signal exposure on brain glucose metabolism. *JAMA* 305(8):808–13.

von Klitzing, L., (1995): Low-frequency pulsed electromagnetic fields influence EEG of man. *Phys Med* XI:77–80.

von Klitzing, L., (2021): Artificial EMG by WLAN-exposure. *J Biostat Biom App* 6:101. ISSN: 2455-765X.

Wada, H., Inagaki, N., Yamatodani, A. and Watanabe, T., (1991): Is the histaminergic neuron system a regulatory center for whole brain activity? *Trends Neurosci* 14(9):415–8.

Wagner, P., Röschke, J., Mann, K., Fell, J., Hiller, W., Frank, C. and Grözinger, M., (2000): Human sleep EEG under the influence of pulsed radio frequency electromagnetic fields. Results from polysomnographies using submaximal high power flux densities. *Neuropsychobiology* 42(4):207–12.

Walleczek, J., (1992): Electromagnetic field effects on cells of the immune system: The role of calcium signaling. *FASEB J* 6(13):3177–85.

Weisenseel, M. H., (1983): Control of differentiation and growth by endogenous electric currents. In W. Hoppe, W. Lohmann, H. Markl and H. Ziegler ed., *Biophysics*. Berlin: Springer–Verlag, pp. 460–5.

Wever, R., (1970): The effects of electric fields on circadian rhythmicity in men. *Life Sci Space Res* 8:177–87.

Wever, R., (1973): Human circadian rhythms under the influence of weak electric fields and the different aspects of these studies. *Int J Biometeorol* 17(3):227–32.

Wever, R., (1979): *The circadian system of man: Results of experiments under temporal isolation.* Springer-Verlag, New York.

WHO (World Health Organization)., (2005): *Electromagnetic fields and public health, electromagnetic hypersensitivity; WHO fact sheet no. 296.* Geneva: World Health Organization.

WHO (World Health Organization)., (2006): *Framework for developing health-based EMF standards.* Geneva: WHO. ISBN: 9241594330. www.who.int/peh-emf/standards/EMF_standards_framework%5b1%5d.pdf.

WHO (World Health Organization)., (2014): *Electromagnetic fields and public health: Mobile phones; fact sheet no. 193.* Geneva: World Health Organization.

Williams, R. A. and Webb, T. S., (1980): Exposure to radiofrequency radiation from an aircraft radar unit. *Aviat Space Environ Med* 51(11):1243–4.

Xiong, L., Sun, C. F., Zhang, J., et al., (2015): Microwave exposure impairs synaptic plasticity in the rat hippocampus and PC12 cells through over-activation of the NMDA receptor signaling pathway. *Biomed Environ Sci* 28(1):13–24.

Yakymenko, I., Tsybulin, O., Sidorik, E., Henshel, D., Kyrylenko, O. and Kyrylenko, S., (2016): Oxidative mechanisms of biological activity of low-intensity radiofrequency radiation. *Electromagn Biol Med* 35(2):186–202.

Yang, S., Chen, Y., Deng, X., et al., (2013): Hemoglobin-induced nitric oxide synthase overexpression and nitric oxide production contribute to blood-brain barrier disruption in the rat. *J Mol Neurosci* 51(2):352–63.

Yang, X. S., He, G. L., Hao, Y. T., Xiao, Y., Chen, C. H., Zhang, G. B. and Yu, Z. P., (2012): Exposure to 2.45 GHz electromagnetic fields elicits an HSP-related stress response in rat hippocampus. *Brain Res Bull* 88(4):371–8.

Yao, L., McCaig, C. D. and Zhao, M., (2009): Electrical signals polarize neuronal organelles, direct neuron migration, and orient cell division. *Hippocampus* 19(9):855–68.

Yenari, M. A., Liu, J., Zheng, Z., Vexler, Z. S., Lee, J. E. and Giffard, R. G., (2005): Antiapoptotic and anti-inflammatory mechanisms of heat-shock protein protection. *Ann N Y Acad Sci* 1053:74–83.

Zalata, A., El-Samanoudy, A. Z., Shaalan, D., El-Baiomy, Y. and Mostafa, T., (2015): In vitro effect of cell phone radiation on motility, DNA fragmentation and clusterin gene expression in human sperm. *Int J Fertil Steril* 9(1):129–36.

Zaret, M. M., (1973): Microwave cataracts. *Med Trial Tech Q* 19(3):246–52.

Zhang, Y., Li, Z., Gao, Y. and Zhang, C., (2015): Effects of fetal microwave radiation exposure on offspring behavior in mice. *J Radiat Res* 56(2):261–8.

Zheng, F., Gao, P., He, M. et al., (2014): Association between mobile phone use and inattention in 7102 Chinese adolescents: A population-based cross-sectional study. *BMC Public Health* 14:1022.

Zheng, F., Gao, P., He, M., et al., (2015): Association between mobile phone use and self-reported well-being in children: A questionnaire-based cross-sectional study in Chongqing, China. *BMJ (Open)* 5(5):e007302.

Zuo, H., Lin, T., Wang, D., et al., (2014): Neural cell apoptosis induced by microwave exre through mitochondria-dependent caspase-3 pathway. *Int J Med Sci* 11(5):426–35.

9 Carcinogenic Effects of Non-thermal Exposure to Wireless Communication Electromagnetic Fields

Igor Yakymenko[1,2] and Oleksandr Tsybulin[3]

[1]Department of Public Health, Kyiv Medical University, Kyiv, Ukraine
[2]Department of Environmental Safety, National University of Food Technologies, Kyiv, Ukraine
[3]European Collegium, Kyiv, Ukraine

CONTENTS

Abstract 370
9.1 Introduction 370
    9.1.1 Man-made EMFs, Non-thermal EMF Exposure, Health Effects 370
    9.1.2 ELF/ULF and RF/Microwave EMFs. Exposure Limits 371
    9.1.3 EMFs from Radars and WC Systems 373
9.2 Man-made EMFs and Carcinogenesis 374
    9.2.1 ELF EMFs and Cancer Promotion in Humans 374
    9.2.2 Radar EMFs and Cancer Promotion in Humans 375
    9.2.3 WC EMFs and Cancer Promotion in Humans 376
    9.2.4 Animal Studies on Cancer Promotion by Man-made EMFs 380
9.3 Man-made EMFs and Markers of Carcinogenesis 381
9.4 Discussion and Conclusions 383
References 385

Keywords: electromagnetic fields, non-ionizing radiation, radio frequency, extremely low frequency, wireless communications, cancer, tumor, risk assessment, safety limits, precautionary principle.

Abbreviations: AAPC: average annual percentage change. APC: annual percentage change. AML: acute myeloid leukemia. BP: 3,4-benzopyrene. CNS: central nervous system. CI: Confidence Interval. DECT: Digitally Enhanced Cordless Telecommunications. DMBA: 7,12-dimethylbenz[a]anthracene. EGF: epidermal growth factor. ELF: Extremely Low Frequency. EMF: electromagnetic field. EHS: electro-hypersensitivity. EMR: electromagnetic radiation. ERK: extracellular-signal-regulated kinase. GSM: Global System for Mobile Telecommunications. HL cancer: hematolymphatic cancer. HR: Hazard Ratio. IARC: International Agency for Research on Cancer. ICNIRP: International Commission on Non-Ionizing Radiation Protection. MT: mobile telephony. MW: microwaves. MMWs: modulated microwaves. NHL: non-Hodgkin lymphoma. ODC: ornithine decarboxylase.

DOI: 10.1201/9781003201052-12

OE: Observed versus Expected Ratio. OR: Odds Ratio. OS: oxidative stress. RADAR: radio detection and ranging. ROS: reactive oxygen species. RF: Radio Frequency. RR: Relative Risk. SAR: Specific Absorption Rate. SIR: Standardized Incidence Ratio. SMR: Standardized Mortality Ratio. UMTS: Universal Mobile Telecommunications System. WC: wireless communications. Wi-Fi: Wireless Fidelity. WHO: World Health Organization. 2G, 3G, 4G, 5G: second, third, fourth, fifth generation of mobile telephony. 8-OH-dG: 8 hydroxy-2-deoxyguanosine.

ABSTRACT

In this chapter, we discuss alarming epidemiological and experimental data on carcinogenic effects of long-term non-thermal exposure to man-made electromagnetic fields (EMFs) and corresponding electromagnetic radiation (EMR), mainly from wireless communication (WC) systems, termed as WC EMFs and WC EMR, respectively. Moreover, since all WC EMFs/EMR include Extremely Low Frequency (ELF) components in the form of pulsations and modulation, the chapter also examines corresponding data from purely ELF man-made EMFs. During the past two decades, a number of scientific reports have revealed that, under certain conditions, non-thermal exposure to WC EMFs/EMR or modulated microwaves (MMWs) can substantially induce cancer progression in humans and animals. The carcinogenic effect of WC EMFs is typically manifested after long-term (usually ≥ 10 years) exposure, e.g., in mobile phone users. Nevertheless, even a year of operation of a powerful base station for mobile telephony (MT) reportedly resulted in a dramatic increase of cancer incidence among the population living nearby. In addition, studies in rodents unveiled a significant increase in carcinogenesis after 17–24 months of MMW exposure both in tumor-prone and intact animals. Data on widely accepted molecular markers of carcinogenesis confirm that exposure to non-thermal levels of MMWs or ELF man-made EMFs can induce tumorigenesis. It is becoming increasingly evident that assessment of biological effects of man-made EMFs/EMR based solely on thermal approach used in recommendations by certain international regulatory agencies, including the International Commission on Non-Ionizing Radiation Protection (ICNIRP), requires urgent and significant re-evaluation. We conclude that available scientific data strongly point to the need for re-elaboration of the current safety limits for man-made EMF exposures. We also emphasize that the everyday exposure of the population to WC EMFs/EMR should be regulated based on the Precautionary Principle, which implies maximum restriction of the risk factor till new, more unambiguous conclusions can be drawn regarding its safety.

9.1 INTRODUCTION

9.1.1 Man-made EMFs, Non-thermal EMF Exposure, Health Effects

Man-made electromagnetic fields (EMFs) and corresponding electromagnetic radiation (EMR), although they are non-ionizing, have become one of the most significant and fastest growing environmental stressors. At first, due to the global electrification more than a century ago, and now due to the intensive development of wireless communication (WC) technologies. All man-made EMFs/EMR are totally polarized and coherent, and thus, they differ from natural EMFs/EMR, which only under few circumstances are partially polarized. Due to their totally polarized and coherent character, man-made EMFs/EMR are more adversely bioactive than natural EMFs (Panagopoulos et al. 2015). The EMFs/EMR emitted by WC, most importantly mobile telephony (MT), antennas, and devices we shall call WC EMFs, and WC EMR. Currently the EMF/EMR levels from man-made sources exceed the levels of natural EMFs in the same band of the electromagnetic spectrum (0–3 × 10^{11} Hz) by thousands of times. Additional development of WC technologies, including the fifth generation (5G) MT system, which is currently being deployed over the world, will only raise this level much further. In this context, the problem of adverse effects of anthropogenic EMFs/EMR on human health and, particularly, the strict assessment of possible carcinogenic effects is extremely important.

In August 2007, an international working group of scientists and public health experts released a report on EMFs and human health (Hardell and Sage 2008). It raised serious concerns about safety limits for public exposure to EMFs from power lines, mobile phones and corresponding antennas, radars, and other sources of man-made EMF exposure in daily life. The authors reviewed more than 2,000 scientific studies and reviews and concluded that the existing public safety limits were inadequate to protect public health. Moreover, a vast number of new, extremely important studies in this field have been published since then. Importantly, the problem, today, is discussed at the highest political levels all over the world. A sound political document is a European Parliament Resolution from April 2, 2009 (www.europarl.europa.eu), in which the direct appeals to activate the research and business strategy for effective restriction of the risks arising from man-made EMFs over the European Union member countries were indicated.

Experts, but not the general public, know that in 2002 the International Agency for Research on Cancer (IARC) of the World Health Organization (WHO) concluded that increased levels of Extremely Low Frequency (ELF) EMFs were associated with childhood leukemia and classified ELF EMFs as possibly carcinogenic to humans (Group 2B) (IARC 2002). Following that, in 2011, the IARC/WHO classified Radio Frequency (RF) EMFs/EMR (actually WC EMFs/EMR) as possibly carcinogenic to humans (Group 2B), mostly due to increased levels of gliomas in "heavy" users of mobile phones (Baan et al. 2011; IARC 2013). And while scientific evidence demands new, even more stringent classification (Miller et al. 2018), few people out of this research area are aware of even these "soft" classifications by the IARC. Therefore, increasing the scientific knowledge on the problem and promoting this knowledge are critically important for educating scientists and the general public and making authorities and governments adopt safer standards and approaches.

In this review, we analyze the scientific data on specific biological effects of man-made non-thermal exposures to ELF, and to WC EMFs/EMR also called here modulated microwaves (MMWs) which always include ELFs in the form of modulation/pulsations/variability (Hyland 2000; 2008; Panagopoulos 2019). Modern WC EMFs/EMR are always emitted in the form of pulsations (digital signals). The RF carrier wave is modulated and contained within the pulses. Both modulation and pulsing frequencies are in most cases ELF. Moreover, there is random variability in the final signal, mainly in the Ultra Low Frequency (ULF) band due to the varying information transmitted by the signal, the varying number of subscribers using the network each moment, and the varying physical parameters of signal transmission, such as position, air conductivity, etc. (Panagopoulos 2019). This is why both ELF and RF/WC EMF studies are examined. More specifically, in this review, we examine epidemiological and experimental studies that deal with potential carcinogenicity of long-term non-thermal man-made EMF exposure from purely ELF or RF/WC systems. By non-thermal EMF exposure, we are referring to usual everyday exposures that do not induce any significant heating in the exposed living tissues (Goodman et al. 1995). Certainly, we are not able to cover all aspects of the problem and, therefore, for additional analysis, we recommend additional reviews (Ahlbom et al. 2004; Breckenkamp et al. 2003; Hardell et al. 2009; Khurana et al. 2009; Kundi 2009; Leszczynski and Xu 2010; Miller et al. 2019; Morgan 2009; Yakymenko and Sidorik 2010; Yakymenko et al. 2011; 2016; Panagopoulos 2019).

9.1.2 ELF/ULF AND RF/MICROWAVE EMFS. EXPOSURE LIMITS

ELF EMFs and RF/microwave EMFs are parts of the electromagnetic spectrum that represent forms of non-ionizing EMFs/EMR produced by electrical and electronic circuits and devices of human technology. Other types of non-ionizing EMFs in our environment are, for example, visible and infrared light, the atmospheric (Schumann) oscillations, or the lower frequency part of ultraviolet radiation. These are natural EMFs that existed throughout the trillions of years of biological evolution in contrast to the man-made EMFs that exist during the past 100–120 years with the global electrification, radars, radio, and television broadcasting, and intensively during the past 30 years with the addition of the modern digital WC technologies.

ELF EMFs occupy the lower part of the electromagnetic spectrum in the frequency range 3–3000 Hz. They are usually associated with the generation, distribution, and use of electric power at the frequencies of 50–60 Hz (called power frequencies) and their harmonics and with the modulation/pulsations of the WC EMFs. The ULF is the lowest part of the spectrum with frequencies 0–3 Hz.

RF is the part of the electromagnetic spectrum with frequencies in the range 300 kHz–300 GHz. The part of RF with the highest frequencies (300 MHz–300 GHz) is referred to as microwaves (MWs). MWs can potentially generate the highest thermal effects (heating) in the absorbing matter, as the energy absorption converted to heat in all materials due to EMF/EMR exposure is proportional to the frequency of the EMF/EMR (Clark et al. 2000).

Any EMF exposure that does not induce any significant heating in the exposed tissues is referred to as non-thermal exposure. RF EMF exposure at intensities up to 1 mW/cm^2, or Specific Absorption Rate (SAR) up to 2 W/kg, for a few minutes (min), are usually non-thermal. Practically, this means that any increase in the temperature of the exposed biological tissues is less than 1°C. This, of course, does not mean that such intensities (up to 1 mW/cm^2) of WC EMFs are safe, as they can induce severe adverse biological and health effects without any significant heating, which are called non-thermal effects and constitute the vast majority of the reported biological effects in the EMF-bioeffects literature. In this chapter, we analyze only the effects of non-thermal EMF exposure.

The international recommendations on safety limits of non-ionizing EMFs/EMR adopted by health authorities and governments are the "Guidelines for Limiting Exposure to Time-Varying Electric, Magnetic, and Electromagnetic Fields (up to 300 GHz)" from a private organization called the International Commission on Non-Ionizing Radiation Protection (ICNIRP 1998). The document directly states that

> "induction of cancer from long-term EMF exposure was not considered to be established, and so these guidelines are based on short-term, immediate health effects such as stimulation of peripheral nerves and muscles, shocks and burns caused by touching conducting objects, and elevated tissue temperatures resulting from absorption of energy during exposure to EMF."

In other words, the ICNIRP limits are based exclusively on thermal effects (heating), and the non-thermal effects are simply ignored.

For ELF EMFs at power line frequencies (50–60 Hz), the ICNIRP (1998) limits for magnetic and electric field intensities were 0.5 mT (5 G) and 10 kV/m for occupational exposure, and 0.1 mT (1 G) and 5 kV/m respectively for the public (ICNIRP 1998). Although these limits were already obsolete since significant adverse biological effects are known to take place at several orders of magnitude weaker ELF EMF exposure (Goodman et al. 1995; Panagopoulos et al. 2021), the ICNIRP revised its limits for EMFs up to 100 kHz in 2010, and the 0.1 mT limit for the public was raised to 0.2 mT (2 G) (ICNIRP 2010).

For RF EMFs, according to the ICNIRP (1998) document, the limits are given in terms of certain parameters of the exposure: 1) The power density or intensity of the incident radiation in W/m^2 (or usually in mW/cm^2, or in µW/cm^2), which indicates the amount of electromagnetic energy incident upon a unit surface per second; and 2) the SAR in W/kg, which indicates the amount of EMR energy absorbed per unit mass of living tissue per second.

The occupational exposure limits in the RF/microwave band according to the ICNIRP (1998) for the incident power density were set to 10–50 W/m^2 (or 1–5 mW/cm^2) average values for 6 min exposure. The corresponding public exposure limits were set to 2–10 W/m^2 (or 0.2–1 mW/cm^2), depending on frequency (averaged over 6 min). For example, for 900 MHz, the ICNIRP (1998) public exposure limit was set to 450 µW/cm^2, and for 2–6 GHz, the corresponding limit was 1 mW/cm^2 (ICNIRP 1998). Surprisingly, the most recent ICNIRP update (ICNIRP 2020) raised the 2–6 GHz 6 min average limit for public exposure, from 1 mW/cm^2 to 4 mW/cm^2, which does not anymore prevent neither the thermal effects.

The SAR limit for the general public for RF EMF exposure in the ICNIRP guidelines is 2 W/kg (for head and trunk). This limit is accepted by the telecommunications industry, and the value of SAR of each mobile phone model must be indicated in the technical specifications of the device. SAR has been seriously questioned as a metric for the assessment of non-thermal biological effects (Panagopoulos et al. 2013). Moreover, only simplified models of the adult human head are currently used by industry for its calculation, while real SAR values depend on geometry, conductivity, and other tissue details, and it was also shown that the SAR is much higher for a child's head than for an adult's head (Christ et al. 2010; de Salles et al. 2006; Gandhi et al. 1996).

Each country has its own national legislation for electromagnetic safety, and national safety limits are different in different countries. Some countries, such as the United States (US) and many European countries have adapted their national EMF/EMR safety limits to the ICNIRP recommendations. Other countries have rather different national limits as compared to the ICNIRP guidelines. For example, for 900 MHz (a carrier frequency used in MT) the safety limits in different countries are in the US and Germany 450 µW/cm^2; in Italy, Russia, Poland, and China 10 µW/cm^2; and in Switzerland 4 µW/cm^2 (Chekhun et al. 2014). As we can see, some countries have significantly more stringent national safety limits than others, staying closer to the Precautionary Principle, which recommends caution, pausing, and review before allowing new technologies that may prove disastrous (Read and O'Riordan 2017).

9.1.3 EMFs from Radars and WC Systems

For a long time in the previous century until about 25 years ago, the main sources of RF EMFs/EMR were radio and television broadcasting systems. In some cases, as in military and aviation, the most powerful local sources of RF EMFs were and still are radars (RADAR: radio detection and ranging). However, the situation has changed dramatically for the general population during the past three decades, and currently, the most prevailing sources of man-made EMF exposures in the human environment are the MT and other WC systems. It is important that both radars and MT/WC systems today use the same microwave part of the RF spectrum, and both use ELF pulsations.

Radar systems are powerful sources of pulsed microwave EMR which mainly expose certain groups of military and service staff and population living nearby. Radars are detection systems which use MW to detect both moving and fixed metallic objects such as aircrafts, ships, missiles, etc. Depending on the tasks, they use different MW frequencies, in most cases 1–12 GHz and pulsing frequencies of 100 Hz–10 kHz (IEEE 2002).

WC systems are undoubtedly the most abundant source of man-made EMFs/EMR in the human environment over the world today. Starting from the first commercial mobile phone networks in Japan, Europe, and the US since 1979–1983, the number of active users of mobile phones has increased globally to more than 5.3 billion (https://www. bankmycell.com). The initial age of youngest users of mobile phones is estimated as 3 years old (Khurana et al., 2009).

WC technology utilizes pulsing microwaves for connection of mobile phones and base antenna stations. A phone is referred to as a "mobile phone" because it is free from wire connection, and it is also referred to as "cell phone" because its technology utilizes a so-called "cellular network". A whole area is covered by many base stations, with each base station antenna operating in one cell (part of the area), and a mobile/cell phone automatically changes stations as it moves from one cell to another.

The first digital system for MT, called Global System for Mobile Telecommunications (GSM), also referred to as second generation (2G) MT system, is still in operation and utilizes carrier frequencies at 900, 1800, or 1900 MHz. This radiation is amplitude and phase modulated, with pulse repetition frequencies at 2, 8, and 217 Hz. The Time Division Multiple Access (TDMA) system is employed. This provides a single pulse from the pulse-packet to each user with a pulse rate about 217 Hz. The third generation (3G) MT system is based on the digital transmission standard called Universal Mobile Telecommunications System (UMTS). In the UMTS, the Code Division Multiple

Access (CDMA) system is employed, according to which a special code is assigned to each user. The carrier frequency in UMTS is between 1920 and 2170 MHz, with pulsing frequencies at 100 and 1500 Hz (Pedersen 1997; Hyland 2000; 2008). The Long Term Evolution (LTE) or fourth generation (4G) MT/WC system is the upgrade to UMTS/CDMA networks with pulsing frequencies at 100 and 1000 Hz. The carrier frequency bands allocated for LTE are different (800, 900, 1800, 2100, 2600 MHz.) in different countries around the world (see Chapter 1, and Ericsson 2009). The fifth generation (5G), the latest MT/WC system which is currently being deployed, uses much higher carrier frequencies (up to 100 GHz) and new supportive technology to enable higher data transmission capacity (Frank 2021). The GSM (2G), UMTS (3G/4G), LTE (4G), and the 5G are cellular MT systems.

The Wireless Fidelity (Wi-Fi) system widely used for wireless connection to the Internet works similarly, with carrier frequency of around 2.45 GHz and 10 Hz pulsations (Belyaev et al. 2016). The cordless domestic (DECT: Digitally Enhanced Cordless Telecommunications) phones are also a significant source of WC EMFs today, combining carrier RF signals of around 1880 MHz with 100 and 200 Hz pulsations (see Chapter 1, and Pedersen 1997). In all WC systems the useful information (sound, picture, video, etc.) is included in the modulation of the RF carrier signal.

It is important to emphasize, again, that all WC technologies utilize modulated and pulsed EMFs/EMR. This actually means that WC EMFs expose biological systems to both RF/MW and ELF EMFs.

9.2 MAN-MADE EMFs AND CARCINOGENESIS

9.2.1 ELF EMFs AND CANCER PROMOTION IN HUMANS

One of the first sound publications about a possible link between anthropogenic non-ionizing EMFs and cancer was published by Wertheimer and Leeper (1979). The authors found association of childhood cancer with "excess of electrical wiring configuration" in Colorado, US in 1976–1977. Children who lived close to high-voltage transmission lines had twice the cases of leukemia (cancer of the blood and bone marrow) and 1.6 times more cases of lymphoma as compared with the control children population. Later, the same authors found less but still significant association between high-electric-current environment and cancer in adults (Wertheimer and Leeper 1982). It was supposed that the ELF (60 Hz) magnetic fields stemming from the high-electric current was the cause of cancer development.

The relative incidence of leukemia, acute myeloid leukemia (AML) and central nervous system (CNS) tumors among workers exposed to ELF EMFs was analyzed in all male industrial workers in Finland aged 25–64 years during 1971–1980 (Juutilainen et al. 1990). The adjusted Relative Risks[*] (RR) with a 95% Confidence Iinterval[†] (CI) in the combined category of "probable" or "possible" exposure, were for all leukemia 1.9 (CI: 1.0–3.5), for AML 1.3 (CI: 0.7–2.3), and for CNS tumors 1.3 (CI: 1.0–1.6), respectively. Thus, the results clearly demonstrated elevated risk among workers exposed to ELF EMFs.

Analysis of mortality during 1990–2008 from different types of hematolymphopoietic cancers in the Swiss National Cohort (3.1 million workers exposed at different levels to ELF EMFs) revealed increased mortality rates in exposed patients with myeloid leukemia (Hazard Ratio[‡] – HR: 1.31, 95% CI: 1.02–1.67), and AML (HR: 1.26, 95% CI: 0.93–1.70). If workers had been highly exposed during their vocational training and at both censuses, the HR increased to 2.24 (95% CI: 0.91–5.53)

---

[*] RR > 1 means the risk is increasing with time.
[†] 95% confidence means 0.95 probability that the cancer is due to the examined parameter. If both numbers of the 95% CI are higher than 1.0, it means that the increasing risk is statistically significant. If one of the numbers of the 95% CI is lower than 1.0 it means the risk elevation is not statistically significant.
[‡] HR is the ratio of the hazard rates corresponding to the conditions described by two levels of an explanatory variable. It is similar with the RR, but the HR represents instantaneous risk over the study time period. HR > 1 means increased risk in exposed population, and 95% CI should be higher than 1.0 (for both limits) in order to be statistically significant.

and 2.75 (95% CI: 1.11–6.83), respectively (Huss et al. 2018). Thus, mortality from AML increased almost threefold in the cohort of most exposed workers to ELF EMFs.

Currently, most meta-analyses of available epidemiological data demonstrate increased carcinogenicity due to ELF EMF exposures. For example, a very recent meta-analysis of studies on childhood cancer due to ELF EMF exposure (33 studies, 186,000 participants) demonstrated a significant association between exposure to ELF EMFs and childhood leukemia, with a dose-response relationship (Seomun et al. 2021).

9.2.2 Radar EMFs and Cancer Promotion in Humans

Radars were the first powerful sources of pulsed microwaves in close proximity to some groups of people, mostly military staff. Early reports on the biological/health effects of radar radiation actually were the first alarming data on possible carcinogenicity of non-thermal man-made EMFs/EMR. Both military and occupational data due to radar operation indicate significant effects of pulsed microwaves on cancer development and other pathological conditions in humans (Puranen and Jokela 1996; Robinette et al. 1980; Szmigielski 1996; Zeeb et al. 2010).

Almost two times more cases of cancer were found among the highly exposed US naval personnel who served during the Korean War (1950–1954) as compared with subjects who had low exposure among 40,000 personnel (Robinette et al. 1980). Death rates for aviation electronic technicians, the group with the highest exposure, were significantly higher than those for the other personnel during the following years up to 1974 (Goldsmith 1997).

Among Polish soldiers (128,000 personnel subjects aged from 20 to 59 years), soldiers of 20–29 years old exposed to radar microwaves from 1970 to 1979 had cancer incidence rates 5.5 times higher than non-exposed soldiers (Szmigielski 1985). The greatest rise of cancer cases was detected in blood-forming organs and lymphatic tissues: The incidence rates were 13.9 times higher for chronic myelocytic leukemia and 8.6 times for myeloblastic leukemia. The level of mortality among all exposed personnel was significantly higher than in unexposed staff. For colorectal cancer the Observed versus Expected ratio[*] (OE) was 3.2 (statistically significant). For cancer of esophagus and stomach, the OE was 3.2 (statistically significant), and for cancer of blood-forming systems and lymphatic tissues, the OE was 6.3 (statistically significant) (Szmigielski 1996).

A substantial increase in cancer incidence was also detected in commercial airline pilots. The Standardized Incidence Ratio[†] (SIR) for malignant melanoma cases was 10.2 (statistically significant) for pilots of commercial airlines in Iceland (Rafnsson et al. 2000). Significantly increased risks of AML (SIR 5.1); skin cancer, excluding melanoma (SIR 3.0); and total cancer (SIR 1.2) were observed also among Danish male jet pilots (Gundestrup and Storm 1999). These data have been explained as a result of excessive exposure to cosmic ionizing radiation or even excessive solar ultraviolet radiation during leisure time. However, analysis of brain cancers among US Air Force personnel has revealed that non-ionizing man-made EMR and particularly pulsed MWs had significant effect on cancer development, with an Odds Ratio[‡] (OR) of 1.38 (statistically significant), whereas ionizing radiation had negative association with cancer cases, with an OR of 0.58 (statistically significant) (Grayson 1996). Also, the Standardized Mortality Ratio[§] (SMR) for brain tumors was 2.1 (statistically significant) among German male cockpit crew members (6,017 people) (Zeeb et al. 2010).

---

[*] OE is the observed rate over the expected rate. If the observed rate is higher than its expected rate (OE >1), then we have increased risk from the event, i.e., a negative effect from the exposure.

[†] SIR is the ratio of the observed number of cancer cases to the expected number of cases.

[‡] The OR is a measure of association between an exposure and an outcome. The OR represents the odds (probability) that an outcome will occur given a particular exposure compared to the odds of the outcome occurring in the absence of that exposure. OR >1 and statistically significant, indicates increased risk of the exposure.

[§] The SMR is the ratio between the observed number of deaths in a study population and the number of deaths that would be expected. When any of these indexes is greater than 1, it indicates an increased risk from the event, i.e., harmful effects from the exposure.

Cancer risk was significantly raised with a risk ratio of 2.2 (statistically significant) among cockpit crew members employed for 30 years, as compared to those employed for less than 10 years. In addition, non-Hodgkin lymphoma (NHL) was also increased with SMR 4.2 (statistically significant) among male cabin crew members (20,757 people). Importantly, any increase in cancers associated with ionizing (cosmic) radiation was not detected in this cohort study.

Six incident cases of testicular cancer occurred within a cohort of 340 police officers between 1979 and 1991 in Seattle, Washington (OE: 6.9; $p < 0.001$) (Davis and Mostofi 1993). Occupational use of hand-held radar was the only common risk factor among all six officers, and all had a routine habit of keeping the emitter ("radar gun") in close proximity to their testicles. Similarly, in Ontario, Canada, risk assessment among police officers exposed to radar devices for speed measurement (1,596 females and 20,601 males) revealed an increased risk among men for testicular cancer (SIR: 1.3) and for melanoma (SIR: 1.45) (Finkelstein 1998).

Other occupational studies revealed the highest risk ratio (2.6) for acute myelogenous leukemia in radio and radar operators among all occupational groups studied (Savitz and Calle 1987). In addition, excessive risk for breast cancer was detected (SIR: 1.5) among Norwegian female radio and telegraph operators (2,619 women) with potential exposure to the corresponding EMF emissions (405 kHz–25 MHz) (Tynes et al. 1996).

In another study, 87 people working with radar (and 150 matched controls) were divided into risk groups according to the various power densities (from 8 µW/cm^2 to 300 µW/cm^2) they were exposed to (Goldsmith 1997). Three specific types of radiation-induced eye-cataracts in persons working with higher power density MW exposure were identified. A dose–response relationship was observed between lens damage and power density level of exposure in the different risk groups. We note that the highest power density in the study (300 µW/cm^2) was significantly below the current ICNIRP (2020) limit e.g. for 2–6 GHz (4000 µW/cm^2).

A statistically significant increase in immature red blood cells among workers exposed to radar radiation was reported (Goldsmith 1997). In addition, the staff exposed to the radar radiation had significantly lower levels of leukocytes and thrombocytes than workers distant from the radar.

In another study (Degrave et al. 2009), the possible association between hematolymphatic (HL) cancers and EMR exposure from radar systems was identified. The results revealed an increased HL cancer mortality rate by 7.2 times (95% CI: 1.09–47.9) among the radar operators compared to a similar non-exposed group of service personnel.

Five cancer cases were reported (RR for all cancers was 8.3, $p < 0.005$) that were linked to occupational radio and radar EMF exposures (Peleg et al. 2018). Three were HL (leukemia, lymphoma, and plasmacytoma).

9.2.3 WC EMFs and Cancer Promotion in Humans

More than 10 years ago, an article in the *New York Times* about mobile (cell) phones (named "cells"), titled "Are Cells the New Cigarettes?" (Dowd 2010), noted:

> "Just as parents now tell their kids that, believe it or not, there was a time when nobody knew that cigarettes were bad for you, those kids may grow up to tell their kids that, believe it or not, there was a time when nobody knew how dangerous it was to hold your phone right next to your head and chat away for hours."

The problem is that mobile (cell) phones, like cordless domestic (DECT) phones, emit modulated and pulsed microwaves very close to our brain while we keep them in contact with the ear during calls. And when this technology was developed, nobody knew the possible effects on our health. But do we know now? Yes, science has known the univocal answer for at least 15 years or more. And this is based not only on data from radar EMF exposure or research with laboratory animals but also on WC EMF exposure on humans. Let us examine the epidemiology.

As early as 2001, a group of Swedish oncologists lead by Prof. Lennart Hardell published an epidemiological study in the *European Journal of Cancer Prevention*, in which they first demonstrated that the use of a mobile phone increased the risk of brain tumors by 2.4 times on the same side of the head where the phone was held (Hardell et al. 2001). Those were histopathologically verified cases on Swedish mobile phone users and a matched control population.

In 2003, Hardell's group reported in the *International Journal of Oncology*, that the risk of acoustic neuroma* on the same side of the head where the phone was used increased in Swedish mobile phone users by 4.4 times (Hardell et al. 2003).

The next year, Hardell's group revealed that, among all analyzed age groups of adults, the most vulnerable was the youngest group, aged 20–29 years. This group had an eightfold increased risk of brain tumor development after 5 years of mobile phone use as compared to non-users of the same age (Hardell et al. 2004). In the same year, another Swedish research team confirmed an almost fourfold increased risk of acoustic neuroma on the side of the head where the phone was normally held after 10 years of use (Lönn et al., 2004). Later, these data were also confirmed by Japanese scientists, which detected a threefold increased risk of acoustic neuroma in Japan with 20 or more minutes of daily cell phone use after 5 years (Sato et al. 2011).

In 2006, the Hardell group published the next results of their epidemiological research, which demonstrated that the risk of glioma, the most aggressive type of brain tumor, increased in Swedes by 5.4 times after 10 years of cell phone use on the same side of the head where the phone was reported to be held (Hardell et al. 2006).

In 2008, Israeli scientists discovered that another vulnerable target for mobile phone radiation is the parotid glands. These salivary glands are situated just near the ear and are exposed to WC EMFs/EMR every time we talk on a mobile phone. The researchers found that those people who talked on a mobile phone 3 hours a day for 5 years had 2.3-fold increased risk of parotid malignant tumor compared to non-users of the same age (Sadetzki et al. 2008). Moreover, they found a fourfold increase in the number of parotid gland cancers over the country from 1980 to 2006, while the other salivary gland cancers had remained stable.

Carcinogenic effects of cell phone radiation were confirmed by the international project known as the Interphone Study. Research teams from 13 countries under the auspice of the World Health Organization (WHO) were involved in the epidemiological study with a total budget of about $25 million. The aim of the study was to assess a possible link between mobile phone radiation and brain tumors in adult human population. Despite fair criticism on the project's design, its financial dependence on the industry, and uncertain final conclusions, the project clearly confirmed a significantly increased risk of brain tumors in heavy users of mobile phones. According to the project findings, people who talked on a mobile phone for 1 hour a day for 4 years had a 3.8–4.8 times increased risk of brain cancers (glioma and meningioma) and 2.8 times increased risk of acoustic neuroma compared to non-users (Cardis et al. 2010; Interphone 2011).

A French national study on mobile phone use and brain tumors (glioma and meningioma) revealed that between 2004 and 2006 the risk of glioma for heavy ($\geq$ 896 h of use during this time) users was OR = 2.54 (95% CI: 1.19–5.41). And a cohort of urban users showed an eightfold increased risk for brain tumors, excluding tumors in the temporal or frontal lobes (OR: 8.2, 95% CI: 1.37–49.07) (Coureau et al. 2014). As already explained, OR = 8.2 means that the risk for these tumors increased under the mobile phone EMF-exposure by 8.2 times compared to the unexposed population,† and this was a statistically significant effect.

Another study found that the patients with glioblastoma multiform (GBM) who had used mobile phones $\leq$ 3 h per day had better survival than those with mobile phone use of $\geq$ 3 h per day

---

* Acoustic neuroma is a benign tumor of the acoustic nerve that connects the ear with the brain. The malady results in chronic pain and hearing loss.
† We should note that the OR does not coincide with the RR, but for large groups of a population with rare pathologies, such as tumors, the values of these two indexes are close.

(Akhavan-Sigari et al. 2014). Glioblastoma, also known as GBM, is the most aggressive type of brain cancer. The authors investigated the p53 mutant gene expression in peripheral and central zones within the tumor and found significantly higher mutant type p53 expression in the peripheral zone of glioblastoma of patients who had used mobile phones ≥ 3 h per day. The protein p53 is an important tumor-suppressor protein that is altered in most cancers. The p53 protein activates various responses, including cell-cycle arrest and apoptosis, which contribute to tumor suppression. Thus, increased level of mutant type p53 expression indicates an increased level of damage and mutation in the corresponding gene and, thus, an increased risk for tumors under the particular exposure.

Hardell and coworkers found the OR for NHL of T-cell,[*] cutaneous cancer, and leukemia types for mobile phone users to be 6.1 (statistically significant), and 5.5 (statistically significant) for cordless phone users (Hardell et al. 2005). An American study found an OR of 1.6 (statistically significant) for NHL in mobile phone users with more than 8 years of usage (Linet et al. 2006). In another study by the Hardell group (Hardell et al., 2020), it was shown that, in the Swedish Cancer Registry, the average annual percentage change (AAPC) of the NHL incidence in men increased from 1970 to 2017 by 1.66% (95% CI: 1.40–1.91%).[†] The authors argued that the increasing incidence of primary CNS lymphoma may be due to the exposure of the population to WC EMFs.

One of the first carcinogenicity studies on WC EMFs analyzed 118 cases of uveal melanoma among people in Germany who were probably or certainly exposed to mobile phone radiation, with 475 controls, and revealed an OR of 4.2 (statistically significant)[‡] (Stang et al. 2001).

Testicular cancer (seminoma) risk was found to have an OR of 1.8 (statistically significant) for men keeping the cell phone during "stand by" mode in ipsilateral trousers pockets (Hardell et al., 2007). The results were based on 542 cases of seminoma in Sweden.

Increasing thyroid cancer incidence has been observed in Sweden during the period 1970–2017 (Carlberg et al. 2020). In women, during this time period, the incidence increased significantly with AAPC 2.13% (95% CI: 1.43–2.83%). The highest increase was found during 2010–2017 with an annual percentage change (APC) of +9.65%, (95% CI: 6.68–12.71%). In men, the AAPC increased from 1970 to 2017 by 1.49% (95% CI: 0.71–2.28%). The highest increase was found between 2001 and 2017 with APC +5.26% (95% CI: 4.05–6.49%). Similar results were found for all Nordic countries based on the NORDCAN database[§] for the period 1970–2016 with APC +5.83% (95% CI: 4.56–7.12%) in women from 2006 to 2016, and APC + 5.48% (95% CI: 3.92–7.06%) in men from 2005 to 2016. The authors attributed the effect to mobile phone use, concluding that WC EMF exposure is a causative factor for the increasing thyroid cancer incidence.

Regarding research on possible carcinogenesis from MT base station antennas, there are also several publications. In the past 30 years, millions of MT base stations have been installed around the world. MT base station antennas emit the same type of pulsing EMFs/EMR with the corresponding mobile phones. Although each base antenna is usually about 100 times more powerful than a mobile phone (Panagopoulos et al. 2010), their distance from the exposed population is significantly greater than that of a mobile phone from the human head/body, resulting, usually, in much lower but continuous exposure. The WHO had suggested a priority to study the effects mainly from mobile phones, while it discouraged studies on the effects of the corresponding base station antennas (with an exception of years 2003–2006, when the WHO recommended studies on possible effects from base antennas as well) (Kundi and Hutter 2009). This is probably the reason for the smaller number of publications regarding exposures from MT base stations than from mobile phones that can be found to date (Abdel-Rassoul et al. 2007; Eger et al. 2004; Gulati et al. 2016;

---

[*] The NHL is divided into different types based on the type of lymphocyte involved: B lymphocytes (B cells) or T lymphocytes (T cells).
[†] The CI here is a range of percentage with 0.95 probability.
[‡] As explained already, for large groups of a population with rare diseases, the OR is close to the RR, while for smaller groups, the OR is higher than the RR.
[§] A database of cancer statistics for the Nordic countries (Denmark, Finland, Iceland, Norway, Sweden, Faroe Islands, and Greenland).

Hutter et al. 2006; Navarro et al. 2003; Santini et al. 2002; Viel et al. 2009; Wolf and Wolf 2007; Zothansiama et al. 2017).

In one of the first such studies, French and Spanish researchers revealed that inhabitants living near MT base stations (up to 300 m) developed significantly higher rates of many subjective health symptoms such as headache, fatigue, sleep disorder, depression, etc., as compared with the matched control population from more distant areas (Santini et al. 2002). In a similar study in Spain (Navarro et al. 2003) on a population living in the vicinity of a MT base station, the results showed significant correlation between the severity of similar symptoms, usually referred to as "microwave sickness" or electro-hypersensitivity (EHS), and the measured power density of the base antenna radiation.

A comparison of cancer cases among people living up to 400 m from MT base station antennas and people living further than 400 m from the antennas from 1994 to 2004 was carried out in Germany (Eger et al. 2004). The total increase in cancer cases among people living near the MT base station antennas compared to the control population was 1.26 times during the first 5-year period (1994–1998) and 3.11 times during the second 5-year period (1999–2004) of the MT antennas operation. Particularly, in the second period, the increase of cancer cases was statistically significant both as compared with the population from more distant areas and with the expected background cancer incidence.

Population (622 individuals) living in the area near (up to 350 m) an MT antenna base station (850 MHz, 1500 watt of full power) for one year of operation, and matched controls (1,222 individuals) from other areas had been compared in Israel (Wolf and Wolf 2007) with respect to cancer incidence. There were 4.15 times more cases of cancer in the antenna base station area than in the other areas. Relative cancer rates for females were 10.5 in the area close to the base station, 0.6 for the more distant (control) areas, and 1 for the whole town. The differences were statistically significant. Considering that such a very significant increase in the number of cancer cases took place in the span of only 1 year, while cancer development in humans usually takes much longer (more than 10–15 years), the authors suggested that MT/WC EMFs could provoke/accelerate latent cases of cancer in the population around the antenna base station. This means that WC EMF exposure may stimulate the faster development of tumors which already were initiated by various carcinogens but were at early stages (and possibly some of them would never develop into clinically recognizable cancers).

There was a significant increase in micronuclei in buccal cells and tail moment value in alkaline comet assay (both are indicators of DNA damages) in subjects exposed to radiation from MT antennas (3.65 ± 2.44 and 6.63 ± 2.32, respectively) compared with control subjects (1.23 ± 0.97 and 0.26 ± 0.27, respectively) (Gulati et al. 2016).

People residing within a perimeter of 80 m of MT base stations had significantly ($p < 0.0001$) higher frequency of micronuclei compared to the control group, residing ≥ 300 m away from the MT base stations. Also, a significant depletion in glutathione concentration ($p < 0.01$), activities of catalase ($p < 0.001$) and superoxide dismutase ($p < 0.001$), and rise in lipid peroxidation in the plasma of exposed individuals were detected compared to controls (Zothansiama et al. 2017).

Brazilian researchers estimated the rate of deaths (2010–2017) from cancer as a result of MT base station WC EMF/EMR exposure, especially for breast, cervix, lung, and esophagus cancers (Rodrigues et al. 2021). In the adjusted analysis, the results showed that the higher the logarithm of base station WC EMF/EMR exposure (sum of the number of base stations multiplied by the exposure time to each one), the higher the cancer mortality rate. The highest adjusted risk was observed for cervix cancer (Rate Ratio 2.18),[*] which means that there was a 2.18 times higher risk of cervix cancer in the highly-exposed population (close to the base stations) compared to the less exposed population.

It is important to emphasize that for short-term mobile phone use, harmful effects were absent or less evident (Morgan 2009). We should keep in mind that cancer development is a slow process, which can last a couple of decades. For example, the radiation released by the nuclear bomb in

---

[*] A Rate Ratio compares the incidence rates or mortality rates between two groups.

Hiroshima induced a peak in cancers among exposed population 30 years after the bombing. In another example, 30 years after the Chernobyl nuclear accident, the Ukrainian population still had a stable increase of thyroid cancer in the affected regions, while the first cases were registered a few years after the accident (Sidorik and Yakymenko 2013). This means that data on increased cancer risk in mobile phone users after 5–10 years of use may be just the precursor to much bigger problems.

9.2.4 Animal Studies on Cancer Promotion by Man-made EMFs

A highly representative study has been carried out at the University of Washington, Seattle, commissioned by the US Air Force (Chou et al. 1992). Experimental rats (100 animals) were exposed to 2450 MHz microwaves with pulsing frequencies at 8 and 800 Hz for 24 months, 21.5 hours per day. Whole-body calorimetry, thermographic analysis, and power-meter analysis indicated that pulsed MWs delivered at 0.144 W to each exposure waveguide resulted in an average SAR of 0.4 W/kg for a 200 g rat. That was a model of long-term exposure of Air Force pilots to pulsed MWs of radar systems. A total of 155 indexes of metabolism were checked out during the study. As a result, the most expressive effect of the long-term irradiation of the animals was a dramatic increase in cancer cases. In total, 3.6 times more cancer cases were detected in irradiated animals than in the matched controls. Lymphoma cases were diagnosed in the irradiated animals 4.5 times more often than in the control group. In addition, benign tumors of the adrenal glands were detected 7 times more often in the irradiated animals than in the controls.

In the next study carried out under the US Air Force contract, 200 female C3H/HeJ mice prone to mammary gland tumors were exposed for 21 months (22 h/day, 7 days/week) to a 435 MHz pulsed EMR (1.0 kHz pulse repetition rate, 1.0 ps pulse width) with an incident power density of 1.0 mW/cm^2 (SAR 0.32 W/kg), while 200 mice were sham-exposed (Toler et al. 1997). Although, under the conditions of this study, the exposure of mice did not affect the incidence of mammary gland tumors compared with the controls, some other tumor cases had increased markedly. For example, bilateral cases of ovarian epithelial stromal tumors increased by 5 times, multiple cases of hepatocellular carcinoma increased 3 times, and total adrenal gland tumor cases increased by 1.63 times.

In the third published study of this series of experiments (Frei et al. 1998), the same type of mammary tumor-prone mice were exposed for 20 months (20 h/day, 7 days/week) to continuous-wave 2450 MHz MWs with SAR ranging from 0.3 to 1 W/kg. A hundred mice were exposed, while another 100 mice were sham-exposed. The result was that the exposed mice had higher level of mammary gland tumors (1.27 times) and a higher total level of all types of tumors (1.38 times) as compared with the sham-exposed animals, but the difference between groups was not statistically significant. Meanwhile, multiple mammary tumor cases occurred in the exposed mice two times more frequently than in the sham-exposed.

In other experiments, mice with a high incidence of spontaneous breast cancer and mice treated with 3,4-benzopyrene (BP) were exposed to continuous-wave 2450 MHz microwaves in an anechoic chamber at 5 or 15 mW/cm^2 (2 hours daily, six sessions per week, for 3 months) (Szmigielski et al. 1982). The exposure to MWs at either 5 or 15 mW/cm^2 (significantly higher power densities than in the other above-described experiments) resulted in acceleration of development of BP-induced skin cancer. MW-exposed mice with a high incidence of spontaneous breast cancer developed breast tumors earlier than the control animals. The authors indicated that the promotion of cancer development and the lowering of natural antineoplastic resistance was similar in mice exposed to MW at 5 mW/cm^2 and in mice stressed by prolonged confinement (2 hours daily, six sessions per week, for 3 months), but the ratio of cancer cases in animals exposed to 15 mW/cm^2 was significantly higher than in animals stressed by confinement.

In another series of experiments, transgenic mice moderately predisposed to develop lymphoma spontaneously were exposed to pulsing EMR of 900 MHz with a pulse repetition frequency of 217 Hz and a pulse duration of 0.6 ms. Incident power densities were 2.6–13 W/m^2 (0.26–1.3 mW/cm^2)

and average SAR 0.13–1.4 W/kg (Repacholi et al. 1997). One group of mice (101 females) was exposed for two 30 min periods per day for up to 18 months. Another group (100 females) was sham-exposed. Lymphoma risk was significantly higher, more than twice, in the exposed mice than in the sham-exposed (OR 2.4; statistically significant). In particular, follicular lymphoma was the major contributor to the increased tumor incidence.

The carcinogenic or co-carcinogenic potential of ELF EMFs was assessed in rats treated with DMBA (7,12-dimethylbenz[a]anthracene) to develop mammary gland cancer (Löscher 2001). It was demonstrated that magnetic field intensities of 50 or 100 µT significantly increased the growth of mammary tumors independently of whether a single dose DMBA was given or repeated doses over a prolonged period. Thus, a significant co-carcinogenic potential of ELF EMFs at environmental intensities (e.g. similar to those in close proximity to high-voltage power lines) was demonstrated.

In long-term experiments, pregnant C57BL/6NCrj mice were exposed to 50 Hz, 500 mG (50 µT) magnetic field for 1 week (12 h per day), and 24 male and 42 female B6C3F1 mice born from them were further exposed for up to 15.5 months (Qi et al. 2015). As a result, the average body weight of exposed male and female offspring was decreased significantly compared to the control offspring, and in the exposed female offspring, the incidence of chronic myeloid leukemia (3 cases in 42 mice, 7%) was significantly higher than in the control group (0%).

The recent study of the US National Toxicology Program (NTP 2018) on mobile phone radiation risks attracted a lot of attention both from experts and from the media. This study was commissioned by the US Department of Health. Despite the fact that they applied some unrealistically high SAR values for mobile phones (up to 9 W/kg) the data on the rats exposed to 1.5 W/kg WC EMR are valuable for the assessment of non-thermal WC EMF exposure effects and are very interesting as well. Rats exposed to 1.5 W/kg in GSM (2G) or CDMA (3G) modes displayed statistically significant increases in the levels of both cancerous and non-cancerous pathologies. About 3.3% of the GSM-exposed male rats (3 out of 90) and 3.3% of the CDMA-exposed female rats (also 3 out of 90) developed malignant glioma (brain cancer), while none of the 180 control animals had it. Interestingly, female rats exposed to higher (thermal) levels of the CDMA mode did not develop brain cancer, similarly with the control animals. In addition, 2% of the male rats exposed to the GSM or the CDMA mode developed malignant schwannoma of the heart, while none of the control animals had it. The cancer induction was accompanied by DNA damage induction in the exposed animals (Smith-Roe et al. 2020). Both types of cancer are very rare, and their development along with the DNA damage in rats under "safe exposure conditions" for humans is very indicative on the actual relevance of the current exposure limits in most countries. Moreover, non-cancer pathologies of prostate gland appeared in 14.4% of the GSM-exposed (1.5 W/kg) male rats compared to 5.5% of the control animals (2.6 times more often). Pituitary gland pathologies were found in 31.1% of the GSM-exposed male rats compared to 19.1% of control animals (1.6 times more often). Cardiomyopathy (a non-cancerous heart pathology) appeared in the exposed female rats (1.5 W/kg GSM) 2.3 times more often than in the control (10% in the exposed group compared to 4.4% in the control) (NTP 2018).

Similarly, an Italian life-span exposure study of rats in a simulated 2G MT EMF also found induction of heart schwannomas and brain glial tumors, confirming the results of the NTP study (Falcioni et al. 2018).

In conclusion, both ELF and WC EMF non-thermal exposures have been proven in model experiments in the laboratory with rats and mice to have a significant carcinogenic potential.

9.3 MAN-MADE EMFs AND MARKERS OF CARCINOGENESIS

Previously, we showed that a vast majority (93%) of experimental studies that examined possible induction of oxidative stress (OS) by modulated/pulsed RF/MW EMFs demonstrated statistically significant OS effects in living cells in different biological models (see Chapter 3, and Yakymenko et al. 2016). For example, a significant increase of nitric oxide (NO•) and other reactive oxygen species (ROS) in cells under non-thermal intensities of WC EMFs has been detected both *in vivo* (Ferreira et al. 2006; Grigoriev et al. 2010; Irmak et al. 2002; Ozguner et al. 2005a; 2005b; Ozgur et al. 2010; Sokolovic et al. 2008) and *in vitro* (Agarwal et al. 2009; De Iuliis et al. 2009; Friedman et al. 2007; Luukkonen et al. 2009; Zmyslony et al. 2004).

It is important to underline that most research on oxidative effects induction has been carried out with modulated/pulsed RF/MW EMFs and, in many cases, with real devices and systems for WC (in most cases mobile phones). This means that these studies are directly related with WC EMF biological/health effects and mechanisms. Moreover, as we discussed before, WC EMFs include both RF/MW as carrier signals and ELF EMFs as modulation/pulsing components. Thus, for the study of the biological/health effects of WC EMFs, experimental data on both types of EMFs (RF and ELF) are not only relevant, but necessary.

The effect of ROS overproduction in living cells under ELF EMF non-thermal exposure, can be explained by the dysfunction (irregular gating) of voltage-gated ion channels in cell membranes caused by this type of exposures (Panagopoulos et al. 2002; 2015; 2021). It is important to underline that this mechanism can be applied to both purely ELF EMFs and to RF/MW EMFs pulsed/modulated by ELF components, which is the case for all WC EMFs.

It was proposed that the action of "mobile phone irradiation" (875 MHz, 0.07 mW/cm^2) on cultured living cells (Rat1 and HeLa) was mediated in the plasma membrane by the NADH oxidase, which can rapidly (within minutes) generate ROS (Friedman et al. 2007). ROS directly stimulate matrix metalloproteinases and allow them to cleave and release heparin-binding epidermal growth factor (EGF). This secreted factor activates the EGF receptor, which, in turn, activates the extracellular-signal-regulated kinase (ERK) cascade, inducing transcription and activating other cellular pathways.

On the other hand, on the model of purified human spermatozoa exposed to simulated GSM mobile phone radiation (1.8 GHz, SAR ranging from 0.4 to 27.5 W/kg), a significant overproduction of ROS in mitochondria and oxidative damage of DNA was detected, along with significant reduction in motility and vitality of spermatozoa (De Iuliis et al. 2009). All observed effects were significantly correlated with non-thermal SAR levels, suggesting that significant effects of the WC EMF exposure occurred under non-thermal mechanisms.

Therefore, WC EMFs can induce cellular OS, which, in turn, can cause DNA damage and cancer stimulation (Halliwell and Whiteman 2004; Valko et al. 2007). Today, it is known that, in addition to damage via OS, ROS in cells can play a role of secondary messengers for certain intracellular signaling cascades which can induce oncogenic transformation (Valko et al. 2006).

DNA damage in cells exposed to non-thermal levels of MW/WC EMFs both *in vivo* and *in vitro* was demonstrated in dozens of independent studies (Ruediger 2009; Panagopoulos 2019). The most frequently applied method for detection of DNA damage after the MW/WC EMF exposure was alkaline comet assay. A statistically significant increase in both single strand and/or double strand breaks of DNA has been detected by various methods in humans (Gandhi and Anita 2005; Yadav and Sharma 2008), animal models (Lai and Singh 1995; 1996; Ferreira et al. 2006; Kesari et al. 2010; Panagopoulos et al. 2010; Panagopoulos 2019), and cell cultures (Lai and Singh 1995; Diem et al. 2005; Paulraj and Behari 2006; Schwarz et al. 2008; Wu et al. 2008) exposed to non-thermal MW/WC EMF intensities.

Among a set of experimental studies on mutagenic effects of non-thermal EMF exposures, particular studies have identified changes in the levels of a specific marker of oxidative damage of DNA, 8-hydroxy-2′-deoxyguanosine (8-OH-dG) (Yokus et al. 2005; 2008; De Iuliis et al. 2009; Xu et al. 2010; Guler et al. 2012; Khalil et al. 2012; Burlaka et al. 2013; Ding et al. 2018; Yakymenko et al. 2018; Alkis et al. 2019). It is important, that both ELF and WC EMFs were demonstrated to cause such effects. For example, the level of 8-OH-dG in human spermatozoa was shown to be significantly increased after *in vitro* exposure to non-thermal levels of simulated GSM 1800 MHz EMF (De Iuliis et al. 2009). Likewise, we demonstrated that the exposure of quail embryos *in ovo* to real-life GSM 900 MHz EMF of 0.25 μW/cm^2 power density for a few days was sufficient for a significant, two- to threefold, increase of 8-OH-dG levels in embryonic cells (Burlaka et al. 2013). Similarly, significant oxidative damage of DNA (8-OH-dG increased levels) was detected in laboratory rats after long-term (100 days or longer) exposure to ELF EMFs of 50 Hz, 100 µT or higher (Yokus et al. 2005; 2008).

As ELF and RF/MW/WC EMFs, especially at non-thermal intensities, do not have enough energy to directly ionize DNA molecules, and as overproduction of ROS in living cells, due to both ELF and WC EMFs exposure, has been reliably documented, it is clear that we have an indirect effect of these types of man-made EMFs on DNA in the form of oxidative damage. The most aggressive form of ROS, which is able to affect the DNA molecule, is the hydroxyl radical (OH•) (Halliwell 2007). Hydroxyl radicals are generated in cells by the Fenton reaction:

$$Fe^{2+} + H_2O_2 \rightarrow Fe^{3+} + OH\bullet + OH^-$$

and by the Haber–Weiss reaction (Valko et al. 2006):

$$O_2\bullet^- + H_2O_2 \rightarrow O_2 + OH\bullet + OH^-$$

The "fuel" for these reactions is the superoxide radical ($O_2\bullet^-$) which can be converted to hydrogen peroxide (H_2O_2) and may be generated in living cells under non-thermal ELF/WC EMF exposure as previously demonstrated (Yakymenko et al. 2016; Panagopoulos et al. 2021).

On the other hand, increased concentration of nitric oxide (NO•) in addition to superoxide radical in EMF-exposed cells can lead to formation of another aggressive form of ROS, peroxynitrite (ONOO⁻), which can also cause DNA damage (Valko et al. 2006).

Thus, due to overproduction of ROS and depletion of the antioxidant system capacity in living cells/organisms under non-thermal ELF or WC EMF exposure, the oxidative damage of DNA is the inevitable result.

Ornithine decarboxylase (ODC) significantly changes its activity under conditions of non-thermal MMW exposure (Byus et al. 1988; Hoyto et al. 2007; Litovitz et al. 1993; Litovitz et al. 1997; Paulraj et al. 1999). This was one of the first markers of carcinogenesis revealed to be activated under non-thermal MMW exposure. ODC is involved in processes of cell growth and differentiation, and its activity is raised in tumor cells. Although overexpression of ODC is not sufficient for transformation of normal cells into tumor cells, an increased activity of this enzyme was shown to promote the development of tumors in pre-tumor cells (Clifford et al. 1995).

9.4 DISCUSSION AND CONCLUSIONS

In this chapter, we presented evidence for carcinogenic effects of ELF and WC EMFs. Both epidemiological and experimental data indicate that at least under certain conditions, long-term non-thermal exposure to WC EMFs or MMWs can lead to tumorigenesis. Supporting evidence comes from statistically significant epidemiological data based either on long-term analysis, e.g., on mortality of the US Navy personnel within 20 years after exposure during the Korean War (Goldsmith 1997), or on a relatively short, 1 year exposure, e.g., from MT base station in Israel (Wolf and Wolf 2007). In the latter case, we fully agree with the authors that WC EMF exposure most likely results in acceleration of preexisting cancer development. It is of note here that the same conclusion was drawn from epidemiological research on fast increase cancer incidence among adult population in Colorado exposed to ELF (60 Hz) EMFs from electric power transmission lines (Wertheimer and Leeper 1982).

Thus, epidemiological data on significant carcinogenic effects of man-made and especially WC EMFs are available. The increased cancer risk due to man-made EMFs found in the studies described in this chapter represents an increased probability for an adverse outcome, which we definitely do not want. Therefore, any unnecessary exposure to man-made and especially WC EMFs should be avoided as much as possible in order to keep the risk as low as possible.

The main shortcoming of the most epidemiological data, both in military and in MT/WC risk assessment studies, is the inevitable lack of a precise measurement of the exposure parameters. We strongly suggest that in forthcoming epidemiological studies an improved assessment of intensity and duration of the EMF exposure should be provided. Good examples of large-scale epidemiological studies employing personal MW dosimeters are the studies carried out in Germany (Heinrich et al. 2010; Milde-Busch et al. 2010; Roosli et al. 2010; Thomas et al. 2010). On the other hand, we also realize that the levels of the EMF exposure in contemporary epidemiological studies, at least in those which deal with MT/WC systems, were well below the official "safety limits" set by national organizations and the ICNIRP (1998; 2010; 2020) recommendations.

Therefore, taking into account the reviewed data, we conclude that the relatively long-term (e.g., 10 years) exposure to MMWs emitted from MT/WC devices operating within "safety limits" set by current regulatory bodies can be considered as a factor for cancer promotion. Indeed, in most of the studies on rodents, the intensity of MMW exposure was appropriately measured, and in the majority of them, it was well below the ICNIRP (1998; 2020) limits. Nevertheless, the majority of these studies demonstrated obvious carcinogenic effects after long-term exposure (up to 24 months). This further emphasizes that at least under certain conditions the exposure to pulsed/modulated MWs with intensities well below the current official "safety limits" can indeed promote cancer development.

In addition, experimental evidence of involvement of typical markers of carcinogenesis, such as overproduction of ROS or formation of 8-OH-dG under non-thermal ELF and MMW exposures, further indicate potential danger of these types of man-made EMFs for human health. It is important to emphasize here that experimental data, especially obtained from *in vitro* studies, often reveal significant effects even after short-term (e.g., only a few minutes) and/or much weaker intensity of exposure to MMWs (by several orders of magnitude lower) than in the ICNIRP recommendations. Taking these data into account, we strongly emphasize that the currently applied "thermal" assessment of potential hazards of man-made EMF exposure is far from being appropriate and safe.

We would also like to underline another important issue for newcomers in this field of research. WC technology constitutes a huge and important part of modern life. Even under increased risks from the WC EMFs, this technology is extremely useful for human progress and prosperity. We should definitely not discuss the issue of how to reject or ban such technology, but we should know about the risks and apply the most safe, scientifically approved standards and approaches. The position of the health authorities and the telecommunications industry itself here is extremely important. Unfortunately, in many cases this position is not as scientifically and socially responsible as it should be.

Big pharma, big tobacco, and big WC industries, as well as the nuclear industry, all have in common a heavy influence on the results of the research they are funding. They need scientific data to provide them with evidence that their products are safe, reliable, and effective, because they are respectful companies and cannot just state something without strong scientific evidence. But the effectiveness/ineffectiveness of pharmaceuticals reported in scientific publications varied by 4 times depending on the presence or absence of financial support from the pharma industry (Lexchin et al. 2003). And this is a relatively soft pressure from the sponsor. Research funded by the tobacco industry demonstrates much stronger bias. As it turned out, scientific studies funded by big tobacco companies have admitted negative effects of passive smoking 88 (!) times less often than independent studies (Barnes and Bero 1998). And then such "reliable scientific data" migrate into the media. Younger people may not be familiar with advertisements such as "More doctors smoke Camel" or "No throat irritation due to smoking", But older people say those advertisements were rather persistent a few decades ago, and all of them were supported by supposedly "strong scientific evidence".

Unfortunately, the WC industry, which produces one of the most developed technologies in our time, is in this line when we are talking about public health risk assessment. Independent studies have demonstrated adverse effects of WC EMFs about 10 times more often than those funded by big WC industry (Huss et al. 2007) For example, almost all governmental or independent studies on ELF EMF risks find either a statistically significant association between electric or magnetic field exposure and childhood leukemia or an elevated risk of at least OR = 1.5, while almost all industry supported studies fail to find any significant or even suggestive association (Carpenter 2019). As Dr. Martin Blank from Columbia University had reasonably explained, one of the most effective strategies of the industry to manage public concern is to compromise undesirable scientific data and to sow doubts in minds through the "alternative findings" (Blank 2014).

Taken together, we may conclude here that, today, there is enough convincing data to assert that the long-term exposure to non-thermal EMF/EMR levels from WC systems can promote cancer development. Therefore, the official recommendations by the ICNIRP, and by many national regulatory bodies for devices emitting WC EMFs, such as all mobile communication systems, must be re-assessed according to the current alarming data, and independent studies on WC EMF risk assessment must be facilitated. Currently, we strongly suggest the wide implementation of the Precautionary Principle (Read and O'Riordan 2017) for everyday WC EMF-exposure, which recommends maximum restriction of any unnecessary exposure.

REFERENCES

Abdel-Rassoul, G., El-Fateh, O. A., Salem, M. A., et al. (2007). Neurobehavioral effects among inhabitants around mobile phone base stations. *Neurotoxicology* 28(2):434–440.

Agarwal, A., Desai, N. R., Makker, K., et al. (2009). Effects of radiofrequency electromagnetic waves (RF-EMW) from cellular phones on human ejaculated semen: An in vitro pilot study. *Fertil Steril* 92(4):1318–1325.

Ahlbom, A., Green, A., Kheifets, L., et al. (2004). Epidemiology of health effects of radiofrequency exposure. *Environ Health Perspect* 112(17):1741–1754.

Akhavan-Sigari, R., Mazloum Farsi Baf, M., Ariabod, V., Rohde, V., Rahighi, S. (2014). Connection between cell phone use, p53 gene expression in different zones of glioblastoma multiforme and survival prognoses. *Rare Tumors* 6(3):5350.

Alkis, M. E., Bilgin, H. M., Akpolat, V., et al. (2019). Effect of 900-, 1800-, and 2100-MHz radiofrequency radiation on DNA and oxidative stress in brain. *Electromagn Biol Med* 38(1):32–47.

Baan, R., Grosse, Y., Lauby-Secretan, B., et al. (2011). Carcinogenicity of radiofrequency electromagnetic fields. *Lancet Oncol* 12(7):624–626.

Barnes, D. E., Bero, L. A. (1998). Why review articles on the health effects of passive smoking reach different conclusions. *JAMA* 279(19):1566–1570.

Belyaev, I., Dean, A., Eger, H., et al. (2016). EUROPAEM EMF Guideline 2016 for the prevention, diagnosis and treatment of EMF-related health problems and illnesses. *Rev Environ Health* 31(3):363–397.

Blank, M. (2014). *Overpowered: The dangers of electromagnetic radiation (EMF) and what you can do about it.* Seven Stories Press, New York.

Breckenkamp, J., Berg, G., Blettner, M. (2003). Biological effects on human health due to radiofrequency/microwave exposure: A synopsis of cohort studies. *Radiat Environ Biophys* 42(3):141–154.

Burlaka, A., Tsybulin, O., Sidorik, E., et al. (2013). Overproduction of free radical species in embryonic cells exposed to low intensity radiofrequency radiation. *Exp Oncol* 35(3):219–225.

Byus, C. V., Kartun, K., Pieper, S., Adey, W. R. (1988). Increased ornithine decarboxylase activity in cultured cells exposed to low energy modulated microwave fields and phorbol ester tumor promoters. *Cancer Res* 48(15):4222–4226.

Cardis, E., Deltour, I., Vrijheid, M. (2010). Brain tumour risk in relation to mobile telephone use: Results of the INTERPHONE international case-control study. *Int J Epidemiol* 39(3):675–694.

Carlberg, M., Koppel, T., Hedendahl, L. K., Hardell, L. (2020). Is the increasing incidence of thyroid cancer in the Nordic countries caused by use of mobile phones? *Int J Environ Res Public Health* 17(23). 9129.

Carpenter, D. O. (2019). Extremely low frequency electromagnetic fields and cancer: How source of funding affects results. *Environ Res* 178:108688.

Chekhun, V., Yakymenko, I., Sidorik, E., et al. (2014). Current state of international and national public safety limits for radiofrequency radiation. *Sci J Minist Health Ukr*, 1 (5):57–64.

Chou, C. K., Guy, A. W., Kunz, L. L., et al. (1992). Long-term, low-level microwave irradiation of rats. *Bioelectromagnetics* 13(6):469–496.

Christ, A., Gosselin, M. C., Christopoulou, M., Kühn, S., Kuster, N. (2010). Age-dependent tissue-specific exposure of cell phone users. *Phys Med Biol* 55(7):1767–1783.

Clark, D. E., Folz, D. C., West, J. K. (2000). Processing materials with microwave energy. *Mater Sci Eng A* 287(2):153–158.

Clifford, A., Morgan, D., Yuspa, S. H., Soler, A. P., Gilmour, S. (1995). Role of ornithine decarboxylase in epidermal tumorigenesis. *Cancer Res* 55(8):1680–1686.

Coureau, G., Bouvier, G., Lebailly, P., et al. (2014). Mobile phone use and brain tumours in the CERENAT case-control study. *Occup Environ Med* 71(7):514–522.

Davis, R. L., Mostofi, F. K. (1993). Cluster of testicular cancer in police officers exposed to hand-held radar. *Am J Ind Med* 24(2):231–233.

De Iuliis, G. N., Newey, R. J., King, B. V., Aitken, R. J. (2009). Mobile phone radiation induces reactive oxygen species production and DNA damage in human spermatozoa in vitro. *PLOS ONE* 4(7):e6446.

De Salles, A. A., Bulla, G., Rodriguez, C. E. (2006). Electromagnetic absorption in the head of adults and children due to mobile phone operation close to the head. *Electromagn Biol Med* 25(4):349–360.

Degrave, E., Meeusen, B., Grivegnee, A. R., Boniol, M., Autier, P. (2009). Causes of death among Belgian professional military radar operators: A 37-year retrospective cohort study. *Int J Cancer* 124(4):945–951.

Diem, E., Schwarz, C., Adlkofer, F., Jahn, O., Rüdiger, H. (2005). Non-thermal DNA breakage by mobile-phone radiation (1800 MHz) in human fibroblasts and in transformed GFSH-R17 rat granulosa cells in vitro. *Mutat Res* 583(2):178–183.

Ding, S.-S., Sun, P., Zhang, Z., et al. (2018). Moderate dose of trolox preventing the deleterious effects of wi-fi radiation on spermatozoa in vitro through reduction of oxidative stress damage. *Chin Med J (Engl)* 131(4):402.

Dowd, M. (2010). Are cells the new cigarettes? *The New York Times*, June 27: 11.

Eger, H., Hagen, K., Lucas, B., et al. (2004). Influence of the proximity of mobile phone base stations on the incidence of cancer. *Environ Med Soc* 17:273–356.

Ericsson. (2009). LTE – An introduction. White Paper.

Falcioni L, Bua L, Tibaldi E, et al. (2018): Report of final results regarding brain and heart tumors in Sprague-Dawley rats exposed from prenatal life until natural death to mobile phone radiofrequency field representative of a 1.8GHz GSM base station environmental emission, *Environmental Research*, 165: 496–503.

Ferreira, A. R., Bonatto, F., de Bittencourt Pasquali, M. A., et al. (2006). Oxidative stress effects on the central nervous system of rats after acute exposure to ultra high frequency electromagnetic fields. *Bioelectromagnetics* 27(6):487–493.

Finkelstein, M. M. (1998). Cancer incidence among Ontario police officers. *Am J Ind Med*, 34(2):157–162.

Frank, J. W. (2021). Electromagnetic fields, 5G and health: What about the precautionary principle? *J Epidemiol Community Health*, 75(6):562–566.

Frei, M. R., Jauchem, J. R., Dusch, S. J., et al. (1998). Chronic, low-level (1.0 W/kg) exposure of mice prone to mammary cancer to 2450 MHz microwaves. *Radiat Res*, 150(5):568–576.

Friedman, J., Kraus, S., Hauptman, Y., Schiff, Y., Seger, R. (2007). Mechanism of short-term ERK activation by electromagnetic fields at mobile phone frequencies. *Biochem J*, 405(3):559–568.

Gandhi, G., Anita. (2005). Genetic damage in mobile phone users: Some preliminary findings. *Indian J Hum Genet* 11(2):99–104.

Gandhi, O., Lazzi, G., Furse, C. (1996). Electromagnetic absorption in the human head and neck for mobile telephones at 835 and 1900 MHz. *IEEE Trans Microw Theor Tech* 44(10):1884–1897.

Goldsmith, J. R. (1997). Epidemiological evidence relevant to radar (microwave) effects. *Environ Health Perspect* 105(6):1579–1587.

Goodman, E. M., Greenebaum, B., Marron, M. T. (1995). Effects of electro-magnetic fields on molecules and cells. *Int Rev Cytol* 158:279–338.

Grayson, J. K. (1996). Radiation exposure, socioeconomic status, and brain tumor risk in the US Air Force: A nested case-control study. *Am J Epidemiol* 143(5):480–486.

Grigoriev, Y. G., Grigoriev, O. A., Ivanov, A. A., et al. (2010). Confirmation studies of Soviet research on immunological effects of microwaves: Russian immunology results. *Bioelectromagnetics* 31(8):589–602.

Gulati, S., Yadav, A., Kumar, N., et al. (2016). Effect of GSTM1 and GSTT1 polymorphisms on genetic damage in humans populations exposed to radiation from mobile towers. *Arch Environ Contam Toxicol* 70(3):615–625.

Guler, G., Tomruk, A., Ozgur, E., et al. (2012). The effect of radiofrequency radiation on DNA and lipid damage in female and male infant rabbits. *Int J Radiat Biol* 88(4):367–373.

Gundestrup, M., Storm, H. H. (1999). Radiation-induced acute myeloid leukaemia and other cancers in commercial jet cockpit crew: A population-based cohort study. *Lancet* 354(9195):2029–2031.

Halliwell, B., Whiteman, M. (2004). Measuring reactive species and oxidative damage in vivo and in cell culture: How should you do it and what do the results mean? *Br J Pharmacol* 142(2):231–255.

Halliwell, B. (2007). Biochemistry of oxidative stress [review]. *Biochem Soc Trans* 35(5):1147–1150.

Hardell, L., Mild, K. H., Påhlson, A., Hallquist, A. (2001). Ionizing radiation, cellular telephones and the risk for brain tumours. *Eur J Cancer Prev* 10(6):523–529.

Hardell, L., Mild, K. H., Carlberg, M. (2003). Further aspects on cellular and cordless telephones and brain tumours. *Int J Oncol* 22(2):399–408.

Hardell, L., Mild, K. H., Carlberg, M., Hallquist, A. (2004). Cellular and cordless telephone use and the association with brain tumors in different age groups. *Arch Environ Health An Int J* 59(3):132–137.

Hardell, L., Eriksson, M., Carlberg, M., Sundström, C., Mild, K. H. (2005). Use of cellular or cordless telephones and the risk for non-Hodgkin's lymphoma. *Int Arch Occup Environ Health* 78(8):625–632.

Hardell, L., Carlberg, M., Mild, K. H. (2006). Pooled analysis of two case–control studies on use of cellular and cordless telephones and the risk for malignant brain tumours diagnosed in 1997–2003. *Int Arch Occup Environ Health* 79(8):630–639.

Hardell, L., Carlberg, M., Ohlson, C. G., et al. (2007). Use of cellular and cordless telephones and risk of testicular cancer. *Int J Androl* 30(2):115–122.

Hardell, L., Sage, C. (2008). Biological effects from electromagnetic field exposure and public exposure standards. *Biomed Pharmacother* 62(2):104–109.

Hardell, L., Carlberg, M., Hansson Mild, K. (2009). Epidemiological evidence for an association between use of wireless phones and tumor diseases. *Pathophysiology* 16(2–3):113–122.

Hardell, L., Carlberg, M., Koppel, T., Nordström, M., Hedendahl, L. K. (2020). Central nervous system lymphoma and radiofrequency radiation - A case report and incidence data in the Swedish cancer register on non-Hodgkin lymphoma. *Med Hypo* 144:110052.

Heinrich, S., Thomas, S., Heumann, C., von Kries, R., Radon, K. (2010). Association between exposure to radiofrequency electromagnetic fields assessed by dosimetry and acute symptoms in children and adolescents: A population based cross-sectional study. *Environ Health* 9:75.

Hoyto, A., Juutilainen, J., Naarala, J. (2007). Ornithine decarboxylase activity is affected in primary astrocytes but not in secondary cell lines exposed to 872 MHz RF radiation. *Int J Radiat Biol* 83(6):367–374.

Huss, A., Egger, M., Hug, K., Huwiler-Müntener, K., Röösli, M. (2007). Source of funding and results of studies of health effects of mobile phone use: Systematic review of experimental studies. *Environ Health Perspect* 115(1):1–4.

Huss, A., Spoerri, A., Egger, M., et al. (2018). Occupational extremely low frequency magnetic fields (ELF-MF) exposure and hematolymphopoietic cancers–Swiss National Cohort analysis and updated meta-analysis. *Environ Res* 164:467–474.

Hutter, H. P., Moshammer, H., Wallner, P., Kundi, M. (2006). Subjective symptoms, sleeping problems, and cognitive performance in subjects living near mobile phone base stations. *Occup Environ Med* 63(5):307–313.

Hyland, G. J. (2000). Physics and biology of mobile telephony. *Lancet* 356(9244):1833–1836.

Hyland, G. J. (2008). Physical basis of adverse and therapeutic effects of low intensity microwave radiation. *Indian J Exp Biol* 46(5):403–419.

IARC. (2002). IARC monographs on the evaluation of carcinogenic risks to humans: Volume 80. In *Non-ionizing radiation, part 1: Static and extremely low-frequency (ELF) electric and magnetic fields*. International Agency for Research on Cancer, Lyon, France. .

IARC, (2013): *Non-ionizing radiation, part 2: Radiofrequency electromagnetic fields*, Vol. 102, International Agency for Research on Cancer, Lyon, France.

ICNIRP. (1998). Guidelines for limiting exposure to time-varying elecrtic, magnetic and electromagnetic fields (up to 300 GHz). *Health Phys* 74(4):494–522.

ICNIRP. (2010). Guidelines for limiting exposure to time-varying electric and magnetic fields (1 Hz to 100 kHz). *Health Phys* 99(6):818–836.

ICNIRP. (2020). Guidelines for limiting exposure to electromagnetic fields (100 kHz to 300 GHz). *Health Phys* 118(5):483–524.

IEEE. (2002). Std 521–2002 standard letter designations for radar-frequency bands. In IEEE Std 521-2002 (Revision of IEEE Std 521-1984).1-10. doi: 10.1109/IEEESTD.2003.94224.

Interphone. (2011). Acoustic neuroma risk in relation to mobile telephone use: Results of the INTERPHONE international case–control study. *Cancer Epidemiol* 35(5):453–464.

Irmak, M. K., Fadillioglu, E., Gulec, M., et al. (2002). Effects of electromagnetic radiation from a cellular telephone on the oxidant and antioxidant levels in rabbits. *Cell Biochem Funct* 20(4):279–283.

Juutilainen, J., Läärä, E., Pukkala, E. (1990). Incidence of leukaemia and brain tumours in Finnish workers exposed to ELF magnetic fields. *Int Arch Occup Environ Health* 62(4):289–293.

Kesari, K. K., Behari, J., Kumar, S. (2010). Mutagenic response of 2.45 GHz radiation exposure on rat brain. *Int J Radiat Biol* 86(4):334–343.

Khalil, A. M., Gagaa, M. H., Alshamali, A. M. (2012). 8-Oxo-7, 8-dihydro-2′-deoxyguanosine as a biomarker of DNA damage by mobile phone radiation. *Hum Exp Toxicol* 31(7):734–740.

Khurana, V. G., Teo, C., Kundi, M., Hardell, L., Carlberg, M. (2009). Cell phones and brain tumors: A review including the long-term epidemiologic data. *Surg Neurol* 72(3):205–215.

Kundi, M. (2009). The controversy about a possible relationship between mobile phone use and cancer. *Environ Health Perspect* 117(3):316–324.

Kundi, M., Hutter, H. P. (2009). Mobile phone base stations-effects on wellbeing and health. *Pathophysiology* 16(2–3):123–135.

Lai, H., Singh, N. P. (1995). Acute low-intensity microwave exposure increases DNA single-strand breaks in rat brain cells. *Bioelectromagnetics* 16(3):207–210.

Lai, H., Singh, N. P. (1996). Single- and double-strand DNA breaks in rat brain cells after acute exposure to radiofrequency electromagnetic radiation. *Int J Radiat Biol* 69(4):513–521.

Leszczynski, D., Xu, Z. (2010). Mobile phone radiation health risk controversy: The reliability and sufficiency of science behind the safety standards. *Health Res Policy Syst* 8(1):2.

Lexchin, J., Bero, L. A., Djulbegovic, B., Clark, O. (2003). Pharmaceutical industry sponsorship and research outcome and quality: Systematic review. *BMJ* 326(7400):1167–1170.

Linet, M. S., Taggart, T., Severson, R. K., et al. (2006). Cellular telephones and non-Hodgkin lymphoma. *Int J Cancer* 119(10):2382–2388.

Litovitz, T. A., Krause, D., Penafiel, M., Elson, E. C., Mullins, J. M. (1993). The role of coherence time in the effect of microwaves on ornithine decarboxylase activity. *Bioelectromagnetics* 14(5):395–403.

Litovitz, T. A., Penafiel, L. M., Farrel, J. M., et al. (1997). Bioeffects induced by exposure to microwaves are mitigated by superposition of ELF noise. *Bioelectromagnetics* 18(6):422–430.

Lönn, S., Ahlbom, A., Hall, P., Feychting, M. (2004). Mobile phone use and the risk of acoustic neuroma. *Epidemiology* 15(6):653–659.

Löscher, W. (2001). Do cocarcinogenic effects of ELF electromagnetic fields require repeated long-term interaction with carcinogens? Characteristics of positive studies using the DMBA breast cancer model in rats. *Bioelectromagnetics* 22(8):603–614.

Luukkonen, J., Hakulinen, P., Maki-Paakkanen, J., Juutilainen, J., Naarala, J. (2009). Enhancement of chemically induced reactive oxygen species production and DNA damage in human SH-SY5Y neuroblastoma cells by 872 MHz radiofrequency radiation. *Mutat Res* 662(1–2):54–58.

Milde-Busch, A., von Kries, R., Thomas, S., et al. (2010). The association between use of electronic media and prevalence of headache in adolescents: Results from a population-based cross-sectional study. *BMC Neurol* 10:12.

Miller AB, Morgan LL, Udasin I, Davis DL, (2018): Cancer epidemiology update, following the 2011 IARC evaluation of radiofrequency electromagnetic fields (Monograph 102), *Environ Res* 167: 673-683, 2018.

Miller, A. B., Sears, M. E., Morgan, L. L., et al. (2019). Risks to health and well-being from radio-frequency radiation emitted by cell phones and other wireless devices. *Front Public Health* 7:223.

Morgan, L. L. (2009). Estimating the risk of brain tumors from cellphone use: Published case-control studies. *Pathophysiology* 16(2–3):137–147.

Navarro, E., Segura, J., Portoles, M., Gómez-Perretta de Mateo, C. (2003). The microwave syndrome: A preliminary study in Spain. *Electromagn Biol Med* 22(2–3):161–169.

NTP. (2018). Toxicology and carcinogenesis studies in B6C3F1/N mice exposed to whole-body radio frequency radiation at a frequency (1,900 MHz) and modulations (GSM and CDMA) used by cell phones. *National Toxicology Program Technical Report Series* (596).

Ozguner, F., Altinbas, A., Ozaydin, M., et al. (2005a). Mobile phone-induced myocardial oxidative stress: Protection by a novel antioxidant agent caffeic acid phenethyl ester. *Toxicol Ind Health* 21(9):223–230.

Ozguner, F., Oktem, F., Ayata, A., Koyu, A., Yilmaz, H. R. (2005b). A novel antioxidant agent caffeic acid phenethyl ester prevents long-term mobile phone exposure-induced renal impairment in rat. Prognostic value of malondialdehyde, N-acetyl-beta-D-glucosaminidase and nitric oxide determination. *Mol Cell Biochem* 277(1–2):73–80.

Ozgur, E., Guler, G., Seyhan, N. (2010). Mobile phone radiation-induced free radical damage in the liver is inhibited by the antioxidants N-acetyl cysteine and epigallocatechin-gallate. *Int J Radiat Biol* 86(11):935–945.

Panagopoulos, D. J., Karabarbounis, A., Margaritis, L. H. (2002). Mechanism for action of electromagnetic fields on cells. *Biochem Biophys Res Commun* 298(1):95–102.

Panagopoulos, D. J., Chavdoula, E. D., Margaritis, L. H. (2010). Bioeffects of mobile telephony radiation in relation to its intensity or distance from the Antenna. *Int J Radiat Biol* 86(5):345–357.

Panagopoulos, D., Johansson, O., Carlo, G. (2013). Evaluation of specific absorption rate as a dosimetric quantity for electromagnetic fields bioeffects. *PLOS ONE* 8(6):e62663.

Panagopoulos, D., Johansson, O., Carlo, G. (2015). Polarization: A key difference between man-made and natural electromagnetic fields, in regard to biological activity. *Sci Rep* 5(1):1–10.

Panagopoulos, D. J. (2019). Comparing DNA damage induced by mobile telephony and other types of man-made electromagnetic fields. *Rev Mutat Res*, 781:53–62.

Panagopoulos, D., Karabarbounis, A., Yakymenko, I., Chrousos, G. P. (2021). Humanmade electromagnetic fields: Ion forcedoscillation and voltagegated ion channel dysfunction, oxidative stress and DNA damage. *Int J Oncol*, 59(5):1–16.

Paulraj, R., Behari, J., Rao, A. R. (1999). Effect of amplitude modulated RF radiation on calcium ion efflux and ODC activity in chronically exposed rat brain. *Indian J Biochem Biophys* 36(5):337–340.

Paulraj, R., Behari, J. (2006). Single strand DNA breaks in rat brain cells exposed to microwave radiation. *Mutat Res* 596(1–2):76–80.

Pedersen, G. F. (1997). Amplitude modulated RF fields stemming from a GSM/DCS-1800 phone. *Wirel Netw* 3(6):489–498.

Peleg, M., Nativ, O., Richter, E. D. (2018). Radio frequency radiation-related cancer: Assessing causation in the occupational/military setting. *Environ Res* 163:123–133.

Puranen, L., Jokela, K. (1996). Radiation hazard assessment of pulsed microwave radars. *J Microw Power Electromagn Energy* 31(3):165–177.

Qi, G., Zuo, X., Zhou, L., et al. (2015). Effects of extremely low-frequency electromagnetic fields (ELF-EMF) exposure on B6C3F1 mice. *Environ Health Prev Med* 20(4):287–293.

Rafnsson, V., Hrafnkelsson, J., Tulinius, H. (2000). Incidence of cancer among commercial airline pilots. *Occup Environ Med* 57(3):175–179.

Read, R., O'Riordan, T. (2017). The precautionary principle under fire. *Environ Sci Policy Sustain Dev* 59(5):4–15.

Repacholi, M. H., Basten, A., Gebski, V., et al. (1997). Lymphomas in E mu-Pim1 transgenic mice exposed to pulsed 900 MHz electromagnetic fields. *Radiat Res* 147(5):631–640.

Robinette, C. D., Silverman, C., Jablon, S. (1980). Effects upon health of occupational exposure to microwave radiation (radar). *Am J Epidemiol*, 112(1):39–53.

Rodrigues, N. C. P., Dode, A. C., de Noronha Andrade, M. K., et al. (2021). The effect of continuous low-intensity exposure to electromagnetic fields from radio base stations to cancer mortality in brazil. *Int J Environ Res Public Health*, 18(3).1229.

Roosli, M., Frei, P., Bolte, J., et al. (2010). Conduct of a personal radiofrequency electromagnetic field measurement study: Proposed study protocol. *Environ Health* 9:23.

Ruediger, H. W. (2009). Genotoxic effects of radiofrequency electromagnetic fields. *Pathophysiology* 16(2–3):89–102.

Sadetzki, S., Chetrit, A., Jarus-Hakak, A., et al. (2008). Cellular phone use and risk of benign and malignant parotid gland tumors–A nationwide case-control study. *Am J Epidemiol* 167(4):457–467.

Santini, R., Santini, P., Danze, J. M., Le Ruz, P., Seigne, M. (2002). Study of the health of people living in the vicinity of mobile phone base stations: 1. Influences of distance and sex. *Pathol Biol (Paris)* 50(6):369–373.

Sato, Y., Akiba, S., Kubo, O., Yamaguchi, N. (2011). A case-case study of mobile phone use and acoustic neuroma risk in Japan. *Bioelectromagnetics* 32(2):85–93.

Savitz, D. A., Calle, E. E. (1987). Leukemia and occupational exposure to electromagnetic fields: Review of epidemiologic surveys. *J Occup Med* 29(1):47–51.

Schwarz, C., Kratochvil, E., Pilger, A., et al. (2008). Radiofrequency electromagnetic fields (UMTS, 1,950 MHz) induce genotoxic effects in vitro in human fibroblasts but not in lymphocytes. *Int Arch Occup Environ Health* 81(6):755–767.

Seomun, G., Lee, J., Park, J. (2021). Exposure to extremely low-frequency magnetic fields and childhood cancer: A systematic review and meta-analysis. *PLOS ONE* 16(5):e0251628.

Sidorik, E., Yakymenko, I. (2013). A brief review on animal research and human health effects following the Chornobyl accident. *Radiat Emerg Med* 2(1):5–13.

Smith-Roe, S. L., Wyde, M. E., Stout, M. D., et al. (2020). Evaluation of the genotoxicity of cell phone radio-frequency radiation in male and female rats and mice following subchronic exposure. *Environ Mol Mutagen* 61(2):276–290.

Sokolovic, D., Djindjic, B., Nikolic, J., et al. (2008). Melatonin reduces oxidative stress induced by chronic exposure of microwave radiation from mobile phones in rat brain. *J Radiat Res (Tokyo)* 49(6):579–586.

Stang, A., Anastassiou, G., Ahrens, W., et al. (2001). The possible role of radiofrequency radiation in the development of uveal melanoma. *Epidemiology* 12(1):7–12.

Szmigielski, S., Szudzinski, A., Pietraszek, A., et al. (1982). Accelerated development of spontaneous and benzopyrene-induced skin cancer in mice exposed to 2450-MHz microwave radiation. *Bioelectromagnetics* 3(2):179–191.

Szmigielski, S. (1985). Polish epidemiological study links RF/MW exposures to cancer. *Microw News* 5(2):1–2.

Szmigielski, S. (1996). Cancer morbidity in subjects occupationally exposed to high frequency (radiofrequency and microwave) electromagnetic radiation. *Sci Total Environ* 180(1):9–17.

Thomas, S., Heinrich, S., Kuhnlein, A., Radon, K. (2010). The association between socioeconomic status and exposure to mobile telecommunication networks in children and adolescents. *Bioelectromagnetics* 31(1):20–27.

Toler, J. C., Shelton, W. W., Frei, M. R., Merritt, J. H., Stedham, M. A. (1997). Long-term, low-level exposure of mice prone to mammary tumors to 435 MHz radiofrequency radiation. *Radiat Res* 148(3):227–234.

Tynes, T., Hannevik, M., Andersen, A., Vistnes, A. I., Haldorsen, T. (1996). Incidence of breast cancer in Norwegian female radio and telegraph operators. *Cancer Causes Control* 7(2):197–204.

Valko, M., Rhodes, C. J., Moncol, J., Izakovic, M., Mazur, M. (2006). Free radicals, metals and antioxidants in oxidative stress-induced cancer. *Chem Biol Interact* 160(1):1–40.

Valko, M., Leibfritz, D., Moncol, J., et al. (2007). Free radicals and antioxidants in normal physiological functions and human disease. *Int J Biochem Cell Biol* 39(1):44–84.

Viel, J.-F., Clerc, S., Barrera, C., et al. (2009). Residential exposure to radiofrequency fields from mobile phone base stations, and broadcast transmitters: A population-based survey with personal meter. *Occup Environ Med* 66(8):550–556.

Wertheimer, N., Leeper, E. (1979). Electrical wiring configurations and childhood cancer. *Am J Epidemiol* 109(3):273–284.

Wertheimer, N., Leeper, E. (1982). Adult cancer related to electrical wires near the home. *Int J Epidemiol* 11(4):345–355.

Wolf, R., Wolf, D. (2007). Increased incidence of cancer near a cell-phone transmitted station. In F. Columbus (Ed.), *Trends in cancer prevention* (pp. 1–8). Nova Science Publishers Inc., New York.

Wu, W., Yao, K., Wang, K. J., et al. (2008). Blocking 1800 MHz mobile phone radiation-induced reactive oxygen species production and DNA damage in lens epithelial cells by noise magnetic fields. *Zhejiang Da Xue Xue Bao Yi Xue Ban*, 37(1):34–38.

Xu, S., Zhou, Z., Zhang, L., et al. (2010). Exposure to 1800 MHz radiofrequency radiation induces oxidative damage to mitochondrial DNA in primary cultured neurons. *Brain Res* 1311:189–196.

Yadav, A. S., Sharma, M. K. (2008). Increased frequency of micronucleated exfoliated cells among humans exposed in vivo to mobile telephone radiations. *Mutat Res* 650(2):175–180.

Yakymenko, I., Sidorik, E. (2010). Risks of carcinogenesis from electromagnetic radiation of mobile telephony devices. *Exp Oncol* 32(2):54–60.

Yakymenko, I., Sidorik, E., Kyrylenko, S., Chekhun, V. (2011). Long-term exposure to microwave radiation provokes cancer growth: Evidences from radars and mobile communication systems. *Exp Oncol* 33(2):62–70.

Yakymenko, I., Tsybulin, O., Sidorik, E., et al. (2016). Oxidative mechanisms of biological activity of low-intensity radiofrequency radiation. *Electromagn Biol Med* 35(2):186–202.

Yakymenko, I., Burlaka, A., Tsybulin, I., et al. (2018). Oxidative and mutagenic effects of low intensity GSM 1800 MHz microwave radiation. *Exp Oncol* 40(4):282–287.

Yokus, B., Cakir, D. U., Akdag, M. Z., Sert, C., Mete, N. (2005). Oxidative DNA damage in rats exposed to extremely low frequency electro magnetic fields. *Free Radic Res* 39(3):317–323.

Yokus, B., Akdag, M. Z., Dasdag, S., Cakir, D. U., Kizil, M. (2008). Extremely low frequency magnetic fields cause oxidative DNA damage in rats. *Int J Radiat Biol* 84(10):789–795.

Zeeb, H., Hammer, G. P., Langner, I., et al. (2010). Cancer mortality among German aircrew: Second follow-up. *Radiat Environ Biophys* 49(2):187–194.

Zmyslony, M., Politanski, P., Rajkowska, E., Szymczak, W., Jajte, J. (2004). Acute exposure to 930 MHz CW electromagnetic radiation in vitro affects reactive oxygen species level in rat lymphocytes treated by iron ions. *Bioelectromagnetics* 25(5):324–328.

Zothansiama, Zosangzuali, M., Lalramdinpuii, M., et al. (2017). Impact of radiofrequency radiation on DNA damage and antioxidants in peripheral blood lymphocytes of humans residing in the vicinity of mobile phone base stations. *Electromagn Biol Med* 36(3):295–305.

Part C

Effects on Wildlife and Environment

10 Effects of Man-made and Especially Wireless Communication Electromagnetic Fields on Wildlife

Alfonso Balmori

Environmental Department of Castilla and León, Valladolid, Spain

CONTENTS

Abstract	394
10.1 Introduction	394
10.2 Effects of Man-made EMFs on Mammals	399
10.2.1 Effects on Physiology, Health, Behavior, and Orientation	400
10.2.2 Effects on Fertility and Reproduction	402
10.2.3 Effects on OS, Genetic Damage, Blood–Brain Barrier, Gene Expression, and Cell Death	403
10.2.4 Carcinogenic Effects	404
10.3 Effects of Man-made EMFs on Birds	404
10.3.1 Effects on Reproduction, Development, Physiology, and Genotoxicity	404
10.3.2 The Decline of Sparrows	407
10.3.3 Effects on Vision, Behavior, Orientation, and Navigation	409
10.4 Effects of Man-made EMFs on Fish	411
10.5 Effects of Man-made EMFs on Amphibians and Reptiles	412
10.6 Effects of Man-made EMFs on Insects	415
10.6.1 The Decline of Insect Populations	415
10.6.2 ELF EMF Studies	415
10.6.3 WC EMF Studies	416
10.6.4 Effects on Orientation, Navigation, and Colony Collapse	418
10.7 Effects of Man-made EMFs on Plants and Trees	419
10.7.1 Effects on Plants	419
10.7.2 Effects on Trees	423
10.8 Discussion	427
10.8.1 General Considerations on the Described Effects	427
10.8.2 Possible Explanations for the Effects on Animals and Plants	429
10.8.3 Where Do We Go from Here?	430
References	431

DOI: 10.1201/9781003201052-14

Keywords: electromagnetic fields; electromagnetic radiation; extremely low frequency; radio frequency; wireless communications; antennas; radars; power lines; birds; insects; fish; mammals; amphibians; reptiles; trees; orientation.

Abbreviations: AChE: acetylcholinesterase. ACTH: adrenocorticotrophin hormone. CAT: catalase. CCD: colony collapse disorder. DECT: Digitally Enhanced Cordless Telecommunications. EEG: electroencephalogram. ELF: Extremely Low Frequency. EMF: electromagnetic field. EMR: electromagnetic radiation. EP: electrical potential. GABA: γ-aminobutyric acid. GPS: global positioning system. GSH: reduced glutathione. GSM: Global System for Mobile Telecommunications. GTEM chamber: gigahertz transverse electromagnetic chamber. IoT: Internet of Things. LPO: lipid peroxidation. LTE: Long Term Evolution. MT: mobile telephony. OS: oxidative stress. RF: Radio Frequency. ROS: reactive oxygen species. SAR: Specific Absorption Rate. SES: seismic electric signals. SOD: superoxide dismutase. TPC: two-pore channel. ULF: Ultra Low Frequency. UMTS: Universal Mobile Telecommunications System. VGCC: voltage-gated calcium channel. VGIC: voltage-gated ion channel. WC: wireless communication. Wi-Fi: Wireless Fidelity. 1G, 2G, 3G, 4G, 5G: first, second, third, fourth, fifth generation of MT/WC.

ABSTRACT

During the past few decades, millions of mobile telephony (MT) base antennas and antennas of other types of wireless communications (WC) have been installed around the world, in cities and in nature, including protected natural areas, in addition to pre-existing antennas (e.g., for television, radio broadcasting, radars, etc.) and high-voltage power lines. Only the aesthetic aspects or urban regulations have been generally considered in this deployment by the responsible authorities, while the biological and environmental impacts of the associated electromagnetic fields (EMFs) and corresponding non-ionizing electromagnetic radiation (EMR) emissions have not been assessed so far. Therefore, the effects on animals (including humans) and plants living around the anthropogenic EMF sources have not been considered. This deficit is particularly concerning because these EMFs/EMR are very different from natural EMFs/EMR, such as light, geomagnetic and geoelectric fields, atmospheric (Schumann) oscillations, or cosmic microwaves, which not only are not dangerous at normal intensities, but, on the contrary, they are vital to the environment and to all forms of life. This chapter reviews the available research on the effects of anthropogenic and especially WC EMFs on wildlife and the natural environment, published mainly during the past 30 years. It includes studies conducted both in the nature and in the laboratory, with vertebrates (mammals, birds, fish, amphibians, and reptiles), invertebrates (mostly insects), plants, and trees. Most of these studies have shown significant detrimental effects of the anthropogenic EMFs on wildlife, at intensities comparable to the current ambient exposure levels, suggesting that we are facing a new environmental pollutant which threatens the health and existence of these species. It is worrying that, despite the accumulating evidence, the people, governments, and even nature conservation organizations are uninformed and unaware of the risks that anthropogenic, and especially WC EMFs pose to the welfare of biodiversity and ultimately to humans.

10.1 INTRODUCTION

During the past five decades, and more intensively since the beginning of this century, many experimental studies and several reviews have been published on the effects of anthropogenic electromagnetic fields (EMFs) and corresponding electromagnetic radiation (EMR) on wildlife, especially on the effects of EMFs/EMR associated with wireless communications (WC), named here WC EMFs or WC EMR. The first review of the WC EMF effects on the living environment, mainly from Russian studies, indicated that non-thermal levels[*] of WC EMFs could cause reversible changes in

---

[*] "Non-thermal" are called the EMF exposures that do not have a high enough intensity (combined with frequency) to cause any significant temperature increase in the exposed living tissues. The vast majority of the studies examined in this chapter refer to non-thermal EMF exposures.

physiological processes in living organisms, animal behavior, and animal spatial orientation. The authors argued that some species (insects, birds, reptiles, etc.) use natural EMFs (such as the geomagnetic and geoelectric fields) for spatial orientation and navigation, particularly during migration. Thus, anthropogenic EMFs, inevitably interfering with natural ones, affect ecosystems, and because wild animal communities are very sensitive to EMFs (both natural and man-made), they could become the live indicators of electromagnetic pollution (Grigoriev et al. 2003).

It should be emphasized that all anthropogenic EMFs are totally polarized and coherent in contrast to natural EMFs (Panagopoulos et al. 2015a). In addition, WC EMFs are pulsed, modulated, highly variable in time, and include a combination of Radio Frequencies (RF)/microwaves* (0.3 MHz–300 GHz), Extremely Low Frequencies (ELF) (3–3000 Hz), and Ultra Low Frequencies (ULF) (0–3 Hz). More specifically, they include RF carrier waves, which are modulated by ELF signals, and they are emitted in the form of on/off pulsations at ELF rates in order to exchange increasing amounts of information from increasing numbers of users. In addition, apart from the regular ELFs (modulation, pulsations), they also include random variability of their final signals, mainly in the ULF band. On the other hand, natural light, atmospheric EMFs (Schumann oscillations), and cosmic microwaves are not polarized, not pulsed, not modulated, retain relatively constant average intensities without sudden variations, and are not dangerous to life but are vital. In those cases that natural EMFs are significantly polarized, such as the geomagnetic and geoelectric fields, they are static (with constant polarity) and not significantly varying. The combination of polarization and intense variability exists only in the man-made EMFs, especially in the WC EMFs, and is considered the reason for their intense adverse bioactivity (see Chapter 1, and Panagopoulos et al. 2015a; 2021; Panagopoulos 2017; 2019). In cases of sudden variations in the geomagnetic, geoelectric, and atmospheric fields during solar storms every few years, health problems in the human population maximize (Dubrov 1978). We have shown that, when certain natural EMFs are polarized to a significant degree [e.g., approximately (~) 70%], such as intense lightning EMFs or seismic electric signals (SES), and transcend certain intensity levels during intense thunderstorms or prior to major earthquakes, they can be sensed by animals and sensitive individuals (Panagopoulos and Balmori 2017; Panagopoulos et al. 2020) and could be used for protection against intense natural phenomena (Varotsos et al. 1993).

WC EMFs/EMR include mobile telephony (MT) EMFs/EMR from mobile phones and corresponding base station antennas (Figure 10.1), and antenna towers with many base antennas (Figure 10.2A and 10.2B), cordless domestic phones (DECT: Digitally Enhanced Cordless Telecommunications), Internet wireless connections (Wi-Fi: Wireless Fidelity), Bluetooth wireless connections among electrical/electronic devices, etc. Wi-Fi signals comprise 2.45 GHz carrier signals and 10 Hz pulsations and are used to connect computers, "smart" mobile phones, and other electronic devices, such as tablets, to the Internet via a nearby device called a wireless router. Bluetooth signals, used to connect electronic devices, comprise very similar characteristics with the Wi-Fi signals (Pedersen 1997; Hyland 2000; 2008; Zhou et al. 2010).

1G, 2G, 3G, 4G, and 5G represent the five generations of MT/WC systems for which "G" stands for generation and the numbers, 1, 2, 3, 4, and 5, represent the generation number. Since the early 1980s, a new generation of MT networks is developed approximately every ten years. 1G were analog phones; 2G is the Global System for Mobile Telecommunications (GSM) with carrier frequencies at 900, 1800, or 1900 MHz, and pulsing frequencies mainly at 2, 8, and 217 Hz; 3G is the Universal Mobile Telecommunications System (UMTS) (also used in 4G), with carrier signals at 1900–2200 MHz, and pulsations at 100 and 1500 Hz; and 4G is the Long Term Evolution (LTE) system (an expansion of UMTS), with various carrier frequencies up to 2600 MHz and, 100 Hz and 1000 Hz main pulsations. 4G provided broadband Internet access, combining usually UMTS for voice and LTE for Internet (see Chapter 1, and Pedersen 1997; Hyland 2000; 2008; Sauter 2011; Shim et al. 2013). 2G–4G are in use today and are digital systems. The 5G system is now under deployment in order to provide even faster data transmission and broadband Internet access and the

---

* Microwaves are the higher part of the RF band with frequencies of 300 MHz–300 GHz.

FIGURE 10.1 Mobile telephony (MT) base station, with three 4G antennas at a 120° angle between each two of them, and a dish antenna for communication among base stations.

so-called Internet of Things (IoT). Some of the technical characteristics of this new 5G technology will be directional beams, thousands of new/additional base antennas everywhere, higher carrier frequencies capable of inducing thermal effects, in addition to the already existing non-thermal, thousands of satellites in the lower atmosphere to complement the land-based antennas, etc. (see Chapter 1, and Singh et al. 2017; Neufeld and Kuster 2018; Agiwal and Jin 2018).

Because all modern WC EMFs contain both RF and ELF/ULF components, in the present review, I shall examine both ELF and RF/microwave studies. A main source of purely ELF continuous-wave (non-pulsed) EMFs are the high-voltage power transmission lines (Figure 10.3), which have long been connected with health problems in humans (specifically with child leukemia) (Wertheimer and Leeper 1979), and, as I shall describe, they are connected with adverse effects on wildlife as well.

FIGURE 10.2A An antenna tower, with two or three MT base stations, several 2G/3G MT/WC antennas, and dish antennas.

Effects of Man-made and Especially WC EMFs on Wildlife 397

FIGURE 10.2B Bigger antenna tower with several MT base stations and many MT/WC antennas, mainly 3G and 4G.

FIGURE 10.3 High-voltage power transmission lines on metallic pylons.

In one of my previous reviews on the impacts of WC EMFs on wildlife, I concluded that anthropogenic EMFs are a form of environmental pollution that could seriously harm wildlife. MT base antennas are continuously exposing/irradiating many species that could suffer long-term consequences, such as a reduction in their natural defenses, deterioration in their health, problems in reproduction, and reduction of their natural territory through habitat deterioration. Further, MT/WC EMFs can induce an aversive behavioral response in many animals, such as rats, bats, and birds, such as sparrows. Therefore, WC EMF pollution constitutes a probable cause for the decline in animal populations and the health deterioration of plants and trees living near MT/WC antennas, and urgent specific measures are necessary to limit these effects (Balmori 2009).

Levitt and Lai (2010) have warned that the anthropogenic EMFs constitute the fastest growing form of environmental pollution. The authors reviewed the existing studies on people living or working near MT/WC and other antennas. They asked for caution regarding the installation of such antennas and explained that the reported symptoms are related with the "microwave sickness", first described in 1978. Moreover, they noted that the described biological effects were induced by WC EMF exposures at levels far below the current exposure limits adopted by governments. It should be noted that "microwave sickness" or "microwave syndrome" was the name given to the reported symptomatology (fatigue, irritability, headache, nausea, appetite loss, insomnia, discomfort, difficulty in concentration, memory loss, etc.) in workers chronically exposed to anthropogenic RF/microwave EMFs at non-thermal levels (Johnson-Liakouris 1998).

The International Commission on Non-Ionizing Radiation Protection (ICNIRP) is a private organization that issues exposure guidelines, which are then adopted by governments, but has been accused of conflicts of interest (Hardell and Carlberg 2020; Hardell et al. 2021). The ICNIRP (2010; 2020) limits are thousands of times above the levels at which effects are recorded for both ELF and RF man-made EMFs and account only for thermal effects, while the vast majority of recorded effects are non-thermal. These existing guidelines for public health protection only consider the effects of acute intense (thermal) exposures and do not protect wildlife from lower level long-term exposures. Most studies addressing the effects on animals and plants have documented effects and responses at exposures well below the ICNIRP limits, and several countries, such as Russia, Italy, Poland, China, and others, have already set much more stringent exposure limits than those proposed by ICNIRP (see e.g., Madjar 2016). It is, therefore, necessary for public authorities and governments to abandon the ICNIRP guidelines and set much stricter ones, based on the sensitive natural biological response, if they want to protect public health, the natural environment, and wildlife.

Cucurachi et al. (2013) presented an extended systematic review of many published studies on the ecological effects of anthropogenic RF EMFs, mostly pulsed or modulated by ELF (thus, WC EMFs) from WC antennas, radars, etc. In about two-thirds of the reviewed studies, a variety of adverse biological effects of WC/RF EMFs were reported at radiation/field intensities that were well below the ICNIRP limits. Low levels of anthropogenic EMFs are currently present anywhere in the natural environment worldwide.

In another previous study (Balmori 2014), I highlighted that, despite the widespread use of WC around the world, governments, and authorities have paid little attention to the harmful effects of WC EMFs on wildlife. I also reviewed the available scientific information on this topic, recommending further studies and making specific research suggestions to confirm or refute specific available experimental results. I argued that more stringent exposure limits should be introduced, since WC technology, offered as "safe" for the environment, represents a significant threat to species.

Halgamuge (2017) reviewed 45 peer-reviewed scientific publications (from 1996 to 2016), describing 169 experimental observations on possible physiological and morphological changes in plants due to WC EMF exposures, mostly from mobile phones, at non-thermal levels. The review demonstrated that a great percentage of these studies (89.9%) showed adverse effects on the plants, suggesting that plants seem to be more responsive to certain frequencies.

Kostoff et al. (2020) reviewed the adverse effects of WC EMFs reported in the biomedical literature. They emphasized that many simulated MT/WC EMF exposures applied in laboratory experiments do not include pulsing, modulation, and variability of the carrier RF signal as happens in real-life WC EMF exposures, and thus, they are unable to identify the more severe adverse effects induced by the real-life WC EMF exposures. That was first identified in a review by Panagopoulos et al. (2015b) in which it was demonstrated that there is a strong discrepancy between experimental studies employing simulated mobile phone exposures with invariable parameters (intensity, frequencies, pulsations) and no modulation or variability of the signals and studies employing real-life mobile phone exposures by commercially available devices. In the first case, about 50% of the studies did not find effects, while in the second case, more than 96% of the studies found effects. The discrepancy was attributed not only to the pulsation and modulation of the real-life WC EMFs but also to the significant role of variability, especially in intensity, which is inherent in all real-life WC EMF exposures (see Chapter 1, and Panagopoulos et al. 2015b; Panagopoulos 2017; 2019).

Moreover, Kostoff et al. (2020) noted that the majority of studies do not account for synergistic adverse effects with other toxic environmental stressors (chemical, biological, etc.), acting jointly with the WC EMFs in real-life conditions. The authors also presented evidence that the 5G MT/WC technology (the newest under deployment WC/MT system) will have adverse systemic effects on the human/animal health and the environment.

In another recent paper, the authors attempted to address the various problems that could arise from the implementation of 5G and the associated IoT designed by the telecommunications industry. They explained that, while technological companies and, unfortunately, the official regulatory bodies advertise the benefits of "smart" buildings and "smart" cities, there is an equal need to understand the demerits behind developing such systems. [In the IoT, electrical/electronic devices equipped with wireless sensors connect and exchange data with other devices and systems over the Internet, and smart buildings are defined as those that utilize artificial intelligence to provide real-time control to users and IoT technology] The authors argued that, although some aspects of "smart" buildings do have tangible benefits, the potential adverse consequences from underdiscussed threats of this technology could undermine the former. According to the same authors, the effects of the increased EMF exposure associated with this "smart" technology include cancer, cardiovascular diseases, DNA and gene alterations, electromagnetic pollution issues relevant to ecology and their negative contribution to the environment. They proposed making the buildings "wired" instead of wireless in order to reduce EMF exposure inside them. They also emphasized that a holistic assessment of all the risks and a comparison with the intended benefits would provide all stakeholders with the real picture, which could be utilized for government policy making or creating worldwide individual awareness. The impacts on human health, environment, and climate change must be regarded as a top priority before the deployment of such a technology on a global scale (Raveendran and Tabet Aoul 2021).

An extended review of the effects of WC EMFs on living beings has been published during the preparation of this chapter (Levitt et al. 2021a; 2021b). The authors conducted a review on several topics, such as how officially accepted exposure standards relate to wildlife, the increasing ambient background levels of anthropogenic EMFs/EMR, the difficulties in assessing ambient exposures, the measured levels, the MT/WC antenna towers (huge installations with many antennas) (Figure 10.2A and 10.2B) in wilderness areas, new technologies, species sensitivity to anthropogenic EMFs, possible mechanisms of animal magnetoreception, other direct (cell membranes, ion channels) and indirect (free radicals, stress proteins, resonance) possible mechanisms of biological significance, the Earth's magnetic field, and a review of studies on effects in animals and plants among other topics. The authors concluded that biological effects can occur at the already existing levels of man-made EMFs in our environment and can affect delicate ecosystems. They suggested that man-made EMFs can be regulated like other pollutants and explained that wildlife loss is often unseen and undocumented until tipping points are reached. Long-term, biologically relevant EMF exposure standards, which currently do not exist, should be set accordingly for wildlife, and environmental laws should be strictly enforced (Levitt et al. 2021a).

This chapter reviews studies on RF/WC and ELF anthropogenic EMF exposures on wildlife and the natural environment (wild animals, trees, plants). As mentioned already, since apart from RF, ELF emissions are also included in WC EMFs, studies on the effects of purely ELF exposures, such as from electric power transmission lines, are equally important with studies on the effects of MT/WC EMFs, and thus, they are also examined in the present review. Finally, because polarized (anthropogenic) EMFs and corresponding EMR are completely interconnected, the terms anthropogenic/man-made EMFs and anthropogenic/man-made EMR are used interchangeably and with the same meaning throughout the chapter.

10.2 EFFECTS OF MAN-MADE EMFs ON MAMMALS

Mammals are probably the most extensively studied animal group in relation to the effects of anthropogenic EMFs, especially rats and mice, which have long been the most common laboratory animals. This is because most of the effects found in them can be carefully extrapolated to the other groups of vertebrates that are less used in experimentation and to humans as well. Several studies have also been carried out on the effects of radiation from MT antennas on cattle and on the alignment of these animals with respect to the geomagnetic field. Studying the effects on livestock is very useful for understanding what happens to wild species that share habitats, although wild mammals, or, e.g., reptiles, in many cases spend more time underground and/or move more, which possibly results in being less exposed to WC EMFs.

10.2.1 Effects on Physiology, Health, Behavior, and Orientation

Animals exposed to WC EMFs can exhibit changes in behavior that disappear when they move away from the antennas (Marks et al. 1995; Loscher and Kas 1998). WC EMFs modify sleep and alter the brain electrical activity (see Chapter 7, and Mann and Roschkle 1996; Kramarenko and Tan 2003; Marino et al. 2003; Mohammed et al. 2013). EMFs from MT base antennas may favor passiveness or aggressiveness in people and animals such as dogs, possibly due to long-term smooth excitability/irritation/inflammation (personal observation).

The pioneering experiments by Delgado (1985) employed ELF EMFs as a non-invasive methodology to study their effects on the central nervous system and in different brain structures. It was found that the excitability of monkeys' central nervous system is influenced by applying EMFs focused on the cerebellum. Frequencies below 50 Hz were found to influence spontaneous behavior in monkeys (Delgado 1985).

In the rat brain, pulsed (with 500 Hz pulse repetition frequency) or continuous-wave 2.45 GHz microwave EMFs have been found to activate endogenous neurotransmitters, such as acetylcholine, involved in many important physiological and behavioral functions. Acetylcholine is synthesized by choline. Sodium-dependent choline uptake was measured in the striatum, frontal cortex, hippocampus,[*] and hypothalamus of rats after acute exposure for 45 minutes (min). The average whole-body Specific Absorption Rate (SAR) was at 0.6 W/kg (corresponding to a power density emitted from a mobile phone at close proximity). Decrease in choline uptake was observed in the frontal cortex after exposure in either exposure conditions (pulsed or continuous-wave). Hippocampal choline uptake was decreased after exposure to pulsed but not to continuous-wave microwaves. Striatal choline uptake was decreased after exposure to either pulsed or continuous-wave. Such effects can have long-term consequences, such as in pain perception, motivational behaviors, and motor functions (Lai et al. 1989).

Exposure to 900 MHz EMR at 1 mW/cm^2 power density for 28 days significantly impaired spatial memory in rats (Tang et al. 2015). Recently, Sultangaliyeva et al. (2020) found that MT EMR affects behavioral reactions of animals, such as the number of horizontal movements, the rate of standing on the hind legs, the number of acts of cleansing (grooming), the number of acts of sniffing, immobility time, and urination.

In an older study, rats avoided pulsed EMR and spent more time in the halves of shuttle boxes that were shielded from the pulsing microwave radiation (Frey and Feld 1975). Bat activity was significantly reduced in habitats exposed to radar EMR (Nicholls and Racey 2007). These results indicate that this radiation could deteriorate the habitats. Navakatikian and Tomashevskaya (1994) described a series of experiments in which they observed disruption of rat behavior (active avoidance of exposure site) induced by microwave EMF exposure at a power density of 0.1 mW/cm^2. In another study, mice exposed to GSM 900 MHz mobile phone radiation expressed a visible individual panic reaction, loss of collective defense reaction, disorientation, and a greater degree of anxiety. In the control (unexposed) group, these behavioral deviations were not observed. The effects in the exposed animals were attributed to disturbance of cognitive functions and disorders in deep subcortical structures (Krstic et al. 2005).

Cortisol is a glucocorticoid steroid hormone, and its release is controlled by the hypothalamic–pituitary–adrenal axis. In response to stress, adrenocorticotrophin hormone (ACTH) from the

---

[*] The hippocampus is a brain structure embedded deep in the temporal lobe.

pituitary gland stimulates the adrenal cortex to secrete cortisol into the bloodstream. Exposure to 915 MHz EMF (whole-body SAR = 4 W/kg) during pregnancy caused a significant elevation in cortisol in the circulating blood and adrenal glands of pregnant rats, which means that it was significantly stressful. However, it seems that the placental barrier protected the fetus from maternal stress, because no significant change in the level of placental blood cortisol was found in the exposed group compared with the sham-exposed (Kim et al. 2021). The levels of γ-aminobutyric acid (GABA) and aspartic acid (amino acid neurotransmitters) in the cortex and hippocampus significantly decreased in adolescent male mice after intense 1.8 GHz simulated MT EMF exposure (for 28 days, 6 hours (h)/day) by a generator at power density 530 μW/cm^2, and these changes were probably involved in increased anxiety (Zhang et al. 2017). It is noteworthy that, even though the power density in this study was relatively high compared to many other experiments, the corresponding ICNIRP (2020) limit (averaged over 30 min and over the whole body) for 1.8 GHz is 900 μW/cm^2 (0.9 mW/cm^2). Moreover, the effects could be even stronger had the authors used a real mobile phone for exposure (Panagopoulos 2019).

A daily 30 min exposure to a pulsing microwave EMF (1.5 GHz, 0.3 mW/cm^2, with 0.12 Hz pulse repetition frequency) for 1 month, produced anxiety and alarm in rabbits (Grigoriev et al. 1995). In another study, male rats were exposed to 2.5 GHz pulsing WC exposure from an installed Wi-Fi device/router (10 Hz pulsations) for a period of 4, 6, and 8 weeks. The Wi-Fi exposure caused a significant increase in anxiety and affected the locomotor function. The authors concluded that long-term exposure to Wi-Fi may lead to neuro-degenerative diseases, as they observed a significant alteration on acetylcholinesterase (AChE) gene expression,[*] and some other neuro-behavioral parameters associated with brain damage (Obajuluwa et al. 2017).

Mobile phone EMF exposure during pregnancy affects the future adult behavior of the exposed embryos. Adult mice which had been exposed during embryonic development were hyperactive and had impaired memory. These behavioral changes were due to altered neuronal developmental, with evidence of neuropathology (Aldad et al. 2012).

Human exposure to mobile phone EMFs prenatally – and, to a lesser degree, postnatally – has been associated with behavioral difficulties in children, such as emotional and hyperactivity problems around the age of school entry. These findings should be of extreme public health concern given the widespread use of this technology, and the fact that many mothers, uneducated on the risks, expose their embryos by using mobile phones during pregnancy (Divan et al. 2008).

Prenatal and early postnatal exposure of rats to 900 MHz WC EMFs, emitted from a MT base station antenna simulator (placed 1 m away from the cages of the exposed subgroups), with an average power density of 0.68 mW/cm^2 (680 μW/cm^2), adversely affected learning and memory in rat pups, with a greater impact when the exposure was during the prenatal period (Azimzadeh and Jelodar 2020a). Children exposed to 154–162 MHz EMR, pulsed at 24.4 Hz, within 20 km radius from a radar station at various power densities (peak power density at 3.7 km distance was 16.4 μW/cm^2) had less developed memory and attention, their reaction time was slower, and the functional state of their neuro-muscular system, assessed by a specifically designed test, was decreased (Kolodynski and Kolodynska 1996).

Animals exposed to man-made EMFs could exhibit health deterioration, such as reproductive and developmental problems, behavioral abnormalities, and vision problems (cataracts) (Marks et al. 1995; Loscher and Kas 1998; Hässig et al. 2012). Man-made EMFs, especially WC EMFs, have been found to weaken animal immune systems (Chou et al. 1992; Novoselova and Fesenko 1998; Galeev 2000) and increase their susceptibility to infectious diseases (Fernie and Bird 2001). Exposure to 900 MHz EMF with an average power density of 0.679 mW/cm^2, disrupted trace element (Fe, Cu, and Zn) homoeostasis in the brain (Azimzadeh and Jelodar 2020b). The reduced body weight found in pregnant rats exposed continuously to 27.12 MHz at 0.1 mW/cm^2 power density during different periods of pregnancy reflects a negative impact on their health, with the exposure duration playing an important

---

[*] Acetylcholinesterase is an enzyme involved in parasympathetic neurotransmission.

role (Tofani et al. 1986). Mice exposed to mobile phone EMF for 2 months gained less weight compared with control mice (Krstic et al. 2005). Another study that investigated the effect of 900 MHz EMF exposure of rat dams during the prenatal period, found that the exposure affected the development and caused pathological changes on the vertebrae of rat pups (Keles 2020). Detrimental effects of WC EMFs on the cornea of rats have also been shown, just like in birds (see Section 10.3) (Balci et al. 2007; Akar et al. 2013; Suetov and Alekperov 2019).

Anthropogenic WC and power-line EMFs affect orientation in birds, mammals, and insects (Vácha et al. 2009; Engels et al. 2014; Balmori 2015; Malkemper et al. 2015), as they inevitably interfere with the natural terrestrial EMFs that allow orientation. Natural and anthropogenic EMFs interfere with orientation mechanisms in cetaceans (aquatic mammals). Perturbations in the Earth's natural EMFs during solar storms are closely related to strandings of grey whales (Granger et al. 2020). Bartos et al. (2019) reported that static magnetic field or weak broadband RF noise (co-existence of ELF was not excluded in the report) could impact the circadian system of cockroaches (*Blattella germanica*).

Alignment of ruminants appeared to be disturbed by high-voltage power lines. Cattle and roe deer resting and grazing in open pastures display very consistent N-S alignment. By contrast, cattle grazing under or in the vicinity (< 150 m) of high-voltage overhead power lines were randomly aligned and this effect of the ELF EMFs on body alignment diminished with increasing distance from the power lines (Burda et al. 2009).

A recent study shows that alignment with the geomagnetic field enhances homing efficiency of hunting dogs. In most cases, dogs start their return with a short (about 20 m long) run (a compass run), mostly performed along the north–south axis irrespectively of the homeward direction (Benediktová et al. 2020).

10.2.2 Effects on Fertility and Reproduction

Magras and Xenos (1997) reported a progressive drop in the number of births in mice exposed continuously to EMFs from a radio-television antenna park (measured power densities 0.168–1.053 μW/cm^2) in Thessaloniki, Greece. The progressive decrease in the number of newborns with increasing time of continuous exposure resulted in permanent sterility of the parental animals. In a review of Russian studies on the effects from various RF antennas at non-thermal intensities, Nikolaevich et al. (2001) reported a decrease in fertility, an increase in deaths after birth in rats, and dystrophic abnormalities in their reproductive organs. Exposure of humans to MT EMFs has been repeatedly reported to have negative effects on their reproductive capacity. Many studies have revealed a decreased number and motility of spermatozoa, induction of oxidative stress (OS), and DNA damage in sperm cells after *in vivo* or *in vitro* mobile phone exposure (see Chapter 6, and Dasdag et al. 1999; Davoudi et al. 2002; Fejes et al. 2005; Wdowiak et al. 2007; Agarwal et al. 2008; De Iuliis et al. 2009).

Pandey et al. (2018) investigated the effect of GSM 900 MHz (2G) MT EMF on mouse germ cell development during spermatogenesis. They found extensive DNA damage in the germ cells of the exposed animals, low sperm count, and sperm head abnormalities. Further biochemical analysis revealed excessive free radical/OS generation, resulting in histological changes in testes and morphological changes in germ cells (Pandey et al. 2018). Long-term exposure of adult male rats to 4G (2 GHz) MT EMF induced cell damage in the testes and corresponding decrease in reproductive capacity (Yu et al. 2020). MT EMFs emitted from GSM and UMTS mobile phones (900 MHz, 1800 MHz, and 2100 MHz) 2 h/day for 6 months, induced cell damage in rat testicular tissue (Alkis et al. 2019b). These findings on mouse and rat spermatogenesis are in complete agreement with the aforementioned damaging effects on human sperm. In a recent review, Gang et al. (2021) indicated that the studies have already extensively investigated and demonstrated the adverse effects of MT/WC EMF/EMR exposure on human/animal spermatogenesis.

Pregnant rats exposed to RF EMF (27.12 MHz, 0.1 mW/cm^2) during different periods of pregnancy, miscarried before the twentieth day of gestation in about 50% of the pregnancies, compared with only a 6% corresponding percentage in the unexposed control pregnant rats. Moreover, 38% of the viable fetuses from the exposed rats had incomplete cranial ossification, compared with less than 6% in the controls (Tofani et al. 1986). Liver damage was induced in rat embryos by combined mobile phone and Wi-Fi radiation exposure from a "smart phone" (adjusted to "talk" and Wi-Fi mode) during gestation. The embryos (and their mothers) were exposed for 6, 12, or 24 h/day for 20 consecutive days during gestation. The prenatal WC EMF exposure lead to OS-mediated hepatotoxicity at levels increasing with increasing daily exposure duration and the embryonic liver damage persisted into puberty in the exposed rats (Tumkaya et al. 2019). The ovarian reserve of the female offspring was diminished with WC EMF exposure from a "smart phone" during pregnancy (Calis et al. 2019). In humans, ELF EMF exposure from power lines, at various electric and magnetic field intensities, depending on the distance, was positively associated with miscarriage risk (Lee et al. 2002).

Moorhouse and Macdonald (2005) found a substantial decline in the number of female water vole population that had been radio-collared (both males and females) for monitoring purposes. The radio-collars were of a type commonly used for small mammals in the United Kingdom (UK) for the study and monitoring of wild animal populations, comprising a small transmitter with battery. Recruits to the radio-tracked population were skewed heavily in favor of males (43:13). This may possibly be interpreted as the female embryos being more vulnerable to the transmitter EMF than the male embryos. Moreover, this effect, along with the reported corneal damage in birds equipped with radio transmitters (see Section 10.3), shows that attaching microwave EMF transmitters to animals in order to study them is actually a bad methodology, as it exposes the animals to genotoxic radiation.

In an older study, continuous exposure of female rats during pregnancy since the time of conception to 27.12 MHz, 100 $\mu W/cm^2$ EMF, resulted in more males born by the exposed dams. Moreover, the exposure resulted in reduced body weight of the dams with the exposure duration playing an important role (Tofani et al. 1986). The results of this study are in agreement with the inadvertent finding of the previous study, possibly suggesting that the female embryos might be more vulnerable to the EMF exposure than the male embryos. This is a significant issue that needs further investigation. Yet such an effect has not been reported by other studies on reproductive effects of man-made EMFs.

10.2.3 Effects on OS, Genetic Damage, Blood–Brain Barrier, Gene Expression, and Cell Death

EMFs emitted by mobile phones induced OS and DNA damage in the frontal lobe of rat brains. Furthermore, 2100 MHz EMF induced DNA single-strand breaks (Alkis et al. 2019a). Ozgur et al. (2013) investigated the effects of prenatal and/or postnatal exposure of rabbits to 1800 MHz GSM MT EMF. Whole-body exposure induced OS and changes in blood chemistry parameters (cholesterol, creatinine, urea, and uric acid).

In another study, the authors examined the impact of a conventional Wi-Fi EMF exposure (2.45 GHz carrier frequency, 10 Hz pulsations) on the liver of female Wistar rats. The animals were exposed for 24 h/day for 40 successive days at an average whole-body SAR 0.01 W/kg. The Wi-Fi exposure induced severe OS in the rat liver, with consequent molecular and functional adverse effects. The severe histological and ultrastructural alterations in the hepatic tissues point to toxic effects induced by the Wi-Fi exposure (Fahmy and Mohammed 2021). Non-thermal Wi-Fi exposure at power levels 3 or 12 W in a gigahertz transverse electromagnetic (GTEM) chamber for 30 min induced hyperplasia of parafollicular cells in the rat thyroid gland (López-Martín et al. 2020). In another study, a simulated GSM signal induced OS and DNA damage in the liver of rats (Alkis et al. 2021). Chronic exposure to mobile phone EMF induced significant increase in hippocampal OS and elevated level of circulatory pro-inflammatory cytokines (Singh et al. 2020). Long-term exposure to GSM or UMTS mobile phone EMF (1800 or 2100 MHz, respectively) caused OS and DNA damage in rat testicular tissue and generated DNA single-strand breaks (Alkis et al. 2019b). EMF exposure to an indoor Wi-Fi device (router), at 10 cm distance, for 4, 6, or 8 weeks, altered hematological profiles and biochemical parameters and induced OS in male albino rats at degrees depending on exposure duration (Bamikole et al. 2019).

Mazor et al. (2008) reported that 72 h *in vitro* exposure of human lymphocyte samples to RF EMF in a waveguide resonator (800 MHz, continuous-wave), induced a non-thermal increase in aneuploidy (cells with abnormal number of chromosomes). Human peripheral blood lymphocytes exposed *in vitro* to UMTS mobile phone EMF (~ 1950 MHz with 100 Hz and 1500 Hz pulsations) for 15 min at 1 cm distance from the handset (29 ± 14 µW/cm^2) during the G$_2$ phase of their cell cycle, exhibited chromatid gaps and breaks at a higher degree than by a very high caffeine dose ~ 290 times above the permissible caffeine dose limit. The power density of the exposure was ~ 136 times below the ICNIRP (2020) limit. The combination of UMTS exposure and caffeine dramatically increased the number of chromatid aberrations in all subjects, and the combined effect increased with increasing exposure duration (5, 15, and 25 min) (Panagopoulos 2020).

MT EMF exposure at levels of mobile phone use has been found to affect the permeability of the blood–brain barrier and, consequently, allow toxic substances to pass from the blood into the brain of rats (Salford et al. 2003; Eberhardt et al. 2008; Tang et al. 2015; Sırav and Seyhan 2016). The results of these studies suggest that MT radiation could cause disruption and increased permeability of the blood–brain barrier under non-thermal exposure levels.

Male Wistar rats exposed to mobile phone EMFs 900, 1800, and 2450 MHz, at SAR 5.84 × 10^{-4} W/kg, 5.94 × 10^{-4} W/kg, and 6.4 × 10^{-4} W/kg, respectively, for 2 h per day for 1-month, 3-month, and 6-month periods, led to significant epigenetic effects that altered gene expression in the hippocampus. The effects increased with increasing exposure duration (Kumar et al. 2021).

There are enzymes called redox proteins that link the two complementary parts (reduction and oxidation) of a redox reaction through an electron transport chain. Biological processes, such as respiration, depend on these enzymes (Armstrong et al. 1988). Cattle exposed to MT emissions (900 MHz) from nearby MT base station antennas displayed changes in their redox proteins and enzyme activities (Hässig et al. 2014). Exposure of rats to 2.6 GHz EMF (21.74 V/m in the RF band) for 30 days (30 min/day, 5 days/week), affected their brain biochemistry and histology and induced structural deformation and apoptosis (Delen et al. 2021).

10.2.4 Carcinogenic Effects

The US National Toxicology Program (NTP 2018) tested the two main modulation types used for MT communication worldwide, CDMA (Code Division Multiple Access) employed in UMTS (3G/4G) and TDMA (Time Division Multiple Access) employed in GSM (2G) in a 2-year rodent cancer bioassay under near-field exposure conditions. The experiments included additional assay for genotoxicity endpoints (Smith-Roe et al. 2020). They found clear evidence of carcinogenic activity and, more specifically, malignant schwannoma of the heart, malignant glioma of the brain, and benign, malignant, or complex (combined) pheochromocytoma of the adrenal medulla. They also found increased DNA damage (measured by the comet assay) in the frontal cortex of male mice, in the leucocytes of female mice, and in the hippocampus of male rats, indicating that MT EMFs could cause DNA damage and consequent carcinogenesis. In a similar large carcinogenicity study, Falcioni et al. (2018) examined far field exposure to GSM 1800 MHz EMF and reported very similar results with the NTP study. Specifically, they found increased incidence of tumors of the brain and heart in the MT EMF-exposed Sprague Dawley rats. Further, these tumors are of the same histotype as those observed in some epidemiological studies on mobile phone users (Hardell et al. 2007).

10.3 EFFECTS OF MAN-MADE EMFs ON BIRDS

10.3.1 Effects on Reproduction, Development, Physiology, and Genotoxicity

The effects of anthropogenic EMFs on birds have been studied for many years, especially in the laboratory but also in the nature. In nature, the EMF sources are in fixed locations, and the researchers approximate the EMF/EMR intensity/power density by the distance from the sources and necessary

Effects of Man-made and Especially WC EMFs on Wildlife 405

FIGURE 10.4A White storks (*Ciconia ciconia*) breeding away from MT antennas and power lines with two chicks.

field/radiation measurements. Each species has its preferences for the space, height, and location of nests and feeding, among other factors, and these preferences may result in different exposure levels, even among species that live in the same area. Man-made EMFs can negatively affect bird populations in places with high electromagnetic pollution. The most vulnerable species are likely to be those that have the habit of feeding, singing, sleeping, or installing nests in high and unprotected places, such as roofs, antennas, or electric current cables/wires.

Researchers have conducted several open-air studies in cities and in nature to investigate the effects of EMF exposure from MT base antennas on birds. In Spain, we investigated effects on the reproduction of the white stork (*Ciconia ciconia*) (Figure 10.4A). The total egg productivity in nests located within 200 m from MT antennas (RF electric field strength ≥ ~2 V/m) was practically half that of those located more than 300 m from the antennas (RF electric field strength < 1 V/m) (Figure 10.5). All nests were located on buildings, and in both cases, we excluded nests situated upon power-line pylons (Figure 10.4B). Furthermore, ~ 40% of the nests located within 200 m from the

FIGURE 10.4B White storks (*Ciconia ciconia*) breeding on power lines with no chicks.

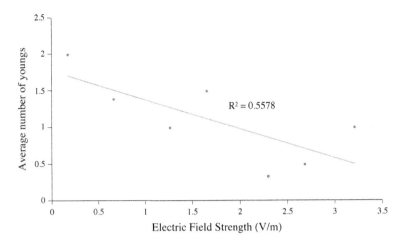

FIGURE 10.5 Inverse relationship between electric field strength (in the RF band) and productivity in white storks. R^2 is the coefficient of determination showing significant correlation (data from Balmori 2005).

antennas never had chicks, while only one nest (3.3%) of those located further than 300 m had no chicks. In nesting sites located within 100 m of one or several MT base antennas, with the main beam of radiation directly impacting the nests, many chicks died from unknown causes. This study was the first to show that EMFs from MT base antennas impact white stork reproduction (Balmori 2005).

In India, researchers observed decreases in the populations of different bird species near MT antenna towers. Most birds were found in the areas with the lowest radiation levels. Avian nests were not detected near MT base station antennas but were found at distances $\geq$ 80 m away from them. The data based on various distances from two MT antenna towers and the four directions of the space that they studied clearly indicate that the occurrence of birds was inversely correlated with the power density (Bhattacharya and Roy 2014). In another review, the same authors discussed the impact of WC EMFs on birds and concluded that, although there is still some uncertainty due to the scarcity of studies on this matter, most of the existing peer-review published literature indicates changes in behavior, adverse effects on physiology and breeding success, and increased mortality (Bhattacharya and Roy 2013).

Other researchers have reported that the total number of birds counted within a 200 m distance from an MT base station was significantly smaller ($n = 94$) than the corresponding number outside the 200 m radius ($n = 171$) in a similar area (Bhat and Singh 2019). In another study in India, the occurrence of birds in exposed and non-exposed zones was 28.08% and 71.92%, respectively (Bose et al. 2020). The findings of these studies are in agreement with my studies (Balmori 2005; 2007).

ELF EMFs from power lines (50–60 Hz) have been found to negatively affect bird reproduction (Doherty and Grubb 1996; Fernie and Reynolds 2005). A study on the diversity and breeding of birds living near power transmission lines and MT antenna towers in Ludhiana, Punjab (India), found low bird numbers and low breeding success near MT antennas and power transmission lines compared with a corresponding unexposed control site (Kler et al. 2018). White storks (*C. ciconia*) breeding directly on pylons of operating power lines had significantly lower breeding success compared with storks breeding on similar pylons of non-operating lines (Vaitkuvienė and Dagys 2014) (see also Figures 10.4A and 10.4B).

In the laboratory, several experimental studies have found significantly increased embryonic death in bird embryos (fertilized eggs) exposed to radiation from mobile phones, accompanied in several cases by OS and genetic damage, which is obviously the cause (Farrel et al. 1997; Youbicier-Simo et al. 1999; Grigoriev 2003; Batellier et al. 2008; Burlaka et al. 2013; Yakymenko et al. 2018). This established effect of the WC/MT EMF exposure also concerns the wild birds living in areas with increased levels of WC EMFs. Moreover, WC EMFs produce non-thermal responses in several

Effects of Man-made and Especially WC EMFs on Wildlife

types of neurons of the nervous system in birds. GSM exposure (900 MHz carrier, pulsed at 217 Hz) induced changes in the activity of more than half of the (*in vitro*) exposed brain neurons, which increased their firing rates by ~ 3.5 times compared to unexposed neurons (Beasond and Semm 2002). As already reported, WC EMFs have also been found to disrupt the blood–brain barrier and damage neurons in the mammalian brain (Salford et al. 2003). It has been documented in a variety of animals and cell types that WC EMF exposure damages biomolecules, particularly DNA (Lai and Singh 1995; 1997; Diem et al. 2005; Panagopoulos et al. 2007a; 2010; Panagopoulos 2012; 2019; Yakymenko et al. 2018) and adversely affects the immune system (Galeev 2000), the reproductive capacity (Magras and Xenos 1997; Fernie et al. 2000; Davoudi et al. 2002; Panagopoulos et al. 2004; 2007b; Fejes et al. 2005), the brain and nervous system (Kramarenko and Tan 2003; Marino et al. 2003; Salford et al. 2003), and the embryonic development, leading to miscarriages (Berman et al. 1990). The outcomes of these effects in birds are neural damage, locomotor defects, and infertility/decreased reproduction (Surendran et al. 2020). Researchers have examined the negative outcomes of WC EMF exposures, especially on humans/mammals, birds, and the environment, and have indicated the urgent need for remediation, emphasizing that it is crucially important that public authorities adopt stricter exposure standards (Surendran et al. 2020).

10.3.2 The Decline of Sparrows

The house sparrow (*Passer domesticus*) lives in urban and suburban habitats and has spread from its original areas in Eurasia to a large number of cities around the world. During the past few decades, there has been a decline in the populations of house sparrows in several European cities (De Laet and Summers-Smith 2007; De Coster et al. 2015; Mohring et al. 2020). Indeed, this decline has been a global phenomenon (Mohring et al. 2020). In India, the number of house sparrows has decreased dramatically in several parts of the country (Singh et al. 2013; Shende and Patil 2015). Despite the many studies published already on this phenomenon, researchers have not yet come to solid conclusions regarding its etiology, although they have proposed several hypotheses to explain this worldwide rapid population decline in urban areas.

Among the most accepted hypotheses are: a) lack of food, particularly insects; b) cleaner streets providing reduced foraging opportunities; c) competition for food from other urban species; d) pollution (air quality); e) increased use of pesticides in parks and gardens; f) increased predation by domestic cats and sparrowhawks (*Accipiter nisus*); g) loss of nesting sites, particularly on the roofs of houses; h) disease transmission; and i) reduction in colony size below some critical value, resulting in the disappearance of the colony as a breeding unit (Crick et al. 2002; Summers-Smith 2003). Moreover, authors have argued that some of the factors that caused the decline in sparrow populations are still present (Crick et al. 2002). A recent study in London revealed that avian malaria (*Plasmodium relictum*) infection is more prevalent in sparrows and could be a factor contributing to the decline in their populations (Dadam et al. 2019). In North America, house sparrow population declines coincide with an increase in populations and expansion of the winter distributions of sparrowhawks. However, the authors do not find a direct connection between the presence of sparrowhawks at count sites and house sparrow population declines (Berigan et al. 2020).

The most complete study to examine this enigma has been recently carried out in Paris. The authors concluded that the decline in house sparrows could be due to several interactive and cumulative effects. This eventuality is worrying because this species, living in cities and peri-urban areas, acts as an urban sentinel, an important bioindicator of the health status of the urban ecosystems (Mohring et al. 2020). Interestingly, while house sparrow populations decline in the cities, another species appeared: Sparrowhawks. But sparrowhawks' (*A. nisus*) first appearance in Paris was in 2008, when house sparrows were already declining. Thus, the appearance of sparrowhawks alone cannot explain the disappearance of the house sparrows. On the other hand, house sparrows do not actually lack nesting sites in the urban areas of Paris. Neither weather fluctuations, nor air pollution are responsible for the current observed decline of house sparrow populations in this city.

Further, house sparrows have declined at all sites, independently of habitat characteristics. The authors noted that the potential influence of other factors has not been adequately assessed: the increasing electromagnetic pollution from man-made EMFs, pollution by artificial light and noise, diseases, and parasites (Mohring et al. 2020).

I have recently argued that the increased levels of man-made EMF exposures and particularly WC EMFs constitute a probable cause for the decline in sparrow populations that should be seriously considered in synergy with the other factors previously proposed (Balmori 2021b).

As it will be discussed next, previous studies indicate that sparrows disappear from areas most contaminated by anthropogenic EMFs, and the house sparrow has been significantly, negatively associated with increasing levels of such EMFs. In addition, many researchers have revealed the negative impact of such EMFs on other species of birds and other animals at field strengths that can be found in the cities. For this reason, the increased levels of anthropogenic EMFs around the world, especially in cities, could certainly be the cause of the sparrow disappearance in combination with other factors.

Everaert and Bauwens (2007) searched for a possible effect of long-term exposure to ambient levels of 2G (GSM) MT radiation from base station antennas on the number of house sparrows during the breeding season in Flanders, Belgium. They sampled 150 locations in six areas to examine small-scale geographic variation in the number of house sparrow males as well as the intensity of EMFs from the MT base stations. At each location, a count of 5 minutes was made of the number of house sparrow males that were singing or were visible within a distance of ~ 30 m. Sightings of birds were done with binoculars. Counts were performed during the morning hours (07:00–11:00), when male house sparrows are most active, on days with favorable weather conditions. The number of house sparrow males at the different locations was inversely and significantly correlated with the power density levels in each of the 900 or 1800 MHz GSM bands and with the total power density of both bands. The inverse correlation was very similar within each of the six study areas, despite differences among the areas in the number of birds and radiation levels. Thus, the authors showed that the number of sparrows correlates (inversely) with WC EMF ambient levels (Everaert and Bauwens 2007).

In another study on sparrows (Balmori and Hallberg 2007), we counted the number of sparrows at 30 places, visited every month for more than 3 years in Valladolid (Spain). The counting was performed between 7:00 and 10:00 by the same ornithologist, following the same protocol. Each counting took place on Sunday because there was less traffic and noise that day. Repeated countings took place between October 2002 and May 2006 in the same selected areas: squares, urban parks, and relatively isolated, tree-lined streets that facilitated the counting process. In each area, we counted all sparrows that were heard or seen, without differentiating the birds by sex or age. In addition, we measured the mean electric field strength (in the band 1 MHz–3 GHz) in V/m (corresponding to power density), with a portable broadband RF meter using a unidirectional antenna. We found a general population decline over time, and a significantly lower bird density in areas with higher radiation levels (Figure 10.6).

Similar studies in India also found that sparrows are disappearing from areas where MT antenna towers have been installed and from cities where electromagnetic pollution is highest. By monthly monitoring in urban and rural areas, researchers found that house sparrow populations are declining in urban areas where MT base antennas are more prevalent compared with rural sites (Shende and Patil 2015). In a 2-year study, the authors selected rural areas where nesting sites, food, roosting sites, and water were plentifully available, and the competition for nesting sites, food, and risk of predation was decreased. In such locations the population should increase, but instead, the population actually decreased. They argued that EMFs from MT base antennas can be the cause because the maximum decrease in the number of nests was found where the number of MT antennas was greater (Singh et al. 2013).

The lack of invertebrate prey used to feed chicks in the nest during the reproductive period has also been suggested as a possible explanation for the decline of house sparrow populations in urban

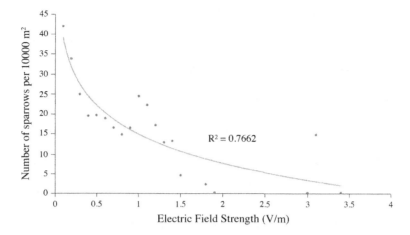

FIGURE 10.6 Mean number of sparrows per 10000 m² inversely correlated with the electric field strength (in the RF band) at different locations (data from Balmori and Hallberg 2007). R^2 is the coefficient of determination showing significant correlation.

centers (Summers-Smith 2003). The availability of key insect prey such as members of aphidoidea, curculionidae, orthoptera, and lepidoptera is crucial for the growth and the development of nestlings (Crick et al. 2002). Numerous studies have shown that electromagnetic pollution, especially from WC/MT EMFs, decreases the number of insects by decreasing their reproduction (Panagopoulos et al. 2004; 2007a; 2007b; Sharma and Kumar 2010; Cucurachi et al. 2013; Odemer and Odemer 2019; Balmori 2021a). Thus, EMF pollution decreases a most important factor for chick sparrow survival, which is their food.

Anthropogenic EMFs is a most plausible factor for the decline of house sparrows because it is the only factor that affects all the other hypotheses that have been proposed so far. Man-made EMFs affect productivity (Doherty and Grubb 1996; Fernie et al. 2000; Balmori 2005) and fertility (Fernie et al. 2000), decrease the number of insects by which chicks feed (Lázaro et al. 2016), cause loss of habitats (bird presence is significantly reduced in habitats exposed to EMFs) (Grigoriev et al. 2003; Balmori 2014), and decrease immunity (Chou et al. 1992; Novoselova and Fesenko 1998; Galeev 2000). A stressed immune system could increase the susceptibility of birds to infectious diseases, bacteria, viruses, and parasites (Fernie and Bird 2001). Moreover, many authors have shown the impact of man-made EMFs on other species of birds and other animals (Balmori 2009; 2014; Levitt and Lai 2010; Kostoff et al. 2020; Levitt et al. 2021a). The results of these studies support the hypothesis that electromagnetic pollution could be responsible, by itself or in conjunction with other factors, for the recent reduction in sparrow populations in the cities. Furthermore, the disappearance of sparrows and the introduction of MT antennas are temporally correlated: The decline of sparrows matches chronologically the deployment of MT/WC networks, especially during the past two to three decades. For these reasons, man-made and, especially WC EMFs, constitute not simply a possible, but a most probable factor for the decline of house sparrows that must be seriously considered in synergy with the other factors proposed so far.

10.3.3 Effects on Vision, Behavior, Orientation, and Navigation

Fritz et al. (2020) described a reintroduction program for northern bald ibises (*Geronticus eremita*), an endangered species, in Europe. The authors of this study documented that birds equipped with solar radio transmitters (RF/GSM transmitters powered by solar light) fixed on their upper-back position, developed unilateral corneal opacity in their eyes after they were equipped with the devices. They did not find any cases in which an affected bird recovered while the transmitter remained; rather, the opacity further advanced and became irreversible. The authors also mentioned a fast recovery of the affected birds after removal or repositioning of the device and verified the unilateral corneal opacity in another similar reintroduction program with the same species carried out in Spain. They also provided information from a Japanese study (Nagata and Yamagish 2013) that was stopped because of the higher mortality in birds equipped with solar-powered GPS (global positioning system) devices on their backs. The northern bald ibis's roosting habit, with the head on the back, suggests a plausible causal relationship between the placement of the GPS device at the upper-back position and unilateral corneal opacity because this roosting position brings the one eye closer to the device than the other for a long time. According to Fritz et al. (2020), a cumulative effect on the nearby eye during roosting phases seems to be the most probable cause for the disease. The authors proposed several possible effects caused by the device to explain the problem and concluded that the most probable explanation for the symptomatology is a repetitive slight temperature rise in the corneal tissue due to EMF exposure by the GSM module of the device. However, my proposal (Balmori 2022) is that these effects do not necessarily have to be thermal, but they can be due to cell death and tissue damage caused by the GSM radiation as shown in Panagopoulos et al. (2007a; 2010), Chavdoula et al. (2010), and several other studies.

There is a long-lived scientific debate regarding the non-thermal effects of man-made EMFs in the eyes (Frey 1985). Exposure to man-made EMFs at levels frequently encountered in daily life has been found to cause damage in the cornea of rats (Akar et al. 2013). The results of another study suggest that mobile phone radiation leads to OS in the corneal and lens tissues (Balci et al. 2007). In rabbits, microwave radiation at power density levels similar to those emitted by a mobile phone, led to adverse dose-dependent changes in the structure of the exposed eyes, including formation of cataract, partial or complete de-epithelialization, stromal edema, endothelial damage and inflammatory infiltration in the cornea, and effusion of proteins in the aqueous humor (Suetov et al. 2019). In Switzerland, a large number of calves were born with eye-cataracts after a MT base station was erected in the vicinity of the barn in a farm. Cataract problems started to appear with a latency of around 12 months after the installation of the base station and disappeared 12 months after its dismantling. Calves born in this farm showed a 3.5-fold higher risk for heavy cataract compared with the Swiss average (Hässig et al. 2012).

Most radio tracking systems involve transmitters tuned to different frequencies (just like different radio stations) that allow individual identification (Mech and Barber 2002). In a previous review of studies that have shown effects on radio-marked animals, I argued that long-term exposure of animals to man-made EMFs from GPS transmitters at non-thermal levels could explain tissue damage and mortality found in such animals (Balmori 2016). I also called for more systematic documentation and publications on the failures and any impairing effects on animals caused by such radio transmitters in order to improve the methods of animal monitoring. Thus, in the analysis of environmental pollutants, the effects of man-made EMFs must be taken into serious consideration. Serious effects may be caused by long-term exposure to relatively low levels of WC EMFs, as found with animals (Hässig et al. 2012), plants (Waldmann-Selsam et al. 2016), and humans (Gómez-Perretta et al. 2013; Rodrigues et al. 2021; López et al. 2021) exposed to MT base station antennas.

Even though radar RF/microwave radiation does not transmit information, and thus, it is not modulated, it is pulsed by on/off pulses repeated at ELF rates. It is not WC EMR, but it does have similarities. It is usually emitted at higher power densities and is considered very hazardous (Puranen and Jokela 1996). A study on the effects of airport radar on birds provided evidence that birds sense the radar emissions, and slight differences in power density and pulse properties may alter bird behavior. The authors found behavioral changes in birds associated with the radar emissions. More specifically, they found that birds moved faster and decreased their vigilance when exposed to the radiation (Sheridan et al. 2015).

Because birds have navigation ability, considered to be related with the sensing of the geomagnetic field, they may perceive man-made magnetic fields as well (Liboff and Jenrow 2000) and actively avoid areas polluted with anthropogenic EMFs (Surendran et al. 2020). Various animal groups are sensitive to ELF EMFs, and many species use the natural electric and magnetic fields for orientation (Kalmijn 1988). It seems that animals can use the direction of the magnetic field as a compass and its intensity as a component of their navigational map (Kirschvink et al. 2001; Johnsen and Lohmann 2005; Hsu et al. 2007; Wiltschko et al. 2007; Ritz et al. 2009; Wajnberg et al. 2010). Applied static magnetic fields affect the circadian rhythms and orientation of insects, and changes in the geomagnetic field intensity affect the immune system of rats (Roman and Tombarkiewicz 2009; Yoshii et al. 2009). Current evidence indicates that exposure to man-made EMFs at levels found in the environment (in urban areas and near MT base stations) could particularly affect the animal ability to orient. Man-made EMFs are found to disrupt the orientation of birds and, therefore, disable their natural compass as long as they are present (Wiltschko et al. 2015). Ritz et al. (2004; 2009) have reported that the orientation of European robins (*Erithacus rubecula*) can be disturbed by man-made magnetic fields. The orientation of migratory birds is disrupted when very weak man-made EMFs (broadband field of 0.1–10 MHz of 85 nT or a 1.3 MHz field of 480 nT) are applied (actually added to the static geomagnetic field of 46000 nT) (Thalau et al. 2006). Engels et al. (2014) have shown that robins are unable to orient in the presence of broadband electromagnetic noise in the range 2 kHz–9 MHz. Therefore, electrosmog interferes with animal orientation and should be restricted by more stringent regulations in order to protect endangered species. Unfortunately, in most of those studies, no attempt was made to determine which specific frequency bands (ELF/RF) were more effective in interfering with bird orientation.

A laboratory study performed on three species of migratory songbirds provided evidence that they all align their bodies parallel to the N–S geomagnetic field axis, and the authors suggested that this behavior is involved in the underlying mechanism for orientation and navigation/migration (Bianco et al. 2019).

In a study performed in nature in eastern Poland, the head and body orientation of the white stork (*C. ciconia*) during incubation was non-random but was modified mainly by wind direction and speed, as well as by the presence of predators. However, white storks in nests located on power-line pylons have been observed to align their bodies parallel to the power lines in a statistically significant degree (Zbyryt et al. 2021).

10.4 EFFECTS OF MAN-MADE EMFs ON FISH

Effects on OS, Behavior, and Orientation

Anthropogenic EMFs represent an important stressor with increasing influence on the marine environment. Thus, the health risks that anthropogenic EMFs pose to aquatic organisms, need to be addressed (Lee and Yang 2014; Balmori 2015). Man-made EMFs may distort or mask vital natural electromagnetic stimuli necessary for many species (Hutchison et al. 2020). Most of the studies that have been carried out with fishes have focused on the EMFs generated by subsea power transmission cables in the aquatic environment (Formicki et al. 2019). Other studies have investigated the effects of mobile phone emissions on locomotor behavior of fishes and other aquatic species (Lee et al. 2015).

Cable deployments in coastal waters are increasing worldwide due to growing demands for electrical power and telecommunications. This phenomenon could alter the local electromagnetic natural stimuli used by electro-magneto-sensitive species (Hutchison et al. 2020). Offshore wind energy is the most developed of the renewable marine energies. The potential interactions between diadromous fishes* of conservation importance and subsea electromagnetic noise from marine renewable energy developments have been studied in the United Kingdom. The authors concluded that it is not yet possible to determine whether effects due to EMF pollution and sound pollution are biologically significant (Gill et al. 2012).

Other authors found a significant behavioral response of certain species, such as the electro-sensitive little skate (*Leucoraja erinacea*) and the presumed magneto-sensitive American lobster

---

* Fishes that spend their lives partially in fresh water and partially in salt water.

(*Homarus americanus*), to the EMFs of subsea, high-voltage cables (Hutchison et al. 2020). The results of this study clearly demonstrated that both skates and lobsters released into the experimental enclosures and video-monitored, changed their behavior in response to the EMFs of an electrical transmission cable buried under the treatment enclosure. There were multiple statistically significant differences in behavioral parameters of the animals due to the EMF exposure. There were significant differences in the total distance travelled by skates, their speed, and their height from the seabed, within a larger than expected area of influence along the cable route (Hutchison et al. 2020).

Some authors have investigated the effects of radiation from mobile phones in the laboratory and have shown changes in the locomotor behavior of fishes. In one of these studies, the researchers investigated the locomotor behavior of the guppy fish (*Poecilia reticulata*) and the zebrafish (*Danio rerio*) during exposure to mobile phone EMF. The exposure caused a change in the trajectory of their movements, and both species stayed as far away as possible from the mobile phone (Lee et al. 2015). It is noteworthy that a similar effect in the locomotor behavior during mobile phone EMF exposure has been recorded in ants (Cammaerts and Johansson 2014).

Another laboratory study with zebrafish showed that mobile phone radiation (GSM 900 MHz) exposure significantly decreased the exploration time, increased the total distance travelled by the exposed animals compared with the non-exposed, and caused a significant impairment of learning and memory. Moreover, the EMF exposure decreased superoxide dismutase (SOD), and catalase (CAT) activities[*], and increased the levels of reduced glutathione (GSH)[†] and lipid peroxidation (LPO),[‡] events related with compromised antioxidant defense (Nirwane et al. 2016). Similar effects on learning and memory are also found for ants (Cammaerts et al. 2012) and mice (Ntzouni et al. 2011).

Alignment constitutes the simplest directional response of living organisms to the Earth's magnetic field (Begall et al. 2013). Fishes show alignment with the Earth's magnetic field, as do amphibians (Phillips et al. 2002), reptiles (Diego-Rasilla et al. 2017), mammals (Begall et al. 2008; 2013), and birds (Bianco et al. 2019). A study found that carp (*Cyprinus carpio*) display a highly significant spontaneous preference to align their bodies along the north–south axis and, in the absence of any other common orientation cues that could explain this directional preference, the authors attributed this alignment to the geomagnetic field lines (Hart et al. 2012). The majority of developing embryos of both migratory and sedentary[§] fish species are aligned along the north–south axis, and it has been hypothesized that magnetoreception came into existence as one of the first senses (Formicki et al. 2019).

10.5 EFFECTS OF MAN-MADE EMFs ON AMPHIBIANS AND REPTILES

There are fewer studies on the effects of man-made EMFs on amphibians and reptiles, but their authors indicate a wide variety of non-thermal effects on these vertebrate groups (Rafati et al. 2015).

Effects on Embryogenesis, Development, Behavior, and Orientation

Several studies on the effects of man-made EMFs on amphibians have been conducted in the laboratory. Earlier studies have reported teratogenic effects due to EMF exposure (Levengood 1969; Grefner et al. 1998). In more recent studies, authors have indicated a wide variety of non-thermal effects of pulsed RF EMFs from various sources, including mobile phone jammers[¶] on several types

---

[*] SOD and CAT are protective antioxidant enzymes involved in critical aspects of health, and their reduced levels indicate compromised antioxidant defense.

[†] The antioxidant enzymes (SOD and CAT) and the GSH scavenge the free radicals in the brain, acting as cellular antioxidants. The increase in GSH levels may be attributed to an adaptive response to ROS production.

[‡] Because lipids are in high concentrations in cellular membranes, the problem of their peroxidation by free radicals leads to variety of functional changes in the membranes, most of which have a negative effect on cellular physiology.

[§] Fish species which, during the harvestable stage, are either immobile on or under the seabed or move in constant physical contact with the seabed or the subsoil.

[¶] Devices used to block mobile phone emissions by interfering with their signals.

of cells, including epithelial, endothelial, epidermal, cardiac muscle cells, and fibroblasts, exposed *in vivo* or *in vitro* (Blank and Goodman 2009; Rafati et al. 2015).

Amphibians can be especially sensitive; for example, the aquatic salamander olm (*Proteus anguinus*) and the Pyrenean salamander (*Euproctus asper*) display an avoidance response to weak sinusoidal 20–30 Hz EMFs at thresholds as low as 10 mV/m, but the sensitivity of the organisms extends over a frequency range from below 0.1 Hz to 1–2 kHz, covering almost the entire ULF and ELF bands (Schlegel 1997). Exposure to pulsing RF EMFs from mobile phone jammers at power densities 30–60 mW/cm^2 (this is more than 100 times stronger than a mobile phone signal at close proximity to the device) has been found to alter the heart rhythm of frogs, possibly due to the effect on the nervous system and the latency period of muscle contractions (Rafati et al. 2015). Static magnetic fields 10^4 G (1 T) have been found to inhibit the early embryonic development in the northern leopard frog (*Rana pipiens*) (Neurath 1968). Common frog tadpoles (*Rana temporaria*) exposed to a 50 Hz EMF developed more slowly and less synchronously than control tadpoles, remained at the early stages for a longer time, and showed increased mortality. In addition, the exposed tadpoles developed allergies and changes in their blood indexes (Grefner et al. 1998).

Non-thermal exposure of frogs (*Xenopus laevis*) to a pulsing 10 MHz EMF (0.73 µT, 219 V/m) resulted in dilated arterioles. The vasodilatory effect on arterioles was optimum when the pulses were applied for half of the total exposure time at 10 kHz repetition frequency (Miura and Okada 1991).

The scientific literature contains plenty of data showing that weak endogenous electric currents consisting of ion flows through cells and tissues play a crucial role in the morphogenesis and embryonic development in all animals (Levin 2003). Externally applied static electric fields of similar intensities with the endogenous electric fields have been found to directly affect the differentiation of some tail structures in amphibians and stimulate tissue regeneration. Alterations of the endogenous electric fields can disrupt the electro-chemical gradient in cells/tissues and the electric signals received by embryonic cells. Cells have been found to respond to externally applied electric fields by changes in their biochemical processes, development, and differentiation. Endogenous electric currents of specific spatial and temporal patterns are often correlated with specific morphogenetic events such as limb bud formation (Borgens 1988; Nuccitelli 1988; Levin 2003). Amphibians may be specifically sensitive to applied EMFs because their skin is always moist and, thus, more conductive (Balmori 2006).

In an early laboratory study, Levengood (1969) exposed embryos (fertilized eggs) of the wood frog (*Rana sylvatica*) and the spotted salamander (*Ambystoma maculatum*) to strong static magnetic fields of 0.6 and 1.8 T during different developmental stages.[*] A short-time exposure (5 min) of the early embryo produced several types of abnormalities: Microcephalia, scoliosis, edema, and retarded growth (Levengood 1969). Several of the treated tadpoles developed teratogenic features, such as severe leg malformations and extra legs as well as a pronounced alteration of histogenesis which took the form of subepidermal blistering and edema (Levengood 1969). Such abnormalities are sporadically observed to occur in nature because of yet unknown reasons (Balmori 2006).

Laboratory experiments have shown that lizards are also sensitive to externally applied EMFs. A study showed that the central bearded dragon (*Pogona vitticeps*), a species of agamid lizard found in arid regions of Australia, is sensitive to ELF EMFs. The animals exposed to a sinusoidal 6 or 8 Hz EMF (peak magnetic and electric field intensities 2.6 µT and 10 V/m) reacted by lifting their tails during the exposure with a significantly greater frequency (average number of tail lifts per individual per day) than the unexposed animals (Nishimura et al. 2010).

In one of my studies (Balmori 2010), common frogs (*R. temporaria*) were exposed to EMFs/EMR from MT base antennas, located at a distance of ~ 140 m, from the egg phase until an advanced phase of tadpole prior to metamorphosis. In the exposed group, the tadpoles displayed low coordination of movements (uncoordinated swimming, rotational behavior, no response to sound stimuli,

---

[*] Note that the static geomagnetic field average intensity is ~ 0.05 mT (Dubrov 1978).

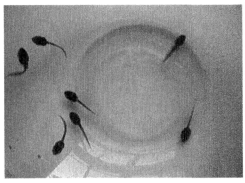 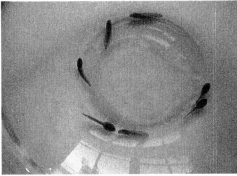

FIGURE 10.7 A picture from the experiment with the tadpoles. On the left, the group that was protected with a Faraday cage. On the right, the exposed group. There is a clear difference in size and appearance of the tadpoles. Although EMF-shielding is not recommended for permanent use (see Panagopoulos and Chroussos 2019), it is very useful for this kind of experiments in open-air studies.

and passivity), asynchronous growth (resulting in both big and small tadpoles), and high mortality (90%). In the control group, the coordination of movements was normal, the development was synchronous, and mortality was 4.2%. The results of this experiment, conducted in real-life conditions in the city of Valladolid (Spain), indicate that tadpoles living near MT antennas and exposed to levels of WC EMFs (1.8–3.5 V/m in the RF band) frequently encountered at distances usually ≤ 100 m, display problematic development and increased mortality (Figure 10.7). In fact, this could be a significant contributing factor for the decline in amphibian populations noted before (Balmori 2006).

Anthropogenic EMFs, especially WC EMFs, affect the immune, nervous, and endocrine systems and operate independently or in cooperation with other factors such as ultraviolet (UV) radiation or chemical pollutants (Walleczek 1992; Johansson 2009; Szmigielski 2013; Lupi et al. 2020; 2021). UV can kill cells by destroying their DNA. Higher levels of photolyase activity would be expected to mediate efficient repair of UV damage to DNA. Death of embryos due to UV radiation in the nature does not occur when the DNA repair mechanisms (such as those associated with the photoreactive enzyme photolyase) are effective (Blaustein et al. 1996). Man-made EMFs affect the immune system, obstructing DNA repair in amphibians and other animals, including humans (Grefner et al. 1998; Hallberg and Johansson 2013).

Magnetic sensing and orientation have been demonstrated in a wide range of animal taxa (Nishimura et al. 2010). While the mechanism of polarized varying electric and/or magnetic fields to initiate biological effects through ion forced oscillation and dysfunction of voltage-gated ion channels (VGICs) is the most plausible and widely recognized (Panagopoulos et al. 2000; 2002; 2015a; 2020; 2021), the specific mechanism of animal orientation with the geomagnetic field is not yet specified.

The hypothesis that animals may sense the geomagnetic field through the possible existence of biogenic magnetite in their brain, which could be reoriented by weak magnetic forces has never been proven, but it is still considered possible (Kirschvink and Gould 1981; Kirschvink 1989; Kirschvink et al. 2001).

It is also hypothesized that photoreceptors specialized for sensing the Earth's magnetic field are present in animals and work via photo-signaling pathways initiated by the so-called radical-pair mechanism (Ritz et al. 2000; Phillips et al. 2010).

The so-called magnetic alignment, an orientation of the body determined by the Earth's magnetic field, is an observed directional response of adaptive significance, but its mechanism or biological function are still unknown (Begall et al. 2013). Many researchers have reported magnetic alignment in insects and several species of vertebrates, including fish, amphibians, reptiles, birds, and mammals. Free-living lacertid lizards exhibit spontaneous alignment behavior, tending to align their body axes with the geomagnetic field lines. Moreover, the orientation of the lizards is found to be significantly correlated with the geomagnetic field intensity at the time of each observation (Diego-Rasilla et al. 2017). Magnetic alignment might assist animals in reading and organizing their mental map of space and encoding their environment. A constant directional reference for spatial orientation might be useful not only for navigation but also for escaping from predators efficiently and many other important functions. For example, the common wall lizard (*Podarcis muralis*) and the Lilford's wall lizard (*Podarcis lilfordi*) maintain magnetic alignment while basking. This alignment might improve their mental map of space to accomplish efficient escape behavior (Diego-Rasilla et al. 2017).

10.6 EFFECTS OF MAN-MADE EMFs ON INSECTS

10.6.1 THE DECLINE OF INSECT POPULATIONS

The biodiversity of insects is threatened worldwide (Hallmann et al. 2017; Vogel 2017; Powney et al. 2019; Sánchez-Bayo and Wyckhuys 2019). Numerous studies have reported a serious decline in insect populations that has occurred in recent decades around the world, and the same is happening with the important group of pollinators (honeybees, butterflies, moths, beetles, bumblebees, solitary bees, hoverflies, and others) which are crucially important animals for the pollination of crops and, thus, for the very existence of plants, trees, animals, and (ultimately) humans (Balmori 2021a). The loss of insect diversity and abundance is expected to provoke a cascade of effects on the food chain and the ecosystem (Hallman et al. 2017; Møller 2019). Insects are specifically sensitive to man-made EMFs, and there is a strong scientific evidence from many laboratory and open-air studies showing that this factor contributes greatly (if not primarily) to the insect decline (Balmori 2006; 2021a).

10.6.2 ELF EMF STUDIES

Exposure to 50–60 Hz ELF power-line EMFs present around overhead power transmission lines has been found to affect the behavior, neuronal and muscular responses, and levels of heat shock proteins in locusts (*Schistocerca gregaria*) (Wyszkowska et al. 2016). Exposure of pupae of the house fly (*Musca domestica*) to 50 Hz EMF caused a significant delay in metamorphosis (Stanojevic et al. 2005). In another study, static electric field exposure induced alterations in CAT and peroxidase (POD) activities and significantly decreased the metabolic rates of the English grain aphid (*Sitobion avenae*). The effects persisted in subsequent generations (Luo et al. 2019).

Panagopoulos et al. (2013) exposed newly eclosed fruit flies (*Drosophila melanogaster*, Oregon R wild type strain) to 50 Hz sinusoidal EMF produced by a crossover coil in the lab at three different magnetic field (and corresponding induced electric field) intensities 0.1, 1.1, and 2.1 mT (0.13, 1.43, and 2.72 V/m). The exposure was continuous for 120 h (5 days) and resulted in statistically significant decreases in reproduction (number of F_1 pupae) by up to 4.3%, depending on field intensities. The decline in reproduction was found to be due to DNA fragmentation and consequent cell death in the reproductive cells. The comparison was made to sham-exposed insects in an identical coil with antiparallel turns in which magnetic and induced electric fields were zero but all other parameters (like temperature, light, etc.) were identical.

Honeybee colonies exposed to high-voltage power lines (765 kV, 60 Hz) at 7 kV/m displayed increased motor activity, increased propolis[*] production, decreased hive weight, queen loss, abnormal production of queen cells,[†] decrease in the colony population, and poor winter survival. When the colonies were exposed to decreasing electric (and magnetic) field intensities (7, 5.5, 4.1, 1.8, and

---

[*] Propolis is a resin-like material made by the bees from a mixture of bee saliva and buds of poplar and cone-bearing trees. The bees use it to seal gaps in the hive to keep out intruders. Propolis also has antimicrobial properties and may protect bees from pathogens.

[†] Queen cells are constructions in which larvae develop and mature into new queens.

0.65–0.85 kV/m) at increasing distances from the power lines, the aforementioned effects attenuated. The groups with the two lowest intensities had normal hive net weight during 25 weeks of exposure duration. The effects became insignificant for distances greater than 23 m from the ground line projection of the power lines. Abnormal propolization of hive entrances did not occur for intensities below 4.1 kV/m. Queen loss occurred in 6 of 7 colonies at 7 kV/m, and in 1 of 7 at 5.5 kV/m, but not for lower intensities. Foraging rates were significantly lower at 7 and 5.5 kV/m (Greenberg et al. 1981). While these effects on bees were observed for a 25-week exposure duration, long-term exposure (years) of human populations to high-voltage power-line fields is connected with childhood leukemia for electric field intensities as low as 10 V/m or distances up to 600 m (Coghill et al. 1996; Kheifets et al. 2010).

A recent study demonstrated that ELF EMFs (50 Hz) from power lines affect honeybee olfactory learning, flight, foraging activity, and feeding, and could represent a prominent environmental stressor for honeybees, potentially reducing their ability to pollinate crops (Shepherd et al. 2018). In Italy, deleterious effects on bees from both pesticides and EMFs from high-voltage power lines have been found. In the EMF-exposed areas, overactivation of all analyzed enzymatic biomarkers (AChE, CAT, glutathione-*S*-transferase, and alkaline phosphatase) was observed at the end of the season (October), representing potential problems for the winter survival of the bees (Lupi et al. 2020; 2021). Recent research showed that high-voltage power-line EMFs disrupt the natural behavioral pattern of the bees with a negative impact on workers related to both the duration of exposure and the field intensities. The authors explained that the behavioral disturbances could reduce the efficiency of bees as pollinators, which would be translated into a decrease in crop production (Migdał et al. 2021).

10.6.3 WC EMF Studies

The effects of non-thermal levels of microwave/WC EMF exposures on insects have been known for decades. Insects seem to perceive WC EMFs as environmental stressors (Yanagawa et al. 2020). WC EMFs emitted by mobile phones have been found to elevate hsp70 (heat shock protein) levels and affecting the reproduction and development of the fruit fly (*Drosophila melanogaster*) (Weisbrot et al. 2003).

A series of studies by Panagopoulos et al. on fruit fly reproduction have shown that the reproductive capacity of fruit flies decreased by 50%–60% after only a few min daily non-thermal exposure to the "talk" ("modulated") or "listening" ("non-modulated") signal of a mobile phone during the first 2–5 days of adult life, with the "talk" signal being significantly more effective due to its higher power density (Panagopoulos et al. 2004). The same authors compared the biological activities between GSM 900 MHz and GSM 1800 MHz and found that both types of radiation significantly decreased the reproductive capacity of fruit flies, with the GSM 900 MHz being even more effective, mainly due to its higher intensity (Panagopoulos et al. 2007a). This non-thermal effect diminished with increasing distance from the device (decreasing intensity) and was provoked by the induction of cell death in reproductive cells due to DNA fragmentation (Panagopoulos et al. 2007b; 2010; Panagopoulos 2012; 2019) as well as damage in the actin cytoskeleton of the ovarian cells (Chavdoula et al. 2010).

Other authors have also observed a statistically significant decrease in the mean fecundity of the fruit fly verifying the aforementioned experiments (Atli and Ünlü 2006; Margaritis et al. 2014), while various problems in certain publications with fruit fly experiments are described in Panagopoulos and Karabarbounis (2020). The mean duration of the pupation stage was extended linearly with increasing duration of exposure to a 10 GHz carrier signal pulsed at 1 kHz at power density 0.0156 W/m^2 (1.56 μW/cm^2), and the mean offspring number was significantly decreased (Atli and Ünlü 2006; 2007).

The increasing use of WC technology in our daily lives certainly affects surrounding animals and has an ecological impact. Genotoxicity and behavioral alterations have recently been detected in *D. melanogaster* for non-thermal exposure to 2.45 GHz radiation produced by a generator (ELF pulsations not reported). The number of wing spots, reflecting genotoxicity, increased with increasing power density (0 µW/cm^2, 100 µW/cm^2, 100 mW/cm^2). Moreover, the exposure induced free radical production and genotoxicity in the insect body along with behavioral changes (Yanagawa et al. 2020).

Increasing evidence indicates that flies and spiders, among other invertebrates, disappear from areas with the highest levels of EMFs from MT antennas. These observations are consistent with numerous laboratory studies showing the negative effects of WC EMFs on reproduction, development, and navigation (Panagopoulos et al. 2004; Balmori 2009; Sharma and Kumar 2010; Lázaro et al. 2016). EMFs from MT antennas have been found to affect the abundance and composition of several species of wild pollinator insects in natural habitats, a phenomenon that could have important ecological and economic impacts and could significantly affect the maintenance of wild plant diversity, crop production, and human welfare (Lázaro et al. 2016).

A recent study has looked at the effects of EMF emissions from an MT base station (GSM 1800) on some insects in Nigeria. EMF levels and numbers of pollinating insects were monitored daily for 22 weeks in five sampling locations (four at different distances around the MT base station and a control location). The abundance, distribution, and diversity of insects were affected by the MT EMFs, showing that most insects favored areas with lower EMF levels (Adelaja et al. 2021).

In another study carried out in nature, the authors examined the impact of exposure to ambient EMF levels from MT base stations for a 48-h period on the reproductive capacity of four different invertebrate species. The measured power density levels were in all cases < 0.00002 µW/cm^2 (a very low EMR level that, indeed, is difficult to find anywhere anymore). They did not find a significant impact on reproductive capacity, which was to be absolutely expected since the duration of exposure was too short and the ambient EMR levels too low for inducing acute effects, but they noted that more attention should be paid to the possible impacts on biodiversity, as exposure to WC EMFs is ubiquitous and continues to increase rapidly over large areas (Vijver et al. 2014).

Intermittent exposure of a few min of ticks (*Dermacentor reticulatus*) to a 900 MHz signal emitted by a horn antenna at approximately 700 µW/m^2 (0.07 µW/cm^2) power density induced an immediate tick locomotor response manifested either as a spasmodic movement of the whole body or spasmodic movement of the first pair of legs. In addition, ticks exhibited overall significantly greater mobility during the exposure (Vargova et al. 2017). It should be noted, for a comparison, that the power density of a mobile phone attached to the head usually exceeds 100 µW/cm^2 (Carpenter 2013; Panagopoulos 2017)

Pereira et al. (2019) demonstrated how low-intensity changes in the normal local geomagnetic field values affect the behavior of workers of three ant species, inducing significant changes in their foraging activities. Other researchers have demonstrated effects of a simulated GSM signal on ant olfactory and visual learning, revealing an impact on their physiology (Cammaerts et al. 2012). In a series of experiments, it was shown that the ants' speed and trajectory of movement were immediately altered by exposure to an active GSM mobile phone. More specifically, when the GSM phone was activated, the ant linear speed statistically decreased, and their angular speed statistically increased, meaning that instead of their normal straight trajectories, the ants started moving in abnormal trajectories in different changing directions (Cammaerts and Johansson 2014). The authors concluded that the MT EMFs affect the ant behavior and physiology, and this represents an additional negative impact of the WC EMFs on insects (Cammaerts et al. 2012).

Lopatina et al. (2019) examined the effects of 24 h exposure to a Wi-Fi router on olfactory excitability, food motivation, and memory in honeybees. The EMF exposure had a significant adverse effect on food excitability and short-term memory. In other studies, only a few bees exposed to DECT radiation from cordless domestic phones returned to the beehive, and the honeycomb weight was lower in the exposed beehives (Stever et al. 2005; Harst et al. 2006).

WC EMFs from active mobile phone handsets had a dramatic impact on bees initiating an alarm behavior as manifested by the worker piping alarm signal, which is an announcement of disturbance of the bee colony (Favre 2011). In another study with GSM (900 MHz) mobile phones, the authors found a significant decline in colony strength and egg-laying rate by the queen. The behavior of exposed foragers was negatively influenced by the exposure: There was neither honey nor pollen in the colony at the end of the experiment (Sharma and Kumar 2010). A more recent study found that GSM mobile phone radiation exposure significantly affected the queen development and caused an increased mortality during pupation, resulting in a reduced hatching rate (Odemer and Odemer 2019).

A study in India conducted with *Apis cerana* honeybee colonies close to an MT base station found that the flight activity and returning of worker honeybees were significantly higher in colonies placed at 500 m than at 100 m from the base station (Taye et al. 2017)

In the laboratory, the concentrations of carbohydrates, proteins, and lipids in the hemolymph of the bee *Apis mellifera* increased under the influence of mobile phone radiation (Kumar et al. 2013). Another study observed an increase in the mortality of *Apis mellifera* after 2 h exposure to 13.56 MHz, 4.04 mV/m EMR, or 868 MHz, 5 V/m EMR produced by generators (Darney et al. 2016).

WC EMF exposure has negative impacts on living organisms; ants react quickly to WC EMFs in their environment, and bees may behave abnormally when exposed to MT EMFs (Cammaerts et al. 2013). According to a review on the effects of man-made EMF pollution on invertebrates, the majority of the studies in each of the categories have shown effects and declines in the populations of key pollinator groups (wild bees, hoverflies, bee flies, beetles, and wasps). These effects have been attributed to WC EMFs and especially emissions from MT base antennas (Friesen and Havas 2020). Considering these studies together, WC EMFs should be considered a main cause for the dramatic decline in insects, acting in synergy with agricultural intensification, pesticides, invasive species, and climate change. Taking into account the benefits insects provide to nature and humankind, the Precautionary Principle should be applied before any new deployment (such as the 5G) is considered (Balmori 2021a).

10.6.4 Effects on Orientation, Navigation, and Colony Collapse

As with birds, man-made EMFs disrupt orientation in insects, as this becomes evident from both laboratory and open-air studies on different invertebrate species (Camlitepe et al. 2005; Balmori 2015).

It is a most plausible hypothesis that orientation and navigation of animals is accomplished by their ability to sense the direction of the geomagnetic field (Kirschvink and Gould 1981; Kirschvink 1989; Kirschvink et al. 2001; Wiltschko et al. 2007; Begall et al. 2008; Engels et al. 2014; Hiscock et al. 2017; Granger et al. 2020). The geomagnetic and geoelectric fields are basically static with ULF/ELF variations (Dubrov 1978). Thus, it is expected that ULF/ELF man-made EMFs distort the local terrestrial fields, disabling animals from using them for orientation and navigation. Man-made EMFs not only distort local terrestrial EMFs but also affect the animal neural system, causing further inability to sense the natural EMFs (Presman 1977; Dubrov 1978; Panagopoulos 2013).

GSM 900 MHz mobile phone radiation has been found to exert a severe impact on ants, especially affecting their visual and olfactory memory, causing the loss of their ability to use the corresponding cues. This finding indicates that MT EMFs have an impact on the orientation and navigation of animals, irrespectively of their interference with the natural EMFs (Cammaerts et al. 2012; Cammaerts and Johansson 2014).

The orientation of the American cockroach (*Periplaneta americana*) was found to be disturbed by weak EMF exposure (1.2 MHz or 2.4MHz) produced by an RF generator. The RF EMF was administered in repeated 5 min on/off pulses, resulting in an additional ULF component of 0.003 Hz. The threshold of the disruptive effect for the 1.2 MHz EMF was found to be between 12 nT and 18 nT, whereas the threshold for the 2.4 MHz EMF was between 18 nT and 44 nT (Vácha et al. 2009).

Adverse biochemical changes and disorientation have been reported for honeybees and other invertebrates (Friesen and Havas 2020). Adult honeybees possess an orientation sense, and reported difficulties in returning back to their hives may also be due to anthropogenic EMFs and their interference with the geomagnetic field (Ferrari 2014).

Regarding the so-called colony collapse disorder (CCD), which has been observed in honeybee colonies around the world over the past 20 years, several authors consider that anthropogenic EMF exposure provides a better explanation compared with other possible causes (Cammaerts et al. 2012). Several authors have warned that the increasing EMF levels produced by mobile phones and MT antennas disturb the navigational skills of honeybees, preventing them from returning to their hives (Warnke 2009). Honeybees are sensitive to pulsed EMFs generated by mobile phones and MT antennas and the observed changes in their behavior could be one explanation for the loss of colonies (Favre 2011). In fact, winter colony losses in the northeast US correlated with the occurrence of geomagnetic field fluctuations (solar storms) are consistent with the observed effects on honeybee behavior and development (Ferrari 2014). In recent years, there has been an important advance in understanding the underlying mechanisms for the orientation of insects and other animals and the biological consequences of its perturbation (Sutherland et al. 2018).

Certainly, apart from orientation problems caused by man-made EMFs, CCD, most importantly, is related to the reproductive decline in all insects due to cell death in their reproductive cells, causing reduction in reproductive capacity, infertility, embryonic mortality, etc., induced by the anthropogenic and, especially WC, EMFs (Panagopoulos et al. 2004; 2007a; 2007b; 2010; 2013; Chavdoula et al. 2010; Sharma and Kumar 2010; Odemer and Odemer 2019). This is also in agreement with the observed mammalian and human infertility, bird embryonic mortality, etc., consistently found to be induced by WC EMF exposures (Magras and Xenos 1997; Wdowiak et al. 2007; Agarwal et al. 2008; De Iuliis et al. 2009; Burlaka et al. 2013; Yakymenko et al. 2018).

10.7 EFFECTS OF MAN-MADE EMFs ON PLANTS AND TREES

The deployment of WC and the exponential increase in the use of mobile phones, which has occurred throughout the world over the past 25–30 years, has increased electromagnetic pollution of ambient anthropogenic EMF levels by several orders of magnitude, especially in inhabited areas (Bandara and Carpenter 2018). The WC EMF effects on wildlife have not been foreseen, especially on species such as trees and plants that, because of their immobility and their proximity in many cases to EMF sources (such as MT and other antennas or high-voltage power lines), they may be subjected to chronic intense exposure (Figure 10.8).

Since the mid-twentieth century, several studies have investigated the effects of anthropogenic EMFs on plants, both in laboratory studies (Kiepenheuer et al. 1949; Brauer 1950; Harte 1950; 1972; Jerman et al. 1998; Lerchl et al. 2000; Sandu et al. 2005; Tkalec et al. 2005; 2009; Roux et al. 2006; 2008; Beaubois et al. 2007; Sharma et al. 2009; Kundu and IEEE 2013; Pesnya and Romanovsky 2013; Cammaerts and Johansson 2015; Grémiaux et al. 2016; Vian et al. 2016) and in nature (Bernatzky 1986; Volkrodt 1987; 1991; Balodis et al. 1996; Selga and Selga 1996; Haggerty 2010). Moreover, during the past two decades, studies have also found worrying effects from MT base antennas on trees (Balmori 2004; Waldmann-Selsam and Eger 2013; Waldmann-Selsam et al. 2016). The effects of WC EMFs on plants and trees depend on the particular species, growth stage, exposure duration, and power density, among other factors (Senavirathna and Takashi 2014; Halgamuge et al. 2015).

10.7.1 Effects on Plants

Almost 40 years ago, Marino et al. (1983) demonstrated that man-made EMFs could affect plants non-thermally by altering their physiology. The authors found that an applied electric field (5 kV/m, 60 Hz) could induce a statistically significant decrease of about 5% in the germination rate of the sunflower. An increasing number of studies have highlighted biological responses and modifications at molecular and whole-organism levels after exposure to WC EMFs (Vian et al. 2016). Man-made EMFs have been found to evoke a multitude of adverse effects in plants that may lead to ecological impacts. Seeds of *Brassicaceae Lepidium sativum* (garden cress) exposed to 0.0070–0.0100 µW/cm^2 WC EMR

FIGURE 10.8 Damaged trees with dry tips in a sensitive species (*Populus nigra*) due to excessive man-made EMF exposure. This usually happens after long-term exposure at levels ≥ 2 V/m electric field strength in the RF band (power density ≥ 1 μW/cm²). The trees are exposed not only to MT antennas (1) but also to a radio broadcasting antenna (2) and power lines (3) simultaneously.

from two MT base station antennas at about 200 meters distance never germinated. When moved away from the MT antennas to a location with significantly lower ambient EMR levels (0.0002–0.0003 µW/cm^2) and similar other conditions, the seeds germinated normally (Cammaerts and Johansson 2015).

The effects of anthropogenic EMFs on plants (like the effects on animals) initiate at the subcellular level, altering the activity of several enzymes, including those related with reactive oxygen species (ROS) production and metabolism. Mobile phone EMF (GSM 900 MHz) exposure has been found to affect the process of rhizogenesis through biochemical alterations that manifest as oxidative damage resulting in root impairment. The exposure enhanced the activities of proteases, polyphenol oxidases, and peroxidases in mung bean hypocotyls. It also enhanced malondialdehyde, hydrogen peroxide, and proline levels, indicating a ROS-mediated oxidative damage in hypocotyls. The effect was confirmed by the upregulation in the activities of antioxidant enzymes (SOD, ascorbate peroxidase, guaiacol peroxidase, CAT, and glutathione reductase) (Singh et al. 2012). The enhanced activity of certain enzymes is a well-known marker of plant responses to various types of environmental stressors. It evokes the expression of specific genes implicated in plant responses to wounds and modifies the growth of the whole plant.

Modulated 1800 MHz EMR produced by a generator with average power density 332 mW/m^2 (33.2 µW/cm^2) was found to affect enzymatic activities and to inhibit the growth of corn (*Zea mays*) (Kumar et al. 2016). Long-term exposure of *Arabidopsis thaliana*[*] to cordless (DECT) phone EMF reduced the number of chloroplasts as well as the stroma thylakoids and photosynthetic pigments (Stefi et al. 2016). Belyavskaya (2001) observed effects on calcium balance and chromatin alterations in meristematic cells of the pea root (*Pisum sativum*) subjected to a condition of decreased geomagnetic field down to the order of nT by use of a "magnetic screen device", producing a static magnetic field in opposite direction to the geomagnetic (~ 0.05 mT), resulting in its attenuation. Mitochondria were the most sensitive organelles under such a condition, and their size and relative volume in cells increased. The effects on the ultrastructure of root cells were due to disruption of various metabolic systems, including the Ca^{2+} homeostasis. These effects recorded in the absence of the natural static geomagnetic field show that the natural terrestrial EMFs are absolutely vital and necessary to life. The adverse effects are induced by the man-made EMFs due to their important differences from the natural EMFs as described in Section 10.1, and more extensively in Chapter 1. Probably, one of the reasons why other planets of our solar system are unfriendly to life, is that contrary to our planet Earth, they lack of a natural static magnetic field similar to the geomagnetic.

Cytotoxic effects detected in plants exposed to mobile phone (pulsed and modulated RF) EMR have clearly shown the capability of this radiation to induce DNA damage in eukaryotic cells and/or disturb the mitotic apparatus, as denoted by micronuclei formation (Gustavino et al. 2016). These results are consistent with similar findings in animals and humans exposed to WC EMFs (Panagopoulos 2019; 2020).

Non-thermal exposure of wheat (*Triticum vulgare*) to a 9.75 GHz EMF has been reported to induce chromosomal aberrations, such as inter-chromosomal bridges, chromosomal fragments, and micro-nuclei formation (Pavel et al. 1998). In other experiments, modulated and non-modulated RF EMF exposure at various non-thermal intensities, induced mitotic aberrations in root meristematic cells of onion (*Allium cepa*). Modulated RF EMF 23 V/m (in the RF band) (~ 140 µW/cm^2) caused strong inhibition of the growth during the whole experiment, and the observed effects were markedly dependent on the field strength (power density) and the modulation applied. The effects of the modulated RF EMF (with ELF modulation) were significantly stronger than the effects of the corresponding non-modulated EMF (Tkalec et al. 2009).

Során et al. (2014) investigated the effects of simulated Wi-Fi and GSM EMF exposures on leaf anatomy, essential oil content, and volatile emissions of *Petroselinum crispum*, *Apium graveolens*, and *Anethum graveolens*. The exposure levels were similar to those measured for high level MT/WC exposure in the open space (100 mW/m^2 or 10 µW/cm^2) and indoor environments' Wi-Fi (70

---

[*] A small plant in the mustard family used as a model system in plant biology.

mW/m^2 or 7 µW/cm^2). The exposures induced thinner cell walls, smaller chloroplasts and mitochondria, and enhanced emissions of volatile compounds. There was a direct relationship between the structural and the chemical modifications of the three plant species studied. These data collectively demonstrate that WC EMFs constitute a stress to the plants. Engelmann et al. (2008), however, found that a simulated UMTS signal at an unusually high average intensity (8 mW/cm^2) induced only a slight effect on the gene activity of plant cells in suspension culture. Only a few genes displayed an altered transcription (mRNA) level after 24 h exposure, and these alterations did not exceed a 2.5-fold reduction or increase in gene expression.

Plants are sensitive and exhibit various responses to both natural and man-made EMFs. Exposure of rose bush plants to 900 MHz EMF produced by a generator (pulsing frequencies were not reported) at electric field intensity in the RF band 5 V/m (corresponding to power density 6.63 µW/cm^2) delayed and significantly reduced the plant growth non-thermally (Grémiaux et al. 2016). Pronounced decline in germination rate and embryonic stem length of wheat (*Triticum aestivum*) was also induced by mobile phone radiation (Hussein and El-Maghraby 2014). In other experiments, exposure of soybean (*Glycine max*) seedlings to mobile phone (pulsed) radiation (GSM 900 MHz), or to continuous-wave radiation (without pulsation) produced by a generator at field intensities in the RF band of 0.56–41 V/m, reduced epicotyl and hypocotyl growth, whereas root outgrowth was stimulated. For GSM exposure with a signal strength of 41 V/m (corresponding to power density ~ 446 µW/cm^2), L1 (epicotyl length between leaves and cotyledon) was reduced by 1.7 mm, whereas L2 (hypocotyl length between cotyledon and start of root and its branches) and L3 (length of root and its branches) did not change. GSM exposure of 5.6 V/m (corresponding to power density ~8.3 µW/cm^2) did not result in changes in outgrowth. For continuous-wave exposure with a signal strength of 41 V/m, only L3 was reduced significantly. Continuous-wave exposure of 5.7 V/m resulted only in a reduction of L2 by 1.3 mm. These findings indicate that the observed effects are mainly dependent on power density/field strength and the ELF pulsation and amplitude modulation of the WC EMFs (Halgamuge et al. 2015).

In experiments carried out in the laboratory testing the effects of static magnetic fields on germination, with seeds of various plants exposed to 0.125 – 0.250 T, the germination speed and the percentage of germinated seeds increased due to the exposure. In growth experiments, the exposed seedlings developed more in length and weight (Martínez et al. 2003). In general, the results showed a stimulating effect of static magnetic fields on the growth and development of plants and an inhibitory effect of a varying magnetic field (50 Hz, 15 µT) (Martínez et al. 2003).

Very low power density (10^{-6} µW/cm^2), non-thermal, wide band (40–78 GHz), modulated (at 10, 100, and 1000 Hz) and non-modulated continuous-wave microwave radiation has been found to affect the response of tobacco plants (*Nicotiana tabacum*) to the tobacco mosaic virus. The ELF modulation (especially the lowest 10 Hz modulation frequency) seemed to be the most bioactive parameter and induced hypersensitive response to the virus compared with the corresponding non-modulated EMR (Betti et al. 2004). The authors noted that their results confirmed the prediction of the mechanism published by Panagopoulos et al. (2002) that the ELF components are the most bioactive parameters of microwave EMFs.

Endogenous electromagnetic signals are important for rapid and "long-distance" communication within a plant (and any living organism). The signals propagate as changes in the plasma membrane electrical potential (EP) of adjacent cells in plants and contain encoded information determined by the shape of the signal. Therefore, if such endogenous signals in plants/organisms are altered by any external factors, they transmit incorrect information, distorting the corresponding biological functions controlled by such signals. Experimental evidence on the effects of man-made EMFs on the electrical signaling in the parrot feather plant (*Myriophyllum aquaticum*) has been provided. The plant was exposed to continuous-wave EMR at 2, 2.5, 3.5, and 5.5 GHz emitted by a signal generator with a maximum intensity 23–25 V/m for every frequency. The effect was recorded as alterations in the EP (in mV) measured between two electrodes inserted in the plant with 7 cm distance between them. The findings confirm the non-thermal effects of man-made EMFs/EMR on the endogenous electrical signaling of plants. The changes persisted after the end of the EMR exposure (Senavirathna and Asaeda 2014).

Singh et al. (2012) investigated the biochemical mechanism of the effect of GSM 900 MHz mobile phone radiation on rhizogenesis in the mung bean (*V. radiata*). The authors concluded that MT radiation affects the process of rhizogenesis through biochemical alterations that manifest as oxidative damage, resulting in root impairment. MT radiation inhibits rhizogenesis through biochemical alterations induced by OS. The results of this research suggest that man-made EMFs have a significant impact on biota and ecosystem functioning (Singh et al. 2012).

Non-thermal (4–10 mW/m² or 0.4–1 µW/cm²) long-term exposure (24 h daily, continuously from sowing to harvesting) to non-modulated carrier signal 915 ± 5 MHz EMR emitted by an RF generator evoked morphophysiological changes in the common bean (*Phaseolus vulgaris*) (Surducan et al. 2020). The exposed plants showed significant morphological modifications, with greater height, less inflorescence, extremely long roots and changes in chlorophyll and carotenoid content, all closely connected with ultrastructural changes in the leaves. There was also an increase in the trichome size and density on the abaxial side of the leaves. Plants physiologically develop protection against the harsh environment and pests by increasing the trichome, but in this case, the EMF was the only stressor. Further, the globular chloroplasts in the exposed samples had an increased number of plastoglobules (lipoprotein particles inside chloroplasts), which is a sign of OS. The authors concluded that the EMF exposure was the primary cause of the ultrastructural alterations and the OS signs. The most affected leaves were the ones exposed to the highest levels of EMR, closer to the source, indicating that the induced damages were dependent on power density. The authors warned that MT evolution to 5G and 6G would add additional concerns regarding the effects of WC EMFs on biological systems (Surducan et al. 2020).

In a study carried out under high-voltage power transmission lines between Austria and the Czech Republic, the authors examined the effect of the power-line EMFs (50 Hz) on wheat and maize crops. There was an average 7% reduction in wheat production in the crops closest to the power-line (2–40 m horizontal distance) during the 5 years of the study investigation, but there was no significant effect on maize production (Soja et al. 2003).

Another study was carried out on aquatic plants exposed to pulsed EMFs from the Skrunda Radio Location Station.* Cultures of duckweed (*Spirodela polyrhiza*) exposed to a power density of 0.1–1.8 µW/cm² had lower longevity, reproductive problems, and morphological and developmental abnormalities compared with a control group of the same species that grew far from the radar (Magone 1996).

In their review on the effects of WC EMFs on plants, Jayasanka and Asaeda (2013) concluded that these effects depend on the plant family and the growth stage as well as on the exposure duration, power density, and frequency. They emphasized that most studies addressing the effects of microwaves on animals and plants have documented effects and responses at exposures well below the ICNIRP limits, and it is, therefore, necessary to reconsider the existing guidelines (Jayasanka and Asaeda 2013). Barman and Bhattacharya (2016) noted that electromagnetism is an essential component of plant life. They concluded that anthropogenic EMFs exert significant stress on plants, as has been reported by many observers throughout the world.

10.7.2 Effects on Trees

Trees have several advantages over animals as experimental subjects: They are continuously exposed with a constant orientation with regard to certain EMF sources (e.g., antennas, radars, etc.) due to their immobility (Vian et al. 2016). In addition, one can easily identify changes over time, such as disturbed growth, dying branches, or premature color change in the leaves. Moreover, the

---

* Early Soviet warning radar system emitting in the band of 156–162 MHz, located in Latvia.

damage to trees is objective and cannot be attributed to psychological or psychosomatic factors (Waldmann-Selsam et al. 2016).

Almost 50 years ago, Canadian researchers observed unpredictable damage in plants exposed to high power density microwave radiation. The authors gave an early warning for urgent attention on commercial equipment producing microwave EMR (Tanner and Romero-Sierra 1974). In the area directly exposed to radiation from the Skrunda Radio Location Station in Latvia, pine trees (*Pinus sylvestris*) displayed less radial growth. This did not occur beyond the area of direct incidence of the electromagnetic waves. In addition, there was a statistically significant inverse correlation between the tree growth and the intensity of the EMR, and it was confirmed that the beginning of the symptoms coincided with the beginning of the radar emissions. Other environmental factors that could have been involved were evaluated and excluded (Balodis et al. 1996). In another study that investigated cellular effects on pine trees irradiated by the same radar, there was an increase in resin production. The authors interpreted this phenomenon as a stress response induced by the radiation, which also explains the accelerated aging and decreased growth of trees subjected to the pulsed radar microwaves. In addition, there was decreased germination of seeds in the pines exposed to the highest EMR levels (Selga and Selga 1996).

Another study reported no effect of long-term exposure (3.5 years) to 2450 MHz EMR generated by a 600 W magnetron source with power densities of 0.007–300 W/m^2 (0.0007–30 mW/cm^2) on leaf density, height growth, and photosynthesis of young spruce and beech trees. However, the same study reported a decrease in the levels of calcium and sulfur in the beech (*Fagus sylvatica*) leaves directly related to the power density of the radiation (Schmutz et al. 1996). Trees near a large powerful radio broadcasting antenna in a forest in Michigan, USA, had grown unusually fast since the antenna was installed in 1986. Forest researchers have attributed this "extra" growth to the EMF emission from the antenna. Different species reacted differently to the radiation: Oak (*Quercus rubra*) and birch trees (*Betula papyrifera*) did not seem to be affected, but the pine trees near the antenna grew more than those further away, and the poplar (*Populus tremuloides*) and maple trees (*Acer rubrum*) grew less near the antenna than those further away. These observations suggest that anthropogenic EMFs can have a complicated influence on the forest (Kiernan 1995). In Ouruhia, New Zealand, in places exposed directly to the main radiation beam from a powerful radio broadcasting antenna, the trees died. According to the observations, the affected trees seemed more vulnerable when they had their roots in the water or were near the river. At the places with the highest EMR levels, the trees appeared more affected, or they were dead (Balmori 2004).

During the Cold War, on the border between East and West Germany, there were numerous radars operating. The areas with damaged forest almost always coincided with those swept by the radar pulsing microwaves. Immediately after dismantling the radars, which had been in operation for two to three decades, there was a visible regeneration and recovery of the forests. In these areas, there was no conventional environmental pollution (Volkrodt 1991). In Canada, there were also devastating effects in forests near radar installations (Volkrodt 1991). In Schwarzenburg, Switzerland, trees near a powerful radio broadcasting antenna grew up not vertically but toward opposite directions from the antenna (Balmori 2004), a curious phenomenon also described by Hertel (1991).

Haggerty (2010) noted that the anthropogenic EMF background exerted a strong adverse effect on the growth rate, anthocyanin production, development, vigor (denoted by total growth length and leaf surface), and reproduction in aspen trees (*P. tremuloides*) and suggested that this could be the underlying factor for the observed aspen population decline. In the trees exposed to higher levels of ambient EMFs, the anthocyanin production was decreased, and the incidence of necrotic leaf tissue was increased compared to shielded trees during the study (Haggerty 2010).

For some years, we have observed, in Spain, a gradual deterioration of trees next to various EMR antennas, especially MT base antennas, and especially in inhabited areas (Figure 10.8). Some trees located inside the main lobe of the radiation display sickly appearance, growth retardation, and increased susceptibility to pests and diseases. In some places, where we have measured levels above 1–2 V/m electric field strength in the RF band (~ 0.2–1 $\mu W/cm^2$ power density) continuously,

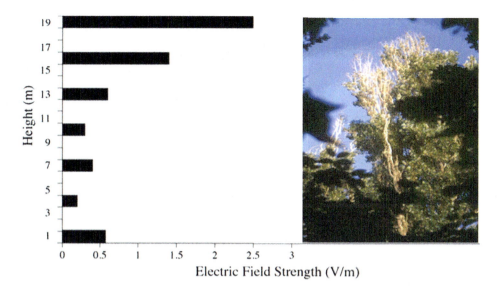

FIGURE 10.9 Electric field intensity (in the RF band) at different heights above the ground (real data measured 100 meters away from a MT base station). Note that the values increase with height as they enter the main emission lobe of the antenna. For this reason, the radiation especially affects the upper parts of the trees (author's data).

the trees display deterioration. In wooded masses, trees generally do not grow above the height of others, and those that stand out usually have dry upper branches (Figure 10.9). There is a varying susceptibility depending on the species. We have observed effects mainly in poplar (*Populus* sp.) and willow trees (*Salix* sp.). Other species such as *Platanus* and privets (*Ligustrum japonicum*) seem more resistant (Balmori 2004).

Waldmann-Selsam et al. (2016) performed a detailed, long-term (2006–2015) nature-monitoring study in the cities of Bamberg and Hallstadt, Germany. We made observations, took photographs of unusual or unexplainable tree damage, and measured EMF levels. The photographs provided a record of the state of the trees to monitor damage and growth over several years. Many trees showed damage patterns that are not usually attributable to diseases (fungi, bacteria, and viruses), or pests (insects and nematodes), or other environmental factors (such as heat, drought, frost, intense sunlight, water excess, compaction of the soil, or air and soil pollutants). In the five most representative species – the Norway maple (*Acer platanoides*), the European hornbeam (*Carpinus betulus*), the linden (*Tilia* sp.), the yew (*Taxus baccata*), and the northern white cedar (*Thuja occidentalis*) – most trees only showed unilateral damage (on one side). Most of these trees had sparse leaves or needles (crown transparency). In many of them, the leaves turned prematurely yellow or brown in June, a phenomenon that always began at the leaf perimeter. In many trees, the leaves fell prematurely or there were dead branches (usually peak branches were dried). Some trees stopped growing in height, while in others, the main guide died. All the needles of *T. baccata* and *T. occidentalis* changed their color to yellow, red, or brown. Generally, in all trees of the study, the damage was higher in the areas with higher EMF levels and occurred, in most cases, on the side where the nearest MT base antenna was located. Moreover, areas with more antennas had higher radiation levels and damaged trees were found most often in such areas. The observations in all the affected trees revealed significant differences between the side facing an MT antenna and the opposite side. The study found that differences in power density between the two sides of the tree corresponded to differences in damage, and the differences in the damage were statistically significant. These results are consistent with the fact that the damage to trees by MT antennas usually starts on one side and extends to the whole tree over time. It should be noted that trees in the same location that were shielded by buildings or other trees were healthy, and all trees in low-radiation areas showed no damage (Waldmann-Selsam et al. 2016).

Forest plants of all species, deciduous and coniferous trees as well as shrubs, are affected by the anthropogenic EMFs, but the differences in susceptibility among different species depend on the specific biological/physical properties of each species and the condition of the trees. It is hypothesized that differences are also related to the electrical conductivity, the tissue (wood) density (fast- or slow-growing species), and the percentage of water in the tissues (Waldmann-Selsam et al. 2016).

Waldmann-Selsam et al. (2016) also documented adverse effects on trees, similar to those described in Bamberg and Hallstadt, in other parts of Germany (Munich, Nürnberg, Erlangen, Bayreuth, Neuburg/Donau, Garmisch-Partenkirchen, Murnau, Stuttgart, Kassel, Fulda, Göttingen, Rhoen Biosphere Reserve, Tegernsee Valley, and in several small towns), as well as in Spain (Valladolid, Salamanca, Madrid, Palencia, and León), Austria (Graz), Belgium (Brussels), and Luxemburg. Each MT antenna can harm many trees, and each tree can be affected by several MT antennas from different base stations. Damaged trees seem to exist around each antenna, and the several millions of MT base stations in the world are potentially exerting a negative impact on the growth and health of millions of trees. This phenomenon may occur in cities (Figure 10.10), forests, and national parks where MT base stations are being installed without prior studies on their environmental impact. For this reason, prior to any new base station implantation, it is essential to assess the environmental impact. Decreased growth and quality of trees/forests due to MT/WC antenna emissions could lead to huge ecological consequences, decreased tree crops and timber productivity in plantations (Waldmann-Selsam et al. 2016).

A recent study in Turkey, investigated the effects around an MT base station on flower, cone yield, and germination percentage in *Pinus brutia*. The base station in the study area was installed in 2005, and the study materials were collected from 27-year-old trees in 2019. The results revealed that the MT base station antennas EMFs significantly reduced the number of flowers and cones

FIGURE 10.10 MT pico-antennas (2G/3G) camouflaged with urban furniture in Valladolid (Spain).

in the trees and that the number of flowers measured in trees at a distance of 800 m from the base station was 11 times greater than at a distance of 100 m, and the number of cones 7 times greater. Moreover, germination percentage among the more distant trees was three times greater than among those closest to the base station. These results show that trees are affected significantly by MT base station antenna EMF emissions (Ozel et al. 2021).

From the above-described studies, it becomes evident that WC EMFs constitute a danger for trees worldwide. Evidence of plant damage due to MT/WC antennas was not taken into account in determining the exposure criteria set by the ICNIRP (2010; 2020) and accepted by most governments. Now that the problem has become evident, the exposure criteria, especially for WC EMFs, should be reconsidered (Waldmann-Selsam et al. 2016). Proper risk assessment of WC EMFs should be undertaken to develop management strategies for reducing this new form of pollution in the natural environment. Further deployment of WC/MT antennas should be stopped and scientific research elucidating the effects on trees under real exposure conditions must continue (Waldmann-Selsam et al. 2016).

10.8 DISCUSSION

10.8.1 General Considerations on the Described Effects

Based on the current scientific knowledge I have examined in this chapter, the reported consequences of anthropogenic EMF exposures on ecosystems and wildlife. It is evident that anthropogenic EMFs should be considered an important factor for the decline in bird and insect populations and the adverse effects on many other species, especially in urban areas where species are subjected to higher exposure levels. Man-made EMF emissions, especially from MT/WC antennas and high-voltage power lines, constitute a toxic stress that chronically affects humans, animals, and plants/trees and threatens life and the environment (Grigoriev et al. 2003; Belyaev 2005; Lai 2005; Balmori 2009; 2014; 2015; 2021a; 2021b; Cucurachi et al. 2013; Pall 2016; Waldmann-Selsam et al. 2016; Lázaro et al. 2016; Panagopoulos 2017; 2019; 2020; Kostoff et al. 2020; Levitt and Lai 2010; 2021a; 2021b).

The described effects have important implications on wildlife, especially in urban and suburban areas, but also in rural, natural, and protected areas where powerful WC antennas and radars have been installed (Bürgi et al. 2014). Even though not all the effects of man-made EMFs on wildlife have been recorded yet, the consequences are already evident and continue because the existing adopted guidelines for public health protection only consider the effects of acute thermal exposures (Hyland 2000; Balmori 2009; 2014; 2015; Yakymenko et al. 2016; Panagopoulos 2019; Panagopoulos et al. 2021; Kostoff et al. 2020; López et al. 2021) and do not protect wildlife from lower level, long-term exposures. EMF safety standards should be based on the more sensitive natural biological response (Blank 2014), which is actually the only scientifically valid criterion for determining EMF-exposure risks, and not on recommendations by a private organization (ICNIRP) which has been accused of conflicts of interest (Hardell and Carlberg 2020; Hardell et al. 2021). The ICNIRP (2010; 2020) limits are thousands of times above the levels at which effects are recorded for both ELF and RF/WC EMFs and account only for thermal effects, while the vast majority of recorded effects (as described also in the present chapter) are non-thermal (i.e., are not accompanied by any significant heating of the exposed living tissues).

Scientific research on the effects of anthropogenic EMFs on wild animals and plants/trees is not as intense as it should be, considering the great impact on the environment in combination with the remarkable expansion of WC technology over the past 30 years. Nevertheless, the available amount of evidence is already huge and indicates that man-made, and especially WC, EMF exposure at levels found in the environment (in urban areas and near MT/WC antennas) impacts the ability of animals to survive, reproduce, orient, and navigate, which can have tremendous adverse consequences on ecosystems.

In the areas most polluted by WC EMFs (within 300–500 m radius from MT antennas, especially in the direction of the main beam), there is a deterioration of the optimal habitat for the permanence of the animals, which can cause the abandonment of the breeding areas or roosts. Taking into account the cumulative effects of the anthropogenic EMFs (Adey 1996; Magras and Xenos 1997; Panagopoulos 2017; 2020; Bamikole et al. 2019; Tumkaya et al. 2019; Kumar et al. 2021), the exposure duration is crucial to assess the effects induced on wildlife. Electromagnetic pollution is not exclusive to urban centers. Indeed, road infrastructure, mountains, or promontories on the coasts frequently house antennas that are installed to cover large territorial areas. The effects of the antennas on the habitats are difficult to quantify, but they could cause their deterioration, generating areas of silence without singing birds or reproductive pairs. The impact on the animal ecosystems is further accentuated by the adverse effects on prey invertebrate (insect) populations and on plants/trees. Such effects reduce the food which is available to birds and other animals.

Certain studies in nature have not found effects on animals induced by the ambient lower-level exposures, especially when the lower levels of ambient EMFs are combined with short-term exposures (e.g., Vijver et al. 2014). Ambient levels of WC EMFs are much lower (usually of the order of 0.0001–0.1 $\mu W/cm^2$) than levels of acute exposure to, e.g., mobile phones in contact with the head/body (usually 100–300 $\mu W/cm^2$). On the other hand, animals and plants/trees in nature can be exposed continuously during their whole lives to the ambient man-made EMFs. This is why it is necessary to apply long-term exposures (e.g., months) when we study the effects from low-intensity man-made EMFs.

There is a variety of recorded non-thermal biological effects by which WC EMFs affect animals. Scientific evidence indicates that prolonged exposure to such EMFs, at levels encountered in the environment, affects the immune system, altering various biological processes (Chou et al. 1992; Novoselova and Fesenko 1998; Galeev 2000; Szmigielski 2013). A stressed immune system may increase the susceptibility of animals to infectious diseases (Fernie and Bird 2001). ELF EMFs from high-voltage power lines (50–60 Hz) decrease reproduction in birds (Doherty and Grubb 1996; Fernie et al. 2000) and insects (Panagopoulos et al. 2013). WC EMFs cause genotoxic effects in cells and animals, decreasing reproduction as well (Lai and Singh 1995; Balode 1996; Diem et al. 2005; Panagopoulos et al. 2007a; 2010; De Iuliis et al. 2009; Chavdoula et al. 2010; Burlaka et al. 2013; Yakymenko et al. 2018). Genotoxic effects in reproductive cells, such as DNA and chromosome damage, imply long-term detrimental effects, as they are related with causation of infertility, birth defects, heritable mutations, and cancer (Gandhi and Singh 2005). Such effects have already been found to be induced by exposure to WC EMFs (Di Carlo et al. 2002; Leszczynski et al. 2002; Panagopoulos et al. 2007a; 2010; De Iuliis et al. 2009; Burlaka et al. 2013; Yakymenko et al. 2018).

MT/WC EMFs have a significant action on the development and incidence of embryonic abnormalities and embryonic death. An increase in mortality and morphological abnormalities, especially in the neural tube, has been recorded in chicken embryos exposed to mobile phone radiation (Youbicier-Simo and Bastide 1999), with genetic factors accounting for the differential susceptibility among individuals (Úbeda et al. 1983; 1994; Farrel et al. 1997). Authors have reported induction of OS, genetic damage, and a high mortality of bird embryos exposed to mobile phone EMFs compared to the control animals (Grigoriev 2003; Batellier et al. 2008; Burlaka et al. 2013; Yakymenko et al. 2018).

An overall literature review by Khan et al. (2018) revealed that altered physiology could be due to disturbed biochemical processes in animals and plants. Their conclusion was based mainly on the increases in OS biomarkers frequently recorded after WC EMF exposure as also documented by Yakymenko et al. (2016). WC EMF exposure causes a number of anomalies at the cytogenetic and genetic levels, such as disturbed mitosis, DNA damage, chromosomal aberrations, and alterations in various metabolites, which have active roles in cellular metabolism (Khan et al. 2018).

WC EMFs may affect animals in at least one of the following ways: a) DNA damage and mutations that may lead to organic disorders, infertility, and other diseases including cancer (Lai and Singh 1995; 1997; Diem et al. 2005; Panagopoulos et al. 2007a; 2010; De Iuliis et al. 2009; Burlaka et al. 2013; Yakymenko et al. 2018); b) damage to the nervous system due to alterations in the brain electrical activity (EEG) (Huber et al. 2000; Mohammed et al. 2013), modification of neuronal response (Beasond and Semm 2002), or disruption/permeabilization of the blood–brain barrier (Salford et al. 2003); c) Alteration of circadian rhythms (sleep–wake cycle) due to interference with the pineal gland and hormonal imbalances (Bartos et al. 2019); d) changes in heart rate and blood pressure (Kim et al. 2021); e) impaired health and immunity toward pathogens, weakness, exhaustion, and developmental problems (Chou et al. 1992; Aldad et al. 2012); f) problems in fertility, embryonic development, and survival of offspring (Balmori 2005; Burlaka et al. 2013; Pandey et al. 2018; Panagopoulos 2017; 2019; Yakymenko et al. 2018); g) genetic malformations, locomotor problems, albinisms and partial melanisms, or promotion of tumors (Balmori 2006); and h) orientation and navigation problems (Engels et al. 2014).

10.8.2 Possible Explanations for the Effects on Animals and Plants

The ampullae of Lorenzini are specialized electrosensory organs that elasmobranch fishes, such as sharks, rays, and skates, use to detect extremely small changes in natural environmental electric fields. The voltage-gated calcium channels (VGCCs) and the big conductance calcium-activated potassium (BK) channels are preferentially expressed by electrosensory cells in the little skate (*L. erinacea*) and functionally couple to mediate electrosensory cell membrane voltage oscillations, which are important for the detection of specific weak electrical signals (Bellono et al. 2017). Application of ELF EMFs (10 Hz, 5 ms pulses) significantly increased intracellular calcium concentrations in glass catfish (*Kryptopterus bicirrhis*), allowing wireless control of cellular function by activation of a specific protein responsive to EMFs (Krishnan et al. 2018). Although such experiments further verify the profound electromagnetic nature of all forms of life (which is already verified by many studies), it should perhaps be noted that the attempt to control cellular functions and living organisms by electromagnetic signals is in opposition with basic bioethical principles.

These findings are clearly related with and can be explained by the biophysical mechanism proposed by Panagopoulos et al. (2000; 2002; 2015a; 2020; 2021), according to which, polarized (anthropogenic) EMFs, especially ELF EMFs such as those of power lines or even more the ELF/ULF pulsation/modulation/variability of the WC EMFs, cause dysfunction of VGICs in the cell membranes, which results in impaired intracellular ion concentrations and cellular function. The same mechanism can provide plausible explanations for at least the majority of the reported effects of man-made EMFs on animals and plants. The polarized and coherent nature of all man-made EMFs and their intensity and frequency in the environment can cause impairment of VGICs in the cell membranes of all cells and, thus, account for the disruptive effects discussed in this chapter (Panagopoulos et al. 2000; 2002; 2015a; 2020; 2021; Pall 2013; Panagopoulos and Balmori 2017), including the deleterious effects on gonads and reproductive systems (De Iuliis et al. 2009; Panagopoulos 2012; Adams et al. 2014; Gang et al. 2021).

While the aforementioned mechanism and the vast majority of the recorded effects on animals and plants are non-thermal (i.e., they are not accompanied by tissue heating), the authors of a recent study argued that the increased carrier frequencies employed in 5G MT become comparable to the body size of insects, which will result in increased energy absorption and consequent tissue heating (Thielens et al. 2018). Because all WC EMFs (including 5G) are based on ELF pulsations, the non-thermal effects will continue to occur, but thermal effects are expected to occur as well, most likely resulting in even more detrimental action of the 5G compared to the previous generations of WC/MT EMFs. This needs to be carefully addressed by the scientific community and the health authorities.

Endogenous electric fields in cells and tissues have long been identified to play a causal role in cellular function, embryonic development, wound healing, etc. Alteration of the endogenous electric currents due to interference with man-made EMFs in the environment may disrupt the electro-chemical gradient and the signals received by cells. Endogenous electric current patterns are often correlated with specific embryonic and morphogenetic events (Hotary and Robinson 1992; Levin 2003). Moreover, available data suggest that non-thermal EMF effects depend on genotype, gender, physiological, and individual factors (Belyaev 2005; Lai 2005). Genomic differences could influence cellular responses to WC EMFs. Data analysis has highlighted a wide inter-individual variability in the responses to EMF exposure (Belyaev 2005). It is expected that each species and each individual show different susceptibility to radiation because vulnerability depends on genetic tendency and the physiological and neurological state of the exposed organism (Farrel et al. 1997; Flipo 1998; Fernie et al. 2000; Hyland 2000; Fernie and Reynolds 2005). Differential susceptibility of each species has also been shown in wild birds exposed to EMFs from high-voltage power lines (Doherty and Grubb 1996).

It has been proposed that man-made EMFs act similarly in plants and animals by impairing the function of the VGICs, especially the VGCCs (Pall 2016). Plasma membrane calcium channels and other VGICs are irregularly activated/inactivated by man-made EMFs in both animals and plants, increasing the intracellular [Ca^{2+}] and altering intracellular ion concentrations, initiating OS, which in turn, leads to DNA damage (Panagopoulos et al. 2002; 2021; Pall 2016). DNA damage, and especially DNA strand breakage, leads to the formation of micronuclei and to chromosomal damage and rearrangements. The two-pore channels (TPC) are additional possible targets in plants for possibly producing such effects. These channels contain a voltage sensor that is activated by partial depolarization of the plasma membrane (similarly with the VGICs) and is predicted to be extremely sensitive to low-intensity non-thermal man-made EMFs because of its structure and physical location in the plasma membrane (Pall 2016).

10.8.3 WHERE DO WE GO FROM HERE?

Anthropogenic, and especially WC, EMFs have become a very common stressor for animal and plant species (Ribeiro-Oliveira 2019). The effects on wildlife from the increasing exposure to man-made EMFs have been identified as an important issue for the global conservation of biological diversity (Sutherland et al. 2018). According to the official document, "Late Lessons of Early Warnings" (European Environment Agency 2013), before the deployment of any new technology such as the 5G, its effects should be clearly assessed (Vanbergen et al. 2019). As explained in this chapter, man-made EMFs, and especially WC EMFs, should be seriously considered as a significant driver for the dramatic decline in insects and other animals, acting in synergy with other environmental factors.

With all these studies published in peer-reviewed, scientific journals, one would think that the governments/authorities have seriously considered the accumulated data. Unfortunately, this is not the case. The European Parliament recently published an official report (Thielens 2021) that attempts to discredit many peer-review studies that have demonstrated adverse effects, without counterarguments and without going through the usual peer-review process. It is clear that the author considered only the possibility of effects through tissue heating (thermal effects) in the current science (in line with the ICNIRP), which is clearly obsolete. Further, the author cited only certain studies and did not consider the vast number of non-thermal effects recorded by hundreds/thousands of scientific studies. It would not be unreasonable to think that this report could be biased to benefit the continuation of the deployment of 5G and to pave the way for telecommunication companies, although this approach is absolutely contrary to the Precautionary Principle of the European Union (Harremoes et al. 2013), the scientific recommendations (Sutherland et al. 2018), and the extended published peer-reviewed, scientific literature described and referenced in this chapter.

The Precautionary Principle is one of the fundamental principles of the European Union, governing policies related to the environment, health, and food safety (Harremoes et al. 2013). This principle enables decision-makers to adopt precautionary measures when the scientific evidence regarding an environmental or human health factor is not certain on its safety. The principle makes clear that the proponent of an activity, rather than the public, should bear the burden to prove that this activity is safe for human health and the environment.

Despite the existing ample and rapidly increasing scientific evidence, no significant progress has been made over all these years, at least at the level of guidelines issued by the responsible authorities and official regulatory bodies. This phenomenon jeopardizes the full functioning of the European democracy. Organizations, such as the ICNIRP, a private organization that issues exposure guidelines, which are then (strangely) adopted by governments, and is accused of having conflicts of interest (Hardell and Carlberg 2020; Hardell et al. 2021), or the national regulatory bodies with the same problems at the official level (Hardell 2021), herald an inevitable and deep environmental/health crisis. Fortunately, there are many more people who are concerned about the risks, especially regarding the pressing implementation of 5G.

Hopefully, this chapter will help official bodies to base their decisions on what science says and not on spurious interests contrary to the maintenance of health and life on our planet. In this way, we would avoid repeating the serious mistakes made with asbestos, tobacco, climate change, and other factors for which the necessary decisions were delayed, with the corresponding irreversible impact on people's health and the environment.

REFERENCES

Adams, J. A., Galloway, T. S., Mondal, D., Esteves, S. C., and Mathews, F. (2014). Effect of mobile telephones on sperm quality: A systematic review and meta-analysis. *Environment International*, 70, 106–112.

Adelaja, O. J., Ande, A. T., Abdulraheem, G. D., Oluwakorode, I. A., Oladipo, O. A., and Oluwajobi, A. O. (2021). Distribution, diversity and abundance of some insects around a telecommunication mast in Ilorin, Kwara State, Nigeria. *Bulletin of the National Research Centre*, 45(1), 1–7.

Adey, W. R. (1996). Bioeffects of mobile communications fileds: Possible mechanisms for cumulative dose. In N. Kuster, Q. Balzano, and J. C. Lin (Eds.), *Mobile communication safety*, pp. 95–131. Chapman & Hall, London.

Agarwal, A., Deepinder, F., Sharma, R. K., Ranga, G., and Li, J. (2008). Effect of cell phone usage on semen analysis in men attending infertility clinic: An observational study. *Fertility and Sterility*, 89(1), 124–128.

Agiwal, M., and Jin, H. (2018). Directional paging for 5G communications based on partitioned user ID. *Sensors*, 18(6), 1845. https://doi.org/10.3390/s18061845.

Akar, A., Karayiğit, M. Ö., Bolat, D., Gültiken, M. E., Yarim, M., and Castellani, G. (2013). Effects of low level electromagnetic field exposure at 2.45 GHz on rat cornea. *International Journal of Radiation Biology*, 89(4), 243–249.

Aldad, T. S., Gan, G., Gao, X. B., and Taylor, H. S. (2012). Fetal radiofrequency radiation exposure from 800–1900 MHz-rated cellular telephones affects neurodevelopment and behavior in mice. *Scientific Reports*, 2, 312.

Alkis, M. E., Bilgin, H. M., Akpolat, V., Dasdag, S., Yegin, K., Yavas, M. C., and Akdag, M. Z. (2019a). Effect of 900-, 1800-, and 2100-MHz radiofrequency radiation on DNA and oxidative stress in brain. *Electromagnetic Biology and Medicine*, 38(1), 32–47.

Alkis, M. E., Akdag, M. Z., Dasdag, S., Yegin, K., and Akpolat, V. (2019b). Single-strand DNA breaks and oxidative changes in rat testes exposed to radiofrequency radiation emitted from cellular phones. *Biotechnology and Biotechnological Equipment*, 33(1), 1733–1740.

Alkis, M. E., Akdag, M. Z., and Dasdag, S. (2021). Effects of low-intensity microwave radiation on oxidant-antioxidant parameters and DNA damage in the liver of rats. *Bioelectromagnetics*, 42(1), 76–85.

Armstrong, F. A., Hill, H. A. O., and Walton, N. J. (1988). Direct electrochemistry of redox proteins. *Accounts of Chemical Research*, 21(11), 407–413.

Atli, E., and Ünlü, H. (2006). The effects of microwave frequency electromagnetic fields on the development of Drosophila melanogaster. *International Journal of Radiation Biology*, 82(6), 435–441.

Atli, E., and Ünlü, H. (2007). The effects of microwave frequency electromagnetic fields on the fecundity of Drosophila melanogaster. *Turkish Journal of Biology*, 31, 1–5.

Azimzadeh, M., and Jelodar, G. (2020a). Prenatal and early postnatal exposure to radiofrequency waves (900 MHz) adversely affects passive avoidance learning and memory. *Toxicology and Industrial Health*, 36(12), 1024–1030.

Azimzadeh, M., and Jelodar, G. (2020b). Trace elements homeostasis in brain exposed to 900 MHz RFW emitted from a BTS-antenna model and the protective role of vitamin E. *Journal of Animal Physiology and Animal Nutrition*, 104(5), 1568–1574.

Balci, M., Devrim, E., and Durak, I. (2007). Effects of mobile phones on oxidant/antioxidant balance in cornea and lens of rats. *Current Eye Research*, 32(1), 21–25.

Balmori, A. (2004). Pueden afectar las microondas pulsadas emitidas por las antenas de telefonía a los árboles y otros vegetales? *Ecosistemas*, 13, 79–87. (available in:https://www.revistaecosistemas.net/index.php/ecosistemas).

Balmori, A. (2005). Possible effects of electromagnetic fields from phone masts on a population of white stork (Ciconia ciconia). *Electromagnetic Biology and Medicine*, 24(2), 109–119.

Balmori, A. (2006). The incidence of electromagnetic pollution on the amphibian decline: Is this an important piece of the puzzle? *Toxicological and Environmental Chemistry*, 88(2), 287–299.

Balmori, A., and Hallberg, Ö. (2007). The urban decline of the house sparrow (Passer domesticus): A possible link with electromagnetic radiation. *Electromagnetic Biology and Medicine*, 26(2), 141–151.

Balmori, A. (2009). Electromagnetic pollution from phone masts: Effects on wildlife. *Pathophysiology*, 16(2–3), 191–199.

Balmori, A. (2010). Mobile phone mast effects on common frog (*Rana temporaria*) tadpoles: The city turned into a laboratory. *Electromagnetic Biology and Medicine*, 29(1–2), 31–35.

Balmori, A. (2014). Electrosmog and species conservation. *Science of the Total Environment*, 496, 314–316.

Balmori, A. (2015). Anthropogenic radiofrequency electromagnetic fields as an emerging threat to wildlife orientation. *Science of the Total Environment*, 518, 58–60.

Balmori, A. (2016). Radiotelemetry and wildlife: Highlighting a gap in the knowledge on radiofrequency radiation effects. *Science of the Total Environment*, 543(A), 662–669.

Balmori, A. (2021a). Electromagnetic radiation as an emerging driver factor for the decline of insects. *Science of the Total Environment*, 767, 144913.

Balmori, A. (2021b). Electromagnetic pollution as a possible explanation for the decline of house sparrows in interaction with other factors. *Birds*, 2(3), 329–337.

Balmori, A. (2022). Corneal opacity in Northern Bald Ibises (*Geronticus eremita*) equipped with radio transmitters. *Electromagnetic Biology and Medicine*. In press. https://doi.org/10.1080/15368378.2022.2046046.

Balode, Z. (1996). Assessment of radio-frequency electromagnetic radiation by the micronucleus test in Bovine peripheral erythrocytes. *Science of the Total Environment*, 180(1), 81–85.

Balodis, V. G., Brumelis, K., Kalviskis, O., Nikodemus, D., Tjarve, V. Z., and Znotia, V. (1996). Does the Skrunda radio location station disminish the radial growth of pine trees? *Science of the Total Environment*, 180(1), 57–64.

Bamikole, A. O., Olukayode, O. A., Obajuluwa, T., Pius, O., Ibidun, O. O., Adewale, F. O., and Adeleke, O. O. (2019). Exposure to a 2.5 GHz non-ionizing electromagnetic field alters hematological profiles, biochemical parameters, and induces oxidative stress in male albino rats. *Biomedical and Environmental Sciences*, 32(11), 860–863.

Bandara, P., and Carpenter, D. O. (2018). Planetary electromagnetic pollution: It is time to assess its impact. *The Lancet Planetary Health*, 2(12), e512–e514.

Barman, P., and Bhattacharya, R. (2016). Impact of electric and magnetic field exposure on young plants-A review. *International Journal of Current Research and Academic Review*, 2(2), 182–192.

Bartos, P., Netusil, R., Slaby, P., Dolezel, D., Ritz, T., and Vacha, M. (2019). Weak radiofrequency fields affect the insect circadian clock. *Journal of the Royal Society Interface*, 16(158), 20190285.

Batellier, F., Couty, I., Picard, D., and Brillard, J. P. (2008). Effects of exposing chicken eggs to a cell phone in "call" position over the entire incubation period. *Theriogenology*, 69(6), 737–745.

Beasond, R. C., and Semm, P. (2002). Responses of neurons to an amplitude modulated microwave stimulus. *Neuroscience Letters*, 333(3), 175–178.

Beaubois, E., Girard, S., Lallechere, S., Davies, E., Paladian, F., Bonnet, P., Ledoit, G., and Vian, A. (2007). Intercellular communication in plants: Evidence for two rapidly transmitted systemic signals generated in response to electromagnetic field stimulation in tomato. *Plant, Cell and Environment*, 30(7), 834–844.

Begall, S., Červený, J., Neef, J., Vojtěch, O., and Burda, H. (2008). Magnetic alignment in grazing and resting cattle and deer. *Proceedings of the National Academy of Sciences of the United States of America*, 105(36), 13451–13455.

Begall, S., Malkemper, E. P., Červený, J., Němec, P., and Burda, H. (2013). Magnetic alignment in mammals and other animals. *Mammalian Biology*, 78(1), 10–20.

Belyaev, I. (2005). Non-thermal biological effects of microwaves. *Microwave Review*, 11(2), 13–29.

Bellono, N. W., Leitch, D. B., and Julius, D. (2017). Molecular basis of ancestral vertebrate electroreception. *Nature*, 543(7645), 391–396.

Belyavskaya, N. A. (2001). Ultrastructure and calcium balance in meristem cells of pea roots exposed to extremely low magnetic fields. *Advances in Space Research*, 28(4), 645–650.

Benediktová, K., Adámková, J., Svoboda, J., Painter, M. S., Bartoš, L., Nováková, P., Hart, V., Phillips, J., and Burda, H. (2020). Magnetic alignment enhances homing efficiency of hunting dogs. *eLife*, 9, e55080.

Berigan, L. A., Greig, E. I., and Bonter, D. N. (2020). Urban house sparrow (Passer domesticus) populations decline in North America. *The Wilson Journal of Ornithology*, 132(2), 248–258.

Berman, E., Chacon, L., House, D., Koch, B. A., Koch, W. E., Leal, J., Wagner, P., Mantiply, E., Martin, A. H., and Martucci, G. I. (1990). Development of chicken embryos in a pulsed magnetic field. *Bioelectromagnetics*, 11(2), 169–187.

Bernatzky, A. (1986). Elektromagnetischer Smog – Feind des Lebens. *Der Naturarzt*, 11, 22–25.

Betti, L., Trebbi, G., Lazzarato, L., Brizzi, M., Calzoni, G. L., Marinelli, F., and Borghini, F. (2004). Nonthermal microwave radiations affect the hypersensitive response of tobacco to tobacco mosaic virus. *Journal of Alternative and Complementary Medicine*, 10(6), 947–957.

Bhat, T. A., and Singh, D. (2019). Effect of mobile tower radiation on avian fauna: A case study of lolab valley, Kupwara Jammu and Kashmir. *JETIR*, 6, 570–576.

Bhattacharya, R., and Roy, R. (2013). Impacts of communication towers on avians: A review. *IJECT*, 4, Issue Spl - 1, 137–139.

Bhattacharya, R., and Roy, R. (2014). Impact of electromagnetic pollution from mobile phone towers on local birds. *International Journal of Innovative Research in Science, Engineering and Technology*, 3, 32–36.

Bianco, G., Köhler, R. C., Ilieva, M., and Åkesson, S. (2019). Magnetic body alignment in migratory songbirds: A computer vision approach. *Journal of Experimental Biology*, 222(5), jeb196469.

Blank, M. (2014). Cell biology and EMF safety standards. *Electromagnetic Biology and Medicine*, 25, 1–3.

Blank, M., and Goodman, R. (2009). Electromagnetic fields stress living cells. *Pathophysiology*, 16(2–3), 71–78. https://doi.org/10.1016/j.pathophys.2009.01.006.

Blaustein, A. R., Hoffman, P. D., Kiesecker, J. M., and Hays, J. B. (1996). DNA repair activity and resistance to solar UV-B radiation in eggs of the red-legged frog. *Conservation Biology*, 10(5), 1398–1402.

Borgens, R. B. (1988). Stimulation of neuronal regeneration and development by steady electrical fields. In S. G. Waxman (Ed.), *Advances in neurology; functional recovery in neurological disease*, 47, 547–564. Raven Press, New York.

Bose, S., Roy, R., Chakraborti, U., Samanta, R., Jana, S., Mondal, T., and Bhattacharya, S. C. R. (2020). Impressions of high frequency radio-waves from cell phone towers on birds: A base-line study. *Journal of Multidisciplinary Research*, 1, 54–62.

Brauer, I. (1950). Experimentelle Untersuchungen über die Wirkung von Meterwellen verschiedener Feldstärke auf das Teilungswachstum der Pflanzen. *Chromosoma*, 3(6), 483–509.

Burda, H., Begall, S., Červený, J., Neef, J., and Němec, P. (2009). Extremely low-frequency electromagnetic fields disrupt magnetic alignment of ruminants. *Proceedings of the National Academy of Sciences of the United States of America*, 106(14), 5708–5713.

Burlaka, A., Tsybulin, O., Sidorik, E., Lukin, S., Polishuk, V., Tsehmistrenko, S., and Yakymenko, I. (2013). Overproduction of free radical species in embryonic cells exposed to low intensity radiofrequency radiation. *Experimental Oncology*, 35(3), 219–225.

Bürgi, A., Scanferla, D., and Lehmann, H. (2014). Time averaged transmitter power and exposure to electromagnetic fields from mobile phone base stations. *International Journal of Environmental Research and Public Health*, 11(8), 8025–8037.

Calis, P., Seymen, M., Soykan, Y., Delen, K., Aral, B. S., Take Kaplanoglu, G., and Karcaaltincaba, D. (2019). Does exposure of smart phones during pregnancy affect the offspring's ovarian reserve? A rat model study. *Fetal and Pediatric Pathology*, 40(2), 142–152.

Cammaerts, M. C., De Doncker, P., Patris, X., Bellens, F., Rachidi, Z., and Cammaerts, D. (2012). GSM 900 MHz radiation inhibits ants' association between food sites and encountered cues. *Electromagnetic Biology and Medicine*, 31(2), 151–165.

Cammaerts, M. C., Rachidi, Z., Bellens, F., and De Doncker, P. (2013). Food collection and response to pheromones in an ant species exposed to electromagnetic radiation. *Electromagnetic Biology and Medicine*, 32(3), 315–332.

Cammaerts, M. C., and Johansson, O. (2014). Ants can be used as bio-indicators to reveal biological effects of electromagnetic waves from some wireless apparatus. *Electromagnetic Biology and Medicine*, 33(4), 282–288.

Cammaerts, M. C., and Johansson, O. (2015). Effect of man-made electromagnetic fields on common Brassicaceae *Lepidium sativum* (cress d'Alinois) seed germination: A preliminary replication study. *Fyton*, 84, 132–137.

Camlitepe, Y., Aksoy, V., Uren, N., Yilmaz, A., and Becenen, I. (2005). An experimental analysis on the magnetic field sensitivity of the black-meadow ant Formica pratensis Retzius (Hymenoptera: Formicidae). *Acta Biologica Hungarica*, 56(3–4), 215–224.

Carpenter, D. O. (2013). Human disease resulting from exposure to electromagnetic fields. *Reviews on Environmental Health*, 28(4), 159–172. https://doi.org/10.1515/reveh-2013-0016.

Chavdoula, E. D., Panagopoulos, D. J., and Margaritis, L. H. (2010). Comparison of biological effects between continuous and intermittent exposure to GSM-900-MHz mobile phone radiation: Detection of apoptotic cell-death features. *Mutation Research/Genetic Toxicology and Environmental Mutagenesis*, 700(1–2), 51–61.

Chou, C. K., Guy, A. W., Kunz, L. L., Johnson, R. B., Crowley, J. J., and Krupp, J. H. (1992). Long-term, low-level microwave irradiation of rats. *Bioelectromagnetics*, 13(6), 469–496.

Coghill, R. W., Steward, J., and Philips, A. (1996). Extra low frequency electric and magnetic fields in the bed place of children diagnosed with leukaemia: A case-control study. *European Journal of Cancer Prevention*, 5(3), 153–158.

Crick, H. Q., Robinson, R. A., Appleton, G. F., Clark, N. A., and Rickard, A. D. (2002). Investigation into the causes of the decline of starlings and house sparrows in Great Britain. BTO Research Report N 290. Department for Environment, Food and Rural Affairs (DEFRA), London.

Cucurachi, S., Tamis, W. L. M., Vijver, M. G., Peijnenburg, W. J. G. M., Bolte, J. F. B., and Snoo, G. R. (2013). A review of the ecological effects of radiofrequency electromagnetic fields (RF-EMF). *Environment International*, 51, 116–140.

Dadam, D., Robinson, R. A., Clements, A., Peach, W. J., Bennett, M., Rowcliffe, J. M., and Cunningham, A. A. (2019). Avian malaria-mediated population decline of a widespread iconic bird species. *Royal Society Open Science*, 6(7), 182197.

Darney, K., Giraudin, A., Joseph, R., Abadie, P., Aupinel, P., Decourtye, A., and Gauthier, M. (2016). Effect of high-frequency radiations on survival of the honeybee (Apis mellifera L.). *Apidologie*, 47(5), 703–710.

Dasdag, S., Ketani, M. A., Akdag, Z., Ersay, A. R., Sar, I., Demirtas, O. C., and Celik, M. S. (1999). Whole body microwave exposure emitted by cellular phones and testicular function of rats. *Urological Research*, 27(3), 219–223. https://doi.org/10.1007/s002400050113.

Davoudi, M., Brössner, C., and Kuber, W. (2002). Der Einfluss elektromagnetischer Wellen auf die Spermienmotilität. *Journal für Urologie und Urogynäkologie*, 9, 18–22.

De Coster, G., De Laet, J., Vangestel, C., Adriaensen, F., and Lens, L. (2015). Citizen science in action—Evidence for long-term, region-wide house sparrow declines in Flanders, Belgium. *Landscape and Urban Planning*, 134, 139–146.

De Iuliis, G. N., Newey, R. J., King, B. V., and Aitken, R. J. (2009). Mobile phone radiation induces reactive oxygen species production and DNA damage in human spermatozoa in vitro. *PLOS ONE*, 4(7), e6446.

De Laet, J., and Summers-Smith, J. D. (2007). The status of the urban house sparrow Passer domesticus in North-Western Europe: A review. *Journal of Ornithology*, 148(2), 275–278.

Delen, K., Sırav, B., Oruç, S., Seymen, C. M., Kuzay, D., Yeğin, K., and Take Kaplanoğlu, G. (2021). Effects of 2600 MHz radiofrequency radiation in brain tissue of male Wistar rats and neuroprotective effects of melatonin. *Bioelectromagnetics*, 42(2), 159–172.

Delgado, J. M. R. (1985). Biological effects of extremely low frequency electromagnetic fields. *Journal of Bioelectricity*, 4(1), 75–91.

Di Carlo, A., White, N., Guo, F., Garrett, P., and Litovitz, T. (2002). Chronic electromagnetic field exposure decreases HSP70 levels and lowers cytoprotection. *Journal of Cellular Biochemistry*, 84(3), 447–454.

Diego-Rasilla, F. J., Pérez-Mellado, V., and Pérez-Cembranos, A. (2017). Spontaneous magnetic alignment behaviour in free-living lizards. *The Science of Nature*, 104(3–4), 13.

Diem, E., Schwarz, C., Adlkofer, F., Jahn, O., and Rudiger, H. (2005). Non-thermal DNA breakage by mobile-phone radiation (1800 MHz) in human fibroblasts and in transformed GFSH-R17 rat granulosa cells in vitro. *Mutation Research*, 583(2), 178–183.

Divan, H. A., Kheifets, L., Obel, C., and Olsen, J. (2008). Prenatal and postnatal exposure to cell phone use and behavioral problems in children. *Epidemiology*, 19(4), 523–529.

Doherty, P. F., and Grubb, T. C. (1996). Effects of high-voltage power lines on birds breeding within the power lines' electromagnetic fields. *Sialia*, 18, 129–134.

Dubrov, A. P. (1978). *The geomagnetic field and life - Geomagnetobiology*. Plenum Press, New York.

Eberhardt, J. L., Persson, B. R., Brun, A. E., Salford, L. G., and Malmgren, L. O. (2008). Blood-brain barrier permeability and nerve cell damage in rat brain 14 and 28 days after exposure to microwaves from GSM mobile phones. *Electromagnetic Biology and Medicine*, 27(3), 215–229.

Engelmann, J. C., Deeken, R., Müller, T., Nimtz, G., Roelfsema, M. R. G., and Hedrich, R. (2008). Is gene activity in plant cells affected by UMTS-irradiation? A whole genome approach. *Advances and Applications in Bioinformatics and Chemistry: AABC*, 1, 71.

Engels, S., Schneider, N. L., Lefeldt, N., Hein, C. M., Zapka, M., Michalik, A., Elbers, D., Kittel, A., Hore, P. J., and Mouritsen, H. (2014). Anthropogenic electromagnetic noise disrupts magnetic compass orientation in a migratory bird. *Nature*. http://doi.org/10.1038/nature13290.

European Environment Agency. (2013). Late lessons from early warnings: Science, precaution, innovation. EEA Report No 1/2013. http://www.eea.europa.

Everaert, J., and Bauwens, D. (2007). A possible effect of electromagnetic radiation from mobile phone base stations on the number of breeding house sparrows (Passer domesticus). *Electromagnetic Biology and Medicine*, 26(1), 63–72.

Fahmy, H. M., and Mohammed, F. F. (2021). Hepatic injury induced by radio frequency waves emitted from conventional wi-fi devices in Wistar rats. *Human and Experimental Toxicology*, 40(1), 136–147.

Falcioni, L., Bua, L., Tibaldi, E., Lauriola, M., De Angelis, L., Gnudi, F. and Belpoggi, F. (2018). Report of final results regarding brain and heart tumors in Sprague-Dawley rats exposed from prenatal life until natural death to mobile phone radiofrequency field representative of a 1.8 GHz GSM base station environmental emission. *Environmental Research*, 165, 496–503.

Farrell, J. M., Litovitz, T. L., Penafiel, M., Montrose, C. J., Doinov, P., Barber, M., and Litovitz, T. A. (1997). The effect of pulsed and sinusoidal magnetic fields on the morphology of developing chick embryos. *Bioelectromagnetics*, 18(6), 431–438.

Favre, D. (2011). Mobile phone-induced honeybee worker piping. *Apidologie*, 42(3), 270–279.

Fejes, I., Za Vaczki, Z., Szollosi, J., Kolosza, R. S., Daru, J., Kovacs, L., and Pa, L. A. (2005). Is there a relationship between cell phone use and semen quality? *Archives of Andrology*, 51, 385–393. https://doi.org/10.1080/014850190924520.

Fernie, K. J., Bird, D. M., Dawson, R. D., and Laguë, P. C. (2000). Effects of electromagnetic fields on the reproductive success of American kestrels. *Physiological and Biochemical Zoology*, 73(1), 60–65.

Fernie, K. J., and Bird, D. M. (2001). Evidence of oxidative stress in American kestrels exposed to electromagnetic fields. *Environmental Research*, 86(2), 198–207.

Fernie, K. J., and Reynolds, S. J. (2005). The effects of electromagnetic fields from power lines on avian reproductive biology and physiology: A review. *Journal of Toxicology and Environmental Health, Part B*, 8(2), 127–140.

Ferrari, T. (2014). Magnets, magnetic field fluctuations and geomagnetic disturbances impair the homing ability of honey bee (Apis m.). *Journal of Apicultural Research*, 53(4), 452–456.

Flipo, D., Fournier, M., Benquet, C., Roux, P., Le Boulaire, C., Pinsky, C., and Krzystyniak, K. (1998). Increased apoptosis, changes in intracellular Ca2+, and functional alterations in lymphocytes and macrophages after in vitro exposure to static magnetic field. *Journal of Toxicology and Environmental Health: Part A*, 54(1), 63–76.

Formicki, K., Korzelecka-Orkisz, A., and Tański, A. (2019). Magnetoreception in fish. *Journal of Fish Biology*, 95(1), 73–91.

Frey, A. H. (1985). Science and standards: Data analysis reveals significant microwave-induced eye damage in humans. *Journal of Microwave Power*, 20(1), 53–55.

Frey, A. H., and Feld, S. R. (1975). Avoidance by rats of illumination with low power nonionizing electromagnetic energy. *Journal of Comparative and Physiological Psychology*, 89(2), 183–188. https://doi.org/10.1037/h0076662.

Friesen, M., and Havas, M. (2020). Effects of non-ionizing electromagnetic pollution on invertebrates, including pollinators such as honey bees: What we know, what we don't know, and what we need to know. In D. Danyluk (Ed.), *Working landscapes: Proceedings of the 12th Prairie conservation and endangered species conference*, pp. 127–138. February 2019. Critical Wildlife Habitat Program, Winnipeg, Manitoba. http://pcesc.ca/media/45404/final-2019-pcesc-proceedings.pdf.

Fritz, J., Eberhard, B., Esterer, C., Goenner, B., Trobe, D., Unsoeld, M., and Scope, A. (2020). Biologging is suspect to cause corneal opacity in two populations of wild living Northern Bald Ibises (Geronticus eremita). *Avian Research*, 11(1), 1–9.

Galeev, A. L. (2000). The effects of microwave radiation from mobile telephones on humans and animals. *Neuroscience and Behavioral Physiology*, 30(2), 187–194.

Gandhi, G., and Singh, P. (2005). Cytogenetic damage in mobile phone users: Preliminary data. *International Journal of Human Genetics*, 5(4), 259–265.

Gang, Y., Bai, Z., Chao, S., Cheng, Q., Wang, G., Tang, Z., and Yang, S. (2021). Current progress on the effect of mobile phone radiation on sperm quality: An updated systematic review and meta-analysis of human and animal studies. *Environmental Pollution*, 282, 116952.

Gill, A. B., Bartlett, M., and Thomsen, F. (2012). Potential interactions between diadromous fishes of UK conservation importance and the electromagnetic fields and subsea noise from marine renewable energy developments. *Journal of Fish Biology*, 81(2), 664–695.

Gómez-Perretta, C., Navarro, E. A., Segura, J., and Portolés, M. (2013). Subjective symptoms related to GSM radiation from mobile phone base stations: A cross-sectional study. *BMJ Open*, 3(12), e003836.

Granger, J., Walkowicz, L., Fitak, R., and Johnsen, S. (2020). Gray whales strand more often on days with increased levels of atmospheric radio-frequency noise. *Current Biology*, 30(4), R155–R156.

Greenberg, B., Bindokas, V. P., and Gauger, J. R. (1981). Biological effects of a 765-kV transmission line: Exposures and thresholds in honeybee colonies. *Bioelectromagnetics*, 2(4), 315–328.

Grémiaux, A., Girard, S., Guérin, V., Lothier, J., Baluška, F., Davies, E., Bonnet, P., and Vian, A. (2016). Low-amplitude, high-frequency electromagnetic field exposure causes delayed and reduced growth in Rosa hybrida. *Journal of Plant Physiology*, 190, 44–53.

Grefner, N. M., Yakovleva, T. L., and Boreysha, I. K. (1998). Effects of electromagnetic radiation on tadpole development in the common frog (Rana temporaria L.). *Russian Journal of Ecology*, 29, 133–134.

Grigoriev, I. U. G., Luk'ianova, S. N., Makarov, V. P., Rynskov, V. V., and Moiseeva, N. V. (1995). Motor activity off rabbits in conditions of chronic low-intensity pulse microwave irradiation. *Radiatsionnaia Biologiiia Radioecologiia*, 35, 29–35.

Grigor'ev, O. A., Bicheldeĭ, E. P., and Merkulo, A. V. (2003). Anthropogenic EMF effects on the condition and function of natural ecosystems. *Radiatsionnaia Biologiia, Radioecologiia*, 43(5), 544–551.

Grigoriew, J. G. (2003). Influence of the electromagnetic field of the mobile phones on chickens embryo, to the evaluation of the dangerousness after the criterion of this mortality. *Radiation Biology*, 5, 541–544.

Gustavino, B., Carboni, G., Petrillo, R., Paoluzzi, G., Santovetti, E., and Rizzoni, M. (2016). Exposure to 915 MHz radiation induces micronuclei in Vicia faba root tips. *Mutagenesis*, 31(2), 187–192.

Haggerty, K. (2010). Adverse influence of radio frequency background on trembling aspen seedlings: Preliminary observations. *International Journal of Forestry Research*, 2010, 836278.

Halgamuge, M. N., Yak, S. K., and Eberhardt, J. L. (2015). Reduced growth of soybean seedlings after exposure to weak microwave radiation from GSM 900 mobile phone and base station. *Bioelectromagnetics*, 36(2), 87–95.

Halgamuge, M. N. (2017). Review: Weak radiofrequency radiation exposure from mobile phone radiation on plants. *Electromagnetic Biology and Medicine*, 36(2), 213–235.

Hallmann, C. A., Sorg, M., Jongejans, E., Siepel, H., Hofland, N., Schwan, H., and Goulson, D. (2017). More than 75 percent decline over 27 years in total flying insect biomass in protected areas. *PLOS ONE*, 12(10), e0185809.

Hallberg, Ö., and Johansson, O. (2013). Increasing melanoma—Too many skin cell damages or too few repairs? *Cancers*, 5(1), 184–204.

Hardell, L., Carlberg, M., Söderqvist, F., Mild, K. H., and Morgan, L. L. (2007). Long-term use of cellular phones and brain tumours: Increased risk associated with use for > or =10 years > or =10 years. *Occupational and Environmental Medicine*, 64(9), 626–632.

Hardell, L., and Carlberg, M. (2020). [Comment] health risks from radiofrequency radiation, including 5G, should be assessed by experts with no conflicts of interest. *Oncology Letters*, 20(4), 1–1.

Hardell, L. (2021). Health council of the Netherlands and evaluation of the fifth generation, 5G, for wireless communication and cancer risks. *World Journal of Clinical Oncology*, 12(6), 393.

Hardell, L., Nilsson, M., Koppel, T., and Carlberg, M. (2021). Aspects on the international commission on non-ionizing radiation protection (ICNIRP) 2020 guidelines on radiofrequency radiation. *Journal of Cancer Science and Clinical Therapeutics*, 5(2), 250–285.

Harremoes, P., Gee, D., MacGarvin, M., Stirling, A., Keys, J., Wynne, B., and Vaz, S. G. (Eds.). (2013). *The precautionary principle in the 20th century: Late lessons from early warnings*. Routledge, London.

Harst, W., Kuhn, J., and Stever, H. (2006). Can electromagnetic exposure cause a change in behaviour? Studying possible non-thermal influences on honeybees—An approach within the framework of educational informatics. *Acta Syst IIAS Int. J*, 6(1), 1–6.

Hart, V., Kušta, T., Němec, P., Bláhová, V., Ježek, M., Nováková, P. and Burda, H. (2012). Magnetic alignment in carps: Evidence from the Czech Christmas fish market. *PLOS ONE*, 7(12), e51100.

Harte, C. (1950). Mutationsauslösung durch Ultrakurzwellen. *Chromosoma*, 3, 140–147.

Harte, C. (1972). Auslösung von Chromosomenmutationen durch Meterwellen in Pollenmutterzellen von Oenothera. *Chromosoma*, 36(4), 329–337.

Hässig, M., Jud, F., and Spiess, B. (2012). Increased occurrence of nuclear cataract in the calf after erection of a mobile phone base station. *Schweizer Archiv fur Tierheilkunde*, 154(2), 82–86.

Hässig, M., Wullschleger, M., Naegeli, H., Kupper, J., Spiess, B., Kuster, N., and Murbach, M. (2014). Influence of non ionizing radiation of base stations on the activity of redox proteins in bovines. *BMC Veterinary Research*, 10(1), 1–11.

Hertel, U. (1991). The forest dies as politicians look on. *Raum y Zeit*, 51(May–junio), 91.

Hiscock, H. G., Mouritsen, H., Manolopoulos, D., and Hore, P. J. (2017). Disruption of magnetic compass orientation in migratory birds by radiofrequency electromagnetic fields. *Biophysical Journal*, 113(7), 1475–1484. https://doi.org/10.1016/j.bpj.2017.07.031.

Hotary, K. B., and Robinson, K. R. (1992). Evidence of a role for endogenous electrical fields in chick embryo development. *Development*, 114(4), 985–996.

Hsu, C. Y., Ko, F. Y., Li, C. W., Fann, K., and Lue, J. T. (2007). Magnetoreception system in honeybees (Apis mellifera). *PLOS ONE*, 2(4), e395. http://doi.org/10.1371/journal.pone.0000395.

Huber, R., Graf, T., Cote, K. A., Wittmann, L., Gallmann, E., Matter, D., Schuderer, J., Kuster, N., Borbely, A. A., and Achermann, P. (2000). Exposure to pulsed high-frequency electromagnetic field during waking affects human sleep EEG. *NeuroReport*, 11(15), 3321–3325.

Hussein, R. A., and El-Maghraby, M. A. (2014). Effect of two brands of cell phone on germination rate and seedling of wheat (Triticum aestivum). *Journal of Environment Pollution and Human Health*, 2, 85–90.

Hutchison, Z. L., Gill, A. B., Sigray, P., He, H., and King, J. W. (2020). Anthropogenic electromagnetic fields (EMF) influence the behaviour of bottom-dwelling marine species. *Scientific Reports*, 10(1), 1–15.

Hyland, G. J. (2000). Physics and biology of mobile telephony. *Lancet*, 356(9244), 1833–1836.

Hyland, G. J. (2008). Physical basis of adverse and therapeutic effects of low intensity microwave radiation. *Indian Journal of Experimental Biology*, 46(5), 403–419.

ICNIRP. (2010). Guidelines for limiting exposure to time-varying electric and magnetic fields (1 Hz to 100 kHz). *Health Physics*, 99(6), 818–836.

ICNIRP. (2020). Guidelines for limiting exposure to electromagnetic fields (100 kHz to 300 GHz). *Health Physics*, 118(5), 483–524.

Jayasanka, S. M. D. H., and Asaeda, T. (2013). The significance of microwaves in the environment and its effect on plants. *Environmental Reviews*, 22(3), 220–228.

Jerman, I., Berden, M., Ruzic, R., and Skarja, M. (1998). Biological effects of TV set electromagnetic fields on the growth of spruce seedlings. *Electromagnetic Biology and Medicine*, 17(1), 31–42.

Johansson, O. (2009). Disturbance of the immune system by electromagnetic fields-A potentially underlying cause for cellular damage and tissue repair reduction which could lead to disease and impairment. *Pathophysiology*, 16(2–3), 157–177.

Johnsen, S., and Lohmann, K. J. (2005). The physics and neurobiology of magnetoreception. *Nature Reviews Neuroscience*, 6(9), 703–712.

Johnson-Liakouris, A. G. (1998). Radiofrequency (RF) sickness in the Lilienfeld study: An effect of modulated microwaves? *Archives of Environmental Health: An International Journal*, 53(3), 236–238.

Kalmijn, A. J. (1988). Detection of weak electric fields. In J. Atema, R.R. Fay, A.N. Popper, W.N. Tavolga (Eds): *Sensory biology of aquatic animals*, pp. 151–186. Springer, New York.

Keles, A. I. (2020). Morphological changes in the vertebrae and central canal of rat pups born after exposure to the electromagnetic field of pregnant rats. *Acta Histochemica*, 122(8), 151652.

Khan, M. D., Ali, S., Azizullah, A., and Shuijin, Z. (2018). Use of various biomarkers to explore the effects of GSM and GSM-like radiations on flowering plants. *Environmental Science and Pollution Research International*, 25(25), 24611–24628.

Kheifets, L., Ahlbom, A., Crespi, C. M., Draper G, Hagihara J, et al. (2010). Pooled analysis of recent studies on magnetic fields and childhood leukaemia. *British Journal of Cancer*, 103(7), 1128–1135.

Kiepenheuer, K. O., Brauer, I., and Harte, C. (1949). Über die Wirkung von Meterwellen auf das Teilungswachstum der Pflanzen. *Naturwissenschaften*, 36(1), 27.

Kiernan, V. (1995). Forest grows tall on radio waves. *New Scientist*, 14, 5.

Kim, H. S., Choi, H. D., Pack, J. K., Kim, N., and Ahn, Y. H. (2021). Biological effects of exposure to a radiofrequency electromagnetic field on the placental barrier in pregnant rats. *Bioelectromagnetics*, 42(3), 191–199.

Kirschvink, J. L., and Gould, J. L. (1981). Biogenic magnetite as a basis for magnetic field sensitivity in animals. *Biological Systems*, 13(3), 181–201.

Kirschvink, J. L. (1989). Magnetite biomineralization and geomagnetic sensitivity in higher animals: An update and recommendations for future study. *Bioelectromagnetics*, 10(3), 239–259.

Kirschvink, J. L., Walker, M. M., and Diebel, C. E. (2001). Magnetite-based magnetoreception. *Current Opinion in Neurobiology*, 11(4), 462–467.

Kler, T. K., Kumar, M., and Vashishat, N. (2018). Effects of electromagnetic radiations on diversity and breeding biology of birds living near power lines and mobile towers at Ludhiana, Punjab. *Journal of Environmental Biology*, 39(2), 247–252.

Kolodynski, A. A., and Kolodynska, V. V. (1996). Motor and psychological functions of school children living in the area of the Skrunda radio location station in Latvia. *Science of the Total Environment*, 180(1), 87–93.

Kostoff, R. N., Heroux, P., Aschner, M., and Tsatsakis, A. (2020). Adverse health effects of 5G mobile networking technology under real-life conditions. *Toxicology Letters*, 323, 35–40.

Kramarenko, A. V., and Tan, U. (2003). Effects of high-frequency electromagnetic fields on human EEG: A brain mapping study. *International Journal of Neuroscience*, 113(7), 1007–1019.

Krishnan, V., Park, S. A., Shin, S. S., Alon, L., Tressler, C. M., Stokes, W., and Pelled, G. (2018). Wireless control of cellular function by activation of a novel protein responsive to electromagnetic fields. *Scientific Reports*, 8(1), 1–12.

Krstic, D. D., Dindic, B. J., Sokolovic, D. T., Markovic, V. V., Petkovic, D. M. & Radic, S. B. (2005, September). The results of experimental exposition of mice by mobile telephones. In TELSIKS 2005-2005 International Conference on Telecommunication in Modern Satellite, Cable and Broadcasting Services (Vol. 2, pp. 465–470). IEEE.

Kumar, N. R., Rana, N., and Kalia, P. (2013). Biochemical changes in haemolymph of Apis mellifera L. drone under the influence of cell phone radiations. *Journal of Applied and Natural Science*, 5(1), 139–141.

Kumar, A., Singh, H. P., Batish, D. R., Kaur, S., and Kohli, R. K. (2016). EMF radiations (1800 MHz) inhibited early seedling growth of maize (Zea mays) involves alterations in starch and sucrose metabolism. *Protoplasma*, 253(4), 1043–1049.

Kumar, R., Deshmukh, P. S., Sharma, S., and Banerjee, B. D. (2021). Effect of mobile phone signal radiation on epigenetic modulation in the hippocampus of Wistar rat. *Environmental Research*, 192, 110297.

Kundu, A., and IEEE. (2013). Specific absorption rate evaluation in apple exposed to RF radiation from GSM mobile towers. In *IEEE applied electromagnetics conference (AEMC)*, pp. 1–2. IEEE. (ISBN 978-1-4799-3266-5). doi: 10.1109/AEMC.2013.7045044.

Lai, H. (2005). Biological effects of radiofrequency electromagnetic field. *Encyclopedia of Biomaterials and Biomedical Engineering*, 10, 1–8.

Lai, H., Carino, M. A., Horita, A., and Guy, A. W. (1989). Acute low-level microwave exposure and central cholinergic activity: A doseresponse study. *Bioelectromagnetics*, 10(2), 203–209.

Lai, H., and Singh, N. P. (1995). Acute low-intensity microwave exposure increases DNA single-strand breaks in rat brain cells. *Bioelectromagnetics*, 16(3), 207–210.

Lai, H., and Singh, N. P. (1997). Acute exposure to a 60 Hz magnetic field increases DNA strand breaks in rat brain cells. *Bioelectromagnetics*, 18(2), 156–165.

Lázaro, A., Chroni, A., Tscheulin, T., Devalez, J., Matsoukas, C., and Petanidou, T. (2016). Electromagnetic radiation of mobile telecommunication antennas affects the abundance and composition of wild pollinators. *Journal of Insect Conservation*, 20(2), 315–324.

Lee, G. M., Neutra, R. R., Hristova, L., Yost, M., and Hiatt, R. A. (2002). A nested case-control study of residential and personal magnetic field measures and miscarriages. *Epidemiology*, 13(1), 21–31.

Lee, W., and Yang, K. L. (2014). Using medaka embryos as a model system to study biological effects of the electromagnetic fields on development and behavior. *Ecotoxicology and Environment Safety*, 108, 187–194.

Lee, D., Lee, J., and Lee, I. (2015). Cell phone-generated radio frequency electromagnetic field effects on the locomotor behaviors of the fishes Poecilia reticulata and Danio rerio. *International Journal of Radiation Biology*, 91(10), 843–850.

Lerchl, D., Lerchl, A., Hantsch, P., Bitz, A., Streckert, J., & Hansen, V. (2000). Studies on the effects of radiofrequency fields on conifers. In Bioelectromagnetics Society Annual Meeting (pp. 9–16).

Leszczynski, D., Joenväärä, S., Reivinen, J., and Kuokka, R. (2002). Non-thermal activation of the hsp27/p38MAPK stress pathway by mobile phone radiation in human endothelial cells: Molecular mechanism for cancer-and blood-brain barrier-related effects. *Differentiation; Research in Biological Diversity*, 70(2–3), 120–129.

Levengood, W. C. (1969). A new teratogenic agent applied to amphibian embryos. *Journal of Embryology and Experimental Morphology*, 21(1), 23–31.

Levin, M. (2003). Bioelectromagnetics in morphogenesis. *Bioelectromagnetics*, 24(5), 295–315.
Levitt, B. B., and Lai, H. (2010). Biological effects from exposure to electromagnetic radiation emitted by cell tower base stations and other antenna arrays. *Environmental Reviews*, 18, 369–395.
Levitt, B. B., Lai, H. C., and Manville, A. M. (2021a). Effects of non-ionizing electromagnetic fields on flora and fauna, part 1. Rising ambient EMF levels in the environment. *Reviews on Environmental Health*, 37(1), 81–122.
Levitt, B. B., Lai, H. C., and Manville, A. M. (2021b). Effects of non-ionizing electromagnetic fields on flora and fauna, part 2 impacts: How species interact with natural and man-made EMF. *Reviews on Environmental Health*, 37(3): 327–406.
Liboff, A. R., and Jenrow, K. A. (2000). New model for the avian magnetic compass. *Bioelectromagnetics*, 21(8), 555–565.
Lopatina, N. G., Zachepilo, T. G., Kamyshev, N. G., Dyuzhikova, N. A., and Serov, I. N. (2019). Effect of non-ionizing electromagnetic radiation on behavior of the honeybee, Apis mellifera L. (Hymenoptera, Apidae). *Entomological Review*, 99(1), 24–29.
López-Martín, E., Jorge-Barreiro, F. J., Relova-Quintero, J. L., Salas-Sánchez, A. A., and Ares-Pena, F. J. (2020). Exposure to 2.45 GHz radiofrequency modulates calcitonin-dependent activity and HSP-90 protein in parafollicular cells of rat thyroid gland. *Tissue and Cell*, 68, 101478.
López, I., Félix, N., Rivera, M., Alonso, A., and Maestú, C. (2021). What is the radiation before 5G? A correlation study between measurements in situ and in real time and epidemiological indicators in Vallecas, Madrid. *Environmental Research*, 194, 110734.
Loscher, W., and Kas, G. (1998). Conspicuous behavioural abnormalities in a dairy cow herd near a TV and radio transmitting antenna. *Practical Veterinary Surgeon*, 29, 437–444.
Luo, K., Luo, C., Li, G., Yao, X., Gao, R., Hu, Z., and Zhao, H. (2019). High-voltage electrostatic field-induced oxidative stress: Characterization of the physiological effects in Sitobion avenae (Hemiptera: Aphididae) across multiple generations. *Bioelectromagnetics*, 40(1), 52–61.
Lupi, D., Tremolada, P., Colombo, M., Giacchini, R., Benocci, R., Parenti, P., Zambon, G., and Vighi, M. (2020). Effects of pesticides and electromagnetic fields on honeybees: A field study using biomarkers. *International Journal of Environmental Research*, 14(1), 107–122.
Lupi, D., Palamara Mesiano, M., Adani, A., Benocci, R., Giacchini, R., Parenti, P. and Tremolada, P. (2021). Combined effects of pesticides and electromagnetic-fields on honeybees: Multi-stress exposure. *Insects*, 12(8), 716.
Madjar, H. M. (2016). Human radio frequency exposure limits: An update of reference levels in Europe, USA, Canada, China, Japan and Korea. *In 2016 international symposium on electromagnetic compatibility-EMC EUROPE* (pp. 467–473). IEEE. doi: 10.1109/EMCEurope.2016.7739164
Magone, I. (1996). The effect of electromagnetic radiation from the Skrunda Radio Location Station on Spirodela polyrhiza (L.) Schleiden cultures. *Science of the Total Environment*, 180(1), 75–80.
Magras, I. N., and Xenos, T. D. (1997). RF radiation-induced changes in the prenatal development of mice. *Bioelectromagnetics*, 18(6), 455–461.
Malkemper, P., Eder, S. H. K., Begall, S., Phillips, J. B., Winklhofer, M., Hart, V., and Burda, H. (2015). Magnetoreception in the wood mouse (Apodemus sylvaticus): Influence of weak frequency-modulated radio frequency fields. *Scientific Reports*, 5, 9917. https://doi.org/10.1038/srep09917.
Mann, K., and Roschkle, J. (1996). Effects of pulsed high-frequency electromagnetic fields on human sleep. *Neuropsychobiology*, 33(1), 41–47.
Margaritis, L. H., Manta, A. K., Kokkaliaris, K. D. et al. (2014). Drosophila oogenesis as a bio-marker responding to EMF sources. *Electromagnetic Biology and Medicine*, 33(3), 165–189.
Marino, A. A., Hart, F. X., and Reichmanism, M. (1983). Weak electric fields affect plant development. *IEEE Transactions on Bio-Medical Engineering*, 12(12), 833–834.
Marino, A. A., Nilsen, E., and Frilot, C. (2003). Nonlinear changes in brain electrical activity due to cell phone radiation. *Bioelectromagnetics*, 24(5), 339–346.
Marks, T. A., Ratke, C. C., and English, W. O. (1995). Strain voltage and developmental, reproductive and other toxicology problems in dogs, cats and cows: A discussion. *Veterinary and Human Toxicology*, 37, 163–172.
Martínez, E., Carbonell, M. V., and Flórez, M. (2003). Estimulación de la germinación y el crecimiento por exposición a campos magnéticos. *Investigación y Ciencia*, 324, 24–28.
Mazor, R., Korenstein-Ilan, A., Barbul, A., Eshet, Y., Shahadi, A., Jerby, E., and Korenstein, R. (2008). Increased levels of numerical chromosome aberrations after in vitro exposure of human peripheral blood lymphocytes to radiofrequency electromagnetic fields for 72 hours. *Radiation Research*, 169(1), 28–37.

Mech, L. D., and Barber, S. M. (2002). *A critique of wildlife radio-tracking and its use in national parks*. A Report to the US National Park Service, Northern Prairie Wildlife Research Center.

Migdał, P., Murawska, A., Bieńkowski, P., Berbeć, E., and Roman, A. (2021). Changes in honeybee behavior parameters under the influence of the E-field at 50 Hz and variable intensity. *Animals*, 11(2), 247.

Miura, M., and Okada, J. (1991). Non-thermal vasodilatation by radio frequency burst-type electromagnetic field radiation in the frog. *Journal of Physiology*, 435, 257–273.

Mohammed, H. S., Fahmy, H. M., Radwan, N. M., and Elsayed, A. A. (2013). Non-thermal continuous and modulated electromagnetic radiation fields effects on sleep EEG of rats. *Journal of Advanced Research*, 4(2), 181–187. https://doi.org/10.1016/j.jare.2012.05.005.

Mohring, B., Henry, P. Y., Jiguet, F., Malher, F., and Angelier, F. (2020). Investigating temporal and spatial correlates of the sharp decline of an urban exploiter bird in a large European city. *Urban Ecosystems*, 24(3), 501–513.

Møller, A. P. (2019). Parallel declines in abundance of insects and insectivorous birds in Denmark over 22 years. *Ecology and Evolution*, 9(11), 6581–6587.

Moorhouse, T. P., and Macdonald, D. W. (2005). Indirect negative impacts of radio-collaring: Sex ratio variation in water voles. *Journal of Applied Ecology*, 42(1), 91.

Nagata, H., and Yamagish, S. (2013). Re-introduction of crested ibis on Sado Island, Japan. In P. S. Soorae (Ed.), *Global re-introduction perspectives: Further case studies from around the globe*, pp. 58–62. Gland and Abu Dhabi: IUCN/SSC Re-introduction Specialist Group and Environment Agency-Abu Dhabi.

National Toxicology Program (NTP), (2018). *Toxicology and carcinogenesis studies of GSM- and CDMA-modulated cell phone radiofrequency radiation at 900 MHz in Sprague Dawley Rats (whole body exposure)*. Technical Report Series No. 595. National Toxicology Program Public Health Service U.S. Department of Health and Human Services ISSN: 2378-8925 Research Triangle Park, North Carolina, USA.

Navakatikian, M. A., and Tomashevskaya, L. A. (1994). Phasic behavioral and endocrine effects of microwaves of nonthermal intensity. In D. O. Carpenter (Ed.), *Biological effects of electric and magnetic fields*, vol. 1, 333–342, Academic Press, San Diego.

Neufeld, E., and Kuster, N. (2018). Systematic derivation of safety limits for time-varying 5G radiofrequency exposure based on analytical models and thermal dose. *Health Physics*, 115(6), 705–711.

Neurath, P. W. (1968). High gradient magnetic fields inhibit embryonic development of frogs. *Nature*, 21(5161), 1358–1359.

Nicholls, B., and Racey, P. A. (2007). Bats avoid radar installations: Could electromagnetic fields deter bats from colliding with wind turbines? *PLOS ONE*, 3(3), e297.

Nikolaevich, N., Igorevna, A., and Vasil, G. (2001). *Influence of highfrequency electromagnetic radiation at non-thermal intensities on the human body (a review of work by Russian and Ukrainian researchers)*. No Place to Hide, 3 Supplement.

Nirwane, A., Sridhar, V., and Majumdar, A. (2016). Neurobehavioural changes and brain oxidative stress induced by acute exposure to GSM900 mobile phone radiations in zebrafish (Danio rerio). *Toxicological Research*, 32(2), 123–132.

Nishimura, T., Okano, H., Tada, H., Nishimura, E., Sugimoto, K., Mohri, K., and Fukushima, M. (2010). Lizards respond to an extremely low-frequency electromagnetic field. *Journal of Experimental Biology*, 213(12), 1985–1990.

Novoselova, E. T., and Fesenko, E. E. (1998). Stimulation of production of tumor necrosis factor by murine macrophages when exposed in vio and in vitro to weak electromagnetic waves in the centimeter range. *Biofizika*, 43(6), 1132–1133.

Ntzouni, M. P., Stamatakis, A., Stylianopoulou, F., and Margaritis, L. H. (2011). Short-term memory in mice is affected by mobile phone radiation. *Pathophysiology*, 18(3), 193–199.

Nuccitelli, R. (1988). Ionic currents in morphogenesis. *Experientia*, 44(8), 657–666.

Obajuluwa, A. O., Akinyemi, A. J., Afolabi, O. B., Adekoya, K., Sanya, J. O., and Ishola, A. O. (2017). Exposure to radio-frequency electromagnetic waves alters acetylcholinesterase gene expression, exploratory and motor coordination-linked behaviour in male rats. *Toxicology Reports*, 4, 530–534.

Odemer, R., and Odemer, F. (2019). Effects of radiofrequency electromagnetic radiation (RF-EMF) on honey bee queen development and mating success. *Science of the Total Environment*, 661, 553–562.

Ozel, H. B., Cetin, M., Sevik, H., Varol, T., Isik, B., and Yaman, B. (2021). The effects of base station as an electromagnetic radiation source on flower and cone yield and germination percentage in Pinus brutia Ten. *Biologia Futura*, 72(3), 359–365..

Ozgur, E., Kismali, G., Guler, G., Akcay, A., Ozkurt, G., Sel, T., and Seyhan, N. (2013). Effects of prenatal and postnatal exposure to GSM-like radiofrequency on blood chemistry and oxidative stress in infant rabbits, an experimental study. *Cell Biochemistry and Biophysics*, 67(2), 743–751.

Pall, M. L. (2013). Electromagnetic fields act via activation of voltage-gated calcium channels to produce beneficial or adverse effects. *Journal of Cellular and Molecular Medicine*, 17(8), 958–965.

Pall, M. (2016). Electromagnetic fields act similarly in plants as in animals: Probable activation of calcium channels via their voltage sensor. *Current Chemical Biology*, 10(1), 74–82.

Panagopoulos, D. J., Messini, N., Karabarbounis, A., Filippetis, A. L., and Margaritis, L. H. (2000). A mechanism for action of oscillating electric fields on cells. *Biochemical and Biophysical Research Communications*, 272(3), 634–640.

Panagopoulos, D. J., Karabarbounis, A., and Margaritis, L. H. (2002). Mechanism for action of electromagnetic fields on cells. *Biochemical and Biophysical Research Communications*, 298(1), 95–102.

Panagopoulos, D. J., Karabarbounis, A., and Margaritis, L. H. (2004). Effect of GSM 900 MHz mobile phone radiation on the reproductive capacity of Drosophila melanogaster. *Electromagnetic Biology and Medicine*, 23(1), 29–43.

Panagopoulos, D. J., Chavdoula, E. D., Nezis, I. P., and Margaritis, L. H. (2007a). Cell death induced by GSM 900-MHz and DCS 1800-MHz mobile telephony radiation. *Mutation Research/Genetic Toxicology and Environmental Mutagenesis*, 626(1–2), 69–78.

Panagopoulos, D. J., Chavdoula, E. D., Karabarbounis, A., and Margaritis, L. H. (2007b). Comparison of bioactivity between GSM 900 MHz and DCS 1800 MHz mobile telephony radiation. *Electromagnetic Biology and Medicine*, 26(1), 33–44.

Panagopoulos, D. J., Chavdoula, E. D., and Margaritis, L. H. (2010). Bioeffects of mobile telephony radiation in relation to its intensity or distance from the antenna. *International Journal of Radiation Biology*, 86(5), 345–357.

Panagopoulos, D. J. (2012). Effect of microwave exposure on the ovarian development of Drosophila melanogaster. *Cell Biochemistry and Biophysics*, 63(2), 121–132.

Panagopoulos, D. J. (2013). Electromagnetic interaction between environmental fields and living systems determines health and well-being. In M.H. Kwang and S.O. Yoon (Eds) *Electromagnetic fields: Principles, engineering applications and biophysical effects*, 99–140. Nova Science Publishers, New York.

Panagopoulos, D. J., Karabarbounis, A., and Lioliousis, C. (2013). ELF alternating magnetic field decreases reproduction by DNA damage induction. *Cell Biochemistry and Biophysics*, 67(2), 703–716.

Panagopoulos, D. J., Johansson, O., and Carlo, G. L. (2015a). Polarization: A key difference between man-made and natural electromagnetic fields, in regard to biological activity. *Scientific Reports*, 5, 14914. https://doi.org/10.1038/srep14914.

Panagopoulos, D. J., Johansson, O., and Carlo, G. L. (2015b). Real versus simulated mobile phone exposures in experimental studies. *BioMed Research International*, 2015, 607053.

Panagopoulos, D. J., and Balmori, A. (2017). On the biophysical mechanism of sensing atmospheric discharges by living organisms. *Science of the Total Environment*, 599, 2026–2034.

Panagopoulos, D. J. (2017). Mobile telephony radiation effects on insect ovarian cells: The necessity for real exposures bioactivity assessment: The key role of polarization, and the ion forced-oscillation mechanism. In C. D. Geddes (Ed.), *Microwave effects on DNA and proteins*. Springer, Cham, Switzerland.

Panagopoulos, D. J. (2019). Comparing DNA damage induced by mobile telephony and other types of man-made electromagnetic fields. *Mutation Research Reviews*, 781, 53–62.

Panagopoulos, D. J. (2020). Comparing chromosome damage induced by mobile telephony radiation and a high caffeine dose: Effect of combination and exposure duration. *General Physiology and Biophysics*, 39(6), 531–544.

Panagopoulos, D. J., and Karabarbounis, A. (2020). Comments on "diverse radiofrequency sensitivity and radiofrequency effects of mobile or cordless phone near fields exposure in Drosophila melanogaster". *Advances in Environmental Studies*, 4(1), 271–276.

Panagopoulos, D. J., Balmori, A., and Chrousos, G. P. (2020). On the biophysical mechanism of sensing upcoming earthquakes by animals. *Science of the Total Environment*, 717, 136989.

Panagopoulos, D. J., Karabarbounis, A., Yakymenko, I., and Chrousos, G. P. (2021). Human-made electromagnetic fields: Ion forced-oscillation and voltage-gated ion channel dysfunction, oxidative stress and DNA damage. *International Journal of Oncology*, 59(5), 1–16.

Pandey, N., and Giri, S. (2018). Melatonin attenuates radiofrequency radiation (900 MHz)-induced oxidative stress, DNA damage and cell cycle arrest in germ cells of male Swiss albino mice. *Toxicology and Industrial Health*, 34(5), 315–327.

Pavel, A., Ungureanu, C. E., Bara, I. I., Gassner, P., and Creanga, D. E. (1998). Cytogenetic changes induced by low-intensity microwaves in the species Triticum aestivum. *Revista Medico-Chirurgicala a Societatii de Medici si Naturalisti din Iasi*, 102(3–4), 89–92.

Pedersen, G. F. (1997). Amplitude modulated RF fields stemming from a GSM/DCS-1800 phone. *Wireless Networks*, 3(6), 489–498.

Pereira, M. C., Guimarães, I. D. C., Acosta-Avalos, D., and Antonialli Junior, W. F. (2019). Can altered magnetic field affect the foraging behaviour of ants? *PLOS ONE*, 14(11), e0225507.

Pesnya, D. S., and Romanovsky, A. V. (2013). Comparison of cytotoxic and genotoxic effects of plutonium-239 alpha particles and mobile phone GSM 900 radiation in the Allium cepa test. *Mutation Research/Genetic Toxicology and Environmental Mutagenesis*, 750(1–2), 27–33.

Phillips, J. B., Borland, S. C., Freake, M. J., Brassart, J., and Kirschvink, J. L. (2002). Fixed-axis' magnetic orientation by an amphibian: Non-shoreward-directed compass orientation, misdirected homing or positioning a magnetite-based map detector in a consistent alignment relative to the magnetic field? *Journal of Experimental Biology*, 205(24), 3903–3914.

Phillips, J. B., Jorge, P. E., and Muheim, R. (2010). Light-dependent magnetic compass orientation in amphibians and insects: Candidate receptors and candidate molecular mechanisms. *Journal of the Royal Society*, 7(Suppl 2), S241–S256.

Powney, G. D., Carvell, C., Edwards, M., Morris, R. K., Roy, H. E., Woodcock, B. A., and Isaac, N. J. (2019). Widespread losses of pollinating insects in Britain. *Nature Communications*, 10(1), 1–6.

Presman, A. S. (1977). *Electromagnetic fields and life*, 3rd ed. Plenum Press, New York.

Puranen, L., and Jokela, K. (1996). Radiation hazard assessment of pulsed microwave radars. *Journal of Microwave Power and Electromagnetic Energy*, 31(3), 165–177.

Radomska, M. M., Horobtsov, I. V., Cherniak, L. M., and Tykhenko, O. M. (2021). The analysis of airports' physical factors impacts on wildlife. *Науковий вісник НЛТУ України*, 31(3), 74–79.

Rafati, A., Rahimi, S., Talebi, A., Soleimani, A., Haghani, M., and Mortazavi, S. M. J. (2015). Exposure to radiofrequency radiation emitted from common mobile phone jammers alters the pattern of muscle contractions: An animal model study. *Journal of Biomedical Physics and Engineering*, 5(3), 133.

Raveendran, R., and Tabet Aoul, K. A. (2021). A meta-integrative qualitative study on the hidden threats of smart buildings/cities and their associated impacts on humans and the environment. *Buildings*, 11(6), 251.

Ribeiro-Oliveira, J. P. (2019). Electromagnetism and plant development: A new unknown in a known world. *Theoretical and Experimental Plant Physiology*, 31(4), 423–427.

Ritz T., Adem S., Schulten K. (2000). A model for photoreceptor-based magnetoreception in birds. *Biophys. J.*, 78, 707–718.

Ritz, T., Thalau, P., Phillips, J. B., Wiltschko, R., and Wiltschko, W. (2004). Resonance effects indicate a radical-pair mechanism for avian magnetic compass. *Nature*, 429(6988), 177–180.

Ritz, T., Wiltschko, R., Hore, P. J., Rodgers, C. T., Stapput, K., Thalau, P., and Wiltschko, W. (2009). Magnetic compass of birds is based on a molecule with optimal directional sensitivity. *Biophysical Journal*, 96(8), 3451–3457.

Rodrigues, N. C. P., Dode, A. C., Andrade, M. K. D. N., O'Dwyer, G., Monteiro, D. L. M., Reis, I. N. C., Frossard, V. C., and Lino, V. T. S. (2021). The effect of continuous low-intensity exposure to electromagnetic fields from radio base stations to cancer mortality in brazil. *International Journal of Environmental Research and Public Health*, 18(3), 1229.

Roman, A., and Tombarkiewicz, B. (2009). Prolonged weakening of the geomagnetic field (GMF) affects the immune system of rats. *Bioelectromagnetics*, 30(1), 21–28.

Roux, D., Vian, A., Girard, S., Bonnet, P., Paladian, F., Davies, E., and Ledoigt, G. (2006). Electromagnetic fields (900 MHz) evoke consistent molecular responses in tomato plants. *Physiologia Plantarum*, 128(2), 283–288.

Roux, D., Vian, A., Girard, S., Bonnet, P., Paladian, F., Davies, E., and Ledoigt, G. (2008). High frequency (900 MHz) low amplitude (5 V m − 1) electromagnetic field: A genuine environmental stimulus that affects transcription, translation, calcium and energy charge in tomato. *Planta*, 227(4), 883–891.

Salford, L. G., Brun, A. E., Eberhardt, J. L., Malmgren, L., and Persson, B. R. (2003). Nerve cell damage in mammalian brain after exposure to microwaves from GSM mobile phones. *Environmental Health Perspectives*, 111(7), 881–883.

Sánchez-Bayo, F., and Wyckhuys, K. A. (2019). Worldwide decline of the entomofauna: A review of its drivers. *Biological Conservation*, 232, 8–27.

Sandu, D. D., Goiceanu, C., Ispas, A., Creanga, I., Miclaus, S., and Creanga, D. E. (2005). A preliminary study on ultra high frequency electromagnetic fields effect on black locust chlorophylls. *Acta Biologica Hungarica*, 56(1–2), 109–117.

Sauter, M. (2011). *From GSM to LTE: An introduction to mobile networks and mobile broadband*. John Wiley & Sons, New Jersey.

Selga, T., and Selga, M. (1996). Response of Pinus sylvestris L. needles to electromagnetic fields: Cytological and ultrastructural aspects. *Science of the Total Environment*, 180(1), 65–73.

Senavirathna, M. D. H. J., and Asaeda, T. (2014). Radio-frequency electromagnetic radiation alters the electric potential of Myriophyllum aquaticum. *Biologia Plantarum*, 58(2), 355–362.

Senavirathna, M. D., and Takashi, A. (2014). The significance of microwaves in the environment and its effect on plants. *Environmental Reviews*, 22(3), 220–228.

Sharma, V. P., Singh, H. P., Kohli, R. K., and Batish, D. R. (2009). Mobile phone radiation inhibits *Vigna radiata* (mung bean) root growth by inducing oxidative stress. *Science of the Total Environment*, 407(21), 5543–5547.

Sharma, V. P., and Kumar, N. R. (2010). Changes in honeybee behaviour and biology under the influence of cellphone radiations. *Current Science*, 98, 1376–1378.

Shepherd, S., Lima, M. A. P., Oliveira, E. E., Sharkh, S. M., Jackson, C. W., and Newland, P. L. (2018). Extremely low frequency electromagnetic fields impair the cognitive and motor abilities of honey bees. *Scientific Reports*, 8(1), 1–9.

Shende, V. A., and Patil, K. G. (2015). Electromagnetic radiations: A possible impact on population of house sparrow (Passer domesticus). *Engineering International*, 3(1), 45–52.

Sheridan, E., Randolet, J., DeVault, T. L., Seamans, T. W., Blackwell, B. F., and Fernández-Juricic, E. (2015). The effects of radar on avian behavior: Implications for wildlife management at airports. *Applied Animal Behaviour Science*, 171, 241–252.

Schlegel, P. A. (1997). Behavioral sensitivity of the European blind cave salamander, Proteus anguinus, and a Pyrenean newt, Euproctus asper, to electrical fields in water. *Brain Behaviour and Evolution*, 49(3), 121–131.

Schmutz, P., Siegenthaler, J., Stäger, C., Tarjan, D., and Bucher, J. B. (1996). Long-term exposure of young spruce and beech trees to 2450 MHz microwave radiation. *Science of the Total Environment*, 180(1), 43–48.

Sharma, V. P., and Kumar, N. R. (2010). Changes in honeybee behaviour and biology under the influence of cellphone radiations. *Current Science*, 98, 1376–1378.

Shim, Y., Lee, I., and Park, S. (2013). The impact of LTE UE on audio devices. *ETRI Journal*, 35(2), 332–335.

Singh, H. P., Sharma, V. P., Batish, D. R., and Kohli, R. K. (2012). Cell phone electromagnetic field radiations affect rhizogenesis through impairment of biochemical processes. *Environmental Monitoring and Assessment*, 184(4), 1813–1821.

Singh, K. V., Gautam, R., Meena, R., Nirala, J. P., Jha, S. K., and Rajamani, P. (2020). Effect of mobile phone radiation on oxidative stress, inflammatory response, and contextual fear memory in Wistar rat. *Environmental Science and Pollution Research International*, 27(16), 19340–19351.

Singh, R., Kour, D. N., Ahmad, F., and Sahi, D. N. (2013). The causes of decline of House Sparrow (Passer domesticus, Linnaeus 1758) in urban and suburban areas of Jammu region, J & K. *Entomology Zoology*, 8, 803–811.

Singh, R. K., Bisht, D., and Prasad, R. C. (2017). Development of 5G mobile network technology and its architecture. *International Journal of Recent Trends in Engineering & Research*, 3(10), 196–201.

Sirav, B., and Seyhan, N. (2016). Effects of GSM modulated radio-frequency electromagnetic radiation on permeability of blood–brain barrier in male & female rats. *Journal of Chemical Neuroanatomy*, 75(B), 123–127.

Szmigielski, S. (2013). Reaction of the immune system to low-level RF/MW exposures. *Science of the Total Environment*, 454–455, 393–400.

Smith-Roe, S. L., Wyde, M. E., Stout, M. D., Winters, J. W., Hobbs, C. A., Shepard, K. G., and Witt, K. L. (2020). Evaluation of the genotoxicity of cell phone radiofrequency radiation in male and female rats and mice following subchronic exposure. *Environmental and Molecular Mutagenesis*, 61(2), 276–290.

Soja, G., Kunsch, B., Gerzabek, M., Reichenauer, T., Soja, A. M., Rippar, G., and Bolhar-Nordenkampf, H. R. (2003). Growth and yield of winter wheat (Triticum aestivum) and corn (Zea mays) near a high voltage transmission line. *Bioelectromagnetics*, 24(2), 91–102.

Soran, M. L., Stan, M., Niinemets, Ü., and Copolovici, L. (2014). Influence of microwave frequency electromagnetic radiation on terpene emission and content in aromatic plants. *Journal of Plant Physiology*, 171(15), 1436–1443.

Stanojević, V., Prolić, Z., Savić, T., Todorović, D., and Janać, B. (2005). Effects of extremely low frequency (50 Hz) magnetic field on development dynamics of the housefly (Musca domestica L.). *Electromagnetic Biology and Medicine*, 24(2), 99–107.

Stefi, A. L., Margaritis, L. H., and Christodoulakis, N. S. (2016). The effect of the non ionizing radiation on cultivated plants of Arabidopsis thaliana (Col.). *Flora*, 223, 114–120.

Stever, J., Kuhn, C., Otten, B., and Wunder, W. H. (2005). *Verhaltensanderungunter Elektromagnetischer Exposition. Pilotstudie, Institut für Mathematik. Arbeitsgruppe, Bildungsinformatik*. Universität Koblenz-Landau, Mainz, Germany.

Suetov, A. A., and Alekperov, S. I. (2019). Acute ocular lesions after exposure to electromagnetic radiation of ultrahigh frequency (an experimental study). *Vestnik Oftalmologii*, 135(4), 41–49.

Sultangaliyeva, I., Beisenova, R., Tazitdinova, R., Abzhalelov, A., and Khanturin, M. (2020). The influence of electromagnetic radiation of cell phones on the behavior of animals. *Veterinary World*, 13(3), 549.

Summers-Smith, J. D. (2003). The decline of the House Sparrow: a review. *Brit. Bird*, 96, 439–446.

Surducan, V., Surducan, E., Neamtu, C., Mot, A. C., and Ciorîţă, A. (2020). Effects of long-term exposure to low-power 915 MHz unmodulated radiation on Phaseolus vulgaris L. *Bioelectromagnetics*, 41(3), 200–212.

Surendran, N. S., Siddiqui, N. A., Mondal, P., and Nandan, A. (2020). Repercussion of electromagnetic radiation from cell towers/mobiles and their impact on migratory birds. In: Siddiqui, N., Tauseef, S., Abbasi, S., Khan, F. (eds) Advances in Air Pollution Profiling and Control. (pp. 193–202), Transactions in Civil and Environmental Engineering. Springer, Singapore. https://doi.org/10.1007/978-981-15-0954-4_12.

Sutherland, W. J., Butchart, S. H., Connor, B., Culshaw, C., Dicks, L. V., Dinsdale, J., and Jiang, Z. (2018). A 2018 horizon scan of emerging issues for global conservation and biological diversity. *Trends in Ecology and Evolution*, 33(1), 47–58.

Tanner, J. A., and Romero-Sierra, C. (1974). Beneficial and harmful growth induced by the action of nonionizing radiation. *Annals of the New York Academy of Sciences*, 238, 171–175.

Tang, J., Zhang, Y., Yang, L., Chen, Q., Tan, L., Zuo, S., Chen, Z., and Zhu, G. (2015). Exposure to 900 MHz electromagnetic fields activates the mkp-1/ERK pathway and causes blood-brain barrier damage and cognitive impairment in rats. *Brain Research*, 1601, 92–101.

Taye, R. R., Deka, M. K., Rahman, A., and Bathari, M. (2017). Effect of electromagnetic radiation of cell phone tower on foraging behaviour of Asiatic honey bee, Apis cerana F.(Hymenoptera: Apidae). *Journal of Entomology and Zoology Studies*, 5, 1527–1529.

Thalau, P., Ritz, T., Burda, H., Wegner, R. E., and Wiltschko, R. (2006). The magnetic compass mechanisms of birds and rodents are based on different physical principles. *Journal of the Royal Society. Interface / the Royal Society*, 3(9), 583–587.

Thielens, A., Bell, D., Mortimore, D. B., Greco, M. K., Martens, L., and Joseph, W. (2018). Exposure of insects to radio-frequency electromagnetic fields from 2 to 120 GHz. *Scientific Reports*, 8(1), 1–10.

Thielens, A., and European Parliamentary Research Service. (2021). Environmental impacts of 5G: A literature review of effects of radio-frequency electromagnetic field exposure of non-human vertebrates, invertebrates and plants. http://www.europarl.europa.eu/stoa (STOA website).

Tkalec, M., Malarić, K., and Pevalek-Kozlina, B. (2005). Influence of 400, 900, and 1900 MHz. Electromagnetic fields on Lemna minor growth and peroxidase activity. *Bioelectromagnetics*, 26(3), 185–193.

Tkalec, M., Malarić, K., Pavlica, M., Pevalek-Kozlina, B., and Vidaković-Cifrek, Ž. (2009). Effects of radio-frequency electromagnetic fields on seed germination and root meristematic cells of Allium cepa L. *Mutation Research/Genetic Toxicology and Environmental Mutagenesis*, 672(2), 76–81.

Tofani, S., Agnesod, G., Ossola, P., Ferrini, S., and Bussi, R. (1986). Effects of continuous low-level exposure to radio-frequency radiation on intrauterine development in rats. *Health Physics*, 51(4), 489–499.

Tumkaya, L., Yilmaz, A., Akyildiz, K., Mercantepe, T., Yazici, Z. A., and Yilmaz, H. (2019). Prenatal effects of a 1,800-MHz electromagnetic field on rat livers. *Cells, Tissues, Organs*, 207(3–4), 187–196.

Úbeda, A. J., Leal, M. A., Trillo, M. A., Jimenez, J., and Delgado, M. R. (1983). Pulse shape of magnetic fields influences chick embryogenesis, *J. Anat.*, 137, 513–536.

Úbeda, A., Trillo, M. A., Chacón, L., Blanco, M. J., and Leal, J. (1994). Chick embryo development can be irreversibly altered by early exposure to weak extremely-low-frequency magnetic fields. *Bioelectromagnetics*, 15(5), 385–398.

Vácha, M., Půžová, T., and Kvíčalová, M. (2009). Radio frequency magnetic fields disrupt magnetoreception in American cockroach. *Journal of Experimental Biology*, 212(21), 3473–3477.

Vaitkuvienė, D., and Dagys, M. (2014). Possible effects of electromagnetic field on White Storks Ciconia ciconia breeding on low-voltage electricity line poles. *Zoology and Ecology*, 24(4), 289–296.

Vanbergen, A. J., Potts, S. G., Vian, A., Malkemper, E. P., Young, J., and Tscheulin, T. (2019). Risk to pollinators from anthropogenic electro-magnetic radiation (EMR): Evidence and knowledge gaps. *Science of the Total Environment*, 695, 133833.

Vargová, B., Kurimský, J., Cimbala, R., Kosterec, M., Majláth, I., Pipová, N., Jankowiak, Ł, and Majláthová, V. (2017). Ticks and radio-frequency signals: Behavioural response of ticks (*Dermacentor reticulatus*) in a 900 MHz electromagnetic field. *Systematic and Applied Acarology*, 22(5), 683–693.

Varotsos, P., Alexopoulos, K., and Lazaridou, M. (1993). Latest aspects of earthquake prediction in Greece based on seismic electric signals, II. *Tectonophysics*, 224, 1–37.

Vian, A., Davies, E., Gendraud, M., and Bonnet, P. (2016). Plant responses to high frequency electromagnetic fields. *BioMed Research International*, 2016, 1830262.

Vijver, M. G., Bolte, J. F., Evans, T. R., Tamis, W. L., Peijnenburg, W. J., Musters, C. J. M., and de Snoo, G. R. (2014). Investigating short-term exposure to electromagnetic fields on reproductive capacity of invertebrates in the field situation. *Electromagnetic Biology and Medicine*, 33(1), 21–28.

Vogel, G. (2017). Where have all the insects gone? *Science*, 356(6338), 576–579.

Volkrodt, W. (1987). Wer ist am Waldsterben Schuld? Mikrowellensmog der Funk- und Nachrichtensysteme. *Raum Zeit*, 26, 53–62.

Volkrodt, W. (1991). Droht den Mikrowellen ein ähnliches Fiasko wie der Atomenergie? *Wetter-Boden-Mensch*, 4, 16–23.

Wajnberg, E., Acosta-Avalos, D., Alves, O. C., de Oliveira, J. F., Srygley, R. B., and Esquivel, D. M. (2010). Magnetoreception in eusocial insects: An update. *Journal of the Royal Society*, 7, S207–S225.

Waldmann-Selsam, C., and Eger, H. (2013). *Baumschäden im Umkreis von Mobilfunksendeanlagen*, 26. Umwelt Medizin Gesellschaft, pp. 198–208.

Waldmann-Selsam, C., Balmori-de la Puente, A., Breunig, H., and Balmori, A. (2016). Radiofrequency radiation injures trees around mobile phone base stations. *Science of the Total Environment*, 572, 554–569.

Walleczek, J. (1992). Electromagnetic field effects on cells of the immune system: The role of calcium signaling. *FASEB Journal*, 6(13), 3177–3185.

Warnke, U. (2009). *Bees, Birds and Mankind: Destroying nature by 'electrosmog' effects of wireless communication technologies. A brochure series by the competence initiative for the protection of humanity, environment and democracy, Kempten*, 1st edn, November 2007, ISBN: 978-3-00-023124-7, English edn, March 2009, pp. 14–33.

Wdowiak, A., Wdowiak, L., and Wiktor, H. (2007). Evaluation of the effect of using mobile phones on male fertility. *Annals of Agricultural and Environmental Medicine*, 14(1), 169–172.

Weisbrot, D., Lin, H., Ye, L., Blank, M., and Goodman, R. (2003). Effects of mobile phone radiation on reproduction and development in Drosophila melanogaster. *Journal of Cellular Biochemistry*, 89(1), 48–55.

Wertheimer, N., and Leeper, E. (1979). Electrical wiring configurations and childhood cancer. *American Journal of Epidemiology*, 109(3), 273–284.

Wiltschko, R., Stapput, K., Ritz, T., Thalau, P., and Wiltschko, W. (2007). Magnetoreception in birds: Different physical processes for two types of directional responses. *HFSP Journal*, 1(1), 41–48.

Wiltschko, R., Thalau, P., Gehring, D., Nießner, C., Ritz, T., and Wiltschko, W. (2015). Magnetoreception in birds: The effect of radio-frequency fields. *Journal of the Royal Society. Interface / the Royal Society*, 12(103), 20141103.

Wyszkowska, J., Shepherd, S., Sharkh, S., Jackson, C. W., and Newland, P. L. (2016). Exposure to extremely low frequency electromagnetic fields alters the behaviour, physiology and stress protein levels of desert locusts. *Scientific Reports*, 6(1), 1–9.

Yakymenko, I., Tsybulin, O., Sidorik, E., Henshel, D., and Kyrylenko, S. (2016). Oxidative mechanisms of biological activity of low-intensity radiofrequency radiation. *Electromagnetic Biology and Medicine*, 35(2), 186–202.

Yakymenko, I., Burlaka, A., Tsybulin, I., Brieieva, I., Buchynska, L., Tsehmistrenko, I., and Chekhun, F. (2018). Oxidative and mutagenic effects of low intensity GSM 1800 MHz microwave radiation. *Experimental Oncology*, 40(4), 282–287.

Yanagawa, A., Tomaru, M., Kajiwara, A., Nakajima, H., Quemener, E. D. L., Steyer, J. P., and Mitani, T. (2020). Impact of 2.45 GHz microwave irradiation on the fruit fly, Drosophila melanogaster. *Insects*, 11(9), 598.

Yoshii, T., Ahmad, M., and Helfrich-Förster, C. (2009). Cryptochrome mediates light-dependent magnetosensitivity of Drosophila's circadian clock. *PLOS Biology*, 7(4), e1000086.

Youbicier-Simo, B. J., and Bastide, M. (1999). Pathological effects induced by embryonic and postnatal exposure to EMFs radiation by cellular mobile phones. *Radiation Protection*, 1, 218–223.

Yu, G., Tang, Z., Chen, H., Chen, Z., Wang, L., Cao, H., and Bai, Z. (2020). Long-term exposure to 4G smartphone radiofrequency electromagnetic radiation diminished male reproductive potential by directly disrupting Spock3–MMP2-BTB axis in the testes of adult rats. *Science of the Total Environment*, 698, 133860.

Zbyryt, A., Jankowiak, L., Jerzak, L., and Tryjanowski, P. (2021). Head and body orientation of the White Stork *Ciconia Ciconia* during incubation: Effect of wind, apex predators and power lines. *Journal of Ornithology*. https://doi.org/10.1007/s10336-021-01920-x.

Zhang, J. P., Zhang, K. Y., Guo, L., Chen, Q. L., Gao, P., Wang, T., Guo, G. Z., and Ding, G. R. (2017). Effects of 1.8 GHz radiofrequency fields on the emotional behavior and spatial memory of adolescent mice. *International Journal of Environmental Research and Public Health*, 14(11), 1344.

Zhou, R., Xiong, Y., Xing, G., Sun, L., and Ma, J. (2010). *ZiFi: Wireless LAN Discovery via ZigBee Interference Signatures MobiCom'10*, September 20–24. Chicago, Illinois.

Part D

Biophysical and Biochemical Mechanisms of Action

11 Mechanism of Ion Forced-Oscillation and Voltage-Gated Ion Channel Dysfunction by Polarized and Coherent Electromagnetic Fields

Dimitris J. Panagopoulos[1,2,3]

[1] National Center for Scientific Research "Demokritos", Athens, Greece
[2] Choremeion Research Laboratory, Medical School, National and Kapodistrian University of Athens, Athens, Greece
[3] Electromagnetic Field-Biophysics Research Laboratory, Athens, Greece

CONTENTS

Abstract 450
11.1 Introduction 450
    11.1.1 The Need for a Plausible Mechanism to Explain Man-made EMF Bioeffects 450
    11.1.2 Historical Background 452
    11.1.3 Mobile Ions and Transmembrane Electric Field 456
    11.1.4 Structure and Function of VGICs 457
11.2 Biophysical Mechanism of Action of Polarized and Coherent EMFs on Cells, Causing VGIC Dysfunction and Disruption of Cell Electrochemical Balance 458
11.3 Discussion 465
    11.3.1 The Explanation of Recorded EMF Bioeffects by the IFO-VGIC Mechanism 465
    11.3.2 The Solution to the Thermal Noise Problem 467
    11.3.3 The Increased Sensitivity of the VGICs to Polarized EMFs 467
    11.3.4 The Explanation of "Windows" and Other Unexplained Phenomena 468
    11.3.5 The Meaning of "Mechanism" in Science 469
    11.3.6 The Explanation of EMF-induced ROS Generation in Cells 470
    11.3.7 Conclusion 470
11.4 Appendix 471
    11.4.1A Calculation of the Damping Coefficient for an Ion in an Aqueous Solution 471
    11.4.1B Calculation of the Damping Coefficient and the Ion Velocity within a Channel Protein 471
    11.4.2 Solution of Equation 11.3 471
    11.4.3 Calculation of the Force on the Voltage Sensors Required to Gate VGICs 473
    11.4.4 Prevalence of the Polarized and Coherent Forced Oscillation of the Ion Against the Greater but Chaotic Thermal Motion 474
References 474

Keywords: electromagnetic fields, ion channels, mobile ions, ion forced oscillation, voltage-gated ion channels, biological effects, mechanism of action.

Abbreviations: B-field: magnetic field. DECT: Digitally Enhanced Cordless Telecommunications. E-field: electric field. EHS: electro-hypersensitivity. ELF: Extremely Low Frequency. EMF: electromagnetic field. ESR: electron spin resonance. ICR: ion cyclotron resonance. IFO: ion forced oscillation. IPR: ion parametric resonance. MT: mobile telephony. OS: oxidative stress. RF: Radio Frequency. ROS: reactive oxygen species. ULF: Ultra Low Frequency. VGCC: voltage-gated calcium channel. VGIC: voltage-gated ion channel. VLF: Very Low Frequency. WC: wireless communication. Wi-Fi: Wireless Fidelity. 2G/3G/4G/5G: second/third/fourth/fifth generation of MT/WC.

ABSTRACT

Exposure of living organisms to man-made electromagnetic fields (EMFs) causes a variety of adverse biological and health effects including oxidative stress (OS), genetic damage, cell death, and cancer, as is today documented by a great number of indisputable scientific studies. How does this happen? Key signaling molecules in all cells are the mobile ions, the concentrations of which control all cellular functions. The mobile ions move in and out of the cells through ion channels. A most important class of ion channels are the voltage-gated ion channels (VGICs) which open or close by polarized forces on the electric charges of their voltage-sensors generated by changes ≥ 30 mV in the membrane voltage. Polarization, coherence, and existence of Extremely Low Frequencies (ELFs) are common features of all man-made EMFs. Polarized and coherent oscillating EMFs force mobile ions to oscillate in parallel and in phase with them. This coordinated oscillation generates electrical forces on neighboring charges. The forces increase with increasing EMF intensity and decreasing EMF frequency. The oscillating ions close to the voltage-sensors of VGICs generate similar forces on them as those generated by 30 mV changes in the membrane voltage, causing irregular opening and closing of the VGICs. Continuance of such a dysfunction disrupts intracellular ionic concentrations, which determine the cell's electrochemical balance and homeostasis. Impairment of this balance triggers overproduction of reactive oxygen species (ROS) in cells which create OS and can damage DNA and other critical biomolecules. Since no convincing corresponding non-thermal mechanism exists for Radio Frequency (RF) EMFs, and because all RF EMFs employed in wireless communications (WC) and other applications are necessarily combined with ELF pulsation, modulation, and random variability, it seems that all non-thermal biological effects of man-made EMFs attributed, until now, to RF EMFs are actually due to their ELF components and can be explained by this mechanism.

11.1 INTRODUCTION

11.1.1 THE NEED FOR A PLAUSIBLE MECHANISM TO EXPLAIN MAN-MADE EMF BIOEFFECTS

A growing number of experimental and epidemiological studies connect man-made electromagnetic field (EMF) and corresponding electromagnetic radiation (EMR) exposures with a plethora of biological and health effects, including oxidative stress (OS), genetic damage, infertility, electro-hypersensitivity (EHS), and cancer. The number of peer-reviewed published studies showing such effects has increased tremendously in recent years, and this scientific evidence is indisputable (see reviews Phillips et al. 2009; Yakymenko et al. 2011; 2016; La Vignera et al. 2012; Cucurachi et al. 2013; Manna and Ghosh 2016; Leach et al. 2018; Miller et al. 2018; 2019; Panagopoulos 2019; Lai 2021). Most studies concern effects induced by man-made EMFs of Extremely Low Frequency (ELF) (3–3000 Hz), such as those from electric power transmission lines (50–60 Hz) and EMFs/EMR emitted by devices and antennas of wireless communications (WC). WC EMFs always employ Radio Frequency (RF) (300 kHz–300 GHz) and more specifically microwave (300 MHz–300 GHz) carrier waves, which are pulsed and modulated at ELF (mostly) and Very Low Frequencies (VLF) (3–30 kHz). Moreover, the final signals always exhibit random variability, mainly in the Ultra Low Frequency (ULF) (0–3 Hz) band (Pedersen 1997; Hyland 2008; Panagopoulos 2019; Pirard and Vatovez). The most important sources of WC EMFs/EMR are the mobile telephony (MT) devices and antennas [specifically second, third, and fourth generation (2G, 3G, 4G) mobile phones and corresponding base station antennas, while the fifth generation (5G) is currently being deployed despite the request by the scientific community for a moratorium (Hardell and Nyberg 2020)], cordless domestic phones (DECT: Digitally Enhanced Cordless Telecommunications), wireless internet routers (Wi-Fi: Wireless Fidelity), "Bluetooth" wireless connections among electronic devices, etc. Since man-made EMF/EMR exposure levels have grown exponentially, especially in the past 25 years, with the massive deployment of WC technologies, there is a need to understand the nature of these technical EMFs that are actually very different from the natural EMFs and cause so many adverse biological and health effects in contrast to the natural ones, which are vital to all living organisms. Moreover, there is a need to understand the mechanism(s) by which these anthropogenic EMFs affect living cells and organisms.

The genetic damage effects are certainly the most serious, as they can lead to a variety of pathologies, including poor health, infertility, and cancer (Panagopoulos et al. 2021). Any damage in the DNA or in any other biological molecule requires the breakage of chemical/electronic bonds in molecules/atoms; in other words, it requires ionization. Visible and infrared natural light do not cause ionization, even though they expose us to higher frequencies and radiation intensities than man-made EMFs in daily life (Panagopoulos et al. 2015a). Man-made EMFs having frequencies up to the low limit of infrared ($0 - 3 \times 10^{11}$ Hz) cannot directly cause ionization, except for very strong field intensities ($\geq 10^6$ V/m) (Francis 1960; Gomer 1961). Such field intensities rarely exist in the environment, apart from close proximity to atmospheric discharges (lightning) or very close proximity to high-voltage power lines and transformers. Therefore, the question is, how can man-made EMFs at environmental intensities be capable of damaging DNA and other biological molecules. Obviously they have the ability to break chemical bonds indirectly through the action of some primary biophysical mechanism(s) and subsequent initiation of intra-cellular biochemical processes. Today we know that the ionization and the corresponding biological damage caused by the man-made EMFs is executed by the reactive oxygen species (ROS) that are generated in the cells due to the man-made EMF exposures (De Iuliis et al. 2009; Campisi et al. 2010; Burlaka et al. 2013; Yakymenko et al. 2016; 2018; Zothansiama et al. 2017; Panagopoulos et al. 2021). Therefore, there must be a unique property of the anthropogenic EMFs that makes them capable of inducing ROS generation and subsequent ionization in living cells in contrast to natural infrared and visible light and other natural EMFs.

It has been shown that this unique property is that all man-made EMFs/EMR are totally polarized and coherent, meaning they possess net electric and magnetic fields, apart from radiation intensity, which exert forces on any electrically charged (or polar) particle/molecule, such as the mobile/dissolved ions and the charged macromolecules that are abundant in all biological systems. Therefore, polarization is a key property of EMFs with respect to their ability to induce biological effects, and all anthropogenic EMFs are totally polarized (and coherent), something that does not occur with natural EMFs, which only in certain cases are locally and partially polarized (Panagopoulos et al. 2015a).

Another key finding is that ELF man-made EMFs are much more bioactive than pure RF man-made EMFs at non-thermal levels. It has been shown by several experimental studies that the most bioactive EMFs are the lower frequency ones (ELF/ULF). In numerous cases of induced biological effects by complex RF EMFs modulated by ELFs, it has been found that the modulation (ELF) and not the carrier (RF) is responsible for the recorded effects. In addition, it has been repeatedly found that pulsed RF EMFs with ELF pulse-repetition rates are more active biologically than continuous-wave (non-pulsed) fields of identical other parameters (Bawin et al. 1975; 1978; Blackman et al. 1980; Delgado 1985; Frei et al. 1988; Bolshakov and Alekseev 1992; Goodman et al. 1995; Penafiel et al.

1997; Huber et al. 2002; Betti et al. 2004; Santini et al. 2005; Höytö et al. 2008; Phillips et al. 2009; Franzellitti et al. 2010; Campisi et al. 2010; Mohammed et al. 2013; Pall 2018; Panagopoulos 2019). These findings are in direct agreement with the mechanism that will be described in this chapter.

Therefore, we do know that polarized and low frequency EMFs are very bioactive, and their action in cells is mediated by the generation of ROS. But how can polarized low frequency EMFs induce ROS generation in living cells? This had not been known until recently (Panagopoulos et al. 2021), despite important steps that had been made. The present chapter describes the biophysical mechanism by which such EMFs, at intensities existing in the environment today, impair cellular function by causing dysfunction of the voltage-gated ion channel (VGIC) proteins in cell plasma membranes and alteration of the normal concentrations of mobile ions in cells, triggering production of ROS which finally damage biological molecules, resulting in the adverse biological and health effects reported in the EMF-bioeffects literature.

11.1.2 Historical Background

A common statement in many articles referring to the biological and health effects of EMFs still is that "there is no generally accepted mechanism" to explain the action of "weak" (non-thermal) EMFs on cells. Before the first publication of the mechanism that is the subject of this chapter (Panagopoulos et al. 2000), several other attempts for the explanation of the biological action of man-made EMFs, in most cases ELF magnetic fields, faced fatal difficulties on important physical issues. Although several of them had considered the cell membrane and the membrane ion-transport proteins as the main sites of interaction, only one of them (Balcavage et al. 1996) focused on the gating (opening or closing) of the VGICs, which are, by far, the most relevant targets for externally applied man-made EMFs in order to induce biological effects. The other attempts simply restricted on studying the motion of ions through such proteins under the influence of an applied EMF (Liboff 1985; 2003a; Liboff and McLeod 1988; Bianco et al. 1988; Zhadin 1998) or the binding of ions on proteins (Chiabrera et al. 1984; Lednev 1991). In addition, none of them (including Balcavage et al.) could calculate damping or restoration forces on the motion of ions within proteins, which is necessary in order to solve any related motion equation, and none of them could solve the "thermal noise" problem, which is briefly that the ions, apart from their velocity due to the applied EMF, have thousands of times greater velocities due to their thermal (Brownian) motion, which would render any other velocity (and corresponding effect) insignificant (Adair 1991a; 1991b).

Liboff's ion cyclotron resonance (ICR) model refers to an ion moving within a channel protein and to the well-established ICR phenomenon in physics, in which a charged particle, in this case, the ion, with a charge q and mass m_i, moves in the presence of a static magnetic field B_o (such as, e.g., the static geomagnetic field which is present anyway). If the ion's velocity u is not parallel to the field lines, the ion will acquire a helical motion of radius r, due to the magnetic force $F = qu \times B_o$ (or circular motion when u is vertical to B_o), which becomes a centripetal force $F = m_i(u^2/r)$ and, in the absence of any other force (such as damping), gives a constant circular frequency to the motion of the ion. The circular frequency of this motion, called ICR frequency will be: $\omega_o = (q/m_i) B_o$ (Alonso and Finn 1967). But while in the cyclotron, we may accept the absence of damping in the ion motion as an approximation, within a channel protein we cannot. Moreover, apart from damping, the ion receives a strong force from the transmembrane electric field. The authors of this model (Liboff 1985; 2003a; Liboff and McLeod 1988) hypothesized that a resonance effect will take place once an additional oscillating EMF of the same frequency as the ICR frequency is applied, and this would facilitate the ion transport through ion channel proteins initiating biological effects. But in the presence of damping, the circular frequency is not constant and, thus, any resonance effect is questionable, if not impossible. Even though it has been claimed that certain resonance effects have been observed in experiments at frequencies close to the ICR frequency (Liboff 2003a), this mechanism faces unresolved issues, such as damping, and the great difficulty (if not impossibility) of cyclotron/helical motion of ions restricted within a channel pore of approximately (~) 3 Å diameter. Certainly, the well-established physical phenomenon of ICR is present due to the geomagnetic field, and we cannot exclude the possibility that it may potentiate effects induced by some other primary mechanism at certain ELF frequencies. Thus, Liboff's suggestion cannot be excluded as a complementary mechanism, as also noted by others (Creasey and Goldberg 2001).

Chiabrera et al. (1984) focused on the magnetic field effect on the binding of ligands (lectins or calcium ions) on receptor proteins, which is a very different and significantly less plausible approach. A quantum mechanical extension of Liboff's ICR model developed (similarly) for binding of calcium ions on calmodulin, called ion parametric resonance (IPR) model (Lednev 1991), assumed ion transitions corresponding to the absorption of photons by man-made EMFs, which is very unlikely especially in the ELF band, as such "photons" do not exist (see Chapter 1, and Panagopoulos 2018). This and the ICR model (Liboff 1985) were found physically impossible in the presence of damping within a protein by a numerical test study with a special computer program (Halgamuge and Abeyrathne 2011). Several other suggestions were alternatives of Liboff's and Lednev's theories (e.g., Blanchard and Blackman 1994; Engstrom 1996).

Other suggestions that were not focusing on ion motion (or binding) in proteins, such as the possible extension of free radical lifetimes (usually referred to as radical pair mechanism) in cells (Brocklehurst and McLauchlan 1996) or the possible existence of biogenic magnetite in animal brains, allowing animals to sense the geomagnetic field and man-made EMFs (Kirschvink 1989), were even more uncertain, in addition to facing important physical inadequacies (Creasey and Goldberg 2001).

The radical-pair mechanism suggests that magnetic fields can affect the recombination of free radicals within cells, extending their lifetimes, resulting in biological effects (Brocklehurst and McLauchlan 1996). The effect of strong static magnetic fields of the order of 1 T on free radical recombination by absorption of microwave radiation, called electron spin resonance (ESR) effect, is well-established in physics (Gautreau and Savin 1978), but it seems unlikely for ULF/ELF/VLF EMFs and magnetic field intensities smaller than ~ 1 mT, as electronic spin recombination takes place on a timescale of a few tens of ns corresponding to a favorable field frequency of ~ 0.1 GHz and normally requires strong magnetic fields as already mentioned (Brocklehurst and McLauchlan 1996). Moreover, no quantitative description has been provided so far for such an effect in biological tissue for ELF EMFs existing in the environment. The increased free radical/ROS concentrations reported after exposures to ELF magnetic fields and arbitrarily attributed to the radical pair effect (Walleczek 1995; Georgiou 2010) are well explained by the mechanism presented next and may well be due to the electric component that always co-exists with the ELF magnetic fields (Panagopoulos et al. 2021).

The suggestion that animals may sense magnetic fields through the existence of biogenic magnetite in their brains, which could be reoriented by magnetic forces (Kirschvink 1989), may also be possible, but a quantitative description of how reorientation of such particles (assuming they exist) could result in cellular changes has never been provided. Actually, animals do not need any magnetite particles or any specific EMF-receptor organs, as has also been suggested (Ritz et al. 2000; Wiltschko et al. 2015), in order to be able to sense the geomagnetic field or other EMFs, as all of their trillions of cells in their bodies are equipped with the most sensitive electromagnetic sensors which are no other than the VGICs.

In 1998, I was in the middle of my PhD work on the biological effects of EMFs, and Drs. A. Filippetis and C. Lioliousis, professors at the physics department in the University of Athens, asked me to prepare an introduction on the possible mechanisms for our next Saturday meeting with a small group of scientists and students (mostly physicists and biologists) interested in the topic. We all knew that this is an unresolved problem. I thought I would just repeat what was told by everyone, that "there is no generally accepted mechanism", I would make brief descriptions of Liboff's, Lednev's, and Balcavage's suggestions, and that would be good enough for the meeting. Then I thought, why not introduce, in a few words, a possible way of resolving the problem, and wrote in a piece of paper the standard equation for a damping forced oscillation of an ion under the influence of an applied polarized oscillating EMF, as this was, in my mind, the most obvious effect that would naturally take place and could possibly cause disruption of cellular function (Alexopoulos 1960; Alonso and Finn 1967). There were unknown parameters in the forces, which is why others (Bianco et al. 1988; Zhadin 1998) could not solve similar equations, such as the damping coefficient (β), the ion's maximum velocity (u_o), and the ion's self-frequency (ω_o). I had no idea how to estimate these parameters, but I thought I would discuss this with the others in the meeting and that would be a good enough introduction to the problem. I did, and while I thought I simply presented something to fill the time, they told me I should try to move forward with this toward a publication. During the next year, I searched the details of the VGICs, their voltage-sensors, looked deeper at the previous suggestions, especially those focusing on ion motion, and calculated the maximum ion velocity within a channel and the damping coefficient from available patch-clamp recording* data (see Appendix 1). Although the equation was still unsolved, I had calculated all forces apart from the restoration force (which depends upon ω_o), and thought I shall submit the paper for publication and see what the reviewers would say. I first submitted the paper to the *Bioelectromagnetics* journal. The reviewers made useful remarks, but the paper was rejected. I made small revisions based on their comments, and moreover, I added equations for the forces exerted by the oscillating ions on the channel sensors of VGICs as Balcavage et al. (1996) had done, but this time, damping and restoration forces were included. I thought I should send it to a higher journal, so I submitted it to the *Biochemical and Biophysical Research Communications* journal where the Balcavage et al. (1996) paper was published. This time, my paper was not rejected. One reviewer (probably someone from the Balcavage group) correctly noted that I should be more specific on the value of ω_o (the ion's self-frequency). If this value were significantly smaller than the frequency of the applied EMF, then the equation could be solved. I ran into science libraries and searched information on the ions' behavior within cells irrespectively of any externally applied EMF. I was surprised to find plenty of information on "spontaneous ionic oscillations". Since such oscillations were recorded and they were "spontaneous", their frequencies should represent the ions' self-frequencies (ω_o)! In some of the publications they had measured the frequencies of the spontaneous oscillations. In all cases, the measured frequencies ranged from 0.016 to 0.2 Hz. Indeed, hundreds or thousands of times smaller than the frequencies of applied man-made EMFs (e.g., the 50–60 Hz of power lines, or the 217 Hz of 2G mobile phone EMF pulsations)! Now the equation could be completely solved. I solved it all the way resulting in bioactive combinations of intensity and frequency for an applied EMF to irregularly gate VGICs. Dr. A. Karabarbounis, professor at the physics department, mentor, co-author, and collaborator (whose help in this mechanism and other studies was invaluable), once again thoroughly checked my equations and calculations. Everything looked correct. The solution of the equation implied that even very weak polarized (man-made) EMFs in the low frequency bands (ULF/ELF/VLF) can irregularly gate VGICs in the cell plasma membrane and, thus, disrupt the cell's electrochemical balance. I resubmitted the revised paper to the journal, and it was immediately accepted for publication in May 2000. Soon after it was published, about a hundred letters came to the department of Cell Biology and Biophysics from all around the world asking for reprints. The secretary of the department said this had never happened before with any other publication.

The next year, Creasey and Goldberg (2001) published a review article on EMF-bioeffect mechanisms, titled: "A New Twist on an Old Mechanism for EMF Bioeffects":

> "While there is no generally accepted mechanism to explain how weak EMFs can produce biological effects, several classes of possible mechanisms have been advanced. All these proposed mechanisms have deficiencies, which is why a theoretical article by a group of researchers at the University of Athens in Greece is of interest. Panagopoulos and colleagues postulate that alternating electric fields vibrate all ions inside and outside cells in tune with the intensity and frequency of the field, and when the displacement

---

* The patch-clamp technique uses a micropipette that is sealed against a small patch of a cell membrane isolating a few VGICs. Then, depolarizing the membrane by electric pulses, they measure the ion current passing through the isolated VGICs (Neher and Sakmann 1992; Stryer 1995).

of this ocean of ions (charged molecules) reaches a critical value it trips the switch for opening or closing the channel in cell membranes through which ions such as calcium or sodium move in or out of cells. This process may either allow or prevent the passage of these critical ions, thereby disrupting the cell's integrity and function in very weak fields, like those associated with power lines. This article reviews three major theories of EMF bioeffects mechanisms and discusses Panagopoulos' new theory."

The other three major theories that were reviewed by Creasey and Goldberg (2001) were Liboff's ICR, the radical pair mechanism, and the biogenic magnetite suggestion. The physical constraints of these theories were analyzed. The article concluded that my suggestion was the only viable one. The situation has not been changed significantly since then.

In October 2000, and while I had not yet completed my PhD, I participated in an international conference in Crete on the biological effects of EMFs in which some of the world's top scientists were among the participants, including Liboff. I attended his presentation, and he attended mine, and we had long and fruitful discussions on the possible mechanisms. He also attended the presentation of my PhD experiments showing reproductive collapse in fruit flies exposed for a few min to mobile phone radiation. Even though our suggestions on the mechanism were different, he was very supportive and told the conference organizer Professor P. Stavroulakis that my theoretical and experimental work was important. I will always be indebted to this great man for the knowledge he shared and his encouragement and generosity. Accordingly, Dr. Stavroulakis asked me to help him with his book (Stavroulakis 2003; Panagopoulos and Margaritis 2003a; 2003b). Liboff was also the first to accept my experiments for publication in the *Electromagnetic Biology and Medicine* journal in which he was the chief editor (Panagopoulos et al. 2004). Another participant of the same conference was Dr. R. Goodman, who has had many significant works published. She was impressed with my fruit fly experiments using the mobile phone itself as the exposure device and was motivated to replicate them, but later she and her group decided to change the experimental procedure, and they found an increase instead of a decrease in reproduction, which is a very different result than of most other similar studies with fruit flies (Weisbrot et al. 2003).

In Panagopoulos et al. (2002) the theory was extended to include oscillating magnetic fields, discuss more extensively the thermal noise problem, and show that many positive experimental and epidemiological findings fit with its predictions. The article was published again in the *Biochemical and Biophysical Research Communications* journal, and one of the reviewers noted that:

> "The theory on the mechanism of forced-vibration of ions on the cell surface is unique and original. In addition, several positive experimental and epidemiological results were proved to be fit with the theory, and this is an evidence for availability of the theory for practical use."

A few days after this paper was published in November 2002, Dr. Neil Cherry, distinguished biophysicist and professor of environmental health at Lincoln University in New Zealand, known also for his integrity, posted a public comment on the EMF-Omega News site regarding the paper, simply saying,

> "Real and honest science says YES."

Sadly, a year later N. Cherry passed away.

The theory was reviewed in Panagopoulos and Margaritis (2003a) in Stavroulakis' book with additional discussion on ion diffusion, Nernst equation, thermal and non-thermal EMF-bioeffects, and other theoretical issues. Liboff (2003b) wrote for this chapter:

> "It is very detailed, summarizing some of the ways that electromagnetic radiation can be biologically interactive with biological systems, including non-thermal possibilities. This reviewer was especially impressed with this work"

The mechanism was reviewed again in Panagopoulos (2011; 2013), and in Panagopoulos et al. (2015a) it was applied to demonstrate the key role of polarization/coherence, a unique property of all man-made

EMFs that makes them very different and much more adversely bioactive than natural EMFs. Then it was reviewed again more extensively in Panagopoulos (2017). In 2017 and 2020, the theory was applied to explain the sensing of upcoming thunderstorms by sensitive individuals, and the sensing of upcoming earthquakes by animals, respectively (Panagopoulos and Balmori 2017; Panagopoulos et al. 2020), phenomena that were known for centuries but had remained unexplained. The most recent update of the theory was in 2021 when it was extended to explain how VGIC dysfunction causes overproduction of ROS, resulting in OS and DNA damage in cells (Panagopoulos et al. 2021). Before getting into the details of the theory, some important related information is reviewed next.

11.1.3 Mobile Ions and Transmembrane Electric Field

On both sides of every cell plasma or internal membrane, there are free/mobile ions (such as K^+, Na^+, Ca^{+2}, Cl^-, etc.), dissolved into the aqueous solutions in and out of the cell, which create a strong electric field that exists between the two sides of all cell membranes, determine the cell volume by controlling the osmotic entrance or exit of water, and play a major role in all metabolic cell processes such as the signaling cascades, cell development, proliferation, differentiation, etc. (Hodgkin and Huxley 1952; Baker et al. 1962; Honig et al. 1986; Hille 1992; Alberts et al. 2002; Stryer 1995; Panagopoulos 2013).

The voltage V across the membrane (called transmembrane voltage) is the total potential difference $V_o - V_i$ between the membrane's external and internal sides, and it is primarily due to concentration differences of positive ions (mainly Ca^{2+}, Na^+, and K^+) between the aqueous solutions outside and inside, and secondarily due to the majority of negative lipids in the membrane's internal layer (Honig et al. 1986; Stryer 1995). The VGICs (passive transport proteins), which are mostly cation channels, and most of all the K^+ channels, are considered the main contributors to the generation and maintenance of the transmembrane voltage, in cooperation with the K^+/Na^+ pump (ATPase), while other ion pumps (active ion transporters) contribute to a smaller degree (Hodgkin and Huxley 1952; Baker et al. 1962; Hille 1992; Alberts et al. 2002; Stryer 1995).

Ion fluxes through cell membranes are caused by forces due to concentration and voltage gradients between the two sides of the membrane. The equilibrium of concentration and voltage gradients between the two sides of the membrane is called electrochemical balance of the membrane. Under equilibrium conditions, the net ion flux through the membrane is zero, and the membrane has a voltage difference $V = V_o - V_i$ between its external and internal sides, with the internal always negative with regard to the external. This voltage across the plasma membrane in animal cells varies between 20 and 200 mV (Hodgkin and Huxley 1952; Baker et al. 1962; Honig et al. 1986; Alberts et al. 2002).

We accept a representative value for the transmembrane voltage $V = 100$ mV. Also accepting that $s = 10^{-8}$ m for the width of the membrane, the intensity of the transmembrane electric field is found to be: $E_m = V/s = \sim 10^7$ V/m.

The contribution of each ion type to the transmembrane voltage under equilibrium conditions (when the net flux of this ion type is zero) is described by the Nernst equation (Alberts et al. 2002):

$$V_o - V_i = -\frac{RT}{zF_c} \ln \frac{C_o}{C_i} \qquad [11.1]$$

where V_o, V_i are the electrical potentials on the external and internal sides of the membrane due to this ion type, R the gas constant, T the temperature in K, z the ion's valence, F_c the Faraday constant, and C_o, C_i the concentrations of this ion type in the external and internal sides of the membrane, respectively, at equilibrium. The total transmembrane voltage is the sum of the contributions from all ion types. [For a more extended analysis of the Nernst equation, see Panagopoulos and Margaritis (2003a)].

In the mechanism presented next (Section 11.2), I consider that, indeed, the primary site of interaction between an externally applied EMF and the cell is the plasma membrane and, specifically, the VGICs. Any externally applied polarized and oscillating electric field will exert parallel oscillating forces on the mobile ions, forcing them to oscillate in parallel lines and in phase with the applied field. I assume, for simplicity, that the oscillating electric field is an alternating (harmonic) one. When the amplitude of the forced oscillation transcends some critical value, the oscillating ions will exert electrical (Coulomb) forces on the voltage sensors of the VGICs, similar to the forces exerted by membrane voltage changes that physiologically gate these channels, resulting in their irregular opening and closing and, thus, in their dysfunction. That would disorder the membrane's electrochemical balance, disrupt the normal ion concentrations across the membrane and in the cell, disrupt the reduction/oxidation (redox) balance (Baker et al. 1962; Hille 1992; Stryer 1995), and consequently, disrupt the function of the whole cell by overproduction of ROS and induction of OS (Panagopoulos et al. 2021).

11.1.4 Structure and Function of VGICs

The majority of cation channels (Ca^{+2}, K^+, Na^+, H^+, etc.) in the membranes of all animal cells are voltage-gated (Alberts et al. 2002; Stryer 1995). The VGICs (also called electro-sensitive channels) convert between open and closed states when the electrostatic force on their voltage sensors, due to transmembrane voltage changes, exceeds some critical value. The voltage sensors of the VGICs are four symmetrically arranged, transmembrane, positively charged α-helices, each one named S4. They occupy the fourth position in a group of six parallel α-helices (S1–S6). The channel consists of four identical such groups in symmetrical positions around the pore of the channel. More specifically, the sensors are positive Lys and Arg amino acids in the S4 helices (Noda et al. 1986; Stühmer et al. 1989; Tytgat et al. 1993; Stryer 1995). The effective (net) charge on each S4 sensor has been calculated to be $q = 1.7\ q_e$, where q_e is the elementary charge (Liman et al. 1991). The positive charges of the S4 sensors are paired with counter charges from adjacent S1–S3 helices so that the net charge in the pore of the channel is zero (Stryer 1995; Groome and Bayless-Edwards 2020). The VGICs open or close by translocation of their S4 helices, responding to the appropriate transmembrane voltage changes (Bezanilla 2000; 2018).

Changes in the transmembrane voltage $dV \geq 30$ mV are normally required to gate electrosensitive channels (change their status from opened to closed and vice-versa) (Noda et al. 1986; Liman et al. 1991). More recent data indicate that there is also an upper limit in the voltage change for the gating, which seems to be ~ 100 mV. Thus, the VGICs respond by changing their status from opened to closed and vice versa to membrane voltage changes dV within a "window" from ~ 30 to ~ 100 mV (Sandipan and Baron 2015; Shi et al. 2020; Villalba-Galea and Chiem 2020). The existence of this "window" for the response of VGICs to changes in the membrane voltage is important, as it probably provides a base for the explanation of the so-called "window effects" reported occasionally in EMF-bioeffect studies for many years, and they are still unexplained (Bawin et al. 1975; 1978; Blackman et al. 1980; Panagopoulos et al. 2010a; Panagopoulos and Margaritis 2010). The S5–S6 helices of the four groups form the walls of the channel pore. Among the S1–S4 α-helices, the S4 are the closest to the pore-forming S5–S6 helices, having less than 1 nm distance from the pore (Tombola et al. 2006; Schmidt and Thomas 2012). The diameters of the α-helices are ~ 3.5 Å, and the diameter of the pore, at its narrowest part in K^+ channels, is ~ 3 Å (Stryer 1995; Miller 2000; MacKinnon 2003; Schmidt and Thomas 2012). The VGICs are ion-specific, and their selectivity depends upon the ion diameter. The ions pass through the pores of VGICs in single file and dehydrated (Palmer 1986; Stryer 1995; Miller 2000). All VGICs of various ion (mostly cation) types (K^+, Na^+, Ca^{2+}, etc.) are very similar in their structure and properties (Stryer 1995).

Several ions may interact simultaneously at any instant with an S4 sensor from a distance of the order of 1 nm, as – except for the ion(s) that may be passing through the pore any moment or are just outside the gate ready to pass – a few more ions are bound (also dehydrated) close to the pore at specific ion-binding sites (e.g., three in potassium channels) (Miller 2000; Zhang et al. 2018). Thus, it is absolutely realistic to consider at least 4 ions in a channel interacting simultaneously with the S4 sensors from ~ 1 nm distance.

Proton voltage-gated channels studied more recently also contain S4 transmembrane α-helices with charged Arg residues as voltage-sensors, similar to the metallic cation channels (De Coursey 2003; Seredenina et al. 2015).

11.2 BIOPHYSICAL MECHANISM OF ACTION OF POLARIZED AND COHERENT EMFs ON CELLS, CAUSING VGIC DYSFUNCTION AND DISRUPTION OF CELL ELECTROCHEMICAL BALANCE

It has been shown that polarized and coherent EMFs, even at very low field intensities in the ULF, ELF, and VLF bands, can cause irregular gating (dysfunction) of electro-sensitive ion channels (VGICs) in the cell membranes through the "ion forced-oscillation mechanism" (Panagopoulos et al. 2000; 2002; 2015a; 2021) with consequent disruption of the cell's electrochemical balance (the electrical and osmotic equilibrium maintained by specific concentrations of all dissolved/mobile ions across all cell membranes according to the Nernst equation (11.1)) (Alberts et al. 2002). Since, as explained, ULF/ELF/VLF components exist also in the complex WC EMFs, this mechanism – which will be thoroughly reviewed next – accounts for the biological effects of the vast majority of man-made (polarized and coherent) EMFs.

The mechanism is based on molecular and physical data and the forces on mobile ions in the vicinity of the voltage sensors of VGICs exerted by an applied polarized oscillating EMF. An oscillating polarized (and coherent) electric field will force mobile ions to oscillate on parallel lines and in phase with the field. This coordinated motion of electrically charged particles exerts Coulomb forces on the voltage sensors, similar to the forces exerted on them by changes in the transmembrane electric field known to physiologically gate these channels, and thus, the channels are gated irregularly by the applied EMF. The forces are proportional to the amplitude of the forced oscillation, and thus, the amplitude is a direct measure of the bioactivity of the applied EMF. It has been shown that the amplitude (bioactivity) is proportional to EMF intensity, inversely proportional to EMF frequency, and doubles for pulsed EMFs. Repeated irregular gating of electro-sensitive ion channels disrupts the cell's electrochemical balance and homeostasis (Baker et al. 1962; Alberts et al. 2002; Stryer 1995), leading to overproduction of ROS/free radicals (Panagopoulos et al. 2021).

The validity of the proposed mechanism has been verified by a numerical test that other previously suggested mechanisms have failed to pass. In this numerical test study by Halgamuge and Abbeyrathne (2011), this theory, Liboff's ICR, and Lednev's IPR theories were tested using a special computer program. The authors put the equations of each theory in the program along with the values of the different parameters (as calculated by Panagopoulos et al. (2000; 2002). The computer was able to verify only this mechanism, while the other two were found impossible in real damping conditions within cells (Halgamuge and Abbeyrathne 2011; Panagopoulos and Karabarbounis 2011).

Since the phenomenon of the ion forced oscillation (IFO) under the influence of a polarized/coherent EMF and its effect on the function of the VGICs are both central in the described mechanism, for simplicity, I shall use the name "IFO-VGIC mechanism".

The net electric field from an infinite number of individual electric pulses of random polarization and/or random phase (as e.g., photons of natural light), tends to zero at any moment (and similarly the net magnetic field):

$$\lim_{n \to \infty} \sum_{i=1}^{n} \vec{E}_i = \vec{E}_1 + \vec{E}_2 + \vec{E}_3 + \ldots + \vec{E}_n = 0 \qquad [11.2]$$

Thus, non-polarized/incoherent EMFs (as e.g., light, cosmic microwaves, etc.) at any radiation intensity cannot cause any parallel/coherent oscillation of charged/polar molecules (Panagopoulos et al. 2015a). By contrast, polarized and coherent (man-made) oscillating EMFs force all charged/polar molecules in biological tissue to oscillate in parallel with their polarization plane and in phase with them. This is crucially important for understanding the mechanism described. Most biological molecules of critical importance such as ions, water molecules, proteins, nucleic acids, lipids, etc. are either polar or carry a net electric charge (Stryer 1995; Alberts et al 2002). The forced oscillation will be most intense on the mobile ions, the smallest charged particles dissolved in large concentrations in the cytosolic and extracellular aqueous solutions in all living cells/tissues, controlling practically all cellular/biological functions (Alberts et al. 2002; Stryer 1995).

Even though all molecules move randomly with much greater velocities/displacements due to thermal energy, this has no biological effect other than increasing tissue temperature. By contrast, a polarized and coherent oscillation of much lower energy than average thermal molecular energy can initiate biological effects as will be shown next (see also Panagopoulos et al. 2000; 2002; 2015a; 2021).

Let us consider four identical mobile ions at distances of the order of 1 nm from the channel sensors (S4) (as explained in Section 11.1.4) and an externally applied oscillating EMF. The average electric (and magnetic) force on each ion due to any non-polarized EMF is zero (Eq. 11.2). On the contrary, the force due to a polarized field with an electrical component E, is $F = Ezq_e$, (zq_e is the ion's electric charge).

In the most usual and simplest case of a sinusoidal alternating electric field of intensity $E = E_o \sin\omega t$, the motion (forced oscillation) of a mobile ion is described by the standard equation from oscillation mechanics (Alexopoulos 1960; Alonso and Finn 1967), which in this case, is written as (Panagopoulos et al. 2000; 2002; 2015a; 2021):

$$m_i \frac{d^2 r}{dt^2} + \beta \frac{dr}{dt} + m_i \omega_o^2 r = E_o z q_e \sin \omega t \qquad [11.3]$$

where m_i is the ion mass, r the ion displacement due to the forced oscillation, z the ion valence (e.g., $z = 1$ for K$^+$, Na$^+$ or $z = 2$ for Ca^{2+} ions), $q_e = 1.6 \times 10^{-19}$ C the elementary charge, β the damping coefficient (being within channels $\beta = \frac{E_m q_e}{u_o} \cong 6.4 \, z \times 10^{-12}$ kg/s, with $E_m \sim 10^7$ V/m the transmembrane electric field, and $u_o = 0.25$ m/s the ion velocity through an open channel; see Appendix, 11.4.1B), $\omega_o = 2\pi\nu_o$ (ν_o the ion's oscillation self-frequency), $\omega = 2\pi\nu$ (ν the frequency of the applied field), and E_o the intensity amplitude of the applied oscillating field. [For detailed calculations of u_o, β, and ω_o, see Appendix (11.4.1 and 11.4.2) and Panagopoulos et al. (2000)].

The right part of Eq. 11.3 is the force on the ion due to the applied electric field. The first term of the left part ($m_i \frac{d^2 r}{dt^2}$) is the resultant force on the ion, the second term ($\beta \frac{dr}{dt}$) is a damping force, and the third term ($m_i \omega_o^2 r$) a restoration force exerted by the medium (Panagopoulos et al. 2000; 2002; 2015a; 2021). Although these are the standard forces in a forced-oscillation equation, their parameters depend on the specific system.

While an oscillating ion close to the S4 sensors exerts gating forces on them, it receives zero reaction, as the S4 charges are paired with counter charges from adjacent helices of the channel (see Section 11.1.4). Eq. 3 is a second-order linear differential equation with constant coefficients, which is solvable once we know the values of the different parameters. The solution is described analytically in the Appendix (11.4.2). The general solution is found to be:

$$r = \frac{E_o z q_e}{\beta \omega} \cos \omega t + \frac{E_o z q_e}{\beta \omega} \qquad [11.4]$$

The constant term $\frac{E_o z q_e}{\beta \omega}$ in the solution represents a constant displacement of the ion and has no effect on the oscillating term $\frac{E_o z q_e}{\beta \omega} \cos\omega t$. This constant displacement ($\frac{E_o z q_e}{\beta \omega}$) represents a jump of the whole oscillation at a distance equal to the amplitude; in other words, it doubles the amplitude

$\frac{E_o z q_e}{\beta \omega}$ of the oscillation at the moment when the field is applied or interrupted. For pulsed fields (such as all WC EMFs), this on/off transition occurs all the time with every repeated pulse. Therefore, pulsed fields are predicted to be twice as bioactive as continuous-wave (non-pulsed) fields of the same other parameters, and this explains a plethora of experimental findings that have shown increased bioactivity of pulsed EMFs compared to non-pulsed, which were previously unexplained (such as Frei et al. 1988; Bolshakov and Alekseev 1992; Huber et al. 2002; Höytö et al. 2008; Franzellitti et al. 2010; Campisi et al. 2010; Mohammed et al. 2013).

Ignoring the constant term in Eq. 11.4, the amplitude of the ion's forced oscillation is:

$$A = \frac{E_o z q_e}{\beta \omega} \quad [11.5]$$

An oscillating ion of charge $z q_e$ (the motion of which is described by Eq. 11.4) close to the S4 sensors of a VGIC exerts a force F on the effective charge q of each S4, described by the Coulomb law: $F = \frac{1}{4\pi\varepsilon\varepsilon_o} \cdot \frac{q \cdot z q_e}{r^2}$, ($r$ here is the distance between the oscillating ion and the S4 sensor[*] accepted to be ~ 1 nm, ε the relative dielectric constant of the medium[†], ε_o the dielectric constant of vacuum, and $\pi = 3.14$). The ion displaced by dr during its oscillation induces an additional force dF on each S4 sensor:

$$dF = -\frac{q \cdot z q_e}{2\pi\varepsilon\varepsilon_o r^3} dr \quad [11.6]$$

While in the case of a random/chaotic movement of the ion due to, e.g., thermal motion, $\lim \sum d\vec{r} = 0$, and thus, $\lim \sum d\vec{F} = 0$, in the case of a coordinated polarized and coherent forced oscillation, the additional force on each S4 sensor from all four ions[‡] (each one displaced by dr) is:

$$dF = -2\frac{q \cdot z q_e}{\pi\varepsilon\varepsilon_o r^3} dr \quad [11.7]$$

The effective (net) charge of each S4 domain is found to be $q = 1.7 \, q_e$ (Liman et al. 1991). The force on this charge generated by a change of 30 mV in the transmembrane voltage required normally to gate the channel is calculated to be [see Appendix (11.4.3) and Panagopoulos et al. (2000)]:

$$dF = 8.16 \times 10^{-13} \text{N}.$$

---

[*] This distance is taken ~1 nm as the distance of the S4 from the channel pore (see Section 11.1.4), and also because the ion concentration in cells is of the order of 1 ion per nm^3 (Alberts et al. 2002).
[†] ε ~ 4 within proteins and ~ 80 in the aqueous solutions (Honig et al. 1986).
[‡] More explanation on the gating ability of the oscillating ions due to the prevalence of their polarized and coherent movement against their greater but chaotic thermal movement is provided in the Appendix (11.4.4).

The displacement of one single-valence ion within the channel vertically to the membrane* corresponding to this minimum force, according to Eq. 11.6 (for $z = 1$, $\varepsilon \cong 4$, and $r \sim 1$ nm) is:

$$dr = 4 \times 10^{-12} \, \text{m}.$$

For four single-valence ions in the channel oscillating in parallel and in phase with an applied polarized (and coherent) oscillating field vertically to the membrane, the minimum displacement is, according to Eq. 11.7, reduced to:

$$dr = 10^{-12} \, \text{m}.$$

The corresponding necessary displacement for ions outside the channel would be 20 times greater due to the corresponding greater dielectric constant of the aqueous solutions.

Thus, a crucial finding has been reached: Any external polarized and coherent oscillating EMF (like all technical/man-made EMFs) able to force mobile ions to oscillate with amplitude

$$\frac{E_o z q_e}{\beta \omega} \geq 10^{-12} \, \text{m} \qquad [11.8]$$

is able to irregularly gate VGICs in cell membranes. For $z = 1$ (e.g., K$^+$ or Na$^+$ ions), and replacing q_e, β by their values in Condition 11.8, we get:

$$E_o \geq 0.25 v \times 10^{-3} \qquad [11.9] \qquad (v \text{ in Hz}, E_o \text{ in V/m})$$

For double-valence cations ($z = 2$) (e.g., Ca^{+2}) the condition becomes:

$$E_o \geq 1.2 v \times 10^{-4} \qquad [11.10] \qquad (v \text{ in Hz}, E_o \text{ in V/m})$$

For pulsed fields (such as all MT/WC fields) the right part of Condition 11.10 is further divided by 2, becoming:

$$E_o \geq 0.6 v \times 10^{-4} \qquad [11.11] \qquad (v \text{ in Hz}, E_o \text{ in V/m})$$

It is clear that the amplitude of the forced oscillation given by Eq. 11.5 is the key parameter to determine the ability of a polarized/coherent EMF to induce biological/health effects. We call it "Bioactivity of the EMF", or "EMF Bioactivity". Thus,

$$\text{EMF-Bioactivity} = \frac{E_o z q_e}{\beta \omega} = k \cdot \frac{E_o}{v} \qquad [11.12]$$

---

* Ion displacements not vertical to the membrane are expected to be less effective, as the forces generated from such displacements are not parallel to those that physiologically gate the VGICs.

where $k = \dfrac{zq_e}{2\pi\beta} = \dfrac{u_o}{2\pi E_m} \cong 4 \times 10^{-9}$ C·s/kg is a constant quantity (depending upon the membrane electric field E_m and the ion's velocity through an open channel u_o), E_o is the intensity amplitude, and ν is the frequency of the applied electric field. We call k the "bioactivity constant".

Thus, we have come to a most reasonable and elegant result, that the bioactivity of a polarized oscillating EMF is proportional to its maximum intensity (E_o) and inversely proportional to its frequency (ν), meaning that lower frequency fields are predicted to be more bioactive than higher frequency fields of the same intensity. Although we got this result considering the most usual/simple case of harmonically oscillating polarized EMFs, it is evident that non-harmonically oscillating polarized fields can also be approximately described in terms of their bioactivity by Eq. 11.12.

For pulsed EMFs with harmonically oscillating carriers, the amplitude doubles with each pulse, and so does the bioactivity:

$$\text{Pulsed EMF-Bioactivity} = 2k \cdot \frac{E_o}{\nu} \quad [11.13]$$

The same mechanism explains the biological action of polarized oscillating magnetic fields as well; if we replace in Eq. 11.3 the electric force $F_E = Ezq_e$, with a magnetic force,

$$F_B = Buzq_e \quad [11.14]$$

exerted on an ion with charge zq_e, moving with velocity u, vertically to the direction of a magnetic field of intensity B (in which case the magnetic force is maximum). In the simplest (and most usual) case of an alternating magnetic field $B = B_o \sin\omega t$ with intensity amplitude B_o and based on the same reasoning as above, we get corresponding bioactivity conditions for an oscillating magnetic field.

For one single-valence ion moving through an open channel vertically to the direction of the applied magnetic field with $u = u_o = 0.25$ m/s (the velocity calculated for ions moving through an open channel) [see Appendix (11.4.1B) and Panagopoulos et al. (2000)] and for the case of a continuous oscillating magnetic field, the corresponding bioactivity condition is:

$$\frac{B_o u_o q_e}{\beta\omega} \geq 4 \times 10^{-12} \text{m} \quad [11.15] \quad (\omega \text{ in rad/s, } u \text{ in m/s, } B_o \text{ in T)},$$

from which, we get:

$$B_o \geq 4 \times 10^{-3} \nu \quad [11.16] \quad (\nu \text{ in Hz, } B_o \text{ in T)},$$

or

$$B_o \geq 4 \times 10^{3} \nu \quad [11.17] \quad (\nu \text{ in Hz, } B_o \text{ in μT)}$$

For double-valence ions the right part of Cond. 11.17 is divided by 2:

$$B_o \geq 2 \times 10^{3} \nu \quad [11.18] \quad (\nu \text{ in Hz, } B_o \text{ in μT)}$$

For double-valence ions and pulsing magnetic fields, the right part of Cond. 11.18 is further divided by 2, and the bioactivity condition becomes:

$$B_o \geq 10^3 \nu \quad [11.19] \quad (\nu \text{ in Hz}, B_o \text{ in } \mu T)$$

It should be noted that, apart from the ion's drift velocity through the channel ($u_o = 0.25$ m/s) that we have accepted as a max velocity, the ion will acquire an additional velocity dr/dt due to the forced oscillation. From Eq. 11.4, we get:

$$\frac{dr}{dt} = -\frac{E_o z q_e}{\beta} \sin \omega t \quad [11.20]$$

(or respectively: $\dfrac{dr}{dt} = -\dfrac{B_o u_o z q_e}{\beta} \sin \omega t$ for a sinusoidal magnetic field).

The corresponding magnetic force due to this additional velocity, $B z q_e (dr/dt)$, is negligible (more than 10^8 times smaller) compared with the damping force β (dr/dt), and thus, it is not taken into account in the corresponding motion equation.

The maximum ($\dfrac{E_o z q_e}{\beta}$ or $\dfrac{B_o u_o z q_e}{\beta}$) of this additional velocity is independent of the frequency of the field (ω) and is much smaller (more than 10^5 times) for usual field intensities than the ion velocity through an open channel $u_o = \dfrac{E_m z q_e}{\beta} \cong 0.25$ m/s (because $E_o \ll E_m$), which in turn, is more than 10^3 times smaller than its corresponding average thermal velocity u_{kT} [see Appendix (11.4.4), and Panagopoulos et al. (2013)]. Thus, the described ion forced oscillation does not add to tissue temperature, and this mechanism is non-thermal in contrast to the known heating ability of the high intensity microwaves (Metaxas 1991). The non-thermal nature of man-made EMF bioeffects, including those of WC EMFs in contrast to high power microwaves has been discussed by others as well (Creasey and Goldberg 2001; Belyaev 2005).

This theory, having resulted in final bioactivity conditions, allows accurate predictions for the bioactivity of various types of EMFs, including man-made EMFs widely present in today's environment. For the sinusoidal alternating (continuous) 50 Hz electric (E) and magnetic (B) fields of high-voltage power lines with intensities of the order of $E \sim 10$ kV/m, and $B \sim 0.1$–1G (or ~ 10–100 μT) at close proximity to such lines, the bioactivity Conditions 11.10 and 11.18 for double valence cations (e.g. Ca^{+2}) give: $E_o \geq 6 \times 10^{-3}$ V/m or $E_o \geq 6$ mV/m (which is satisfied by more than 10^6 times), and $B_o \geq 10^5$ μT, which is not satisfied, showing that the recorded effects from high-voltage power lines are due to the electric rather than the magnetic component of the resultant EMF in contrast to what is usually considered. Thus, the electric component of power line EMFs is certainly capable of inducing biological and health effects in living organisms according to the presented mechanism, not only for the high intensities of the high-voltage power lines, but even for intensities down to 1–10 V/m that exist in most homes and work places, with the 10 V/m being associated with cancer (Coghill et al. 1996), and down to threshold intensities of a few mV/m as found in experimental studies to induce biological effects (McLeod et al. 1987; Cleary et al. 1988). These experimental and epidemiological results had not been explained before by any other suggested mechanism, as no other mechanism has resulted in final bioactivity conditions.

For the pulsing ELF E and B fields of MT/WC EMFs with a pulse repetition frequency ~ 100 Hz (3G/4G MT, DECT), $E \sim 10$ V/m, and $B \sim 1$ mG (or ~ 0.1 μT) (see Chapter 1, and Pedersen 1997; Panagopoulos 2020) the bioactivity Conditions 11.11 and 11.19, respectively, give: $E_o \geq 6 \times 10^{-3}$ V/m or $E_o \geq 6$ mV/m which is satisfied by more than 10^3 times, and $B_o \geq 10^5$ µT, which is not satisfied for direct action, but it may be satisfied by the magnetically induced electric field, which is significant in this case due to the very short rise/fall times of the pulses (see calculations in Panagopoulos et al. 2015a). Similar results are obtained for the 217 Hz pulsing E/B fields of 2G MT (Pedersen 1997; Panagopoulos et al. 2010a). Thus, the electric component of the ELF MT/WC pulsations emitted by mobile/cordless phones and corresponding antennas is certainly capable of inducing biological and health effects in living organisms according to the presented mechanism, even for intensities 1,000 times smaller than those of mobile and cordless phones in contact with the devices, as is the case of a few meters distance from such devices or a few hundred meters from corresponding MT base station antennas. This verifies the results of experimental studies with mobile phones and MT antennas inducing OS, DNA damage, infertility, and other effects (Ji et al. 2004; Panagopoulos 2011; 2017; 2019; 2020; La Vignera et al. 2012; Burlaka et al. 2013; Yakymenko et al. 2016; 2018; Gulati et al. 2016; Zothansiama et al. 2017), and epidemiological studies with mobile/cordless phones and MT antennas inducing EHS and cancer (Hardell et al. 2007; 2009; Miller et al. 2018; 2019; López et al. 2021).

For Wi-Fi and Bluetooth wireless connections with a pulsing frequency of around 10 Hz, $E \sim 1$ V/m, and $B \sim 0.1$ mG (or ~ 0.01 µT) (Zhou et al. 2010) the bioactivity Conditions 11.11 and 11.19, respectively, give: $E_o \geq 0.6 \times 10^{-3}$ V/m or $E_o \geq 0.6$ mV/m, which is satisfied by more than 10^3 times, and $B_o \geq 10^4$ µT, which is not satisfied for direct action. Again, the electric component of the ELF pulsations emitted by Wi-Fi routers and Bluetooth devices is certainly capable of inducing biological and health effects in living organisms according to the presented mechanism, even for intensities 1,000 times smaller than those in close proximity to such devices (and, thus, present in all modern home and working environments with operating Wi-Fi routers and Bluetooth).

The above numerical examples clearly demonstrate that the electric field is the bioactive component of an EMF rather than the magnetic, and it is the ELF components rather than the RF carriers of the WC EMFs responsible for the effects, in contrast to what has been considered before by studies and health agencies (IARC 2002; 2013). The magnetically induced electric field can also be bioactive in the case of the sudden ELF pulses of WC signals with short rise/fall times (Panagopoulos et al. 2015a).

The bioactivity Conditions 11.10 and 11.11 for continuous and pulsed electric fields, respectively, are depicted in **Figure 11.1**. The space above Line 1 (including the line) represents the bioactive

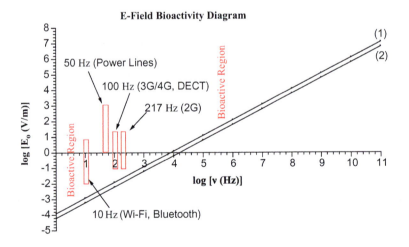

FIGURE 11.1 E-field bioactivity diagram showing the bioactive combinations of electric field intensity and frequency capable of inducing biological/health effects according to the IFO-VGIC mechanism. The ELF electric components of power line EMFs, 2G/3G/4G MT, DECT, Wi-Fi, and Bluetooth EMFs lie within the bioactive region (above the lines). Line 1 refers to continuous (non-pulsed) fields, such as those from power lines (Condition 11.10), while Line 2 refers to pulsed fields, such as the ELF pulsations of WC EMFs (Condition 11.11).

combinations of intensity amplitude (E_o) and frequency (ν) that fulfill Condition 11.10 for continuous fields, and the space above Line 2 (including the line) represents the bioactive combinations that fulfill Condition 11.11 for pulsed fields. The ELF electric fields of power lines, 2G/3G/4G MT, DECT, WiFi, and Bluetooth lie within the bioactive region predicted by this theory, explaining the great number of experimental and epidemiological findings that have shown adverse effects from such exposures.

11.3 DISCUSSION

11.3.1 THE EXPLANATION OF RECORDED EMF BIOEFFECTS BY THE IFO-VGIC MECHANISM

According to the IFO-VGIC mechanism described above and in Panagopoulos et al. (2000; 2002; 2015a; 2021), man-made (polarized and coherent) ULF/ELF/VLF EMFs or the ULF/ELF/VLF modulation/pulsing/variability components of modern RF/WC EMFs can alter the concentrations of mobile ions in living cells by causing dysfunction (irregular gating) of VGICs in cell membranes. As recently documented (Panagopoulos et al. 2021), this leads immediately to OS in the cells by ROS (over)production in the cytosol and/or the mitochondria, which can damage DNA and other biological molecules when cells are unable to reinstate electrochemical balance (the normal intracellular ionic concentrations).

The mechanism shows that the bioactivity of polarized/coherent EMFs is proportional to their intensity, inversely proportional to their frequency, and doubles for pulsed fields, meaning that the ULF/ELF/VLF EMFs and even more the pulsing RF EMFs with ELF pulsations, such as all WC EMFs, are predicted to be the most bioactive. This explains the recorded effects of purely ULF/ELF/VLF man-made EMFs and those of modulated/pulsing/variable RF EMFs (see reviews Goodman et al. 1995; Santini et al. 2005; Balmori 2006; 2010; Phillips et al. 2009; La Vignera et al. 2012; Cucurachi et al. 2013; Yakymenko et al. 2016; Manna and Ghosh 2016; Leach et al. 2018; Pall 2018; Panagopoulos 2019; Lai 2021).

As emphasized in the introduction of this chapter (see Section 11.1.1), all types of RF exposure from all types of antennas and WC devices (named WC EMFs) necessarily combine RF carrier signals with ELF/ULF/VLF components in the form of pulsing, modulation, and random variability. The RF carrier signal alone does not contain information. The information is always contained in the ELF signals that modulate the RF (Panagopoulos 2019). This is the reason why pure RF EMFs have very few (if any) technological applications. Significant experimental evidence shows that the bioactive parameters in a complex RF/WC signal are its ELF components and that non-modulated and non-pulsed RF signals alone do not usually induce biological effects (Bawin et al. 1975; 1978; Blackman et al. 1980; Frei et al. 1988; Bolshakov and Alekseev 1992; Penafiel et al. 1997; Huber et al. 2002; Betti et al. 2004; Höytö et al. 2008; Franzellitti et al. 2010; Campisi et al. 2010; Mohammed et al. 2013) (see review in Chapter 1, and in Panagopoulos 2019), other than heating when they possess high enough frequency and intensity (Metaxas 1991; Creasey and Goldberg 2001; Belyaev 2005; Panagopoulos et al. 2021). Therefore, the IFO-VGIC mechanism explains all these experimental findings and suggests that the vast majority of non-thermal effects attributed, until now, to RF EMFs are actually due to their ELF/ULF/VLF components, and many studies and health authorities have been looking in the wrong direction trying to explain the WC EMF bioeffects.

It has been claimed that the ELF components of complex RF-ELF EMFs of WC need to be "demodulated" in order to be sensed by living organisms (Sheppard et al. 2008). "Demodulated" or not, the fact is that the ELF components of modulated/pulsed WC signals can be directly sensed by both ELF meters/spectrum analyzers and the living organisms (see examples of ELF measurements of WC signals in Chapter 1, and in Pedersen 1997; Zhou et al. 2010; Panagopoulos et al. 2010a; Panagopoulos 2020).

The presented IFO-VGIC mechanism and the provided numerical examples show that it is the direct ELF electric fields (and the magnetically induced electric fields when they are greater than the minimum bioactive values predicted by this theory, as usually happens in cases of sudden pulses), not the direct magnetic fields themselves, that are the bioactive components in contrast to what has been considered before by studies and health agencies (IARC 2002; 2013) and in agreement with early experimental findings (Liburdy 1992). Although electric fields are less penetrating in living tissue than magnetic fields, penetration depends upon the inverse square root of frequency, and thus, ELF electric fields are significantly penetrating. Penetration depends also upon the inverse square root of the medium conductivity. More specifically, the "penetration depth" δ for plane electromagnetic waves within a material (defined as the depth at which the wave is damped to $1/e = 0.368$ of its initial amplitude) is given by the equation: $\delta = \sqrt{\dfrac{2}{\mu\omega\sigma}}$ (where $\omega = 2\pi\nu$, ν the frequency of the wave, and μ, σ the magnetic permeability and specific conductivity of the material respectively) (Ludwig 1974; Jackson 1975). Even though sea water is much more conductive than living tissue, ELF electromagnetic waves (thus, both the electric and the magnetic parts of the waves) penetrate several meters into sea water, accommodating communications with submarines (Barr et al. 2000). Moreover, it is known that isolated tissues respond to externally applied pulsed or sinusoidal ELF electric fields at very low thresholds (~ 10^{-3} V/m) similar to those predicted by this theory (McLeod et al. 1987; Cleary et al. 1988; Lee et al. 1993). This evidence shows that ELF electric fields penetrate enough to induce effects in living tissue, even at very low field intensities. Finally, skin cells, nerve terminals, eyes, and organs close to the surface, such as the brain, heart, testes, etc., are directly exposed to externally applied EMFs. For all these reasons, no distinction has been made between externally applied ELF electric fields and internally induced ones.

Liburdy (1992), in a series of important experiments, had demonstrated that a) an increase in intracellular calcium in rat lymphocytes occurred only 100 s after the beginning of exposure to a 60 Hz EMF (22 mT, 0.17 V/m); b) the increase was not due to release from intracellular calcium stores but due to the influx of extracellular calcium, therefore, identifying the calcium ion channels, most of which are voltage-gated and, thus, voltage-gated calcium channels (VGCCs), as the cites of interaction; and c) it is the electric (including the magnetically-induced electric) rather than the magnetic part of an EMF responsible for the effects. Liburdy's important findings are fully explained by the described biophysical mechanism. It was surprising to come across a note of the Office of Research Integrity (1999) accusing Liburdy for "scientific misconduct in biomedical research by intentionally falsifying and fabricating data and claims about the purported cellular effects of electric and magnetic fields" in his study (!), as the findings and conclusions of this paper are, today, verified by both the described mechanism and a great number of published peer-reviewed experimental studies showing alterations in intracellular calcium and other ion concentrations due to dysfunction of VGICs/VGCCs caused by ELF EMFs (Walleczek 1992; Goodman et al. 1995; Piacentini et al. 2008; Pall 2013; Cecchetto et al. 2020; Zheng et al. 2021; Bertagna et al. 2021; Panagopoulos et al. 2021).

The IFO-VGIC mechanism was described by realistic equations for the forces exerted on mobile ions in the vicinity of the voltage sensors of VGICs in cell membranes by externally applied manmade (polarized) EMFs. The theory resulted in bioactivity conditions connecting the intensity of an applied polarized EMF with its frequency. The bioactivity Conditions 11.9–11.11 and 11.17–11.19 provide the bioactive intensity-frequency combinations for continuous-wave and pulsed electric and magnetic fields. The final numbers explain almost all the experimental and epidemiological findings connecting biological/health effects with man-made EMF exposures. The minimum bioactive EMF intensities predicted by the presented theory (down to 10^{-4} V/m for ELF/ULF EMFs) are consistent with the experimental and epidemiological findings for the threshold ELF electric field intensities found to induce biological effects or to be associated with cancer (McLeod et al. 1987; Cleary et al. 1988; Lee et al. 1993; Coghill et al. 1996; Miller et al. 1996). The IFO-VGIC mechanism shows that the exposure limits set by the private organization called the International Commission on Non-Ionizing Radiation Protection (ICNIRP) (1998; 2010; 2020) and adopted by governments, being millions of times higher than the minimum intensities predicted by this theory, are totally inadequate to protect human health and the environment.

11.3.2 THE SOLUTION TO THE THERMAL NOISE PROBLEM

The velocity acquired by an ion due to its forced oscillation (given by Eq. 11.20) is normally more than 10^5 times smaller than the maximum ion velocity through an open channel u_o, which in turn, is more than 10^3 times smaller than the average thermal velocity (Appendix 11.4.4), and this shows that this mechanism is absolutely non-thermal, contrary to the ICNIRP's (1998; 2010; 2020) claims that the only established EMF effect is heating. Even though the ion velocity due to its forced oscillation is so much smaller than its velocity due to thermal motion, it prevails in its ability to gate VGICs (see Appendix 11.4.4), and thus, Adair's (1991a; 1991b) argument that the ion thermal motion would mask any other motion of the ion within a cell cannot stand in this case. Adair did not consider ion movement within a channel, which is necessarily confined by a mean free path of ~ 10^{-11} m and, most importantly, did not consider polarized and coherent movement of the ions due to the applied EMF in a single direction and in phase, in contrast to the chaotic thermal movement which breaks into ~ (4×) 9300 mean free paths, each one in a different direction resulting in zero net force on the channel sensor (Appendix 11.4.4). Indeed, if the S4 sensors responded to the chaotic non-polarized forces due to the ions' thermal motion, the channels would be gated chaotically all the time, the cell's electrochemical balance and homeostasis would be continuously disrupted, and no cell/organism would survive.

11.3.3 THE INCREASED SENSITIVITY OF THE VGICS TO POLARIZED EMFS

The IFO-VGIC mechanism shows that the VGICs are very sensitive to polarized and coherent (man-made) ULF/ELF EMFs with threshold intensities below 10^{-4} V/m (**Figure 11.1**) due to the phenomenon of the forced oscillation of the ions in the vicinity of the voltage sensors of the VGICs, which explains the recorded sensitivity to man-made EMFs without the need of any amplification mechanisms or any specific EMF-receptor organs in animals as was thought before by others (Ritz et al. 2000; Mathie et al. 2003; Wiltschko et al. 2015). For example, the sensing of the geomagnetic field by animals can be explained by the same mechanism, according to the fact that, in each cell, those VGICs with optimum orientation (e.g., vertical to the geomagnetic field lines at opposite sides of the cell surface) will be most activated, while those VGICs with poorer orientation (e.g., parallel to the geomagnetic field lines) will not be activated. This immediately provides each cell of an organism with a sense of orientation in the geomagnetic field. Certainly, some organisms can be more electrosensitive than others, but this can simply be due to higher percentages of (and possibly more specialized) electrosensitive ion channels in their cells. Therefore, as already noted in the introduction of this chapter (Section 11.1.2) all cells in all animals (and other living organisms) are equipped with the most sensitive electromagnetic detectors: The VGICs in their plasma (and internal) membranes.

As made clear by the description of the mechanism and was previously emphasized (Panagopoulos et al. 2015a), the VGICs are not gated by externally applied EMFs via forces exerted directly on their channel sensors – as that would require very strong polarized EMFs of the order of the trans-membrane physiological fields (~ 10^7 V/m) – but via the forced oscillation of mobile ions close to their sensors. The Coulomb forces dF exerted on the S4 sensors depend upon the inverse third power of the distance ($1/r^3$) between the oscillating ions and the S4 (Eqs. 11.6 and 11.7), while the direct forces from an applied field E on the S4 depend upon the first power of the field ($F = E \cdot q$ or $dF = dE \cdot q$). Thus, a smaller distance between two electric charges is more crucial in generating forces than a greater external field.

Although the mechanism was first published in 2000 (Panagopoulos et al. 2000) based on the available data on the structure and function of the VGICs, newer details on the roles of S1–S6 helices, channel structure, relaxation, hysteresis, and gating have not refuted, but verified and extended that knowledge (Tombola et al. 2006; Schmidt and Thomas 2012; Zhang et al. 2018; Sandipan and Baron 2015; Groome and Bayless-Edwards 2020; Shi et al. 2020; Villalba-Galea and Chiem 2020).

11.3.4 The Explanation of "Windows" and Other Unexplained Phenomena

What is more difficult to explain is the existence of non-linear phenomena, such as the increased bioactivity "windows" reported occasionally in the EMF-bioeffects literature, in which certain effects are intensified within certain values of an EMF-exposure parameter (intensity in most cases, or frequency) (Goodman et al. 1995; Bawin et al. 1975; 1978; Blackman et al. 1980; Panagopoulos et al. 2010a; Panagopoulos and Margaritis 2010). The existence of "windows" shows that the response of living cells/organisms to EMFs is not generally proportional to the above EMF parameters (intensity, frequency). Non-linear responses of living cells have not been explored in depth, and it will take many years until they are. Nevertheless, a probable explanation of observed intensity "windows", according to the described mechanism, is that they can be due to a corresponding "window" in the membrane gating voltage change (Panagopoulos and Margaritis 2010). Although that was suggested as a hypothesis in Panagopoulos and Margaritis (2010), indeed, such a "window" seems to exist. The VGICs respond to membrane voltage changes from ~ 30 mV (min) to ~ 100 mV (max) where the conductivity of the channel saturates (Sandipan and Baron 2015; Villalba-Galea and Chiem 2020). Therefore, it seems that the so far unexplained window effects observed occasionally in the EMF-bioeffects literature may now be explained for the first time according to the IFO-VGIC mechanism. Although such an explanation is very simple and reasonable as an idea, in practice, there are several parameters that need to be considered for an accurate prediction/explanation of the window effects which make such a prediction difficult. These include the direction of the externally applied E-field and the number of VGICs oriented in parallel to this direction (optimum orientation). In Panagopoulos and Margaritis (2010) we had calculated the membrane voltage-change limits that corresponded to an observed window of increased DNA fragmentation in fruit fly ovarian cells attributed to the 217 Hz pulsations of 2G mobile phone EMFs (Panagopoulos et al. 2010a). The calculated voltage-change limits (90–180 mV) corresponded to the forces on each S4 sensor exerted by one single-valence ion in a channel. For four ions interacting simultaneously with the S4, which is the most realistic case, these forces are divided by 4, and the above transmembrane voltage-change window is reduced to 22.5–45 mV, which is finally translated to 30 – 45 mV (since the voltage-change must be $\geq$ 30 mV). This calculated window lies within the 30–100 mV transmembrane voltage-change limits that VGICs respond to, and thus, it is a reasonable result. Finally, the calculated voltage change refers to the channels (VGICs) that are oriented parallel to the direction of the applied E-field (optimum orientation). VGICs in other orientations would expectedly require greater gating forces from the oscillating ions in order to contribute to the above observed window. Therefore, it seems that the "window effects" in the EMF-bioeffects literature may be explained according to the IFO-VGIC mechanism, even though their accurate prediction is complicated because of unknown variables.

Another effect not included in the bioactivity Equations 11.12 and 11.13 is the increased bioactivity of highly and unpredictably varying exposures as those from WC devices (mobile phones, Wi-Fi, etc.) and corresponding antennas (Panagopoulos et al. 2015b; 2021; Panagopoulos 2017; 2019). The described mechanism results in accurate predictions when the applied EMFs have constant parameters (intensity, frequency, etc.). When the parameters are highly and unpredictably variable, the mechanism – and any possible mechanism – can only estimate effects according to the average or the maximum values of the varying EMF parameters. Finally, the bioactivity equations include field (and tissue) parameters and not exposure variables, such as exposure duration, intermittence, etc., which are also very important (Ivancsits et al. 2002; 2003; Diem et al. 2005; Chavdoula et al. 2010; Panagopoulos 2017; 2020). One way to include such parameters (variability, exposure duration, etc.) is to multiply the right parts of Equations 11.12 and 11.13 by certain coefficient(s) which would be estimated experimentally. This and a more accurate explanation of the window effects could be subjects for future developments of the theory.

Other natural phenomena that had remained unexplained for centuries (and more) have been the sensing of upcoming thunderstorms by sensitive individuals and the sensing of upcoming earthquakes by animals. Both phenomena were recently precisely explained for the first time by the

IFO-VGIC mechanism as being due to the partially polarized natural ULF/ELF EMF emissions associated with these phenomena (Panagopoulos and Balmori 2017; Panagopoulos et al. 2020).

11.3.5 The Meaning of "Mechanism" in Science

Any theory on a mechanism in science (particularly in physics) must be based on simple and reasonable postulates and must, necessarily, be expressed quantitatively (by solvable equations that result in numbers). The values of the different parameters in the equations must be based on physical/molecular data. The presented biophysical theory is the only one that fulfills these criteria for a theory on a mechanism in the case of EMF-induced bioeffects. Qualitative descriptions alone or incomplete quantitative descriptions based on incomplete or unsolvable equations do not, in fact, constitute a "theory" or a "mechanism". The EMF-bioeffects literature is full of uncertain suggestions based on qualitative descriptions and/or unsolved general equations. Such examples are the radical-pair mechanism, the biogenic magnetite suggestion, the IPR mechanism discussed in the introduction (see Section 11.1.2), or the Fröhlich (1968) suggestion for a supposedly possible non-thermal action of purely RF EMFs. Several publications referring to EMF-bioeffect mechanisms still look at complicated and impossible mechanistic schemes (such as, e.g., general quantum mechanical considerations) instead of looking at the simple and obvious. In this way, a confusion perpetuates with the industry, ICNIRP, and health authorities refuting the vast number of experimental studies showing, e.g., genotoxic effects, EHS, infertility, cancer, etc., just because "there is no generally accepted mechanism" to explain these effects.

As described in the introduction of this chapter (Section 11.1.2), previous important attempts on mechanisms focusing on ions moving inside membrane channels or other proteins (Liboff 1985; Bianco et al. 1988; Balcavage et al. 1996; Zhadin 1998) were not successful, mainly because 1) They had not taken into account damping and restoration forces (Liboff 1985; Balcavage et al. 1996) or did not calculate them (Bianco et al. 1988; Zhadin 1998). The difficulty was related to calculating their parameters such as β, u_o, and ω_o. 2) They did not consider coordinated motion of several ions oscillating in parallel and in phase due to polarization and coherence, exerting constructive forces on channel sensors which prevail against the greater but chaotic forces due to the ions' random thermal motions. 3) They focused on magnetic fields or magnetically induced electric fields and ignored externally applied electric fields, which eventually, seem to be the bioactive ones. 4) They did not result in specific intensities and frequencies predicted to affect cells. 5) Apart from Balcavage et al. (1996), they did not focus on the gating of VGICs, which is, by far, the most probable event to initiate biological effects, but simply on the motion of ions within channels/proteins. The fact that Balcavage et al. (1996) were the first to focus on this obvious scenario makes their study an exceptional contribution despite its described limitations.

What has been called by Pall (2018) "VGCC activation mechanism" since 2018, and presented as his own discovery, is none other than the mechanism presented here. A commentary paper/letter to the editor was published on this major ethical issue (Panagopoulos 2021). Pall also spoke of "direct forces" exerted by the EMF on the channel sensors, which are "amplified" by the cell membranes and are independent of the field's frequency so that purely microwave EMFs (without ELF/ULF/VLF components) can allegedly cause dysfunction (or "activation" as he called it) of VGICs/VGCCs through the action of their "highly penetrating" magnetic parts. The impossibility of such claims was also explained in Panagopoulos (2021). Earlier, Pall (2013) in a review of experimental EMF studies with ion channel blockers, noted the unique role of VGICs (and especially VGCCs, as the studies involved calcium ion channel blockers) in the recorded bioeffects, which was an important observation in complete agreement with Liburdy (1992) and the IFO-VGIC mechanism. Moreover, he had expressed his support on this mechanism by a public comment (see Panagopoulos 2017; 2021):

> "The argument that has been made by the advocates of the current safety standards is that the low intensity, non-thermal EMFs produce only very weak forces on charged groups, weaker than those due

to thermal motions at body temperature. They argue, therefore that any effect would be no more than effects produced all the time spontaneously in the body. That physics argument has been disproven by Panagopoulos and his colleagues when they published two biophysical studies in 2000 and 2002. The problem with that can be seen when one looks at the voltage-gated calcium channels (VGCCs). The VGCCs have a specific number of charged amino acid residues each of which has a role in the opening and closing of the channels. Each of these are pushed by weak forces, acting in the same direction when a change in the electrical potential across the plasma membrane opens the channel. In much the same way, weak forces acting in the same direction produced by the EMFs should be able to open the channel as well. Whereas thermal motions act randomly in three dimensions and will therefore only extremely rarely be all acting in the same direction, the forces produced by these fields like the forces produced by changes in plasma membrane electrical potential, do act coordinately in the same direction and can, therefore open these channels. This was the insight that led Panagopoulos to formulate his "Ion Forced-Oscillation" theory and it is, in my view, a brilliant insight! The whole basis of the heating/thermal/SAR paradigm of action of these fields is entirely based on the claim that "there is no biophysically viable mechanism for the action of these weak non-thermal or micro-thermal fields and that claim was shown by Panagopoulos to be wrong and the empirical evidence shows that Panagopoulos is right. In addition to that there are literally thousands of studies that falsify the heating/thermal/SAR paradigm. This is THE best example I have seen of a clearly strongly supported paradigm shift within the last 50 years."

I would welcome Pall and others to work with us on the details and implications of this mechanism or on its experimental verification, like in Pall (2013). Claiming someone else's work that is already published, or neglecting a work that would greatly help in explaining important effects, cannot stand for long. Apart from Pall, others have also attempted to either present this mechanism as their own suggestion or downgrade it, and this has occurred also with other studies. This has given us the opportunity to write several commentary articles which are very useful for discussing details of the studies and various important scientific and methodological issues (Panagopoulos et al. 2010b; 2016; Panagopoulos and Karabarbounis 2011; 2020; Panagopoulos 2021).

11.3.6 The Explanation of EMF-induced ROS Generation in Cells

In Panagopoulos et al. (2021) it was documented how the impaired function of VGICs in the membranes of living cells triggers (over)production of free radicals/ROS. This is of great importance because the electrochemical imbalance caused in the cell by the VGIC dysfunction is translated into specific biological and health effects (OS, DNA damage, infertility, cancer, etc.). ROS, such as the hydroxyl radical (OH•) produced by hydrogen peroxide (H_2O_2) via the Fenton reaction and peroxynitrite (ONOO-) produced by nitric oxide (NO•), had already been accused for the observed damages induced by man-made EMFs (De Iuliis et al. 2009; Burlaka et al. 2013; Pall 2013; Yakymenko et al. 2016; Panagopoulos et al. 2021). The hydroxyl radical and peroxynitrite are considered the main damaging ROS for DNA and other critical biological molecules (Coggle 1983; Hall and Giaccia 2006; Pacher et al. 2007). Furthermore, it has been explained how unrepaired/misrepaired DNA lesions/damages such as strand breaks, covalent bond breakage, or nucleotide base damages, lead to cell senescence, cell death, or mutations and related pathologies including cancer (Panagopoulos et al. 2021). All these are now described in the next (final) chapter of this book.

11.3.7 Conclusion

In conclusion, the IFO-VGIC mechanism described in this chapter and the ROS production that is triggered by the VGIC dysfunction in the cells (which is the subject of the next chapter in this book) explain, for the first time, the vast number of experimental and epidemiological findings connecting man-made EMF exposure with OS, DNA damage, and related pathologies, such as EHS, infertility, cancer, neuro-degenerative diseases, etc. Hopefully, the presented mechanism will set

EMF-bioeffect research on a new base in which the "argument" of "no generally accepted mechanism" will no longer be used as a relevant argument to justify the continuation of unrestricted use of man-made EMF technology without considering the consequences in health and environment. Otherwise, those who insist on such an argument and neglect this mechanism should bear the burden of specifically proving that this mechanism is wrong. Neglecting this mechanism and looking at other uncertain and impossible mechanistic schemes should not anymore be accepted as a responsible scientific position.

11.4 APPENDIX

11.4.1A Calculation of the Damping Coefficient for an Ion in an Aqueous Solution

In this case, the damping coefficient β depends upon the internal friction coefficient (viscosity) n of the medium (the cytoplasm or the extracellular aqueous solution) and the radius a of the (hypothetically spherical) ion $\beta = 6\pi n a$. Assuming the viscosity is that of water at 37°C, $n = 7 \times 10^{-4}$ kg/m·s, and the ion's radius is (Na$^+$ radius = 0.98×10^{-10} m) $a \cong 10^{-10}$ m, we get: $\beta \cong 1.32 \times 10^{-12}$ kg/s.

11.4.1B Calculation of the Damping Coefficient and the Ion Velocity within a Channel Protein

It has been found by conductivity measurements with the "patch-clamp" technique that the intensity of ionic electric current through an open sodium channel is $\sim 4 \times 10^{-12}$ A, when the transmembrane voltage is 100 mV. Since the electric charge of each Na$^+$ ion is 1.6×10^{-19} C, that means $\sim 2.5 \times 10^7$ Na$^+$ ions per s pass through an open sodium channel (Neher and Sakmann 1992; Stryer 1995). Accepting that the channel's length is equal to the membrane's width of $\sim 10^{-8}$ m, and that the ions move through the channel in single file (Palmer 1986; Stryer 1995), we get that the transit time of every Na$^+$ ion through an open Na$^+$ channel is of the order of $\sim 0.4 \times 10^{-7}$ s, and thus, the ion velocity through the channel is: $u = 2.5 \times 10^7 \times 10^{-8}$ m/s, or $u_o = \mathbf{0.25}$ **m/s**. In such a case, on every Na$^+$ ion only the force from the transmembrane electric field $E_m z q_e$ (with $E_m \approx 10^7$ V/m, $z = 1$) and the damping force $-\beta u$ are exerted. As the ion accelerates, these forces become equal: $\beta u_o = E_m q_e \Rightarrow \beta = \dfrac{E_m q_e}{u_o}$

$\Rightarrow \beta \cong 6.4 \times 10^{-12}$ kg/s (Panagopoulos et al. 2000). As we can see, the damping coefficient within a channel is 4.8 times greater than the damping coefficient in the cytoplasm for the same ion type, which is a very reasonable result. For a z-valence ion, $\beta = \dfrac{E_m z q_e}{u_o} \cong 6.4z \times 10^{-12}$ kg/s.

Here, we have two important findings: 1) The maximum velocity of an ion through an open ion channel (~ 0.25 m/s), which is an upper limit for any ion velocity in cells and, thus, as explained in Chapter 1, represents a biophysical constant; and 2) the damping coefficient for the ion motion through an open ion channel ($\sim 6.4z \times 10^{-12}$ kg/s). Both of these parameters are necessary for the solution of the motion Eq. 11.3. Moreover, they provide the measure of two important physical phenomena taking place in living cells: The ion transit velocity through an open channel (and an upper limit for any ion velocity in cells), and the damping force exerted on ions passing through ion channels.

11.4.2 Solution of Equation 11.3

Equation 11.3 is a second order linear differential equation with constant coefficients (Bear 1962; Bronson 1973; Stephenson 1973), and has a particular solution of the type:

$$r_p = A_1 \cos \omega t + A_2 \sin \omega t$$

or
$$r_p = A\cos(\omega t - \varphi) \qquad [11.21]$$

where A is the amplitude of the forced oscillation and φ is the phase difference between the forced oscillation and the externally applied periodical force.

From Eqs 11.3 and 11.21, after operations, (method of undetermined coefficients), we get:

$$A = \frac{E_o z q_e}{\sqrt{m_i^2(\omega^2 - \omega_o^2)^2 + \beta^2 \omega^2}} \qquad [11.22]$$

and
$$\tan\varphi = \frac{m_i(\omega^2 - \omega_o^2)}{\beta\omega} \qquad [11.23]$$

The general solution of Eq. 11.3, will be the sum of the particular solution (11.21) and the general solution r_o of the corresponding homogeneous differential equation:

$$r = r_p + r_o \qquad [11.24]$$

The corresponding homogeneous equation is:

$$m_i \frac{d^2 r}{dt^2} + \beta \frac{dr}{dt} + m_i \omega_o^2 r = 0 \qquad [11.25]$$

and its general solution is of the type (Bronson 1973):

$$r_o = c_1 e^{\xi_1 t} + c_2 e^{\xi_2 t} \qquad [11.26]$$

where c_1, c_2 are constants that can be calculated from the initial conditions, if we apply them to the general solution, and ξ_1, ξ_2 are the roots of the corresponding "characteristic equation" of Eq. 11.25:

$$m_i \xi^2 + \beta \xi + m_i \omega_o^2 = 0 \qquad [11.27]$$

The discriminant of Eq. 11.27 is: $\Delta = \beta^2 - 4m_i^2 \omega_o^2$, and the roots are:

$$\xi_{1,2} = \frac{-\beta \pm \sqrt{\beta^2 - 4m_i^2 \omega_o^2}}{2m_i} \qquad [11.28]$$

The only unknown parameter is now ω_o. It has been revealed that the mobile ions in cells oscillate spontaneously at low frequencies (ULF), and this is reflected as a slow variation in the plasma membrane potential/voltage. All the experimental evidence on cytosolic free ion spontaneous oscillations, as well as membrane potential spontaneous oscillations, in many different types of cells, shows that the frequencies of such oscillations do not transcend the value of 1 Hz. Most of these oscillations display frequencies ranging from 0.016 to 0.2 Hz, (Ueda et al. 1986; Berridge 1988; Berridge and Galione 1988; Gray 1988; Tsunoda 1990; Furuya et al. 1993). It is most reasonable to accept that the frequencies of these spontaneous ionic oscillations represent the ions' self-frequencies $v_o = \omega_o/2\pi$. This is a most crucial observation allowing us to solve Eq. 11.3 all the way.

Thus, for a typical ion, e.g., Na^+ with $m_i = 3.8 \times 10^{-26}$ kg and $\beta \sim 10^{-12}$ kg/s, we get that:

$$\beta \gg 2 m_i \omega_o \qquad [\text{Eq. 11.29}]$$

Then, $\Delta > 0$ and the roots ξ_1, ξ_2 of Eq. 11.27 in this case will be real, unequal, and negative (the quantities: β, m_i, and ω_o are positive).

According to above values for m_i, β, ω_o and ω, the amount $m_i^2(\omega^2 - \omega_o^2)^2$ in Eq. 11.22, is negligible compared to the amount $\beta^2\omega^2$. Thus, the amplitude A in the particular solution is:

$$A = \frac{E_o z q_e}{\beta\omega} \quad [11.30]$$

Moreover, the amount $m_i(\omega^2-\omega_o^2)$ is very small compared to the amount $\beta\omega$, and from Eq. 11.23, we get: $tan\varphi \cong 0$. In addition, $tan\varphi \geq 0$ for $\omega \geq \omega_o$, and thus, $\varphi \cong 0$. Therefore, the particular solution (Eq. 11.21) becomes:

$$r_p = \frac{E_o z q_e}{\beta\omega} cos\omega t \quad [11.31]$$

Substituting in Eq. 11.24 r_o from Eq. 11.26 and r_p from Eq. 11.31, the general solution of Eq. 11.3 becomes:

$$r = \frac{E_o z q_e}{\beta\omega} cos\omega t + c_1 e^{\xi_1 t} + c_2 e^{\xi_2 t} \quad [11.32]$$

For a representative value of $v_o = 0.1$ Hz, from Eq. 11.28, we find that: $\xi_1 = -1.5 \times 10^{-14}$ s^{-1} $\cong 0$ and $\xi_2 \cong -2.63 \times 10^{13}$ s^{-1}.

Accepting as initial conditions (for $t = 0$) $r_{t=0} = 0$ and $(dr/dt)_{t=0} = u_o$, from Eq. 11.32, we get: $c_1 + c_2 = -\frac{E_o z q_e}{\beta\omega}$ and $c_1 \xi_1 + c_2 \xi_2 = u_o$, from which we get:

$$c_1 \cong -\frac{E_o z q_e}{\beta\omega} \text{ and } c_2 \cong -u_o 4 \times 10^{-14}$$

For any reasonable value of u_o, (the maximum ion velocity due to the forced oscillation)* $c_2 e^{\xi_2 t} \cong 0$, and ignoring the minus sign for c_1, we get:

$$r_o \cong \frac{E_o z q_e}{\beta\omega} \quad [11.33] \quad (r_o \text{ in m}, E_o \text{ in V·m}^{-1}, \omega \text{ in Hz}),$$

[and thus, Zhadin's (1998) postulate that $r_o = 0$ was not correct].

Substituting in Eq. 11.32, we finally get the general solution of Eq. 11.3:

$$r = \frac{E_o z q_e}{\beta\omega} cos\omega t + \frac{E_o z q_e}{\beta\omega}$$

11.4.3 CALCULATION OF THE FORCE ON THE VOLTAGE SENSORS REQUIRED TO GATE VGICS

The intensity of the transmembrane electric field is: $E_m = \frac{V}{s}$

Moreover, $E_m = \frac{F}{q}$

where F, in this case, is the force acting on an S4 sensor and q is the effective charge of the S4, which has been estimated to be (Liman et al. 1991): $q = 1.7\ q_e$

---

* An upper limit for u_o is the value 0.25 m/s that we calculated for Na$^+$ ions passing through an open Na$^+$ channel (Appendix 11.4.1B). In fact, the max velocity is ~ 10^5 times smaller (see discussion in Eq. 11.20).

Equating the right parts in the above equations for the transmembrane electric field we get:

$$\frac{F}{q} = \frac{V}{s} \Rightarrow F = \frac{V}{s}q$$

Differentiating with respect to V, we get: $dF = dV \frac{q}{s}$

Substituting $dV = 30$ mV, $s = 10^{-8}$ m, and $q = 1.7\, q_e$, we get: $dF = 8.16 \times 10^{-13}$ N.

This is the minimum force on each S4 voltage sensor of a VGIC (corresponding to the minimum transmembrane voltage change of 30 mV) normally required to interconvert the channel between closed and open states. A similar calculation was first performed by Balcavage et al. (1996), resulting in a slightly different value for dF. It is evident that these forces on the S4 sensors that physiologically gate VGICs are parallel to the channel direction/orientation (i.e., vertical to the membrane).

11.4.4 Prevalence of the Polarized and Coherent Forced Oscillation of the Ion Against the Greater but Chaotic Thermal Motion

Even if we consider only one ion interacting with an S4 sensor, this ion moving with a drift velocity of $u = 0.25$ m/s, (see Appendix 11.4.1B) within an open channel, it needs a time interval, $dt = dr/u \cong 1.6 \times 10^{-11}$ s, in order to be displaced at the distance $dr = 4 \times 10^{-12}$ m and generate the force on the voltage sensor required to gate the channel, according to Eq. 11.6. During this time interval, the ion will also be displaced at a total distance* of $dr_{kT} = u_{kT}\, dt = 9.3 \times 10^{-9}$ m due to its thermal velocity

$u_{kT} = \sqrt{\dfrac{3kT}{m_i}}$ ($\cong 0.58 \times 10^3$ m/s) for human body temperature 37°C or 310 K ($k = 1.381 \times 10^{-23}$ J·K^{-1} the Boltzmann constant, $m_i \cong 3.8 \times 10^{-26}$ kg the ion mass for Na$^+$) (Panagopoulos et al. 2013). The mean free path of the ions in the aqueous solutions around the membrane is $\sim 10^{-10}$ m (Chiabrera et al. 1994) and within a channel (with a pore diameter $\sim 3 \times 10^{-10}$ m) is $\sim 10^{-11}$ m (Panagopoulos et al. 2002). Therefore, the ion within the channel and during the above time interval dt will cover because of its thermal activity $\sim 9.3 \times 10^2 = 9300$ mean free paths, each one in a different direction, resulting in mutually destructive/canceling forces on the channel sensors, while at the same time, the ion's displacement due to the external field ($dr = 4 \times 10^{-12}$ m) is in a certain direction, exerting on each S4 sensor a constant force of constant direction. If the polarization of the external electric field is parallel to the channel direction/orientation, the condition is optimum.

If we consider several (at least four as explained) ions interacting simultaneously with an S4 sensor, which is the realistic case, then the effect of the external field is multiplied by the number of ions becoming even more significant, whereas the effect of their chaotic thermal motions is further divided by the number of ions becoming even more negligible.

REFERENCES

Adair RK, (1991a): Biological effects on the cellular level of electric field pulses, *Health Physics*, 61(3), 395–399.

Adair RK, (1991b): Constraints on biological effects of weak extremely-low-frequency electromagnetic fields, *Physical Review: Part A*, 43(2), 1039–1048.

Alberts B, Johnson A, Lewis J, Raff M, Roberts K, Walter P, (2002): *Membrane transport of small molecules and the electrical properties of membranes - molecular biology of the cell*, Garland Science, New York.

---

* The formula for this thermal displacement is $dr_{kT} = \left(\sqrt{\dfrac{3kT}{m_i}}\right) dt$, which corrects the corresponding formula given before (Panagopoulos et al 2000; 2002; Panagopoulos and Margaritis 2003a) as X_{kT} (taken from Adair 1991a) that had resulted in significantly smaller displacement.

Alexopoulos CD, (1960): Mechanics and Acoustics, Athens (Αλεξόπουλος ΚΔ, Μηχανική-Ακουστική, Αθήνα, 1960).
Alonso M, Finn EJ, (1967): *Fundamental university physics*, Vol. 1, Mechanics and Thermodynamics, Addison-Wesley.
Baker PF, Hodgkin AL, Shaw TL, (1962): The effects of changes in internal ionic concentration on the electrical properties of perfused giant axons, *Journal of Physiology*, 164, 355–374.
Balcavage WX, Alvager T, Swez J, Goff CW, Fox MT, Abdullyava S, King MW, (1996): A mechanism for action of extremely low frequency electromagnetic fields on biological systems, *Biochemical and Biophysical Research Communications*, 222(2), 374–378.
Balmori A, (2006): The incidence of electromagnetic pollution on the amphibian decline: Is this an important piece of the puzzle?, *Toxicological and Environmental Chemistry*, 88(2), 287–299.
Balmori A, (2010): Mobile phone mast effects on common frog (Rana temporaria) tadpoles: The city turned into a laboratory, *Electromagnetic Biology and Medicine*, 29(1–2), 31–35.
Barr R, Llanwyn Jones D, Rodger CJ, (2000): ELF and VLF radio waves, *Journal of Atmospheric and Solar-Terrestrial Physics*, 62(17–18), 1689–1718.
Bawin SM, Kaczmarek LK, Adey WR, (1975): Effects of modulated VHF fields, on the central nervous system, *Annals of the New York Academy of Sciences*, 247, 74–81.
Bawin SM, Adey WR, Sabbot IM, (1978): Ionic factors in release of $^{45}Ca^{2+}$ from chick cerebral tissue by electromagnetic fields, *Proceedings of the National Academy of Sciences of the UnitedStates of America*, 75(12), 6314–6318.
Bear HS, (1962): *Differential equations*, Addison Wesley, Reading, Mass., USA.
Belyaev I, (2005): Non-thermal biological effects of microwaves, *Microwave Review*, 11(2), 13–29.
Berridge MJ, (1988): Inositol triphosphate-induced membrane potential oscillations in Xenopus oocytes, *Journal of Physiology*, 403, 589–599.
Berridge MJ, Galione A, (1988): Cytosolic calcium oscillators, *FASEB Journal*, 2(15), 3074–3082.
Bertagna F, Lewis R, Silva SRP, McFadden J, Jeevaratnam K, (2021): Effects of electromagnetic fields on neuronal ion channels: A systematic review, *Annals of the New York Academy of Sciences*, 1499(1), 82–103.
Betti L, Trebbi G, Lazzarato L, Brizzi M, Calzoni GL, Marinelli F, Nani D, Borghini F, (2004): Nonthermal microwave radiations affect the hypersensitive response of tobacco to tobacco mosaic virus, *Journal of Alternative and Complementary Medicine*, 10(6), 947–957.
Bezanilla F, (2000): The voltage sensor in voltage-dependent ion channels, *Physiological Reviews*, 80(2), 555–592.
Bezanilla F, (2018): Gating currents, *Journal of General Physiology*, 150(7), 911–932.
Bianco B, Chiabrera A, Morro A, Parodi M, (1988): Effects of magnetic exposure on ions in electric fields, *Ferroelectrics*, 86(1), 159–168.
Blackman CF, Benane SG, Elder JA, House DE, Lampe JA, Faulk JM, (1980): Induction of calcium-ion efflux from brain tissue by radiofrequency radiation: Effect of sample number and modulation frequency on the power-density window, *Bioelectromagnetics*, 1(1), 35–43.
Blanchard JP, Blackman CF, (1994): Clarification and application of an ion parametric resonance model for magnetic field interactions with biological systems, *Bioelectromagnetics*, 15(3), 217–238.
Bolshakov MA, Alekseev SI, (1992): Bursting responses of Lymnea neurons to microwave radiation, *Bioelectromagnetics*, 13(2), 119–129.
Brocklehurst B, McLauchlan KA, (1996): Free radical mechanism for the effects of environmental electromagnetic fields on biological systems, *International Journal of Radiation Biology*, 69(1), 3–24.
Bronson R, (1973): *Differential equations*, McGraw-Hill, New York.
Burlaka A, Tsybulin O, Sidorik E, Lukin S, Polishuk V, Tsehmistrenko S, Yakymenko I, (2013): Overproduction of free radical species in embryonic cells exposed to low intensity radiofrequency radiation, *Experimental Oncology*, 35(3), 219–225.
Campisi A, Gulino M, Acquaviva R, Bellia P, Raciti G, Grasso R, Musumeci F, Vanella A, Triglia A, (2010): Reactive oxygen species levels and DNA fragmentation on astrocytes in primary culture after acute exposure to low intensity microwave electromagnetic field, *Neuroscience Letters*, 473(1), 52–55.
Cecchetto C, Maschietto M, Boccaccio P, Vassanelli S, (2020): Electromagnetic field affects the voltage-dependent potassium channel Kv1.3, *Electromagnetic Biology and Medicine*, 39(4), 316–322.
Chavdoula ED, Panagopoulos DJ, Margaritis LH, (2010): Comparison of biological effects between continuous and intermittent exposure to GSM-900 MHz mobile phone radiation: Detection of apoptotic cell death features, *Mutation Research*, 700(1–2), 51–61.
Chiabrera A, Grattarola M, Viviani R, (1984): Interaction between electromagnetic fields and cells: Microelectrophoretic effect on ligands and surface receptors, *Bioelectrornagnetics*, 5(2), 173–191.

Chiabrera A, Bianco B, Moggia E, Tommasi T, (1994): Interaction mechanism between electromagnetic fields and ion absorption: Endogenous forces and collision frequency, *Bioelectrochemistry and Bioenergetics*, 35(1–2), 33–37.

Cleary SF, Liu LM, Graham R, Diegelmann RF, (1988): Modulation of tendon fibroplasia by exogenous electric currents, *Bioelectromagnetics*, 9(2), 183–194.

Coggle JE, (1983): *Biological effects of radiation*, Taylor & Francis, London.

Coghill RW, Steward J, Philips A, (1996): Extra low frequency electric and magnetic fields in the bed place of children diagnosed with leukaemia: A case-control study, *European Journal of Cancer Prevention*, 5(3), 153–158.

Creasey WA, Goldberg RB, (2001): A new twist on an old mechanism for EMF bioeffects?, *EMF Health Report*, 9(2), 1–11.

Cucurachi S, Tamis WL, Vijver MG, Peijnenburg WJ, Bolte JF, de Snoo GR, (2013): A review of the ecological effects of radiofrequency electromagnetic fields (RF-EMF), *Environment International*, 51, 116–140.

De Coursey TE, (2003): Interactions between NADPH oxidase and voltage-gated proton channels: Why electron transport depends on proton transport, *FEBS Letters*, 555(1), 57–61.

De Iuliis GN, Newey RJ, King BV, Aitken RJ, (2009): Mobile phone radiation induces reactive oxygen species production and DNA damage in human spermatozoa in vitro, *PLOS ONE*, 4(7), e6446.

Delgado JMR, (1985): Biological effects of extremely low frequency electromagnetic fields, *Journal of Bioelectricity*, 4(1), 75–91.

Diem E, Schwarz C, Adlkofer F, Jahn O, Rudiger H, (2005): Non-thermal DNA breakage by mobile-phone radiation (1800 MHz) in human fibroblasts and in transformed GFSH-R17 rat granulosa cells in vitro, *Mutation Research*, 583(2), 178–183.

Engstrom S, (1996): Dynamic properties of Lednev's parametric resonance mechanism, *Bioelectromagnetics*, 17(1), 58–70.

Francis G, (1960): *Ionization phenomena in gases*, Butterworths Scientific Publications, London.

Franzellitti S, Valbonesi P, Ciancaglini N, Biondi C, Contin A, Bersani F, Fabbri E, (2010): Transient DNA damage induced by high-frequency electromagnetic fields (GSM 1.8 GHz) in the human trophoblast HTR-8/SVneo cell line evaluated with the alkaline comet assay, *Mutation Research*, 683(1–2), 35–42.

Frei M, Jauchem J, Heinmets F, (1988): Physiological effects of 2.8 GHz radio-frequency radiation: A comparison of pulsed and continuous-wave radiation, *Journal of Microwave Power and Electromagnetic Energy*, 23(2), 2.

Fröhlich H, (1968): Long-range coherence and energy storage in biological systems, *International Journal of Quantum Chemistry*, 2(5), 641–649.

Furuya K, Enomoto K, Yamagishi S, (1993): Spontaneous calcium oscillations and mechanically and chemically induced calcium responses in mammary epithelial cells, *Pflügers Archiv: European Journal of Physiology*, 422(4), 295–304.

Gautreau R, Savin W, (1978): *Theory and problems of modern physics*, McGraw-Hill, New York.

Georgiou CD, (2010): Oxidative stress-induced biological damage by low-level EMFs: Mechanism of free radical pair electron spin-polarization and biochemical amplification. In Giuliani L, Soffrittis M (Eds.), *Non-thermal effects and mechanisms of interaction between electromagnetic fields and living matter*, 63–113.

Gomer R, (1961): *Field emission and field ionization*, Harvard University Press, Cambridge, Mass., USA.

Goodman EM, Greenebaum B, Marron MT, (1995): Effects of electro-magnetic fields on molecules and cells, *International Review of Cytology*, 158, 279–338.

Gray PTA, (1988): Oscillations of free cytosolic calcium evoked by cholinergic and catecholaminergic agonists in rat parotid acinar cells, *Journal of Physiology*, 406, 35–53.

Groome JR, Bayless-Edwards L, (2020): Roles for counter charge in the voltage sensor domain of ion channels, *Frontiers in Pharmacology*, 11, 160. https://doi.org/10.3389/fphar.2020.0016.

Gulati S, Yadav A, Kumar N, Kanupriya, Aggarwal NK, Kumar R, Gupta R, (2016): Effect of GSTM1 and GSTT1 polymorphisms on genetic damage in humans populations exposed to radiation from mobile towers, *Archives of Environmental Contamination and Toxicology*, 70(3), 615–625.

Halgamuge MN, Abeyrathne CD, (2011): Behavior of charged particles in a biological cell exposed to AC-DC electromagnetic fields, *Environmental Engineering Science*, 28(1), 1–11.

Hall EJ, Giaccia AJ, (2006): *Radiobiology for the radiologist*, Lippincott Williams & Wilkins, Philadelphia.

Hardell L, Carlberg M, Söderqvist F, Mild KH, Morgan LL, (2007): Long-term use of cellular phones and brain tumours: Increased risk associated with use for > or =10 years, *Occupational and Environmental Medicine*, 64(9), 626–632.

Hardell L, Carlberg M, Hansson Mild K, (2009): Epidemiological evidence for an association between use of wireless phones and tumor diseases, *Pathophysiology*, 16(2–3), 113–122.

Hardell L, Nyberg R, (2020): Appeals that matter or not on a moratorium on the deployment of the fifth generation, 5G, for microwave radiation, *Molecular and Clinical Oncology*. https://doi.org/10.3892/mco.2020.1984.

Hille B, (1992): *Ionic channels of excitable membranes*, 2nd ed., Sinauer, Sunderland.

Hodgkin AL, Huxley AF, (1952): A quantitative description of membrane current and its application to conduction and excitation in nerve, *Journal of Physiology*, 117(4), 500–544.

Honig BH, Hubbell WL, Flewelling RF, (1986): Electrostatic interactions in membranes and proteins, *Annual Review of Biophysics and Biophysical Chemistry*, 15, 163–193.

Höytö A, Luukkonen J, Juutilainen J, Naarala J, (2008): Proliferation, oxidative stress and cell death in cells exposed to 872 MHz radiofrequency radiation and oxidants, *Radiation Research*, 170(2), 235–243.

Huber R, Treyer V, Borbély AA, Schuderer J, Gottselig JM, Landolt HP, Werth E, Berthold T, Kuster N, Buck A, Achermann P, (2002): Electromagnetic fields, such as those from mobile phones, alter regional cerebral blood flow and sleep and waking EEG, *Journal of Sleep Research*, 11(4), 289–295.

Hyland GJ, (2008): Physical basis of adverse and therapeutic effects of low intensity microwave radiation, *Indian Journal of Experimental Biology*, 46(5), 403–419.

IARC, (2002): Non-ionizing radiation, part 1: Static and extremely low-frequency (ELF) electric and magnetic fields, Vol. 80, World Health Organization, Lyon, France.

IARC, (2013): Non-ionizing radiation, part 2: Radiofrequency electromagnetic fields (Vol. 102), World Health Organization, Lyon, France.

ICNIRP, (1998): Guidelines for limiting exposure to time-varying electric, magnetic and electromagnetic fields (up to 300GHz), *Health Physics*, 74, 494–522.

ICNIRP, (2010): Guidelines for limiting exposure to time-varying electric and magnetic fields (1Hz to 100 kHz), *Health Physics*, 99(6), 818–836.

ICNIRP, (2020): Guidelines for limiting exposure to electromagnetic fields (100 kHz to 300 GHz), *Health Physics* [Published ahead of print].

Ivancsits S, Diem E, Pilger A, Rüdiger HW, Jahn O, (2002): Induction of DNA strand breaks by intermittent exposure to extremely-low-frequency electromagnetic fields in human diploid fibroblasts, *Mutation Research*, 519(1–2), 1–13.

Ivancsits S, Diem E, Jahn O, Rüdiger HW, (2003): Intermittent extremely low frequency electromagnetic fields cause DNA damage in a dose-dependent way, *International Archives of Occupational and Environmental Health*, 76(6), 431–436.

Jackson JD, (1975): *Classical electrodynamics*, John Wiley & Sons, Inc., New York.

Ji S, Oh E, Sul D, Choi JW, Park H, Lee E, (2004): DNA damage of lymphocytes in volunteers after 4 hours use of mobile phone, *Journal of Preventive Medicine and Public Health*, 37(4), 373–380.

Kirschvink JL, (1989): Magnetite biomineralization and geomagnetic sensitivity in higher animals: An update and recommendations for future study, *Bioelectromagnetics*, 10(3), 239–259.

Lai H, (2021): Genetic effects of non-ionizing electromagnetic fields, *Electromagnetic Biology and Medicine*, 40(2), 264–273.

La Vignera S, Condorelli RA, Vicardi E, D'Agata R, Calogero AE, (2012): Effects of the exposure to mobile phones on male reproduction: A review of the literature, *Journal of Andrology*, 33(3), 350–356.

Leach V, Weller S, Redmayne M, (2018): A novel database of bio-effects from non-ionizing radiation, *Reviews on Environmental Health*, 33(3), 1–8.

Lednev VV, (1991): Possible mechanism for the influence of weak magnetic fields on biological systems, *Bioelectromagnetics*, 12(2), 71–75.

Lee RC, Canaday DJ, Doong H, (1993): A review of the biophysical basis for the clinical application of electric fields in soft-tissue repair, *Journal of Burn Care and Rehabilitation*, 14(3), 319–335.

Liboff AR, (1985): Cyclotron resonance in membrane transport. In Chiabrera A, Nicolini C, Schwan HP (Eds.), *Interactions between electromagnetic fields and cells*, Plenum Press, London, 281–296.

Liboff AR, McLeod BR, (1988): Kinetics of channelized membrane ions in magnetic fields, *Bioelectromagnetics*, 9(1), 39–51.

Liboff AR, (2003a): Ion cyclotron resonance in biological systems: Experimental evidence. In Stavroulakis P (Ed.), *Biological effects of electromagnetic fields*, Springer, Berlin, 76–113.

Liboff AR, (2003b): Biological effects of electromagnetic fields (book review), *Electromagnetic Biology and Medicine*, 22(1), 85–86.

Liburdy RP, (1992): Calcium signaling in lymphocytes and ELF fields: Evidence for an electric field metric and a site of interaction involving the calcium ion channel, *FEBS Letters*, 301(1), 53–59.

Liman ER, Hess P, Weaver F, Koren G, (1991): Voltage sensing residues in the S4 region of a mammalian K+ channel, *Nature*, 353(6346), 752–756.

López I, Félix N, Rivera M, Alonso A, Maestú C, (2021): What is the radiation before 5G? A correlation study between measurements in situ and in real time and epidemiological indicators in Vallecas, Madrid, *Environmental Research*, 194, 110734.

Ludwig HW, (1974): Electric and magnetic field strengths in the open and in shielded rooms in the ULF to LF zone. In Persinger MA (Ed.), *ELF and VLF electromagnetic fields*, Plenum Press, New York, 35–80.

MacKinnon R, (2003): Potassium channels, *FEBS Letters*, 555(1), 62–65.

Manna D, Ghosh R, (2016): Effect of radiofrequency radiation in cultured mammalian cells: A review, *Electromagnetic Biology and Medicine*, 35(3), 265–301.

Mathie A, Kennard LE, Veale EL, (2003): Neuronal ion channels and their sensitivity to extremely low frequency weak electric field effects, *Radiation Protection Dosimetry*, 106(4), 311–316.

McLeod KJ, Lee RC, Ehrlich HP, (1987): Frequency dependence of electric field modulation of fibroblast protein synthesis, *Science*, 236(4807), 1465–1469.

Metaxas AC, (1991): Microwave heating, *Power Engineering Journal*, 5(5), 237–247.

Miller AB, To T, Agnew DA, Wall C, Green LM, (1996): Leukemia following occupational exposure to 60-Hz electric and magnetic fields among Ontario electric utility workers, *American Journal of Epidemiology*, 144(2), 150–160.

Miller AB, Morgan LL, Udasin I, Davis DL, (2018): Cancer epidemiology update, following the 2011 IARC evaluation of radiofrequency electromagnetic fields (Monograph 102), *Environmental Research*, 167, 673–683.

Miller AB, Sears ME, Morgan LL, Davis DL, Hardell L, Oremus M, and Soskolne CL, (2019): Risks to health and well-being from radio-frequency radiation emitted by cell phones and other wireless devices, *Frontiers in Public Health*, 7, 223. https://doi.org/10.3389/fpubh.2019.00223.

Miller C, (2000): An overview of the potassium channel family, *Genome Biology*, 1(4).

Mohammed HS, Fahmy HM, Radwan NM, Elsayed AA, (2013): Non-thermal continuous and modulated electromagnetic radiation fields effects on sleep EEG of rats, *Journal of Advanced Research*, 4(2), 181–187.

Neher E, Sakmann B, (1992): The patch clamp technique, *Scientific American*, 266(3), 28–35.

Noda M, Ikeda T, Kayano T, Suzuki H, Takeshima H, Kurasaki M, Takahashi H, and Numa S, (1986): Existence of distinct sodium channel messenger RNAs in rat brain, *Nature*, 320(6058), 188–192.

Office of Research Integrity, (1999): Findings of scientific misconduct, *NIH Guide for Grants and Contracts*, 1999(June 18), 1. PMID: 12458593; PMCID: PMC4259611.

Pacher P, Beckman JS, Liaudet L, (2007): Nitric oxide and peroxynitrite in health and disease, *Physiological Reviews*, 87(1), 315–424.

Pall ML, (2013): Electromagnetic fields act via activation of voltage-gated calcium channels to produce beneficial or adverse effects, *Journal of Cellular and Molecular Medicine*, 17(8), 958–965.

Pall ML, (2018): Wi-fi is an important threat to human health, *Environmental Research*, 164, 405–416.

Palmer LG, (1986): *New insights into cell and membrane transport processes*. In Poste G, Crooke ST (Eds.), Plenum Press, New York, 331.

Panagopoulos DJ, Messini N, Karabarbounis A, Filippetis AL, Margaritis LH, (2000): A mechanism for action of oscillating electric fields on cells, *Biochemical and Biophysical Research Communications*, 272(3), 634–640.

Panagopoulos DJ, Karabarbounis A, Margaritis LH, (2002): Mechanism for action of electromagnetic fields on cells, *Biochemical and Biophysical Research Communications*, 298(1), 95–102.

Panagopoulos DJ, Margaritis LH, (2003a): Theoretical considerations for the biological effects of electromagnetic fields. In Stavroulakis P (Ed.), *Biological effects of electromagnetic fields*, Springer, Berlin, 5–33.

Panagopoulos DJ, Margaritis LH, (2003b): Effects of electromagnetic fields on the reproductive capacity of Drosophila melanogaster. In Stavroulakis P (Ed.), *Biological effects of electromagnetic fields*, Springer, Berlin, 545–578.

Panagopoulos DJ, Karabarbounis A, Margaritis LH, (2004): Effect of GSM 900-MHz mobile phone radiation on the reproductive capacity of Drosophila melanogaster, *Electromagnetic Biology and Medicine*, 23(1), 29–43.

Panagopoulos DJ, Margaritis LH, (2010): The identification of an intensity "window" on the bioeffects of mobile telephony radiation, *International Journal of Radiation Biology*, 86(5), 358–366.

Panagopoulos DJ, Chavdoula ED, Margaritis LH, (2010a): Bioeffects of mobile telephony radiation in relation to its intensity or distance from the antenna, *International Journal of Radiation Biology*, 86(5), 345–357.

Panagopoulos DJ, Karabarbounis A, Margaritis LH, (2010b): Comment on mechanism for alternating electric fields induced-effects on cytosolic calcium, *Chinese Physics Letters*, 27(4), 1.

Panagopoulos DJ, (2011): Biological impacts, action mechanisms, dosimetry and protection issues of mobile telephony radiation. In Barnes MC, Meyers NP (Eds.), *Mobile phones: Technology, networks and user issues*, Nova Science Publishers, Inc., New York, 1–54.

Panagopoulos DJ, Karabarbounis A, (2011): Comments on study of charged particle's behavior in a biological cell exposed to AC-DC electromagnetic fields, and on comparison between two models of interaction between electric and magnetic fields and proteins in cell membranes, *Environmental Engineering Science*, 28(10), 749–751.

Panagopoulos DJ, (2013): Electromagnetic interaction between environmental fields and living systems determines health and well-being. In MH Kwang and SO Yoon (Eds), *Electromagnetic fields: Principles, engineering applications and biophysical effects*, Nova Science Publishers, New York, 87–130.

Panagopoulos DJ, Johansson O, Carlo GL, (2013): Evaluation of specific absorption rate as a dosimetric quantity for electromagnetic fields bioeffects, *PLOS ONE*, 8(6), e62663. https://doi.org/10.1371/journal.pone.0062663.

Panagopoulos DJ, Johansson O, Carlo GL, (2015a): Polarization: A key difference between man-made and natural electromagnetic fields, in regard to biological activity, *Scientific Reports*, 5, 14914. https://doi.org/10.1038/srep14914.

Panagopoulos DJ, Johansson O, Carlo GL, (2015b): Real versus simulated mobile phone exposures in experimental studies, *BioMed Research International*, 2015, 607053.

Panagopoulos DJ, Cammaerts MC, Favre D, Balmori A, (2016): Comments on environmental impact of radiofrequency fields from mobile phone base stations, *Critical Reviews in Environmental Science and Technology*, 46(9), 885–903.

Panagopoulos DJ, (2017): Mobile telephony radiation effects on insect ovarian cells. The necessity for real exposures bioactivity assessment. The key role of polarization, and the "Ion forced-oscillation mechanism". In Geddes CD (Ed.), *Microwave effects on DNA and proteins*, Springer, Cham, Switzerland, 1–48.

Panagopoulos DJ, Balmori A, (2017): On the biophysical mechanism of sensing atmospheric discharges by living organisms, *Science of the Total Environment*, 599–600, 2026–2034.

Panagopoulos DJ, (2018): Man-made electromagnetic radiation is not quantized. In Reimer A (Ed.), *Horizons in world physics*, Vol. 296, Nova Science Publishers, New York, 1–57.

Panagopoulos DJ, (2019): Comparing DNA damage induced by mobile telephony and other types of man-made electromagnetic fields, *Mutation Research Reviews*, 781, 53–62.

Panagopoulos DJ, (2020): Comparing chromosome damage induced by mobile telephony radiation and a high caffeine dose: Effect of combination and exposure duration, *General Physiology and Biophysics*, 39(6), 531–544.

Panagopoulos DJ, Balmori A, Chrousos GP, (2020): On the biophysical mechanism of sensing upcoming earthquakes by animals, *Science of the Total Environment*, 717, 136989.

Panagopoulos DJ, Karabarbounis A, (2020): Comments on "diverse radiofrequency sensitivity and radiofrequency effects of mobile or cordless phone near fields exposure in Drosophila melanogaster", *Advances in Environmental Studies*, 4(1), 271–276.

Panagopoulos DJ, Karabarbounis A, Yakymenko I, Chrousos GP, (2021): Mechanism of DNA damage induced by human-made electromagnetic fields, *International Journal of Oncology*, 59(5), 92.

Panagopoulos DJ, (2021): Comments on Pall's "Millimeter (MM) wave and microwave frequency radiation produce deeply penetrating effects: The biology and the physics", *Reviews on Environmental Health*. https://doi.org/10.1515/REVEH-2021-0090.

Pedersen GF, (1997): Amplitude modulated RF fields stemming from a GSM/DCS-1800 phone, *Wireless Networks*, 3(6), 489–498.

Penafiel LM, Litovitz T, Krause D, Desta A, Mullins JM, (1997): Role of modulation on the effects of microwaves on ornithine decarboxylase activity in L929 cells, *Bioelectromagnetics*, 18(2), 132–141.

Phillips JL, Singh NP, Lai H, (2009): Electromagnetic fields and DNA damage, *Pathophysiology*, 16(2–3), 79–88.

Piacentini R, Ripoli C, Mezzogori D, Azzena GB, Grassi C, (2008): Extremely low-frequency electromagnetic fields promote in vitro neurogenesis via upregulation of Ca_v1-channel activity, *Journal of Cellular Physiology*, 215(1), 129–139.

Pirard W, Vatovez B: *Study of pulsed character of radiation emitted by wireless telecommunication systems*, Institut scientifique de service public, Liège, Belgium. https://www.issep.be/wp-content/uploads/7IWSBEEMF_B-Vatovez_W-Pirard.pdf.

Ritz T, Adem S, Schulten K, (2000): A model for photoreceptor-based magnetoreception in birds, *Biophysical Journal*, 78(2), 707–718.

Sandipan C, Baron C, (2015): Basic mechanisms of voltage sensing. In Zheng J, Trudeau MC (Eds.), *Handbook of ion channels*, CRC Press, London, 25–39.

Santini MT, Ferrante A, Rainaldi G, Indovina P, Indovina PL, (2005): Extremely low frequency (ELF) magnetic fields and apoptosis: A review, *International Journal of Radiation Biology*, 81(1), 1–11.

Schmidt WF, Thomas CG, (2012): More precise model of α-helix and transmembrane a-helical peptide backbone structure, *Journal of Biophysical Chemistry*, 3(4), 295–303.

Seredenina T, Demaurex N, Krause KH, (2015): Voltage-gated proton channels as novel drug targets: From NADPH oxidase regulation to sperm biology, *Antioxidants & Redox Signaling*, 23(5), 490–513.

Sheppard AR, Swicord ML, Balzano Q, (2008): Quantitative evaluations of mechanisms of radiofrequency interactions with biological molecules and processes, *Health Physics*, 93(4), 365–396.

Shi YP, Thouta S, Claydon TW, (2020): Modulation of hERG K+ channel deactivation by voltage sensor relaxation, *Frontiers in Pharmacology*, 11, 139. https://doi.org/10.3389/fphar.2020.00139.

Stavroulakis P, (2003): *Biological effects of electromagnetic fields*, Springer, Berlin.

Stephenson G, (1973): *Mathematical methods for science students*, 2nd ed., Longman group, London.

Stryer L, (1995): *Biochemistry*, Freeman, New York.

Stühmer W, Conti F, Suzuki H, Wang X, Noda M, Yahagi N, Kubo H, Numa S, (1989): Structural parts involved in activation and inactivation of the sodium channel, *Nature*, 339(6226), 597–603.

Tombola F, Pathak MM, Isacoff EY, (2006): How does voltage open an ion channel?, *Annual Review of Cell and Developmental Biology*, 22, 23–52.

Tsunoda Y, (1990): Cytosolic free calcium spiking affected by intracellular pH change, *Experimental Cell Research*, 188(2), 294–301.

Tytgat J, Nakazawa K, Gross A, Hess P, (1993): Pursuing the voltage sensor of a voltage-gated mammalian potassium channel, *Journal of Biological Chemistry*, 268(32), 23777–23779.

Ueda S, Oiki S, Okada Y, (1986): Oscillations of cytoplasmic concentrations of Ca^{2+} and K^+ in fused L cells, *Journal of Membrane Biology*, 91(1), 65–72.

Villalba-Galea CA, Chiem AT, (2020): Hysteretic behavior in voltage-gated channels, *Frontiers in Pharmacology*, 11, 579596. https://doi.org/10.3389/fphar.2020.579596.

Walleczek J, (1992): Electromagnetic field effects on cells of the immune system: The role of calcium signaling, *FASEB Journal*, 6(13), 3177–3185.

Walleczek J, (1995): Magnetokinetic effects on radical pairs: A paradigm for magnetic field interactions with biological systems at lower than thermal energy, *Advances in Chemotherapy*, 250, 395–420.

Weisbrot D, Lin H, Ye L, Blank M, Goodman R, (2003): Effects of mobile phone radiation on reproduction and development in Drosophila melanogaster, *Journal of Cellular Biochemistry*, 89(1), 48–55.

Wiltschko R, Thalau P, Gehring D, Nießner C, Ritz T, Wiltschko W, (2015): Magnetoreception in birds: The effect of radio-frequency fields, *Journal of the Royal Society. Interface*, 12(103), 20141103.

Yakymenko I, Sidorik E, Kyrylenko S, Chekhun V, (2011): Long-term exposure to microwave radiation provokes cancer growth: Evidences from radars and mobile communication systems, *Experimental Oncology*, 33(2), 62–70.

Yakymenko I, Tsybulin O, Sidorik E, Henshel D, Kyrylenko O, Kyrylenko S, (2016): Oxidative mechanisms of biological activity of low-intensity radiofrequency radiation, *Electromagnetic Biology and Medicine*, 35(2), 186–202.

Yakymenko I, Burlaka A, Tsybulin I, Brieieva I, Buchynska L, Tsehmistrenko I, Chekhun F, (2018): Oxidative and mutagenic effects of low intensity GSM 1800 MHz microwave radiation, *Experimental Oncology*, 40(4), 282–287.

Zhadin MN, (1998): Combined action of static and alternating magnetic fields on ion motion in a macromolecule: Theoretical aspects, *Bioelectromagnetics*, 19(5), 279.

Zhang XC, Yang H, Liu Z, Sun F, (2018): Thermodynamics of voltage-gated ion channels, *Biophysics Reports*, 4(6), 300–319.

Zheng Y, Xia P, Dong L, Tian L, Xiong C, (2021): Effects of modulation on sodium and potassium channel currents by extremely low frequency electromagnetic fields stimulation on hippocampal CA1 pyramidal cells, *Electromagnetic Biology and Medicine*, 17, 1–12.

Zhou R, Xiong Y, Xing G, Sun L, Ma J, (2010): ZiFi: Wireless LAN discovery via ZigBee interference signatures, *MobiCom'10*, September 20–24, Chicago.

Zothansiama, Zosangzuali M, Lalramdinpuii M, Jagetia GC, (2017): Impact of radiofrequency radiation on DNA damage and antioxidants in peripheral blood lymphocytes of humans residing in the vicinity of mobile phone base stations, *Electromagnetic Biology and Medicine*, 36(3), 295–305.

12 Electromagnetic Field-induced Dysfunction of Voltage-Gated Ion Channels, Oxidative Stress, DNA Damage, and Related Pathologies

Dimitris J. Panagopoulos[1,2,3], Igor Yakymenko[4,5], and George P. Chrousos[6]

[1]National Center for Scientific Research "Demokritos", Athens, Greece

[2]Choremeion Research Laboratory, Medical School, National and Kapodistrian University of Athens, Athens, Greece

[3]Electromagnetic Field-Biophysics Research Laboratory, Athens, Greece

[4]Department of Public Health, Kyiv Medical University, Kyiv, Ukraine

[5]Department of Environmental Safety, National University of Food Technologies, Kyiv, Ukraine

[6]University Research Institute of Maternal and Child Health and Precision Medicine, and UNESCO Chair on Adolescent Health Care, National and Kapodistrian University of Athens, Medical School, Aghia Sophia Children's Hospital, Athens, Greece

CONTENTS

Abstract 482
12.1 Introduction 482
    12.1.1 Experimental and Epidemiological Findings Connect Exposure of Living Organisms to ELF or WC EMFs with OS, Genetic Damage, Infertility, and Cancer 482
    12.1.2 Negative Findings and Real-Life Versus Simulated WC EMF Exposures 485
    12.1.3 DNA Damage and Its Consequences 485
12.2 EMF Exposure and Dysfunction of Voltage-Gated Ion Channels 486
12.3 Biochemical Processes Activated by Dysfunction of VGICs, Leading to OS, and DNA Damage 487
    12.3.1 Dysfunction of VGICs and OS 487
    12.3.2 OS and DNA Damage 491
12.4 Discussion 492
    12.4.1 Man-made EMFs Damage DNA by Causing VGIC Dysfunction and Triggering ROS Production in Cells 492
    12.4.2 Possible Outcomes of DNA Damage and Repair Capability of Cells 494
    12.4.3 EMFs and Therapeutic Effects 496
    12.4.4 Conclusions 497
References 497

DOI: 10.1201/9781003201052-17

Keywords: electromagnetic fields, voltage-gated ion channels, oxidative stress, free radicals, ROS, DNA damage, cell death, infertility, cancer

Abbreviations: ATP: adenosine triphosphate. DECT: Digitally Enhanced Cordless Telecommunications. EHS: electro-hypersensitivity. ELF: Extremely Low Frequency. EMF: electromagnetic field. EMR: electromagnetic radiation. ETC: electron transport chain. IFO: ion forced oscillation. MT: mobile telephony. NADH: nicotinamide adenine dinucleotide. NADPH: nicotinamide adenine dinucleotide phosphate. NOS: nitric oxide synthase. OS: oxidative stress. RF: Radio Frequency. ROS: reactive oxygen species. SOD: superoxide dismutase. VGCC: voltage-gated calcium channel. VGIC: voltage-gated ion channel. VLF: Very Low Frequency. WC: wireless communication. Wi-Fi: Wireless Fidelity. ULF: Ultra Low Frequency. 2G, 3G, 4G, 5G: second, third, fourth, fifth generation of MT/WC.

ABSTRACT

A plethora of studies show that exposure of living organisms to man-made polarized and coherent electromagnetic fields (EMFs), especially in the Extremely Low Frequency (ELF) and the microwave/Radio Frequency (RF) bands, may lead to oxidative stress (OS) and DNA damage. DNA damage is associated with mutations, cell senescence, cell death, infertility, and other pathologies, including cancer. ELF EMF exposures from high-voltage power lines and complex "RF" EMF exposures from wireless communication (WC) antennas/devices have been associated with increased cancer risk. Almost all man-made microwave/RF EMFs, and especially those employed in WC, are combined with ELF components in the form of modulation, pulsation, and random variability. Thus, in addition to polarization/coherence, the existence of ELFs is a common feature of almost all man-made EMFs. Polarized/coherent ELF EMFs are predicted to induce dysfunction of voltage-gated ion channels (VGICs) in cell membranes through the ion forced oscillation mechanism, and this has been verified by many experimental studies. Dysfunction of VGICs disrupts intracellular concentrations of critical ions, such as calcium, sodium, potassium, etc. This condition initiates biochemical processes leading to OS by reactive oxygen species (ROS) overproduction. Such processes include a) increased calcium signaling, leading to nitric oxide (NO•) overproduction by the nitric oxide synthases (NOS) in various locations in the cell, and superoxide anion ($O_2•^-$) overproduction in the mitochondria; b) activation of NADPH/NADH oxidase in the plasma membrane, leading to increased production of $O_2•^-$; and c) dysfunction of the Na^+/K^+ pump (ATPase) in the plasma and internal cell membranes, triggering mitochondrial ROS production. At least these processes may result in excessive OS, leading to DNA damage and related diseases, including infertility and cancer. Thus, it seems that there is a plausible explanation for the genetic damage and related effects found to be induced by man-made EMF exposures as reported by many experimental and epidemiological studies.

12.1 INTRODUCTION

12.1.1 Experimental and Epidemiological Findings Connect Exposure of Living Organisms to ELF or WC EMFs with OS, Genetic Damage, Infertility, and Cancer

A great number of experimental findings connect exposures of experimental animals or cells to man-made electromagnetic fields (EMFs) and corresponding electromagnetic radiation (EMR) of Extremely Low Frequency (ELF) (3–3000 Hz) or EMFs/EMR emitted by wireless communications (WC), with oxidative stress (OS), genetic damage/alterations (DNA damage, chromosome damage, mutations, etc.), cell senescence, cell death, and related effects (see reviews Goodman et al. 1995; Santini et al. 2005; Phillips et al. 2009; Yakymenko et al. 2016;

Panagopoulos 2019a). WC EMFs from mobile phones and corresponding base station mobile telephony (MT) antennas, cordless domestic phones (DECT: Digitally Enhanced Cordless Telecommunications), internet (Wi-Fi: Wireless Fidelity) or "Bluetooth" wireless connections, etc., always combine Radio Frequency (RF)/microwave (300 kHz–300 GHz) carrier signals with ELF/VLF (Very Low Frequency) (3–30 kHz) pulsation and modulation and Ultra Low Frequency (ULF) (0–3 Hz) random variability of the final signal (Panagopoulos 2019a; 2020; Panagopoulos et al. 2021).

The number of *in vivo* or *in vitro* experimental laboratory studies showing genetic damage and related effects induced by man-made ELF or RF (combined with ULF/ELF/VLF) EMFs on a variety of organisms/cell types under different experimental conditions is rapidly increasing, especially in recent years (such studies in chronological order are by Delgado 1985; Garaj-Vrhovac et al. 1990; 1991; 1992; Ma and Chu 1993; Sarkar et al. 1994; Lai and Singh 1995; 1996; 1997; Svedenstal et al. 1999; Koana et al. 2001; Ivancsits et al. 2002; 2003; Mausset-Bonnefont et al. 2004; Ji et al. 2004; Winker et al. 2005; Diem et al. 2005; Hong et al. 2005; Belyaev et al. 2005; Markova et al. 2005; Aitken et al. 2005; Nikolova et al. 2005; Zhang et al. 2006; Lixia et al. 2006; Ferreira et al. 2006; Panagopoulos et al. 2007a; Yan et al. 2007; Yao et al. 2008; Yadav and Sharma 2008; Sokolovic et al. 2008; Lee et al. 2008; De Iuliis et al. 2009; Agarwal et al. 2009; Mailankot et al. 2009; Luukkonen et al. 2009; Panagopoulos et al. 2010; Chavdoula et al. 2010; Guler et al. 2010; Tomruk et al. 2010; Franzellitti et al. 2010; Campisi et al. 2010; Panagopoulos 2012; Panagopoulos et al. 2013; Liu et al. 2013; Burlaka et al. 2013; Pesnya and Romanovsky 2013; Mihai et al. 2014; Daroit et al. 2015; Banerjee et al. 2016; D'Silva et al. 2017; Yakymenko et al. 2018; Panagopoulos 2019b; 2020; Yuan et al. 2020; Sharma et al. 2020). If we add studies that found induction of OS in cells due to reactive oxygen species (ROS) generation and/or cell senescence, the list becomes much longer (Yakymenko et al. 2016; Shahin et al. 2017; Choi et al. 2020; Kim et al. 2021). Yakymenko et al. (2016) reviewed 100 experimental studies that examined OS in cells exposed to WC EMFs, from which, 93 found increased OS due to ROS overproduction (more recent review in Chapter 3).

Several of the above findings involve OS, DNA damage, and consequent cell death in reproductive cells of various animals, resulting in decreased reproduction/infertility or embryonic death. Especially the effects of pulsing WC EMFs on the DNA of reproductive cells reported by different studies on a variety of animals (Aitken et al. 2005; Panagopoulos et al. 2007a; 2010; 2013; Yan et al. 2007; De Iuliis et al. 2009; Chavdoula et al. 2010; Panagopoulos 2012; Burlaka et al. 2013; Yakymenko et al. 2018) display a remarkable similarity and explain other findings that connect WC EMF exposures with insect, bird, and mammalian (including human) infertility (Ma and Chu 1993; Magras and Xenos 1997; Panagopoulos et al. 2004; 2007b; Wdowiak et al. 2007; Agarwal et al. 2008; Gul et al. 2009; Sharma and Kumar 2010; La Vignera et al. 2012; Shahin et al. 2017), or declines in bird and insect populations (especially bees) during the past 17 years (Balmori 2005; Balmori and Hallberg 2007; Everaert and Bauwens 2007; Bacandritsos et al. 2010; Cucurachi et al. 2013). Significant decrease in reproduction (egg laying or embryonic death) after exposure to MT radiation was identically observed in fruit flies (Panagopoulos et al. 2004; 2007a; 2007b; 2010), chicken or quail embryos (Batellier et al. 2008; Burlaka et al. 2013; Yakymenko et al. 2018), birds (Balmori 2005; Balmori and Hallberg 2007; Everaert and Bauwens 2007), and bees (Sharma and Kumar 2010). Similar effects are reported for amphibians (Balmori 2006; 2010), rats and mice (Yan et al. 2007; Gul et al. 2009; Shahin et al. 2017), and human sperm (decreased number and motility of spermatozoa) (Wdowiak et al. 2007; Agarwal et al. 2008). These remarkably similar findings in different animals by different research groups can be explained by the observed cell death in reproductive cells or embryonic death after DNA damage, as seen in fruit fly ovarian cells (Panagopoulos et al. 2007a; 2010; Chavdoula et al. 2010; Panagopoulos 2012), human sperm cells (De Iuliis et al. 2009), mouse or rat sperm cells (Aitken et al. 2005; Yan et al. 2007), and quail embryos (Burlaka et al. 2013; Yakymenko et al. 2018). Decreased reproduction after DNA damage and cell death in reproductive cells or embryonic death induced by purely ELF EMF exposures have also been reported (Ma and Chu 1993; Hong et al. 2005; Panagopoulos et al. 2013).

At the same time, epidemiological/statistical studies increasingly link man-made EMF exposures with health problems, genetic damage, and cancer in human populations. More specifically, ELF EMFs from power lines and high-voltage transformers (mainly 50–60 Hz plus additional ELF frequencies due to harmonics, noise, discharges, etc.) are linked with childhood leukemia and other cancer types (Werthheimer and Leeper 1979; Savitz et al. 1988; Coleman et al. 1989; Feychting and Ahlbom 1993; 1994; Miller et al. 1996; Coghill et al. 1996; Villeneuve et al. 2000a; 2000b; Ahlbom et al. 2000; Greenland et al. 2000; Draper et al. 2005; Kheifets et al. 2010) for magnetic field intensities as low as 2 mG (0.2 µT) (Feychting and Ahlbom 1994; Kheifets et al. 2010) or distances from power lines of up to 600 m (Draper et al. 2005) and electric field intensities as low as 10 V/m (Coghill et al. 1996). RF exposures from various antennas, always including ELF components, especially radio broadcasting and MT antennas, are linked to various forms of cancer. Hallberg and Johansson (2002) found a connection between skin cancer (melanoma) incidence in humans and residential exposure to radio broadcasting antennas, while two more recent studies found significantly increased genetic damage in peripheral blood lymphocytes of people residing in the vicinity of MT base antennas (Gulati et al. 2016; Zothansiama et al. 2017). During the past 15 years, epidemiological studies find an increasing association between mobile or cordless phone use and brain tumors in humans (Hardell et al. 2007; 2009; 2013a; 2013b; 2013c; Khurana et al. 2009; Hardell and Carlberg 2009; Wang and Guo 2016; Carlberg and Hardell 2017; Hardell 2018; Momoli et al. 2017; Miller et al. 2018; 2019). A very recent study found the cancer incidence among people residing close to MT base antennas increased by 10 times compared to the cancer incidence among the general population (Lopez et al. 2021). Moreover, during the past 20 years, other epidemiological/statistical studies find association between exposure to MT base station antennas and devices with reported symptoms of unwellness referred to as "microwave syndrome", or "electro-hypersensitivity" (EHS). The symptoms include headaches, fatigue, sleep disorders, etc. (Santini et al. 2002; Navarro et al. 2003; Salama et al. 2004; Hutter et al. 2006; Abdel-Rassoul et al. 2007; Blettner et al. 2009; Viel et al. 2009; Kundi and Hutter 2009; Lopez et al. 2021). A high percentage (~ 80%) of EHS self-reporting patients were recently found with increased OS in their peripheral blood (Irigaray et al. 2018).

Induction of cancer in experimental animals by long-term MT exposures containing ELF pulsations has also been reported (Tillmann et al. 2010; Lerchl et al. 2015). A recent study of the US National Toxicology Program (NTP) found that rats exposed in the near-field of simulated second (2G) or third generation (3G) MT emissions for 2 years, 9 hours (h) per day, developed brain cancer (glioma) and heart cancer (malignant schwannoma) for both lower and higher radiation levels than the officially accepted limits (NTP 2018). Moreover, the study found significantly increased DNA damage (strand breaks) in the brains of exposed animals (Smith-Roe et al. 2019), confirming that DNA damage is closely related to carcinogenesis. An Italian life-span exposure study of rats to a simulated 2G MT far-field EMF also found induction of heart schwannomas and brain glial tumors, confirming the results of the NTP study (Falcioni et al. 2018).

As already mentioned, a review of studies involving exposures to complex RF EMFs with ELF pulsation/modulation revealed that 93% of them reported induction of OS/ROS overproduction in biological systems (see Chapter 3, and Yakymenko et al. 2016). Most common ROS found after EMF exposures are superoxide anion radical ($O_2^{\bullet-}$), which may be transformed to the most damaging hydroxyl radical, ($OH\bullet$), and nitric oxide ($NO\bullet$), which may be transformed to peroxynitrite ($ONOO^-$), which is also a very damaging non-radical ROS.

The findings on animal carcinogenicity, along with the epidemiological cancer findings in humans, the DNA damage and OS findings, and the adverse effects on reproduction due to DNA damage in ovarian or sperm cells or embryonic death, point toward the same direction: Man-made EMF exposures induce OS and DNA damage, which may lead to reproductive declines, cancer, and other related pathologies. It is important to note that the exposure levels in the vast majority of all the above studies were significantly below the officially accepted exposure limits for ELF and RF EMFs, which were set by a private organization called the International

Commission on Non-Ionizing Radiation Protection (ICNIRP) to prevent discharges on humans in the case of ELF and heating of living tissues in the case of RF EMFs (ICNIRP 1998; 2010; 2020; Hardell and Carlberg 2021).

The International Agency for Research on Cancer (IARC), which is a branch of the World Health Organization (WHO), has already classified since long time both ELF and "RF" (in fact WC) EMFs as possibly carcinogenic to humans (Group 2B) (IARC 2002; 2013; Baan et al. 2011). Based on additional scientific evidence after the 2011 IARC classification for "RF" EMFs, several studies have suggested that "RF"/WC EMFs should be re-evaluated and classified as probably carcinogenic (Group 2A) or carcinogenic (Group 1) to humans (Hardell et al. 2013c; Miller et al. 2018; Hardell and Nyberg 2020; Hardell and Carlberg 2020).

12.1.2 Negative Findings and Real-Life Versus Simulated WC EMF Exposures

At the same time, several other studies have reported no effects of ELF or RF EMFs in all of the above end-points (see reviews in Goodman et al. 1995; IARC 2002; 2013; Panagopoulos et al. 2004; 2013; 2015a; Santini et al. 2005; Phillips et al. 2009; Manna and Ghosh 2016; Yakymenko et al. 2016; Verschaeve 2017; Leach et al. 2018; Wood et al. 2021, and Lerchl et al. 2020; Schuermann et al. 2020), especially studies that employed simulated MT/WC exposures from generators with invariable parameters and no modulation. By contrast, more than 95% of the studies that employed real-life MT/WC exposures, from commercially available devices (mobile/cordless phones, Wi-Fi, etc.) with high signal variability, find effects (Panagopoulos et al. 2015a; Panagopoulos 2017; 2019a; Leach et al. 2018). Some studies can be misleading in the way they present information. For example, Schuermann et al. (2020) in their experiments applied fixed (invariable) pulsations emitted by a generator, which they described as "modulation". Although a fixed pulsation is very different and significantly less bioactive than pulsations and modulation of real-life signals emitted by WC devices/antennas, they claimed that they applied modulated signals and found no effects. A fixed pulsation used in simulated WC signals is invariable in intensity, repetition frequency, and other parameters. By contrast, in the real signals, the pulsations are not fixed in repetition frequency or intensity, and the modulation of the signal is unpredictable, as it depends on the transmitted information which is variable each moment. Moreover, a fixed pulsation is not "modulation" because it does not carry any information (and consequent variability) and cannot be regarded as such (Panagopoulos et al. 2015a; Panagopoulos 2017; 2019a).

Regardless of real-life or simulated exposures, the majority of experimental studies (more than 70%) both in the "RF" (combined with ELF) and purely ELF bands do find effects (Manna and Ghosh 2016; Yakymenko et al. 2016; Panagopoulos 2017; 2019a; Leach et al. 2018). The vast majority of reported effects in all the aforementioned studies, induced by ELF or complex RF (containing ELF) EMFs, were not accompanied by any significant heating of the exposed living tissues and are, thus, categorized as non-thermal effects. Finally, as already emphasized, in the vast majority of studies that employed EMF exposures characterized simply as "RF", the ELF/ULF components were present.

12.1.3 DNA Damage and Its Consequences

DNA damage is connected with cell senescence (cell aging and loss of replicative capacity), cell death, infertility, neuro-degenerative diseases, aging of an organism, and carcinogenesis. DNA damaging events take place anytime in the cells of any living organism due to a variety of events (such as exposure to ultraviolet radiation, natural radioactivity, or cytotoxic chemicals), but efficient DNA repair mechanisms have evolved during biological evolution to provide protection. Damage in the DNA is any modification in a nucleotide base, sugar (deoxyribose), a break in a covalent bond between deoxyribose and nucleotide base, or a break in a phosphodiester bond in one or both strands (Ames 1989; Lieber 1998; Hanahan and Weinberg 2000; 2011; Zglimicki et al. 2005; Helleday et al. 2007; Phillips et al. 2009; Lahtz and Pfeifer 2011; Shah et al. 2012; Bernstein et al.

2013; Yao and Dai 2014; Rodgers and McVey 2016; Basu 2018). In every case, a damage or chemical alteration in the DNA or other biological molecule involves breakage of chemical bonds, in other words, ionization.

Replication of damaged (or inaccurately repaired) DNA which has not been repaired or blocked during the cell division cycle, or any damage that may occur after the cell has passed the checkpoint in the G_2 phase of the cell division cycle, and the cell proceeds to mitosis, replicating itself with the damaged DNA, can lead to gene mutations, which may give rise to altered proteins with defective functions. Mutations in oncogenes, tumor-suppressor genes, or genes that control the cell division cycle can generate a clonal cell population with a distinct ability to proliferate. Many such events, which may accumulate over a long period of time in cases of chronic exposure to carcinogens, can lead to genomic instability and cancer (Hanahan and Weinberg 2000; 2011; Yao and Dai 2014; Rodgers and McVey 2016; Basu 2018; Panagopoulos 2019b; 2020). Epigenetic effects, such as altered gene expression through various mechanisms, may also lead to cellular dysfunction and carcinogenesis (Hanahan and Weinberg 2000; 2011; Lahtz and Pfeifer 2011; Panagopoulos et al. 2021).

The time length of cancer development (latency period) after irreparable DNA damage may be many years depending on the organism and the type of cancer. The latency period for gliomas (a type of brain cancer) is usually longer than 20 years in humans (Nadler and Zurbenko 2014). This probably explains why only relatively recent epidemiological studies performed during the past ~ 15 years have started showing an association between mobile phone use and cancer (Hardell et al. 2007), whereas cancer linked to high-voltage power transmission lines, which are several decades older than MT/WC, was indicated long before (Wertheimer and Leeper 1979).

A crucial factor for the survival and health of any living organism in the presence of environmental insults is the repair capability of its cells and the general health of the organism.

12.2 EMF EXPOSURE AND DYSFUNCTION OF VOLTAGE-GATED ION CHANNELS

As indicated by the long list of experimental and epidemiological/statistical studies referenced above, man-made EMF exposures are connected with OS, genetic damage, infertility, and cancer. Man-made EMFs with frequencies up to the low limit of infrared ($0-3\times10^{11}$ Hz) examined here cannot directly break chemical bonds and cause ionization, except for very strong field intensities ($\geq 10^6$ V/m) (Francis 1960; Gomer 1961). Such field intensities are rarely present in the environment, apart from very close proximity to high-voltage power lines and transformers, or very close to atmospheric discharges (lightning). How then are man-made EMFs at environmental intensities capable of damaging DNA and other biological molecules? Obviously they have the ability of breaking chemical bonds indirectly through the action of some primary biophysical mechanism(s) and subsequent initiation of intracellular biochemical processes. Since both DNA damage and OS are frequently found separately or combined after WC/ELF EMF exposures, and since OS may, indeed, lead to breakage of chemical bonds and DNA damage, it is reasonable to look for mechanistic links between EMF exposures and OS.

Visible and infrared natural light cannot break chemical bonds, even though they expose us with higher frequencies and radiation intensities than man-made EMFs in daily life (Panagopoulos et al. 2015b). Then what is the unique property of the man-made EMFs that makes them capable of inducing adverse biological/health effects in contrast to natural light? It has been shown in Panagopoulos et al. (2015b) and in the first chapter of this book, that man-made EMFs/EMR are totally polarized and coherent, while natural light is not. That means they possess net electric and magnetic field intensities – apart from radiation intensity – which exert forces on every electrically charged (or polar) particle/molecule, such as mobile/dissolved ions and charged/polar macromolecules in any biological system (Panagopoulos et al. 2015b).

A biophysical mechanism for triggering biological effects was originally described in Panagopoulos et al. (2000; 2002; 2015b; 2020; 2021) and extensively reviewed in the previous chapter. It explains, theoretically, how polarized and coherent EMFs, even at very low field intensities in the ULF/ELF/ VLF bands, can cause irregular gating (dysfunction) of electrosensitive or voltage-gated ion channels (VGICs) in cell membranes with consequent disruption of a cell's electrochemical balance (the equilibrium between concentration and voltage gradients maintained by specific concentrations of all dissolved/mobile ions across all cell membranes according to the Nernst equation) (see Chapter 11, and Stryer 1996; Alberts et al. 2002; Panagopoulos et al. 2000; Panagopoulos and Margaritis 2003). Because ELF/ULF/VLF EMFs are basic components of the WC EMFs, this mechanism accounts for the biological effects of the vast majority of man-made (polarized and coherent) EMFs.

Irregular gating of VGICs in cell membranes induced by man-made EMFs is, today, verified by many experimental studies (Liburdy 1992; Piacentini et al. 2008; El-Swefy et al. 2008; Cecchetto et al. 2020; Zheng et al. 2021) and reviews (Walleczek 1992; Pall 2013; Bertagna et al. 2021). Short-term exposures (10–30 min) to ELF EMFs of 0.5–2 mT are found to gate (open or close) sodium and potassium voltage-gated channels in rat neurons and modulate the corresponding ion currents, as shown by whole-cell patch-clamp recordings (Zheng et al. 2021). Exposure of Shaker cells (involved in the immune and nervous systems of animals/humans) to ELF EMF of 0.27–0.90 mT induced immediate (within seconds) changes in potassium ion currents, as was also shown by whole-cell patch-clamp recordings. The changes lasted for several minutes (min) after the end of the exposures (Cecchetto et al. 2020). Exposure of mouse brain neural cells to 1 mT 50 Hz ELF EMF induced changes in the ion flows of voltage-gated calcium channels (VGCCs) and in the corresponding neuronal markers. The changes ceased when the calcium channel blocker nifedipine was added in the cell cultures (Piacentini et al. 2008). Calcium channel blockade by amlodipine was found to alleviate brain injury in rats induced by long-term exposure to MT EMFs (El-Swefy et al. 2008).

Blank and Goodman (2009; 2011) noted that both ELF and "RF" (actually WC) EMFs produce similar effects, especially in inducing synthesis of stress proteins in cells very rapidly (within a few min). In their effort to explain the similarity of effects between ELF and RF/WC EMF exposures, they hypothesized that DNA functions as a "fractal antenna" (a combined ELF and RF antenna), and the EMFs (in both ELF and RF bands) interact directly with electrons in the DNA molecule. We consider it unlikely that DNA has components in its structure serving as "antennas", especially RF antennas, as RF EMFs are totally artificial and, thus, new in the course of biological evolution. The cosmic "microwaves" are not actually microwaves but infrared radiation shifted toward lower (microwave) frequencies due to the cosmic expansion and the Doppler effect (see Chapter 1, and Panagopoulos 2018). What we consider a plausible explanation of the common ELF and RF/ WC EMF effect is that "RF" EMFs affect cells not by their carrier (RF) frequencies but by their ELF components of pulsing and modulation. This simple and obvious explanation escaped not only Blank and Goodman (2009; 2011) but also most EMF-bioeffect researchers (Panagopoulos 2021; Panagopoulos et al. 2021). It is now revealed that actually only the ELF EMFs are those that induce the non-thermal biological effects, and they do not act directly on DNA but indirectly through VGIC dysfunction and consequent induction of OS (Panagopoulos et al. 2021).

The biochemical processes initiated in living cells due to the dysfunction of the VGICs in their cell membranes are the subject of the next section in the present chapter. We shall examine whether and how such a dysfunction may result in OS, genetic damage, and related pathologies, providing an explanation for the plethora of findings reviewed above.

12.3 BIOCHEMICAL PROCESSES ACTIVATED BY DYSFUNCTION OF VGICs, LEADING TO OS, AND DNA DAMAGE

12.3.1 Dysfunction of VGICs and OS

The biophysical mechanism for irregular gating and consequent dysfunction of VGICs in cell membranes by man-made (polarized and coherent) ELF/WC EMFs is described in the previous chapter and originally in Panagopoulos et al. (2000; 2002; 2015b; 2020; 2021). Since the phenomenon of the ion forced oscillation (IFO) under the influence of a polarized EMF and its effect on the function of the VGICs are both central in this mechanism, for simplicity, we shall use the name "IFO-VGIC mechanism" introduced in the previous chapter.

The VGICs (also called electrosensitive ion channels) are the most important and abundant type of ion channels in all animal cells (Alberts et al. 1994; 2002; Stryer 1995).

Any dysfunction in ion channels will affect the intracellular ionic concentrations, disrupting the cell's electrochemical balance and homeostasis and the intracellular redox status. We shall show how this can trigger OS by ROS overproduction and subsequent DNA damage. It is known that the intracellular redox status (the concentrations of reducing and oxidizing molecules, which is an index of ROS content in the cell) can trigger gating (opening or closing) of Ca^{2+}, Na^+, and K^+ channels in order to reinstate homeostasis (Akbarali 2014). Inversely, the function of these channels determines the redox status and the cell's electrochemical balance (Kourie 1998; Panagopoulos et al. 2000).

The large amounts of energy in the form of adenosine triphosphate (ATP) required for the maintenance of a cell's aerobic life are generated predominantly by oxidative phosphorylation in the mitochondria at the expense of molecular oxygen. A small amount (~ 2%) of oxygen during this process is converted to superoxide radical, hydrogen peroxide, and related ROS. Even though ROS play critical roles in the maintenance of life in living cells, they can also damage biological molecules (such as lipids, proteins, RNA, and DNA), causing various diseases when they are in excess (Alberts et al. 1994; Stryer 1996; Margaritis 1996; Inoue et al. 2003).

Most (not all) ROS are free radicals. Free radicals are highly unstable molecules containing an unpaired electron, which is denoted by a dot (•), and possessing a very strong tendency to chemically react with surrounding molecules and/or with each other in order to couple their unpaired electron, balance the electron spins, and become stable. This is why they have extremely short lifetimes. Most ROS react rapidly with surrounding biomolecules, including DNA, breaking chemical bonds and causing chemical alterations (Inoue et al. 2003; Lushchak et al. 2014). DNA damage induced by OS/ROS leading to mutations and disease has been well documented (Cooke et al. 2003; Barzilai and Yamamoto 2004), and similarly well documented is the increased production of ROS in living cells after EMF exposures (see Chapter 3, and Yakymenko et al. 2016). In the present chapter, we shall show how this is initiated by dysfunction of VGICs.

Many studies have found a connection between voltage-gated calcium, potassium, sodium, and chloride channels with induction of OS and related pathologies (Batcioglu et al. 2012; Pall 2013; Ramírez et al. 2016; O'Hare Doig et al. 2017). Indeed, it is repeatedly shown that dysfunction of VGICs induced by EMF exposures may trigger immediate ROS generation in the cells, and the effect is significantly diminished by use of ion channel blockers (Piacentini et al. 2008; El-Swefy et al. 2008; Pilla 2012; Pall 2013).

Two important free radical ROS found after man-made EMF exposures are superoxide anion ($O_2^{\bullet-}$) and nitric oxide (NO•) (Yakymenko et al. 2016; 2018). The superoxide anion free radical may be converted into hydroxyl radical (OH•) or react with nitric oxide to form peroxynitrite (ONOO-). Both final products (hydroxyl radical and peroxynitrite) are very reactive ROS with many critical biological molecules, especially DNA, as shown before (Ischiropoulos et al. 1992; Pall 2013; Yakymenko et al. 2016; 2018) and discussed below. But how can the initial ROS ($O_2^{\bullet-}$ and NO•) be produced by dysfunction of VGICs? This has been a missing link until recently (Panagopoulos et al. 2021). We have identified several pathways for this.

A. Calcium Signaling, Nitric Oxide Synthases (NOS), and Mitochondrial ROS Production: The effect of man-made EMFs, especially ELF EMFs, on calcium concentrations in exposed cells and the unique role of the VGCCs in EMF-induced bioeffects has been well-documented for long time (Bawin et al. 1975; 1978; Bawin and Adey 1976; Walleczek 1992; Liburdy 1992; Goodman et al. 1995; Barbier et al. 1996; Morgado-Valle et al. 1998; Gobba et al. 2003; Grassi et al. 2004; Lisi et al. 2006; Jeong et al. 2006; Marchionni et al. 2006; Piacentini et al. 2008; El-Swefy et al. 2008; Pall 2013; Bertagna et al. 2021) and explained by the IFO-VGIC mechanism (Panagopoulos et al. 2000; 2002; 2015b; 2020; 2021).

Dysfunction of VGCCs will cause alterations in the intracellular calcium concentrations, which in turn, will affect key cellular signaling pathways known as the Ca^{2+} signaling system. The effect of ELF EMF, or RF EMF amplitude-modulated by ELF signals, on intracellular calcium concentrations has been shown since the early years of bioelectromagnetic research (Bawin et al. 1975; 1978; Bawin and Adey 1976; Lin-Liu and Adey 1982; Blackman et al. 1980; Dutta et al. 1984). Walleczek (1992) reviewed many studies showing effects of ELF EMFs on cells of the immune system, revealing the critical role of intracellular calcium. But until that time, the site of interaction of EMFs with cells was unknown, even though the facts were pointing toward the calcium ion channels in the cell membranes as a most reasonable explanation. At the same time, Liburdy (1992) in a series of experiments, showed that calcium influx in lymphocytes, which occurred within minutes after the onset of ELF EMF exposures, was due to the interaction of the EMF with the calcium channels in the cell plasma membranes (most of which are voltage-gated) and not due to release from intracellular stores. Today, it has been revealed that not only the VGCCs, but all VGICs are the sites of the action (rather than interaction) of man-made EMFs on cells (Panagopoulos et al. 2000; 2002; 2015b; 2020; 2021).

Ca^{2+} is a most important signaling factor in living cells, regulating many cell functions, including cell proliferation, differentiation, and apoptosis (Brookes et al. 2004; Lang et al. 2005; Görlach et al. 2015; Lombardi et al. 2019). The ROS regulatory system is closely linked to the Ca^{2+} signaling system and is affected by changes in intracellular Ca^{2+} concentrations (Görlach et al. 2015). Dysfunction of voltage-gated Ca^{2+} channels in the plasma or the mitochondrial membranes can lead to critical changes in cytosolic and/or mitochondrial concentrations of Ca^{2+} ions, triggering regulatory mechanisms crucial for several key cell functions. Changes in intra-cellular Ca^{2+} concentration (such as those following EMF exposures) are connected with modifications in cellular activity, including cytotoxicity (Walleczek 1992; Kourie 1998; Brookes et al. 2004; Akbarali 2014; Ramirez et al. 2016; O'Hare Doig et al. 2017).

Increased levels of intracellular Ca^{2+} stimulate nitric oxide free radical (NO•) synthesis by the nitric oxide synthases (NOS) in various locations of the cell. Several NOS enzymes have been described, such as the neuronal NOS, or the endothelial NOS which are plasma membrane enzymes. Their activation seems to be dependent on intracellular calcium levels and calmodulin. NO• exists everywhere within and between cells in all vertebrates as an intercellular messenger, and certain studies suggest that it is not particularly toxic by itself. A single NO• molecule can readily move between cells many times within its life span (Ischiropoulos et al. 1992; Pacher et al. 2007). Increases in Ca^{2+} and NO• levels in cells have been found to be triggered very rapidly (within few seconds) by EMF exposure (Pilla 2012), and the induced DNA damage (especially by peroxynitrite) can be blocked by NOS inhibitors (Lai and Singh 2004) and antioxidants (Moon et al. 2006; Sakihama et al. 2012; Pall 2013).

A major source of ROS in all cells is the mitochondrion (Stryer 1995; Inoue et al. 2003; Yakymenko et al. 2016; Schuermann and Mevissen 2021). Voltage-gated anion channels in the outer membrane of the mitochondria regulate Ca^{2+} entry into the intermembrane space and in the matrix, which is crucial for mitochondrial ROS production. Increased levels of Ca^{2+} in the mitochondria stimulate superoxide anion radical (O_2•$^-$) synthesis by the electron transport chain (ETC). NO• inhibits complex IV of the ETC. Inhibition of the ETC causes release of free electrons, increasing the generation of O_2•$^-$ (Inoue et al. 2003; Brookes et al. 2004). ROS overproduction in the mitochondria may damage DNA both in the mitochondria and the nucleus and may initiate a signaling cascade leading to apoptosis. Proper apoptosis regulation is crucial for anticancer control. By contrast, excessive apoptosis, induced by increased ROS levels, has been linked to inflammatory diseases and cancer (Lowe and Lin 2000).

Increased levels of intracellular Ca^{2+}, in some cases, are associated with increased apoptosis, probably due to activation of Ca^{2+} dependent DNase I (Nitahara et al. 1998). This may be an alternative pathway for DNA damage and related pathologies.

B. NADPH/NADH Oxidase and ROS Production: Apart from the effect of EMFs on metallic cation voltage-gated channels (Ca^{2+}, Na^+, K^+, etc.), proton (H^+) voltage-gated channels will be affected as well, as they operate in a very similar fashion (De Coursey 2003; Seredenina et al. 2015). This, in turn, will affect the function of NADPH/NADH oxidases, plasma membrane enzymes found in abundance in all cells, which normally generate ROS for the elimination of invading microorganisms. It has been argued that the NAD(P)H oxidase is one enzyme with different affinities for two substrates: nicotinamide adenine dinucleotide phosphate (NADPH) and nicotinamide adenine dinucleotide (NADH), with the NADPH substrate as the most common (Iverson et al. 1977; Gamaley et al. 1995; Azumi et al. 1999; Zalba et al. 2000; Li and Shah 2001; Bedard and Krause 2007; Panday et al. 2015). Yet, it seems that the NADH oxidase and the NADPH oxidase are two distinct systems (Friedman et al. 2007; Low et al. 2012). The NADPH/NADH oxidases catalyze the production of superoxide anion free radical by transferring electrons to oxygen from NAD(P)H according to the reaction:

$$NAD(P)H + 2O_2 \rightarrow NAD(P)^+ + 2O_2^{\bullet-} + H^+ \qquad [12.1]$$

Thus, they generate an electron flux for the reduction of extracellular O_2 to superoxide anion radical ($O_2^{\bullet-}$). The activity of NADPH/NADH oxidase is strongly connected with H^+ channels, and it may even act directly as a H^+ voltage-gated channel due to its gp91^{phox} transmembrane subunit (Henderson 2001; Musset et al. 2009).

The NADPH/NADH oxidase is also activated by cytosolic Ca^{2+} and possesses a Ca^{2+}-binding site apart from its H^+ voltage-gated channel (gp91^{phox} transmembrane domain) (Panday et al. 2015). Thus, perturbation of intracellular concentrations of either H^+ or Ca^{2+}, after irregular gating of their voltage-gated channels, will affect the function of NADPH/NADH oxidase and trigger irregular ROS (over)production.

The NADPH/NADH oxidase has been reasonably suggested as a primary target of man-made EMF exposures on living cells. Friedman et al. (2007) found rapid ROS production by NADH oxidase in cultured cells after a few min EMF exposure.

What is common in the NADPH/NADH oxidase and in the mitochondria in generating ROS is the occurring electron flux, which is called ETC in the mitochondria. In both the NADPH/NADH oxidase electron flux, and the mitochondrial ETC, the generated ROS is superoxide anion radical ($O_2^{\bullet-}$), which may finally result, as already mentioned, either in peroxynitrite (ONOO-), after reaction with nitric oxide, or in the most aggressive hydroxyl radical (OH•), after conversion to hydrogen peroxide.

C. Na$^+$/K$^+$-ATPase Signaling and ROS Production: Apart from calcium, potassium (K^+) channels are also involved in the activation of apoptosis (Lang et al. 2005), and voltage-gated Ca^{2+} and K^+ channels have been associated with cell proliferation and carcinogenesis (Becchetti 2011). Thus, cytosolic concentrations of Ca^{2+} and K^+ free ions play major roles in cellular function and metabolism.

In addition, voltage-gated calcium and potassium channels play important roles in iron entry into the cells (Gaasch et al. 2007; Chattipakorn et al. 2011; Salsbury et al. 2014). Iron catalyzes the production of the hydroxyl free radical (OH•) via the Fenton reaction and, thus, impaired function of these channels can promote cellular toxicity.

Dysfunction of Na$^+$, K$^+$, Mg^{2+}, and Ca^{2+} voltage-gated channels will affect the function of the Na$^+$/K$^+$ pump (ATPase), and Ca^{2+} pumps in the plasma membranes of all cells. The ion pumps (active ion transporters) across all cell membranes in coordination with the ion channels (passive ion transporters) determine the membrane voltage, the cell's volume, and the cell's electrochemical balance and homeostasis and contribute to the control of the redox status. In addition to its role as an ion pump, Na$^+$/K$^+$-ATPase operates as a signal regulator, transducing signals from the plasma membrane to the intracellular organelles (Stryer 1995; Alberts et al. 1994; 2002). It has long been shown that the activity of the Na$^+$/K$^+$-ATPase is affected by ELF EMFs (Serpersu and Tsong 1984; Blank and Soo 1990; Walleczek 1992). It has also been shown that changes in the activity of this enzyme are linked to ROS production by the mitochondria, and in turn, increased mitochondrial ROS production stimulates the signaling function of the enzyme forming a positive amplification feedback loop (Zhang et al. 2008). This was recently verified experimentally in primary cultures of cardiac myocytes. The Na^+/K^+-ATPase became a target for ROS-initiated signaling, and in turn, stimulation of the Na^+/K^+-ATPase by ROS led to further increase in ROS production by the mitochondria (Pratt et al. 2018). Thus, changes in the activity of the Na^+/K^+-ATPase due to EMF exposures can stimulate ROS production by the mitochondria, and the process can be amplified by increasing ROS levels.

12.3.2 OS AND DNA DAMAGE

Therefore, we have described how the dysfunction of VGICs gives rise to the generation of initial ROS, such as NO• and O_2•−, by certain ROS sources in the cell, such as the NOS, the mitochondria, the NADPH/NADH oxidase, and the Na^+/K^+-ATPase that may act as a trigger to stimulate ROS production by the mitochondria. The initial ROS (O_2•− and NO•) give rise to other even more potent ROS, peroxynitrite and hydroxyl radical, which are finally those that mainly damage DNA.

A. Peroxynitrite: Increased concentrations of NO• and O_2•− within a cell lead to peroxynitrite ($ONOO^-$) overproduction after reaction among the two initial ROS, as follows:

$$NO\bullet + O_2\bullet^- \rightarrow ONOO^- \qquad [12.2]$$

Peroxynitrite is a strong non-radical ROS which can damage critical molecules including DNA (Valko et al. 2007). It seems that most of the cytotoxicity attributed to NO• is rather due to its product peroxynitrite. Both nitric oxide and peroxynitrite can be diffused anywhere within the cell and, thus, act directly on DNA or other molecules. The effects of peroxynitrite on DNA include base and sugar oxidative modifications and DNA single-strand breaks (Burney et al. 1999; Pacher et al. 2007). Peroxynitrite has a half-life ~ 10–20 ms, which is sufficient to cross biological membranes, diffuse at distances equal to one or two cell diameters, and allow significant interactions with most critical biomolecules (Pacher et al. 2007). Of the four DNA bases, guanine is the most reactive (Yu et al. 2005). The formation of DNA single-strand breaks represents a critical aspect of peroxynitrite-mediated cytotoxicity (Pacher et al. 2007). Induction of single-strand DNA breaks by peroxynitrite has been demonstrated by several studies, and the effect can be prevented by use of calcium channel blockers and antioxidants (Szabo and Baehrle 2005; Moon et al. 2006; Sakihama et al. 2012; Pall 2013). Pall (2013), in his review of EMF-bioeffects studies with calcium channel blockers, noted a connection between EMF-induced dysfunction of VGCCs and nitric oxide (NO•)/peroxynitrite ($ONOO^-$) overproduction. Peroxynitrite can damage lipids, DNA, and proteins either by direct oxidative action or by indirect radical-mediated processes. Such changes trigger cellular responses ranging from subtle modulations of cell signaling to serious oxidative damage, committing cells to apoptosis or necrosis.

B. Hydroxyl Radical: The superoxide anion radical (O_2•−), produced by the mitochondria or the NADPH/NADH oxidase, can be catalyzed by superoxide dismutase (SOD) enzymes in the cytosol or the mitochondria and convert to hydrogen peroxide (H_2O_2) (Valko et al. 2007; Gamaley et al. 1995; DeCoursey et al. 2003) according to the reaction:

$$2O_2\bullet^- + 2H^+ \xrightarrow{SOD} H_2O_2 + O_2 \qquad [12.3]$$

H_2O_2 is a critical molecule in oxidative damage, as it can move to any intracellular site (including the nucleus), where it can be converted via the Fenton reaction (see below) to the most potent hydroxyl radical (OH•), which can damage any biological molecule, including DNA (Balasubramanian et al. 1998; Cadet et al. 1999; Cadet and Wagner 2013; Halliwell 2007; Tsunoda et al. 2010).

The hydroxyl radical (OH•) is considered the most potent oxidant of DNA. The main mechanism for OH• production involves the iron-catalyzed conversion of H_2O_2 via the Fenton reaction (Fenton 1894): Fe^{2+} is oxidized by H_2O_2 to Fe^{3+}, producing a OH• radical and a hydroxide ion (OH^-)

(Eq. 12.4). Fe^{3+} is then reduced back to Fe^{2+} by another molecule of H_2O_2, producing a hydroperoxyl radical and a proton (Eq. 12.5).

$$Fe^{2+} + H_2O_2 \rightarrow Fe^{3+} + OH\bullet + OH^- \qquad [12.4]$$

$$Fe^{3+} + H_2O_2 \rightarrow Fe^{2+} + HOO\bullet + H^+ \qquad [12.5]$$

The net effect is the conversion of two hydrogen peroxide molecules to produce two different oxygen-radical species, with water ($H^+ + OH^-$) as a byproduct:

$$2\ H_2O_2 \rightarrow OH\bullet + HOO\bullet + H_2O \qquad [12.6]$$

Another way for OH• production is the Haber-Weiss reaction (Valko et al. 2006):

$$O_2\bullet^- + H_2O_2 \rightarrow O_2 + OH\bullet + OH^- \qquad [12.7]$$

The OH• radical reacts with any biological molecule in its immediate environment, including DNA. One example is breaking macromolecules (R-R or R-H) or subtracting atoms from them (such as the various hydrogen atoms of the deoxyribose molecule) by breakage of covalent bonds. This results in chemical alterations of the macromolecules and production of new free radicals (R• or RO•):

$$R\text{-}R + OH\bullet \rightarrow ROH + R\bullet \qquad [12.8]$$

$$RH + OH\bullet \rightarrow R\bullet + H_2O \qquad [12.9]\ \text{or}$$

$$RH + OH\bullet \rightarrow RO\bullet + H_2 \qquad [12.10]$$

The new free radicals will further react with other molecules and with each other, resulting in additional chemical alterations.

Therefore, it clearly comes that irregular gating of VGICs in the plasma or the intracellular membranes by man-made EMF exposures will most likely trigger ROS overproduction, OS, and consequent cellular damage. Although plenty of data connecting impaired ion channel function and induction of cell death or cancer have been available for a long time (Lang et al. 2005; Becchetti 2011), the connection between dysfunction of VGICs and OS (Pall 2013; Batcioglu et al. 2012; Akbarali 2014; Ramírez et al. 2016; O'Hare Doig et al. 2017; Panagopoulos et al. 2021) leading to DNA damage has not gained the necessary attention.

In conclusion, we have a clear sequence of events starting from irregular gating of VGICs by man-made EMFs up to DNA damage and related pathologies, including infertility and carcinogenesis. Figure 12.1 provides a schematic representation of the above-described biochemical processes initiated by EMF-induced dysfunction of VGICs and resulting in DNA damage.

12.4 DISCUSSION

12.4.1 Man-made EMFs Damage DNA by Causing VGIC Dysfunction and Triggering ROS Production in Cells

In this chapter, we reviewed experimental and epidemiological studies that have researched a possible connection between man-made EMF exposure and genetic damage (DNA or chromosome damage), including related effects (cell death, OS, etc.) and related pathologies such as infertility and cancer. We conclude that it is well documented that both purely ELF and WC/RF (containing ELF) man-made EMFs induce OS and genetic damage, which can lead to related pathologies, such as infertility and cancer, in both humans and animals (see also Chapters 3, 4, 5, 6).

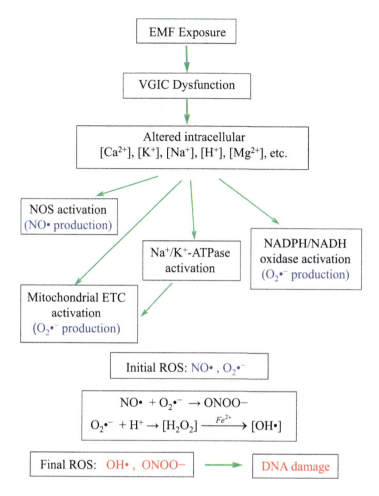

FIGURE 12.1 Schematic diagram of the biochemical processes initiated in living cells after dysfunction of VGICs induced by man-made EMF exposure.

We have explained that man-made EMFs are totally polarized and coherent, meaning that they possess electric and magnetic parts that oscillate on single directions and in phase (in the case of oscillating/varying fields) (see also Chapter 1). This condition induces parallel and coherent forced oscillations of the mobile ions and other charged/polar molecules in living tissues. The IFO-VGIC mechanism has described how such oscillations cause the dysfunction of VGICs in the membranes of all cells resulting in altered intra-cellular ionic concentrations (see previous chapter, and Panagopoulos et al. 2000; 2002; 2015b; 2020; 2021).

Most importantly, in this chapter, we have examined plausible biochemical processes initiated by the altered intracellular concentrations of critical ions such as Ca^{2+}, Na^+, K^+, H^+, etc., leading to ROS generation and induction of OS in all living cells. Indeed, prolonged dysfunction of VGICs caused by man-made EMF exposure alters the intracellular ionic concentrations and disrupts the cell's electrochemical balance and homeostasis (Alberts et al. 1994; 2002; Stryer 1995; Panagopoulos et al. 2000; Akbarali 2014; Ramirez et al. 2016; Panagopoulos et al. 2021). This leads to immediate release of two initial ROS, the nitric oxide radical (NO•) and the superoxide anion radical (O_2•−), produced by certain ROS sources in the cell. Such a condition brings the cell under OS. The initial ROS can be easily converted to peroxynitrite ($ONOO^-$) or the most potent hydroxyl radical (OH•), respectively. These "final" ROS* can damage DNA when cells are unable to reinstate

---

* By "final" ROS we mean those ROS that may finally damage DNA and other critical biomolecules.

electrochemical balance (normal intracellular ionic concentrations). Consequently, DNA damage can lead to reproductive disabilities, organic/neuro-degenerative diseases and corresponding disorders, aging, genetic alterations, and cancer.

According to the available data, it is estimated that about two-thirds of the DNA damages caused by ionizing radiation are due to the hydroxyl radical. Although OH• can only diffuse at distances comparable to the length of a macromolecule due to its extremely short lifetime (of the order of 10^{-9}–10^{-4} s depending on presence of other molecules), it can be formed by H_2O_2 at any location within the cell (including the nucleus) and act instantly upon DNA or other biological molecules. H_2O_2 can move to any intracellular site. As for nitric oxide and peroxynitrite, they can both be diffused anywhere in the cell, including the nucleus and, thus, directly damage DNA and other critical biomolecules (Coggle 1983; Hall and Giaccia 2006; Pacher et al. 2007).

Sources of ROS production after man-made EMF exposure are the ETC in the mitochondria, the electron flux in the NADPH/NADH oxidase in the plasma membrane, the Na^+/K^+-ATPase in the plasma and intracellular membranes which trigger ROS production by the mitochondria, and the NOS enzymes at different locations in the cell. All four sources are affected by the intracellular concentrations of cations such as Ca^{2+}, K^+, Na^+, H^+, and most cation channels are voltage-gated (Stryer 1995; Alberts et al. 1994; 2002).

Although we do not yet know many of the details of the ion signaling that triggers ROS generation by the above sources, we do know that the triggering involves changes in the intracellular concentrations of these and other ions. Since man-made EMFs, indeed, have the ability to cause dysfunction of VGICs, it seems that the basic parts of the entire process leading to DNA damage and related pathologies are for the first time identified.

When overproduction of ROS in a cell overloads the capacity of its antioxidant system, the cell/organism is under OS. A sustained or repeated such condition leads to DNA damage. We discussed how unrepaired/misrepaired DNA lesions, such as strand breaks, covalent bond breakage, or nucleotide base and sugar damages, can lead to cell senescence, cell death (apoptosis/necrosis), or mutations and related pathologies such as aging, infertility, neuro-degenerative diseases, and cancer (Halliwell 2007; Valko et al. 2007; Phillips et al. 2009; Görlach et al. 2015; Panagopoulos et al. 2021).

The dysfunction of VGICs caused by man-made EMF exposure and leading to OS can also explain EHS, since EHS is accompanied by OS (De Luca et al. 2014; Irigaray et al. 2018), and in fact it may be due to chronic OS. The pathophysiological changes in the central nervous system observed to accompany EHS (see Chapter 8) can be explained by the fact that neural cells have a higher percentages of VGICs, since VGICs (and specifically sodium and potassium voltage-gated channels) are the major mediators for the transmission of the nerve impulses (Alberts et al. 1994; 2002).

12.4.2 Possible Outcomes of DNA Damage and Repair Capability of Cells

When a cell's genomic DNA is irreparably damaged for any reason, or the damage is inaccurately repaired, the following outcomes are possible.

1) The cell dies (necrosis) or is led to suicide (induced apoptosis). In the case of cell types with the ability to proliferate, the organism compensates for their loss by creating new cells, practically with no adverse consequences apart from energy consumption, which may lead to accelerated aging when such events occur at a high rate. In the case of differentiated reproductive cells (oocytes, spermatocytes), this may lead to reproductive difficulties, while in the case of reproductive stem cells (oogonia, spermatogonia), this may lead to permanent infertility. In case of cell types that do not have the ability to proliferate, such as neural cells or chondrocytes, the loss of a significant number of cells will probably result in inability of certain tissues/organs to operate normally (organic diseases). In the case of neural cells, this may lead to neuro-degenerative diseases such as Alzheimer's or Parkinson's.

EMF-induced VGIC Dysfunction, OS, DNA Damage, Related Pathologies

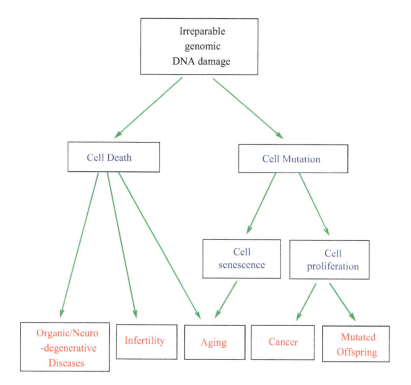

FIGURE 12.2 Schematic diagram of the possible consequences of irreparable damage in a cell's genomic DNA.

2) The cell does not die but survives with modified DNA. In the case of somatic cells that proliferate, the cell with the modified genome will reproduce itself. Even though the organism may recognize such mutant cells as foreign and try to isolate them and remove them, mutant cells strive to survive and may start proliferating uncontrollably, initiating cancer. In the case of reproductive cells (oocytes, spermatocytes), this may lead to mutated new organisms, which may be problematic in many ways, or cancer-prone. In both cases (somatic or reproductive cells), cell senescence is an alternative pathway for deactivating genetically defective cells. Thus, cells with irreparably damaged genomic DNA will result in cell senescence, cell death, cancer, or mutated offspring, depending on cell type and specific biological/environmental conditions (see also Panagopoulos et al. 2021). The described possible consequences of irreparable damage in a cell's genomic DNA are graphically represented in Figure 12.2.

Even though effective mechanisms have evolved in all animals/cells for repairing DNA damage induced by environmental stressors, it is very different when the damaging events are isolated or random (e.g., radioactive particles or gamma photons of cosmic/natural radioactivity, magnetic storms, sporadic x-ray diagnostic exposures, etc.) than persistent/repeated exposures to cytotoxic agents, even when these agents are significantly weaker. Man-made EMF exposures, and especially the most detrimental ones from WC antennas/devices and high-voltage electric power transmission lines, have become a new reality in modern life, exposing billions of people on a daily basis (Panagopoulos 2019a). Even though they are significantly less cytotoxic than radioactivity or certain toxic chemicals, they represent an evolutionary novel and most persistent daily cytotoxic agent, against which any repair mechanisms may not be efficient enough.

Previously existing cytotoxic agents, such as natural (non-polarized and incoherent) ionizing EMR (ultraviolet, gamma), or isolated hazardous events that expose us randomly, have allowed throughout the billions of years of biological evolution the development of protective mechanisms. But when an organism is constantly under OS due to repeated daily exposure to man-made (totally polarized and coherent) EMFs, no pre-existing protective mechanism can be effective enough. Especially in individuals who are already genetically or epigenetically compromised. Such a compromise serving as a "first hit" renders these individuals specifically vulnerable to the EMF-OS-DNA damage cascade, which serves as a "second hit", as originally suggested by Knudson in his famed two-hit hypothesis for carcinogenesis (Knudson 1985; Jones and Laird 1999). Granted, that up to 15% of cancer patients have innate vulnerability to cancer, the exposure to man-made EMFs should be taken under serious consideration.

Therefore, a crucial factor for good health is minimizing any genotoxic environmental stressors, such as toxic chemicals, pesticides, and in particular man-made EMFs by prudent avoidance of exposure to such stressors. Practical ways for minimizing man-made EMF exposure in daily life are provided in Chapter 5 (Discussion) and in Panagopoulos and Chrousos (2019). Another crucial factor for cell/organism integrity and resistance against any external stressor, once the various environmental pollutants are prudently minimized, is the repair capability of cells in response to DNA damage. The threshold of damage above which it becomes irreparable depends on cell type and the health and status of the organism. An organism with poor health and/or under chronic stress and inflammation is expected to have decreased DNA repair capability and increased cancer risk. Therefore, good physical and mental health and minimization/avoidance of stressors in everyday life is key to increased repair capability, immunity, and protection against any disease.

12.4.3 EMFs and Therapeutic Effects

Apart from the detrimental effects of man-made EMF exposures through the mechanism of ROS generation described in this chapter, therapeutic effects have also been reported to be induced by pulsing ELF EMFs, especially in bone fracture healing (Pilla 1993; 2003; Ryabi 1998; Chalidis et al. 2011; Pall 2013; Wade 2013; Daish et al. 2018). Altered intracellular calcium levels have also been reported to accompany the therapeutic effects, and the same biophysical mechanism of induced VGIC gating seems to be involved in both the detrimental and therapeutic effects of man-made EMFs (Pall 2013; Wade 2013; Daish et al. 2018). We believe a critical point to determine whether an effect on VGIC gating would be detrimental or beneficial is whether the EMF exposure mimics exposure by natural EMFs. Once we know that the most bioactive EMFs are the ULF/ELF ones, the critical issue for an applied ULF/ELF EMF is whether its included frequencies (and other parameters such as waveform, polarity, etc.) reinforce or cancel the endogenous electrical activity of the cells which is responsible for the specific therapeutic action (Panagopoulos 2013; Panagopoulos and Chrousos 2019). For example, the atmospheric "Schumann" electromagnetic resonances are found to be vital and necessary for any living organism on Earth, exciting the brain activity of all humans/animals and controlling circadian rhythms in coordination with the natural 24-h light periodicity (Wever 1970; 1973; 1974; 1979; Persinger 1974; Cherry 2003; Panagopoulos 2013; Panagopoulos and Chrousos 2019). Not only their basic frequency (7.83 Hz) but also their harmonics are detected in the human/animal brain activity, and the physical parameters of electromagnetic brain activity and lightning display remarkable similarities (Persinger 2012; 2014). Thus, we suggest that the therapeutic effects of pulsed EMFs might be most evident at pulsing frequencies coinciding with the Schumann frequencies. But there are also important differences to be considered: All anthropogenic EMFs are totally polarized and coherent, something that does not occur with the natural EMFs, which are only partially polarized on certain occasions (Panagopoulos et al. 2015b). This seems to be the reason why the vast majority of man-made EMF bioeffects are detrimental and not beneficial.

12.4.4 CONCLUSIONS

The present chapter, in combination with the IFO-VGIC mechanism for impairment of VGICs (described in the previous chapter), provides a complete and precise biophysical/biochemical picture to explain the great number of experimental and epidemiological findings connecting man-made EMF exposures with OS, DNA damage, and related pathologies, such as poor health, EHS, infertility, cancer, neuro-degenerative diseases, etc. Even though many more details of this picture need to be illuminated, we believe its basic scheme is revealed already.

Thus, the long existing experimental and epidemiological findings connecting exposure to man-made EMFs and DNA damage, infertility, and cancer are now explained by the induced dysfunction of VGICs and the consequent induction of OS/ROS overproduction in the exposed cells. We hope this novel finding provides a basis for future research and the establishment of biologically relevant EMF exposure guidelines.

Prudent avoidance and minimization of anthropogenic EMF exposures and other environmental stressors in everyday life, and preservation of good physical and mental health, is crucially important under any circumstances.

REFERENCES

Abdel-Rassoul G, El-Fateh OA, Salem MA, Michael A, Farahat F, El-Batanouny M, Salem E, (2007): Neurobehavioral effects among inhabitants around mobile phone base stations. *Neurotoxicology*, 28(2):434–440.

Agarwal A, Deepinder F, Sharma RK, Ranga G, Li J, (2008): Effect of cell phone usage on semen analysis in men attending infertility clinic: An observational study. *Fertil Steril*, 89(1):124–128.

Agarwal A, Desai NR, Makker K, Varghese A, Mouradi R, Sabanegh E, Sharma R, (2009): Effects of radiofrequency electromagnetic waves (RF-EMW) from cellular phones on human ejaculated semen: An in vitro pilot study. *Fertil Steril*, 92(4):1318–1325.

Ahlbom A, Day N, Feychting M, Roman E, Skinner J, Dockerty J, Linet M, McBride M, Michaelis J, Olsen JH, Tynes T, Verkasalo PK, (2000): A pooled analysis of magnetic fields and childhood leukaemia. *Br J Cancer*, 83(5):692–698.

Aitken RJ, Bennetts LE, Sawyer D, Wiklendt AM, King BV, (2005): Impact of radio frequency electromagnetic radiation on DNA integrity in the male germline. *Int J Androl*, 28(3):171–179.

Akbarali HI, (2014): Oxidative stress and ion channels. In Lahers I (Ed.), *Systems biology of free radicals and antioxidants*, Springer, Berlin, Heidelberg, 355–373.

Alberts B, Bray D, Lewis J, Raff M, Roberts K, Watson JD, (1994): *Molecular Biology of the Cell*, Garland Publishing, Inc., N.Y., USA.

Alberts B, Johnson A, Lewis J, Raff M, Roberts K, Walter P, (2002): *Membrane transport of small molecules and the electrical properties of membranes-molecular biology of the cell*, Garland Science, New York.

Ames BN, (1989): Endogenous DNA damage as related to cancer and aging. *Mutat Res*, 214(1):41–46.

Azumi H, Inoue N, Takeshita S, Rikitake Y, Kawashima S, Hayashi Y, Itoh H, Yokoyama M, (1999): Expression of NADH/NADPH oxidase p22phox in human coronary arteries. *Circulation*, 100(14):1494–1498. https://doi.org/10.1161/01.cir.100.14.1494.

Baan R, Grosse Y, Lauby-Secretan B, El Ghissassi F, Bouvard V, Benbrahim-Tallaa L, Guha N, Islami F, Galichet L, Straif K; WHO International Agency for Research on Cancer Monograph Working Group, (2011): Carcinogenicity of radiofrequency electromagnetic fields. *Lancet Oncol*, 12(7):624–626.

Bacandritsos N, Granato A, Budge G, Papanastasiou I, Roinioti E, Caldon M, Falcaro C, Gallina A, Mutinelli F, (2010): Sudden deaths and colony population decline in Greek honey bee colonies. *J Invertebr Pathol*, 105(3):335–340.

Balasubramanian B, Pogozelski WK, Tullius TD, (1998): DNA strand breaking by hydroxyl radical is governed by the accessible surface area of the hydrogen atom of the DNA backbone. *PNAS*, 95(17):9738–9743.

Balmori A, (2005): Possible effects of electromagnetic fields from phone masts on a population of White Stork (Ciconia ciconia). *Electromagn Biol Med*, 24(2):109–119.

Balmori A, (2006): The incidence of electromagnetic pollution on the amphibian decline: Is this an important piece of the puzzle? *Toxicol Environ Chem*, 88(2):287–299.

Balmori A, Hallberg O, (2007): The urban decline of the house sparrow (Passer domesticus): A possible link with electromagnetic radiation. *Electromagn Biol Med*, 26(2):141–151.

Balmori A, (2010): Mobile phone mast effects on common frog (Rana temporaria) tadpoles: The city turned into a laboratory. *Electromagn Biol Med*, 29(1–2):31–35.

Banerjee S, Singh NN, Sreedhar G, Mukherjee S, (2016): Analysis of the genotoxic effects of mobile phone radiation using buccal micronucleus assay: A comparative evaluation. *J Clin Diagn Res*, 10(3):ZC82-5. https://doi.org/10.7860/JCDR/2016/17592.7505.

Barzilai A, Yamamoto K, (2004): DNA damage responses to oxidative stress. *DNA Repair (Amst)*, 3(8–9):1109–1115.

Barbier E, Vetret B, Dufy B, (1996): Stimulation of Ca^{2+} influx in rat pituitary cells under exposure to a 50 Hz magnetic field. *Bioelectromagnetics*, 17:303–11.

Basu AK, (2018): DNA damage, mutagenesis and cancer. *Int J Mol Sci*, 19(4):970.

Batcioglu K, Uyumlu AB, Satilmis B, Yildirim B, Yucel N, Demirtas H, Onkal R, Guzel RM, Djamgoz MB, (2012): Oxidative stress in the in vivo DMBA rat model of breast cancer: Suppression by a voltage-gated sodium channel inhibitor (RS100642). *Basic Clin Pharmacol Toxicol*, 111(2):137–141.

Batellier F, Couty I, Picard D, Brillard JP, (2008): Effects of exposing chicken eggs to a cell phone in "call" position over the entire incubation period. *Theriogenology*, 69(6):737–745.

Bawin SM, Kaczmarek LK, Adey WR, (1975): Effects of modulated VHF fields, on the central nervous system. *Ann N Y Acad Sci*, 247:74–81.

Bawin SM, Adey WR, (1976): Sensitivity of calcium binding in cerebral tissue to weak environmental electric fields oscillating at low frequency. *Proc Natl Acad Sci U S A*, 73(6):1999–2003.

Bawin SM, Adey WR, Sabbot IM, (1978): Ionic factors in release of $^{45}Ca^{2+}$ from chick cerebral tissue by electromagnetic fields. *Proc Natl Acad Sci U S A*, 75(12):6314–6318.

Becchetti A, (2011): Ion channels and transporters in cancer. 1. Ion channels and cell proliferation in cancer. *Am J Physiol Cell Physiol*, 301(2):C255–C265.

Bedard K, Krause KH, (2007): The NOX family of ROS-generating NADPH oxidases: Physiology and pathophysiology. *Physiol Rev*, 87(1):245–313. https://doi.org/10.1152/physrev.00044.2005.

Belyaev IY, Hillert L, Protopopova M, Tamm C, Malmgren LO, Persson BR, Selivanova G, Harms-Ringdahl M, (2005): 915 MHz microwaves and 50 Hz magnetic field affect chromatin conformation and 53BP1 foci in human lymphocytes from hypersensitive and healthy persons. *Bioelectromagnetics*, 26(3):173–184.

Bernstein C, Prasad AR, Nfonsam V, Bernstein H, (2013): DNA damage, DNA repair and cancer. In Clarc C (Ed.), *New research directions in DNA repair*, InTech, Rijeka, 413–465.

Bertagna F, Lewis R, Silva SRP, McFadden J, Jeevaratnam K, (2021): Effects of electromagnetic fields on neuronal ion channels: A systematic review. *Ann N Y Acad Sci*, 1499(1):82–103.

Blackman CF, Benane SG, Elder JA, House DE, Lampe JA, Faulk JM, (1980): Induction of calcium-ion efflux from brain tissue by radiofrequency radiation: Effect of sample number and modulation frequency on the power-density window. *Bioelectromagnetics*, 1(1):35–43.

Blank M, and Soo L, (1990): Ion activation of the Na,K-ATPase in alternating currents. *Bioelectrochem Bioenerg*, 24:51–61.

Blank M, Goodman R, (2009): Electromagnetic fields stress living cells. *Pathophysiology*, 16(2–3):71–78.

Blank M, Goodman R, (2011): DNA is a fractal antenna in electromagnetic fields. *Int J Radiat Biol*, 87(4):409–415.

Blettner M, Schlehofer B, Breckenkamp J, Kowall B, Schmiedel S, Reis U, Potthoff P, Schüz J, Berg-Beckhoff G, (2009): Mobile phone base stations and adverse health effects: Phase 1 of a population-based, cross-sectional study in Germany. *Occup Environ Med*, 66(2):118–123.

Brookes PS, Yoon Y, Robotham JL, Anders MW, Sheu SS, (2004): Calcium, ATP, and ROS: A mitochondrial love-hate triangle. *Am J Physiol Cell Physiol*, 287(4):C817–C833.

Burlaka A, Tsybulin O, Sidorik E, Lukin S, Polishuk V, Tsehmistrenko S, Yakymenko I, (2013): Overproduction of free radical species in embryonic cells exposed to low intensity radiofrequency radiation. *Exp Oncol*, 35(3):219–225.

Burney S, Caulfield JL, Niles JC, Wishnok JS, Tannenbaum SR, (1999): The chemistry of DNA damage from nitric oxide and peroxynitrite. *Mutat Res*, 424(1–2):37–49.

Cadet J, Delatour T, Douki T, Gasparutto D, Pouget JP, Ravanat JL, Sauvaigo S, (1999): Hydroxyl radicals and DNA base damage. *Mutat Res*, 424(1–2):9–21.

Cadet J, Wagner JR, (2013): DNA base damage by reactive oxygen species, oxidizing agents, and UV radiation. *Cold Spring Harb Perspect Biol*, 5(2):a012559.

Campisi A, Gulino M, Acquaviva R, Bellia P, Raciti G, Grasso R, Musumeci F, Vanella A, Triglia A, (2010): Reactive oxygen species levels and DNA fragmentation on astrocytes in primary culture after acute exposure to low intensity microwave electromagnetic field. *Neurosci Lett*, 473(1):52–55.

Carlberg M, Hardell L, (2017): Evaluation of mobile phone and cordless phone use and glioma risk using the Bradford Hill viewpoints from 1965 on association or causation. *BioMed Res Int*, 2017:9218486. https://doi.org/10.1155/2017/9218486.

Cecchetto C, Maschietto M, Boccaccio P, Vassanelli S, (2020): Electromagnetic field affects the voltage-dependent potassium channel Kv1.3. *Electromagn Biol Med*, 39(4):316–322.

Chattipakorn N, Kumfu S, Fucharoen S, Chattipakorn S, (2011): Calcium channels and iron uptake into the heart. *World J Cardiol*, 3(7):215–218.

Chalidis B, Sachinis N, Assiotis A, Maccauro G, (2011): Stimulation of bone formation and fracture healing with pulsed electromagnetic fields: Biologic responses and clinical implications. *Int J Immunopathol Pharmacol*, 24(1):17020.

Chavdoula ED, Panagopoulos DJ, Margaritis LH, (2010): Comparison of biological effects between continuous and intermittent exposure to GSM-900 MHz mobile phone radiation: Detection of apoptotic cell death features. *Mutat Res*, 700(1–2):51–61.

Cherry NJ, (2003): Human intelligence: The brain, an electromagnetic system synchronised by the Schumann Resonance signal. *Med Hypo*, 60(6):843–844.

Choi J, Min K, Jeon S, Kim N, Pack JK, Song K, (2020): Continuous exposure to 1.7 GHz LTE electromagnetic fields increases intracellular reactive oxygen species to decrease human cell proliferation and induce senescence. *Sci Rep*, 10(1):9238. https://doi.org/10.1038/s41598-020-65732-4.

Coggle JE, (1983): *Biological effects of radiation*, Taylor & Francis, London.

Coghill RW, Steward J, Philips A, (1996): Extra low frequency electric and magnetic fields in the bed place of children diagnosed with leukaemia: A case-control study. *Eur J Cancer Prev*, 5(3):153–158.

Coleman MP, Bell CM, Taylor HL, Primic-Zakelj M, (1989): Leukaemia and residence near electricity transmission equipment: A case-control study. *Br J Cancer*, 60(5):793–798.

Cooke MS, Evans MD, Dizdaroglu M, Lunec J, (2003): Oxidative DNA damage: Mechanisms, mutation, and disease. *FASEB J*, 17(10):1195–1214.

Cucurachi S, Tamis WL, Vijver MG, Peijnenburg WJ, Bolte JF, de Snoo GR, (2013): A review of the ecological effects of radiofrequency electromagnetic fields (RF-EMF). *Environ Int*, 51:116–140.

Daish C, Blanchard R, Fox K, Pivonka P, Pirogova E, (2018): The application of pulsed electromagnetic fields (PEMFs) for bone fracture repair: Past and perspective findings. *Ann Biomed Eng*, 46(4):525–542.

Daroit NB, Visioli F, Magnusson AS, Vieira GR, Rados PV, (2015): Cell phone radiation effects on cytogenetic abnormalities of oral mucosal cells. *Braz Oral Res*, 29:1–8.

De Coursey TE, (2003): Interactions between NADPH oxidase and voltage-gated proton channels: Why electron transport depends on proton transport. *FEBS Lett*, 555(1):57–61.

De Coursey T, Morgan D, Cherny V, (2003): The voltage dependence of NADPH oxidase reveals why phagocytes need proton channels. *Nature*, 422(6931):531–534.

De Iuliis GN, Newey RJ, King BV, Aitken RJ, (2009): Mobile phone radiation induces reactive oxygen species production and DNA damage in human spermatozoa in vitro. *PLOS ONE*, 4(7):e6446.

Delgado JMR, (1985): Biological effects of extremely low frequency electromagnetic fields. *J Bioelectr*, 4(1):75–91.

De Luca C, Thai JC, Raskovic D, et al., (2014): Metabolic and genetic screening of electromagnetic hypersensitive subjects as a feasible tool for diagnostics and intervention. *Mediators Inflamm*, 2014: 924184

Diem E, Schwarz C, Adlkofer F, Jahn O, Rudiger H, (2005): Non-thermal DNA breakage by mobile-phone radiation (1800 MHz) in human fibroblasts and in transformed GFSH-R17 rat granulosa cells in vitro. *Mutat Res*, 583(2):178–183.

Draper G, Vincent T, Kroll ME, Swanson J, (2005): Childhood cancer in relation to distance from high voltage power lines in England and Wales: A case-control study. *BMJ*, 330(7503):1290.

D'Silva MH, Swer RT, Anbalagan J, Rajesh B, (2017): Effect of radiofrequency radiation emitted from 2G and 3G cell phone on developing liver of chick embryo - A comparative study. *J Clin Diagn Res*, 11(7):5–9.

Dutta SK, Subramaniam A, Ghosh B, Parshad R, (1984): Microwave radiation - Induced calcium ion efflux from human neuroblastoma cells in culture. *Bioelectromagnetics*, 5(1):71–78.

El-Swefy S, Soliman H, Huessein M, (2008): Calcium channel blockade alleviates brain injury induced by long-term exposure to an electromagnetic field, *J Appl Biomed*, 6:153–163.

Everaert J, Bauwens D, (2007): A possible effect of electromagnetic radiation from mobile phone base stations on the number of breeding house sparrows (Passer domesticus). *Electromagn Biol Med*, 26(1):63–72.

Falcioni L, Bua L, Tibaldi E, Lauriola M, De Angelis L, Gnudi F, Mandrioli D, Manservigi M, Manservisi F, Manzoli I, Menghetti I, Montella R, Panzacchi S, Sgargi D, Strollo V, Vornoli A, Belpoggi F, (2018): Report of final results regarding brain and heart tumors in Sprague-Dawley rats exposed from prenatal life until natural death to mobile phone radiofrequency field representative of a 1.8 GHz GSM base station environmental emission. *Environ Res*, 165:496–503.

Fenton HJH, (1894): LXXIII.—Oxidation of tartaric acid in presence of iron. *J Chem Soc Trans*, 65:899–911.

Ferreira AR, Knakievicz T, Pasquali MA, Gelain DP, Dal-Pizzol F, Fernández CE, de Salles AA, Ferreira HB, Moreira JC, (2006): Ultra high frequency-electromagnetic field irradiation during pregnancy leads to an increase in erythrocytes micronuclei incidence in rat offspring. *Life Sci*, 80:43–50.

Feychting M, Ahlbom A, (1993): Magnetic fields and cancer in children residing near Swedish high - Voltage power lines. *Am J Epidemiol*, 138(7):467–81.

Feychting M, Ahlbom A, (1994): Magnetic fields, leukemia and central nervous system tumors in Swedish adults residing near high - Voltage power lines. *Epidemiology*, 5(5):501–509.

Francis G, (1960): *Ionization phenomena in gases*, Butterworths Scientific Publications, London.

Franzellitti S, Valbonesi P, Ciancaglini N, Biondi C, Contin A, Bersani F, Fabbri E, (2010): Transient DNA damage induced by high-frequency electromagnetic fields (GSM 1.8 GHz) in the human trophoblast HTR-8/SVneo cell line evaluated with the alkaline comet assay. *Mutat Res*, 683(1–2):35–42.

Friedman J, Kraus S, Hauptman Y, Schiff Y, Seger R, (2007): Mechanism of short-term ERK activation by electromagnetic fields at mobile phone frequencies. *Biochem J*, 405(3):559–568.

Gaasch JA, Geldenhuys WJ, Lockman PR, Allen DD, Van der Schyf CJ, (2007): Voltage-gated calcium channels provide an alternate route for iron uptake in neuronal cell cultures. *Neurochem Res*, 32(10):1686–1693.

Gamaley I, Augsten K, Berg H, (1995): Electrostimulation of macrophage NADPH oxidase by modulated high-frequency electromagnetic fields. *Bioelectrochem Bioenerg*, 38(2):415–418.

Garaj-Vrhovac V, Horvat D, Koren Z, (1990): The effect of microwave radiation on the cell genome. *Mutat Res*, 243(2):87–93.

Garaj-Vrhovac V, Horvat D, Koren Z, (1991): The relationship between colony-forming ability, chromosome aberrations and incidence of micronuclei in V79 Chinese hamster cells exposed to microwave radiation. *Mutat Res*, 263(3):143–149.

Garaj-Vrhovac V, Fucić A, Horvat D, (1992): The correlation between the frequency of micronuclei and specific chromosome aberrations in human lymphocytes exposed to microwave radiation in vitro. *Mutat Res*, 281(3):181–186.

Gobba F, Malagoli D, Ottaviani E, (2003): Effects of 50 Hz magnetic fields on fMLP-induced shape changes in invertebrate immunocytes: the role of calcium ion channels. *Bioelectromagnetics*, 24:277–82.

Görlach A, Bertram K, Hudecova S, Krizanova O, (2015): Calcium and ROS: A mutual interplay. *Redox Biol*, 6:260–271.

Gomer R, (1961): *Field emission and field ionization*, Harvard University Press, Cambridge, Massachusetts.

Goodman EM, Greenebaum B, Marron MT, (1995): Effects of electro-magnetic fields on molecules and cells. *Int Rev Cytol*, 158:279–338.

Grassi C, D'Ascenzo M, Torsello A, Martinotti G, Wolf F, Cittadini A, Azzena GB, (2004): Effects of 50 Hz electromagnetic fields on voltage-gated Ca^{2+} channels and their role in modulation of neuroendocrine cell proliferation and death. *Cell Calcium*, 35(4):307–315.

Greenland S, Sheppard AR, Kaune WT, Poole C, Kelsh MA, (2000): A pooled analysis of magnetic fields, wire codes, and childhood leukemia. Childhood leukemia-EMF Study Group. *Epidemiology*, 11(6):624–634.

Gul A, Celebi H, Uğraş S, (2009): The effects of microwave emitted by cellular phones on ovarian follicles in rats. *Arch Gynecol Obstet*, 280(5):729–733.

Gulati S, Yadav A, Kumar N, Kanupriya, Aggarwal NK, Kumar R, Gupta R, (2016): Effect of GSTM1 and GSTT1 polymorphisms on genetic damage in humans populations exposed to radiation From mobile towers. *Arch Environ Contam Toxicol*, 70(3):615–625.

Guler G, Tomruk A, Ozgur E, Seyhan N, (2010): The effect of radiofrequency radiation on DNA and lipid damage in non-pregnant and pregnant rabbits and their newborns. *Gen Physiol Biophys*, 29(1):59–66.

Hallberg O, Johansson O, (2002): Melanoma incidence and frequency modulation (FM) broadcasting. *Arch Environ Health*, 57(1):32–40.

Hall EJ, Giaccia AJ, (2006): *Radiobiology for the radiologist*, Lippincott Williams & Wilkins, Philadelphia.

Halliwell B, (2007): Biochemistry of oxidative stress. *Biochem Soc Trans*, 35(5):1147–1150.

Hanahan D, Weinberg RA, (2000): The hallmarks of cancer. *Cell*, 100(1):57–70.

Hanahan D, Weinberg RA, (2011): The hallmarks of cancer: The next generation. *Cell*, 144(5):646–674.

Hardell L, Carlberg M, Söderqvist F, Mild KH, Morgan LL, (2007): Long-term use of cellular phones and brain tumours: Increased risk associated with use for > or =10 years. *Occup Environ Med*, 64(9):626–632.

Hardell L, Carlberg M, (2009): Mobile phones, cordless phones and the risk for brain tumours. *Int J Oncol*, 35(1):5–17.

Hardell L, Carlberg M, Söderqvist F, Mild KH, (2013a): Pooled analysis of case-control studies on acoustic neuroma diagnosed 1997–2003 and 2007–2009 and use of mobile and cordless phones. *Int J Oncol*, 43(4):1036–1044.

Hardell L, Carlberg M, Söderqvist F, Mild KH, (2013b): Case-control study of the association between malignant brain tumours diagnosed between 2007 and 2009 and mobile and cordless phone use. *Int J Oncol*, 43(6):1833–1845.

Hardell L, Carlberg M, Hansson Mild K, (2013c): Use of mobile phones and cordless phones is associated with increased risk for glioma and acoustic neuroma. *Pathophysiology*, 20(2):85–110.

Hardell L, (2018): Effects of mobile phones on children's and adolescents' health: A commentary. *Child Dev*, 89(1):137–140.

Hardell L, Nyberg R, (2020): Appeals that matter or not on a moratorium on the deployment of the fifth generation, 5G, for microwave radiation. *Mol Clin Oncol*. https://doi.org/10.3892/mco.2020.1984.

Hardell L, Carlberg M, (2020): Health risks from radiofrequency radiation, including 5G, should be assessed by experts with no conflicts of interest. *Oncol Lett*, 20(4):15.

Hardell L, Carlberg M, (2021): Lost opportunities for cancer prevention: Historical evidence on early warnings with emphasis on radiofrequency radiation. *Rev Environ Health*. https://doi.org/10.1515/reveh-2020-0168.

Helleday T, Loc J, van Gentd DC, Engelward BP, (2007): DNA double-strand break repair: From mechanistic understanding to cancer treatment. *DNA Repair*, 6(7):923–935.

Henderson LM, (2001): NADPH oxidase subunit gp91phox: A proton pathway. *Protoplasma*, 217(1–3):37–42.

Hong R, Zhang Y, Liu Y, Weng EQ, (2005): Effects of extremely low frequency electromagnetic fields on DNA of testicular cells and sperm chromatin structure in mice. *Zhonghua Lao Dong Wei Sheng Zhi Ye Bing Za Zhi*, 23(6):414–417.

Hutter H-P, Moshammer H, Wallner P, Kundi M, (2006): Subjective symptoms, sleeping problems, and cognitive performance in subjects living near mobile phone base stations. *Occup Environ Med*, 63(5):307–313.

IARC, (2002): *Non-ionizing radiation, part 1: Static and extremely low-frequency (ELF) electric and magnetic fields*, Vol. 80, International Agency for Research on Cancer, Lyon, France.

IARC, (2013): *Non-ionizing radiation, part 2: Radiofrequency electromagnetic fields*, Vol. 102, International Agency for Research on Cancer, Lyon, France.

ICNIRP, (1998): Guidelines for limiting exposure to time-varying electric, magnetic, and electromagnetic fields (up to 300 GHz). *Health Phys*, 74(4):494–521.

ICNIRP, (2010): Guidelines for limiting exposure to time-varying electric and magnetic fields (1 Hz to 100 kHz). *Health Phys*, 99(6):818–836.

ICNIRP, (2020): Guidelines for limiting exposure to electromagnetic fields (100 kHz to 300 GHz). *Health Phys*, 118(5):483–524.

Inoue M, Sato EF, Nishikawa M, Park AM, Kira Y, Imada I, Utsumi K, (2003): Mitochondrial generation of reactive oxygen species and its role in aerobic life. *Curr Med Chem*, 10(23):2495–2505.

Ischiropoulos H, Zhu L, Beckman JS, (1992): Peroxynitrite formation from macrophage-derived nitric oxide. *Arch Biochem Biophys*, 298(2):446–451.

Irigaray P, Caccamo D, Belpomme D, (2018): Oxidative stress in electrohypersensitivity self-reporting patients: Results of a prospective *in vivo* investigation with comprehensive molecular analysis. *Int J Mol Med*, 42(4):1885–1898.

Ivancsits S, Diem E, Pilger A, Rüdiger HW, Jahn O, (2002): Induction of DNA strand breaks by intermittent exposure to extremely-low-frequency electromagnetic fields in human diploid fibroblasts. *Mutat Res*, 519(1–2):1–13.

Ivancsits S, Diem E, Jahn O, Rüdiger HW, (2003): Intermittent extremely low frequency electromagnetic fields cause DNA damage in a dose-dependent way. *Int Arch Occup Environ Health*, 76(6):431–436.

Iverson D, De Chatelet LR, Spitznagel JK, Wang P, (1977): Comparison of NADH and NADPH oxidase activities in granules isolated from human polymorphonuclear leukocytes with a fluorometric assay. *J Clin Invest*, 59(2):282–290. https://doi.org/10.1172/JCI108639.

Jeong JH, Kum C, Choi HJ, et al. (2006): Extremely low frequency magnetic field induces hyperalgesia in mice modulated by nitric oxide synthesis. *Life Sci*, 78:1407–12.

Ji S, Oh E, Sul D, Choi JW, Park H, Lee E, (2004): DNA damage of lymphocytes in volunteers after 4 hours use of mobile phone. *J Prev Med Public Health*, 37(4):373–380.

Jones PA, Laird PW, (1999): Cancer epigenetics comes of age. *Nat Genet*, 21(2):163–167.

Kim JH, Jeon S, Choi HD, Lee JH, Bae JS, Kim N, Kim HG, Kim KB, Kim HR, (2021): Exposure to long-term evolution radiofrequency electromagnetic fields decreases neuroblastoma cell proliferation via Akt/mTOR-mediated cellular senescence. *J Toxicol Environ Health A*:1–12. https://doi.org/10.1080/15287394.2021.1944944.

Kheifets L, Ahlbom A, Crespi CM, Draper G, Hagihara J, Lowenthal RM, Mezei G, Oksuzyan S, Schüz J, Swanson J, Tittarelli A, Vinceti M, Wunsch Filho V, (2010): Pooled analysis of recent studies on magnetic fields and childhood leukaemia. *Br J Cancer*, 103(7):1128–1135. Erratum in: *Br J Cancer*, (2011), 104(1):228.

Khurana VG, Teo C, Kundi M, Hardell L, Carlberg M, (2009): Cell phones and brain tumors: A review including the long-term epidemiologic data. *Surg Neurol*, 72(3):205–214.

Knudson AG, (1985): Hereditary cancer, oncogenes, and antioncogenes. *Cancer Res*, 45(4):1437–1443.

Koana T, Okada MO, Takashima Y, Ikehata M, Miyakoshi J, (2001): Involvement of eddy currents in the mutagenicity of ELF magnetic fields. *Mutat Res*, 476(1–2):55–62.

Kourie JI, (1998): Interaction of reactive oxygen species with ion transport mechanisms. *Am J Physiol*, 275(1) (Cell Physiol. 44):C1–C24.

Kundi M, Hutter HP, (2009): Mobile phone base stations-effects on wellbeing and health. *Pathophysiology*, 16(2–3):123–135.

Lahtz C, Pfeifer GP, (2011): Epigenetic changes of DNA repair genes in cancer. Review. *J Mol Cell Biol*, 3(1):51–58.

Lai H, Singh NP, (1995): Acute low-intensity microwave exposure increases DNA single-strand breaks in rat brain cells. *Bioelectromagnetics*, 16(3):207–210.

Lai H, Singh NP, (1996): Single- and double-strand DNA breaks in rat brain cells after acute exposure to radiofrequency electromagnetic radiation. *Int J Radiat Biol*, 69(4):513–521.

Lai H, Singh NP, (1997): Acute exposure to a 60 Hz magnetic field increases DNA strand breaks in rat brain cells. *Bioelectromagnetics*, 18(2):156–165.

Lai H, Singh NP, (2004): Magnetic-field-induced DNA strand breaks in brain cells of the rat. *Environ Health Perspect*, 112(6):687–694.

Lang F, Föller M, Lang KS, Lang PA, Ritter M, Gulbins E, Vereninov A, Huber SM, (2005): Ion channels in cell proliferation and apoptotic cell death. *J Membr Biol*, 205(3):147–157.

La Vignera S, Condorelli RA, Vicardi E, D'Agata R, Calogero AE, (2012): Effects of the exposure to mobile phones on male reproduction: A review of the literature. *J Androl*, 33(3):350–356.

Leach V, Weller S, Redmayne M, (2018): A novel database of bio-effects from non-ionizing radiation. *Rev Environ Health*, 33(3): 1–8.

Lee KS, Choi JS, Hong SY, Son TH, Yu K, (2008): Mobile phone electromagnetic radiation activates MAPK signaling and regulates viability in Drosophila. *Bioelectromagnetics*, 29(5):371–379.

Lerchl A, Klose M, Grote K, Wilhelm AF, Spathmann O, Fiedler T, Streckert J, Hansen V, Clemens M, (2015): Tumor promotion by exposure to radiofrequency electromagnetic fields below exposure limits for humans. *Biochem Biophys Res Commun*, 459(4):585–590.

Lerchl A, Klose M, Drees K, (2020): No increased DNA damage observed in the brain, liver, and lung of fetal mice treated With ethylnitrosourea and exposed to UMTS radiofrequency electromagnetic fields. *Bioelectromagnetics*, 41(8):611–616.

Li JM, Shah AM, (2001): Differential NADPH- versus NADH-dependent superoxide production by phagocyte-type endothelial cell NADPH oxidase. *Cardiovasc Res*, 52(3):477–486. https://doi.org/10.1016/s0008-6363(01)00407-2.

Liburdy RP, (1992): Calcium signaling in lymphocytes and ELF fields. *FEBS Lett*, 301(1):53–59.

Lieber MR, (1998): Pathological and physiological double-strand breaks: Roles in cancer, aging, and the immune system. *Am J Pathol*, 153(5):1323–1332.

Lin-Liu S, Adey WR, (1982): Low frequency amplitude modulated microwave fields change calcium efflux rates from synaptosomes. *Bioelectromagnetics*, 3(3):309–322.

Lisi A, Ledda M, Rosola E, et al. (2006): Extremely low frequency electromagnetic field exposure promotes differentiation of pituitary corticotrope-derived AtT20 D16V cells. *Bioelectromagnetics*, 27:641–51.

Liu C, Gao P, Xu SC, Wang Y, Chen CH, He MD, Yu ZP, Zhang L, Zhou Z, (2013): Mobile phone radiation induces mode-dependent DNA damage in a mouse spermatocyte-derived cell line: A protective role of melatonin. *Int J Radiat Biol*, 89(11):993–1001.

Lixia S, Yao K, Kaijun W, Deqiang L, Huajun H, Xiangwei G, Baohong W, Wei Z, Jianling L, Wei W, (2006): Effects of 1.8 GHz radiofrequency field on DNA damage and expression of heat shock protein 70 in human lens epithelial cells. *Mutat Res*, 602(1–2):135–142.

Lombardi AA, Gibb AA, Arif E, Kolmetzky DW, Tomar D, Luongo TS, Jadiya P, Murray EK, Lorkiewicz PK, Hajnóczky G, Murphy E, Arany ZP, Kelly DP, Margulies KB, Hill BG, Elrod JW, (2019): Mitochondrial calcium exchange links metabolism with the epigenome to control cellular differentiation. *Nat Commun*, 10(1): 1–17.

López I, Félix N, Rivera M, Alonso A, Maestú C, (2021): What is the radiation before 5G? A correlation study between measurements in situ and in real time and epidemiological indicators in Vallecas, Madrid. *Environ Res*, 194:110734.

Low H, Crane FL, Morre DJ, (2012): Putting together a plasma membrane NADH oxidase: a tale of three laboratories. *Int J Biochem Cell Biol*, 44(11): 1834–1838

Lowe SW, Lin AW, (2000): Apoptosis in cancer. *Carcinogenesis*, 21(3):485–495.

Lushchak VI, (2014): Free radicals, reactive oxygen species, oxidative stress and its classification. *Chem Biol Interact*, 224:164–175.

Luukkonen J, Hakulinen P, Mäki-Paakkanen J, Juutilainen J, Naarala J, (2009): Enhancement of chemically induced reactive oxygen species production and DNA damage in human SH-SY5Y neuroblastoma cells by 872 MHz radiofrequency radiation. *Mutat Res*, 662(1–2):54–58.

Ma TH, Chu KC, (1993): Effect of the extremely low frequency (ELF) electromagnetic field (EMF) on developing embryos of the fruit fly (Drosophila melanogaster L.). *Mutat Res*, 303(1):35–39.

Magras IN, Xenos TD, (1997): RF radiation-induced changes in the prenatal development of mice. *Bioelectromagnetics*, 18(6):455–461.

Mailankot M, Kunnath AP, Jayalekshmi H, Koduru B, Valsalan R, (2009): Radio frequency electromagnetic radiation (RF-EMR) from GSM (0.9/1.8 GHz) mobile phones induces oxidative stress and reduces sperm motility in rats. *Clin (Sao Paulo)*, 64(6):561–565.

Manna D, Ghosh R, (2016): Effect of radiofrequency radiation in cultured mammalian cells: A review. *Electromagn Biol Med*, 35(3):265–301.

Marchionni I, Paffi A, Pellegrino M, et al. (2006): Comparison between low-level 50 Hz and 900 MHz electromagnetic stimulation on single channel ionic currents and on firing frequency in dorsal root ganglion isolated neurons. *Biochim Biophys Acta*, 1758:597–605.

Margaritis LH, (1996): Cell Biology, Litsas medical publications, Athens [Μαργαρίτης ΛΧ, Κυτταρική Βιολογία, Ιατρικές Εκδόσεις Λίτσας, Γ΄έκδοση, Αθήα]

Markova E, Hillert L, Malmgren L, Persson BR, Belyaev IY, (2005): Microwaves from GSM mobile telephones affect 53BP1 and gamma-H2AX foci in human lymphocytes from hypersensitive and healthy persons. *Environ Health Perspect*, 113(9):1172–1177.

Mausset-Bonnefont AL, Hirbec H, Bonnefont X, Privat A, Vignon J, de Sèze R, (2004): Acute exposure to GSM 900-MHz electromagnetic fields induces glial reactivity and biochemical modifications in the rat brain. *Neurobiol Dis*, 17(3):445–454.

Mihai CT, Rotinberg P, Brinza F, Vochita G (2014): Extremely low-frequency electromagneticfields cause DNA strand breaks in normal cells. *J Environ Health Sci Eng*, 12(1):15.

Miller A, To T, Agnew DA, Wall C, Green LM, (1996): Leukemia following occupational exposure to 60-Hz electric and magnetic fields among Ontario electric utility workers. *Am J Epidemiol*, 144(2):150–160.

Miller AB, Morgan LL, Udasin I, Davis DL, (2018): Cancer epidemiology update, following the 2011 IARC evaluation of radiofrequency electromagnetic fields (Monograph 102). *Environ Res*, 167:673–683.

Miller AB, Sears ME, Morgan LL, Davis DL, Hardell L, Oremus M, Soskolne CL, (2019): Risks to health and well-being from radio-frequency radiation emitted by cell phones and other wireless devices. *Front Public Health*, 7:223. https://doi.org/10.3389/fpubh.2019.00223.

Momoli F, Siemiatycki J, McBride ML, Parent MÉ, Richardson L, Bedard D, Platt R, Vrijheid M, Cardis E, Krewski D, (2017): Probabilistic multiple-bias modeling applied to the Canadian data from the interphone study of mobile phone use and risk of glioma, meningioma, acoustic neuroma, and parotid gland tumors. *Am J Epidemiol*, 186(7):885–893.

Moon HK, Yang ES, Park JW, (2006): Protection of peroxynitrite-induced DNA damage by dietary antioxidants. *Arch Pharm Res*, 29(3):213–217.

Morgado-Valle C, Verdugo-Dıaz L, Garcıa DE, et al. (1998): The role of voltage-gated Ca^{2+} channels in neurite growth of cultured chromaffin cells induced by extremely low frequency (ELF) magnetic field stimulation. *Cell Tissue Res*, 291:217–30.

Musset B, Cherny VV, Morgan D, DeCoursey TE, (2009): The intimate and mysterious relationship between proton channels and NADPH oxidase. *FEBS Lett*, 583(1):7–12.

Nadler DL, Zurbenko IG, (2014): Estimating cancer latency times using a Weibull model. *Adv Epidemiol*, 2014, 746769.

Navarro A, Segura J, Portolés M, Gómez-Perretta de Mateo C, (2003): The microwave syndrome: A preliminary study in Spain. *Electromagn Biol Med*, 22(2–3):161–169.

Nikolova T, Czyz J, Rolletschek A, Blyszczuk P, Fuchs J, Jovtchev G, Schuderer J, Kuster N, Wobus AM, (2005): Electromagnetic fields affect transcript levels of apoptosis-related genes in embryonic stem cell-derived neural progenitor cells. *FASEB J*, 19(12):1686–1688.

Nitahara JA, Cheng W, Liu Y, Li B, Leri A, Li P, Mogul D, Gambert SR, Kajstura J, Anversa P, (1998): Intracellular calcium, DNase activity and myocyte apoptosis in aging Fischer 344 rats. *J Mol Cell Cardiol*, 30(3):519–535.

NTP (National Toxicology Program), (2018): Toxicology and Carcinogenesis studies in Hsd: Sprague Dawley SD rats exposed to whole-body Radio Frequency radiation at a frequency (900 MHz) and modulations (GSM and CDMA) used by cell phones, NTP TR 595, Department of Health and Human Services, USA.

O'Hare Doig RL, Chiha W, Giacci MK, Yates NJ, Bartlett CA, Smith NM, Hodgetts SI, Harvey AR, Fitzgerald M, (2017): Specific ion channels contribute to key elements of pathology during secondary degeneration following neurotrauma. *BMC Neurosci*, 18(1):62.

Pacher P, Beckman JS, Liaudet L, (2007): Nitric oxide and peroxynitrite in health and disease. *Physiol Rev*, 87(1):315–424.

Pall ML, (2013): Electromagnetic fields act via activation of voltage-gated calcium channels to produce beneficial or adverse effects. *J Cell Mol Med*, 17(8):958–965.

Panagopoulos DJ, Messini N, Karabarbounis A, Filippetis AL, Margaritis LH, (2000): A mechanism for action of oscillating electric fields on cells. *Biochem Biophys Res Commun*, 272(3):634–640.

Panagopoulos DJ, Karabarbounis A, Margaritis LH, (2002): Mechanism for action of electromagnetic fields on cells. *Biochem Biophys Res Commun*, 298(1):95–102.

Panagopoulos DJ, Margaritis LH, (2003): Theoretical considerations for the biological effects of electromagnetic fields. In Stavroulakis P (Ed.), *Biological effects of electromagnetic fields*, Springer, Berlin Heidelberg, 5–33.

Panagopoulos DJ, Karabarbounis A, Margaritis LH, (2004): Effect of GSM 900-MHz mobile phone radiation on the reproductive capacity of Drosophila melanogaster. *Electromagn Biol Med*, 23(1):29–43.

Panagopoulos DJ, Chavdoula ED, Nezis IP, Margaritis LH, (2007a): Cell death induced by GSM 900 MHz and DCS 1800 MHz mobile telephony radiation. *Mutat Res*, 626(1–2):69–78.

Panagopoulos DJ, Chavdoula ED, Karabarbounis A, Margaritis LH, (2007b): Comparison of bioactivity between GSM 900 MHz and DCS 1800 MHz mobile telephony radiation. *Electromagn Biol Med*, 26(1):33–44.

Panagopoulos DJ, Chavdoula ED, Margaritis LH, (2010): Bioeffects of mobile telephony radiation in relation to its intensity or distance from the antenna. *Int J Radiat Biol*, 86(5):345–357.

Panagopoulos DJ, (2012): Effect of microwave exposure on the ovarian development of Drosophila melanogaster. *Cell Biochem Biophys*, 63(2):121–132.

Panagopoulos DJ, (2013): Electromagnetic interaction between environmental fields and living systems determines health and well-being. In Kwang M-H, and Yoon S-O (Eds.), *Electromagnetic fields: Principles, engineering applications and biophysical effects*, Nova Science Publishers, New York, 87–130.

Panagopoulos DJ, Karabarbounis A, Lioliousis C, (2013): ELF alternating magnetic field decreases reproduction by DNA damage induction. *Cell Biochem Biophys*, 67(2):703–716.

Panagopoulos DJ, Johansson O, Carlo GL, (2015a): Real versus simulated mobile phone exposures in experimental studies. *BioMed Res Int*, 2015:607053.

Panagopoulos DJ, Johansson O, Carlo GL, (2015b): Polarization: A key difference between man-made and natural electromagnetic fields, in regard to biological activity. *Sci Rep*, 5:14914. https://doi.org/10.1038/srep14914.

Panagopoulos DJ, (2017): Mobile telephony radiation effects on insect ovarian cells. The necessity for real exposures bioactivity assessment. The key role of polarization, and the ion forced-oscillation mechanism. In Geddes CD (Ed.), *Microwave effects on DNA and proteins*, Springer, Cham, Switzerland, 1–48.

Panagopoulos DJ, (2018): Man-made electromagnetic radiation is not quantized. In Reimer A (Ed.), *Horizons in World Physics*. Vol. 296, Nova Science Publishers, New York, 1–57.

Panagopoulos DJ, (2019a): Comparing DNA damage induced by mobile telephony and other types of man-made electromagnetic fields. *Mutat Res Rev Mutat Res*, 781:53–62.

Panagopoulos DJ, (2019b): Chromosome damage in human cells induced by UMTS mobile telephony radiation. *Gen Physiol Biophys*, 38(5):445–454.

Panagopoulos DJ, Chrousos GP, (2019): Shielding methods and products against man-made electromagnetic fields: Protection versus risk. *Sci Total Environ*, 667C:255–262.

Panagopoulos DJ, (2020): Comparing chromosome damage induced by mobile telephony radiation and a high caffeine dose: Effect of combination and exposure duration. *Gen Physiol Biophys*, 39(6):531–544.

Panagopoulos DJ, Balmori A, Chrousos GP, (2020): On the biophysical mechanism of sensing upcoming earthquakes by animals. *Sci Total Environ*, 717:136989.

Panagopoulos DJ, Karabarbounis A, Yakymenko I, Chrousos GP, (2021): Human-made electromagnetic fields: Ion forced-oscillation and voltage-gated ion channel dysfunction, oxidative stress and DNA damage. *Int J Oncol*, 59(5):92.

Panday A, Sahoo MK, Osorio D, Batra S, (2015): NADPH oxidases: An overview from structure to innate immunity-associated pathologies. *Cell Mol Immunol*, 12(1):5–23.

Persinger MA, (1974): *ELF and VLF electromagnetic fields*, Plenum Press, New York.

Persinger MA, (2012): Brain electromagnetic activity and lightning: Potentially congruent scale - Invariant quantitative properties. *Front Integr Neurosci*, 6(19):1–7.

Persinger MA, (2014): Schumann resonance frequencies found within quantitative electroencephalographic activity: Implications for earth-brain interactions. *Int Lett Chem Phys Astron*, 11(1):24–32.

Pesnya DS, Romanovsky AV, (2013): Comparison of cytotoxic and genotoxic effects of plutonium-239 alpha particles and mobile phone GSM 900 radiation in the Allium cepa test. *Mutat Res*, 750(1–2):27–33.

Phillips JL, Singh NP, Lai H, (2009): Electromagnetic fields and DNA damage. *Pathophysiology*, 16(2–3):79–88.

Piacentini R, Ripoli C, Mezzogori D, Azzena GB, Grassi C, (2008): Extremely low-frequency electromagnetic fields promote in vitro neurogenesis via upregulation of Ca_v1-channel activity. *J Cell Physiol*, 215(1):129–139.

Pilla AA, (1993): State of the art in electromagnetic therapeutics. In Blank M (Ed.), *Electricity and magnetism in biology and medicine*, San Francisco Press Inc., 17–22.

Pilla AA, (2003): Weak time-varying and static magnetic fields: From mechanisms to therapeutic applications. In Stavroulakis P (Ed.), *Biological effects of electromagnetic fields*, Springer, Berlin, Heidelberg, 34–75.

Pilla AA, (2012): Electromagnetic fields instantaneously modulate nitric oxide signaling in challenged biological systems. *Biochem Biophys Res Commun*, 426(3):330–333.

Pratt RD, Brickman CR, Cottrill CL, Shapiro JI, Liu J, (2018): The Na/K-ATPase signaling: From specific ligands to general reactive oxygen species. *Int J Mol Sci*, 19(9):2600.

Ramírez A, Vázquez-Sánchez AY, Carrión-Robalino N, Camacho J, (2016): Ion channels and oxidative stress as a potential link for the diagnosis or treatment of liver diseases. *Oxid Med Cell Longev*, 2016:3928714.

Rodgers K, McVey M, (2016): Error-prone repair of DNA double-strand breaks. *J Cell Physiol*, 231(1):15–24.

Ryabi JT, (1998): Clinical effects of electromagnetic fields on fracture healing. *Clin Orthop Relat Res*, 355(Suppl. l):S205–S215.

Sakihama Y, Maeda M, Hashimoto M, Tahara S, Hashidoko Y, (2012): Beetroot betalain inhibits peroxynitrite-mediated tyrosine nitration and DNA strand damage. *Free Radic Res*, 46(1):93–99.

Salama OE, Abou El Naga RM, (2004): Cellular phones: Are they detrimental? *J Egypt Public Health Assoc*, 79(3–4):197–223.

Salsbury G, Cambridge EL, McIntyre Z, Arends MJ, Karp NA, Isherwood C, Shannon C, Hooks Y, Sanger Mouse Genetics Project, Ramirez-Solis R, Adams DJ, White JK, Speak AO, (2014): Disruption of the potassium channel regulatory subunit KCNE2 causes iron-deficient anemia. *Exp Hematol*, 42(12):1053–1058.

Santini MT, Ferrante A, Rainaldi G, Indovina P, Indovina PL, (2005): Extremely low frequency (ELF) magnetic fields and apoptosis: A review. *Int J Radiat Biol*, 81(1):1–11.

Santini R, Santini P, Danze JM, Le Ruz P, Seigne M, (2002): Study of the health of people living in the vicinity of mobile phone base stations: I. Influences of distance and sex. *Pathol Biol (Paris)*, 50(6):369–373.

Sarkar S, Ali S, Behari J, (1994): Effect of low power microwave on the mouse genome: A direct DNA analysis. *Mutat Res*, 320(1–2):141–147.

Savitz DA, Wachtel H, Barnes F, John EM, Tvrdik JG, (1988): Case-control study of childhood cancer and exposure to 60 Hz magnetic fields. *Am J Epidemiol*, 128(1):21–38.

Schuermann D, Ziemann C, Barekati Z, Capstick M, Oertel A, Focke F, Murbach M, Kuster N, Dasenbrock C, Schär P, (2020): Assessment of genotoxicity in human cells exposed to modulated electromagnetic fields of wireless communication devices. *Genes*, 11(4):347. https://doi.org/10.3390/genes11040347.

Schuermann D, Mevissen M, (2021): Manmade electromagnetic fields and oxidative stress—Biological effects and consequences for health. *Int J Mol Sci*, 22(7):3772.

Serpersu EH, and Tsong TY, (1984): Activation of electrogenic Rb transport of (Na/K)-ATPase by an electric field. *J Biol Chem*, 259:7155–7162.

Seredenina T, Demaurex N, Krause KH, (2015): Voltage-gated proton channels as novel drug targets: From NADPH oxidase regulation to sperm biology. *Antioxid Redox Signal*, 23(5):490–513.

Shah DJ, Sachs RK, Wilson DJ, (2012): Radiation-induced cancer: A modern view. *Br J Radiol*, 85(1020):1166–1173.

Shahin S, Singh SP, Chaturvedi CM, (2017): Mobile phone (1800 MHz) radiation impairs female reproduction in mice, Mus musculus, through stress induced inhibition of ovarian and uterine activity. *Reprod Toxicol*, 73:41–60. https://doi.org/10.1016/j.reprotox.2017.08.001.

Sharma A, Shrivastava S, Shukla S, (2020): Exposure of radiofrequency electromagnetic radiation on biochemical and pathological alterations. *Neurol India*, 68(5):1092–1100. https://doi.org/10.4103/0028-3886.294554.

Sharma VP, Kumar NR, (2010): Changes in honeybee behaviour and biology under the influence of cellphone radiations. *Curr Sci*, 98:1376–1378.

Smith-Roe SL, Wyde ME, Stout MD, Winters JW, Hobbs CA, Shepard KG, Green AS, Kissling GE, Shockley KR, Tice RR, Bucher JR, Witt KL, (2019): Evaluation of the genotoxicity of cell phone radiofrequency radiation in male and female rats and mice following subchronic exposure. *Environ Mol Mutagen*, 1(2):276–290.

Sokolovic D, Djindjic B, Nikolic J, Bjelakovic G, Pavlovic D, Kocic G, Krstic D, Cvetkovic T, Pavlovic V, (2008): Melatonin reduces oxidative stress induced by chronic exposure of microwave radiation from mobile phones in rat brain. *J Radiat Res (Tokyo)*, 49(6):579–586.

Stryer L, (1995): *Biochemistry*. Freeman, New York.

Svedenstal BM, Johanson KJ, Mild KH, (1999): DNA damage induced in brain cells of CBA mice exposed to magnetic fields. *In Vivo*, 13(6):551–552.

Szabo G, Baehrle S, (2005): Role of nitrosative stress and poly(ADP-ribose) polymerase activation in myocardial reperfusion injury. *Curr Vasc Pharmacol*, 3(3):215–220.

Tillmann T, Ernst H, Streckert J, Zhouc Y, Taugner F, Hansen V, Dasenbrock C, (2010): Indication of cocarcinogenic potential of chronic UMTS-modulated radiofrequency exposure in an ethylnitrosourea mouse model. *Int J Radiat Biol*, 86(7):529–541.

Tomruk A, Guler G, Dincel AS, (2010): The influence of 1800 MHz GSM-like signals on hepatic oxidative DNA and lipid damage in nonpregnant, pregnant, and newly born rabbits. *Cell Biochem Biophys*, 56(1):39–47.

Tsunoda M, Sakaue T, Naito S, Sunami T, Abe N, Ueno Y, Matsuda A, Takénaka A, (2010): Insights into the structures of DNA damaged by hydroxyl radical: Crystal structures of DNA duplexes containing 5-formyluracil. *J Nucleic Acids*, 2010:107289.

Valko M, Rhodes CJ, Moncol J, Izakovic M, Mazur M, (2006): Free radicals, metals and antioxidants in oxidative stress-induced cancer. *Chem Biol Interact*, 160(1):1–40.

Valko M, Leibfritz D, Moncol J, Cronin MT, Mazur M, Telser J, (2007): Free radicals and antioxidants in normal physiological functions and human disease. *Int J Biochem Cell Biol*, 39(1):44–84.

Verschaeve L, (2017): Misleading scientific papers on health effects from wireless communication devices. In Geddes CD (Ed.), *Microwave effects on DNA and proteins*, Springer, Cham, Switzerland.

Viel JF, Clerc S, Barrera C, Rymzhanova R, Moissonnier M, Hours M, Cardis E, (2009): Residential exposure to radiofrequency fields from mobile phone base stations, and broadcast transmitters: A population-based survey with personal meter. *Occup Environ Med*, 66(8):550–556.

Villeneuve PJ, Agnew DA, Miller AB, Corey PN, (2000a): Non-Hodgkin's lymphoma among electric utility workers in Ontario: The evaluation of alternate indices of exposure to 60 Hz electric and magnetic fields. *Occup Environ Med*, 57(4):249–257.

Villeneuve PJ, Agnew DA, Miller AB, Corey PN, Purdham JT, (2000b): Leukemia in electric utility workers: The evaluation of alternative indices of exposure to 60 Hz electric and magnetic fields. *Am J Ind Med*, 37(6):607–617.

Wade B, (2013): A review of pulsed electromagnetic field (PEMF) mechanisms at a cellular level: A rationale for clinical use. *Am J Health Res*, 1(3):51–55.

Walleczek J, (1992): Electromagnetic field effects on cells of the immune system: The role of calcium signaling. *FASEB J*, 6(13):3177–3185.

Wang Y, Guo X, (2016): Meta-analysis of association between mobile phone use and glioma risk. *J Cancer Res Ther*, 12(Suppl):C298–C300. https://doi.org/10.4103/0973-1482.200759.

Wdowiak A, Wdowiak L, Wiktor H, (2007): Evaluation of the effect of using mobile phones on male fertility. *Ann Agric Environ Med*, 14(1):169–172.

Wertheimer N, Leeper E, (1979): Electrical wiring configurations and childhood cancer. *Am J Epidemiol*, 109(3):273–284.

Wever R, (1970): The effects of electric fields on circadian rhythmicity in men. *Life Sci Space Res*, 8:177–187.

Wever R, (1973): Human circadian rhythms under the influence of weak electric fields and the different aspects of these studies. *Int J Biometeorol*, 17(3):227–232.

Wever R, (1974): ELF effects on human circadian rhythms. In Persinger MA (Ed.), *ELF and VLF electromagnetic fields*, Plenum Press, New York.

Wever R, (1979): *The circadian system of man: Results of experiments under temporal isolation*, Springer-Verlag, New York.

Winker R, Ivancsits S, Pilger A, Adlkofer F, Rüdiger HW, (2005): Chromosomal damage in human diploid fibroblasts by intermittent exposure to extremely low-frequency electromagnetic fields. *Mutat Res*, 585(1–2):43–49.

Wood A, Mate R, Karipidis K, (2021): Meta-analysis of in vitro and in vivo studies of the biological effects of low-level millimetre waves. *J Expo Sci Environ Epidemiol*. https://doi.org/10.1038/s41370-021-00307-7.

Yadav AS, Sharma MK, (2008): Increased frequency of micronucleated exfoliated cells among humans exposed in vivo to mobile telephone radiations. *Mutat Res*, 650(2):175–180.

Yakymenko I, Tsybulin O, Sidorik E, Henshel D, Kyrylenko O, Kyrylenko S, (2016): Oxidative mechanisms of biological activity of low-intensity radiofrequency radiation. *Electromagn Biol Med*, 35(2):186–202.

Yakymenko I, Burlaka A, Tsybulin I, Brieieva I, Buchynska L, Tsehmistrenko I, Chekhun F, (2018): Oxidative and mutagenic effects of low intensity GSM 1800 MHz microwave radiation. *Exp Oncol*, 40(4):282–287.

Yan JG, Agresti M, Bruce T, Yan YH, Granlund A, Matloub HS, (2007): Effects of cellular phone emissions on sperm motility in rats. *Fertil Steril*, 88(4):957–964.

Yao K, Wu W, Wang K, Ni S, Ye P, Yu Y, Ye J, Sun L, (2008): Electromagnetic noise inhibits radiofrequency radiation-induced DNA damage and reactive oxygen species increase in human lens epithelial cells. *Mol Vis*, 19(14):964–969.

Yao Y, Dai W, (2014): Genomic instability and cancer. *J Carcinog Mutagen*, 5. https://doi.org/10.4172/2157-2518.1000165.

Yu H, Venkatarangan L, Wishnok JS, and Tannenbaum SR, (2005): Quantitation of Four Guanine Oxidation Products from Reaction of DNA with Varying Doses of Peroxynitrite. *Chem Res Toxicol*, 18(12):1849–1857.

Yuan LQ, Wang C, Lu DF, Zhao XD, Tan LH, Chen X, (2020): Induction of apoptosis and ferroptosis by a tumor suppressing magnetic field through ROS-mediated DNA damage. *Aging (Albany NY)*. 2020 8;12(4):3662–3681. https://doi.org/10.18632/aging.102836.

Zalba G, Beaumont FJ, San José G, Fortuño A, Fortuño MA, Etayo JC, Díez J, (2000): Vascular NADH/NADPH oxidase is involved in enhanced superoxide production in spontaneously hypertensive rats. *Hypertension*, 35(5):1055–1061. https://doi.org/10.1161/01.hyp.35.5.1055.

Zglinicki T, Saretzki G, Ladhoff J, d'Adda di Fagagna F, Jackson SP, (2005): Human cell senescence as a DNA damage response. *Mech Ageing Dev*, 126(1):111–117.

Zhang DY, Xu ZP, Chiang H, Lu DQ, Zeng QL, (2006): Effects of GSM 1800 MHz radiofrequency electromagnetic fields on DNA damage in Chinese hamster lung cells. *Zhonghua Yu Fang Yi Xue Za Zhi*, 40(3):149–152.

Zhang L, Zhang Z, Guo H, Wang Y, (2008): Na+/K+-ATPase-mediated signal transduction and Na+/K+-ATPase regulation. *Fundam Clin Pharmacol*, 22(6):615–621.

Zheng Y, Xia P, Dong L, Tian L, Xiong C, (2021): Effects of modulation on sodium and potassium channel currents by extremely low frequency electromagnetic fields stimulation on hippocampal CA1 pyramidal cells. *Electromagn Biol Med*, 17:1–12.

Zothansiama, Zosangzuali M, Lalramdinpuii M, Jagetia GC, (2017): Impact of radiofrequency radiation on DNA damage and antioxidants in peripheral blood lymphocytes of humans residing in the vicinity of mobile phone base stations. *Electromagn Biol Med*, 36(3):295–305.

Index

A

AAPC, *see* average annual percentage change
abdominal pain, 330
aberrant cells, 143, 144
aberration, 198
abortion, 141
absorbed power, 46
accelerated aging, 424
accelerating electric charges, 20
accelerating free electrons, 54
acceleration of free electrons/ions, 61
acceleration of pre-existed cancer development, 383
accumulative damage, 258
acentric fragments, 147, 151
acetylcholine, 400
acetylcholinesterase (AChE), 148, 401
achromatic lesion, 198, 202; *see also* chromatid gaps
acoustic neuroma, 81, 122, 377
actin cytoskeleton, 103, 150, 236, 416
actin cytoskeleton damage, 120
actin cytoskeleton disorganization, 193
actin microfilaments, 103
action potential, 277
active ion transporters, 456
acute myelogenous leukemia, 376
acute myeloid leukaemia (AML), 374, 375
adaptive beam-forming, 39
Adaptive Power Control, 39
addiction disorders, 140
addictive behavior, 84
adenocarcinoma cells, 152
adenosine triphosphate-ATP, 488
adjoint operators, 58
adrenal glands benign tumors, 380
adrenal gland tumor, 380
adrenaline (AD), 323
adrenergic nerve endings, 323
adrenocorticotrophin hormone (ACTH), 400
advanced oxidation protein products (AOPP), 148
adverse effects, 141, 169
adverse health effects, 304
aerobic life, 488
agamid lizard, 413
aging, 485, 494
air conductivity, 371
air-force pilots, 380
airplane mode, 211
airport radars, 410
air-tube headset, 211
alarm behaviour, 417
albinism, 429
alcohol, 89
aldehyde dehydrogenase, 256
alertness, 280
alignment, 412
alignment with the geomagnetic field, 402
allergic hypersensitivity, 122

alpha band, 280
alpha rhythm changes, 283
alpha waves, 279
α-helice diameter, 457
α-helices, 457
altered gene expression, 404
alternating electric current, 1
Alzheimer's disease, 300, 315, 316, 327, 345, 494
ambient levels of WC EMFs/EMR, 100
amenorrhea, 229
amino-acids, 457
amlodipine, 487
amphibians, 80, 222, 412, 483
amplification mechanisms, 467
amplitude, 30
amplitude (of the ion's forced-oscillation), 460
amplitude modulation (AM), 21, 37, 103, 146, 373
amplitude of resultant wave, 33
amplitude of the forced-oscillation, 461, 472
ampullae of Lorenzini, 429
amyloid plaque formation, 223
amyotrophic lateral sclerosis, 316
analogue modulation, 37
analysis of variance test, 196
anaphase, 198
anatennae towers, 395
androgens, 248
anechoic chamber, 380
aneuploidy, 143, 229, 250, 404
animal carcinogenicity, 83, 484
animal cells, 210
animal EEG, 287
animal monitoring, 410
animal orientation, 411, 414
animal sample size, 227
annual percentage change (APC) for cancer, 378
Anomalous Viscosity Time Dependencies (AVTD), 143
anovulation, 229
ant behaviour, 417
antenna arrays, 34, 39, 66, 87
antennae emission spectra, 60
antennae spectra, 26, 54
antennae towers, 399
antenna park, 20
antennas, 20, 211, 371, 464, 484
anthropogenic electromagnetic fields, 3, 8, 20, 232, 343, 345, 394, 397, 398, 409, 427, 451; *see also* man-made EMFs
anthropogenic EMF exposures, 331, 332, 399
anthropogenic non-ionizing EMFs, 374
anticancer control, 489
antigen challenges, 122
antigenic determinant, 313
antigonadotrophic control, 248
anti-histaminics, 344
anti-inflammatory response, 312
antimycin A, 243
antineoplastic resistance, 380

antioxidant capacity, 249
antioxidant enzyme activity, 237, 255
antioxidant enzymes, 104, 119, 421
antioxidant factors, 172
antioxidant proteins, 252
antioxidants, 99, 119, 142, 148, 171, 257, 258, 344, 491
antioxidant systems, 251, 383, 494
antioxidative enzymes, 319
antioxidative non-enzymatic proteins, 319
antrum formation, 237
ants, 412, 417
anxiety, 121, 140, 305
aphid, 415
aphidoidea, 409
apoptosis, 110, 147–149, 151, 152, 195, 254, 315, 378, 404, 489, 491
apoptosis-related proteins, 149
apoptotic cells, 146, 241
apoptotic factors, 171
apoptotic genes, 172
apoptotic index, 148
apoptotic markers, 251
appetite loss, 304
aquatic organisms, 411
aquatic plants, 423
arrhythmia, 330
arterial pressure instability, 330
arthralgia, 310
artificial electromagnetic fields (EMFs), 20, 299; see also man-made EMFs
artificial intelligence, 399
asbestos, 90
As Low As Reasonably Achievable (ALARA) practice, 172
aspartic acid, 401
aspen trees, 424
assisted reproductive technology (ART), 222, 231
asynchronous growth, 414
atmospheric discharges, 3, 28, 54, 61, 200
atmospheric electromagnetic fields, 2
atmospheric electromagnetic oscillations, 280
atmospheric (Schumann) oscillation, 371
atomic clocks, 59
atomic transitions, 59
atoms, 61
ATP production, 249
ATP synthase beta subunit (ASBS) protein, 241
atretic follicles, 237
attention, 85, 279, 285
attention deficiency, 310
attention deficit hyper-activity disorder (ADHD), 85
attention/memory loss, 335
attention task performance, 336
attractus, 198
auditory memory tasks, 285
Australian Radiation Protection and Nuclear Safety Agency, 64
autoantibodies, 313
autoimmune response, 316
autonomic nervous system, 333
average annual percentage change (AAPC), 378
average annual percentage change (AAPC) for cancer, 378
average radiation intensity, 31
average thermal velocity, 463, 467

avian malaria, 407
aviation electronic technicians, 375
avoidance response, 413

B

balance disorder, 310, 328
Balcavage, 453
baro-reflex activity, 336
basal body temperature (BBT), 229
base antennas, 3, 7, 38, 98, 191
base station antennas, 139, 373, 395, 408
base station Mobile Telephony (MT) antennas, 483
bat activity, 400
BBB, see blood-brain barrier
BBB disruption/opening, 322, 330
BBB permeability, 311
BBF regional changes, 327
Bcl-2 protein, 257
beacons, 39
beamforming antennas, 227
bean, 423
bean hypocotyls, 421
beech trees, 424
bees, 416, 417, 483
behavior, 400
behavioral changes, 417
behavioral changes in birds, 410
behavioral difficulties in children, 401
behavioral disorder, 85
behavioral parameters, 412
behavioral reactions, 400
Belousov-Zhabotinski reaction, 103
beneficial effect, 121
benign tumor, 82
beta band, 283
beta rhythm, 280
β-globin locus, 149
β-lactoglobulin, 103
bias, 384
big conductance calcium-activated potassium (BK) channels, 429
binary phase-shift keying, 37
binding protein, 143
bioactive parameters of EMFs, 209
bioactive region, 464, 465
bioactivity, 208, 210
bioactivity condition, 462–464
bioactivity constant, 462
bioactivity diagram, 464
bioactivity of electromagnetic field, 3, 4, 36
bioactivity of polarized/coherent EMFs, 465
biochemical mechanism, 301
biochemical mechanism of action, 8
bioelectromagnetics, 7
biogenic amines, 324
biogenic magnetite, 340, 414, 453
biological alterations, 313
biological damage, 451
biological effects, 85, 141, 372, 450
biological evolution, 3, 209, 371, 487
biological/health effects, 5, 9
biological response, 427
biomarkers, 312, 318, 322

Index

biophysical mechanism, 280, 290, 429, 451, 466
biophysical mechanism of action, 8, 301
biophysical model, 336
birch trees, 424
bird embryos, 406, 428
bird reproduction, 406
birds, 80, 222, 483
birth defects, 232
black body, 45
black body radiation, 54
blood-brain barrier (BBB), 85, 141, 315, 407, 429
blood-forming organs, 375
blood leukocytes
blood plasma, 153
blood pressure variability, 333
blood sample, 143, 202
blood-testis barrier, 240
blood toxic substances, 315
blood viscosity, 336
Bluetooth, 3, 38, 39, 80, 211, 464, 483
Bluetooth connection, 21
Bluetooth wireless connections, 395
body temperature, 323
body weight, 381
Bohr, 26, 52
bone fracture healing, 22, 496
bone marrow, 147, 149, 153
bone regeneration, 22
bone restoring surgery, 344
Born, 26, 57
bound charged microparticles, 61
bound microparticles, 52
bound state, 52
brain, 323
brain activity, 4, 85, 285, 496
brain angioscan, 325
brain blood flow (BBF) resistance, 325
brain blood flow (BBF) velocity, 325
brain cancers, 82, 83, 141, 169, 375, 381, 484
brain capillary endothelial cells, 315
brain cells, 152, 153
brain-computer interface (BCI), 279
brain computerized tomography (CT), 325
brain cortex, 149
brain damage, 401
brain electrical activity, 4, 400, 429
brain function, 248
brain hypoxia, 330
brain magnetic resonance imaging (MRI), 325
brain malignancies, 224
brain mast cells, 330
brain neural cells, 487
brain neuroinflammation, 331
brain neurological disorder, 323
brain neurological syndrome, 330
brain neurologic pathological disorder, 346
brain neurons, 407
brain pathological alterations, 330
brain-related neurologic disorder, 324
brain response, 336
brain tumors, 81, 82, 122, 141, 377, 484
brain vascular insufficiency, 335
brain vascularization, 336
brain waves, 139, 278
breaks, 198, 203; *see also* chromatid breaks
breast cancer, 82, 122, 376, 379
breast tumors, 380
broadband RF meter, 408
bromic acid, 103
buccal cells, 379
buccal mucosa cells, 144, 145
building sickness syndrome, 303
burning sensation, 305

C

Ca^{2+} pumps, 490
cabin crew members, 376
caffeic acid phenethyl ester, 119
caffeine, 143, 195, 197, 202, 203, 206, 211
caffeine dose, 206
caffeine single dose limit, 208
calcium channel blockers, 487, 491
calcium-dependent signaling cascades, 102
calcium influx, 489
calcium ion channels, 466, 489
calcium ions, 453
calcium signaling, 247, 488
calmodulin, 453, 489
calorimetry law, 48
cancer, 4, 7, 99, 141, 171, 191, 347, 372, 374, 375, 380–382, 428, 450, 484, 489, 494, 495
cancer development, 375, 379, 384, 385, 486
cancer epidemiology, 80
cancer in adults, 374
cancer incidence, 379
cancer incidence rates, 375
cancer mortality rate, 379
cancer of esophagus and stomach, 375
cancer of the central nervous system, 122
cancer risk, 376
capacitive coupling, 50
capacitor, 196
capillary BBF, 327
capsulo-thalamic area, 327
carcinogenesis, 378, 404, 484, 485, 496
carcinogenesis markers, 381, 384
carcinogenic effects, 122, 370, 377, 379, 384
carcinogenic effects of ELF and WC EMFs, 383
carcinogenicity, 371
carcinogenic potential, 381
carcinogenic risk, 241
cardiovascular dysfunction, 304
cardiovascular effects, 140
cardiovascular symptoms, 328, 330
carps, 412
carrier frequency, 41, 100, 225, 373
carrier RF signal, 398
carrier signals, 299
carrier waves, 21, 139, 280
case-control, 88
case-control studies, 81
caspase activation, 110
caspases, 147, 151, 152
catalase (CAT), 99, 142, 148, 150, 153, 171, 254, 318, 412
catalase activity, 144, 145, 241, 379
cataract, 410
catecholamines, 288

catfish, 429
cation channels, 457
cations, 247
cattle, 400, 404
causality, 55, 56
causality criteria, 331
cavity resonator, 58
CDMA, see Code Division Multiple Access
cell cycle arrest, 152, 378
cell damage, 402
cell death, 4, 6, 110, 141, 147, 150, 171, 223, 415, 416, 482, 483, 485, 494
cell division, 152
cell division cycle, 192, 194, 207, 486
cell electrochemical balance, 290, 336
cell growth, 122
cell membrane, 5, 192, 458
cell membrane electric field, 28, 61
cell membrane permeability, 141
cell phone, 80, 373
cell (digital mobile) phone, 80
cell phone radiation, 377
cell plasma membrane cell membrane, 452
cell proliferation, 148, 489
cell proliferation rate, 152
cell senescence, 171, 482, 485, 494
cell towers, 88, 89
cellular effects, 169, 209
cellular metabolism, 102
cellular OS, 223
cellular signaling, 120
cellular stress, 312
cell volume, 456
central bearded dragon, 413
Central Brain Tumor Registry of the U.S., 82
central nervous system (CNS), 281, 311, 315, 323, 400
central nervous system (CNS) autonomic effects, 304
central nervous system (CNS) tumors, 374
centripetal force, 452
cerebellum, 153
cerebral atrophy, 345
cerebral BBF, 327
cerebral biomarkers, 330
cerebral blood flow, 284, 289
cerebral cortex, 323
cerebral cortical activity, 280
cerebral imaging, 330
cerebral neurotransmitters, 330
cerebral vascular hypoperfusion, 330
cerebrospinal fluid, 323
cervical mucus observation, 229
cervix cancer, 379
cetaceans, 402
changes in blood chemistry, 403
channel conductivity, 468
channel length, 471
Channel pore, 457
channel pore diameter, 457
channel selectivity, 457
channel sensors of VGICs, 454
chaperone functions, 312
characteristic equation, 472
charged macromolecules, 451
charged/polar macromolecules, 486

checkpoint kinase protein, 151
checkpoints, 153, 200, 205, 486
chelation, 344
chemical bond, 486
chemical exposure, 315
chemical tolerance threshold, 335
Chernobyl nuclear accident, 380
Cherry Neil, 455
chick embryos, 149
childhood cancer, 374
childhood leukemia, 84, 122, 371, 375, 385, 416, 484
child leukaemia, 396
Chinese hamster lung cells, 145, 152
chlorophyll, 423
chloroplasts, 421, 423
choline, 400
choriogenesis, 193
choroid plexus, 153
chromatid aberrations, 151
chromatid breaks, 143, 144, 147, 198, 202, 206
chromatid breaks and gaps, 145
chromatid gaps, 143, 144, 198, 202, 206
chromatid-type aberrations, 143, 195, 198, 202, 203, 206
chromatid width, 198
chromatin alterations, 343
chromatin condensation, 143, 150, 193
chromosomal aberrations, 142, 146, 202, 421
chromosomal damage, 198, 202, 207, 210
chromosomal instability, 343
chromosomal sensitivity, 197
chromosomal translocations, 149
chromosome aberrations, 143, 148, 151
chromosome breaks, 147, 151
chromosome damage, 482
chromosome exchanges, 143
chromosome gaps, 151
chromosomes, 119
chronic fatigue, 335
chronic fatigue syndrome, 303
chronic health effects, 86
chronic myelocytic leukemia, 375
chronic OS, 258
chronic (EHS) symptoms, 335
circadian rhythmicity, 289
circadian rhythms, 411, 429, 496
circadian system, 402
circular frequency, 26, 30
circularly polarized fields/waves, 27
circulatory system, 247
classical electromagnetism, 26, 53
clinical syndrome, 311
CNS, see central nervous system
CNS-related tolerance threshold, 345
co-carcinogenicity, 83
cocarcinogenic potential, 381
cockpit crew members, 375
Code Division Multiple Access (CDMA), 39, 100, 140, 373
coenzyme Q10 (CoQ10), 318
coffee consumption, 206
coffee drinking, 210
cognition, 85, 323
cognitive decline, 315
cognitive deficiency, 305, 306
cognitive functions, 85, 140, 279

cognitive performance, 336
cognitive processes, 279, 285
coherence, 23, 208
coherent electromagnetic field, 1, 9
coherent polarized fields/waves, 29
coherent superimposed waves/fields, 34
cohorts of population, 227
cohort studies, 82, 88
cohort study, 81
coils, 196
colcemid, 198
collective intrinsic modes, 103
colony collapse disorder (CCD), 419
colony population, 415
colorectal cancer, 375
combination of polarization and intense variability, 395
combination of polarization and variability, 36
combined exposure, 211
combined mobile phone and Wi-Fi radiation, 403
comet assay, 119, 144, 148–150, 152, 239, 241, 379, 382
commercial airline pilots cancer incidence, 375
commercially available devices, 398
commutation relations, 58
compass run, 402
compensatory mechanism, 236
complex II of the ETC, 256
complex III of the ETC, 243
complex EMF, 21
complexity of the signals, 210
complex RF signal, 24
complex RF/WC signal, 465
complex term, 57
complex WC EMFs, 24
Compton effect, 62
computed tomography (CT), 279
concentration deficiency, 310
concentration loss, 304, 305
concentration problems, 140
concentration voltage gradient, 456
Confidence Interval (CI), 81
conflicts of interest, 7, 142, 289, 398
conjugate, 57
conjugated dienes, 251
constructive interference, 32
constructive or destructive interference, 29
continuous "classical" waves, 67
continuous ELF electric field exposure, 333
continuous EMF exposure, 306
continuous emission, 53, 54
continuous emission spectra, 54
continuous exposure, 145
continuous fields, 465
continuous (non-pulsed) fields, 460
continuous-wave EMFs, 287
continuous-wave (non-pulsed) fields, 451
continuous-wave microwaves, 380
continuous-wave radiation, 422
continuous waves, 6, 25, 51, 53, 60, 62, 146
contralateral use, 81, 141
coordinated motion (of electrically charged particles), 458
cordless domestic phones, 3, 25, 37, 80, 81, 98, 280, 374, 376, 395, 483
cordless phones, 122
cordless phone use, 378

corneal opacity, 409
cornea of rats, 410
corn growth, 421
corpus luteum, 237
correlation, 379
cortex, 153
cortical neurons, 146
corticosterone, 288
cortisol, 400
cosmic expansion, 53, 487
cosmic ionizing radiation, 375
cosmic microwave radiation, 53
cosmic microwaves, 2, 3, 26, 395, 487
co-stress condition, 207
Coulomb forces, 458
Coulomb law, 460
counter charges, 457, 459
covalent bond, 485, 492
COVID-19 viral infection, 316
cows, 147
creatinine, 313, 316
crop production, 416, 417
cross-sectional study, 88, 286
cryogenics, 59
cryptochroms, 340
cumulative effects, 236, 287, 288, 428
curculionidae, 409
cutaneous cancer, 378
cutaneous lesions, 302, 310
cyclin, 151
cyclotron, 452
cystoblast, 193
cytochrome c, 110, 257
cytochrome oxidase, 103
cytokines, 152
cytoplasmic vacuolization, 149
cytoskeleton, 103
cytotoxic agent, 200
cytotoxic chemicals, 193, 200, 205, 211, 485

D

damping, 452, 453
damping coefficient, 45, 454, 459, 471
damping force, 459
death rate, 375
De Broglie, L, 55
decline in amphibian populations, 414
decline in bird and insect populations, 427
decline in insect populations, 415
decline of house sparrows, 409
declines in animal populations, 397
declines in bird and insect populations, 483
decreased germination, 424
decreased growth, 424
DECT, see Digitally Enhanced Cordless Telecommunications
DECT phones, 38, 39, 80, 140
DECT radiation, 417
default mode network (DMN), 327
default mode network hyper-connectivity, 327
dehydration, 193, 200, 205, 211
delayed effects, 288
delta band, 283
delta waves, 278, 279

dementia, 315
demodulation, 64
demodulation of ELF components, 465
dendrites, 276
dental amalgam fillings, 344
deoxyribose, 485, 492
depression, 99, 121, 140, 345, 379
depression tendency, 310
depressive tendency, 305, 328, 335
destructive interference, 33
development, 413, 416
developmental capacity, 141
developmental stages, 201, 205
developmental toxicity, 232
diagnostic criteria, 300
dialed mode, 146
dialing mode, 146
dicentric chromosomes, 144, 147, 151
dichloro-diphenyl-trichloroethane (DDT), 80
dielectric constant, 460
dielectric loss factor, 46
diethyl ether, 195
differences between natural and man-made EMFs/EMR, 60, 67
differential equation, 459, 471
differentiation, 122, 413, 489
Digitally Enhanced Cordless Telecommunications (DECT), 37, 80, 98, 374, 395
digital modulation, 37
digital WC, 21
Dirac, PAM, 26, 56
direct electric current, 1
direct ELF electric fields, 465
directional beam forming, 100
directional beams, 396
Dirichlet conditions, 55
Discontinuous Transmission Mode (DTX), 38, 42
discriminant, 472
disease, 209
disease biomarkers, 311–313
disrupted hormone signaling, 222
dizziness, 305, 310
DNA base damage, 146
DNA bases, 491
DNA breaks, 249
DNA damage, 4, 7, 83, 84, 122, 143–146, 148–153, 172, 191, 207, 210, 223, 239, 241, 301, 381, 382, 402–404, 421, 428, 430, 482–485, 492
DNA damage in the brain cells, 233
DNA double-strand breaks (DNA DSBs), 145, 147, 149–153, 382
DNA fragmentation, 103, 119, 144–146, 149–151, 153, 193, 196, 198–200, 205, 206, 227, 232, 236, 241, 244, 415, 416
DNA integrity, 241
DNA inversion, 149
DNA methylation, 152
DNA oxidation, 241
DNA oxidation and fragmentation, 256
DNA oxidative damage, 120, 244
DNA polymerase, 153
DNA-protein cross-links, 152
DNA repair, 171, 236
DNA repair inhibition, 343

DNA repair mechanisms, 414, 485
DNA repair pathways, 152
DNA replication, 313
DNase, 489
DNA single-strand breaks (DNA SSBs), 143–145, 147, 150–152, 382, 491
DNA strand breaks, 146, 148, 151, 152, 341
DNA synthesis, 153
DNA synthesis rate, 6
domestic cordless phones, 139
domestic telephones, 211
dopamine (DA), 323
Doppler effect, 53
dose-dependent, 207
dosimeters, 384
double strand breaks, 195
driverless vehicles, 87
Drosophila, 153
Drosophila melanogaster, 119, 150, 192, 193, 236, 415, 416
Drosophila oogenesis, 200
DTX, *see* Discontinuous Transmission Mode
duckweed, 423
duration of mobile phone use, 226
duration of use, 81
dysesthesia, 304, 310
dysfunction/irregular gating of VGICs, 107, 465
dysfunction of VGICs, 429, 487
dysregulated meiosis, 223
dysthermia, 310
dystrophic abnormaities, 402

E

ear canal, 144
ear heat, 310
early oogenesis, 193
early oogenesis checkpoint, 193
ear problems, 302
Earth, 9
earthing, 344
earthquakes, 395
earth's magnetic field, 399
ECG, 336
ecological consequences, 426
ecosystems, 99, 415, 427
eectroencephalography (EEG), 276
EEG alterations, 336
EEG changes, 283, 284
EEG power, 283
EEG rhythms, 278
EEG signals, 278
effective charge, 457
effective charge of S4, 460
effects of anthropogenic EMFs on birds, 404
egg chamber cells, 153, 201, 205
egg chambers, 193, 199, 236
egg laying, 418, 483
egg-productivity in nests, 405
eggs, 240
egg-shell, 193
EHS, *see* electro-hypersensitivity
EHS as a brain disorder, 324
EHS-associated symptom occurrence, 331
EHS biomarkers, 329

EHS causal origin, 329, 331
EHS causation, 341
EHS cause, 345
EHS clinical diagnostic criteria, 328
EHS clinical symptoms, 309
EHS definition, 345
EHS diagnosis, 328, 329
EHS diagnostic criteria, 347
EHS etiology, 331
EHS female predominance, 308
EHS gender-related genetic susceptibility, 306
EHS genesis, 331, 343, 347
EHS genetically susceptible people, 332
EHS intolerable symptoms, 332
EHS/MCS combined syndrome, 310
EHS occurrence percentages, 345
EHS pathogenesis and etiology, 329
EHS pathophysiological mechanism, 336
EHS patients, 318, 323
EHS patients tolerance threshold, 336
EHS persisting symptoms, 332
EHS provocation studies, 334
EHS self-reported, 306, 309
EHS self-reported patients, 302, 303, 306
EHS symptomatic picture studies, 306
EHS temporary symptoms, 332
EHS treatment, 343
8-hydroxy-2'-deoxyguanosine (8-OH-dG), 99, 120, 145, 243, 382
8-OH-dG DNA adducts, 147–150, 152, 153
8-OH-dG formation, 384
Einstein, 26
elastic wave, 30
electrical and electronic circuits, 371
electric charge, 1
electric component of the ELF MT/WC pulsations, 464
electric current density, 47
electric devices, 140
electric/electronic circuit, 20
electric field, 99
electric force, 462
electric power transmission lines, 140, 383
electric properties of human blood, 336
electric skin potential (ESP), 333
electrochemical balance, 66, 102, 456–458, 487, 488, 493
electrochemical balance in a cell, 454
electrochemical homeostasis, 171
electrochemical imbalance, 470
electroencephalogram (EEG), 333
electroencephalogram (EEG) signal, 277
electro-hypersensitivity (EHS), 4, 86, 121, 140, 300, 304, 305, 315, 318, 322, 332, 335, 341, 379, 450, 484
electromagnetic beam, 31
electromagnetic biology, 102
electromagnetic detectors, 467
electromagnetic fields (EMFs), 1, 19, 80, 139, 221
electromagnetic noise, 411
electromagnetic oscillation circuit, 20
electromagnetic pollution, 347, 395, 408, 409
electromagnetic radiation (EMR), 1, 19, 80, 98, 139, 221, 299, 370, 394
electromagnetic sensors, 192, 453
electromagnetic spectrum, 2, 370, 371
electromagnetic wave, 1, 30

electromagnetic wave velocity, 20
electromyogram (EMG), 333
electron beam, 58
electron flux, 490
electronic matter-waves, 55
electronic spin recombination, 453
electron leakage from complex III of the ETC, 249
electron magnetic moment, 59
electron paramagnetic resonance (EPR) spin-trapping, 254
electrons, 61
electron spin recombination, 103
electron spin resonance (ESR), 53, 59
electron spin resonance (ESR) effect, 453
electron spins, 488
electron transfer, 103
electron transport chain (ETC), 108, 252, 253, 404, 489, 490
electron transport chain (ETC) inhibitor, 243
electrophilic aldehyde formation, 223
electrosensitive channels, 457
electro-sensitive ion channels, 458, 467, 488
electrosmog, 411
elementary charge, 457, 459
elementary particles, 52
ELF, see Extremely Low Frequency
ELF components, 465
ELF electric fields, 466
ELF (60 Hz) electric fields, 84
ELF EMF non-thermal exposure, 381
ELF EMFs, 99, 104, 107, 120, 121, 153–168, 232, 305, 371, 374, 382, 383, 416
ELF EMFs carcinogenicity, 84
ELF magnetic field, 151
ELF (60 Hz) magnetic fields, 84, 374
ELF modulation, 422
ELF modulation/pulsation
ELF pulsation/modulation, 287, 290
ELF pulsations, 23, 284, 285, 429
ELF pulsations, modulations and variability components, 240
ELF pulses, 282
ELF pulsing/modulation, 289
ELF pulsing/modulation components, 169
ELF/ULF component, 4
elliptically polarized fields/waves, 27
embryo development, 223, 231, 232
embryonic cells, 121, 383
embryonic death, 6, 406, 483
embryonic development, 121, 233, 407, 413, 429
embryonic stem cells, 239
embryos, 483
EMF-bioactivity, 211, 461
EMF bioeffects, 142
EMF-bioeffects literature, 372
EMF-dosimetry, 50
EMF emission pattern, 37
EMF exposure, 300, 315
EMF-exposure avoidance, 343, 344
EMF exposure limits, 208
EMF-exposure provocation studies, 303
EMF-induced brain injury, 323
EMF-induced developmental disruption, 233
EMF-induced neurotransmitter changes, 324
EMF intensity, 306, 323

EMF Portal, 88
EMF public exposure safety limits, 371
EMF quantization, 58
EMF-receptor organs, 453, 467
EMF-related immune deficiency, 316
EMF-sensors, 336
EMF-shielding, 344
EMF-tolerance threshold, 335
emission spectra, 52
emotional disorder, 84
emotivity, 310
endangered species, 411
endogenous electrical activity, 496
endogenous electric currents, 4, 35, 413, 429
endogenous electric currents in cells and tissues, 22
endogenous electric fields, 413, 429
endogenous electromagnetic signals, 422
endogenous physiological electric fields, 342
endogenous physiological electromagnetic field, 4
endogenous signals, 422
endometrial degeneration, 237
endometrial fibroids, 229
endometrial tissue, 229
endometriosis, 229
endometritis, 255
endometrium, 254
endoplasmic reticulum, 192
endothelial cells, 255
energy absorption, 84
energy conservation law, 56
energy density, 30, 62
energy loss, 62
energy metabolism, 110
environment, 8
environmental/health crisis, 431
environmental impact, 426
environmental intolerance, 303
environmental intolerance-related disorder, 304
environmental intolerance-related illnesses, 311
environmental intolerance-related neurological syndrome, 347
environmental pollution, 140, 397
environmental sensitivity-related diseases, 303
environmental stressor brain effect model, 330
environmental stressors, 210, 221, 370, 496
enzymatic biomarkers, 416
enzyme activities, 257
eosinophils, 254
epicotyl length, 422
epidemiological study, 122, 141, 384, 484
epididymal sperm, 149
epididymal spermatozoa, 146, 239
epididymis, 231, 240, 242
epigenetic alterations, 341
epigenetic effects, 404, 486
epigenetic mechanisms, 343
epilepsy, 286
epileptic patients, 286, 287
epithelial, 193
epithelial cells, 153, 193
epitope, 313
erectile dysfunction, 230
esophagus cancer, 379
ESR, *see* electron spin resonance

estrogen levels, 255
estrous cycle, 236, 238
estrus, 141
ETC, *see* electron transport chain
ETC complex I, 108
ETC complex II, 108
ETC complex III, 108
ethanol oxidation, 120
etoposide, 193, 200
Euler formula, 55
European Parliamentary Research Service (EPRS), 64
European Parliament Resolution, April 2, 2009, 371
European Union, 430
event-related desynchronization (ERD), 285
event-related synchronization (ERS), 285
evoked electrical potentials, 336
evoked potential, 103
excitation, 102
excitation/de-excitation, 27, 28, 61
experimental rats, 380
exposure assessment, 88
exposure chamber, 7
exposure criteria, 427
exposure duration, 6, 20, 142–145, 147, 194, 208, 227, 236, 284, 306, 323
exposure guidelines, 5
exposure limit for WC EMFs, 7
exposure limits, 87, 101, 210, 228, 398
exposure time, 145, 151
extracellular current, 278
Extremely Low Frequency (ELF), 2, 3, 21, 80, 139, 190, 234, 235, 280, 299, 395
Extremely Low Frequency (ELF) EMFs, 98, 371
Extremely Low Frequency (ELF) modulation and pulsation, 221
eye cataract, 305
eye-closed condition, 285
eye-closed waking state, 279
eye-open condition, 285
eye problems, 304

F

Facebook, 172
Faraday cage, 226, 345
Faraday constant, 456
far-field, 192
fatigue, 86, 99, 121, 140, 304, 310, 379, 484
fecundity, 150, 194, 222, 416
female fertility, 222
female hormone production, 231
female infertility, 229
female reproductive capacity, 236
Fenton reaction, 103, 120, 383
fermented papaya preparation (FPP), 344
ferritin, 103
ferrozine, 103
fertility, 205, 222, 402
fertility decline, 222
fertility drugs, 231
fertilization, 149
fertilization potential, 231
fertilized eggs, 231, 413
fetal anomalies, 232

Index

fetal growth, 141, 233
fetal organogenesis, 232
fetal period, 153, 315
Feynman, R.P., 26
fibroblasts, 103
fibromyalgia, 303
field intensity, 29, 32, 451
field/radiation intensity, 208
field strength, 151
Fifth Generation (5G), 3, 4, 7, 24, 34, 39, 45, 46, 64–66, 86, 90, 100, 139, 169, 206, 207, 225, 227, 281, 345, 370, 374, 395, 399, 429, 430, 451
50 Hz alternating EMF, 200
50 Hz alternating MF, 196
50 Hz MFs, 205
financial support, 384
Finite Difference Time Domain, (FDTD), 49
First Generation (1G), 3, 140, 281, 395
Fischer rats, 148
5G New Radio (NR), 191
5-hydoxytryptamine (5-HT), 323
fluorescein dUTP, 196
fluorescence, 201
fluorescence microscopy, 196
foci method, 195
follicle apoptosis, 254
follicle cells, 150, 193, 200, 236
follicle stimulating hormone (FSH), 141, 238
follicular lymphoma, 381
Food and Drug Administration (FDA), 311
food chain, 415
foraging, 416
forced-oscillation amplitude, 458
forced-oscillation equation, 459
forced-oscillation of an ion, 453
forced-oscillation of charged/polar particles, 32
forced-oscillation of ions, 467
forced oscillation of mobile ions, 290
forced-oscillation/rotation, 43
forced-oscillations, 23, 29, 49, 61, 62, 493
forced-vibration of ions, 455
forced-vibration/oscillation of free ions, 102
force on each S4 voltage sensor, 474
force on effective charge, 460
force on S4 sensor, 460
force on the voltage-sensors, 473
forest plants, 426
formaldehyde, 84
Fourier amplitudes, 58
Fourier integral/transform, 55
Fourier series, 26, 55, 57
Fourier theorem, 57
Fourier theory, 55
Fourth Generation (4G), 3, 36, 37, 39, 100, 140, 281, 374, 395
fractal antenna, 487
frame repetition, 39
frames, 39, 139
free electron acceleration, 20
free electrons, 20, 27, 53, 61, 139
free radical lifetimes, 103, 453
free radical pair, 103
free radical production, 247, 248
free radical recombination, 453

free radicals, 5, 99, 152, 470, 488
free-radical theory of ageing, 250
frequency, 1, 20, 30, 100
frequency modulation (FM), 21, 37
frequency-shift keying, 37
friction of oscillating molecules, 45
frogs, 413
frog tadpoles, 413
Fröhlich, J, 469
fruit fly, 150, 192, 193, 200, 415, 416, 483
fruit fly oogenesis, 205
fruit fly ovarian cells, 103, 193
fruit fly reproduction, 236
fruit fly studies, 194
functional MRI (fMRI), 327
functional symptoms, 332
funding, 84, 142

G

G0 phase, 152, 194
G1 phase, 152
G2 assay, 195
G2 checkpoint, 195
G2 phase, 143, 192, 195
gallic acid, 119
gametes, 240
gamma band, 283
gamma radiation, 3, 84, 247
gamma rays, 25
gamma rhythm, 280
γ-Aminobutyric Acid (GABA), 401
γH2AX foci, 143, 151, 152
γH2AX foci formation, 145
gaps, 198, 203; *see also* chromatid gaps
garlic, 119
gas constant, 456
gastric cancer cells, 152
gastrointestinal tract, 323
Gauss, (G), 100
Gaussian Minimum Shift Keying (GMSK), 37, 39
gene activation, 103
gene expression, 85, 152, 343, 486
gene expression changes, 315
gene mutation, 486
general public exposure, 101
general solution, 472, 473
generators, 169
genes polymorphism, 145
genetic damage, 9, 99, 145, 202, 343, 406, 450, 482–484
genetic polymorphism, 324
genetic predisposition, 325
genomic DNA, 192
genomic instability, 486
genotoxic, 201
genotoxic assault, 227
genotoxic effects, 141, 154–169, 171, 191, 241, 428
genotoxicity, 141, 169, 417
genotoxicity mechanism of man-made EMFs, 171
genotyping of drug metabolism-related enzymes, 324
geoelectric, 61
geoelectric field, 3, 22, 28, 34, 418
geomagnetic, 61
geomagnetic field, 3, 22, 28, 34, 400, 410, 418, 421, 452, 467

geomagnetic field axis, 411
geomagnetic field intensity, 415
geomagnetic field lines, 412
geomagnetic pulsations, 34, 209
germarium, 193, 200
germarium checkpoint, 193
germ cell development, 242
germ cells, 99, 193, 222, 255, 258
germinal epithelium, 240
germination, 426
germination rate, 419, 422
germline cyst, 193
germline perturbations, 223
gestation, 148, 149
Giemsa solution, 198
gigahertz transverse electro-magnetic (GTEM) chamber, 403
ginkgo biloba, 344
glial tumors, 83
glioblastoma, 82, 83, 378
glioblastoma multiforme (GBM), 141, 377
glioma, 81, 83, 371, 377, 381, 404, 484
global positioning system (GPS) devices, 410
Global System for Mobile Telecommunication (GSM), 36–38, 42, 100, 140, 149, 191, 198, 373, 395, 404
glucose-6-phosphate dehydrogenase, 148
glucose metabolism activity, 336
glutathione (GSH), 148, 149, 151, 171, 318, 344
glutathione concentration, 379
glutathione peroxidase (GPx), 99, 142, 318
glutathione reductase (GR), 318
glutathione-*S*-transferase (GST), 142, 148, 171, 318
GMSK, *see* Gaussian Minimum Shift Keying
gonadotropins, 248
GPS transmitters, 410
GPx, 148, 153, 171, 318
GPx activity, 254, 255
Graafian follicles, 254
green tea, 119
grounding, 344
growth factor-induced signaling, 120
growth rate, 84
GSH, *see* glutathione
GSH concentration, 144
GSM, *see* Global System for Mobile Telecommunication
GSM "basic," 145
GSM "talk," 145
GSSG, *see* oxidized glutathione
GST, *see* glutathione-*S*-transferase
guanine, 491
Guglielmo, Marconi, 138
guppy fish, 412

H

Haber-Weiss reaction, 120, 383, 492
habitat deterioration, 397
hair follicle cells, 84, 144
hair root cells, 144
Hamiltonian, 58
hand-held radar, 376
hand rigidity, 140
Hardell, Lennart, 377
harmonic oscillators, 59

harmonic plane wave, 29
harmonic wave, 57
harmonic wave function, 56, 67
Hazard Ratio (HR), 374
headache, 86, 99, 121, 122, 140, 302, 304–306, 310, 379, 484
health burden, 222
health deterioration, 397
health effects, 99, 121, 141, 210, 300, 301, 450
health impacts, 257
health of the offspring, 257
health problems, 211, 484
health risks, 207
heart activity, 4
heart cancer, 484
heart rate, 429
heart rate variability (HRV), 330, 333, 335
heart rhythm, 413
heat, 193
heating, 45, 62, 141, 200
heat shock proteins (HSPs), 153, 233, 244, 312, 315, 343, 416
heat stress, 245
heavy users of mobile phones, 377
Heisenberg, 26, 56
HeLa cells, 151
hematolymphatic (HL) cancers, 376
hematolymphopoietic cancers, 374
hemizona assay, 246
hemoglobin, 321
hemoglobin deoxygenation, 336
hepatic tissue destruction, 232
hepatocellular carcinoma, 380
hepatocytes, 149
heritable mutations, 205, 428
Herz, Heinrich, 20
heterochromatin, 145
high-sensitivity C Reactive Protein (hs-CRP), 311, 313
high-voltage power lines, 100, 122, 198, 200, 211, 416, 463
high-voltage power transmission lines, 396
high-voltage transformers, 484
high voltage transmission lines, 374
hippocampal neurons, 149
hippocampus, 147, 149, 153, 278, 323, 341, 400
Hippocrates, 332
Hiroshima nuclear bomb, 379
histamine, 122, 311, 315, 330
histamine release, 330
histamine release in peripheral blood, 330
histological changes in testes, 402
histone kinase, 241
homeostasis, 488, 493
homogeneous differential equation, 472
honeybee, 415
honeybee colonies, 419
hormesis effect, 121
hormonal control for reproductive function, 248
hormonal disorders, 99, 140
hormonal imbalances, 122, 429
hormone regulation
house fly, 415
house sparrow, 407
HPBLs, *see* human peripheral blood lymphocytes
hs-CRP, *see* high-sensitivity C Reactive Protein

Index

human amniocytes, 145
human amniotic cells, 151
human amniotic epithelial cells, 145
human body infrared radiation, 23
human buccal epithelial cells, 145
human carcinogen, 80
human cells, 210
human Chorionic Gonadotropin (hCG) levels, 231
human diploid skin fibroblasts, 145, 151
human fertility, 228
human fibroblasts, 151, 239
human infertility, 224
human lens epithelial cells, 145
human lung fibroblasts, 151
human lymphocytes, 151, 202
human-made EMFs, 23
human melanocytes, 151
human monocytes, 151
human peripheral blood lymphocytes (HPBLs), 142, 150, 194, 202, 206
human primary fibroblasts, 151
human reproductive system, 223
human skeletal muscle cells, 151
human skin fibroblasts, 145
human sperm, 85, 402, 483
human spermatozoa, 108, 120, 145, 239, 244, 382
human technology, 371
human tissue density, 48
human trophoblast cells, 23
human umbilical vein endothelial cells, 145
Huygens principle, 29
hydrogen peroxide, 142, 251, 383, 491
hydroperoxides, 251
hydroxyl free radical, 102, 171
hydroxyl radical, 120, 142, 383, 484, 488, 491, 494
hyperactivity, 85, 233, 309
hyperacusis, 310
hyperdiploid cells, 144
hyperfine splitting, 59
hypocotyl length, 422
hypodiploid cells, 144
hypoperfusion, 316
hypothalamus, 248, 323
hypoxia up-regulated protein 1 precursor (HYOU1), 241
hysterosalpingogram (HSG), 229

I

IARC, *see* International Agency for Research on Cancer
ICNIRP, *see* International Commission on Non-Ionizing Radiation Protection
ICNIRP exposure limits, 5
ICNIRP guidelines, 228, 373, 398
ICNIRP limits, 207, 210, 372, 423
ICNIRP recommendations, 373
ICR frequency, 452
idiopathic environmental intolerance (IEI), 301, 303, 335
idiopathic infertility, 222
IEI, *see* idiopathic environmental intolerance
IFO-VGIC mechanism, 458, 465–467, 488, 497
IgE, *see* immunoglobulin E
imaginary unit, 55
imaging techniques, 323, 346
immediate memory loss, 304, 305

immune function, 87
immune system, 4, 141, 316, 401, 407, 409, 411, 414, 428, 487, 489
immunity, 496
immunoglobulin E, 122, 311
immunoglobulin G, 313
immunosuppression, 343
impacts on reproduction and development, 232
impacts on reproductive systems, 227
impaired immunity, 429
impaired memory, 233
impedance spectroscopy, 336
incident EMF, 66
incident EMR, 225
incident field, 51
incident radiation, 101
incident radiation/field, 50
incident radiation/field intensity, 67
incident radiation intensity, 248
increase in cancer cases, 379
independent studies, 384, 385
indirect non-thermal mechanisms, 247
industrial workers, 374
industry, 377
industry supported studies, 385
infectious diseases, 409
infertility, 4, 9, 85, 228, 248, 428, 450, 483, 485, 494
inflammation, 121, 300, 311
inflammatory amplification, 330
inflammatory cells, 331
inflammatory diseases, 489
inflammatory mediator, 322, 330
inflammatory response, 330
inflorescence, 423
information-carrying electromagnetic waves, 21
information carrying EMFs, 36
information signal, 21
infrared, 2
infrared band, 45
infrared light, 371
initial conditions, 472
initial ROS, 491
inner mitochondrial membrane, 252
insect decline, 415, 418
insecticide, 80
insects, 80, 222
insomnia, 86, 140, 304, 335
inspiration/expiration ratio, 336
Institute of Electrical and Electronic Engineers (IEEE), 87
insulin, 103
insulin signaling, 120
intensification of genotoxic effects, 171
intensity, 100
intensity measurement, 50
intensity of radiation/field, 226
intensity of the incident radiation, 372
intensity or frequency "windows," 51
intensity variations, 3
intensity windows, 468
interaction of EMFs/EMR with matter, 61
intercellular tight junctions, 315
interictal EEG spiking activity, 286
interleukin, 148
intermembrane space, 489

intermittence of EMF exposure, 25
intermittence of exposure, 5
intermittent exposure, 145, 146, 150, 151, 236
internal desynchronization, 344
internal electric field, 47
International Agency for Research on Cancer (IARC), 4, 24, 80, 84, 191, 299, 371, 485
International Commission on Non-Ionizing Radiation Protection (ICNIRP), 5, 24, 80, 89, 101, 139, 191, 280, 289, 300, 372, 384, 398, 427, 484
International Journal of Oncology, 377
internet, 138, 172, 211, 395
Internet of Things (IoT), 7, 87, 396, 399
internet wireless connections, 395
Interphone study, 81, 377
interstitial tissue, 248
intolerance to man-made EMFs, 336
intracellular calcium, 466, 489
intracellular ionic concentrations, 22, 35, 171, 488, 493
intracellular organelles, 192
intracytoplasmic sperm injection (ICSI), 231
intrauterine insemination, 231
intrinsic oscillatory electrical activity, 278
in utero exposure, 233
invariable signals, 169
in vitro fertilization (IVF), 231
ion and protein transportation, 248
ion-binding sites
ion channel blockers, 285, 488
ion channel gating, 209
ion channels, 44, 488, 490
ion concentrations, 429, 430
ion currents, 487
ion cyclotron resonance (ICR), 452
ion diameter, 457
ion diffusion, 455
ion displacement, 459, 461
ion flows, 413
Ion fluxes, 456
ion forced-oscillation (IFO), 414, 458, 488
ion forced-oscillation (IFO) mechanism, 24, 66, 102, 458
ion forced-oscillation (IFO) velocity, 463, 467
ionic channels, 285
ionic clusters, 102
ionic concentrations, 66, 192, 209
ionic electric current, 471
ionic oscillations, 54, 61
ionization, 5, 25, 61, 100, 102, 139, 451, 486
ionizing radiation, 25, 61, 172, 375
ion kinetic energy, 44
ion mass, 459
ion mean free path, 474
ionosphere, 20
ion parametric resonance (IPR) model, 453
ion pumps, 490
ions, 61
ion self-frequency, 472
ion's maximum velocity, 454
ion's self-frequency, 454
ion transitions, 453
ion transport, 452
ion-transport protein, 452
ion valence, 459
ion velocities though opened channels, 65

ion velocity, 44, 459
ion velocity due to the forced-oscillation, 473
ion velocity through a channel, 471
IoT, *see* Internet of Things
ipsilateral phone use, 81
ipsilateral use, 141
iron chelation, 103
irregular gating, 487
irregular gating (of ion channels), 458
irregular gating of electro-sensitive ion channels, 171
irregular gating of VGICs, 487
irritability, 305, 310
irritation, 122
ischemic-hypoxic damage, 330
isochromatid break, 146
isolated tissues, 466

J

Japanese quail embryos, 150
Jordan, 57

K

karyorrhexis, 149
kinetic energy, 65
kinetic energy of forced-oscillation, 43
Kirchhoff's law, 46
Kirchhoff's theorem, 45; *see also* Kirchhoff's law
Klystron, 58
K^+/Na^+ pump, 456
Korean war, 375, 383
Kronecker's delta function, 58

L

labile emotional behavior, 304
lability, 323
laboratory animals, 299
laboratory rats, 120, 383
lactate dehydrogenase (LDH), 148
lactate dehydrogenase isoenzyme, 241
Laplace operator, 56
lasers, 28
latency, 81
latency period, 486
latency time, 83
late oogenesis, 193
L-carnitine, 119
LDH, *see* lactate dehydrogenase
lead, 80, 85
learning, 279
learning and memory, 288, 412
learning and memory capacity, 323
learning and memory in rat, 401
learning and memory performance, 323
lectins, 453
LED, *see* light emitting diode
lens damages, 376
lepidoptera, 409
leukemia, 374, 376, 378
leukocytes, 149, 153
Leydig cells, 225, 248
LF, 2

Index

Liboff, 452
Liburdy, 466
life-cycle, 193
ligand, 453
ligand-gated ion channels, 277
light emitting diode (LED), 119
light interference, 29
light microscopy, 198
lightning, 28, 496
lightning discharges, 28, 196
lightning EMFs, 395
light periodicity, 496
light velocity, 20
linearly polarized field/wave, 27, 28
line spectra, 52
lipid peroxidation (LPO), 99, 104, 142, 144, 145, 147, 148, 150–153, 171, 223, 233, 241, 249, 254, 255, 257, 318, 379, 412
listening mode, 42
liver, 149
liver damage, 403
liver dysfunction, 311
living tissue, 50
lizards, 413
lobster, 411
locally polarized field, 28
locomotor behavior, 412
locomotor function, 401
locomotor problems, 429
locomotor response, 417
locust, 415
long-term EMF exposure, 372
Long Term Evolution (LTE), 37, 39, 42, 100, 140, 191, 374, 395
long-term exposure limit, 208
long-term exposures, 86, 210, 288, 416, 428
long-term irradiation, 380
loss of immediate memory, 310
low grade inflammation, 313
LPO, *see* lipid peroxidation
LTE, *see* Long Term Evolution
lung cancer, 379
lung cancer cells, 152
luteal cells, 254
luteinizing hormone (LH), 229, 238, 248
lymphatic tissues, 375
lymphocytes, 197
lymphoma, 122, 374, 376, 380

M

macrophage cells, 151
macrophages, 152, 254
magnetic alignment, 414
magnetically induced electric field, 464
magnetic field (MF), 99, 194, 198, 381
magnetic force, 452, 462
magnetic induction, 196
magnetic nanoparticles, 8
magnetic permeability, 466
magnetic resonance imaging (MRI), 143, 278, 279
magnetic storms, 22, 34, 35, 209
magnetogenetics, 8
magnetoreception, 399, 412
magnetosomes, 340
magnetron, 58, 424
maize, 423
male and female germ cells, 247
male fertility, 85, 231, 238, 257
male germline, 223
male infertility, 230
male reproductive system, 224, 226, 247
malignant brain tumors, 141
malignant melanoma, 375
malignant schwannoma, 404
malignant transformation, 122
malondialdehyde (MDA), 99, 104, 241, 254, 318
mammals, 399
mammary gland cancer, 381
mammary gland tumors, 380
mammary tumor-prone mice, 380
mammary tumors, 381
Manganese Superoxide Dismutase (Mn-SOD), 145
man-made chemicals, 304, 335
man-made electromagnetic fields, 2, 4, 98, 299, 370
man-made ELF EMF, 122
man-made EMF bioeffects, 241
man-made EMF emissions, 427
man-made EMF exposures, 171, 256, 323, 342, 384
man-made EMF interaction with VGICs, 342
man-made EMFs, 1, 3, 5, 19, 20, 22, 34, 139, 140, 169, 200, 210, 211, 232, 239, 243, 300, 304, 331, 335, 371, 374, 408, 418, 450, 451, 486
man-made EMR, 1, 61
man-made non-thermal EMF exposure, 123
many-electron atom, 55
MAPK, *see* mitogen-activated protein kinase
maple trees, 424
marine environment, 411
marketed environmental chemicals, 300
marrow, 84
maser, 58, 59
mast cell degranulation, 311
mast cells, 122, 305, 315, 330
mastocytes, 330
mathematical "quantization" of EMF/EMR, 56, 67
matrix, 489
matter waves, 55
maturation processes, 250
mature epididymal spermatozoa, 256
mature spermatozoa, 241
mature spermatozoon, 239
maximum ion velocity through a channel, 471
maximum ion velocity through an open channel, 467
Maxwell's equations, 20, 30, 57
MCS, *see* multiple chemical sensitivity
MCS patients, 318, 335
MCS self-reported, 308
MDA, *see* malondialdehyde
MDA levels, 251
mean free path, 467
mechanism of action, 458
mechanism of action for man-made EMFs, 246
mechanism of heating biological tissues, 45
median age for final birth, 229
meiotic metaphase, 149
meiotic processes, 250
melanism, 429

melanoma, 122, 376, 484
melatonin, 119, 241, 248, 251, 289, 313, 316
membrane ion channels, 192
membrane potential spontaneous oscillations, 472
membrane voltage, 209, 243
membrane voltage oscillations, 429
memory, 279, 323, 330
memory impairment, 140
meningioma, 82, 377
menstrual cycle, 229
meta-analysis, 82
meta-analysis of studies on childhood cancer, 375
metabolic changes, 99
metallic conductor, 20
metal objects, 227
metal shielding, 8, 345
metaphase, 195, 198, 202
method of undetermined coefficients, 472
methodological shortcomings in EHS provocation studies, 332
MF, see magnetic field
MF exposure, 199
MF intensity, 200
mice, 149, 153, 241, 380, 400, 483
microcephalia, 413
micro-heating, 246
micronuclei, 142, 144–149, 151, 153
micronuclei formation, 119, 227, 254, 421
micronuclei frequency, 379
micro-thermal impacts, 225
microtubules, 103
microwave band, 139
microwave carrier signals, 221
microwave exposure, 304
microwave frequency, 206
Microwave Generators, 58
microwave heating, 24, 45, 65, 206
microwave oven, 21, 45, 102, 141, 247
microwave photons, 26, 58
microwave radiation, 410
microwaves, 45, 98
microwaves (MWs), 2, 3, 20, 100, 299, 372, 395
microwave sickness, 86, 140, 379, 398
microwave syndrome, 304, 332, 341, 398, 484
microwave transitions in atoms, 59
middle cerebral arteries, 325
mid-luteal phase, 229
mid-oogenesis, 193, 200
mid-oogenesis checkpoint, 193
migratory songbirds, 411
millimeter waves, 86, 225
MIMO, see Multiple-Input Multiple-Output
miscarriage, 231
miscarriage risk, 403
misinformation, 9
misleading information, 7
mitochondria, 108, 123, 146, 192, 249, 256, 382, 421, 488
mitochondrial collapse, 223
mitochondrial DNA damage, 149
mitochondrial genome, 241
mitochondrial membrane electrical potential, 252
mitochondrial membrane potential, 108, 149, 250
mitochondrial respiration, 249
mitochondrial ROS, 241, 244

mitochondrion, 240, 489
mitogen-activated protein kinase (MAPK), 152
mitosis, 151, 194, 343
mitotic cell division, 192
mitotic cycle, 143
mitotic division, 239
mitotic index, 148, 153
mm-waves, 4, 46, 65
MMWs, see modulated microwaves
Mn-SOD, see Manganese Superoxide Dismutase
mobile/cordless phones, 464
mobile/dissolved ions, 451, 486
mobile ions, 62, 276, 456, 458, 465
mobile ("free") ions, 43
mobile phone, 3, 23, 120, 138, 143, 146, 149, 191, 202, 255, 286, 302, 305, 373, 377, 416, 417
mobile phone EMF exposure during pregnancy, 401
mobile phone EMFs, 108, 145, 200, 323
mobile phone exposures, 232, 402
mobile phone jammers, 412
mobile phone radiation, 108, 199, 236, 382, 400, 412, 422, 455
mobile phones, 25, 37, 38, 80, 81, 98, 122, 139, 169, 202, 211, 222, 225, 280, 301, 371, 382, 395, 483
mobile (cell) phones, 376
mobile phone use, 141, 172, 210, 233, 239, 245, 377, 378, 486
mobile phone users, 144, 371
mobile telephony (MT), 3, 20, 38, 99, 138, 190, 370, 373, 395, 451
mobile telephony base station, 305
Mobile Telephony radiation, 83
modulated and pulsed microwaves, 376
modulated and pulsed RF EMFs, 65
modulated EMFs, 23
modulated microwaves (MMWs), 371
modulated/pulsed RF/MW EMFs, 381
modulated RF EMFs, 122
modulated wave, 3
modulation, 9, 21, 24, 36, 37, 63, 80, 104, 208, 280, 371, 374, 395, 421, 451, 485
modulation and pulsing frequencies, 371
modulation field/wave, 21
modulation of electromagnetic field, 3, 4
modulation/pulsing components, 382
modulation/pulsing/variability components, 465
molecular/atomic/nuclear spectra, 52, 61
molecular biomarkers, 300
molecular oscillations, 43
molecules, 61
monkeys, 400
monoamine neurotransmitters, 323
mood, 323
mood disturbance, 305
morphogenesis, 413
morphogenetic events, 413
morphological changes in germ cells, 402
morphophysiological changes, 423
mortality, 209, 230, 375, 406, 413, 414, 418, 428
mortality in birds, 410
mortality rate, 376
motion equation, 452
motor activity, 330
motor functions, 280

Index

motor skills, 85
mouse and rat spermatogenesis, 402
mouse embryos, 153
mouse follicles, 141, 237
mouse germ cell development, 402
mouse spermatocytes, 146, 152
mouse spermatogonia, 146
moving charge interaction, 103
M phase, 143, 195
MRI, see magnetic resonance imaging
mRNA, 152
MT, see mobile telephony
MT antennae base station, 379
MT antennas, 425
MT antenna towers, 90, 406
MT base antennas, 37, 88, 89, 405, 408
MT base station antennas, 122, 305, 378, 410
MT base stations, 85, 122, 144, 383, 417
MT EMFs, 169, 198, 200, 402
multi-frame repetition frequency, 38
multiple births, 231
multiple chemical sensitivity (MCS), 300, 303, 304, 315, 335
Multiple-Input Multiple-Output (MIMO), 39, 100
multiplexing, 38, 39, 100, 140, 190
multi-stream transmission, 39
mung bean, 423
muscle and nerve stimulation, 140
muscle contractions, 413
muscular spasm, 328
musculoskeletal discomfort, 304
mutagenic effects, 119, 171
mutant gene expression, 378
mutated offspring, 495
mutations, 149, 151, 428, 482, 494
MW radiation, 102
MWs, see microwaves
myalgia, 310, 328
myelin, 313
myelinating glia, 313
myelin P_0 autoantibodies, 316
myelin protein zero (myelin P_0), 313
myelin sheath, 82
myeloblastic leukemia, 375
myeloid leukemia, 374, 381
myometrial degeneration, 238
myometrium, 254

N

NADH, see nicotinamide adenine dinucleotide
NADPH, see nicotinamide adenine dinucleotide phosphate
NADPH/NADH oxidase, 107, 108, 122, 382, 490
Na^+/K^+-ATPase, 103, 490
Na^+/K^+ pump, 490; see also Na^+/K^+-ATPase
National Toxicology Program (NTP), 191, 484
natural ambient fields, 209
natural electromagnetic field, 4
natural electromagnetic stimuli, 411
natural EMFs, 3, 22, 23, 34, 169, 370, 371, 395, 418
natural EMFs/EMR, 342
natural environment, 8, 9
natural radioactivity, 485
nausea, 330
navigation, 395, 410

near-field, 50, 191
necrosis, 491
negative findings, 142, 164–168
negative reports, 169
neonates, 153
nephroblastoma cells, 152
Nernst equation, 455, 456, 487
nerve impulses, 63, 139, 323
nerve regeneration, 22
nerve sheath tumor, 82
nerve sprouting, 22
nervous system, 304, 407, 413, 429, 487
nestlings, 409
neural system, 418
neural tube, 428
neurasthenia, 305
neuroblastoma cells, 151
neurodegenerative diseases, 316, 401, 485, 494
neurodegenerative disorders, 327, 345
neurodevelopment, 85
neuro-epithelial brain cancer, 82
neuroepithelial brain tumors, 141
neurogenic inflammation, 304
neuro-inflammation, 330
neurological somatic disease, 341
neurological syndrome, 306
neurologic disorder, 310
neuronal activity, 276
neuronal apoptosis, 312
neuronal axon, 277
neuronal destruction, 341
neuronal membrane, 276
neuronal terminals, 277
neurons, 341
neuropsychiatric effects, 85
neurotransmitter levels, 288
neurotransmitters, 277, 287, 323, 400, 401
neutrophils, 152
newborn rabbits, 147
newborn rats, 148
New Radio (NR), 3, 39
NFκB, see nuclear factor kappa B
NFκB-related proteins, 152
NHL, see non-Hodgkin lymphoma
nick translation, 153
nicotinamide adenine dinucleotide (NADH), 490
nicotinamide adenine dinucleotide phosphate (NADPH), 490
nifedipine, 487
95% Confidence Interval, 374
nitric oxide, 104, 119, 150, 311, 381, 383, 484, 488, 491
nitric oxide synthases (NOS), 119, 254, 311, 489
nitrogen oxide, 108, 254
nitrosative stress, 237
nitroso-oxidative stress, 318
nitrotyrosine (NTT), 311, 316, 318
NMR, see nuclear magnetic resonance
nocebo effect, 341
non-carcinogenic health effects, 121
non-enzymatic protein biomarkers, 320
non-Hodgkin lymphoma (NHL), 84, 376
non-ionizing, 99, 139
non-ionizing band, 27
non-ionizing EMFs, 247

non-ionizing EMFs/EMR, 101, 371
non-ionizing EMR, 22
non-ionizing energies, 240
non-ionizing radiation, 25
non-linear phenomena, 468
non-polarized field, 27, 28
non-pregnant rabbits, 233
non-radical ROS, 491
non-rapid eye movement (NREM) sleep, 278
non-thermal biological effects, 102, 487
non-thermal effects, 4, 5, 23, 24, 65, 86, 104, 141, 169, 206, 207, 227, 280, 290, 372, 398, 412, 429, 430, 465, 485
non-thermal effects of man-made EMFs in the eyes, 410
non-thermal EMF exposures, 99, 100, 371
non-thermal EMR, 100
non-thermal exposures, 43, 104, 372
non-thermal intensities of WC EMFs, 382
non-thermal man-made EMF exposure, 120, 121
non-thermal mechanisms, 239, 247, 282, 382
non-thermal microwave effects, 65
non-thermal modes of action, 225
non-thermal pathways, 247
non-thermal WC EMF exposure, 312, 381
norAD, *see* noradrenaline
noradrenaline (norAD), 323
NORDCAN database, 378
northern bald ibises, 409
NOS, *see* nitric oxide synthases
NR, *see* New Radio
NREM, *see* non-rapid eye movement sleep
NREM sleep, 285
NTP, *see* National Toxicology Program
NTT, *see* nitrotyrosine
nuclear factor kappa B (NFκB), 121
nuclear magnetic moment, 59
nuclear magnetic resonance (NMR), 53, 59
nuclear β-globin locus, 241
nucleomegaly, 149
nucleons, 61
nucleotide base, 485
nucleus, 192
number of oocytes, 229
numerical examples, 464
numerical modeling, 49
numerical test study, 458
nurse cells, 150, 193, 200, 236

O

oak trees, 424
objective somatic disorders, 310
Observed to Expected ratio (OE), 375
occupational exposure, 101
occupational exposure limits, 372
occupational radio and radar EMF exposure, 376
occupational studies, 376
occurrence of birds, 406
Oceania Radiofrequency Science Advisory Association, 88
ocular disturbance, 304
ODC, *see* ornithine decarboxylase
Odds Ratio (OR), 81, 375
OE, *see* Observed to Expected ratio; Odds Ratio
oedema, 413

Office of Research Integrity, 466
offshore wind energy, 411
offspring, 240, 381
offspring health, 223
offspring number, 416
Ohm's law, 47
oligomenorrhea, 229
oncogene, 486
oncogenic transformation, 382
on-off keying, 37
on/off pulsations, 280
on/off pulses, 21, 139
on/off transition, 460
onset and removal of EMF exposure, 64
oocyte degeneration, 237
oocyte loss, 229
oocytes, 150, 193, 200, 223, 236, 250
oogenesis, 150, 153, 192, 193, 236
opportunistic infections, 343
optical fibers, 90
optic chiasm, 22
Oregon R wild type strain, 192
organic diseases, 494
organic disorders, 428
orientation, 411, 412, 467
orientation and navigation, 418
orientation of migratory birds, 411
ornithine decarboxylase (ODC), 103, 122, 153, 383
orthoptera, 409
OS, *see* oxidative stress
OS-associated biomarkers, 319
OS biomarkers, 122
oscillating electric field, 1
oscillating magnetic field, 1
OSI, *see* oxidative stress index
OS indexes, 104
OS indicators, 99
OS in EHS patients, 318
osmotic equilibrium, 458
OS pathways, 241
OS response, 257
otalgia, 310
ovarian cells, 198, 205, 483
ovarian epithelial stromal tumor, 380
ovarian growth, 150
ovarian histopathological changes, 254
ovarian hyperstimulation syndrome, 231
ovarian pool of oocytes, 229
ovarian reserve, 403
ovarian ROS content, 254
ovarian steroids, 238
ovaries, 236, 248, 252
ovarioles, 193, 200, 236
ovary size, 141
oviposition, 236
ovulation, 231
ovulation prediction, 229
ovulatory function, 229
oxidative damage, 148, 383, 421, 423
oxidative DNA damage, 104, 146, 233, 240, 254, 382
oxidative effects, 104
oxidative/nitrosative stress, 311, 313
oxidative phosphorylation, 249, 252, 488
oxidative sperm DNA damage, 244

Index

oxidative stress (OS), 4, 5, 7, 84, 85, 99, 104–108, 110–119, 121, 122, 140, 141, 151, 152, 171, 192, 194, 223, 225, 227, 232, 237, 239, 241, 249, 252, 255, 258, 301, 315, 318, 381, 382, 402, 403, 406, 428, 430, 450, 457, 482–484, 494, 497
oxidative stress index (OSI), 252
oxidized glutathione (GSSG), 142, 148, 318
oxygen, 488

P

pain, 323
pancreatic cancer cells, 152
pandemic, 3
Parkinson's disease, 494
parotid glands, 377
parotid malignant tumor, 377
PARP, *see* Poly-ADP Ribose-Polymerase
parrot feather plant, 422
partially polarized light, 29
particular solution, 471, 473
passive transport proteins, 456
patch-clamp, 43
patch clamp ion current recordings, 285
patch-clamp recordings, 454, 487
patch-clamp technique, 471
paternal DNA, 239
paternal genome, 238
paternal genomic damage, 244, 257
paternal stressors, 231
pathological disorder, 311
pathological syndrome, 304
pathophysiological effects, 299
Pauli, 57
PCD, *see* programmed cell death
PEF, *see* pulsed electric field
pelvic inflammatory disease, 229
penetration (of electric and magnetic fields into living tissue), 466
penetration depth (for plane electromagnetic waves within a material), 466
penetration of electromagnetic waves, 20
perception, 280
pericytes, 315
peripheral blood, 153, 311, 315
peripheral blood biomarkers, 313, 346
peripheral blood erythrocytes, 147
peripheral blood lymphocytes, 484
Peripheral Nervous System (PNS), 313
perivascular astrocytes, 312
perivascular tissue, 330
permeability of the blood–brain barrier, 248, 404
permissible single dose, 206
peroxidation, 102
peroxynitrite, 120, 171, 254, 311, 383, 484, 488, 491
Persian gulf war syndrome, 303
pesticides, 407, 416
PET, *see* positron emission tomography
PHA, *see* phytohaemagglutinin
phagocytes, 108
phagocytosis, 193
pharma industry, 384
phase, 1
phased arrays, 100
phase difference, 27, 28, 30, 472
phase modulated, 21
phase modulation (PM), 37, 373
phasic relative amplitude, 285
pheochromocytoma cells, 149, 404
phosphodiester bond, 485
photoelectron, 54
photolyase, 414
photomultiplier, 54
photonic energies, 25, 26, 51
photons, 20, 26, 51, 61, 67
photoreceptors, 414
photosynthesis, 424
photosynthetic pigments, 421
physical parameters of EMF exposure, 208
physical parameters of signal transmission, 371
phytohaemagglutinin (PHA), 197
pineal gland, 289, 323, 429
pine trees, 424
Pinus brutia, 99
pituitary, 248
pituitary gland, 381
pituitary tumors, 82
Planck formula, 26
Planck's constant, 26
plane harmonic wave, 55
plane waves, 27; *see also* linearly polarized field/wave
plant damage, 427
plants, 80, 419
plasmacytoma, 376
plasma membrane, 456
PM, *see* phase modulation
polarization, 3, 23, 28, 60, 63, 208, 451
polarization combined with variability, 209
polarization plane, 1, 27
polarized and coherent ELF EMFs, 102
polarized and coherent EMFs/EMR, 342
polarized and coherent movement, 467
polarized/coherent EMFs, 458
polarized electromagnetic field, 1
polarized EMR, 9
polarized oscillating EMF, 454
polar molecules, 43
police officers, 376
pollinator insects, 417
pollinators, 415
Poly-ADP Ribose-Polymerase (PARP), 152
polycystic ovary syndrome, 229
polymorphonucleocyte, 254
polyploid cell, 146
poor nutrition, 193, 200
poplar trees, 424, 425
population decline, 407, 408
positive findings, 142, 169
positron emission tomography (PET), 279, 325
post-ganglionic sympathetic neurons, 323
postnatal development, 252
postnatal period, 315
postsynaptic potential, 277
power density, 30, 100, 101, 139, 142, 145, 225, 284, 372, 376, 379, 406
power density levels, 408
power density limits, 210, 228
power frequency, 3

power line EMFs, 415, 423
power lines, 371, 406, 484
power transmission lines, 406
Poynting vector, 30
Precautionary Principle, 7, 89, 100, 102, 172, 347, 373, 385, 418, 430
preclinical animal studies, 231
pregnancy, 85, 89, 141
pregnant mice, 381
pregnant rabbits, 147, 233
pregnant rats, 148, 255, 401, 402
pregnant women, 141, 233
preimplantation stage, 232
premature labor, 231
premature ovarian insufficiency, 229
prenatal development, 252
prenatal exposure, 85
preseismic pulsations, 61
preterm delivery, 141
primordial follicles, 141, 237
progesterone, 141, 229, 237, 248, 255
programmed cell death (PCD), 193
pro-inflammatory cytokines, 403
prolactin, 255
prophase, 198
propolis, 415
prostate, 241
prostate cancer, 123
prostate gland, 381
protective measures, 347
protective mechanisms, 496
proteinase-K, 152
protein conformation, 103
protein conformational abnormalities, 341
protein damage, 4
protein homeostasis, 229
protein oxidation, 104
protein S100B, 312, 316
protein synthesis, 85, 343
protein synthesis rate, 6
proteolytic pathways, 229
proteomic analysis, 241
proteomic and epigenomic changes, 231
proteostasis, 244
proton flow, 249
proton (H^+) voltage-gated channels, 489
provocation studies, 300, 302, 332
proximity to EMF-source, 236
psychiatric brain disease, 335
psychiatric disorder, 345
psychiatric symptoms, 305
psychological distress, 305
psychoneurologic pathologies, 300
public exposure limits, 372
public health, 8
public health measures, 348
public health protection, 210
public health risk assessment, 384
public health risks, 86
pulsatility index, 325
pulsation, 9, 24, 36, 38, 104, 485
pulsation and modulation of the WC EMFs, 225
pulsation electromagnetic field, 4
pulsations, 100, 140, 371, 395

pulsed electric field (PEF), 194, 198, 200, 205
pulsed EMF-bioactivity, 462
pulsed EMF exposure, 306
pulsed EMFs, 23, 148
pulsed EMR, 380
pulsed fields, 465
pulsed microwave EMF exposure, 333
pulsed microwave EMR, 373
pulsed microwaves, 375
pulsed/modulated EMFs, 287
pulsed/modulated MWs, 384
pulsed MWs, 375
pulsed MWs of radar systems, 380
pulsed RF EMFs, 451
pulsed wave, 3
pulse rate, 100
pulse repetition, 147
pulse repetition frequencies, 373
pulse ("frame") repetition frequency, 100
pulse repetition rate, 38
pulsing, 21, 63, 80, 208
pulsing, modulation, and variability, 398
pulsing electromagnetic field, 3
pulsing ELF E and B fields of MT/WC EMFs, 463
pulsing ELF electric field exposure, 333
pulsing EMF exposures, 255
pulsing EMFs/EMR, 378
pulsing EMR, 380
pulsing frequencies, 41, 100
pulsing microwave EMF, 401
pulsing microwave radiation, 400
pulsing microwaves, 373
pulsing RF EMR, 282
pulsing WC EMFs, 223
pulsometric index, 327
pupation, 416
pupillary light reflex, 333, 335
p-values, 196, 198
pyramidal neuronal cells, 277

Q

QED, *see* quantum electrodynamics
QEM, *see* quantum electromagnetism
quail embryos, 108, 120, 382
quanta, 26
quantized emission, 52, 53
quantized energy levels, 61
quantized transitions, 52, 54
quantum bits (qubits), 59
quantum computers, 59
quantum electrodynamics (QED), 26
quantum electromagnetism (QEM), 26
quantum microwave emitters, 60
queen cells, 415
queen development, 418
queen loss, 415

R

rabbits, 401
radar microwaves, 375
radar MW frequencies, 373
radar operators, 376

radar pulsing frequencies, 373
radar radiation, 375
radar RF/microwave radiation, 410
radars, 21, 139, 144, 371, 373, 375, 376, 424
radiation dose, 84
radiation-induced eye-cataract, 376
radiation intensity, 29, 30, 32
radiation pattern, 37
radical pair mechanism, 414, 453
radio and radar operators, 376
radio and telegraph operators, 376
radio and television broadcasting, 139, 373
radio broadcasting, 424
radio-collars, 403
Radio Frequencies (RF), 2, 3, 80, 98, 139, 190, 221, 234, 235, 280, 299, 395
Radio-Frequency band, 20
Radio Frequency EMFs/EMR, 371
radio-marked animals, 410
radio tracking systems, 410
Ramazzini Institute, 83
random ELF/ULF variability, 42
random variability, 9, 21, 41, 63, 221, 280, 395
random variability in the final signal, 371
rapid eye movement (REM) sleep, 278
rat and mouse carcinogenicity studies, 83
rat astrocytes, 145, 146
rat behaviour, 400
rat brain, 147, 148, 400, 403
rat embryos, 403
rat endometrium, 255
rate of energy absorbed, 225
rat granulosa cells, 145, 151, 239
rat liver, 148, 403
rat lymphocytes, 103, 466
rats, 147, 381, 400, 483, 484
rats and mice, 399
rat spermatozoa, 147
Rayleigh law, 20
RBC, see red blood cells
reaction time, 285
reactive oxygen species (ROS), 5, 99, 104, 107, 120–122, 140–142, 146, 148, 149, 151, 152, 171, 172, 192, 223, 232, 238, 241, 249, 252, 255, 256, 311, 343, 381, 382, 451, 457, 470, 483, 484, 488
real EMF/EMR exposures, 284
real-life conditions, 414
real-life EMF exposure, 7
real-life EMFs, 24
real-life emissions, 64
real-life exposure conditions, 210
real-life exposures, 210, 485
real-life mobile phone exposures, 284, 398
real-life MT EMFs, 200, 205–207
real-life MT/WC EMFs, 211
real-life WC EMF exposures, 120, 142, 398
real-life WC EMFs, 8, 169, 191, 226
real-life WC exposures, 43
real signals, 226
recall bias, 82
receptor proteins, 453
rectal temperature, 225
red blood cells (RBC), 318, 321
redox balance, 457

redox homeostasis, 249
redox imbalance, 257
redox proteins, 404
redox status, 488
reduced body weight, 403
reduced glutathione (GSH), 104, 142, 318, 412
reduced ovarian follicle counts, 237
reference pulsations, 39
reference signals, 140
relative permittivity, 30
Relative Risk (RR), 374
REM, see rapid eye movement sleep
REM sleep, 287
repair capability, 486, 496
repair mechanisms, 195
reproduction, 141, 207, 232, 234, 235, 409, 415, 416, 428, 483
reproductive biology, 221, 224
reproductive capacity, 402, 407, 416
reproductive cells, 415, 483
reproductive damage, 241
reproductive decline, 150, 207, 419, 484
reproductive endocrine system, 248
reproductive health, 85, 221, 255
reproductive tract, 222
reptiles, 412
research funding, 384
resin production, 424
resonance, 452
responses to EMF-exposure, 335
restoration force, 454, 459
retarded growth, 413
retrograde ejaculation, 230
RF, see Radio Frequencies
RF carrier signal, 169
RF carrier waves, 371, 395
RF EMFs, 153–168, 373
RF EMFs/EMR, 100
RF EMF sources, 225
RF/microwave EMFs, 371
RF/MW carrier signals, 382
RF/MW EMFs, 104, 107
RF/MW/WC EMFs, 383
rhizogenesis, 421, 423
ribosome, 192
ring chromosomes, 144, 147, 151
risk assessment, 427
risk ratio, 376
robins, 411
rodents, 83, 222
Romberg sign, 310
root impairment, 421, 423
root meristematic cells, 421
root outgrowth, 422
ROS, see reactive oxygen species
rose bush, 422
ROS formation, 145
ROS generation, 244
ROS levels, 241
rosmarinic acid, 119
ROS-mediated DNA damage, 257
ROS overproduction, 382–384, 465
ROS pathways, 243
ROS production, 232, 247, 249, 252, 336

ROS sources, 107, 491, 494
RPMI medium, 197
RR, see Relative Risk
ruminants, 402

S

S4 helices, 457
S4 sensor, 457, 474
safety limits, 207, 211, 372, 373, 384
salamander, 413
salivary glands, 377
salivary gland tumors, 82
sample size, 286
SAR, see Specific Absorption Rate
SAR estimation, 49
SAR general public limit, 373
SAR limit, 228
satellites, 7, 396
scalp, 277
scattered EMR, 20
scattering, 61
scattering of electromagnetic waves, 20
scattering particles, 20
SCBF, see skin capillary blood flow
SCENIHR, see Scientific Committee on Emerging and Newly Identified Health Risks
Schroedinger, E, 55, 67
Schumann electromagnetic resonances, 8, 344
Schumann generator, 8
Schumann oscillations, 22, 26, 28, 395; see also Schumann resonances
Schumann resonances, 2, 3, 22, 54, 61, 206, 280, 496
Schwann cells, 82, 313
Schwannoma, 82, 83, 141, 484
schwannoma of the heart, 381
Scientific Committee on Emerging and Newly Identified Health Risks (SCENIHR), 228
SCN, see suprachiasmatic nucleus
scoliosis, 413
screening intervention, 324
scrotal hyperthermia, 225
scrotum, 239, 240, 247
SD, see Standard Deviation
sea water (conductivity), 466
Second Generation (2G), 3, 37, 38, 100, 140, 149, 191, 281, 373, 381, 395
sedentary fish, 412
seismic electric signals (SES), 28, 395
seizure activity, 287
selenium, 110, 119
self-frequency, 459
self-reported symptoms, 332
semen, 141
semen parameters, 222, 250
semen quality, 230, 238
semen volume, 85
seminiferous tubules, 225, 240, 248
seminoma, 122, 378
sensing of upcoming earthquakes, 456
sensing of upcoming thunderstorms, 456
sensitive biological conditions, 210
sensitivity, 205
sensitivity to stressor, 202

serotonin, 288, 323
serum melatonin, 254
serum testosterone, 147
SES, see seismic electric signals
sferics, 28, 200
sham-exposed, 207
short-term exposure, 210
short-term exposure limit, 208
short-term health effects, 372
shrubs, 426
SICD, see stress-induced cell death
signal generators, 38
signaling cascades, 382
signaling hormones, 248
signal reception, 37
signal variability, 142
similarities among humans and animals, 192
simulated EMF-exposure, 7
simulated EMFs, 7
simulated exposures, 285, 485
simulated mobile phone exposures, 398
simulated MT EMFs, 25, 205
simulated MT/WC EMF exposures, 398
simulated MT/WC EMFs, 211
simulated signals, 8, 42, 210; see also simulated EMFs
simulated systems, 240
simulated WC EMFs, 24, 142, 191
simulated WC EMF signals, 284
simulated WC signals, 43
single photons, 54
sinusoidal 50 Hz magnetic fields, 84
sinusoidal alternating electric field, 459
sinusoidal magnetic field, 150
SIR, see Standardized Incidence Ratio
sister chromatid exchanges, 152
6-Hydroxymelatonin Sulfate (6-OHMS), 313, 316
6-OHMS/creatinine ratio, 313
skate, 411, 429
skin cancer, 375, 380, 484
skin capillary blood flow (SCBF), 333
skin conductivity, 333
skin irritation, 99, 140
skin lesions, 310, 328
skin problems, 305
skin rashes, 305
skin symptoms, 305
Skrunda radio, 424
Skrunda radio station, 147, 423
skull, 84, 277
sleep, 323
sleep and waking EEG, 289
sleep cycle, 330
sleep disorder, 122, 379, 484
sleep disturbances, 305, 306, 310, 316, 335
sleep EEG, 281, 285
sleep EEG pattern, 287
sleep problems, 140
slow brain potentials, 336
slow-wave sleep (SWS), 279
small non-coding RNA (sncRNA), 231
SMART, see somatic mutation and recombination test
"smart" buildings, 399
"smart" cities, 399
"smart" mobile phone, 21

Index 529

"smart phone," 403
"smart phone" devices, 223
smoking, 90
SMR, *see* Standardizing Mortality Ratio
sncRNA, *see* small non-coding RNA
SOD, *see* superoxide dismutase
SOD activity, 144
solar activity, 34, 209
solar EMR intensity, 23
solar light, 3
solar radio transmitters, 409
solar storms, 395, 402, 419
solar ultraviolet radiation, 375
somatic cells, 193, 258, 495
somatic mutation and recombination test (SMART), 119
Sommerfeld, A, 52
sonohysterogram, 229
soybeans, 422
sparrowhawk, 407
sparrow populations, 408
spatial memory, 85
spatial orientation, 395
speaking mode, 42
Specific Absorption Rate (SAR), 5, 24, 25, 46, 66, 100, 101, 139, 169, 225, 248, 372, 373, 400
specific activity, 321
specific conductivity, 46, 47, 466
speed measurement radar, 376
sperm, 147, 223
sperm acrosome, 246
spermatocytes, 149
spermatogenesis, 247, 402
spermatogenesis disruption, 225
spermatozoa, 149, 222, 225, 242, 402, 483
spermatozoon, 240
sperm cells, 147, 256, 483
sperm concentration, 85, 230
sperm count, 85, 231, 402
sperm DNA, 238
sperm DNA damage, 239
sperm DNA fragmentation, 85, 245
sperm DNA integrity, 257
sperm head, 240, 246
sperm head abnormalities, 402
sperm maturation, 231
sperm mitochondria, 243
sperm morphology, 239
sperm motility, 141, 231, 239
sperm motility and viability parameters, 244
sperm parameters, 230
sperm quality, 230
sperm quantity and quality, 225
sperm selection barriers, 240
sperm viability, 141
sperm vitality, 231
S-phase, 152, 239
spindle frequencies, 285
spindle oscillatory frequencies, 284
spleen, 149
spontaneous abortion, 232
spontaneous breast cancer, 380
spontaneous intracellular ionic oscillations, 139
spontaneous ionic oscillations, 4, 454, 472
Sprague Dawley rats, 84, 147, 148, 152, 153, 240, 254

spruce, 424
SSB, *see* Synchronization Signal Blocks
stages of oogenesis, 193
standard deviation, 203
Standard Deviation (SD), 196, 199, 202
Standardized Incidence Ratio (SIR), 375
Standardizing Mortality Ratio (SMR), 375
standing interference, 29
starvation, 193, 201, 205, 211
static electric field, 1
static magnetic fields, 1, 411, 422
stationary energy levels, 52
stationary waves, 28
staurosporine, 193, 200
stem cells, 193, 233
sterility, 402
Stern-Gerlach experiment, 53, 59
steroidogenic enzyme activity, 237
strand breaks, 484
strandings of grey whales, 402
stress, 121, 140, 193, 305, 400
stress-induced cell death (SICD), 193, 200
stressors, 200
stress proteins, 103, 487
stress response, 424
stromal cells, 255
Student's t-test, 198
sub-frame, 39
sub-infrared, 2
sub-infrared frequencies, 1
subjective health symptoms, 379
submarines, 466
subsea power transmission cables, 411
substance P, 322
sugar, 485
suicidal ideation, 310
sunflower, 419
superconductive conditions, 59
superconductive/cryogenic conditions, 59
superoxide, 108
superoxide anion, 254, 311, 488
superoxide anion free radical, 249, 252
superoxide anion radical, 484, 491
superoxide dismutase (SOD), 99, 142, 148–150, 153, 171, 254, 318, 412, 491
superoxide dismutase activity, 379
superoxide free radical, 150, 238
superoxide radical, 108, 119, 142, 383
superposition of fields, 27
superposition of non-polarized electromagnetic waves/fields, 34
superposition of non-polarized EMR/EMFs, 31
superposition of polarized electromagnetic waves/fields, 34
superposition of vectors, 31
superposition of waves, 32
suprachiasmatic nucleus (SCN), 22
surface potential, 277
survival, 415
Swiss albino mice, 149, 153
SWS, *see* slow-wave sleep
symmetrical interchanges, 147
sympathetic nerve activity, 330
symptomatic picture, 309
synapse, 277

synaptic cleft, 277
synaptic plasticity, 330
Synchronization Signal Blocks (SSB), 39
synchronization signals, 39, 140
synergistic adverse effects, 399
systemic disorder, 346
systemic effects, 87, 193
systemic inflammation, 313

T

talk mode, 150, 200, 202
"talk" signal, 146
TAS, *see* total antioxidant status
TBARs, *see* thiobarbituric acid-reactive substances
T-cell NHL, 378
TDMA, *see* Time Division Multiple Access
TDU, *see* transcranial Doppler ultrasound
technical EMF, 20; *see also* man-made EMFs
television (TV), 305
temperature of exposed biological tissues, 100
temporal lobes, 327
10-20 system, 277
tenosynovitis, 140
teratogenic effects, 412
teratogenic features, 413
terminal deletion, 198, 202; *see also* chromatid breaks
terminal deoxynucleotide transferase dUTP nick end labeling (TUNEL) assay, 120, 144, 147, 150, 153, 196, 205, 236
terminal transferase, 196
Terrestrial Trunked Radio (TETRA), 41
Tesla (T), 99
testes, 148, 149, 153, 242, 247, 248
testicular cancer, 239, 376, 378
testicular development, 231
testicular proteome, 85
testicular tissue, 403
testis temperature, 225
test mobile phones, 7, 104, 148
testosterone, 148, 240, 248
testosterone level, 230
test phones, 24, 143
TETRA, *see* Terrestrial Trunked Radio
tetraploid cells, 147
thalamocortical neurons, 278
therapeutic effects, 496
thermal effects, 5, 24, 65, 100, 102, 141, 169, 206, 207, 247, 289, 396, 398, 429
thermal (heating) effects, 90, 99, 280, 372
thermal energy, 44, 49, 65, 459
thermal exposures, 45
thermal mechanism, 290
thermal movement, 467
thermal noise problem, 452, 467
thermal velocity, 44, 474
theta band, 283
theta rhythm, 278
thiobarbituric acid-reactive substances (TBARs), 99, 318
Third Generation (3G), 3, 36, 39, 100, 140, 149, 281, 373, 381, 395
Thomas Young, 29
3,4-Benzopyrene (BP), 380
3G/4G, 191

3G (UMTS) mobile phone, 206
three-phase electric power transmission lines, 28
threshold ELF electric field intensities, 466
threshold EMF/EMR intensity levels, 5
threshold intensities, 467
thumb weakness, 140
thunderstorms, 28, 200, 395
thyroid cancer, 378
thyroid dysfunction, 311
thyroid gland, 403
ticks, 417
TILT, *see* toxicant-induced loss of tolerance
Time Division Duplex, 39
Time Division Multiple Access (TDMA), 37, 38, 100, 140, 373
time-finite emission, 61
time slot, 39
tinnitus, 99, 140, 305, 310
tiredness, 305
tissue conductivity, 47, 48
tissue density, 47
tissue heating, 23, 80
tissue hypopulsation, 327
tissue regeneration, 413
tissue specific heat, 47, 48
tobacco, 89
tobacco industry, 384
tobacco mosaic virus, 422
tobacco plants, 422
tolerance threshold, 303
TOS, *see* total oxidant status
total antioxidant status (TAS), 252
total oxidant status (TOS), 252, 255
toxicant-induced loss of tolerance (TILT), 303
toxico-biological effects, 331
toxicological effects, 299
TPC, *see* two-pore channels
trace element homoeostasis, 401
transcranial Doppler ultrasound (TDU), 325
transcription, 422
transformer, 140
transgenerational impacts, 256
transgenic mice, 380
transient receptor potential subfamily V member 1 cation channel (TRPV1), 255
transition, 52, 61
transitions (molecular/atomic/nuclear), 28, 52
transit time (of an ion through an open channel), 471
translocation of S4 helices, 457
transmembrane electric field, 34, 452, 456, 458, 459
transmembrane electric field intensity, 473
transmembrane voltage, 35, 209, 456
transmembrane voltage changes, 457
tree damage, 425
trees, 80, 419
trichome, 423
triggers of EHS clinical symptoms, 331
TRPV1, *see* transient receptor potential subfamily V member 1 cation channel
tubal obstruction, 229
tubal patency, 229
tubulin, 250
tumorigenesis, 383
tumor necrosis factor, 148

Index

tumors of parotid glands, 122
tumor-suppressor gene, 486
tumor suppressor p53-binding protein, 150
tumor-suppressor protein, 378
TUNEL, *see* terminal deoxynucleotide transferase dUTP nick end labeling assay
TUNEL-negative, 200
TUNEL-positive, 200
TUNEL-positive signal, 198
two-hit hypothesis, 496
two-pore channels (TPC), 430

U

UCTS, *see* ultrasonic cerebral tomosphygmography
ULF, *see* Ultra Low Frequencies
ULF/ELF, 4
Ultra Low Frequencies (ULF), 2, 21, 139, 191, 221, 280, 395
Ultra Low Frequency band, 371
Ultra Low Frequency EMFs, 98
ultrasonic cerebral tomosphygmography (UCTS), 327
ultraviolet radiation, 371, 485
ultraviolet (UV) radiation, 414
UMTS, *see* Universal Mobile Telecommunication System
UMTS EMF, 203
UMTS exposure, 202
undescended testicles, 230
undesirable scientific data, 385
unidirectional antenna, 408
United States National Toxicology Program (NTP 2018), 404
Universal Mobile Telecommunication System (UMTS), 36, 39, 42, 100, 140, 149, 191, 373, 395, 404
unpaired electron, 488
urban ecosystems, 407
urine biomarkers, 346
urine samples, 316
US Air Force personnel, 375
USA National Toxicology Program (NTP), 381
USA naval personnel, 375
US Department of Health, 381
US Federal Communications Commission (FCC), 87
US National Institute of Environmental Health Sciences (NIEHS), 305
US National Institute of Health (NIH), 311
U.S. National Toxicology Program (NTP), 83
US Navy personnel mortality, 383
uterine fibroids, 229
uterine hormone levels, 255
uterine OS, 255
UV, *see* ultraviolet radiation
uveal melanoma, 378

V

vacuolation, 254
vacuum permittivity, 30
vacuum ultraviolet, 25
vagal nerve stimulation, 336
variability, 24, 36, 63, 208, 210, 371
variability of electromagnetic field, 3, 4
variability of EMF-exposure, 211
variability of signals, 210, 225
varicocele, 230

vasodilatory effect, 413
VDT, *see* visual display terminals
vector "mode" functions, 57
vector potential, 57, 58
velocity of electromagnetic wave, 31, 57
velocity of light, 1
Very Low Frequency (VLF), 2, 4, 21, 234, 235
vestibular/hearing nerve, 81
vestibular Schwannoma, 82
VGCC activation mechanism, 469
VGCCs, *see* voltage-gated calcium channels
VGIC dysfunction, 9, 171, 172, 249, 470, 493, 497
VGIC function, 457
VGIC mechanism, 346
VGICs, *see* voltage-gated ion channels
VGIC structure, 457
video games, 172
vimentin, 103
virgin female flies, 150
viscosity, 471
visible light, 371
vision problems, 401
visual display terminals (VDT), 302, 304, 305
visual performance, 304
visual working memory tasks, 285
vitamin C, 99, 119, 251
vitamin C and vitamin E, 255
vitamin D, 311, 315, 344
vitamin E, 99, 119, 251
vitellogenesis, 193
VLF, *see* Very Low Frequency
Voice over LTE (VoLTE), 39, 191
voltage-gated anion channels, 489
voltage-gated calcium channels (VGCCs), 66, 209, 247, 256, 290, 429, 430, 453, 456, 458, 466, 487, 491, 492
voltage-gated ion channels (VGICs), 5, 35, 102, 171, 192, 277, 336, 382, 414, 452, 467, 487
voltage gradient, 456
voltage sensors, 171, 457, 467
voltage-sensors of VGICs, 458
VoLTE, *see* Voice over LTE

W

waking and sleep EEG, 284
waking EEG, 285
water decomposition, 102
water vole, 403
wave equation, 57
waveform, 36
wave-function, 55
waveguide, 226
wave intensity, 30
wavelength, 20, 30
wave number, 30
wave-packet, 28, 52
wave penetrative capacity, 248
wave velocity, 20
WC, *see* wireless communication technologies
WC antennas, 169
WC devices, 169, 280
WCDs, *see* wireless communication devices
WC EMF ambient levels, 408

WC EMF effects, 172, 224
WC EMF-emissions, 42
WC EMF exposures, 123, 169, 233, 238, 249, 284, 379, 398
WC EMF-induced dysfunction of VGICs, 256
WC EMF-induced OS, 243, 254
WC EMF non-thermal exposure, 381
WC EMF pollution, 397
WC EMFs, *see* wireless communication EMFs
WC EMFs/EMR, 83, 84, 86, 90, 280, 371, 377
WC EMR, 80, 85, 299, 370, 394
WC industry, 384
WC networks, 232
WC systems, 373
WC technologies, 172, 225, 230, 331, 373, 451
wellbeing, 8
wheat, 421–423
white rabbit, 147
white stork, 405, 406
white stork orientation, 411
white stork reproduction, 406
WHO, *see* World Health Organization
whole blood, 197
Wi-Fi, *see* Wireless Fidelity
Wi-Fi EMF, 403
Wi-Fi enabled laptop computer, 223
Wi-Fi router, 25, 417
wild animals, 299
wild birds, 406
wildlife, 394, 399, 427
willow trees, 425
Wilson, W, 52
window, 150
window effects, 457, 468
window of increased DNA fragmentation, 468
windows of increased bioactivity, 468
wing spots, 417
wire connections, 211
wired connection, 8
wired internet, 211

wireless communication devices (WCDs), 80, 89, 90, 221, 225, 230
wireless communication EMFs (WC EMFs), 9, 23, 36, 38, 42, 63, 98, 103, 104, 107, 139, 140, 153–169, 190, 209, 210, 222, 232, 234, 235, 240, 248, 252, 255–257, 283, 299, 345, 370, 378, 382, 384, 394, 396, 400, 406, 408, 409, 416, 419, 423, 427, 429, 450, 465, 483
wireless communication (WC) technologies, 3, 19, 80, 98, 138, 139, 299, 370, 394
wireless control of cellular function, 429
Wireless Fidelity (Wi-Fi), 3, 21, 38, 39, 80, 89, 98, 139, 141, 211, 233, 255, 302, 374, 395, 401, 464, 483
Wireless Local Area Network (WLAN), 21, 39, 302; *see also* Wi-Fi
wireless router, 395
Wistar rats, 85, 147, 148, 241, 252
worker piping alarm signal, 417
workers exposed to radar radiation, 376
World Health Organization (WHO), 4, 89, 300, 331, 371, 377, 378, 485
wound healing, 22, 429
wound repair, 121

X

xenobiotic-metabolizing enzymes, 324
x-rays, 25, 247
x-ray spectra, 53

Y

youngest mobile phone users, 373

Z

zebrafish, 412
zinc, 119, 344
zona pellucida, 238, 246, 254